D1334452

MidKent College
LEARNING RESOURCE CENTRE
Medway Camp

Class N

MW
MKC EDU
TP 08004898

CAMBRIDGE STUDIES IN
ADVANCED MATHEMATICS 59

Practical Foundations of Mathematics

3C 00152 8395

Cambridge Studies in Advanced Mathematics

Editorial Board: W. Fulton, D.J.H. Garling, T. tom Dieck, P. Walters

3017535

Practical Foundations
of Mathematics

PAUL TAYLOR

CAMBRIDGE
UNIVERSITY PRESS

UNIVERSITIES AT MEDWAY
2 0 FEB 2009
DRILLHALL LIBRARY

PUBLISHED BY THE PRESS SYNDICATE OF THE UNIVERSITY OF CAMBRIDGE
The Pitt Building, Trumpington Street, Cambridge, United Kingdom

CAMBRIDGE UNIVERSITY PRESS
The Edinburgh Building, Cambridge CB2 2RU, UK   http://www.cup.cam.ac.uk
40 West 20th Street, New York, NY 10011-4211, USA   http://www.cup.org
10 Stamford Road, Oakleigh, Melbourne 3166, Australia

© Paul Taylor, 1999

This book is in copyright. Subject to statutory exception
and to the provisions of relevant collective licensing arrangements,
no reproduction of any part may take place without
the written permission of Cambridge University Press.

First published 1999

Typeset in Computer Modern 10/12.6pt using TeX.

Commentary on this book (bibliographical information, answers
to some exercises, readers' remarks and corrections) may be found
at http://www.dcs.qmw.ac.uk/~pt/Practical_Foundations/

Library of Congress Cataloguing-in-Publication Data
Taylor, Paul, 1960–
Practical Foundations of Mathematics / Paul Taylor.
xii+572pp. 23cm.
Includes bibliographical references (pp. 530–552) and index.
ISBN 0 521 63107 6 (hc.)
1. Mathematics. I. Title.
QA39.2.T413   1999
510–dc21   98–39472 CIP

ISBN 0 521 63107 6 hardback

Transferred to digital printing 2003

Contents

Introduction

FOUNDATIONS have acquired a bad name amongst mathematicians, because of the reductionist claim analogous to saying that the atomic chemistry of carbon, hydrogen, oxygen and nitrogen is enough to understand biology. Worse than this: whereas these elements are known with no question to be fundamental to life, the membership relation and the Sheffer stroke have no similar status in mathematics.

Our subject should be concerned with the basic idioms of argument and construction in mathematics and programming, and seek to explain these (as fundamental physics does) in terms of more general and basic phenomena. This is "discrete math for grown-ups".

A moderate form of the logicist thesis is established in the tradition from Weierstrass to Bourbaki, that mathematical treatments consist of the manipulation of assertions built using $\wedge$, $\vee$, $\Rightarrow$, $\forall$ and $\exists$. We shall show how the way in which *mathematicians* (and programmers) — rather than logicians — conduct such discussions really does correspond to a certain semi-formal system of proof in (intuitionistic) predicate calculus. The working mathematician who is aware of this correspondence will be more likely to make valid arguments, that others are able to follow. Automated deduction is still in its infancy, but such awareness may also be expected to help with computer-assisted construction, verification and dissemination of proofs.

One of the more absurd claims of extreme logicism was the reduction of the natural numbers to the predicate calculus. Now we have a richer view of what constitutes logic, based on a powerful *analogy between types and propositions*. In classical logic, as in classical physics, particles enact a logical script, but neither they nor the stage on which they perform are permanently altered by the experience. In the modern view, matter and its activity are created together, and are interchangeable (the observer also affects the experiment by the strength of the meta-logic).

This analogy, which also makes algebra, induction and recursion part of logic, is a structural part of this book, in that we always treat the simpler propositional or order-theoretic version of a phenomenon as well as the type or categorical form.

Besides this and the classical symmetry between $\wedge$ and $\vee$ and between $\forall$ and $\exists$, the modern rules of logic exhibit one between introduction (proof, element) and elimination (consequence, function). These rules are part of an even wider picture, being examples of *adjunctions*.

This suggests a new understanding of foundations apart from the mere codification of mathematics in logical scripture. When the connectives and quantifiers have been characterised as (universal) properties of mathematical structures, we can ask what *other* structures admit these properties. Doing this for coproducts in particular reveals rather a lot of the elementary theory of algebra and topology. We also look for function spaces and universal quantifiers among topological spaces.

A large part of this book is category theory, but that is because for many applications this seems to me to be the most efficient heuristic tool for investigating structure, and comparing it in different examples. Plainly we all write mathematics in a symbolic fashion, so there is a need for fluent translations that render symbols and diagrams as part of the same language. However, it is not enough to see recursion as an example of the adjoint functor theorem, or the propositions-as-types analogy as a reflection of bicategories. We must also *contrast* the examples and give a *full* categorical account of symbolic notions like structural recursion.

You should not regard this book as yet another foundational prescription. I have deliberately not given any special status to any particular formal system, whether ancient or modern, because I regard them as the vehicles of meaning, not its cargo. I actually believe the (moderate) logicist thesis less than most mathematicians do. This book is not intended to be synthetic, but analytic — to ask what mathematics requires of logic, with a view to starting again from scratch [Tay98].

Advice to the reader. Technical books are never written and seldom read sequentially. Of course you have to know what a category is to tackle Chapter V, but otherwise it is supposed to be possible to read any of at least the first six chapters on the basis of general mathematical experience alone; the people listed below were given *individual* chapters to read partly in order to ensure this. There is more continuity between sections, but again, if you get stuck, *move on to the next one*, as secondary material is included at the end of some sections (and subsections). The book is thoroughly indexed and cross-referenced to take you as quickly as possible to specific topics; when you have found what you want, the cross-references should be ignored.

The occasional anecdotes are not meant to be authoritative history. They are there to remind us that *mathematics is a human activity*, which is always done in some social and historical context, and to encourage the reader to trace its roots. The dates emphasise quite how late logic arrived on the mathematical scene. The footnotes are for colleagues, not general readers, and there are theses to be written *à propos* of the exercises.

The first three chapters should be accessible to final year undergraduates in mathematics and informatics; lecturers will be able to select appropriate sections themselves, but should warn students about parenthetical material. Most of the book is addressed to graduate students; Section 6.4 and the last two chapters are research material. I hope, however, that every reader will find interesting topics throughout the book.

Chapters IV, V and VII provide a course on category theory. Chapter III is valuable as a prelude since it contains many of the results (the adjoint functor theorem, for example) in much simpler form.

Sections 1.1–5 (not necessarily in that order), 2.3, 2.4, 2.7, 2.8, 3.1–5, 4.1–5 and 4.7 provide a course on the semantics of the λ-calculus.

For imperative languages, take Sections 1.4, 1.5, 4.1–6, 5.3, 5.5 and 6.4.

An advanced course on type theory would use Chapter IV as a basis, followed by Chapters V and IX with Sections 7.6, 7.7 to give semantic and syntactic points of view on similar issues.

Chapter VI and Section 9.6 discuss topics in symbolic logic using the methods category theory.

Acknowledgements. Pierre Ageron, Lars Birkedal, Andreas Blass, Ronnie Brown, Gian-Luca Cattani, Michel Chaudron, Thierry Coquand, Robert Dawson, Luis Dominguez, Peter Dybjer, Susan Eisenbach, Fabio Gadducci, Gillian Hill, Martin Hyland, Samin Ishtiaq, Achim Jung, Stefan Kahrs, Jürgen Koslowski, Steve Lack, Jim Lambek, Charles Matthews, Paddy McCrudden, James Molony, Edmund Robinson, Pino Rosolini, Martin Sadler, Andrea Schalk, Alan Sexton, Thomas Streicher, Charles Wells, Graham White, Andrew Wilson and Todd Wilson took the trouble to read a chapter or more of the draft and made detailed criticisms of it. Mike Barr, Peter Freyd, Peter Johnstone, Andy Pitts and Phil Scott have also patiently given illuminating answers to many stupid questions, and I learnt the box method from Krysia Broda. These people's remarks have often resulted in substantial rewriting of the text, but even those comments that I chose not to use were illuminating, and I hope to publish some of them as a Web "companion" to the book.

When I began this book in 1991, I was a member of a lively research group led by Samson Abramsky at Imperial College. Those who were there, including Roy Crole, Simon Gay, Radha Jagadeesan, Achim Jung, Yves Lafont, François Lamarche, Ian Mackie, Raja Nagarajan, Luke Ong, Duško Pavlović, Christian Retoré, Leopoldo Román, Mark Ryan, Mike Smyth and Steve Vickers, provided much stimulation. But in 1996 the management saw fit first to deprive me of my office, and later to withhold my EPSRC salary for not being in that office. I am deeply indebted to everyone at Queen Mary and Westfield College for supporting me at this distressing time: besides being a friendlier place all round, QMW has a much healthier working environment. I have since learned a lot from Richard Bornat, Peter Burton, Keith Clarke, Adam Eppendahl, Peter Landin, Peter O'Hearn, David Pym, Edmund Robinson and Graham White.

Jim Lambek was the first of my senior colleagues to express appreciation of this work. At many times when I might otherwise have given up, Carolyn Brown, Adam Eppendahl, Pino Rosolini and Graham White told me repeatedly that it was worth the effort.

Since this is my first and a very personal book, I would also like to record my appreciation of those who have taught me and encouraged my career in mathematics, beginning with my parents, Brenda and Ced(ric) Taylor. Ruth Horner of Stoke Poges county primary school, Buckinghamshire; Christian Puritz, Bert Scott, Doris Wilson and the late Henry Talbot at the Royal Grammar School, High Wycombe; Béla Bollobás, Andrew Casson and Pelham Wilson when I was an undergraduate at Trinity.

Typography. I composed this book using Emacs and typeset it all myself in TeX: I cannot conceive of doing research without these two programs, and all of the software that I have used is public domain (for a long time Mark Dawson kept this going for me). The commutative diagram, proof tree, proof box and design macros are my own. I would like to thank Vera Brice and Leslie Robinson of the London College of Printing for their suggestions and an interesting course in book design. Without Peter Jackson's eagle eye and Roger Astley's patient guidance at Cambridge University Press, however, there would be far more errors and idiosyncracies than you see here now.

Financial. From 1992 to 1997 I held an Advanced Research Fellowship awarded by the Engineering and Physical Sciences Research Council. I also took part in the European Union ESPRIT project *Categorical Logic in Computer Science* (1989–95).

I

First Order Reasoning

HOW DO WE BEGIN to lay the foundations of a palace which is already more than 3600 years old? Alan Turing [Tur35] identified what is perhaps the one point of agreement between the Rhind papyrus and our own time, that a "computer" (*i.e.* a mathematician performing a calculation) puts marks on a page and *transforms* them in some way. Even in its most naïve form, Mathematics is not passive: we recite the multiplication table, *transforming* 7 × 8 into 56, and later find out that $x \times (y + z)$ may be replaced by $(x \times y) + (x \times z)$. We say that these pairs are respectively *equal*, meaning that they denote the same objects in "reality", even though they are written in different ways.

During the process of calculation (from Latin *calx*, a pebble) there are intermediate forms with no directly explicable meaning: accountants refer to "net" and "gross" amounts, and to "pre-tax" profits, in an attempt to give them one. The remarkable feature of mathematics is that we may suspend belief like this, and yet rely on the results of a lengthy calculation, even when it has been delegated to a computer.

The notation of elementary school arithmetic, which nowadays everyone takes for granted, took centuries to develop. There was an intermediate stage called *syncopation*, using abbreviations for the words for addition, square, root, *etc.* For example Rafael Bombelli (*c.* 1560) would write[1]

R. c. L. 2 p. di m. 11 ⌡   for our $\sqrt[3]{2 + 11i}$.

Many professional mathematicians to this day use the quantifiers $(\forall, \exists)$ in a similar fashion,

$$\exists \delta > 0 \text{ s.t. } |f(x) - f(x_0)| < \varepsilon \text{ if } |x - x_0| < \delta, \text{ for all } \varepsilon > 0,$$

in spite of the efforts of Gottlob Frege, Giuseppe Peano and Bertrand Russell to reduce mathematics to logic.

The logical calculus is easier to execute than any of the techniques of mathematics itself, yet only in 1934 did Gerhard Gentzen set it out in a natural way. Even now, mathematics students are expected to learn

---

[1]Radice cubica ligata [*i.e.* bracketed as far as ⌡] 2 più di meno [his name for $+i$] 11.

complicated $(\varepsilon{-}\delta)$-proofs in analysis with no help in understanding the logical structure of the arguments. Examiners fully deserve the garbage that they get in return.

In Sections 1.4 and 1.5 natural deduction is introduced in a way which has been taught successfully to first year informatics undergraduates, for whom reasoning about the elementary details of their programs is a more pressing concern. Section 1.6 shows how formal methods correspond to *carefully* written proofs in the vernacular of mathematics and Section 1.7 formalises the way in which routine proofs are found, maybe by machine.

The manipulation of sets and relations more familiar to mathematics students is treated in Section 1.3 and Chapter II. If you are doubtful of the need for formal logic then I suggest reading the later chapters first.

In the spirit of starting out "from nothing", the first two sections discuss the behaviour of purely symbolic manipulation. Nevertheless, the ideas which they introduce will play a substantial role in the rest of the book. We have, for example, to learn the difference between *object-language* and *meta-language*. Logic is primarily a meta-theory, and historically it has sought its object-language in some strange places: for medieval logicians, the motivation was theology. Today it is mathematics and programming, but in the first instance we shall apply logic to formal manipulation itself.

The final section discusses classical and intuitionistic logic, and why we intend to use the latter. This chapter and the next raise some of the questions of logic. The rest of the book uses modern tools to illuminate a few of those issues.

1.1   SUBSTITUTION

Logic as the foundations of mathematics turns mathematics on itself, but there is *content* even in the study of *form*.

We begin by considering how expressions must transform if they are to have and retain meaning, without being too specific about that meaning. Think of this as learning how to use a balance, ruler and stop-watch in preparation for studying physics. To those whose background is in pure mathematics, the explicit manipulation of syntax and the distinction between object-language and meta-language may be unfamiliar, but they are the stuff of logic and informatics. We shall not be too formal: formal rigour can only be tested dynamically, and those whose business it is to implement formal languages on computers are already well aware of many of the *actual* difficulties from a practical point of view.

The mathematical expressions which we write are graphically of many forms, not necessarily one-dimensional, and are often subject to complex rules of well-formedness. It suffices at first to think of strings composed of constants $(0, 1, ..., \pi, etc.)$ and operation-symbols $(+, \times, \sqrt{}, \sin)$.

Given an expression of this kind, we can evaluate it, but that's the end of the story. The main preoccupation of algebra before the nineteenth century was the solution of equations.

Variables. Algebraic techniques typically involve the manipulation of expressions in which the unknown is represented by a *place-holder*. This may be a letter, or (as often in category theory) an anonymous symbol such as $(-)$ or $(=)$. The transformations must be such that they would remain valid if the place were filled by any particular quantity. They must not depend on its being a variable-name rather than a quantity, so we cannot add the variables b and e to obtain g ("$2 + 5 = 7$") because this is unlikely to continue to make sense when other values are put for the variables. Variables were used in this way by Aristotle.

The symbols are called parameters, constants, *etc.* depending on "how" variable they are taken to be in the context. Usage might describe x as a variable, p and q as parameters and 3 as a constant, but they may change their status during the course of an argument. This is reflected in the history of the cubic equation (Example 4.3.4): Gerolamo Cardano, like the author of the Rhind papyrus, demonstrated his techniques by solving specific but typical numerical examples such as $x^3 + 6x = 20$, but François Viète was the first to use letters for the *coefficients*. (The use of xyz for variables and abc for coefficients is due to René Descartes.)

After we have solved $x^3 = 3px + 2q$ for x in terms of p and q, we may consider how the form of the solution depends on them, in particular on whether $p^3 \geqslant q^2$. In a more general discussion, what were constants before are subject to investigation and become variables. This changing **demarcation** between what is considered constant or variable will be reflected in the *binding* of variables by quantifiers, and in the boxes used in Section 1.5.

The places p and q also differ in that they stand for arbitrary ($\forall$) quantities, whilst x is intended to be filled by individual ($\exists$), though as yet unknown, ones, namely the solutions of the equation. This difference is represented formally by the two *quantifiers*: $\forall p, q. \exists x. x^3 = 3px + 2q$.

Structural recursion.

REMARK 1·1·1: Although we write algebraic expressions and programs as linear strings of symbols, it is better to think of them as **trees**, in which the connectives are nodes and the constants and variables are leaves. Each operation-symbol corresponds to a certain species of node, with a sub-tree[2] for each argument; we speak of **nullary**, **unary**, **binary**, **ternary** and in general **k-ary** or **multiary** operations if they have 0, 1, 2, 3 or k arguments. For example Cardano's first cubic equation might be written

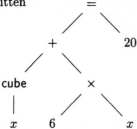

Transformation of expressions becomes surgery on trees. The analysis of a tree begins with the top node, the "outermost" operation-symbol, which is (confusingly) called the **root**. It is important to be able to recognise the root of an expression written in the familiar linear fashion; if it is a unary symbol such as a quantifier ($\forall$, $\exists$) it is usually at the front (a *prefix*), but binary ones ($+$, $\times$, $\vee$, $\wedge$, $\Rightarrow$) are traditionally written between their arguments (*infix*). Like unwrapping a parcel, you must remove the *outermost* operation-symbol *first*.

Expressions are typically stored in a computer by representing each node as a *record*, stating which operation-symbol, constant or variable it is, and with pointers to the records which represent the arguments or branches. Concretely, the pointer is the machine address of the record. The whole expression is denoted by the address of its root record, so the outermost operation-symbol is accessible im*mediate*ly, *i.e.* without inter*mediate* processing.

The leaves of an expression-tree are its variables. The simplest operation on trees is substitution of a term for a variable. A copy of the expression-tree of the term is made for every occurrence of the variable, and attached to the tree in its place. If there were many occurrences, the term and its own variables would be proliferated, but if there were none the latter would disappear. We call this a **direct transformation**.

The application of a single operation-symbol such as $+$ to its family of arguments is a special case of substitution: $3 + 5$ is obtained by substituting 3 for x and 5 for y in $x + y$. Conversely, it can be useful to think of any expression-tree (with variables scattered amongst its leaves) as a

---

[2]We shall use a hyphen (sub-thing) for syntactic or meta-language notions such as sub-expressions. A subthing (without the hyphen) is the source of an inclusion map, in the semantics, for example a subgroup.

generalised operation-symbol; since it replicates the algebraic language by adding in definable operations, the collection of all trees, substitutable for each other's leaves, is called the **clone**.

REMARK 1·1·2: An expression may also be regarded as a **function** of certain variables, written $f(x, y, \ldots)$. Beware that "$(x, y, \ldots)$" signifies *neither* that x, y, etc. actually occur in the expression (if they don't then f is a **constant function**), nor that these exhaust the variables which do occur (others may be parameters). We shall therefore describe this and $f(a, b, \ldots)$, the result of substituting certain terms for the variables, as the **informal notation** for functions.

There are also **indirect transformations** which re-structure the whole tree from top to bottom. For each kind of node, we have to specify the resulting node and how its sub-trees are formed from the original ones and their transformed versions. If you have a mathematical background, but haven't previously thought about syntactic processes, then symbolic differentiation will perhaps be the most familiar example. The most complex case is multiplication.

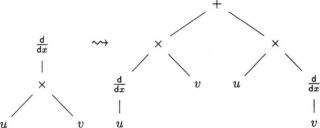

The tree has grown both horizontally and vertically, but notice that all occurrences of $\frac{d}{dx}$ are now closer to the leaves, and this observation is just what we need to see that the recursion terminates: the motto is **divide and conquer**. We shall study recursion in Chapters II and VI.

These symbols u and v are meant to stand for sub-trees such as $x^3 + 6x$ or y, rather than values: we have to use unknowns representing trees in order to apply mathematics to the discussion. It is important to understand that u is a different kind of variable from x. In the differential calculus an approximately analogous distinction is made between dependent and independent variables; this is needed to explain what we mean by $\frac{d}{dx}t$, where the variable x may "occur" in t. In terms of formal languages, t is (a variable in the) **meta-notation**, *i.e.* a **meta-variable**. We must be prepared for the possibility that t does actually stand for a variable, in which case the manipulation usually behaves in one way for the special variable x and in another for other variables.

Equipped with some meta-notation, it is more convenient to change back from trees to the linear notation; then the rules of differentiation are

$$\frac{d}{dx}c \quad\;\; = \quad 0 \qquad\qquad\qquad\qquad\qquad c \text{ is a constant}$$

$$\frac{d}{dx}x \quad\;\; = \quad 1$$

$$\frac{d}{dx}y \quad\;\; = \quad 0 \qquad\qquad\qquad\qquad x, y \text{ distinct variables}^3$$

$$\frac{d}{dx}(u+v) \quad = \quad \left(\tfrac{d}{dx}u\right) + \left(\tfrac{d}{dx}v\right)$$

$$\frac{d}{dx}(u \times v) \quad = \quad \left(\left(\tfrac{d}{dx}u\right) \times v\right) + \left(u \times \left(\tfrac{d}{dx}v\right)\right)$$

$$\frac{d}{dx}(\sin u) \quad = \quad (\cos u) \times \left(\tfrac{d}{dx}u\right) \qquad\qquad \text{and so on.}$$

Terms and substitution. Following these examples we can give

DEFINITION 1·1·3: A **term** is either

(a) a variable, such as x,

(b) a constant symbol, such as c, or

(c) $r(u, v, \ldots)$, *i.e.* an operation-symbol r that is applied to a sequence of sub-terms u, v, ...

Then the set $\mathsf{FV}(t)$ of variables in a term t is defined by

$$\mathsf{FV}(x) \qquad\qquad = \quad \{x\}$$

$$\mathsf{FV}(c) \qquad\qquad = \quad \varnothing$$

$$\mathsf{FV}(r(u, v, \ldots)) \quad = \quad \mathsf{FV}(u) \cup \mathsf{FV}(v) \cup \cdots$$

Beware that FV is not an operation-symbol, since it does depend on the identity of the variables. It's a meta-operation like $\frac{d}{dx}$.

Substitution is, for us, the most important use of structural recursion. We shall study the term a by exploiting its effect by substitution for variables in other terms, which must follow through all of the transformations of expressions. This obligation on each principle of logic gives rise to many curious (and easily overlooked) phenomena.

DEFINITION 1·1·4: By structural induction on t we define **substitution**, $[a/x]^* t$, of a term a for a variable x possibly occurring in t, as follows:

$$[a/x]^* c \qquad\qquad = \quad c \qquad\qquad\qquad c \text{ is a constant}$$

$$[a/x]^* x \qquad\qquad = \quad a \qquad\qquad\qquad a \text{ actually gets used}$$

$$[a/x]^* y \qquad\qquad = \quad y \qquad\qquad\qquad x, y \text{ distinct variables}$$

$$[a/x]^* r(u, v, \ldots) \quad = \quad r\big([a/x]^* u, [a/x]^* v, \ldots\big)$$

$$\qquad\qquad\qquad\qquad\qquad \text{for each operation-symbol } r.$$

---

[3]This is of course partial differentiation, in which x and y are treated as independent variables. Alternatively, y is a *parameter*. This example is given in preparation for studying substitution, in which we shall indeed treat x and y as independent variables.

Our notation for substitution, $[a/x]^*t$, has a star, the real meaning of which will become clear in Chapter VIII. For the moment, it is a reminder that $[a/x]^*t$ is not an expression starting with a square bracket but a modified form of t, having the same first symbol as it has, unless this is x, in which case $[a/x]^*t$ starts with the first symbol of a. Other forms of this notation are to be found in the literature, including $t[a/x]$ from typography and $t[x := a]$ from programming.

The next result, the **Substitution Lemma**, is the key to our semantic understanding of syntax in Chapters IV and VIII.

LEMMA 1·1·5: Let a, b and t be expressions, and x, y distinct variables such that $x, y \notin \mathsf{FV}(a)$ and $y \notin \mathsf{FV}(b)$. Then

$$
\begin{aligned}
[b/y]^*a &= a \\
[a/x]^*[b/y]^*t &= [[a/x]^*b/y][a/x]^*t.
\end{aligned}
$$

PROOF: We use structural induction on the term (a or t), considering each of the cases c, x, y, z (*i.e.* another distinct variable) and $r(u, v, \ldots)$, using the induction hypothesis for the last. The first result, with $[a/x]^*b$ instead of b, is needed to prove the second in the case where $t = x$. □

When both $x \notin \mathsf{FV}(b)$ and $y \notin \mathsf{FV}(a)$, the operations commute,

$$[a/x]^*[b/y]^*t = [b/y]^*[a/x]^*t,$$

justifying **simultaneous substitution**, for which we write

$$[a/x, b/y]^*t \quad \text{or} \quad [\underset{\rightarrow}{a}/\underset{\rightarrow}{x}]^*t.$$

The general form in the lemma, where we allowed x to occur in b, applies when variables do interfere: as in quantum mechanics, the failure of commutation shows that something is *happening*.

Quantification. There are expressions which have logical values as well as those with numerical meaning. We shall say **formula** for a logical expression and **term** for other kinds. Equations ($a = b$) and inequalities ($a \leqslant b$) are (atomic) formulae; we might think of $=$ as a binary operation-symbol whose result is a truth-value. There is an algebra of formulae *à la* Boole, whose operation-symbols include $\wedge$, $\vee$ and $\Rightarrow$.

Using the informal notation, a formula such as $\phi[x]$ containing variables is known as a **predicate**, x being the **subject**. If it has no variables it is called a **proposition**, **sentence** or **closed formula**. Like terms with variables, predicates may be understood as equations to be "solved", defining the class of things which satisfy them. The formation of such a class from a predicate is known as **comprehension** (Definition 2.2.3).

A predicate $\phi[x]$ may also be intended to make a general assertion, *i.e.* of every instance $\phi[a]$ obtained by substituting a term a for x. In this case it is called a **scheme**. It may be turned into a single assertion by **universal quantification**: $\forall x.\ \phi[x]$. Similarly, we may assert that the predicate *quâ* equation has some solution or **witness** by writing $\exists x.\ \phi[x]$, with the **existential quantifier**.

Bound variables. Quantified variables, like those in a definite integral,

$$\int_0^{2\pi} \sin x\ \mathrm{d}x \qquad \sum_{x=1}^{10} x^2 \qquad \exists x.\ \phi[x] \qquad \forall x.\ \phi[x] \qquad \lambda x.\ p(x)$$

have a special status. They are no longer available for substitution, and can be replaced by any other variable not already in use: $\exists x.\ \phi[x]$ is the same as $\exists y.\ \phi[y]$. In this new role, x is called a **bound variable**, where before it was *free*. This distinction was first clarified by Giuseppe Peano (1897). He invented the symbols $\in$, $\exists$ and many others in logic, and a language called *Latino sine flexione* or *Interlingua* in which to write his papers. The twin $\forall$ was added by Gerhard Gentzen in 1935.

Variables *link* together occurrences which are intended to be the same thing, and substitution specifies what that thing is. Variable-binding operations isolate this link from the outside world. The linking may be represented on paper by drawing lines instead (such as in Exercise 1.23) or in a machine by assigning to the variable an address that can *only* be accessed *via* the binding node.

DEFINITION 1·1·6:   The expressions $\exists x.\ \phi[x]$ and $\exists y.\ \phi[y]$ are said to be α-**equivalent**. In the latter, y has been substituted for each occurrence of x inside the quantifier, which now ranges over y. Although we often emphasise the dynamic nature of calculation, we shall treat these expressions as *the same*: there is no preferential choice of going from one to the other. Technically this presents no problems since it is decidable whether two given strings of symbols are α-equivalent. Nikolas de Bruijn devised a method of eliminating (the choice of names of) bound variables in favour of something more canonical (Exercise 2.24), but we shall not use it, as it is less readable by humans.

EXAMPLES 1·1·7:   The predicate $\exists x{:}\mathrm{N}.\ y = x^2$ says whether y is a square number. We may substitute a value (3) or another expression $(w + z)$ for y, to obtain the (false) proposition $\exists x{:}\mathrm{N}.\ 3 = x^2$ or binary predicate $\exists x{:}\mathrm{N}.\ (w + z) = x^2$. But we may not substitute a term involving x for y: $\exists x{:}\mathrm{N}.\ x = x^2$ is a true proposition, which is not a substitution instance of the original formula. Likewise, α-equivalence lets us substitute a new variable for the bound x: $\exists w{:}\mathrm{N}.\ y = w^2$ is the same predicate, but $\exists 3{:}\mathrm{N}.\ y = 3^2$ and $\exists (w + z){:}\mathrm{N}.\ y = (w + z)^2$ are nonsense, whilst $\exists y{:}\mathrm{N}.\ y = y^2$ is again a true proposition.

These difficulties can be overcome by the **strong variable convention**:

CONVENTION 1·1·8: We never use the same variable name twice in an expression. Where circumstances bring together two expressions with common variable names, we use α-equivalent forms instead.

Surely only an undisciplined programmer who couldn't be bothered to make a good choice of names would do otherwise? But no, although $\exists$ and $\forall$ respect it, β-reduction in the λ-calculus may cause duplication of sub-terms, and hence bound variables, so it may later substitute one abstraction within the scope of the other (Exercise 2.25). The convention is therefore a very naïve one, unsuitable for computation. (A weaker convention will be needed in Section 4.3, and variables must be repeated for different reasons in Sections 9.3 and 9.4.)

DEFINITION 1·1·9: When we use variable-binding operators we must therefore have available an **inexhaustible supply of variables**. This means that, whatever (finite) set of variables we're already using, we can always find a new one, *i.e.* one which is different from the rest. Exercise 6.50 examines what is needed to mechanise this. Some accounts of formal methods specify that there are *countably* many variables — even number them as $x_1, x_2, x_3, \ldots$ or $x, x', x'', \ldots$ — but one thing the formalism must not allow is to treat x_n as an expression containing a variable n.

In fact the x_1 written here is a *meta*-variable which stands for whatever actual variable may be desired in the application; it *may* stand for the same variable as x_2 unless we say otherwise. The difference between object-language and meta-language is important, but making it explicit can be taken too far. In this book we make make no systematic distinction between object-variables and the meta-variables used as placeholders for them, or, rather, we shall use x, y, z, x_1, x_2, *etc.* as *examples* of variable names (*cf.* Cardano's *examples* of equations).

In the Substitution Lemma and elsewhere we do need to know whether or not two symbols are meant to be *the same variable* (*intensional* or *by mention* equality); we write $x \not\equiv y$ if they have to be different symbols. Even in this case they may be assigned the same *value* (*extensional* or *by use* equality). On the other hand, we sometimes also need to constrain values, for example to say that $x \neq 0$ as a precondition for division.

Substitution and variable-binding. Bound variables complicate the formal definition of substitution, and it must be given simultaneously with the definitions of free variables and of α-equivalence. As before, we do this by structural recursion on the main term. If this is new to you then you should work through these definitions and prove the Substitution Lemma for yourself, maybe using the tree notation.

DEFINITION 1·1·10: By simultaneous structural recursion on t we define (i) the set, $\mathsf{FV}(t)$, of **free variables**, (ii) **substitution**, $[a/x]^*t$, of a term a for a variable x possibly occurring in it and (iii) α-**equivalence** with another term. To Definitions 1.1.3 and 1.1.4 we add that

(a) the constant c and variable x are each α-equivalent to themselves, but not to anything else;

(b) operation-symbols respect α-equivalence: if u is α-equivalent to u' and v to v', *etc.*, then $r(u, v, \ldots)$ is α-equivalent to $r(u', v', \ldots)$);

(c) if t is $\exists x.\, p$ then it is α-equivalent (to itself and) to any $\exists y.\, ([y/x]^*p)$ as long as $y \notin \mathsf{FV}(p)$, and similarly for $\forall$.

Hence

(d) if t is a quantified formula then (by the previous part, and using the inexhaustible supply of variables) we may assume that it is $\exists y.\, p$ where y is not the same variable as x, *and y does not occur in a*; then $\mathsf{FV}(t) = \mathsf{FV}(p) \setminus \{y\}$ and $[a/x]^*\exists y.\, p = \exists y.\, [a/x]^*p$. (Similarly with $\forall$.) In this sense, quantification respects α-equivalence.

Notice that α-substitution for $y \in \mathsf{FV}(a)$ may be needed within t in order to define $[a/x]^*t$. (Some authors say that "a is *free for x in t*" if there are no clashes between bound variables of t and free variables of a.) The Substitution Lemma remains valid, and, as a corollary, substitution respects α-equivalence.

Recall that the Lemma mentioned expressions in which the variable x *must not* occur freely, and we shall put the same condition on certain formulae in the rules for proving $\forall x.\, \phi[x]$ and using $\exists x.\, \phi[x]$ in Section 1.5. When this condition arises, it often means that at some point the same expression is also used *as if* the variable x really *did* occur in it.

NOTATION 1·1·11: For any term t with $x \notin \mathsf{FV}(t)$, we shall write

$$\hat{x}^*t$$

for the term t in which the variable x *is* considered to occur ("invisibly"). Like $[a/x]^*$, $\hat{x}^*$ is meta-notation, though in this case it does not actually alter the term. The use of $\hat{x}^*$ presupposes $x \notin \mathsf{FV}(t)$, *i.e.* that the term t came from a world in which x did not occur. In particular, $\hat{x}^*x$ and $\hat{x}^*\hat{x}^*t$ are not well formed terms.

The Notation conveys more than a negative listing of the free variables: it says that t is being *imported* into a world in which x is defined. Although this distinction seems trivial, making it and the passage back and forth between these two worlds explicit will help us considerably to understand the quantifiers in Chapter IX. What $\hat{x}^*$ means will become clearer when we use it in Sections 1.5, 2.3 and 4.3. The hat has been adopted from the abbreviation $1, \ldots, \hat{i}, \ldots, n$ for the sequence with i omitted in the theory of matrices and their determinants.

We define $\mathsf{FV}(\hat{x}^*t) = \mathsf{FV}(t) \cup \{x\}$.

PROPOSITION 1·1·12: (**Extended Substitution Lemma**) Let a, b, t, u and v be expressions, and x and y be distinct variables such that $x \notin \mathsf{FV}(a, u)$ and $y \notin \mathsf{FV}(a, b, u, v)$. Then

$$
\begin{aligned}
[a/x]^*[b/y]^*t &= \big[[a/x]^*b/y\big]^*[a/x]^*t \\
\hat{x}^*(\hat{y}^*u) &= \hat{y}^*(\hat{x}^*u) \\
[b/y]^*\hat{y}^*v &= v \\
[a/x]^*\hat{y}^*v &= \hat{y}^*[a/x]^*v \\
[x/y]^*\hat{x}^*[y/x]^*\hat{y}^*v &= v
\end{aligned}
$$

Moreover if t' is α-equivalent to t, u' to u and a' to a, then $[a'/x]^*t'$ and $\hat{x}^*u'$ are α-equivalent to $[a/x]^*t$ and $\hat{x}^*u$ respectively.

PROOF: To the proof of Lemma 1.1.5 we add the case $\exists z. \, s$, in which, without loss of generality (by α-equivalence), the bound variable is not x or y and does not occur in a or b. Then $[a/x]^*$, $[b/y]^*$, $\hat{x}^*$ and $\hat{y}^*$ commute with $\exists z$, and the induction hypothesis applies. Since they also commute with $[w/z]^*\hat{w}^*$, the bound variable z can be renamed. □

The Extended Substitution Lemma will be our point of departure for the semantic treatment of syntax in Sections 4.3 and 8.2. Algebras are the subject of Section 4.6, and *term* algebras of Chapter VI, which considers structural recursion in a novel and sophisticated way.

1.2   DENOTATION AND DESCRIPTION

The basic understanding of operations in an arithmetic expression is that they *transform* their arguments into results, *e.g.*

$$\sin 150° \rightsquigarrow \tfrac{1}{2}.$$

The two sides are not simply *equal*: we have to do a calculation, or at least consult a dusty table of trigonometric functions. In algebra we say

$$u \times (v + w) = (u \times v) + (u \times w);$$

one side may get us closer to the desired result than the other, but which it is depends on the situation.

EXAMPLE 1·2·1: A **rational number** may be represented as a fraction n/d with $d \neq 0$ in many ways. The synonyms are related, written

$$\langle n, d \rangle \sim \langle n', d' \rangle, \quad \text{if} \quad nd' = n'd.$$

There is a normal form, obtained by Euclid's algorithm (Example 6.4.3), in which n and d have no common factor. But there is no need to reduce everything to normal form at each step of a calculation, since the arithmetic operations respect the equivalence relation.

Similarly, a (positive, zero or negative) integer z may be coded as the set of pairs $\langle p, n \rangle$ of (zero or positive) natural numbers subject to

$$\langle p, n \rangle \sim \langle p', n' \rangle \quad \text{if} \quad p + n' = p' + n,$$

so $p - n = z$. The "obvious" way of coding integers, as absolute values together with signs, needs a case analysis to define addition. This is avoided by using equivalence classes, and we see that it is not necessary to use normal forms; for example you may have *both* savings *and* a loan with a particular bank.

In these two examples it is necessary to check that

$$x \sim x' \wedge y \sim y' \Rightarrow r(x, y) \sim r(x', y')$$

for each arithmetic operation $r = +, \times, -$, *etc.*, to show that these are well defined for the *values* (equivalence classes) of rationals and integers.

Platonism and Formalism. As there are syntactically different terms which are synonyms for the same value, the expressions themselves can never capture exactly what mathematical *values* are.

Various mathematical philosophies offer different views about what these eternal values are meant to be. One view is that the constants are tokens for real things like sheep and pebbles (Exercise 1.1) and the operations combine them. Our investigations are merely passive observations of an eternal and unchanging world, in which there is a true answer to every question, even though we may be (provably) unable to find it. This is known as **Platonism**.[4]

---

[4] I feel that this word does an injustice to Plato. In *The Republic* he argued that the objects of our experience are shadows of their pure Forms. For example a table is a shadow of the Form of Tableness. This seems remarkably close to a modern idea of genericity which, for example, we shall need in Section 1.5. The naïve view which we call Platonism derives more from the way in which the Theory of Forms is expressed — the Form of Tableness somehow inhabits a particular place in the Heavens — than from the theory itself.

Haskell Curry [Cur63, page 8] uses the word *contensivism* for this philosophy; the term *Platonism* is due to Paul Bernays [Ber35]. It was Frank Ramsey, a student of Russell, who described formalist mathematics as a "game" (1926), but he went on to say that *meta*mathematics, consisting of real assertions about which formulae can or cannot be proved, is *not* meaningless.

Although most mathematicians habitually think and speak in Platonist language, this philosophy is ridiculously naïve. It brings mathematics down to an experimental science, in which we can only infer laws such as the associativity of addition by *scientific induction* from the cases which have been observed. How can we know when all of the laws have been codified? Have they been asserted in excessive generality? We don't know. Bertrand Russell noticed that Gottlob Frege's use of comprehension was too general, and Kurt Gödel showed that Russell's own system (or anything like it) could express certain true facts about itself which it could not prove, unless it was inconsistent.

Formalism denies absolute being: only the symbols themselves exist, and nothing else. The notion of value must then be a derived one: it is defined as that which is common to all of the other expressions which are directly or indirectly related to the given one. What is common is simply their totality: the equivalence class.

Formalism is regarded by some as nihilism: if mathematics is just a game with symbols and arbitrary rules, what's the point of playing it? Chess is also a game with arbitrary rules played by many mathematicians, but to a chess master it has latent structure, a semantics. We hope to show that the rules of logic are not so arbitrary as those of chess, and exhibit many of the symmetries which mathematicians find beautiful in the world.

So long as we fix our sights on a reasonably small fragment of logic, there is a synthesis between the two points of view: out of the formalism itself we construct a world which has exactly the required properties. The free algebra for an algebraic theory is an example of such a world; it satisfies just those equations which are provable. Joachim Lambek has argued [LS86, p. 123] that it is possible to reconcile the Platonist and Formalist points of view.

Laws as reduction rules. Using the tree notation, the distributive law, seen as a way of transforming or computing with expressions, is

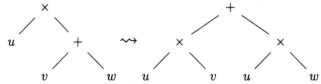

We distinguish between *equations*, which may or may not hold, and *laws*, which hold because (as in the legal sense) we choose to enforce them. In the λ-calculus the term **δ-rule** is used, and occasionally we shall call an *ad hoc* imposed equation a *relation* ("*regulation*"), though this term is normally reserved for another sense. Beware that the word *law* is also used elsewhere for what we call an *operation*, such as multiplication; this sense is archaic in English but current in French (*loi de multiplication*).

DEFINITION 1·2·2:   In a **law** the variables on the left hand side name sub-trees in a pattern to be matched. The variables on the right say where to copy these sub-trees (maybe several times, maybe not at all), so every variable occurring on the right must also occur on the left.[5]

Any (sub-)expression which matches the pattern on the left is called a **redex** (**red**ucible **ex**pression), and it **reduces** to the substituted expression on the right. A term may have many redexes in it.

The result of a "reduction" may be a *longer* expression than the first, and there may indeed be infinite computations. A sequence of terms in which each is obtained by replacing a redex in the previous one by the term to which it reduces is called a **reduction path**. We write $u \leadsto v$ if (u and v are identical expressions or) there is a reduction path whose first and last terms are u and v respectively.

A term with no redexes is said to be **irreducible**, but note that this only means that we have reached a dead end: by going backwards and following another reduction path some different result might be obtained.

Equivalence relations. For an arbitrary set of laws we have no guide to say whether a reduction gets nearer to a "result" or makes matters worse. In this case the equivalence class is the only notion of static value which we can give to an expression, and there is no systematic way of determining whether two expressions are equal.

DEFINITION 1·2·3:   **Reduction**, $\leadsto$, is the sparsest binary relation which contains reduction of redexes, respects substitution and is **reflexive** and **transitive**:

$$\frac{u \leadsto v}{[a/x]^* u \leadsto [a/x]^* v} \text{ pre-subs} \qquad \frac{}{u \leadsto u} \text{ reflex}$$

$$\frac{a \leadsto b}{[a/x]^* t \leadsto [b/x]^* t} \text{ post-subs} \qquad \frac{u \leadsto v \quad v \leadsto w}{u \leadsto w} \text{ trans}$$

Some authors, particularly when studying term-rewriting, use $\leadsto^*$ for the reflexive–transitive closure of $\leadsto$, after Russell's R_*. Another notation for this is $\twoheadrightarrow$, but we shall not use either $\leadsto^*$ or $\twoheadrightarrow$ in this book.

---

[5]If such a variable is repeated, the meaning of the pattern is that the corresponding sub-trees must be the same, for example in the η-rule for pairs, $\langle \pi_0(z), \pi_1(z) \rangle \leadsto z$ (Remark 2.2.2). The possibility that the copies of this sub-tree may undergo independent reductions may lead to failure of confluence.

Equivalence, $\sim$, is the least relation which is **symmetric** as well:

$$\frac{u \sim v}{v \sim u} \ \textsf{symm}$$

Any relation which is reflexive, symmetric and transitive is called an **equivalence relation**. These are essentially Euclid's Common Notions.

(We have just started using a way of expressing conditional assertions, *i.e.* rules of deduction, which will be very useful throughout the book. The ruled line means that whenever what is written above holds then what is written below follows, for the reason on the right.)

The notion of reduction *path* is an example of the **transitive closure** of a relation; it is reflexive if we include the trivial path (the equality or identity relation). Although any relation can be made symmetric by adding its converse, *i.e.* making the arrows bi-directional, this may destroy transitivity. To form the equivalence closure we need

LEMMA 1·2·4: Let $\rightarrow$ be any reflexive–transitive relation, viewed as an **oriented graph**. Then two nodes u and v are related by the smallest equivalence relation containing $\rightarrow$ iff there is a **zig-zag**, *i.e.* a sequence $u = u_0, u_1, u_2, \ldots, u_{2n} = v$ where

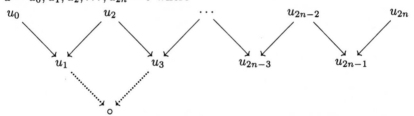

We say that u and v are in the same **connected component**, whatever the orientations of the arrows involved. $\qquad\square$

We shall discuss the transitive closure in Sections 3.8 and 6.4.

In a formal system of expressions and laws with no underlying Platonist meaning, it may be that there are so many indirect laws that the terms which we have chosen to name the numbers and truth values all turn out to be provably equal. This is known as **algebraic inconsistency**.

Confluence. This definition captures the **Church–Rosser Theorem**, which showed that the pure λ-calculus is consistent (Fact 2.3.3).

DEFINITION 1·2·5: A system of reduction rules is said to be **confluent** if, whenever there are reduction *paths* $t \rightsquigarrow u$ and $t \rightsquigarrow v$, there is some term w with reduction paths $u \rightsquigarrow w$ and $v \rightsquigarrow w$. Other names for this property are **diamond** and **amalgamation**.

What we prove in practice is **local confluence**, that any two *one-step* reductions may be brought back together. If it only takes one step on each side to do this then the following result still holds. Usually more than one step is needed, so the paths to be reconciled may just get longer.

Confluence is a property of the *presentation* of a system of rules. Donald Knuth and Peter Bendix [KB70] showed how a system of algebraic laws may *sometimes* be turned into a confluent system of reduction rules.

LEMMA 1·2·6: Suppose u and v are equivalent terms with respect to a transitive confluent relation $\rightsquigarrow$. Then $u \rightsquigarrow w$ and $v \rightsquigarrow w$ for some term w. In particular for each term there is *at most one* irreducible form to which it is equivalent, and distinct irreducible forms are inequivalent, so we call them **normal**.

PROOF: Confluence changes each "zag-zig" to a "zig-zag", as shown with dotted arrows in the diagram accompanying Lemma 1.2.4.   □

EXAMPLE 1·2·7: Soldiers in Lineland[6] face towards the right if they have even rank $(0, 2, 4, ...)$, whilst those of odd rank face left. A sequence of soldiers, which we write with semi-colons between them, is called a *parade* if each can see only those facing in the same direction and of the same or lower rank. Adjacent soldiers annihilate if they are facing each other and their rank differs by exactly 1 (*e.g.* $2 ; 3$ or $6 ; 5$). Otherwise, if a junior is facing an adjacent senior of either parity then they change places, but the senior loses two grades (*e.g.* $4 ; 7 \rightsquigarrow 5 ; 4$, $6 ; 8 \rightsquigarrow 6 ; 6$, $3 ; 1 \rightsquigarrow 1 ; 1$).

Since the total rank decreases, any conflict of forces terminates, and the result is a parade. An algebraist might take the parades alone as the elements of the structure, and seek to show that (;) is an associative operation. The term-rewriting approach would show that the reduction rules are confluent. In fact *exactly the same calculations* are needed either way, and we deduce that any conflict has a foregone conclusion. Hence parades form a monoid whose unit is the empty sequence; we shall use it in Example 7.1.9.   □

DEFINITION 1·2·8: A system of reduction rules is **weakly normalising** if every term has *some* reduction path which leads to an irreducible form. It is **strongly normalising** if there is no infinite reduction path.[7]

THEOREM 1·2·9: In a normalising confluent system of reduction rules, each term is equivalent to exactly one normal form, and *can* be reduced

---

[6] After *Flatland* by Edwin Abbott Abbott, 1884.

[7] That is, (the opposite of) the reduction relation is well founded: see Section 2.5.

to it. If the system is *strongly* normalising then *local* confluence suffices, and *every* reduction path leads to the normal form.   □

Normal forms, where they exist, provide a versatile tool. Suppose that some formula is provable in a logic whose proofs have this property; then (without being given a proof) we know that any such proof must be of a particular form, involving certain other formulae which are also provable. This observation can be used to prove a powerful result about $\lor$ and $\exists$ in intuitionistic logic (Remark 2.4.9 and Section 7.7).

There are idioms in the English language which exploit uniqueness, such as that provided by confluence.

Theory of Descriptions. In the vernacular we speak of *the* Moon, using the **definite article**, because Earth has only one. By contrast, Ganymede is **a** satellite of Jupiter: the **indefinite article** is used since there are several others. Similarly, when there is *exactly one* term v which is normal and equivalent to u, we call it *the* normal form of u.

DEFINITION 1·2·10: Let $\phi[x]$ be a predicate of one argument. Then

(a) ϕ is a **description** if it is satisfied by at most one thing. The best way to say this is that *any* ($\forall$) *two solutions are equal* or, in symbols,

$$\forall x, y. \ \frac{\phi[x] \qquad \phi[y]}{x = y} \ \text{descr}$$

(b) a description ϕ **denotes** if something ($\exists$) satisfies it:

$$\exists x. \ \phi[x].$$

In this case we may speak of *the* solution, *i.e. what* the description describes or denotes. The notation

$$\imath x. \ \phi[x]$$

is sometimes used for this solution. We write $\exists! x. \ \phi[x]$ for unique existence, the conjunction of $\exists x. \ \phi[x]$ and $\forall x, y. \ \phi[x] \land \phi[y] \Rightarrow x = y$. Some equivalent forms will be given in Exercise 1.8.

It is convenient to extend usage to descriptions which are not known to denote, that is, to a general situation in which we would recognise a widget if we met one and know that there cannot be two which are different, but where there need not actually be any. Then we may say

"the widget is grue"

to mean that, should we ever find ourselves in possession of a widget, it will necessarily be found to be grue, where grueness ($\psi[x]$) is any predicate, not necessarily another description. In symbols,

$$\forall x. \ \phi[x] \Rightarrow \psi[x] \qquad \text{where } \forall x, y. \ \phi[x] \land \phi[y] \Rightarrow x = y.$$

As a further abuse,

"the widget (if it exists) is unique"

means that the predicate ϕ characterising widgets is a description.

"The" and "unique" must be used with predicates: it is meaningless to say that a *thing* is unique, whatever abuses advertisers may make of the English[8] language. It is similarly wrong to introduce *putative* things such as unicorns and then treat existence as a *property* of them.

LEMMA 1·2·11: Let ϕ be a description and ψ any predicate. Then

if $\exists x.\ \phi[x] \wedge \psi[x]$ then $\forall x.\ \phi[x] \Rightarrow \psi[x]$

and conversely if ϕ denotes; in the latter case we write

$\psi\big[\imath x.\ \phi[x]\big].$

That is, if we have a grue widget, then we know that all widgets are necessarily grue.

PROOF:

$$\cfrac{\psi[x] \qquad \cfrac{\phi[x] \qquad \phi[y]}{x = y}\ \text{descr}}{\psi[y]}\ \text{subs} \qquad \qquad \square$$

The $\imath$-**calculus** notation had been in use in the late nineteenth century, but it was Bertrand Russell who sorted out its meaning. A description such as "the author of *Waverley*" is not a name — it is *incomplete*, having no meaning until it is embedded in a sentence such as "Scott is the author of *Waverley*" — and may be contingent ("the President"). Unfortunately some logic texts to this day assign an arbitrary value such as zero to the two meaningless cases, leading to nonsense like saying that

the unicorn *is* the author of *Principia Mathematica*

is *true* since $0 = 0$ (and this book had two authors).

Synonyms. The theory of descriptions puts a premium on uniqueness, but how can we reconcile this with the stress we put before on the *many* concrete forms which a mathematical object such as 5 may take? One way is to appoint either the equivalence class (*all* quintuples) or a normal form ($\{0, 1, 2, 3, 4\}$) to be what the object *is*. The former is objectionable

---

[8]For the benefit of those readers whose mother tongue is (perhaps Serbo-Croat or Polish but) not English, it should be stressed that we're discussing issues of logic, not English grammar, in which the word *unique* does behave as an adjective.

In fact existence and uniqueness *are* predicates, in second order logic: they are properties of predicates, not of things (Section 2.8).

Giuseppe Peano wrote ιa for our $\{a\}$ and $\bar{\iota}X = a$ if $X = \iota a$, *i.e.* X is a description. Russell and Whitehead turned $\bar{\iota}$ into $\imath$. (ι is the Greek letter iota.)

because it introduces extraneous material (the protons in a Boron atom, as well as the fingers on one hand), and indeed a proper class of it, whilst Example 1.2.1 shows that the latter is unnecessary. As Pál Halmos [Hal60] points out, we look to physics for the way standards are defined: when we want to measure a kilogram, we *compare* it with a particular platinum–iridium bar in France. See [Ben64] for discussion.

The deeper we dig into foundations, to find out what things *do*, the less sure we are about what they *are*: we must be content with knowing how to exchange *equals for equals*.

DEFINITION 1·2·12:   A binary relation $\sim$ is a **congruence** if it is an equivalence relation (reflexive, symmetric and transitive) and permits the substitution of equals for equals in any predicate ϕ:

$$\frac{\phi[a] \qquad a \sim b}{\phi[b]} \text{ subs}$$

If $\sim$ is the equivalence relation generated by reduction ($\leadsto$), it suffices to verify this rule for $a \leadsto b$ and $b \leadsto a$ (with $w = \phi[x]$, this is post-subs in Definition 1.2.3), and this property is known as **subject reduction**.

The doctrine of interchangeability does not allow us to test for equality *in substance*, and so the "=" sign in Definition 1.2.10 and the proof of Lemma 1.2.11 must be replaced by congruence.

So any two things satisfying the description have to be *congruent* rather than equal. Conversely, the congruence law says that anything which is equivalent to something satisfying the description also satisfies it.

A description specifies *uniquely up to interchangeability*.

The idea is that two things are *the same* if they have all properties in common, *cf.* Leibniz' principle, Proposition 2.8.7. Finally, testing a property for any one representative suffices, by the congruence law again, to prove it for all of them.

For terms denoting *individual* values these remarks are maybe academic, but significant technical issues arise when we turn to *structures*. Then the congruence is not just a property but a *method* of passing from one representation to another and back (*isomorphism*). Now the admissible properties are those which are invariant with respect to isomorphism. We often say things like "the product is unique" or "the quotient is compact", to be understood in the above senses.

Since a structure may have non-trivial isomorphisms *with itself*, few interesting properties survive indiscriminate isomorphisms; for example all of the points on a sphere become identified. There are two isomorphisms between my shoes and my feet, but choosing the wrong one is rather painful! What isomorphisms there are with my *socks* is a significant one geometrically, as a non-trivial *group* is involved (Example 6.6.7). In order to express what we have to say about a structure we must *follow the parts of the structure through the transformations.*

Therefore mathematical objects are defined **up to interchangeability**, but we must pay attention to the **means of exchange**: isomorphisms of objects and equivalences of categories (Definition 4.8.9). Opinions differ on how to handle this issue, for example Michael Makkai [Mak96] has taken an extreme semantic point of view in which equality of objects is unthinkable. The syntactic position is that we only deal with *names* of types, which can therefore be compared. In Section 7.6 we shall show that this conflict can be resolved, by using interchangeability of *categories* to restore equality of *objects.*

We shall explore the relationship between the formal and vernacular ways of expressing mathematics further in Section 1.6. The idea of interchangeability and the means of exchange will be used in Section 4.4. Algebras with laws are the subject of Section 4.6, and Section 5.6 treats equivalence relations. Now we turn to *parametric* descriptions.

1.3   Functions and Relations

For Leonhard Euler (1748) and most mathematicians up to the end of the nineteenth century, a function was an expression formed using the arithmetical operations and transcendental operations such as log. The modern *infor*matician would take a similar view, but would be more precise about the method of formation (algorithm). Two such functions are equal if this can be shown from the laws they are known to obey.

However, during the twentieth century mathematics students have been taught that a function is a set of input–output pairs. The only condition is that for each input value there *exists*, somehow, an output value, which is *unique*. This is the **graph** of the function: plotting output values in the y-direction against arguments along the x-axis, forgetting the algorithm. Now two functions are equal if they have the same output value for each input. (This definition was proposed by Peter Lejeune Dirichlet in 1829, but until 1870 it was thought to be far too general to be useful.)

These definitions capture the *intension* and the effect (*extension*) of a function. Evaluation takes us from the first to the second, but it doesn't say what non-terminating programs do during their execution, and can't distinguish between algorithms for the same function. But each view is both pragmatic and entrenched, so how can this basic philosophical clash ever be resolved? Chapter IV begins the construction of semantic models which recapture the intension extensionally, as part of our reconciliation of Formalism and Platonism.

DEFINITION 1·3·1:

(a) A *binary relation* is a predicate[9] in two variables x and y; we shall write it variously as $R[x, y]$, $x \overset{R}{\mapsto} y$, $R : x \mapsto y$ or xRy.

Such a relation is called

(b) a *functional* or *single-valued relation* or a *partial function* if for each x it is a *description* of y (Definition 1.2.10(a)), *i.e.* for all y_1 and y_2,

$$\frac{x \overset{R}{\mapsto} y_1 \qquad x \overset{R}{\mapsto} y_2}{y_1 = y_2} \text{ func}$$

(c) a *total functional relation*, or just a *function*, if also for each x, it *denotes*, *i.e.* there is in fact some y with $x \overset{R}{\mapsto} y$ (Definition 1.2.10(b)).

Functional relations are more familiarly called f instead of R, and in this case we write "$f(a) = b$" for $f : a \mapsto b$ or $b = \imath y. a \overset{R}{\mapsto} y$. The notation of function-application, like the definite article, implicitly means that the result is uniquely determined and (usually) that it exists.

A relation which satisfies the existence but not necessarily uniqueness axiom for a function is said to be *entire*. Nothing really remains of the functional idea, but the axiom of choice (Definition 1.8.8) says that such a relation, considered as a set of pairs, contains a function.

On the other hand, single-valuedness alone *is* important. It is neither possible nor desirable to require all programs to terminate, but those of mathematical interest can typically be calculated in some manifestly deterministic way. For term-rewriting systems (including λ-calculi), confluence is a commoner property than normalisation. So *partial* functions are the norm, and will be considered in Sections 3.3, 5.3, 6.3 and 6.4.

REMARK 1·3·2: When equality has to be weakened to interchangeability, the functional property becomes that $x \sim x' \Rightarrow f(x) \sim f(x')$ or

$$(x \overset{R}{\mapsto} y) \wedge (x \sim x') \wedge (x' \overset{R}{\mapsto} y') \Rightarrow (y \sim y'),$$

i.e. functions preserve congruence (the means of exchange).

---

[9]You may say "a set of pairs" if you wish, but this wouldn't make any difference, as Definition 2.2.3 makes "subset" no more nor less than a *synonym* for predicate or relation. Keeping this in mind will help you to understand the remarks about subsets where intuitionism seems to differ from classical preconceptions. As a predicate can only be given using the calculus of the next section, both notions of function are, for us, syntactic; two functions are equal if there is a syntactic proof which says so.

Arity, source and target. In this book we shall take the view that

> for each operation or predicate,
> some things may meaningfully be its subject,
> but applying it to anything else yields nonsense.

Although this seems like common sense, it surprises me how readily this principle is dropped when people try to reason about language.

In Chapter II we shall provide ways of forming new types, such as $\mathcal{P}(X)$, $X \times Y$, Y^X and $\mathsf{List}(X)$, but for the time being they are fixed in advance; in this case we often say "sort" instead of (base) type.

NOTATION 1·3·3:   We write $x : X$ and $c : X$ to express the syntactic information that the variable x or constant c is declared to have type X. For each operation-symbol r we must specify not only the type of its result but also those of each argument. We sum up this information as

$$X_1, X_2, \ldots \vdash r : Y$$

and then we impose the ***type discipline***:

> we may only form the expression $r(u_1, u_2, \ldots)$
> if $u_1 : X_1$, $u_2 : X_2$, ...
> and then, by definition, $r(u_1, u_2, \ldots) : Y$.

The list $\underset{\sim}{X}$ of input types is called the ***arity*** of r. Types, like predicates (Definition 1.2.12), must be invariant under ***subject reduction***:

> if $u : X$ and $u \rightsquigarrow v$ (or $u \sim v$) then $v : X$.

Type information can be presented in a graphically immediate way by means of "commutative" diagrams, which we introduce in Section 4.2.

The symbol $\in$ is often used instead of the colon, but this can lead to confusion with the axiom of comprehension (Definition 2.2.3), *i.e.* that the value x satisfies the predicate defining a *sub*set X (Exercise 1.12).

NOTATION 1·3·4:   The types of the variables in Definition 1.3.1 are called the ***source*** $x : X$ and ***target*** $y : Y$. We regard them as an inseparable part of the definition, and indicate them by arrows:

(a) $R : X \rightharpoonup Y$ for a binary relation (this symbol is new),

(b) $R : X \rightharpoonup Y$ for a functional relation (partial function), and

(c) $R : X \to Y$ for a total functional relation (function).

The words ***domain*** and ***codomain*** are more usual in category theory, but we shall avoid them because of confusion with Section 3.4. We also avoid the word *range* because usage is ambiguous as to whether it means Y or the set of outputs which actually arise from some input,

$$\{y \mid \exists x.\ x \overset{R}{\mapsto} y\},$$

which we call the **image**. Again, the word *image* is sometimes used for the value of a function at a particular element, but we shall always use it in the above way as the *collection* of values taken over a set, ignoring repetition. We shall use the word **range** in another sense, for the type of the bound variable of a quantifier ($\forall x$, $\exists x$). An **endofunction** is one whose source and target are the same, *i.e.* a "loop" $\circlearrowleft$.

Semantics.

REMARK 1·3·5: Besides notation and discipline, types also *internalise* values, which need not have names. For example there are (in a classical understanding) far more irrational numbers than we can name in finitely many symbols, but a function "on $\mathbb{R}$" is meant to be defined for all numbers, not just those with names. Even for the natural numbers, where each value does have a name, the symbol $\mathbb{N}$ brings the **completed infinity** of numbers into the discussion.

Theorem 1.2.9 relates terms to total functional relations:

LEMMA 1·3·6: Let t be a term of type Y with a free variable x of type X. Then[10]

$$x \overset{R}{\mapsto} y \overset{\text{def}}{\iff} y = t$$

is a total functional relation from X to Y.

PROOF: The notation is deceptively simple, so we must first clarify its meaning. The term "t" belongs to the intensional syntax and as such may involve the variable x, which is also understood syntactically. The other graphical symbols belong to the extensional semantics. Therefore to interpret the formula "$y = t$" we must convert t from the syntax to the semantics, by substituting a term representing the value x for the variable x wherever it occurs in t, and then evaluating the result.

We may regard the types as the sets of values, where the values may be equivalence classes of terms, or normal forms. If y_1 and y_2 are two values which are both equal to the value of t then they must be equal to each other (it is the confluence property that allows us to use normal forms here instead of equivalence classes), so R is functional. It is total because the value of t itself witnesses $\exists y.\ R : x \mapsto y$, although we may choose to say instead that only those equivalence classes which have normal forms are to be treated as "defined" values. □

---

[10]This is our first encounter with the covariant regular representation, otherwise known as the Yoneda Lemma.

Substitution of another term for a variable in a term is given by

DEFINITION 1·3·7:   Let $R : X \rightharpoonup Y$ and $S : Y \rightharpoonup Z$. Then the **relational composition** is given by

$$x \xrightarrow{R;S} z \text{ or } x \xrightarrow{SoR} z \text{ if } \exists y.\, x \xrightarrow{R} y \wedge y \xrightarrow{S} z.$$

We use these two notations synonymously. The semicolon was used in this sense, for left-to-right composition, by Ernst Schröder in 1895. Today it is used for sequential composition in imperative programming languages (Definition 4.3.1). The **identity relation** id_X is the same as equality on X. It is also called the **diagonal relation** (Δ) because when its values are written out in a square table the entries on the diagonal are true and the others false (Exercise 1.18).

LEMMA 1·3·8:   If R and S are the (total functional) relations which correspond to terms $v : Y$ and $w : Z$, each having one free variable $x : X$ and $y : Y$ respectively, then $S \circ R$ corresponds to $[v/y]^* w$. Also, the diagonal relation corresponds to the variable $x : X$ considered as a term.

PROOF: $t = [v/y]^* w$ iff $\exists y.\, t = w \wedge y = v$.   □

Composition preserves functionality and totality, but we postpone the proof to Lemma 1.6.6 for reasons of exposition.

Relational calculus. The definition of a (total) functional relation is not symmetrical in X and Y, so we can ask what happens if we interchange the roles of the variables in the conditions. Of course what we are then considering is

DEFINITION 1·3·9:   The **converse relation** has $y \xrightarrow{R^{\mathrm{op}}} x \iff x \xrightarrow{R} y$. Its source is now Y and its target X.

DEFINITION 1·3·10:   A function or functional relation R is

(a) **injective** or **1-1** if $x_1 \xrightarrow{R} y \wedge x_2 \xrightarrow{R} y \Rightarrow x_1 = x_2$, i.e. R^{op} is also functional; we write $R : X \hookrightarrow Y$ or $R : X \rightarrowtail Y$;

(b) **surjective** or **onto** if $\forall y.\, \exists x.\, x \xrightarrow{R} y$, i.e. R^{op} is entire; we write $R : X \twoheadrightarrow Y$;

(c) **bijective** if both R and R^{op} are total functional relations.

These properties are examined further in Exercises 1.14–1.16.

Bijectivity can be characterised purely in terms of composition:

LEMMA 1·3·11: The following are equivalent for a function $f : X \to Y$:

(a) f, considered as a functional relation, is bijective;

(b) f is (total,) injective and surjective;

(c) there is a function $g : Y \to X$ such that $g \circ f = \mathsf{id}_X$ and $f \circ g = \mathsf{id}_Y$.

Moreover in the last case g, which we call the **inverse**, f^{-1}, is unique and is given by f^{op}. When f has an inverse we call it an **isomorphism** and write $f : X \cong Y$. (An isomorphism whose source and target are the same type is called an **automorphism** of that type.) Beware that, when there is other structure, a bijection is not necessarily an isomorphism (Example 3.1.6(e)).   □

There is a common situation in which just one of these laws holds:

DEFINITION 1·3·12: An endofunction $e : X \to X$ is called **idempotent** if $e \circ e = e$. In this case, x is in the image of e iff it is fixed by e.

$$
A = \{x \mid e(x) = x\} \underset{q}{\overset{i}{\underset{\longleftarrow}{\lhook\joinrel\longrightarrow}}} X
$$

The inclusion i into and the surjection q onto the set of such points are said to **split** the idempotent; they satisfy $i \,;\, q = \mathsf{id}_A$ and $q \,;\, i = e$. The functions i and q are called respectively **split mono** and **split epi**. The set A is said to be a **retract** of X (sometimes written $A \lhd X$) and i a **section** or **pre-inverse** of q.

Chapter II will begin the study of types, concentrating on functions in Section 2.3. Composition is the basis of category theory, beginning in Chapter IV; in particular Remark 4.4.7 considers isomorphisms. The relational calculus will be discussed further in Sections 3.8, 5.8 and 6.4. We shall now turn to the symbols $\Rightarrow$, $\wedge$, $\forall$ and $\exists$ which we have just started using.

1.4   DIRECT REASONING

To the first approximation, a proof is a sequence of assertions, each justified by the earlier ones. We don't actually perceive the *truth* of the later assertions by *themselves*, but only accept them because, like accountancy, they follow necessarily from what's gone before. David Hilbert proposed that this style be used systematically in mathematical logic. We shall need to modify it in the next section, but those proofs which can be written like this we call **direct deductions**.

The style has been traditional[11] in editions of Euclid's *Elements* since the invention of printing:

> First we write down the hypotheses, each on a separate line and annotated as such. Then we write any formulae we wish to derive from them (and from formulae which have been derived already), noting the names of the rules and hypotheses used.

Each step is introduced by ∴ (***therefore***) and its reason by ∵ (***because***). The *reason* cites one of a small number of **rules of inference** and some previously asserted formulae (the **premises** of the rule). In presenting the rules we shall employ the ruled line notation used in Definition 1.2.3.

Each one of the rules makes an appearance somewhere in ancient or medieval logic, long before Boole. So they all have Latin names, which shed no light at all on their structure (even if you can read Latin). As Russell commented, "Mathematics and Logic, historically speaking, have been entirely different studies. Mathematics has been connected with Science, Logic with Greek." The only Latin name in common use in mathematics is **modus ponens**, which we call $(\Rightarrow \mathcal{E})$. The symbolic names reflect the symmetries of logic.

The language of predicate calculus.

DEFINITION 1·4·1:   ***Formulae*** are built up as follows:

(a) ***Atomic predicates*** $\rho[u_1, \ldots, u_k]$ where u_i is a term of type X_i. The number k is called the **arity**. Predicates for which $k = 0$ are usually called **propositions**; in particular there are constants

$\top$ (true)   and $\bot$ (false).

If $k \geqslant 2$ we speak of **relations**, as in the last section. For example we have the binary relations of equality, $u = v$, order, $u \leqslant v$, and membership, $u \in w$ (Definition 2.2.5), where $u, v : X$ and $w : \mathcal{P}(X)$.

(b) If ϕ and ψ are formulae then so are

$\phi \wedge \psi$ (and), $\phi \vee \psi$ (or)   and $\phi \Rightarrow \psi$ (implies);

$\neg \phi$ (not) and $\phi \Leftrightarrow \psi$ (equivalent) are abbreviations for $\phi \Rightarrow \bot$ and $(\phi \Rightarrow \psi) \wedge (\psi \Rightarrow \phi)$ respectively. The arrow should be read "in so far as ϕ (the **antecedent**) holds, then so does ψ (the **consequent**)". These symbols are collectively known as **connectives**.

---

[11]Notable examples are Pierre Hérigone's highly symbolic *Cursus Mathematicus* in Latin and French (1644), and Thomas Brancker's English translation of Johann Rahn's Dutch *Algebra* "much altered and augmented" by John Pell (1668). Maximus Planudes, who worked in Constantinople in the first half of the fourteenth century, put each step of the manipulation on a separate line in his edition of Diophantus' *Arithmetica*, whereas Diophantus himself had written the equations in running text.

(c) If ϕ is a formula and x a variable of type X then

$$\exists x{:}X.\ \phi\ (\textbf{\textit{for some}}) \qquad \text{and} \qquad \forall x{:}X.\ \phi\ (\textbf{\textit{for all}})$$

are formulae. The symbols $\forall$ and $\exists$, which are called **quantifiers**, bind variables in the way that we described in Definition 1.1.6ff.

The phrase "there exists" will be discussed in Remarks 1.6.2(f) and 1.6.5.

The direct logical rules.

DEFINITION 1·4·2: The **direct logical rules** are

$$\frac{}{\top}\,\top\mathcal{I} \qquad \frac{\phi\wedge\psi}{\phi}\,\wedge\mathcal{E}0 \qquad \frac{\phi\wedge\psi}{\psi}\,\wedge\mathcal{E}1 \qquad \frac{\forall x.\ \phi[x]}{\phi[a]}\,\forall\mathcal{E}$$

$$\frac{\bot}{\vartheta}\,\bot\mathcal{E} \qquad \frac{\phi}{\phi\vee\psi}\,\vee\mathcal{I}0 \qquad \frac{\psi}{\phi\vee\psi}\,\vee\mathcal{I}1 \qquad \frac{\phi[a]}{\exists x.\ \phi[x]}\,\exists\mathcal{I}$$

In the last column, x is a variable (which, by α-equivalence, may be assumed not to occur elsewhere) and a any term of the same type. This is called an **instance** of $\forall x.\ \phi[x]$ or a **witness** of $\exists x.\ \phi[x]$. Using the substitution notation (Definition 1.1.10ff), these two rules become

$$\frac{\forall x.\ \phi}{[a/x]^*\phi}\,\forall\mathcal{E} \qquad \frac{[a/x]^*\phi}{\exists x.\ \phi}\,\exists\mathcal{I}$$

to which we shall return in Remark 1.5.5.

REMARK 1·4·3: The names of the rules specify the connective involved — that is, the *outermost* connective (Remark 1.1.1), as the formulae ϕ and ψ may themselves involve connectives. The $\mathcal{I}$ stands for an **introduction rule**, *i.e.* where the formula in the conclusion is obtained from that in the premise by *adding* the connective in question. Similarly $\mathcal{E}$ indicates an **elimination rule**, where the connective has been *deleted*. We employ the introduction rule to give a *reason* for the formula, and the elimination rule to derive *consequences*[12] from it.

Notice how the elimination rules for $\wedge$ and $\forall$ mirror the introduction rules for $\vee$ and $\exists$. ($\top\mathcal{I}$) and ($\bot\mathcal{E}$) and, to some extent, the other rules also match up, although the duality is seen more strictly in classical logic, Theorem 1.8.3. The other three direct logical rules each have two premises, so cannot have mirror images.

$$\frac{\phi \quad \psi}{\phi\wedge\psi}\,\wedge\mathcal{I} \qquad \frac{\phi \quad \phi\Rightarrow\psi}{\psi}\,\Rightarrow\mathcal{E} \qquad \frac{\phi \quad \neg\phi}{\bot}\,\neg\mathcal{E}$$

For negation, ($\neg\mathcal{E}$) is the special case of ($\Rightarrow\mathcal{E}$) where $\psi = \bot$.

---

[12]For us, a "consequence" is the result of a *proof*, but some authors use this word for the semantic notion which we call *entailment*, Remark 1.6.13.

REMARK 1·4·4:   The $(\bot\mathcal{E})$-rule comes as a surprise to novices. It is like playing the *joker* in a game of cards: $\bot$ stands for any formula you like. The strategy usually disregards this card (*minimal logic* leaves it out), but it can sometimes save the day, *e.g.* in Remark 1.6.9(a).

REMARK 1·4·5:   There are two $(\wedge\mathcal{E})$- and $(\vee\mathcal{I})$-rules, whilst $(\wedge\mathcal{I})$ has two premises and $(\vee\mathcal{E})$ has two sub-boxes (Definition 1.5.1), since these connectives are binary. The nullary $(\top, \bot)$ and infinitary $(\forall, \exists)$ versions follow the same pattern: in particular, there is *no* $(\top\mathcal{E})$- or $(\bot\mathcal{I})$-rule.

Since the $(\wedge\mathcal{I})$, $(\Rightarrow\mathcal{E})$- and $(\neg\mathcal{E})$-rules, together with transitivity (1.2.3), description (1.2.10(a)) and congruence (1.2.12) have two premises, the ancestry of a deduction is not linear but tree-like, *e.g.* Lemma 1.2.11. Proofs, like expressions, involve binary operation-symbols, which we shall come to recognise as pairing for $(\wedge\mathcal{I})$ and evaluation for $(\Rightarrow\mathcal{E})$.

Whilst the tree style shows more clearly the roles played by individual hypotheses in a deduction, it can repeat large parts of the text when a derived formula (*lemma*) is used more than once as a subsequent premise, as in induction (Remark 2.5.12). Big sub-expressions also get repeated in algebraic manipulation, but this can be avoided by use of declarative programming (Definition 1.6.8 and Section 4.3).

The provability relation. The presentation of a proof as a chain or tree of assertions is very convenient when the aim is to show that some result in mathematics is true. But from the point of view of logic *per se*, we need a notation which says that "there is a proof of ϑ from hypotheses $\phi_1,\ldots,\phi_n$". This list is to be understood conjunctively $(\phi_1 \wedge \cdots \wedge \phi_n)$; it is called the **context** and will be denoted by Γ. (Such contexts must not be confused with "context-free languages", Example 4.6.3(d).) The provability assertion is written $\Gamma \vdash \vartheta$ and is called a **sequent**.

REMARK 1·4·6:   Provability is an *inequality* $\phi \vdash \vartheta$ on formulae, whereas reduction rules (Section 1.2) defined when two expressions were *equal*. Ernst Schröder observed that the inequality is more natural in logic. Formulae are then equal if both $\phi \vdash \vartheta$ and $\vartheta \vdash \phi$, abbreviated to $\phi \dashv\vdash \vartheta$.

Beware that $\phi_1, \phi_2 \vdash \vartheta_1, \vartheta_2$ means $\phi_1 \wedge \phi_2 \vdash \vartheta_1 \vee \vartheta_2$ in Gentzen's *classical* sequent calculus, but for us *the comma means conjunction on both sides*.

The structure consisting of inter-provability classes (equivalence classes under $\dashv\vdash$) of formulae is called the **Lindenbaum algebra**.[13] In a much

---

[13]Alfred Tarski may well have formulated the idea first, but he went on to contribute plenty of other ideas to logic after the war, whereas Adolf Lindenbaum died in Białystok concentration camp in 1941. The Lindenbaum algebra is the second occurrence of the regular representation, *alias* the Yoneda Lemma.

more general form (the classifying category or category of contexts and substitutions, Section 4.3) it will be the major object of study in this book; the weakening rule will play a crucial role in our construction.

Sequent presentation. Whereas the rules of natural deduction say

> if ϕ is true then ϑ is also true,

those of the sequent calculus have the form

> if ϕ is provable from Γ then ϑ is provable from Δ,

where Γ and Δ are contexts.

DEFINITION 1·4·7: The provability relation, written using the turnstile $\vdash$ (which comes from Frege, 1879), is generated by three classes of rules:

(a) the structural rules, which govern the way the formulae move around the proof,

(b) the logical rules, which determine the way in which the connectives and quantifiers behave, and

(c) the non-logical rules, which relate to symbols in the object-language.

For the logical rules we may make other distinctions: between the direct rules we have given and the indirect rules of the next section (using temporary hypotheses, which must be delimited somehow), and between introduction and elimination rules.

DEFINITION 1·4·8: The **structural rules** are the **identity axiom, cut, exchange, weakening** and **contraction**:

$$\frac{}{\phi \vdash \phi}\ \text{identity} \qquad \frac{\Gamma \vdash \alpha \quad \alpha, \Psi \vdash \vartheta}{\Gamma, \Psi \vdash \vartheta}\ \text{cut} \qquad \frac{\Gamma, \Psi \vdash \vartheta}{\Gamma, \alpha, \Psi \vdash \vartheta}\ \text{weaken}$$

$$\frac{\Gamma, \alpha, \alpha, \Psi \vdash \vartheta}{\Gamma, \alpha, \Psi \vdash \vartheta}\ \text{contract} \qquad \frac{\Gamma, \alpha, \beta, \Psi \vdash \vartheta}{\Gamma, \beta, \alpha, \Psi \vdash \vartheta}\ \text{exchange}$$

The identity axiom and cut rule show straight away that the $\vdash$ relation is reflexive and transitive (a *preorder*, Definition 3.1.1).

It is precisely the exchange and contraction rules which enable us to treat Γ as a *set* instead of a *list* (indeed in Section 8.2 we shall discuss a similar set of rules for type theory in which exchange is not generally valid, and there Γ must be considered as a list). The cut rule allows us to delete those hypotheses which are derivable from others; the exchange and contraction rules mean that the context of its conclusion is just $\Gamma \cup \Psi$.

The weakening rule says that hypotheses may be added to the context. If we allow weakening by an arbitrary set of hypotheses we can give a meaning to an infinite context: something may be proved in it iff it may be proved in some finite subset (though we shall not take this up).

REMARK 1·4·9:   There are several ways of presenting the direct logical rules in sequent form, *cf.* operation-symbols applied to either variables or terms (Remark 1.1.2). For example ($\wedge\mathcal{E}0$) may be written as any of

$$\phi \wedge \psi \vdash \phi \qquad \frac{\Gamma \vdash \phi \wedge \psi}{\Gamma \vdash \phi} \qquad \frac{\Gamma, \phi \vdash \vartheta}{\Gamma, \phi \wedge \psi \vdash \vartheta}$$

The second and third are obtained from the first by using Cut on the conclusion and premise respectively. The moral of this is that, even when components of two systems perform the same function, the equivalence may rely on more primitive conditions.

Gerhard Gentzen [Gen35] used the third form in his sequent rules for intuitionistic logic. His Hauptsatz (German: main theorem) was that anything that can be proved in the sequent calculus is provable without the Cut rule: cut-free proofs are normal forms. All of his rules apart from cut have the **sub-formula property**: the formulae used in the premises are sub-formulae of those in the conclusion, so *controlling the search for a proof*. This proves consistency, but by using an induction principle which is stronger than the calculi under study.

Cut is redundant in Gentzen's calculus because he saturated the other rules with respect to it. This makes it a very cumbersome notation for justifying theorems, since the context must be copied from one line to the next, usually without change.

1.5   PROOF BOXES

The direct rules cannot prove $\phi \Rightarrow \psi$ or $\forall x.\ \phi[x]$: the best they can offer is a hypothetical argument or a *scheme* of proofs. Nor can they exploit $\exists x.\ \phi[x]$, although they can of course make use of $\phi[a]$ once a is given.

The ($\forall\mathcal{E}$)-rule says that all instances $\phi[a]$ obtainable by substituting terms hold, *and it says no more than this*. But by $\forall x{:}X.\ \phi[x]$ we intend $\phi[x]$ to be true throughout the type X (*cf.* Remark 1.3.5), not just for those values a with names which can be used for substitution.

Universal quantification over numbers (and not just a scheme[14] ranging over numbers) is needed to formulate the principle of induction,

$$\vartheta[0] \wedge \left(\underline{\forall n{:}\underline{\mathbb{N}}.\ \vartheta[n]} \Rightarrow \vartheta[n+1] \right) \Rightarrow \forall n{:}\mathbb{N}.\ \vartheta[n].$$

---

[14]It *is* still a scheme, in that the predicate ϑ has not been specified (Section 2.8): the point is that the number n must be quantified.

Remark 2.5.12 illustrates the role of the part of the proof corresponding to the nested $\forall$, as a repeating feature of a tree. The type X of the variable (the **range** of quantification) is essential to understanding the quantified formula $\forall x : X.\ \phi[x]$.

The indirect rules. To prove $\phi \Rightarrow \psi$ we must *temporarily* assume ϕ, and deduce ψ from that. Then we are (by definition) able to deduce $\phi \Rightarrow \psi$ without assumption. The additional hypothesis ϕ, and anything we deduced from it, are no longer part of the context for $\phi \Rightarrow \psi$, and so our notation must provide a way of "shielding" them. The demarcation between facts, assumptions and hypotheses (page 3) has *changed*.

The word *indirect* has been borrowed from the euphemism *indirect proof* for excluded middle (Definition 1.8.1), because in these cases at certain stages in the deduction we make assertions which are subsequently withdrawn. We saw similar transformations of expressions as trees on pages 4–5. Dag Prawitz [Pra65] called them *proper rules* and *improper rules*.

DEFINITION 1·5·1: The boxed or **indirect logical rules** are

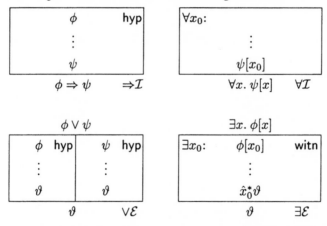

where x_0 is not free in ϑ. $(\neg\mathcal{I})$ is a special case of $(\Rightarrow\mathcal{I})$ when $\psi = \bot$.

REMARK 1·5·2: The formally similar rules for $\forall$ and $\Rightarrow$, and those for $\exists$ and $\wedge$, are combined idiomatically into the **guarded quantifiers**

$$\forall x.\ \phi[x] \Rightarrow \psi[x] \quad \text{and} \quad \exists x.\ \phi[x] \wedge \psi[x],$$

where $\phi[x]$ is essentially the "type" of the bound variable (Exercise 2.17 and Remark 3.8.13(b)). The box for $(\forall\Rightarrow\mathcal{I})$ states both the variable and the formula at the top; $(\forall\Rightarrow\mathcal{E})$ requires an instance a for which $\phi[a]$ has been proved.

Contexts and sequents. The ellipses ($\cdots$) may be avoided.

DEFINITION 1·5·3: The corresponding *sequent rules* are

$$\frac{\Gamma, \phi \vdash \psi}{\Gamma \vdash \phi \Rightarrow \psi} \Rightarrow\!\mathcal{I} \qquad\qquad \frac{\Gamma, \phi \vdash \vartheta \qquad \Gamma, \psi \vdash \vartheta}{\Gamma, \phi \vee \psi \vdash \vartheta} \vee\mathcal{E}$$

$$\frac{\Gamma, x : X \vdash \psi[x]}{\Gamma \vdash \forall x.\, \psi[x]} \forall\mathcal{I} \qquad\qquad \frac{x \notin \mathsf{FV}(\vartheta) \qquad \Gamma, x : X, \phi[x] \vdash \vartheta}{\Gamma, \exists x.\, \phi[x] \vdash \vartheta} \exists\mathcal{E}$$

The translation from box to sequent proofs is as follows. The proof of a formula within a box depends on hypotheses, some of which lie outside (at the top of the main proof) and some inside (as temporary hypotheses for the pending box rule). By weakening, we may as well import the entire external proof. But then the box is irrelevant, and we have a proof of the assertion on the top line of one of the rules shown. The corresponding assertion below the line is that the qualified formula exported from the box has been proved from the original hypotheses.

What happens to the variables (which, as we said, must be accounted for) in these translations? Many authors simply ignore them, so are unable to deal with empty types (see below). Careful presentations add them as subscripts to the turnstile $\vdash$, or leave them on the left hand side, with their types, where they belong. Indeed the parallel treatment of propositions and types will be very important for us, so we revise

DEFINITION 1·5·4: A *context* is a list of formulae *and* typed variables such that the free variables of each formula in the list occur earlier than the formula itself (*cf.* the comments before Remark 1.4.6).

REMARK 1·5·5: As $x \notin \mathsf{FV}(\vartheta, \Gamma)$, we may write the quantifier rules as

$$\frac{\phi \vdash \hat{x}^*\vartheta}{\exists x.\, \phi \vdash \quad \vartheta} \qquad\qquad \frac{\hat{x}^*\gamma \vdash \quad \vartheta}{\gamma \vdash \forall x.\, \vartheta}$$

using Notation 1.1.11, where $\gamma = \bigwedge \Gamma$. The double line means that the deduction goes both up and down. These symmetrical situations are known as *adjunctions*, written $\exists x \dashv \hat{x}^*$ and $\hat{x}^* \dashv \forall x$, and will be discussed fully in Sections 3.6–3.9. They will be the basis of the categorical analysis of quantifiers in the final chapter of the book.

Hypotheses and generic elements. The ($\forall\mathcal{I}$)-rule uses a temporary quantity x_0, and the ($\Rightarrow\!\mathcal{I}$)-rule involves a temporary hypothesis. They must both be shielded from the rest of the proof.

REMARK 1·5·6: If the *generic element* x_0 in an ($\forall\mathcal{I}$)-box were allowed to occur *above* the box, *i.e.* in some external hypothesis, then that would impose some condition on it which would restrict the generality of the terms which could be substituted for it.

Inside a box, x_0 is a constant, just like 0 and π. In particular, its presence asserts the non-trivial hypothesis (inside) that its type is non-empty.

If the generic element were allowed to occur *below* the box, *i.e.* in some external conclusion, then that would amount to an assumption that some value for x_0 actually exists, in other words that its type is non-empty. The scoping rules prevent us from deducing the existence of a white unicorn from the well known fact that all unicorns are white:

$$\forall x.\ \phi[x] \vdash \phi[x] \vdash \exists x.\ \phi[x].$$

It is forbidden because x in the middle formula is not an admissible term of the object-language (the constant-free language of unicorns).

This explains the perhaps unexpected result of quantifying over the empty type $X = \emptyset$, *cf.* $(\bot\mathcal{E})$, Remark 1.4.4:

REMARK 1·5·7: $\forall x{:}\emptyset.\ \phi[x]$ is *true*, and $\exists x{:}\emptyset.\ \phi[x]$ is *false*, for any ϕ.

Why do we not say that $\forall x.\ \phi[x] \Rightarrow \exists x.\ \phi[x]$, making $\forall x{:}\emptyset.\ \phi[x]$ false? Like omitting zero from arithmetic, to allow this would add an unnatural complication to the $(\forall\mathcal{I})$-rule: it would be necessary to prove $\exists x.\ \phi[x]$ too, and Remark 1.5.5 would be spoiled. But whether X has an element or is empty may be the question at issue, as Felix Hausdorff, an opponent of intuitionism, stressed. For example we might write $x : X$ for a typical odd perfect number while investigating whether there are any.

REMARK 1·5·8: Inside the box the hypothesis ϕ is a *fact* (*cf.* x_0 being a constant). We think of the box as that *part* of the world where ϕ is true.

Alfred Tarski showed how to interpret intuitionistic propositions as open subsets of topological spaces in 1935. Marshall Stone found representations of Boolean algebras using compact Hausdorff totally disconnected spaces (Section 5.4). Helena Rasiowa and Roman Sikorski [RS63] explored the connections from classical and intuitionistic logic to lattice theory and topology (but dismissed intuitionistic logic as useless). Steven Vickers [Vic88] has explained how "observable" properties of programs, *i.e.* those which (if they're true) we can see after a finite wait, may be seen as *open* subsets. Non-termination is not an observable property, even when it is obvious from the source code of the program.

REMARK 1·5·9: Making the hypothesis ϕ stronger restricts the part of the world on which it is true, with the effect that the conclusion ψ is easier to prove, so *the implication $\phi \Rightarrow \psi$ is weaker*.

Therefore the hypothesis plays a *negative* or *contravariant* role.

Repeating this, ϕ becomes *positive* (*covariant*) in $(\phi \Rightarrow \psi) \Rightarrow \vartheta$, but not in $\phi \Rightarrow (\psi \Rightarrow \vartheta)$, so the sign of the influence of ϕ depends on whether it lies behind an odd or even number of implications.

This negative behaviour explains why politicians are unwilling to discuss "hypothetical" situations. For example, a socialist may believe that "high unemployment causes damage to Society", but is formally unable to prove this, because to do so would involve the temporary advocacy or implementation of an undesirable policy.

What appears to be a negation in the vernacular may sometimes really have a constructive meaning. For example the criminal law, however draconian, does not *prevent* murder or robbery: it provides a *translation* which turns a *proof* of an offense into punishment.

β-reduction of proofs. The operational meaning of generic elements is demonstrated by substituting actual values. The detour *via* the generic element (introduction rule) and the actual value (elimination rule) can be short-circuited.

REMARK 1·5·10:   If we have an indirect proof which depends on the hypothesis ϕ or the generic element $x : X$, together with an actual proof of ϕ or a term $a : X$, we may substitute a for x and remove the box. So a proof of $\forall x.\ \phi[x] \Rightarrow \vartheta[x]$ is a *template* for the proofs of $\vartheta[a]$ for any required a, in which we have to insert a in the holes.

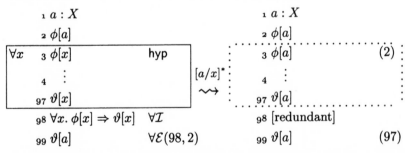

There is a similar translation for $(\exists \mathcal{I})$ followed by $(\exists \mathcal{E})$, which we use in reverse in Definition 1.6.8. If $(\vee \mathcal{I})$ is followed by $(\vee \mathcal{E})$ then one of the two sub-boxes may be omitted. Similarly, if $(\wedge \mathcal{I})$ is expressed as a tree then one of the branches disappears if $(\wedge \mathcal{E})$ follows; in the box style the redundancy is not so obvious.

Section 2.3 presents the rules for types analogous to the ones set out here for propositions. Induction uses nested quantifiers (Section 2.6), so provides examples of proofs using boxes. Modal logic, the fragment of the predicate calculus in which formulae may have at most one free variable, is discussed in Section 3.8.

1.6   FORMAL AND IDIOMATIC PROOF

Most mathematical texts do not use the formal rules of logic which we have given, except as objects of discussion in the study of logic itself. We have nevertheless been brought up to consider mathematical proof as rigorous, and (by degree level, at least) the student has gained a feel for when a proof has all of its *i*s dotted and its *t*s crossed. So where do formal methods come in?

We shall show how to translate the formal rules into the **vernacular** (English, French, *etc.* prose) and back. But it is important to understand that this translation cannot be a precise one like those amongst logical, programming and categorical languages which we give elsewhere in the book. This is because the meaning of a vernacular sentence (if it has one) is not given structurally in terms of the component parts, but depends heavily on unstated contextual information. Sentences which a classical grammarian would parse alike may have very different logical meanings.

The art of translation between human languages lies in *idiom*: particular phrases in the two languages match, not literally, but as a whole. The theory of descriptions (Definition 1.2.10) has already illustrated this.

Direct reasoning.

REMARK 1·6·1: The simplest idiom is the equational one:

$$a = b = c = d = \cdots$$

where the symmetric and transitive laws of equality are being elided.

Little more needs to be said for the other direct rules: the formal style places each assertion below the previous one, spelling out which rule justifies the step. By contrast, for reasons of space, it is usual to present a long argument as running text, divided into paragraphs and sentences to indicate the milestones in our progress. Arguments of this form are hardly literary prose, and we save *hence, thence* and *whence* from the grammatical graveyard simply to avoid the monotony of *therefore*.

Proof boxes delimited by keywords. The vernacular has its own ways of accommodating simple departures from direct logic.

REMARK 1·6·2:

(a) "***Put*** x" indicates a substitution, such as an instance of a universal formula (the substitution used in an $(\forall \mathcal{E})$-rule, Definition 1.4.2 and

Remark 1.5.2) or a declaration (Definition 1.6.8). This associates a *specific* value or expression with the name x.

(b) "***Let*** x" introduces a fresh variable, opening an $(\forall\mathcal{I})$-box. *No* value in particular is given to x — it is generic[15] — until a β-reduction (Remark 1.5.10) annihilates the $(\forall\mathcal{I})$ and corresponding $(\forall\mathcal{E})$.

(c) "***Suppose*** ϕ" opens an $(\Rightarrow\mathcal{I})$-box with a hypothesis.

(d) "***Thus***" (in this way) closes these boxes.

(e) "***If*** ϕ **then**... . ***Otherwise***" delimit $(\vee\mathcal{E})$-boxes. In programming languages "**fi**" or "**endif**" closes these boxes, drawing attention to the common conclusion, but "in either case" is English idiom.

But the most interesting idioms are those corresponding to $\exists$:

(f) "***There exists*** x such that $\phi[x]$" has two linguistic functions: *both* asserting $\exists x.\, \phi[x]$ ("for *some* x") *and* opening the $(\exists\mathcal{E})$-box which makes use of something satisfying ϕ. The same symbol x is both the bound variable and the temporary witness (Remark 1.6.5ff).

Boxes can be avoided by dividing the presentation of a topic into lemmas, each of which deals with a single box. The idiomatic proof of a lemma with $\forall x.\, \vartheta[x]$ as its conclusion or $\exists y.\, \phi[y]$ as its hypothesis would not bother to state the quantified formulae or use any kind of box: the variables are simply global to the proof. The same applies to the proof of $\exists z.\, \vartheta[z]$: the witness z has to be introduced at some point and its properties developed. So only when the proof is complex, with heavily nested $(\Rightarrow\mathcal{I})$- and $(\forall\mathcal{I})$-rules and no natural way of packaging the parts, must we use a formal style to make the argument clear.

Although nested boxes can be handled by additional lemmas, *they make us deaf to anything logic may have to say about a problem*: for example in $(\varepsilon{-}\delta)$-proofs in analysis, information (the degree δ of approximation) *flows backwards* from output to input. Induction, with its nested $\forall$, involves nested $(\forall\mathcal{E})$-boxes, of which Section 2.6 gives examples.

It is easy to translate formal proofs mechanically into the vernacular, though some creativity is needed to make them readable, *cf.* word by word translations from a foreign language using a dictionary. The other way is much more difficult — it would be quicker to reconstruct the proof from scratch (as I often find myself). This raises questions about the usefulness of proofs in printed journals in future.

---

[15]Grammatically, the indeterminacy is expressed by the subjunctive, which is formed by "let" in English. This is shown more clearly in the French "soit" (which is also used for case analysis). Programming languages and mathematical usage often use "let" where we have "put", but we prefer to maintain the distinction between them. Note that "put" is related to substitution or cut and "let" to variables and weakening, *cf.* Notation 1.1.11.

Often the conceptual structure of an argument may already be present in a simpler[16] version of the result, the more substantial one involving some difficult calculation. If the former, which *explains the theorem*, had been laid out, the readers could have supplied or omitted the details of the calculation for themselves.

Importing and exporting formulae. One should think of a box as a separate logical world, interacting with our own only across a membrane, which allows any hypothesis to enter from above (unless, of course, this means taking it out of another nested box).

A formula within a box which has no proof there is a hypothesis; this is replaced by evidence if the formula is imported from outside (β-reduction, Remark 1.5.10). Indeed, the weakening rule specifically allows this for external hypotheses, and nothing stops us from repeating parts of the development; so we may as well import the conclusion instead. The weakening rule for variables is given in Remark 2.3.8.

These are exactly the rules for *scoping* in block-structured programs.

LEMMA 1·6·3: In the case of the $\vee$ we have the *distributive law*

$$\gamma \wedge (\phi \vee \psi) \dashv\vdash (\gamma \wedge \phi) \vee (\gamma \wedge \psi)$$

and, for $\exists$, the so-called *Frobenius law*

$$\gamma \wedge (\exists x.\ \phi[x]) \dashv\vdash \exists x.\ (\gamma \wedge \phi[x]).$$

PROOF: To show LHS $\vdash$ RHS we need to import γ.

	₁ γ			γ				
	₂ $\exists x.\ \phi[x]$			$\phi \vee \psi$				
$\exists x$	₃ $\phi[x]$	hyp		ϕ	hyp	ψ		hyp
	₄ γ	(1)		γ	(1)	γ		(1)
	₅ $\gamma \wedge \phi[x]$	$\wedge\mathcal{I}$		$\gamma \wedge \phi$	$\wedge\mathcal{I}$	$\gamma \wedge \psi$		$\wedge\mathcal{I}$
	₆ $\gamma \wedge \exists y.\ \phi[y]$	$\exists\mathcal{I}$		$\cdots \vee \cdots$	$\vee\mathcal{I}0$	$\cdots \vee \cdots$		$\vee\mathcal{I}1$
	₇ $\exists y.\ \gamma \wedge \phi[y]$	$\exists\mathcal{E}$		$(\gamma \wedge \phi) \vee (\gamma \wedge \psi)$				$\vee\mathcal{E}$ $\square$

In general, the ability to use formulae from earlier in the proof is known as *referential transparency*, it was discussed by Willard Quine for

---

[16]Maybe this is an appropriate formal meaning to give to the word *trivial*: something which should have been learned in an earlier course. (The *trivium* was the first part of the medieval syllabus, the second being the *quadrivium*.) The treatment of posets in Chapter III was designed to help the later category theory in this way.

natural language [Qui60]. For us, it is a manifestation of *invariance under substitution*. This phenomenon will arise

(a) for the λ-calculus as the naturality equation (Definition 4.7.2(c));

(b) for conditionals as stability under pullback (Definition 5.5.1);

(c) for composition of relations (Lemma 5.8.6) and for the existential quantifier (Remark 9.3.7) in the same way;

(d) and for recursion over $\mathbb{N}$ as a product with Γ (Remark 6.1.6).

(e) The effect on $\Rightarrow$ and $\forall$ will only be apparent in Section 9.4, where it is the Beck–Chevalley condition.

REMARK 1·6·4: The box rules allow us to export ϑ from an $(\Rightarrow\mathcal{I})$-box in the form $\phi \Rightarrow \vartheta$, and $\vartheta[x]$ from $(\forall\mathcal{I})$ as $\forall x.\ \vartheta[x]$. Although we presented these rules with just one such formula, and wrote it on the last line of the box, in fact *any number* of formulae may be exported from *any line* of the box, if they are appropriately qualified (by $\forall x$ or $\phi \Rightarrow$). A formula may be exported unaltered from an $(\vee\mathcal{E})$-box *if it occurs on both sides*.

Open-ended boxes. The $(\vee\mathcal{E})$-rule provides a proof in each of the two cases, without prejudice as to which holds. Similarly the $(\exists\mathcal{E})$-rule gives a demonstration in terms of an unspecified witness. Dependence on the alternative or witness means that a box is needed, but we shall now show that the $(\exists\mathcal{E})$-box is open-ended below (we have just seen that all boxes are open above). This rule is the least well understood in the practice of mathematics, although it has a bearing on the use of structure such as products (Remark 4.5.12) and the meaning of finiteness (Remark 6.6.5).

REMARK 1·6·5: Since the conclusion (ϑ) is arbitrary, the closing of the box may be postponed indefinitely, *i.e.* until the end of the enclosing box or proof. This is because any ϑ' which we deduce from ϑ outside the box (necessarily not mentioning the witness) may equally be deduced inside and exported as the conclusion instead of ϑ.

So the box need not be closed at all.

Some unfinished business from Section 1.3 provides an example.

LEMMA 1·6·6:   Composition preserves functionality and totality.

PROOF:

$\forall x$	1	$x \overset{f;g}{\longmapsto} z_1, z_2$	hyp	$\forall x$		
$\forall z_1$	2	$\exists y_1. \; x \overset{f}{\mapsto} y_1 \overset{g}{\mapsto} z_1$	def($f;g$)		$\exists y. \; x \overset{f}{\mapsto} y$	total(f)
$\forall z_2$	3	$\exists y_2. \; x \overset{f}{\mapsto} y_2 \overset{g}{\mapsto} z_2$	def($f;g$)	$\exists y$	$x \overset{f}{\mapsto} y$	witn(2)
$\exists y_1$	4	$x \overset{f}{\mapsto} y_1 \overset{g}{\mapsto} z_1$	witn(2)		$\exists z. \; y \overset{g}{\mapsto} z$	total(g)
$\exists y_2$	5	$x \overset{f}{\mapsto} y_2 \overset{g}{\mapsto} z_2$	witn(3)	$\exists z$	$y \overset{g}{\mapsto} z$	witn(4)
	6	$y_1 = y_2$	func(f)		$\exists y. \; x \overset{f}{\mapsto} y \overset{g}{\mapsto} z$	$\exists \mathcal{I}$
	7	$y_1 \overset{g}{\mapsto} z_2$	subs		$x \overset{f;g}{\longmapsto} z$	def($f;g$)
	8	$z_1 = z_2$	func(g)		$\exists z. \; x \overset{f;g}{\longmapsto} z$	$\exists \mathcal{I}$

Notice that, in the proof on the right, the natural end of the ($\exists y$)-box comes *before* that of ($\exists z$). □

The properties of the ($\exists \mathcal{E}$)-box make it notationally redundant, and explain the idiomatic phrase "there exists". However, the conclusions of such an argument cannot be exported from *enclosing* boxes, unless the witness is unique ($\phi[x]$ is a description, Definition 1.2.10), in which case a function-symbol may be introduced.

Enlarging the box as much as possible is appropriate for the existential quantifier. The conclusion of ($\vee \mathcal{E}$) is also indeterminate, but this is normally exploited in such a way as to *shorten* the proof, by closing the box as soon as the alternatives can be reunited. Remark 2.3.13 sets out the *continuation* rules which are needed in type theory to handle the open-endedness, and Example 7.2.6 explains it categorically.

REMARK 1·6·7:   Notice that the ($\exists \mathcal{E}$)-rule alone — one of the two halves of the meaning of the quantifier — suffices to give a formal justification of the introduction of an *unspecified* witness for any existentially quantified statement, and the continued use of this witness until the end of the enclosing box. It is quite unnecessary to postulate, as Hilbert and later Bourbaki [Bou57] did, a global process which *selects* such witnesses, and indeed to do so amounts to the axiom of choice (Definition 1.8.8).

Declaration and assignment.

DEFINITION 1·6·8:   A very useful application of the ($\exists \mathcal{E}$)-box and its open-endedness is the **definition** or **declaration**. For any well formed term t we may always introduce a fragment of proof of the form

$$
\begin{array}{lll}
& t = t & \text{refl} \\
& \exists x.\ x = t & \exists\mathcal{I} \\
\hline
\exists x & x = t & \text{witn} \\
& \quad\vdots &
\end{array}
$$

After this, as far as the end of the next enclosing box, x and t may be used interchangeably. As t is a term, *i.e.* a long sequence of symbols, x is an **abbreviation** for it, but any conclusion we reach concerning x may be translated into one about t and exported from the box.

Since the box is open-ended, we don't bother to write it, and condense the three steps above to "**put** $x = t$". Although *after* the declaration the relationship between t and x is symmetrical, *during* the defining step they play different roles; some authors indicate this by writing $x := t$. The symbol x is called the **definiendum** or **head** and the term t the **definiens** or **body**. Similarly when t is a proposition we say "if" rather than "iff" or "if and only if" in a declaration.

Notice that any variables or hypotheses in the context are parameters or preconditions, so declarations cannot be exported from enclosing boxes.

REMARK 1·6·9:   The phrase **without loss of generality (wlog)** is a variant of declaration which is analogous to **assignment** in imperative programming languages (Remark 4.3.3). For example,

(a) To show that there is at most one integer satisfying a given property we take x and y having it and may suppose without loss of generality that $x < y$ to derive a contradiction, because the argument from $y < x$ is similar. By the trichotomy law $\forall x, y.\ x < y \lor x = y \lor x > y$ we are left with $x = y$; this does not use excluded middle because, by $(\bot\mathcal{E})$, we can deduce $x = y$ in each of the three cases.

(b) To solve the general cubic equation (Example 4.3.4) we may assume without loss of generality that it is $x^3 - 3px - 2q = 0$, because it is easy to turn the general problem into this form and the result back.

(c) To solve Newton's equations for the motion of the Earth about the Sun, we may assume without loss of generality that the origin is at their centre of mass and motion takes place in the xy-plane.

These assumptions are *desirable* because the problem becomes simpler. They are *permissible* because the general problem may be transformed into the special case and its solution back again. One ought to take care to distinguish the general and special cases by different notation, such as x and x', but commonly this is not done. Nevertheless, when we say "without loss of generality" we must always state the two-way translation involved (the *means of exchange*).

Alternative methods. Until Frege, Aristotle's syllogisms (Exercise 1.25) were the standard treatment of the quantifiers, and they are still taught to unfortunate philosophy students, despite significant advances in logic in both ancient times and the twelfth to fourteenth centuries.

Gottlob Frege was by far the best logician between the Renaissance and the First World War. The distinction in Section 1.2 between an expression and its value is due to him. In 1879, while others such as de Morgan and Schröder were still battling with the propositional connectives, he developed a modern theory of quantifiers, understood bound variables and used second order logic with confidence. His *verbal* explanations are crystal clear, but his space-consuming notation (*Begriffsschrift*, concept-writing) would have consigned his work to oblivion but for Russell's attention. It must have caused nightmares for his printer, but he argued (in a letter to Peano, who printed his own books) that this was not the important consideration: one ought to take advantage of the second dimension to explain mathematics. We use Frege's theory of sequences in Sections 3.8 and 6.4.

REMARK 1·6·10: Hilbert's logic, like that of Frege and Russell, had steps like

$$\text{if } \bigwedge \Gamma \Rightarrow \phi \text{ is true then } \bigwedge \Delta \Rightarrow \vartheta \text{ is also true,}$$

and axioms (instead of rules) such as

$$(\phi \Rightarrow \vartheta) \Rightarrow ((\psi \Rightarrow \vartheta) \Rightarrow ((\phi \vee \psi) \Rightarrow \vartheta)).$$

Although no boxes are needed, the ($\Rightarrow \mathcal{E}$)-rule is unavoidable, as Charles Dodgson (Lewis Carroll) demonstrated in a discussion between Achilles and the Tortoise; this is reprinted with many pieces in a similar *genre* in [Hof79, pp. 43–5]. *Implicational logic* is an interesting fragment of Hilbert's logic; see Exercise 1.23 and Example 2.4.2.

All of the assertions in a Hilbert-style proof are *facts*, whereas idiomatic arguments in mathematics use *hypotheses*. The first formal account of *natural* deduction, together with its equivalence with the Hilbert style, was given by Stanisław Jaśkowski [Jás34], based on ideas of Jan Łukasiewicz. It treated not only implication and conjunction but also the universal quantifier (and substitution), noting and correctly handling the problem of empty domains (Remark 1.5.6).

Gerhard Gentzen [Gen35], a student of Hilbert, treated the connectives individually (whereas previous authors had defined some in terms of others), recognising the symmetry in their rules. He gave translations amongst natural deduction (NK), sequent calculus (LK) and Hilbert-style (LHK) classical and intuitionistic (NJ, LJ, LHJ) logic.

Frederic Fitch wrote the first textbook [Fit52] to make routine use of natural deduction using our proof boxes, *modulo* some syntactic sugar, and Nikolas de Bruijn developed the notation for AUTOMATH.

REMARK 1·6·11: Gentzen used the tree notation (Remark 1.4.5) for his natural deduction. This style remains the prevalent one amongst logicians, but it is highly unsatisfactory for the indirect rules, especially for extended proofs. The formulae at the leaves (with no rule above them) are hypotheses; when $\phi \Rightarrow \vartheta$ has been deduced from ϑ, ϕ need no longer be a hypothesis and so may be *discharged*. The $(\vee\mathcal{E})$-rule is similar. We have done this by closing a box, but it is traditionally indicated by striking through the formula:

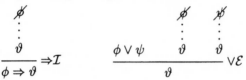

When several of these rules have been used, we cannot see which of them resulted in the discharge without some additional *ad hoc* labelling. Moreover only *some* occurrences of ϕ need be discharged (a "parcel" of them according to [GLT89]), and the notation does not account for free variables. The box method shows these things naturally, and ensures that the introduction and discharge of hypotheses are properly nested.

Model theory. Surely the simplest interpretation of proof is that

ϕ is true, therefore ϑ is true.

However, the verity of a formula such as

$\forall n \geqslant 2.\ \exists p, q \text{ prime. } 2n = p + q$

cannot be perceived by any mortal from its sub-formulae or instances. The truth-values interpretation of $\forall n$ in this formula has been verified up to large values, but a *proof*, if there is one, must be finitary and introduce this quantifier by considering a *generic* n, *i.e.* using the $(\forall\mathcal{I})$-rule.

REMARK 1·6·12: A formula is not true or false *of itself*, but only when interpreted in a *model* $\mathcal{M}$. This specifies

(a) the individuals which are denoted by the constants, and over which the variables range,

(b) for each relation-symbol and tuple of individuals, whether or not this instance of the relation holds, and

(c) for each operation-symbol and tuple of individuals, what individual is the result of the operation.

The meaning of general terms and formulae is defined by structural recursion; this is straightforward for the connectives $\land$ and $\lor$, and we write $\mathcal{M} \vDash \phi$ if ϕ is valid in the interpretation. However, a quantified formula $\forall x.\ \phi[x]$ is valid if $\phi[a]$ holds for each individual a, which is an *infinitary* condition, so the naïve meta-language is no longer adequate. The arity of $\forall x$ is the semantic object which interprets the type of the syntactic variable x. We shall consider infinitary operations, *i.e.* whose arities are objects of the world under mathematical study, in Section 6.1. The quantification over individuals is then performed by the $(\forall \mathcal{I})$-rule in the meta-logic, as we can never escape completely from finitary proof.

Having set this up, one can show that *each of the logical rules preserves validity*, *i.e.* whenever the premises are true, so is the conclusion. This is called **soundness**. To an algebraist, it says that the interpretation is a *homomorphism* for $\land$, $\forall$, *etc.* By structural induction on a proof $\Gamma \vdash \phi$, if its hypotheses are valid ($\mathcal{M} \vDash \Gamma$) then so is its conclusion ($\mathcal{M} \vDash \phi$).

REMARK 1·6·13: On the other hand, suppose that *every* model $\mathcal{M}$ for which $\mathcal{M} \vDash \Gamma$ *also* satisfies ϕ. In this case we say that Γ (semantically) *entails* ϕ and write $\Gamma \vDash \phi$. As we have said, when there is a proof $\Gamma \vdash \phi$, then $\Gamma \vDash \phi$. But as the notions of proof and validity have been defined independently, $\Gamma \vDash \phi$ may perhaps happen without a proof, either if ϕ is supposed to be true but we have forgotten to state some rule of deduction needed to prove it, or if it ought to be false but our class of models is too poor to furnish a counterexample. The proof theory is said to be **complete** for its semantics if proof and truth do coincide. Kurt Gödel showed that *first order logic* (what we have considered so far) is complete, but second or higher order logic (Section 2.8) is not.

Both of these famous theorems raise a number of deep questions, many of which are beyond the scope and viewpoint of this book. As we have noted, there was a tendency in traditional logic to study propositions, regarding variables, individuals and terms as secondary. To correct this attitude we must add more detail, and begin with the algebraic fragment alone. The connectives and quantifiers are added one by one, so we shall not reach the classic model-theoretic results. On the other hand, models in the old framework had to be discrete sets, whereas for us they may perhaps be topological spaces, or come from some more exotic world. In fact it is possible to fashion such worlds out of the syntactic theories themselves, and they contain *generic* models. Conversely, Section 7.6 shows how to design your syntax to fit your semantics.

Model theory, with the completeness theorem at its heart, is a "three sides of a square" approach to logic: given the axioms for, say, groups, instead of deducing theorems directly from the axioms, it seeks to find out what is true about the proper class of models. We aim to do things directly, in particular recognising the Lindenbaum algebra or "classifying category" of a theory as a *syntactic* construction.

REMARK 1·6·14: I have seen logic introduced to first year undergraduates both in the form of truth-assignments and using proof boxes, and firmly believe that the box method is preferable. As this section has shown, it is a formal version of the way mathematicians (and informaticians) *actually* reason, even if they *claim* to use Boolean algebra when asked.

If you want to teach both interpretations, they have to be shown to be equivalent. The formal soundness of $(\vee\mathcal{E})$ and $(\Rightarrow\mathcal{I})$ is more difficult to explain than one might suppose, as it depends on *hypotheses*, *i.e.* the notion of scope (*cf.* proof boxes). Soundness *for a whole proof* makes use of structural induction, which, whilst they should learn it during the first year, is unfamiliar to students just out of school. Although they need to be aware of the truth-values interpretation, together with the statement and explanation of soundness and completeness, this should be given *at the end* of a first logic course, after the students have learned to construct some actual proofs.

Vernacular idioms for induction will be discussed in Sections 2.5–2.8 and 3.7–3.9, and declarative programming in Section 4.3.

1.7   AUTOMATED DEDUCTION

There are mechanised approaches to algebra, combinatorics, numerical analysis, *etc.*, but this book is not about those subjects, it is about logic: all we can discuss in this section is the choreography of $\forall$, $\exists$, $\wedge$, $\vee$ and $\Rightarrow$, and reinforce the importance of the scope of variables and hypotheses. We cannot do the creative part of mathematics, because the solution to a problem in number theory, for example, may be "simply" an ingenious observation about $\zeta(s) = \sum_{n=1}^{\infty} n^{-s}$ or counting points in a cube in $\mathbb{R}^n$.

The steps which can be automated are the **obvious** ones, in a technical sense: this literally means "in the way" in Latin. It is obvious how to go through a foreign airport, not because you know it intimately, but because there are signs telling you where to turn whenever you need them (you hope). This is also known as **exam technique**: write down and exploit what you already know. Whereas the box or sequent rules of predicate calculus from the previous section are the *laws* of the game of proof, the heuristics[17] are hints on the *tactics*. This section is based on teaching first year informatics students to construct proofs *on paper*. Of course this will also give some idea of how to write a program to do it, but the strategy for making choices when backtracking is needed raises issues far outside the scope of this book [Pau92].

George Polya [Pol45] and Imre Lakatos [Lak63] gave two classic accounts of heuristics in mathematics, using Euclidean geometry for examples. Polya's advice — make a plan and carry it out, compare your problem with known theorems, *etc.* — is extremely valuable to help students of mathematics (and professionals) get past the blank sheet of paper, but treats more strategic aspects of proof than we can. An early theorem prover was based on his methods of drawing diagrams and formulating conjectures and counterexamples; that this seems odd now shows both the sophistication of modern proof theory and perhaps also the danger of isolation from the traditional instincts of mathematicians.

Nikolas de Bruijn's AUTOMATH project (late 1960s) set out to codify existing mathematical arguments, rather than to find new theorems, and this remains the research objective of automated reasoning. Johan van Benthem Jutting (1977) translated Edmund Landau's book *Foundations of Analysis* into AUTOMATH and analysed the ratio by which the text is magnified, which was approximately constant from beginning to end. Similar work has been done for other areas of mathematics.

There are certain dangers inherent in the formalisation of mathematics. Systems of axioms acquire a certain sanctity with age, and in the *how* of churning out theorems we forget *why* we were studying these conditions in the first place. Computer languages suffer far more from this problem: nobody would claim any intrinsic merit for FORTRAN or HTML, but sheer weight of existing code keeps them in use. Through the need for a standard — *any* standard — a similar disaster could befall mathematics if set theory were chosen. As with any programming, and also with the verification of programs, far more detail is required than is customary in mathematics. G. H. Hardy (1940) claimed that there is no permanent place in the world for ugly mathematics, but I have never seen a program which is not ugly. Even when the mathematical context and formal language are clear, we should not perpetuate *old* proofs but instead look for new and more perspicuous ones.

Theoretical basis. By Gentzen's cut-elimination theorem, supposing that $\Gamma \vdash \vartheta$ is provable, we need only look for a cut-free proof. We can say something about the final lines of such a proof (in sequent calculus)

---

[17]Greek, $\varepsilon \dot{v} \rho \iota \sigma \kappa \omega$, I find, of which the past tense ($\eta \dot{v} \eta \kappa \alpha$) is Archimedes' "eureka".

from the structure of the rules other than cut, and in particular the sub-formula property (Remark 1.4.9).

Although we must *read* a finished proof from top to bottom, the *search* for and creation of the proof are not so direct. (The commonest misconception about mathematicians amongst the general population is that we act like robots when trying to solve problems.) By the nature of cut-elimination, the heuristics are in fact **goal-driven**: they proceed mainly in the opposite direction from the reading of the completed proof.

For certain fragments of logic, if there is any proof of $\Gamma \vdash \vartheta$ then there is one obtainable by means of the following heuristics. Conversely, if we fail, by completeness (Remark 1.6.13) there is a counterexample, which can be obtained from the trace of our proof-attempts.

FACT 1·7·1:   **Hereditary Harrop formulae** are the *definite* formulae γ and *goals* ϑ respectively defined by the grammar

$$\gamma \quad ::= \quad \alpha \mid (\vartheta \Rightarrow \alpha) \mid \forall x.\, \gamma \mid \gamma_1 \wedge \gamma_2$$
$$\vartheta \quad ::= \quad \alpha \mid (\gamma \Rightarrow \vartheta) \mid \forall x.\, \vartheta \mid \vartheta_1 \wedge \vartheta_2 \mid \exists x.\, \vartheta \mid \vartheta_1 \vee \vartheta_2 \mid \top,$$

where α is atomic. If Γ is a list of definite formulae and ϑ is a goal formula for which $\Gamma \vdash \vartheta$ is provable, then it has a **uniform proof**, *i.e.* one in which each sequent $\Delta \vdash \phi$ with ϕ non-atomic is deduced *only* by means of the introduction rule for the outermost connective of ϕ [MNPS91]. □

So we construct the proof in reverse by dismantling the goal formula.

Resolution. When the goal is an atomic formula, logical manipulation has nothing to say, and we have to make use of the **database**, *i.e.* the axioms Γ given in the problem. These are written at the top of the page, numbering the lines from 1 and giving the justification for each line as "data". The desired conclusion(s) or **goals** ϑ are written at the bottom, numbering the lines backwards from 99 and giving no reason (yet). We shall progressively add more lines 2, 3, ... and 98, 97, ... and also fill in reasons; the lines which have no reason so far are called **pending goals**.

REMARK 1·7·2:   If the atomic formula ϑ is a goal and the axiom

$$\phi_1 \wedge \phi_2 \wedge \ldots \wedge \phi_k \Rightarrow \vartheta$$

is in the database, then the problem is reduced to proving each of ϕ_1, ϕ_2, ..., ϕ_k. The idea of logic *programming* is that this is a *procedure* which defines ϑ (the **head**) in terms of the ϕ_i (the **body**), and the notation is reversed (sometimes using ":=" for $\Leftarrow$):

$$\vartheta \Leftarrow \phi_1,\ \phi_2,\ \ldots,\ \phi_k.$$

In order to answer the query ϑ, we regard the database as a program and call a procedure whose head is ϑ, which calls sub-procedures; if the original call returns successfully then ϑ has been proved, *i.e.* the answer to the query is yes. Notice that while the search for a proof is in progress there may be several pending goals, to be taken conjunctively, just as the database may consist of several hypotheses.

REMARK 1·7·3: The program is non-deterministic, because there may be several procedures for ϑ: if using one of them fails to find a proof, we **backtrack** and try another. To do this by hand, place a new sheet of tracing paper over the proof so far each time you have to make a choice; then if the choice is wrong you can discard the working which *depended* on it and return to the immediately preceding state. Only the last choice is discarded: earlier ones may still be viable until all possibilities at this stage have been exhausted. This means that the choices in the search form a nested system in the heuristics, but this is independent of the nested contexts (boxes) in the completed proof.

REMARK 1·7·4: Similarly, a predicate in the database of the form

$$\forall \underline{x}, \underline{y}. \left(\exists \underline{z}. \, \phi_1[\underline{x}, \underline{y}, \underline{z}] \wedge \cdots \wedge \phi_k[\underline{x}, \underline{y}, \underline{z}] \right) \Rightarrow \vartheta[\underline{c}, \underline{x}]$$

is written as the procedure

$$\vartheta[\underline{c}, \underline{x}] \Leftarrow \phi_1[\underline{x}, \underline{y}, \underline{z}], \ldots, \phi_k[\underline{x}, \underline{y}, \underline{z}].$$

By convention, *the whole formula* is universally quantified over all the free variables, which is the same as saying that it is a *scheme* for the closed formulae obtained by substituting terms for variables. A formula such as this, in which the sub-goals ϕ_i are also atomic predicates, is called a (*positive* or *definite*) **Horn clause** of arity k. (Recall from Definition 1.4.1(a) that the atomic predicates $\phi_i[\underline{x}]$ also have arity — the length of the sequence $\underline{x}$ — but this is independent of the arity of the clause, *i.e.* the number of atomic formulae it contains.)

Suppose that we want to use this Horn clause to prove (solve the query) $\vartheta[\underline{a}, \underline{b}]$. By $(\forall \mathcal{E})$ we put $\underline{a}$ for $\underline{x}$, and by $(\Rightarrow \mathcal{E})$ we have to prove $\phi[\underline{b}, \underline{d}, \underline{e}]$ and *match* $\underline{a} = \underline{c}$, substituting suitable terms $\underline{d}$ and $\underline{e}$ for $\underline{y}$ and $\underline{z}$.

EXAMPLE 1·7·5: A database application might have axioms such as

train [London, Bristol, £40]	$\Leftarrow$	
train [London, York, £40]	$\Leftarrow$	
train [London, Paris, £100]	$\Leftarrow$	
train [Paris, Nice, £80]	$\Leftarrow$	
train $[x, y, u]$	$\Leftarrow$	train $[y, x, u]$
journey $[x, y, u]$	$\Leftarrow$	train $[x, y, u]$
journey $[x, z, u + v]$	$\Leftarrow$	journey $[x, y, u]$, journey $[y, z, v]$.

Then for the query journey [Nice, Bristol, u] we expect not only a proof that one can go from Nice to Bristol by rail, but *also the route and cost.*

So when we assert $\exists x.\ \vartheta[x]$ we give a *definite answer* as to what x is — these substitutions are the *result* of the computation.

REMARK 1·7·6:   John Robinson showed how to do this by **resolution** (1965). Gentzen's Hauptsatz cannot eliminate cuts when axioms are used, and resolution deals with those that remain. It involves substitution of terms for variables, but each resolution step only gives partial information about what has to be substituted: the constraints which fully determine the value may come from quite different parts of the proof (execution of the program).

We can use *declarations* (Definition 1.6.8) to record this information:

1	$\forall \underset{\sim}{x}, \underset{\sim}{y}.\ \phi[\underset{\sim}{x}, \underset{\sim}{y}] \Rightarrow \vartheta[\underset{\sim}{c}, \underset{\sim}{x}]$		1	**put** $x_0 = ?$	
2	**put** $\underset{\sim}{y}_0 = ?$		2	$\vdots$	
3	$\vdots$		97	$\vdots$	
96	$\vdots$		98	$\vartheta[x_0]$	(pending)
97	$\phi[\underset{\sim}{b}, \underset{\sim}{y}_0]$	(pending)	99	$\exists x.\ \vartheta[x_0]$	$\exists\mathcal{I}(98)$
98	$\underset{\sim}{a} = \underset{\sim}{c}$	(pending)			
99	$\vartheta[\underset{\sim}{a}, \underset{\sim}{b}]$	$\forall\mathcal{E}(1, 97, 2, 98)$			

Here $\phi[\underset{\sim}{b}, y_0]$ is a new goal, to be satisfied by further resolution, as we have done with ϑ. The partial proof on the right illustrates the similar way in which existential goals are handled.

The equations $\underset{\sim}{a} = \underset{\sim}{c}$ are also new goals. If these terms are simply names for individuals (London, York, *etc.*) and *there are no axioms to say that individuals with different names can be equal* then we can see immediately whether or not the equations hold. If not, this attempt at resolution fails and we backtrack to find another one. In practice this is done by database-searching techniques.

The programming language PROLOG does resolution and unification. Despite its name, it does not in fact deal with the logical connectives and quantifiers, but what we shall come to call the algebraic fragment (although this will not look like algebra until Section 4.6). The denotational semantics, based on the work of Jacques Herbrand (1930), will be discussed in Sections 3.7 and 3.9.

Unification. Goals involving function-symbols need another technique, called unification. How to *do* unification is easy: the difficult part is to see what it *means*. The functions in question are those whose values might be enumerated in a database, such as mother_of, not arithmetic.

REMARK 1·7·7: A goal of the form $r(y) = r(v)$, where r is an operation-symbol *for which no laws are known*, can only follow by substitution:

$$\frac{u_1 = v_1 \quad \cdots \quad u_k = v_k}{r(u_1, \ldots, u_k) = r(v_1, \ldots, v_k)}$$

with the new goals $u_1 = v_1$, ..., $u_k = v_k$.

This does not mean that every function is injective. We want to carry on building the *logical structure* of a proof, possibly without knowing what terms serve as the subjects of predicates. We postpone filling in these terms, and then try to do so as non-specifically as possible, using only the building blocks we already have in the term calculus of the object-language. The possibility that two terms might denote the same thing is only considered if the terms themselves were formed in the same way.

The point is that

(a) if we have no information about r, then the only hope we have of proving that $r(y) = r(v)$ is by first showing $u_i = v_i$ ($1 \leqslant i \leqslant k$). This step can be built into the proof-layout we have given, by treating $u_i = v_i$ as new goals and giving "substitution" as the reason for the line $r(y) = r(v)$.

So

(b) a match between terms $r(u_1, u_2) = r(v_1, v_2)$ having the same outermost operation creates a new equation for *each* argument, $u_1 = v_1$ and $u_2 = v_2$; despite this proliferation the algorithm does terminate because u_i and v_i are all shorter than $r(y)$ and $r(v)$;

(c) there is no way to prove $r(y) = s(v)$ if r and s are different operation-symbols satisfying no laws; this is called a ***clash***;

(d) if r does satisfy other axioms or laws, unification may be of no help; at any rate it may involve backtracking, as for example with concatenation of lists, where a division must somehow be chosen;

(e) a goal of the form $x_0 = u$, where x_0 is an indeterminate and u a term *in which x_0 does not occur*, forms part of the solution of the unification problem, and completes the unfinished declaration (**put** $x_0 = ?$) in Remark 1.7.6;

(f) an equation such as $x_0 = r(x_0)$, in which x_0 does occur on the right, cannot be satisfied by substitution of a term for x_0 (try it!); this necessitates the ***occurs-check***;

(g) the other axioms for a congruence (Definition 1.2.3) are also applicable: if $u = v$ and $v = w$ are goals then so is $u = w$, and these may match, clash, form part of the solution or fail the occurs-check;

(h) the heuristic applies only to goals, not to hypotheses — it exploits $u = v \Rightarrow r(u) = r(v)$ without asserting the converse.

Eventually, if neither type of failure (clash or occurrence) happens, the system of equations will be saturated, *i.e.* none of these rules will expand it further. Then some of the indeterminates will be expressed as terms, possibly involving the others.

REMARK 1·7·8: Some of the indeterminates may be independent, for example y_0 is arbitrary in the equation $x_0 = r(y_0)$. The full solution to the unification problem is not unique, since we may put anything we please for y_0. However, the solution in which y_0 is left as we have it is the *most general unifier* in that

(a) it is itself a solution,

(b) every other solution is obtained by substituting terms into it, and

(c) any such substitution is a solution.

We can in fact eliminate the confusion of working backwards from goals, and reduce unification to a kind of algebra. A theory with operation-symbols of various arities (numbers of arguments) but *no laws* is called a *free theory*, and unification is the study of its *free* or *term model*. We shall take this up formally in Chapter VI and return to unification in Section 6.5, where we shall see that Remark 1.7.7(g) can be simplified.

Unification in theories with laws is more difficult. It is possible to handle commutativity and associativity, at the cost of uniqueness: there is now a family of maximally general unifiers. Unification under the distributive law would give a uniform way to solve Diophantine equations, but Yuri Matajasivič showed that this is undecidable (1970). Gérard Huet showed that unification in higher order λ-calculi is also undecidable [Hue73].

Box-proof heuristics. Now we turn to the logical symbols themselves. The following methods belong in a course on the predicate calculus: it is probably better to teach resolution quite separately. Unless we say otherwise, any boxes are drawn as large as possible, extending from the end of the database to the first pending goal.

REMARK 1·7·9: The uniform proof (Fact 1.7.1) is found as follows:

(a) Any formula $\phi \wedge \psi$ as a goal or hypothesis may be replaced by ϕ and ψ as two formulae. Similarly $\top$ may be ignored altogether.

1	$\phi \wedge \psi$	data	97	ϕ	(pending)
2	ϕ	$\wedge \mathcal{E}0(1)$	98	ψ	(pending)
3	ψ	$\wedge \mathcal{E}1(1)$	99	$\phi \wedge \psi$	$\wedge \mathcal{I}(97, 98)$

(b) To prove the goal $\forall x. \phi[x] \Rightarrow \vartheta[x]$, we open an $(\forall \mathcal{I})$-box with new variable x, hypothesis $\phi[x]$ and conclusion $\vartheta[x]$. Having now filled in the immediate proof-step which justifies the first goal, albeit without any reason for $\vartheta[x]$ so far, we are excused from considering this goal again by the annotation $(\Rightarrow\mathcal{I})$ or $(\forall\mathcal{I})$ on line 99. The goals $\neg\phi$ and $\phi \Leftrightarrow \vartheta$ are handled in a similar way.

$$
\begin{array}{ll}
\forall x \quad {}_1 \; \phi[x] & \text{hyp} \\
\qquad {}_2 \; \vdots & \\
\qquad {}_{98} \; \vartheta[x] & \text{(pending)} \\
\end{array}
$$
$$
{}_{99} \; \forall x. \; \phi[x] \Rightarrow \vartheta[x] \quad \forall\mathcal{I}
$$

$$
\begin{array}{ll}
{}_1 \; \exists x. \; \phi[x] & \text{data} \\
\exists x \quad {}_2 \; \phi[x] & \text{witn} \\
\qquad {}_3 \; \vdots & \\
\end{array}
$$

(c) The behaviour of $\exists x. \; \phi[x]$ as a hypothesis $(\exists\mathcal{E})$ mirrors that of $\forall$ as a goal, since $\big(\exists x. \; \phi[x]\big) \Rightarrow \vartheta \equiv \forall x. \big(\phi[x] \Rightarrow \vartheta\big)$. Recall, however, that the $(\exists\mathcal{E})$-box is open-ended below (Remark 1.6.5), so as long as the variable x does not occur elsewhere, we can simply add $\phi[x]$ to the data without a box. It is to our advantage to do this as soon as possible, because there may be many things satisfying ϕ, and it could be relevant later that the same one plays two or more different roles in the argument, although to say that we "choose" a witness does not mean that an actual individual is selected (Remark 1.6.7). The original axiom $\exists x. \; \phi[x]$ will not be needed again.

Subject to scoping of variables, these boxes may be nested in any order and so may be taken together in a single step.

(d) A goal $\exists x. \; \vartheta[x]$ can only be deduced from $\vartheta[x_0]$, using $(\exists\mathcal{I})$, where x_0 is a term to be found by resolution (Remark 1.7.6).

(e) If $\phi \vee \psi$ is in the database then an $(\vee\mathcal{E})$-box is opened below it. Each half of the box now has its own copy of the database (with ϕ or ψ respectively replacing $\phi \vee \psi$) and goals. As this step may lead to duplication of the proof, we prefer to do it as late as possible.

The remaining heuristics involve backtracking (Remark 1.7.3).

(f) We use resolution (Remark 1.7.6) to prove an atomic goal $\vartheta[q]$ using $\forall x, y. \; \phi[x, y] \Rightarrow \vartheta[x]$ from the database $(\forall\Rightarrow\mathcal{E})$.

Notice that $(\forall\mathcal{I})$ and $(\exists\mathcal{E})$ mirror each other, but $(\exists\mathcal{I})$ and $(\forall\mathcal{E})$ do not. This is because a goal requires just one proof, whereas a hypothesis may be employed any number of times, or not at all. (Linear logic analyses the reuse of hypotheses, but we shall not consider it in this book.)

(g) If $\perp$ is in the database, all goals are immediately satisfied $(\perp\mathcal{E})$. More generally, if $\forall x. \; \neg\phi[x]$ is in the database then we may replace it by **put** $x_0 = ?$ and all outstanding goals by $\phi[x_0]$, using $\perp$ as a "joker" (Remark 1.4.4).

(h) If $\vartheta_0 \vee \vartheta_1$ is a goal then we seek first a proof of ϑ_0, and then (if that fails) a proof of ϑ_1.

REMARK 1·7·10: During resolution, we used declarations (**put** x_0 = ?) to introduce indeterminates. This was done to allow us to continue building the *logical structure* of the proof without specifying certain of its details. When we have obtained a valid proof, complete apart from the occurrences of an indeterminate x_0, we have to find a term which can be substituted *everywhere* for it. This term must satisfy any equations in which x_0 occurs, *irrespective of how they are nested within the proof box*, so the unification problem cuts across the scoping structure of the proof. Nevertheless, the term must still be well formed at the point of the declaration: the variables belonging to nested $(\forall\mathcal{I})$- and $(\exists\mathcal{E})$-boxes must not be free in it.

REMARK 1·7·11: We have gone beyond Fact 1.7.1 by discussing axioms of the form $\exists x.\ \phi[x]$, $\phi \vee \psi$ and $\neg\phi$. These are not *definite* in the sense of Example 1.7.5, because when $\exists x.\ \phi[x]$ is used the program cannot provide an *answer* for x, and x may remain free in any other answers it gives. Similarly *which* of ϕ or ψ holds in $\phi \vee \psi$ is indeterminate. If the joker $(\bot\mathcal{E})$ is used to prove $\exists y.\ \vartheta[y]$, again we have no idea what y is.

1.8   CLASSICAL AND INTUITIONISTIC LOGIC

Mathematical reasoning as commonly practised makes use of two other logical principles, excluded middle and the axiom of choice, which we shall not use in our main development. *Classical predicate calculus* consists of the rules we have given together with excluded middle.

Excluded middle. "Every proposition is either true or false."

DEFINITION 1·8·1:   ***Excluded middle*** may be expressed as either of

$$\frac{\neg\neg\phi}{\phi}\ \neg\neg \qquad \frac{}{\phi \vee \neg\phi}\ \text{EM}$$

which are equivalent in the sense that

$$\left(\forall\alpha.\ \neg\neg\alpha \Rightarrow \alpha\right) \dashv\vdash \left(\forall\beta.\ \beta \vee \neg\beta\right).$$

However, an individual formula may be $\neg\neg$-closed (satisfy the rule on the left) without being ***decidable*** or ***complemented*** (as on the right).

In saying that *we shall not use* excluded middle, beware that we are not *affirming its negation*, $\neg(\phi \vee \neg\phi) \equiv \neg\phi \wedge \neg\neg\phi$, which is falsity. If we are able to prove neither ϕ nor $\neg\phi$ then *we remain silent about them*.

Of course there are instances of case analysis even in intuitionism, in particular the properties of finite sets. A recurrent example will be *parsing* of terms in free algebras, for example a list *either* is empty *or* has a head (first element) and tail (Section 2.7).

Idiomatically, $(\neg\mathcal{I})$ and $(\vee\mathcal{E})$ are incorporated to give indirect rules

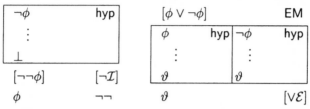

known as **reductio ad absurdum** and **tertium non datur** (Latin: the third is not given) respectively. Some Real Mathematicians use the former habitually, starting *every* proof with "suppose not" ("are you calling me a liar?") without even attempting a positive approach. Many treatments of logic itself (rather than difficult applications) regrettably do not *prove* things, but refute their negations instead.

Certain subjects, notably linear algebra, really have no idea of negation:

EXAMPLE 1·8·2:   Let M be an invertible matrix. To show that $M\mathbf{x} = \mathbf{u}$ has only one solution, it is quite unnecessary to assume that $\mathbf{a} \neq \mathbf{b}$ are *different* solutions, as the proof naturally leads to $\mathbf{a} = \mathbf{b}$ *without the aid of the hypothesis*. It is futile to obtain a contradiction to the hypothesis, *i.e.* $\neg\neg(\mathbf{a} = \mathbf{b})$, and then deduce $\mathbf{a} = \mathbf{b}$ after all.

See Remarks 2.5.7 and 3.7.12 for how this arises in induction.

THEOREM 1·8·3:   Negation is a **duality** in classical logic, interchanging $\top$ with $\bot$, $\wedge$ with $\vee$ and $\forall$ with $\exists$. That is,

$$\neg(\alpha \vee \beta) \quad \dashv\vdash \quad (\neg\alpha) \wedge (\neg\beta) \qquad \neg\exists x.\, \phi[x] \quad \dashv\vdash \quad \forall x.\, \neg\phi[x]$$
$$\neg(\alpha \wedge \beta) \quad \dashv\vdash \quad (\neg\alpha) \vee (\neg\beta) \qquad \neg\forall x.\, \phi[x] \quad \dashv\vdash \quad \exists x.\, \neg\phi[x].$$

The equations on the top row are valid without excluded middle. Those in the left column are called **de Morgan's laws**, although they were well known to William of Ockham. Excluded middle also makes any implication $\alpha \Rightarrow \beta$ equivalent to its **contrapositive**, $\neg\beta \Rightarrow \neg\alpha$. $\square$

REMARK 1·8·4:   In classical propositional calculus with n propositional variables (and no quantifiers), we may enumerate the 2^n cases where each of them is true or false. Such a listing of the values of a formula is called a **truth table**. Each line of the table which has the value $\top$ may be read as a conjunction of possibly negated atomic propositions, and the whole table as the disjunction of these lines; this is the **disjunctive normal form** and is classically equivalent to the given formula. So if two formulae have the same truth table they are classically inter-provable. In other words, this calculus is **complete**: anything which is true in all models (*i.e.* all 2^n cases) is provable.

ϕ	ψ	$\neg\phi$	$\phi\wedge\psi$	$\phi\vee\psi$	$\phi\Rightarrow\psi$
$\top$	$\top$	$\bot$	$\top$	$\top$	$\top$
$\bot$	$\top$	$\top$	$\bot$	$\top$	$\top$
$\top$	$\bot$		$\bot$	$\top$	$\bot$
$\bot$	$\bot$		$\bot$	$\bot$	$\top$

Venn diagrams, in which overlapping circles represent propositional variables, are a popular way of illustrating classical propositional logic, although they are misleading when one of the regions turns out to be empty. They were invented in 1764 by Johann Lambert, better known for his work on light intensity; he also proved the irrationality of π and introduced the hyperbolic functions sinh, cosh, *etc.*

The truth table approach to logic is pedagogically not so simple as is claimed, as the **material implication** $(\bot \Rightarrow \phi) = \top$ is an obstacle right at the start, which has confused every generation of students since ancient times. It took deeper insight into logic to discover the analogy between $(\Rightarrow\mathcal{I})$ and defining functions or sub-routines (Section 2.4), but once pointed out it is completely natural, and the $(\Rightarrow\mathcal{I})$-rule is easily grasped. This analogy is a pearl of modern logic, and I believe students should be allowed to glimpse it, rather than have the prejudices of classical logic reinforced. The material implication does feature in proof theory as the $(\bot\mathcal{E})$-rule, but we have already called this the joker of logic (Remark 1.4.4).

The Sheffer stroke. Whereas in art and in other parts of mathematics symmetry is considered beautiful, many logic texts use the de Morgan duality to eradicate half of the calculus in a quite arbitrary way. Instead of presenting $\forall$ and $\wedge$ with their own natural properties, they are treated as mere abbreviations for $\neg\exists\neg$ and $\neg(\neg\alpha \vee \neg\beta)$, or *vice versa*. In our intuition $\wedge$ and $\vee$ are twins, as are $\forall$ and $\exists$, so why should one of them be treated as a second class citizen in the logical world?

Gottfried Ploucquet (1764) discovered that a single binary connective actually suffices; the operation $\alpha \mid \beta \equiv \neg(\alpha \wedge \beta)$ is commonly known as the Sheffer stroke. Using $\forall x. \neg(\phi[x] \wedge \psi[x])$, Moses Schönfinkel was able to dispose of variables and quantifiers, reducing the whole logical calculus to just one symbol (plus brackets). His paper survives as an important part of the literature because his combinators have types which correspond to the structural rules (Example 2.4.2 and Exercise 2.26ff).

These operations "simplify" logic in the same way that it would simplify chemistry if (after the discovery of oxygen, carbon, hydrogen, *etc.*) *sea water* had been chosen as the one primitive substance ("element"), on the grounds that the usual 93 can all be obtained from it.

Although this nihilist tendency contributed to the failure of mainstream logic to observe the propositions as types analogy (which is very clear in intuitionism), or to recognise the quantifiers as adjoints, the Sheffer stroke is the building block of digital electronics, where it is called **nand**.

REMARK 1·8·5: When a small current is passed between the base and the emitter of a transistor, a larger current flows from the collector to the emitter. This effect is used in analogue circuits for amplification, but it performs an essentially *negating*[18] digital operation: a high voltage on the base (relative to the emitter) causes a low voltage on the collector and *vice versa*.

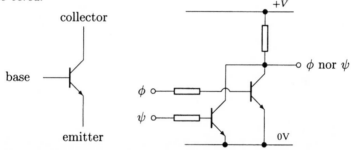

The example shows **nor**, $\neg(\phi \vee \psi)$, which may be used in a similar way to nand, so positive operators are made by concatenating such circuits.

Intuitionism. Excluded middle seems obvious, until you think about the reason for believing it, and see that this begs the question. It requires mathematicians to be omniscient, a power no other scientist would claim. Those who feel obliged to justify this insist on classifying structures up to isomorphism before using them to solve real problems.

William of Ockham (*c.* 1320), in whose name Henry Sheffer wielded his stroke, considered propositions about the future with an "indeterminate" value which even God does not yet know to be either true or false, in connection with the problem of the Free Will of potential sinners. Even Aristotle had his doubts about excluded middle, which is more properly attributed to the Stoics, according to Jan Lukasiewicz.

---

[18] A similar positive effect would be obtained by reckoning voltages relative to the collector, but this would be less reliable as a way of building circuits.

Modern transistors, embedded in integrated circuits, are activated by a potential difference, with negligible current, and so avoid the need for resistors.

This Remark is dedicated to my father, Ced Taylor, who taught me about logic gates and Fourier transforms in the context of electronics long before I knew about propositional calculus or trigonometry.

REMARK 1·8·6:   The modern critique of classical omniscience in analysis was formulated by Jan Brouwer (1907). Given a sequence whose ultimate behaviour is an unsolved problem (for which he took questions about patterns of digits in the decimal expansion of π), he constructed counterexamples to several of the major assumptions and theorems of analysis. Suppose for example that all *known* terms of (a_n) are zero. Although the real number $u = \sum_n a_n 2^{-n}$ is well defined, it is not known whether $u = 0$ or $u > 0$, so we cannot find even the first digit of $1 - u$ in the usual decimal expansion. He showed that continuous functions need not be Lebesgue-integrable, or have local maxima on $[0, 1]$. On the other hand, all intuitionistically definable functions $\mathbb{R} \to \mathbb{R}$ are continuous.

With hindsight, it is unfortunate that Brouwer chose analysis to attack. As we shall demonstrate throughout this book, logic and algebra can be presented intuitionistically, with no noticeable inconvenience, whereas most of the usual properties of the real numbers rely on excluded middle. For this reason we will not attempt to cover constructive analysis, but Errett Bishop [BB85] gave an account which is very much in our spirit: it makes such alterations to the definitions as are required, and gets on with proving the traditional theorems as well as can be done, discarding naïve formulations rather than dwelling on their counterexamples.

Every reform provokes reaction. Although Hilbert's influence is to be found behind almost every pre-war revolution in logic, he also made such comments as "no one shall drive us from the paradise that Cantor created for us" and his Platonist battle-cry, "Wir müssen wissen, wir werden wissen" (we must know, we shall know). Later he claimed that "to prohibit existence statements and the principle of excluded middle is tantamount to relinquishing the science of mathematics altogether".

Rather more of mathematics than Hilbert's "wretched remnants" has now been developed intuitionistically. Often, however, we shall find it convenient to assume excluded middle in order to give some simple introductory examples of concepts, particularly when the most familiar form occurs in the context of $\mathbb{R}$. There are also habits of language ("$\mathbb{N}$ has no proper subalgebra") which one is reluctant to give up: they are to be understood as *idioms*. The reader for whom constructivity is essential will be able to recognise the cases where these conventions apply.

Andrei Kolmogorov (1925) devised a translation of classical proofs into intuitionistic ones, which is often attributed to Kurt Gödel.

FACT 1·8·7:   Define $\phi^{\neg\neg}$ by structural recursion on predicates as follows:

$$\bot^{\neg\neg} = \bot \qquad\qquad \top^{\neg\neg} = \top$$
$$(\phi \wedge \psi)^{\neg\neg} \quad = \quad (\phi^{\neg\neg}) \wedge (\psi^{\neg\neg})$$

$$(\phi \vee \psi)^{\neg\neg} \quad = \quad \neg\neg\big((\phi^{\neg\neg}) \vee (\psi^{\neg\neg})\big)$$
$$(\phi \Rightarrow \psi)^{\neg\neg} \quad = \quad \neg\neg\big((\phi^{\neg\neg}) \Rightarrow (\psi^{\neg\neg})\big)$$
$$(\forall x.\ \phi[x])^{\neg\neg} \quad = \quad \neg\neg\forall x.\ (\phi[x]^{\neg\neg})$$
$$(\exists x.\ \phi[x])^{\neg\neg} \quad = \quad \neg\neg\exists x.\ (\phi[x]^{\neg\neg})$$

If $\Gamma \vdash \phi$ is provable classically, then $\Gamma^{\neg\neg} \vdash \phi^{\neg\neg}$ has an intuitionistic proof, *i.e.* without using excluded middle. In particular, intuitionistic logic does not save us from any inconsistency (the ability to prove $\bot$) which might arise classically. $\qquad\qquad\Box$

The axiom of choice. The increasingly abstract form of late nineteenth century mathematics led to the use of infinite families of choices, often with no conscious understanding that a new logical principle was involved. Giuseppe Peano did formulate, and reject, such an axiom in 1890, and Charles Sanders Peirce gave the following definition in 1893, but it was Ernst Zermelo who first recognised how widely it had already been used. His 1904 proof of the well-ordering principle (Proposition 6.7.13 and Exercise 6.53) attracted vehement opposition — at least, judged by the contemporary standards of courtesy, so much higher than today. It was in order to formulate his response that he found his famous axioms of set theory, which we shall discuss in Remark 2.2.9.

DEFINITION 1·8·8: The *axiom of choice* says that

> any entire relation $R : X \nrightarrow Y$ (Definition 1.3.1(c))
> contains a total functional relation $f : X \to Y$,
> *i.e.* $\forall x.\ x \overset{R}{\mapsto} f(x)$.

Exercises 1.38 and 2.15 were how Burali-Forti and Zermelo formulated it; the former is more convenient for category theory, and Radu Diaconescu showed that it implies excluded middle (1975, Exercise 2.16). Russell and Whitehead used a "multiplicative axiom", but this is only meaningful in the context of a much stronger principle (Proposition 9.6.13). Well-ordering was later supplanted in algebra by maximality properties such as Zorn's Lemma (Exercise 3.16), actually due to Kazimierz Kuratowski.

In the first use of Zermelo's axioms, Georg Hamel showed that $\mathbb{R}$ has a basis as a vector space over $\mathbb{Q}$. Tychonov's theorem (that a product of compact spaces is compact) and many other famous results have also been shown to be equivalent to it.

Not all of the consequences of Choice are benign, for instance it allows us to define non-measurable sets, with the bizarre corollary (due to Felix Hausdorff) that a sphere can be decomposed into two or more spheres, each congruent to the first. The moral of this is that if we allow the

Angels to employ brute force, then the Devil will make use of it too. Even Zermelo and many of the enthusiasts for Choice considered it appropriate to indicate when results depend on it.

When Hilbert gave his basis theorem using Choice, Paul Gordon (Emmy Noether's thesis adviser) said of it, "Das ist nicht Mathematik, das ist Theologie", having worked on the subject for twenty years using what we would now call constructive mathematics. Although we, on the cusp of the millennium, now reject Choice, it was the way forward at the *start* of the twentieth century: it stimulated research throughout mathematics, notably in the Polish school, which we have to thank for numerous ideas in logic and general topology mentioned in this book [Moo82, McC67].

Zermelo conjectured that Choice was independent of his other axioms, and Abraham Fraenkel devised models with permutable ur-elements in order to prove this. Kurt Gödel (1938) showed how to cut down a model of Zermelo's axioms to the "constructible sets", for which Choice is provable. However, it was 1963 when Paul Cohen found a model of Zermelo's other axioms in which Choice fails.

The *axiom* of choice is typically not needed in the concrete cases, because their own structure provides some way of making the selections (we shall indicate real uses of Choice by the capital letter). Often it is used to extend a property of some familiar structures to the generality of an abstract axiomatisation, but even then the need for Choice may be more a feature of that particular *formulation* than of the *actual* mathematical structures. For example, Peter Johnstone [Joh82] showed that, by a conceptual change from points to open sets, Tychonov's Theorem could be proved without Choice. It is for infinitary algebra in Sections 5.6 and 6.2 that this axiom will be most missed in this book. We respond to this difficulty by examining what infinitary operations are of interest in practice.

In the countable case the following assumption, formulated by Paul Bernays, is often more directly applicable.

DEFINITION 1·8·9: The axiom of **dependent choice** says that

any entire relation $X \overset{R}{\leftharpoonup} X$ with an element $x_0 \in X$
contains an (ω-)sequence,
i.e. a function $x_{(-)} : \mathbb{N} \to X$, such that $\forall n.\ x_n \overset{R}{\mapsto} x_{n+1}$.

If R is a function then this is primitive recursion over $\mathbb{N}$ (Remark 2.7.7). Similarly, we get Dependent Choice by repeatedly applying the choice function to the seed. König's Lemma (Corollary 2.5.10) is a widely used form of Dependent Choice throughout informatics and combinatorics. Dependent Choice does not imply excluded middle or *vice versa*.

Logic in a topos. For Jan Brouwer and his student Arend Heyting, Intuitionism was a profound philosophy of mathematics [Hey56, Dum77, Man98], but like increasingly many logicians, we shall use the word *intuitionistic* simply to mean that we do not use excluded middle. This abstinence is nowadays very important in category theory, not because of any philosophical conviction on the part of categorists (indeed most of them still use excluded middle as readily as the Real Mathematician does), but because it is the internal logic of the kind of world (a *topos*) which most naturally axiomatises the familiar mathematical universe.

Joachim Lambek and Philip Scott [LS86] show that the so-called "term model" of the language of mathematics, also known as the "free topos", may be viewed as a preferred world, but Gödel's incompleteness theorem shows that the term model of classical mathematics won't do.

Category theory provides the technology for creating new worlds. This is quite simple so long as we do not require them to be classical. Why should we want such worlds, though? One application is to provide the generic objects (*cf.* those used in proof boxes) in such a way that we may reason with them in the ordinary way in their own worlds, and then instantiate them. Arguments about "generic points" have been used in geometry since Giuseppe Veronese (1891), but the logic is unsound if, for example, the equality of two such points is required to be decidable.

Worlds have also been created with convenient but exotic properties. In synthetic differential geometry [Koc81] *all* functions on $\mathbb{R}$ are continuous, as Jan Brouwer said, and it is legitimate to use infinitesimals in the differential calculus. More recently, synthetic domain theory has similarly postulated that all functions are to be *computable*.

Excluded middle was traditionally regarded as a true fact about the *real* world, so in order to investigate intuitionistic logic it was necessary to build *fictional* worlds where excluded middle does not hold. The point of view of this book is that these worlds are not exotic but quite normal, and their logic is perfectly typical, just as algebraic extensions of the rationals have come to be seen as ordinary number domains with straightforward arithmetic structure.

It is not necessary to know in advance how to construct such worlds from classical ones before learning how to reason in them. Indeed excluded middle is like the fear of water: it's easier to learn to swim as a small child, before anyone's told you that it's difficult.

Whatever your philosophical standpoint may be, intuitionism forces you to write mathematics much more cleanly, and to understand much more deeply how it works. Proof by refutation runs backwards, and so the argument gets tangled. The constructive character of intuitionism is really due to the strong analogy with type theory, which does not extend to excluded middle. We describe it in Section 2.4, making use of another idea due to Kolmogorov.

EXERCISES I

1. When Bo Peep got too many sheep to see where each one was throughout the day, she found a stick or a pebble for each individual sheep and moved them from a pile outside the pen to another inside, or *vice versa*, as the corresponding sheep went in or out. Then one evening there was a storm, and the sheep came home too quickly for her to find the proper objects, so for each sheep coming in she just moved *any* one object. She moved all of the objects, but she was still worried about the wolf. By the next morning she had satisfied herself that the less careful method of reckoning was sufficient. Explain her reasoning *without the aid of numbers.*

2. For each of the connectives and quantifiers, give the phrases in English and any other language you know which usually express them. Point out any ambiguities, and how they are resolved in everyday usage. Now give in logical notation the literal and intended meanings of the following, choosing appropriate abbreviations for the atomic predicates:

>All farmers don't have cows.
>The library has some books by Russell and Whitehead.
>All passes must be shown at the gate.
>Dogs must be carried on the escalator.
>You hit me and I'll hit you.

3. The following equations are familiar in elementary algebra. Which of $\exists x$, $\forall x$ and $\{x \mid \cdots\}$ are understood?

$$(x+y)^2 = x^2 + 2xy + y^2 \quad ax^2 + bx + c = 0 \quad x^2 + y^2 = 1$$

4. Show that reductions in the Lineland Army (Example 1.2.7) are locally confluent. Equivalently, show that (;) is associative.

5. A **Turing machine** [Tur35] consists of a *head*, which may be in any of a finite number of *states*, and a *tape* which extends infinitely in both directions and is divided into *cells* (indexed by $\mathbb{Z}$), each of which contains one *symbol* from a finite *alphabet*. All but finitely many cells contain the *blank* symbol. For each state in which the head may be, and for each symbol which may be written in the cell currently being read, there is specified a new state, a new symbol and a direction of motion (left or right). Show how to express a Turing machine as a rewrite

system. [Hint: the root has the state and the current symbol as two of its arguments; the other two are the left and right parts of the tape, which must be expressed using the blank and one binary operation.] Since computation can only proceed in one place, the head has been called the **von Neumann bottleneck.**

6. Express confluence (Definition 1.2.5) as a formula involving ($\forall$, $\wedge$, $\Rightarrow$, $\rightsquigarrow$ and) $\exists$. Using proof boxes — in particular the ($\exists\mathcal{E}$)-rule — show that R satisfies the property iff $R^{op} ; R \subset R ; R^{op}$.

7. Prove Theorem 1.2.9 about local confluence.

8. Show that the formulae

$$\exists x. \left(\phi[x] \wedge \forall y. (\phi[y] \Rightarrow x = y)\right)$$

$$\exists x. \forall y. (\phi[y] \Leftrightarrow x = y)$$

$$(\exists x. \phi[x]) \wedge (\forall x. \forall y. \phi[x] \wedge \phi[y] \Rightarrow x = y)$$

are inter-provable. Use them to derive idiomatic proof-box rules for $\exists!$.

9. Suppose that "x is a widget" and "x is a gadget" are descriptions. Show that: the widget is the gadget iff the gadget is the widget.

10. Consider the formulae

(a) $\forall x, y, z. \, z = y \vee y = z \vee z = x$;

(b) $\neg\neg\forall x, y, z. \, z = y \vee y = z \vee z = x$;

(c) $\forall x, y, z. (\neg\neg z = y) \vee (\neg\neg y = z) \vee (\neg\neg z = x)$;

(d) $\forall x, y, z. \, \neg\neg(z = y \vee y = z \vee z = x)$.

Show that (a) is the strongest and (d) the weakest, and that any other formula obtained by inserting $\neg\neg$ (other than between $\forall x$, $\forall y$ and $\forall z$) is equivalent to one of these. Restate the weakest using $\wedge$ instead.

11. Devise formulae similar to those of the previous exercise to say that ϕ has exactly three, four, ..., solutions. By adjoining a condition that ϕ and ψ have no common solutions (Example 2.1.7), interpret the equations $1 + 1 = 2$, $1 + 2 = 3$ and $2 + 2 = 4$ and prove them using the box method. ($1 + 1 = 2$ is proved in this sense on page 360 of *Principia Mathematica*.)

12. What, if anything, do the negations of $x \in \mathsf{FV}(t)$ (Definition 1.1.3), $x : X$ (Notation 1.3.3) and $x \in U$ (Definition 2.2.3) mean?

13. Show that $f(x) = f(y)$ defines an equivalence relation on X, where $f : X \to Y$ is any function.

14. Show that a relation $R : X \nrightarrow Y$ is total and functional iff the composite $R \hookrightarrow X \times Y \xrightarrow{\pi_0} X$ is bijective. So functions $X \to Y$ correspond to **sections** i of π_0, *i.e.* such that $i ; \pi_0 = \mathrm{id}$ (Definition 1.3.12).

15. List the sixteen cases where the functional, total (entire), injective and surjective conditions do and do not hold for a binary relation. For each case give an example and, where possible, a name and notation; three less familiar ones represent overlap, subquotient and its converse.

16. Show that a relation $R : X \rightharpoonup Y$ is functional iff $R \circ R^{op} \subset$ id, total iff also id $\subset R^{op} \circ R$, injective iff $R^{op} \circ R \subset$ id and surjective iff id $\subset R \circ R^{op}$. Hence prove Lemma 1.3.11. Show also that a function f is injective iff it is a **monomorphism**: $g_1 \, ; f = g_2 \, ; f \Rightarrow g_1 = g_2$, and surjective iff it is an **epimorphism**: $f \, ; g_1 = f \, ; g_2 \Rightarrow g_1 = g_2$; *cf.* Proposition 5.2.2(d).

17. Describe the sixteen cases for a binary (endo)relation where reflexivity, symmetry, transitivity and functionality do and do not hold, noting those for which idempotence necessarily holds.

18. Let $R : \mathbf{n} \rightharpoonup \mathbf{m}$ be a decidable relation between two finite sets. Write $\overline{R}$ for the $(n \times m)$ matrix with 1 in the (i, j)-position if iRj and 0 otherwise. Compare the matrix product $\overline{R} \cdot \overline{S}$ with the relational composition $\overline{R \, ; S}$.

19. Show that the sequent $\Gamma \vdash \vartheta$ is provable in the sequent calculus iff ϑ is provable from hypotheses Γ in the box style.

20. Describe the introduction and elimination rules for $\neg$ and $\Leftrightarrow$ by adapting those for $\Rightarrow$ and $\wedge$. In each case prove that the derived rules are equivalent, and describe the verbal mathematical idioms which express them.

21. Show that

(a) $\vee$, $\wedge$ and $\Leftrightarrow$ are commutative and idempotent;

(b) $\vee$ and $\wedge$ are associative;

(c) the boxes for $(\forall\mathcal{I})$, $(\Rightarrow\mathcal{I})$ and $(\exists\mathcal{E})$ may be interchanged.

What can be done with $(\wedge\mathcal{I})$ and $(\vee\mathcal{E})$?

We tend to abuse $\Leftrightarrow$ transitively, so that $\phi \Leftrightarrow \psi \Leftrightarrow \chi$ means $(\phi \Leftrightarrow \psi) \wedge (\psi \Leftrightarrow \chi)$, but why is this an abuse?

22. Prove the following by the box method. Make it clear where any formulae are imported into boxes (Lemma 1.6.3).

$(\alpha \wedge \beta) \Rightarrow \gamma$	$\dashv\vdash$ $\alpha \Rightarrow (\beta \Rightarrow \gamma)$	*Currying*
$\exists x. \, [\psi \wedge \phi[x]]$	$\dashv\vdash$ $\psi \wedge \exists x. \, \phi[x]$	*Frobenius law*
$(\alpha \vee \beta) \wedge \phi$	$\dashv\vdash$ $(\alpha \wedge \phi) \vee (\beta \wedge \phi)$	*distributivity*
$(\alpha \wedge \beta) \vee \phi$	$\dashv\vdash$ $(\alpha \vee \phi) \wedge (\beta \vee \phi)$	*codistributivity*
$\phi \Rightarrow (\alpha \wedge \beta)$	$\dashv\vdash$ $(\phi \Rightarrow \alpha) \wedge (\phi \Rightarrow \beta)$	
$(\alpha \vee \beta) \Rightarrow \phi$	$\dashv\vdash$ $(\alpha \Rightarrow \phi) \wedge (\beta \Rightarrow \phi)$	

$$(\exists x.\ \phi[x]) \Rightarrow \psi \quad \dashv\vdash \quad \forall x.\ (\phi[x] \Rightarrow \psi)$$

$$\neg\neg\neg\phi \quad\quad\quad \dashv\vdash \quad \neg\phi$$

where $x \notin \mathsf{FV}(\psi)$.

23. In *linear implicational logic* the only structural rules are identity, cut and exchange (*not weakening and contraction*). Contexts are then bags (unordered lists), not sets. There is only one connective ($\Rightarrow$, but it is usually written $\multimap$), obeying the sequent form of ($\Rightarrow\mathcal{I}$) and

$$\frac{\Gamma \vdash \phi \quad\quad \Delta \vdash \phi \multimap \psi}{\Gamma, \Delta \vdash \psi}\ \multimap\!\mathcal{E}$$

Using ideas of double-entry bookkeeping, develop a box style of proof (like that in Section 1.5) which is sound for this logic. [Hint: the reasons or credit column must cite the two formulae which are used in each ($\multimap\mathcal{E}$), and there must also be a debit column which records where the present formula is used.] Replace the cross-references by arrows (*proof nets*) and $\multimap$ by a ternary node with one outgoing and two incoming arrows. Show that the boxes are then redundant, and investigate the dynamical behaviour when ($\multimap\mathcal{E}$) and ($\multimap\mathcal{I}$) meet.

24. (Only for those who already know linear logic.) Extend the proof box method to the connectives $\otimes$, $\oplus$ and $\&$, giving the translations into and from the sequent, λ-calculus and proof net approaches.

25. Writing (in modern notation)

$$(A_{ij}) \quad \forall x.\ \phi_i[x] \Rightarrow \phi_j[x] \quad\quad\quad (E_{ij}) \quad \forall x.\ \phi_i[x] \Rightarrow \neg\phi_j[x]$$
$$(I_{ij}) \quad\ \exists x.\ \phi_i[x] \wedge \phi_j[x] \quad\quad\quad (O_{ij}) \quad \exists x.\ \phi_i[x] \wedge \neg\phi_j[x],$$

prove Aristotle's *syllogisms*,

$$A_{12}, A_{23} \vdash A_{13} \quad\quad A_{12}, E_{23} \vdash E_{13} \quad\quad I_{12}, A_{23} \vdash I_{13}$$
$$I_{12}, E_{23} \vdash O_{13} \quad\quad O_{12}, A_{32} \vdash O_{13} \quad\quad A_{21}, O_{23} \vdash O_{13}$$

26. Write the direct and indirect rules of propositional logic in the style of Hilbert (Remark 1.6.10). Using steps of the form $\forall \underline{x}.\ \bigwedge \Gamma \Rightarrow \phi$, extend this to the quantifiers (this wasn't how Hilbert did it).

27. Formulate and prove the soundness of the rules of natural deduction with respect to a truth values semantics.

28. An assertion is *immediate* if it is the conclusion of an applicable rule, *i.e.* it may be deduced with no mediating argument. Give examples of mathematical arguments which are obvious but not immediate and *vice versa* (page 44), and also *trivial* topics (page 37).

29. Using Remark 1.6.2, translate the proof of Lemma 1.6.6 (that composition preserves functionality and totality) into the vernacular.

30. Using the (ε–δ) definition of continuity $\mathbb{R} \to \mathbb{R}$, show as a proof box that the composite $g \circ f$ of two such functions is continuous.

31. Why may we assume in Remark 1.7.9(c) that a hypothesis of the form $\exists x.\ \phi[x]$ is used exactly once in a proof, *i.e.* with just one witness?

32. Show how to extend Fact 1.7.1 to $\phi \vee \psi$ and $\exists x.\ \phi[x]$ in the database, though the proof is no longer uniform. This requires us to use $\phi \vee \psi$ as *soon* as possible, contrary to Remark 1.7.9(e); can our Remark be justified?

33. Devise heuristics which use axioms of the forms $\xi \Rightarrow (\phi \vee \psi)$ and $\xi \Rightarrow \exists x.\ \phi[x]$. Why must the new goal ξ be put at line 50 and new data ($\phi \vee \psi$ or $\exists x.\ \phi[x]$) at line 51, instead of their usual places at the bottom and top of the box? Give examples of theorems (in analysis, for example) whose statements are of this form. What are the idioms for dividing up arguments in which such theorems are used?

34. Show that ϕ is decidable iff $\phi \vee \neg\phi$ is $\neg\neg$-closed, and then ϕ itself is also $\neg\neg$-closed. Show that de Morgan's law gives $(\neg\phi) \vee (\neg\neg\phi)$, so under this assumption every $\neg\neg$-closed formula is decidable.

35. Write out the truth tables for the two sides of de Morgan's laws and the distributive laws. Show that if two propositional formulae have the same truth table then they are classically inter-provable.

36. Express each of $\neg$, $\vee$, $\wedge$ and $\Rightarrow$ in terms of either nand (the Sheffer stroke) or nor.

37. Write down the formulae for the sum and carry bits in the binary addition of two single-bit numbers ($0 + 0 = 00$, $0 + 1 = 01$, $1 + 0 = 01$ and $1 + 1 = 10$). By expressing them in terms of *nor*, give a circuit for a **half adder**. Show how to add two n-bit binary numbers; why is the half adder so called?

38. Show that every surjective function $p : X \twoheadrightarrow Y$ has a section $i : Y \hookrightarrow X$ such that $i\,;p = \mathrm{id}_Y$ iff the axiom of choice (Definition 1.8.8) holds. [Hint: *cf.* Exercise 1.14.]

39. Let ω be any proposition whose truth-value you know but which others may dispute, such as "July is in the winter". Suppose that Ω, the type of truth values, consists of true, false, ω and $\neg\omega$ (with apologies to tropical readers). This supposition is consistent with most of pure mathematics. Explain how *it is still the case* that Ω is the two-element set $\{\top, \bot\}$. What element of this set is ω? Where would you have to be to observe Ω as a four-element lattice?

Types and Induction

E VERY MATHEMATICIAN'S toolbox contains tuples and subsets for making ideal elements, and proofs by induction. In this chapter we bring together the traditional techniques which form the received view of the foundations of twentieth century mathematics. Afterwards they will be dismantled and reconsidered in the light of later algebraic experience.

At the beginning of his career, Georg Cantor investigated sets of points of discontinuity which functions could have whilst still admitting Fourier representations. He also gave a construction of the real numbers from the rationals, and showed that there are a lot more reals than rationals (Hermann Weyl later reproached analysts for decomposing the continuum into single points). Cantor was led to considering *abstract* sets, forming hierarchies under constructions such as the set of all subsets.

There are historical parallels between mathematics and programming in the development of types. Cantor was concerned with the magnitudes of sets, whereas FORTRAN distinguished between integer and real data types because they have different storage requirements. (Linear logic shows that resource analysis continues to be a fruitful idea.) Both started from the integers and real numbers alone. Bertrand Russell formulated his theory of types as a way of avoiding the vicious circles which he saw as the root of the paradoxes of set theory. On the other hand, the one lesson which the software industry has learned from informatics is that the type discipline catches a very large proportion of errors and thereby makes programs more reliable. Early calculi provided a static universe in advance, but modern type theories and programming languages create new types dynamically from old ones.

What is an abstract set? Some accounts of set theory claim that it is a voluntary conspiracy of its elements, coming together arbitrarily from independent sources (the *inductive conception*). But this conflicts with mathematical practice, and has little backing even in philosophical tradition. Plato held that members of a class are images of a Form; in practice, we conceive of the Form and *certain* of its instances first. The totality is only a semantic afterthought (and the instances are usually not themselves Forms). Indeed, from Zeno's time, points in geometry *lay on* lines but did not *constitute* them.

Gottlob Frege defined sets by comprehension of predicates, which at first he allowed to take *anything* as their subjects. Russell's famous $\{x \mid x \notin x\}$ showed that things couldn't be done quite so naïvely, so instead we select the elements *from an already given ambient set.*

For us, types are not imposed afterwards to constrain the size of the world, but are a precondition of meaning. In elementary trigonometry sin is thought of as applying to *angles* only, which are only reduced to real numbers by choosing a unit of measurement. Physical quantities may only be added or tested for equality if they measure the same thing (length, mass, energy, electric charge, *etc.*) in the same units; sometimes laws of mechanics can be guessed by this dimensional analysis alone, or from a scale model which preserves the dimensionless part (such as the Reynolds number in fluid mechanics).

More complex types are formed by processes, such as the powerset, like those generating terms and logical formulae. The establishment of certain standard abstract methods of construction made it possible to state and prove results of a generality that would not have been considered in the nineteenth century. As Michael Barr [BW85, p. 88] put it, "The idea of constructing a quotient space without having to have the ambient space including it, for example, was made possible by the introduction of set theory, in particular by the advent of the rather dubious idea that a set can be an element of another set. There is probably nothing in the introduction of topos theory as foundations more radical than that."

On the other hand, the importance of the quotient operation is such that *it* should perhaps be taken as primitive instead. Modern type theory builds hierarchies as Cantor and Zermelo did, but using simpler ways of forming types, such as the product, sum and set of functions. These correspond very directly to conjunction, disjunction and implication of propositions, an analogy which will be an important guiding principle for the rest of the book. We shall find that the structure of the types is characterised, not by their set-theoretic incarnations, but by certain operations, such as projection and evaluation maps, which build terms of that type and take them apart. In particular the λ-calculus handles the terms arising from the function-type.

The second half of the chapter is devoted to induction and recursion. Sections 2.5–2.6 discuss well founded relations, a notion of induction which also comes from the set-theoretic tradition, but we shall motivate it instead from the problem of proving correctness and termination of a wide class of recursive programs. There are classical idioms of induction based on minimal counterexamples and descending sequences, but we shall show how they can often be made intuitionistic. More complicated inductive arguments can be justified by constructing recursion measures using lexicographic products and other methods.

For programming (and foundations), structural recursion over lists, trees and languages is more important. In Section 2.7 we treat lists and Peano induction over the natural numbers in a similar fashion to the function-type. The last section treats second and higher order logic.

2.1 CONSTRUCTING THE NUMBER SYSTEMS

The growth of algebra from the sixteenth to the nineteenth century made the idea of number more and more general, apparently demanding ever greater acts of faith in the existence and meaningfulness of irrational, negative and complex quantities. Then in 1833 William Rowan Hamilton showed how complex numbers ($\mathbb{C}$) could be *defined* as pairs of real numbers, and the arithmetic operations by formulae involving these pairs. Ten years later he discovered a similar system of rules with four real components, the quaternions.

The rationals ($\mathbb{Q}$) may also be represented in the familiar way as pairs of integers ($\mathbb{Z}$), although now there are many pairs representing each rational (Example 1.2.1), and the positive and negative integers may be obtained from the natural numbers ($\mathbb{N}$) in a similar way. This leaves the construction of the reals ($\mathbb{R}$) from the rationals.

The real numbers. The course of the foundations of mathematics in the twentieth century was set on 24 November 1858, when Richard Dedekind first had to teach the elements of the differential calculus, and felt more keenly than before the lack of a really scientific foundation for analysis. In discussing the approach of a variable magnitude to a fixed limiting value, he had to resort to geometric evidences. Observing how a point divides a line into two parts, he was led to what he saw as the essence of continuity:

REMARK 2·1·1: If all points of the straight line fall into two classes such that every point of the first class lies to the left of every point of the second class, then there exists one and only one point which produces this severing of the straight line into two portions.

In [Ded72] he used these **Dedekind cuts** of the set of rational numbers to define real numbers, and went on to develop their arithmetic and analysis. By way of an example, he proved "for the first time" that $\sqrt{2} \times \sqrt{3} = \sqrt{6}$. There is one slight difficulty, in that each rational number gives rise to two cuts, depending on whether it is itself assigned to the lower or upper part — we shall say *neither*.

A real number is then a *pair* of *subsets* of $\mathbb{Q}$, and, from the universe of all pairs of subsets $(L, U \subset \mathbb{Q})$, the collection $\mathbb{R}_D$ of (Dedekind) reals is the *subset* consisting of those satisfying a certain property, namely

$$\forall x.\, \neg(x \in L \wedge x \in U)$$
$$\wedge\, \forall x, y.\, y \in L \wedge x < y \Rightarrow x \in L$$
$$\wedge\, \forall x, y.\, y \in U \wedge x > y \Rightarrow x \in U$$
$$\wedge\, \forall x.\, x \in U \Rightarrow \exists y.\, y \in U \wedge y < x$$
$$\wedge\, \forall x.\, x \in L \Rightarrow \exists y.\, y \in L \wedge y > x$$
$$\wedge\, \forall \varepsilon > 0.\, \exists x, y.\, x \in L \wedge y \in U \wedge y - x < \varepsilon.$$

To do this we have used the **cartesian product** (collecting all pairs), the **powerset** (collecting all subsets), and **comprehension** (forming a subset by selecting those elements which satisfy a particular property, for example the circle $S^1 \subset \mathbb{R}^2$ considered as the set of solutions of $x^2 + y^2 = 1$).

Georg Cantor (1872) gave another construction of $\mathbb{R}$ based on the idea of the convergence of sequences, such as the decimal expansion of π. First we must explain how a sequence may be *abstractly* convergent without having a limit point which is known in advance.

DEFINITION 2·1·2: A **sequence** in a set X is a function $a_{(-)} : \mathbb{N} \to X$. This is called a **Cauchy sequence** (in $X = \mathbb{Q}$ or $\mathbb{R}$) if

$$\forall \varepsilon > 0.\, \exists N.\, \forall n, m > N.\, |a_n - a_m| < \varepsilon.$$

If $\forall \varepsilon > 0.\, \exists N.\, \forall n, m > N.\, |a_n - b_m| < \varepsilon$ then the sequences (a_n) and (b_n) are **equivalent**, and $\mathbb{R}_C$ consists of the equivalence classes.

The Dedekind and Cantor constructions, which are equivalent in classical logic, are developed and related in Exercises 2.2–2.11.

EXAMPLES 2·1·3: Other familiar systems represented by subsets.

(a) A point in **projective *n*-space** is a line through the origin in $\mathbb{R}^{n+1}$, *i.e.* an equivalence class of $\mathbb{R}^{n+1} \setminus \{0\}$ with respect to the relation that $(x_0, \ldots, x_n) \sim (y_0, \ldots, y_n)$ if for some $k \neq 0$, $x_0 = k\, y_0$, ..., $x_n = k\, y_n$. The $(n-1)$-plane at infinity consists of those classes for which the co-ordinate x_0 is zero: no infinite co-ordinates are needed.

(b) As $6 = (1 + \sqrt{-5})(1 - \sqrt{-5}) = 3 \cdot 2$, unique factorisation fails in $R = \mathbb{Z}[\sqrt{-5}]$. To remedy this, Ernst Kummer (1846) introduced **ideal numbers**, which are subsets $I \subset R$ closed under addition and under multiplication by elements of R. An ordinary number $r \in R$ is represented by its set of multiples, and in general we define

$$(r_1, \ldots, r_n) = \{a_1 r_1 + \cdots + a_n r_n \mid a_i \in R\}.$$

The product IJ is the ideal *generated* by $\{ij \mid i \in I, j \in J\}$ and then the prime factorisation of 6 is $(1 + \sqrt{-5}, 2)^2 (1 + \sqrt{-5}, 3)(1 - \sqrt{-5}, 3)$.

Functions and equivalence classes. The Cantor construction of the reals adds two further operations, but these may themselves be defined in terms of the product, powerset and comprehension. The idea in both cases is to internalise the definitions of equivalence relation and function from Sections 1.2 and 1.3 respectively. The connection between the exponential and the set of functions, defined using input–output pairs, was also first made by Cantor (but for cardinal arithmetic, in 1895).

EXAMPLE 2·1·4: For sets X and Y, the function-type is constructed as $Y^X = \{f : \mathcal{P}(X \times Y) \mid \psi[f]\}$, where $\psi[f]$ is (*cf.* Definition 1.3.1)

$$\forall x. \, \forall y_1, y_2. \, \langle x, y_1 \rangle \in f \wedge \langle x, y_2 \rangle \in f \Rightarrow y_1 = y_2$$
$$\wedge \, \forall x. \, \exists y. \, \langle x, y \rangle \in f.$$

Any actual function $p : X \to Y$ is represented by $\{\langle x, p(x) \rangle \mid x \in X\}$. Conversely, given $f \in Y^X$ and $a : X$, from $\psi[f]$ we have

$$\exists! y {:} Y. \, \langle a, y \rangle \in f, \qquad i.e. \, \langle a, y \rangle \in f \Leftrightarrow y = f(a),$$

so $\langle a, y \rangle \in f$ is a *description* (Definition 1.2.10) of the result, called $f(a)$. But in order to understand function-types properly, the **evaluation** operation $\mathrm{ev} : (f, a) \mapsto f(a)$ must be studied in its own right. $\qquad \square$

EXAMPLE 2·1·5: Let $\sim$ be an equivalence relation (Definition 1.2.3) on a set X. Then the *quotient* is the set $X/\!\!\sim \; = \{U \subset X \mid \vartheta[U]\}$ of **equivalence classes**, where $\vartheta[U]$ is

$$\forall x, y. \, x \in U \wedge x \sim y \Rightarrow y \in U$$
$$\wedge \, \forall x, y. \, x \in U \wedge y \in U \Rightarrow x \sim y$$
$$\wedge \, \exists x. \, x \in U.$$

For $x \in X$, we write $[x] = \{y \mid x \sim y\} \in X/\!\!\sim$. The union of these subsets is X, they are inhabited, and if any two them overlap at all then they coincide (classically, we would say that they are non-empty, and either disjoint or equal); such a family of subsets is called a **partition**.

Let $f : X \to \Theta$ be a function such that

$$\forall x, y. \, x \sim y \implies f(x) = f(y);$$

then for any $U \subset X$ that satisfies $\vartheta[U]$, the formula

$$\exists! z. \, \exists x. \, x \in U \wedge z = f(x)$$

provides a *description* of a value $z \in \Theta$, which we call $p(U)$,

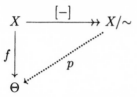

thereby defining a function $p : X/\!\!\sim \ \to \Theta$.   □

Unions and intersections. Having shown that the more powerful operations can be reduced to the product, powerset and comprehension, we complete the picture by treating the simpler ones in the same way. (See also Proposition 2.8.6 for the logical connectives.)

EXAMPLE 2·1·6: Let a be an element of a set X, and $U, V \subset X$ be the subsets characterised by predicates $\phi[x]$ and $\psi[x]$ respectively. Then

(a) the **singleton**, $\{a\}$, is characterised by the predicate $x = a$ in x,

(b) the **union**, $U \cup V$, is characterised by $\phi[x] \lor \psi[x]$,

(c) in particular $\{a, b\} = \{a\} \cup \{b\} = \{x \mid x = a \lor x = b\}$,

(d) the **intersection**, $U \cap V$, is characterised by $\phi[x] \land \psi[x]$, and

(e) the **difference**, $U \smallsetminus V$, by $\phi[x] \land \neg\psi[x]$.

(f) Given excluded middle, $X \smallsetminus V$ is the **complement** of V in X:
$$(X \smallsetminus V) \cap V = \varnothing \qquad (X \smallsetminus V) \cup V = X.$$ □

The operations we have just described form subsets of an ambient set X. The *disjoint* union, like the product, function-type and quotient, forms a *new* set. It can also be constructed using products, powersets and comprehension, though the following construction may be unfamiliar as it is not the same as that used in set theory. (The common set-theoretic construction is not valid in the axiomatisation of the next section.)

EXAMPLE 2·1·7: If X and Y are sets then their **sum** or **disjoint union** is $X + Y = \{\langle U, V \rangle : \mathcal{P}(X) \times \mathcal{P}(Y) \mid \phi[U, V]\}$, where $\phi[U, V]$ says that U and V have exactly one element altogether, *i.e.*

$$[(\exists x.\, x \in U) \lor (\exists y.\, y \in V)]$$
$$\land\ \forall x_1, x_2.\ x_1 \in U \land x_2 \in U \Rightarrow x_1 = x_2$$
$$\land\ \forall y_1, y_2.\ y_1 \in V \land y_2 \in V \Rightarrow y_1 = y_2$$
$$\land\ \forall x, y.\ \neg(x \in U \land y \in V).$$

For $a \in X$ or $b \in Y$ we write $\nu_0(a) = \langle \{a\}, \varnothing \rangle$ and $\nu_1(b) = \langle \varnothing, \{b\} \rangle$.

Conversely, Exercise 2.13 shows that every element of $X + Y$ is of one or other of these forms, but not both — yet another *description*. (Exercise 1.11 was based on a similar idea.) *Case analysis* may be used on such a value: for any two functions $f : X \to \Theta$ and $g : Y \to \Theta$, there is a unique function $p : X + Y \to \Theta$ such that

$$p(\nu_0(a)) = f(a) \qquad p(\nu_1(b)) = g(b),$$

and we write $[f, g]$ for p. □

Singletons and the empty set. As with the union and disjoint union, there is a conceptual difference between the singleton as a free-standing *set* (which we call **1**) and the singleton *subset* consisting of a particular element of a given set. Up to interchangeability (unique isomorphism) there is only one singleton *set*, and only one empty set. The two notions of singleton are related by the correspondence between elements of any set X (and so its singleton subsets) and functions $\mathbf{1} \to X$.

EXAMPLES 2·1·8:

(a) On any set X we have the constantly true and false predicates, $\top$ and $\bot$, which characterise $X \subset X$ and the **empty set** $\emptyset \subset X$; any other subset lies between these. In particular, $\emptyset \cong \{x : \mathbf{1} \mid \bot\}$.

(b) The only subset of $\emptyset$ is itself, because the true and false predicates coincide (Exercise 2.29). Hence $\mathcal{P}(\emptyset)$ has exactly one element, $\emptyset$. We shall write $\mathbf{1} = \{\star\}$ for the **singleton**, since it is clearer to make its element anonymous than to insist on $\mathbf{1} = \mathcal{P}(\emptyset) = \{\emptyset\}$. □

The symbol $\emptyset$ appears to be a 1950s variant of zero, having nothing to do with Latin O, Greek ϕ or Danish/Norwegian Ø.

The roles of $\emptyset$ and $\mathbf{1}$ are summed up by

PROPOSITION 2·1·9: Let X be any set. Then

$$\begin{array}{ccccccc} X \times \emptyset & \cong & \emptyset & \quad & X \times \mathbf{1} & \cong & X \\ X^{\emptyset} & \cong & \mathbf{1} & \quad & X^{\mathbf{1}} & \cong & X \qquad \mathbf{1}^{X} \cong \mathbf{1} \end{array}$$

(a) there is a unique relation $\emptyset \rightharpoondown X$, namely the empty or constantly false one, and this is a function $\emptyset \to X$;

(b) any function $X \to \emptyset$ is bijective;

(c) for any sets X and Y, there is a unique bijection

$$\{x : X \mid \bot\} \cong \{y : Y \mid \bot\},$$

so we are justified in using $\emptyset$ for the smallest subset of *any* set;

(d) total functions $\mathbf{1} \equiv \{\star\} \to X$ are of the form $\star \mapsto a$, so correspond bijectively to elements $a \in X$;

(e) there is a unique total function $X \to \mathbf{1} \equiv \{\star\}$, namely $x \mapsto \star$. □

There is one further major construction of this kind which we shall do, namely that of the free algebra for a free theory (Proposition 6.1.11), but now we turn to the axiomatisation of these operations.

2.2   SETS (ZERMELO TYPE THEORY)

These methods of construction were first set out as a basis of set theory by Ernst Zermelo in 1908. The subsequent work sought to formalise them in terms of a notion of membership in which any entity in the universe may serve either as an element or as a set, and where it is legitimate to ask of any two entities whether one bears this relation to the other.

We shall make a *distinction* between elements and sets, though in such a formalism it is usual to refer to *terms* and *types* as we did in Section 1.3. We shall also modify what Zermelo did very slightly, taking the cartesian product $X \times Y$ as a primitive instead of the unordered pair $\{X, Y\}$, and the singleton instead of the empty set (*cf.* Examples 2.1.8).

Our system conforms very closely to the way mathematical constructions have *actually* been formulated in the twentieth century. The claim that set theory provides the foundations of mathematics is only justified *via* an encoding of this system, and not directly. It is, or at least it should be, surprising that it took 60 years to arrive at an axiomatisation which is, after all, pretty much as Zermelo did it in the first place.

The study of sheaf theory by the Grothendieck school unintentionally wrested foundations from the set-theorists, though it was Bill Lawvere who saw that logic could be done in these new worlds (toposes). The formulation of languages for such reasoning was undertaken by Bénabou, Coste, Fourman, Joyal and Mitchell; although they called it "set theory", they were in fact developing the type theory below and in Chapter V. For a detailed account of the modern system and its history, see [LS86].

DEFINITION 2·2·1:   *Zermelo type theory* consists of:

(a) the *singleton*, **1**, which has just one element, called $\star$;

(b) the *cartesian product*, $X \times Y$, whose elements are *ordered pairs* $\langle a, b \rangle$, where $a \in X$ and $b \in Y$, whenever X and Y are sets; Remark 2.2.2 gives the full definition of the product and associated pairing and projection operations;

(c) the set U, whenever X is a set and $U \subset X$ a subset of it;

(d) the *powerset*, $\mathcal{P}(X)$, whose elements are the subsets of a set X;

(e) and the set $\mathbb{N}$ of natural numbers (Section 2.7).

Singleton and product. We shall give introduction and elimination rules for the types announced in this Definition, as we did in Sections 1.4–1.5 for the predicate calculus. There we were really only interested in the *fact* that various propositions could be proved, but now we want to say that certain terms do or do not denote the same value, so we must give reduction rules relating them (Section 1.2).

REMARK 2·2·2: In the colon notation (1.3.3), the product satisfies

$$\frac{a:X \qquad b:Y}{\langle a,b\rangle : X\times Y}\times\mathcal{I} \qquad \frac{f:X\times Y}{\pi_0(f):X}\times\mathcal{E}0 \qquad \frac{f:X\times Y}{\pi_1(f):Y}\times\mathcal{E}1$$

$$\frac{a=a':X \quad b=b':Y}{\langle a,b\rangle = \langle a',b'\rangle} \qquad \frac{f=f':X\times Y}{\pi_0(f)=\pi_0(f')} \qquad \frac{f=f':X\times Y}{\pi_1(f)=\pi_1(f')}$$

$$\frac{f:X\times Y}{\langle\pi_0(f),\pi_1(f)\rangle = f}\times\eta \qquad \frac{a:X \quad b:Y}{\pi_0\langle a,b\rangle = a}\times\beta 0 \qquad \frac{a:X \quad b:Y}{\pi_1\langle a,b\rangle = b}\times\beta 1$$

The rules have been named in the same way as in Remark 1.4.3: the introduction rule ($\times\mathcal{I}$) creates a term of product type (the pair), which is used by the elimination rules, and the β- and η-rules cut out detours. The equality rules say that substitution interacts with π_0, π_1 and $\langle\ ,\ \rangle$ in the same way that it does with operation-symbols.

The other eight rules make ($\times\mathcal{I}$) into a two-way correspondence:

$$\frac{a=\pi_0(f):X \qquad b=\pi_1(f):Y}{f=\langle a,b\rangle : X\times Y}$$

There are two elimination rules and two β-rules: in the ternary case there would be three, so in the nullary case (the singleton) there is no elimination rule at all. ($\times\mathcal{I}$) has two premises, so ($1\mathcal{I}$) has none.

$$\frac{}{\star : 1}\,1\mathcal{I} \qquad\qquad \frac{f:1}{\star = f}\,1\eta$$

Compare these with the rules for descriptions (Definition 1.2.10ff) and conjunction (Definition 1.4.2ff).

Pairing is the generic binary operation, coding many arguments as one,

$$r(a,b) = \bar{r}(\langle a,b\rangle),$$

so the product type is used in the semantics of algebra (Section 4.6), but we choose not to use it for the syntax. Pairing is sometimes claimed to make variables redundant, but it does so by replacing *named* variables with *numbered* projection functions. In a product of many factors (as for record types in programming languages) we want to name, not number, the fields. [Pit95] treats the product carefully in its own right.

Comprehension and powerset. As we have rejected the "inductive conception" of a set as a voluntary conspiracy of its elements, we are left with the problem of defining what a "subset" is. We shall do this in terms of *predicates* (Definition 1.4.1): a subset of X is *by definition* the same thing as a predicate with a variable of type X. By the same convention, we may treat a k-ary relation as either a predicate in k variables or a subset of the k-fold cartesian product, *cf.* Definition 1.3.1(a).

DEFINITION 2·2·3:   The **axiom of comprehension** says that

> if X is a set, and $\phi[x]$ is a predicate on X,
> then $\{x : X \mid \phi[x]\}$ is also a set.

The syntax for comprehension, like quantification, *binds* the variable x, so it is subject to α-equivalence (Definition 1.1.6). Since it is therefore a context-changing operation, box or sequent methods similar to those of Section 1.5 are needed to formalise it properly. We shall not in fact do this until Section 9.5, because it is preferable to introduce these methods for the function-type instead, as we do in the next section.

The elements of this new set (or, as we prefer to say following Notation 1.3.3, the terms of the new *type*) are given by the two-way rule,

$$\frac{a : X \qquad \phi[a]}{a : \{x : X \mid \phi[x]\}}$$

which we may read downwards as an introduction rule ($\{\}\mathcal{I}$) and upwards as two elimination rules, rather similar to those for the product type. The β- and η-rules say that the term a stays the same; this is because $\phi[a]$, unlike the type Y in the product, has no associated term.

REMARK 2·2·4:   Notice that the term a has both the ambient set X and the subset $\{x : X \mid \phi[x]\}$ as its type, so two occurrences of the same term may have different types. In particular, by one of the elimination rules, any term of the subtype acquires the wider type. This defines an injective function, which is called the **inclusion**:

$$\{x : X \mid \phi[x]\} \hookrightarrow X.$$

In category theory we *define* subsets as injective functions (Section 5.2).

DEFINITION 2·2·5:   For a predicate $\phi[x]$ on X, we write $\{x : X \mid \phi[x]\}$ not only for the new *set* defined above by comprehension, but also for an *element* of the **powerset** $\mathcal{P}(X)$; this is the introduction rule ($\mathcal{PI}$). The elimination rule ($\mathcal{PE}$) provides the binary **membership relation**,

$$(\in_X) : X \rightharpoonup \mathcal{P}(X) \qquad \text{for } each \text{ type } X.$$

Our $\in_X$ is typed as shown, whereas in set theory there is a single $\in$ relation for the whole class of sets. As we mentioned in Notation 1.3.3 and Exercise 1.12, there is a distinction between

(a) $a : X$, the statement in the meta-language that a term a has type X,

(b) and $a \in_X U$, which says that the term a of type X satisfies the predicate $\phi[x]$ (in the object-language) defining $U = \{x : X \mid \phi[x]\}$.

The use of $\in$ in (a) is a rather ingrained habit, to which we shall often revert since the colon is not altogether a satisfactory alternative: if it could be stripped of its set-theoretic confusions, a symbol derived from the Italian *è* (is) would be entirely reasonable, whereas the colon is punctuation. Nor do we often bother to write the subscript. We shall, however, be careful to write $\forall x{:}X.\ \phi$ for quantifiers, reserving $\in$ for the **guarded quantifiers**, so

$$\forall x \in U.\,\phi[x] \quad \text{means} \quad \forall x{:}X.\ x \in U \Rightarrow \phi[x]$$
$$\exists x \in U.\,\phi[x] \quad \text{means} \quad \exists x{:}X.\ x \in U \wedge \phi[x].$$

The ambiguous notation makes the β-rule for powerset look the same as the introduction and elimination rules for comprehension together:

$$\phi[a] \Leftrightarrow a \in \{x : X \mid \phi[x]\}. \tag{$\mathcal{P}\beta$}$$

Regarding them as elements rather than types, we have to say when two subsets $U, V : \mathcal{P}(X)$ are equal:

$$U = V \quad \text{if} \quad \forall x{:}X.\ x \in U \Leftrightarrow x \in V,$$

i.e. the predicates defining these subsets are inter-provable. Like $\Leftrightarrow$ in logic, but unlike equality in arithmetic, we have to give *two* arguments to show that subsets are equal, one in each direction: $U \subset V$ and $U \supset V$ (*cf.* Exercise 2.18). Finally, the $(\mathcal{P}\eta)$-rule is $U = \{x : X \mid x \in U\}$.

We shall study the powerset in Sections 2.8, 5.2 and 9.5.

Notation.

REMARK 2·2·6: The symbols $\leqslant$ and $<$ for the reflexive and irreflexive orders on $\mathbb{N}$ were used for inclusion of subsets in the nineteenth century (and are still used for subgroups), but Ernst Schröder introduced $\Subset$ and $\subset$. Many authors use $\subset$ for *strict* containment, and $\subseteq$ for the non-strict version, but strict inclusion is neither primitive (constructively) nor particularly useful: if $U \subset V$ but $V \not\subset U$ then the latter fact has to be proved — and should be stated — separately. Indeed Louis Couturat rejected the symbol $\Subset$ in 1905 "parce qu'il est complexe, tandis que la relation d'inclusion est simple". The analogy with arithmetic is bogus (Section 3.1): $\subset$ is syntactic sugar for $\vdash$, $\Rightarrow$ or $\hookrightarrow$. In these cases, rightly, no notation has been invented for the strict versions, or any resolution made into strict and equal. So we use $\subset$ in the *non-strict* sense.

A conflicting notation survives in philosophy as $\supset$ for implication (and is also used by some modern authors in type theory to avoid overloading the arrow notations). It is actually older: Joseph Gergonne introduced C for *contient* and $\supset$ for its converse in 1817, and these symbols were used by Peano and by Russell and Whitehead. (In fact Russell and Whitehead *also* used $\subset$ for containment in our sense.)

REMARK 2·2·7:   A common abuse of the subset-forming notation (which we have already committed) is to put a *term* in place of the *variable*:

$$\{f(x,y) \mid \phi[x,y]\} \text{ means } \{z \mid \exists x,y.\, \phi[x,y] \wedge z = f(x,y)\}.$$

Which variables x, y, ..., are deemed to be bound in this notation? This is not made clear — informally, we write "$x \in X, y \in Y, \phi[x,y]$" to indicate what we mean. (In fact this is the same abuse of notation which we ridiculed in Examples 1.1.7.) As an important special case, we often want to apply a function "in parallel" to the elements of a subset, obtaining a (sub)set of results. In this case we write, as in Notation 1.3.4,

$$f_!(U) \stackrel{\text{def}}{=} \{f(x) \mid x \in U\} = \{y \mid \exists x.\, x \in U \wedge y = f(x)\}$$

for the image (see also Remark 3.8.13(b)); in particular $f_!(\{x\}) = \{f(x)\}$. Notice that the extended use of the notation for comprehension, and in particular the image, disguise an existential quantifier, which we shall discuss in Section 2.4.

Another, perhaps unfamiliar, special case is when there is a constant on the left of the divider, or maybe no variables in the expression at all:

$$\{\star \mid \phi\}.$$

(Classically, we would say that this is $\{\star\}$ if ϕ is true and $\varnothing$ otherwise.) In this way, *propositions* correspond to subsets of the singleton, and to elements of $\mathcal{P}(\mathbf{1})$.

Parametric sets. In algebra and the predicate calculus we used terms and formulae containing variables, but the types of the variables were fixed in advance. In Zermelo type theory, by contrast, the comprehension operation $\{x \mid \phi\}$ is not required to bind *all* of the free variables of the formula ϕ. Those which remain free become the free variables of a *type-expression*, for example,

$$\mathsf{Factors}[x] = \{y \mid \exists z.\, x = yz\}.$$

The "arguments" of a **dependent type** $Y[x]$ will be enclosed in *square* brackets, as we have already done for predicates.   This is an *informal notation* like $f(x)$ for functions (Remark 1.1.2). Of course each argument $x : X$ has its own type, but x is not *itself* a type (although in Section 2.8 we shall briefly discuss an extension in which there are type variables and quantification over them). These phenomena are called *polymorphism*, because the same type-expression may be instantiated in many ways.

As $\mathsf{Factors}[x]$ may be used as the type of another variable y in terms, formulae and other type-expressions, we must modify Definition 1.5.4.

DEFINITION 2·2·8: For each typed variable $y : Y$ in a ***context***, the free variables $\underset{\sim}{x}$ of the type-expression $Y[\underset{\sim}{x}]$ must occur earlier in the list than the variable y itself.

This and Remark 2.2.4 make Zermelo type theory very complicated.

Fortunately, Exercise 2.17 shows that it is possible to rewrite any type-expression (using any of the constructors, including comprehension) as a subset of a type defined using $\mathbf{1}$, $\times$, $\mathcal{P}$ and $\mathbb{N}$ alone. So comprehension may be postponed and used just once. Variables may be taken to range over types from this simpler class, where each formula or term is ***guarded*** (Remark 1.5.2) by the predicate defining the subset. Types and terms may once more be treated separately, and the exchange rule allows us to disregard the order of the variables. This system is studied in [LS86].

Comprehension-free types are, however, somewhat artificial and do not allow us to speak directly of functions, real numbers or equivalence classes. But in practice the difficulty which we mentioned does not arise unless we make actual use of types containing non-trivial dependency. The notation for dependent types is not straightforward, but as they are important to the *practical* foundations of mathematics and interpreted in many semantic models we devote the final two chapters to them.

Historical comments. The foregoing motivation is a fiction, in terms of history. Ernst Zermelo had been enticed from applied mathematics into foundations by Hilbert. He was interested in cardinal arithmetic and in particular the well-ordering property (Proposition 6.7.13), which Cantor had assumed but was unable to prove. The effort of formalising his proof of this led Zermelo to a usable system of foundations, which brought set theory to what was arguably its perigee.

REMARK 2·2·9: Zermelo's axioms [Zer08b] were, in modern notation,

(a) Bestimmtheit (literally *definiteness*, but usually known in English as ***extensionality***): if $\forall z. z \in x \Leftrightarrow z \in y$ then $x = y$;

(b) Elementarmengen (basic sets): $\varnothing$, $\{x\}$, $\{x, y\}$ are sets if x and y are;

(c) Aussonderung (comprehension) for *definit* properties (see below);

(d) Potenzmenge (power set);

(e) Vereinigung (union): $\{z \mid \exists y. z \in y \in x\}$ is a set if x is;

(f) Auswahl (Choice): see Exercise 2.15; and

(g) Unendlichkeit (infinity, $\mathbb{N}$ is a set): see Exercise 2.47.

Real numbers and "ideal" algebraic numbers were both constructed as sets, so it was reasonable at the time to treat individuals and collections in the same way. Nowadays we are used to mutual recursive definitions of several distinct syntactic classes (such as commands and expressions in programs), and it is preferable to do this for terms, types and predicates.

The (now archaic) German word Aussonderung means "sorting out" in the sense of discarding what is not wanted. No single English word seems to fit as well, but it is often translated as *separation*, comprehension being reserved for the unbounded way of forming sets which brought Gottlob Frege's system down. But as the word "separation" has wider and more natural uses in, for example, topology, it seems better to re-employ the term whose meaning would be immediately recognisable to non-logicians.

The *definit* properties sparked a new controversy. Again people were forced to think, now about the formulation of the sentential calculus, and the outcome was first order model theory. (Zermelo's vagueness is now perhaps an advantage, since we may (remove Choice and) substitute intuitionistic, classical or some other calculus for his missing definition.)

Extensionality defines equality between individuals, but it also imposes an absolute notion of interchangeability for types, where subsequent experience has taught us that specified isomorphisms should play this role (Definition 1.2.12ff). This first, seemingly innocuous, axiom has some rather bizarre results, particularly for unions and intersections. If the grandchildren z belong to some known set w, then Zermelo's union $\{z \mid \exists y.\ z \in y \in x\}$ is simply the union in the lattice $\mathcal{P}(w)$, which is given by the existential quantification of a family of predicates in our notation. However, if $x = \{y_1, y_2\}$, where y_1 and y_2 are sets given *independently*, they may suddenly be found to overlap. For the more natural *disjoint* union (Example 2.1.7), explicit coding must be used to distinguish the elements of the two sets. (The overlapping union is not definable using Definition 2.2.1, but see [Tay96a] and [Tay96b].)

Even as an equality test, extensionality is highly recursive, although a further axiom (Foundation) is needed to justify the recursion. Dana Scott (1966) showed that it is essential to giving the axiom of replacement its power: *without* extensionality, ZF (see below) is provably consistent in Zermelo type theory (*cf.* Example 7.1.6(g) and Exercise 9.62).

More recently, this idea has arisen in process algebra as **bisimulation**, and can in fact be seen as a notion of *co*-induction [Acz88, Hen88], *cf.* Exercise 3.53, Example 6.3.3 and Remark 6.7.14.

It is easy to find a bijection between $\mathbb{Q}$ and $\mathbb{N}$ (we say that $\mathbb{Q}$ and $\mathbb{Z}$ are **countable**), but Cantor showed that there is none between $\mathbb{R}$ and $\mathbb{N}$. He found his now well known diagonalisation argument in 1891, but had a much prettier proof using intervals rather than decimal expansions in 1873. This began the theory of cardinality. But his next discovery, that there *is* a bijection between $\mathbb{R}^2$ and $\mathbb{R}$, showed at its conception in 1877 that the attempt to classify infinite sets up to (not necessarily continuous) bijection is powerless to define dimension and hence make the distinctions which are important in mathematics. (See the remarks after Proposition 9.6.4 for how cardinals ought to be interpreted.)

As Emile Borel stressed in 1908, the important observation about $\mathbb{Q}$ is that there is an *effective coding*, not anything to do with its "size".

In the same year as Zermelo, Bertrand Russell gave a theory of "ramified" types, later developed in [RW13]. Leon Chwistek (1923–5) and Frank Ramsey (1927) showed how to eliminate ramification, giving a theory (historically known as *simple type theory*) which is essentially equivalent to Zermelo's without the comprehension scheme. Versions of the systems of Frege, Russell and Whitehead, Zermelo, Ramsey and Quine are compared in [Hat82]; for a more philosophical survey, see [Bla33].

Thoralf Skolem, like Russell, set out to deal with the impredicativity questions which had been raised by Poincaré and Weyl (see the end of this chapter). He recognised the difference between the mathematical statement that $\mathcal{P}(\mathbb{N})$ is uncountable, and the *meta*mathematical one that its terms are recursively enumerable. Skolem is best remembered for formalising first order logic and establishing its model theory. As this was, for a long time, the only available tool for the logical analysis of mathematical theories, set theory became the study of axiomatisations of a *first order* $\in$ relation. Since the predicates belonged to the metalanguage, comprehension was turned into an infinite *scheme* of axioms, and, by the Löwenheim–Skolem theorem (Remark 2.8.1), set theory has countable models which contain uncountable cardinalities [Sko22].

This paradox was repugnant to Zermelo, who correctly said that it revealed the limitations of first order logic, not of set theory. Indeed, turning comprehension into salami doesn't explain the type-constructors that actually occur in mathematics (but see Definition 5.2.10). After his treatment by the subsequently dominant tradition in set theory, Zermelo would, I believe, readily forgive my putting *ordered* pairs into his system.

A trick was found for coding the ordered pairs which mathematics needs in terms of the unordered ones Zermelo provided (Exercise 2.19). Like the Sheffer stroke (page 54), this only obscures matters. (Those who like to argue *from authority* should note that [Bou57], whilst indulging in obfuscatory reductionism of its own (Remark 1.6.7), treats pairing as primitive.) Set theory is sometimes called the "machine code" of mathematics, and this comment is supposed to justify the pair formula.

Since mathematicians have traditionally regarded their Platonist world as unchanging, it is understandable for them to think like this, but I find it shocking to hear it from informaticians. What use are the Z80 machine code programs I wrote around 1980? Modularity and portability are the programming values which we should be teaching.

Why Zermelo made the choice he did between ordered and unordered pairs is perhaps worthy of historical study, though it seems unlikely that it actually occurred to him that he was making any significant decision at all. Irredundancy was considered important at the time, probably because Hilbert had only recently found the final settlement of the axiomatics behind Euclid's parallel postulate. In fact the cartesian product had yet to gain the importance which it has now; for example [Die88] remarks that the first mention of the product of two abstract topological spaces was also in 1908.

In 1922 Abraham Fraenkel and Thoralf Skolem added another type-forming operation: the axiom-scheme of **replacement**. Owing to its obscure formulation, use of Replacement is widely overlooked, but it is *incredibly* powerful: Richard Montague (1966) showed that it can prove the consistency, not only of Zermelo set theory itself, but of the extension of this by any single theorem of ZF. We shall try to see what Replacement means in the final section.

The last of the Zermelo–Fraenkel axioms is **foundation**, which says that we may use induction on the membership relation (Definition 6.7.5). On the other hand, it also says that everything in mathematics belongs to the set-theoretic hierarchy, contradicting the intuition that, for example, the group A_5 has elements but not elements of elements. (This group has familiar representations as the even permutations of five objects and the symmetries of a dodecahedron, as well as of two different projective spaces.) The axiom of foundation makes the hierarchy *rigid*, whereas objects of mathematical interest typically have lots of automorphisms. Stone's representation of Boolean algebras as lattices of clopen subsets also gives an example where there is no preferred view in which either "points" belong to "neighbourhoods" or *vice versa*.

After Zermelo, more axioms were added to set theory in an endeavour to formulate the strongest system that would remain consistent, in which everything anyone could possibly want would be provable, and the model would be unique. We mention some of these in Sections 4.1 and 9.6.

We shall take the opposite strategy: by restricting the hypotheses of an argument to *just* what is needed, we are better able to understand how it works, and we are led to generalisations and novel applications. When the box or sequent rules for the correct management of the free variables of terms and comprehension types are taken into account, Zermelo type theory becomes far too complicated to study in one go, and is arguably too powerful for actual mathematics and programming.

2.3 Sums, Products and Function-Types

We begin our reassessment of the constructions needed in mathematics by considering the function-type and sum. The system of types which is freely generated by the binary symbols $\times$, $\to$ and $+$ from $\varnothing$, **1** and a given collection of **base types** or **sorts** will be called **simple type theory**. Applied to sets, $\varnothing$, **1**, $\times$, $+$ and $\to$ refer to the constructions of Section 2.1. Recall that these were accompanied by operations such as π_i, ν_i and ev, which we shall now discuss. Later we shall show how certain other mathematical objects admit directly analogous structures.

Function (λ) abstraction. Sections 1.1–1.3 discussed how functions *act*, but they must also be considered as *entities in themselves*. Early in the history of the integral calculus problems arose in which the unknown was a function *as a whole*, rather than its value at particular or even all points: the Sun's light takes that path through the variable density of the atmosphere which minimises the time of travel; the motion of a stretched string depends on its initial displacement along its whole length.

Remark 2·3·1: In order to consider a function *per se*, we must first identify which of the unknowns in an expression p are inputs. **Lambda abstraction**, $\lambda \underline{x}$, does this, thereby *binding* these variables (Definition 1.1.6). Sometimes the function already has a name, such as sin or sine, but the squaring function can only be written as ^{2} or, now, as $\lambda x.\, x^2$. Since λ is clearer than the *informal* notation $p(x)$ of Remark 1.1.2 as a way of distinguishing the inputs from other variables, we treat all variables, whether free or bound, as *part of the expression p*.

Given a function f and an argument a, the one may be **applied** to the other and **evaluated** to a result. This is usually written fa without brackets, in which the juxtaposition denotes a *formal* operation of application; as we shall need to study this operation in its own right, we shall sometimes write $\mathsf{ev}(f, a)$ instead. The result of the evaluation, which was written $p(a)$ informally, is $[a/x]^* p$. The passage (using substitution)

$$\text{from } (\lambda x.\, p)a \quad \text{to } [a/x]^* p$$

is called β-**reduction**. See Definition 1.2.2 for related terminology.

We also identify $\lambda x.\, \sin x$ with $\sin$, and more generally

$$\lambda x.\, fx \quad \text{with} \quad f, \quad \text{where } x \notin \mathsf{FV}(f).$$

This is, crudely speaking, a converse to the β-rule (where an abstraction is applied) because it says that when you abstract the application of a function you get the function back. It is called η-***reduction***. The interpretation of η as a computation step is not so clear as for β (arguably it should go from f to $\lambda x.\, fx$), but it will be needed (as an equality) in Lemma 4.7.5ff to obtain an exact match between the intensional and extensional notions of function.

Besides the type-theoretic rules, we also intend that the constants may have their own laws, or δ-rules, as they are known in the λ-calculus.

For us, all terms are typed: if x and p have types X and Y, the abstraction $\lambda x.\, p$ has type $X \to Y \equiv Y^X$. Notice that the two reduction rules preserve type, *i.e.* they obey *subject reduction*, Definition 1.2.12.

The type $X \to (Y \to Z)$ or $(Z^Y)^X \cong Z^{(X \times Y)}$ is that of a function of two arguments. This trick — of using λ-abstraction to supply multiple arguments one by one to a function — is called ***Currying***, though it had been observed by Moses Schönfinkel and was implicit in Frege's work.

CONVENTION 2·3·2: It is customary to omit the brackets as follows:

$$
\begin{aligned}
X_1 \to \cdots \to X_n \to Y &\equiv X_1 \to (\cdots \to (X_n \to Y)\cdots) \\
fa_1 a_2 \cdots a_{n-1} a_n &\equiv ((\cdots((fa_1)a_2)\cdots a_{n-1})a_n) \\
\lambda x_1 x_2 \cdots x_{n-1} x_n.\, p &\equiv \lambda x_1.\,(\lambda x_2.\,(\cdots.(\lambda x_n.\, p)\cdots)).
\end{aligned}
$$

The notation is further abbreviated to $\underline{X} \to Y$, $f\underline{a}$ and $\lambda \underline{x}.\, p$ respectively. As this deals with many-argument functions (in fact in a rather useful way), many authors omit pairing from the calculus. By contrast, the type $(X \to Y) \to Z$ is that of what is sometimes called a functional, *i.e.* a function whose argument is itself a function, and so is more complex.

Normalisation. A distinction is made between $(\lambda x.\, p)a$ and $[a/x]^* p$, and between application and evaluation, since the result of a β-reduction is frequently a longer expression than the first: it may contain more λs and so more opportunities for further reduction than the original term. So *strategies* for β-reduction are an important topic of study in themselves.

FACT 2·3·3:   The ***Church–Rosser Theorem*** says that the pure λ-calculus (without δ-rules) is confluent (Definition 1.2.5). The simply typed pure λ-calculus is also strongly normalising (Definition 1.2.8). $\square$

The Church–Rosser Theorem relies too much on intricacies of syntax to be appropriate for this book: see, *e.g.*, [Bar81], [LS86] and [Bar92] for a detailed treatment. The result is valid in many different calculi — including the untyped λ-calculus, in which the normalisation theorems fail — but unfortunately breaks down in some variations which seem semantically benign, such as the untyped calculus with surjective pairing. Without the type discipline (Notation 1.3.3) there need be no normal form to which a term reduces. For example the term $(\lambda x.\, xx)(\lambda x.\, xx)$ reduces to itself, and there are much worse phenomena.

The depth of bracket nesting, with the above Convention for omitting them, considered as a notion of type complexity, can be used to prove weak normalisation: see Example 2.6.4. It follows from Fact 2.3.3 and Theorem 1.2.9 that normal forms exist and are unique. They are characterised in Exercise 2.23, and used in Theorem 7.6.15. Section 7.7 shows in another way that every term is provably equal to a normal form.

REMARK 2·3·4: When the λ-calculus is used as a programming language [Plo77], it is usual to forbid β-reduction under λ, as $\lambda x.\, p$ is regarded as an as yet passive fragment of code, which is only activated when it is applied to some argument. Then there is a choice whether

(a) to reduce $(\lambda x.\, p)a$ straight away (***call by name***), so avoiding the perhaps unnecessary risk of evaluating an undefined argument, or

(b) to wait until a has itself been normalised (***call by value***), so that this is not done repeatedly if x occurs several times in p.

Contexts for the λ-calculus. The variable governed by λ-abstraction, like that bound by the quantifiers $\exists$ and $\forall$, is generic, so we use boxes to delimit it. As in Section 1.5, the variable cannot be used outside the box, because to do so would restrict its generality and prejudice the question of whether its type is empty.

DEFINITION 2·3·5: The box rules for λ-abstraction and application are

$$\begin{array}{|ll|}\hline x:X & \text{hyp}\\ \vdots & \\ p:Y & \\\hline\end{array}$$
$$(\lambda x{:}X.\, p):X\to Y \quad \to\!\mathcal{I}$$

$$\frac{f:X\to Y \quad a:X}{fa:Y}\to\!\mathcal{E}$$

cf. the syntax for *procedures* with *formal parameters* in programming languages. Inside the box we may form and manipulate terms algebraically, treating x as a constant. Its operational meaning is given by the $(\to\!\mathcal{E})$- and β-rules, where an actual parameter a is substituted throughout for the formal x (*cf.* Remark 1.5.10).

NOTATION 2·3·6:   As for the predicate calculus (Definition 1.5.3), the *abstract* study of the λ-calculus needs a sequent form in which

$$\Gamma \vdash t : X \text{ means that } \begin{cases} t \text{ is a well formed term of type } X \\ \text{given the typed variables in } \Gamma. \end{cases}$$

In the simply typed λ-calculus, contexts consist only of typed variables, as there are no propositional hypotheses.

DEFINITION 2·3·7:   The sequent forms of the rules for λ-terms are

$$\frac{\Gamma, x : X \vdash p : Y}{\Gamma \vdash \lambda x{:}X.\ p : X \to Y} \to\mathcal{I} \qquad \frac{\Gamma \vdash f : X \to Y \quad \Gamma \vdash a : X}{\Gamma \vdash fa : Y} \to\mathcal{E}$$

$$\frac{\Gamma, x : X \vdash p = q : Y}{\Gamma \vdash (\lambda x.\ p) = (\lambda x.\ q)} \lambda{=} \qquad \frac{\Gamma \vdash f = g : X \to Y \quad \Gamma \vdash a = b : X}{\Gamma \vdash fa = gb : Y} \text{ ev=}$$

$$\frac{\Gamma \vdash a : X \quad \Gamma, x : X \vdash p : Y}{\Gamma \vdash (\lambda x.\ p)a \rightsquigarrow [a/x]^* p : Y} \to\beta \qquad \frac{\Gamma \vdash f : X \to Y \quad x \notin \Gamma}{\Gamma \vdash (\lambda x{:}X.\ fx) = f} \to\eta$$

The other five rules make ($\to\mathcal{I}$) into a two-way adjoint correspondence:

$$\frac{\Gamma, x : X \vdash p = fx : Y}{\Gamma \vdash f = \lambda x{:}X.\ p : X \to Y}$$

REMARK 2·3·8:   We need structural rules for terms as well as formulae (Definition 1.4.8). In fact these ought to have been given for the ($\forall\mathcal{I}$)- and ($\exists\mathcal{E}$)-boxes (Definition 1.5.1), in order to import terms.

$$\frac{}{x : X \vdash x : X} \text{ identity} \qquad \frac{\Gamma \vdash a : X \quad \Gamma, x : X, \Psi \vdash t : Z}{\Gamma, \Psi \vdash [a/x]^* t : Z} \text{ cut}$$

$$\frac{\Gamma, \Psi \vdash t : Z}{\Gamma, x : X, \Psi \vdash \hat{x}^* t : Z} \text{ weaken} \qquad \frac{\Gamma, x : X, y : X, \Psi \vdash t : Z}{\Gamma, x : X, \Psi \vdash [x/y]^* t : Z} \text{ contract}$$

$$\frac{\Gamma, x : X, y : Y, \Psi \vdash t : Z}{\Gamma, y : Y, x : X, \Psi \vdash t : Z} \text{ exchange}$$

In the informal notation, the effect of cut, weakening and contraction is $t(x) \mapsto t(a)$, $t \mapsto t$ and $t(x, y) \mapsto t(x, x)$ respectively. The Ψ has been included to emphasise that these things can happen anywhere in the context, not just at the end of the list.

REMARK 2·3·9:   There is a superficial similarity between the β- and cut rules. To see the β-rule in action in a symbolic idiom we need to use general expressions for the body p and argument a of the function, *cf.* the various ways of expressing conjunction as a sequent rule in Remark 1.4.9; Example 7.2.7 gives a diagrammatic version with variables instead. Cut makes these substitutions, and this explains the likeness.

The substitution rule (Definition 1.1.10(d)) relates cut to $(\to\mathcal{I})$:

$$\frac{\Gamma \vdash c : Z \qquad \Gamma, z : Z, x : X \vdash p : Y}{\Gamma \vdash [c/z]^* \lambda x.\, p = \lambda x.\, [c/z]^* p : X \to Y} \qquad \frac{\Gamma, x : X \vdash p : Y}{\Gamma, z : Z \vdash \hat{z}^* \lambda x.\, p = \lambda x.\, \hat{z}^* p}$$

The side-conditions $z \not\equiv x \notin \mathsf{FV}(c)$ are forced automatically by the need for the contexts to be well formed. It is these rules which allow any term definable outside the box to be imported, according to the scoping rules familiar in programming languages as block structure, *cf.* Lemma 1.6.3. Remark 9.4.3 discusses the substitution rule further and relates it to the Beck–Chevalley condition in category theory.

The sum type. The rules for sum *largely* mirror those for product. The pair $\langle a, b \rangle$ for the product is supplied by the data, and the program uses π_i to extract what it needs. For the sum, the *program* provides a pair $[f, g]$ of options, from which the input makes a selection using ν_i. There is an ultimate result type Θ to mirror the context of parameters Γ. This symmetry is much more (arguably too) obvious in the categorical presentation (Sections 5.3–5.5); it is really spoilt by the asymmetry of the term calculus, in which the input but not the output may involve parameters. The defects require new variable-binding operations, and rules to handle substitution and its dual notion of continuation.

These rules are rather technical and will not be relevant until Section 5.3, so you should skip the rest of this section unless you are already very familiar with the typed λ-calculus expressed in a contextual style.

REMARK 2·3·10: The sum type has inclusions ν_0 and ν_1,

$$\frac{a : X}{\nu_0(a) : X + Y} +\mathcal{I}0 \qquad \frac{b : Y}{\nu_1(b) : X + Y} +\mathcal{I}1$$

which are operation-symbols and so obey the obvious equality rules. The elimination rule for sums is *case analysis*. Its sequent form is

$$\frac{\Gamma, x : X \vdash f : \Theta \qquad \Gamma, y : Y \vdash g : \Theta}{\Gamma, z : X + Y \vdash [\nu_0 x.\, f(x),\ \nu_1 y.\, g(y)](z) : \Theta} +\mathcal{E}$$

or, by using cut (Remark 2.3.8, *cf.* conjunction in Remark 1.4.9),

$$\frac{\Gamma, x : X \vdash f : \Theta \qquad \Gamma, y : Y \vdash g : \Theta \qquad \Gamma \vdash c : X + Y}{\Gamma \vdash [\nu_0 x.\, f(x),\ \nu_1 y.\, g(y)](c) : \Theta} +\mathcal{E}$$

There is no syntax for the term below the line that is as natural and concise as λ, or that is universally agreed. In programming languages it is written in a form such as on the left below, where each branch offers a *pattern* $\nu_0(x)$ or $\nu_1(y)$ against which to *match* the data c.

switch c **into** $c : X + Y$

     case $\nu_0(x)$:

$x : X$	hyp	$y : Y$	hyp
$\vdots$		$\vdots$	
$f(x) : \Theta$		$g(y) : \Theta$	

$\qquad\qquad\qquad f(x)$

    case $\nu_1(y)$:

$\qquad\qquad\qquad g(y)$

        end $\qquad\qquad \big[\nu_0 x.\, f(x),\ \nu_1 y.\, g(y)\big](c) : \Theta \qquad +\mathcal{E}$

Notice that the variables x and y are local to (bound by) their respective branches. The letter ν in the *introduction* rules behaves as an operation-symbol, and is applied to a term a; in the *elimination* rule it is a variable binder like λ. The β-rules give the meaning of this construct, deleting an elimination rule following an introduction. When the pattern $\nu_0(x)$ meets the data $\nu_0(a)$, the value a is substituted for x in $f(x)$.

$$\big[\nu_0 x.\, f(x),\ \nu_1 y.\, g(y)\big](\nu_0(a)) \rightsquigarrow f(a) \quad \text{or} \quad [f,g] \circ \nu_0 = f$$
$$\big[\nu_0 x.\, f(x),\ \nu_1 y.\, g(y)\big](\nu_1(b)) \rightsquigarrow g(b) \quad \text{or} \quad [f,g] \circ \nu_1 = g.$$

The η-rule says that if the two branches contain the same code p then the switch is redundant: $\big[\nu_0 x.\, p(\nu_0(x)),\ \nu_1 y.\, p(\nu_1(y))\big](c) = p(c).$

There is also an equality rule similar to $(\lambda{=})$ in Definition 2.3.7.

REMARK 2·3·11:   Just as pairing handles many-argument operations uniformly, so λ-abstraction avoids the need for other variable binders. Thus $(+\mathcal{E})$ can be put in a form which doesn't itself alter the context:

$$\dfrac{\dfrac{\Gamma, x : X \vdash f : \Theta}{\Gamma \vdash \lambda x.\, f(x) : \Theta^X}\to\mathcal{I} \quad \dfrac{\Gamma, y : Y \vdash g : \Theta}{\Gamma \vdash \lambda y.\, g(y) : \Theta^Y}\to\mathcal{I}}{\dfrac{\Gamma \vdash \big\langle \lambda x.\, f(x),\, \lambda y.\, g(y) \big\rangle : \Theta^X \times \Theta^Y}{\Gamma \vdash \big[\nu_0 x.\, f(x),\, \nu_1 y.\, g(y)\big](c) : \Theta} \times\mathcal{I} \quad \Gamma \vdash c : X + Y}$$

The $[\ ,\ ]$ notation is at least standard usage in category theory,[1] where the rules can be summed up as the two-way adjoint correspondence

$$\dfrac{\Gamma \vdash f = p \circ \nu_0 : X \to \Theta \qquad \Gamma \vdash g = p \circ \nu_1 : Y \to \Theta}{\Gamma \vdash p = [f,g] : X + Y \to \Theta}$$

so $\Theta^{(X+Y)} \cong \Theta^X \times \Theta^Y.$

REMARK 2·3·12:   Terms may be imported into the boxes, with the effect of substituting for free variables within the binding. The rule for this is exactly analogous to that in Definition 1.1.10(d),

$$[a/w]^*[f,g](c) = \big[[a/w]^*f,\ [a/w]^*g\big](c) \qquad \hat{w}^*[f,g] = [\hat{w}^*f, \hat{w}^*g].$$

---

[1]Freyd and Scedrov [FS90] write $\binom{f}{g}$, adopting the matrix notation from linear algebra. As the square brackets are rather overloaded, this is a better notation.

Semantically, this results in the **distributive law** (*cf.* Lemma 1.6.3),

$$(\Gamma \times X) + (\Gamma \times Y) \cong \Gamma \times (X + Y).$$

REMARK 2·3·13: Remark 1.6.5 showed that the $(\exists \mathcal{E})$-box is essentially open-ended below: any phrase of proof not involving the bound variable can be moved in or out. The $(+\mathcal{E})$-rule has the same property, but now that we are discussing significant terms rather than anonymous proofs, we must state another law, called a **continuation rule** or **commuting conversion**. This says that moving the continuation $\jmath$ into or out of the box has no effect,

$$\jmath \circ [f, g] = [\jmath \circ f, \jmath \circ g],$$

so the η-rule can be expressed as $[\nu_0, \nu_1] = \mathsf{id}$. Continuation is dual to substitution, and is explained in category theory by postcomposition.

As for the function-type, these reduction rules interact, and questions of confluence and normalisation have to be studied. For example we would like to know that *every* definable closed term of sum type is provably equal to either $\nu_0(a)$ or $\nu_1(b)$. This will be considered in Section 7.7.

REMARK 2·3·14: Since there are two binary introduction rules, there is no nullary one (and so no β-rule either), and as the binary elimination rule has two cases (premises), the nullary one has none.

$$\frac{\rule{3cm}{0.4pt}}{\Gamma, x : \oslash \vdash \star : \Theta} \oslash \mathcal{E}$$

The elimination rule provides a function $\oslash \to \Theta$ for each type Θ, and the corresponding equality rule says that this is unique. The $(\oslash \eta)$ and continuation rules mean that it is the identity on $\oslash$ itself, and preserved by any function. The substitution rule gives Proposition 2.1.9: any $X \to \oslash$ is invertible, $\Theta^{\oslash} \cong \mathbf{1}$ and $\Gamma \times \oslash \cong \oslash$.

Contexts will be developed in Section 4.3 and Chapter VIII. Sections 4.5–4.7 give a categorical account of products and function-types, applying the former to universal algebra. We discuss the binary sum further in Section 5.3ff, along with the **if then else fi** programming construct. Section 9.3 treats the infinitary (dependent type) analogue, $\Sigma x.Y[x]$.

2.4 PROPOSITIONS AS TYPES

Although the predicate calculus underlies Zermelo type theory, it is always foolish to assert that one piece of mathematics is more basic than another, because a slight change of perspective overturns any such rigid orders. Indeed Jan Brouwer (1907) considered that logical steps rest on mathematical constructions. One of the most powerful ideas in logic and informatics in recent years has been the analogy between

conjunction ($\wedge$, $\top$)   and   product types ($\times$, **1**)

implication ($\Rightarrow$)   and   function-types ($\rightarrow$)

disjunction ($\vee$, $\perp$)   and   sum types ($+$, $\varnothing$)

which puts propositions and types on a par. This is sometimes called the **Curry–Howard isomorphism**, as William Howard (1968) identified it in Haskell Curry's work (1958), although Nikolas de Bruijn, Joachim Lambek, Hans Läuchli and Bill Lawvere also deserve credit for it in the late 1960s. The idea was developed by Dana Scott to give substance to Brouwer's intuition, and rather more extensively by Per Martin-Löf.

Formulae correspond to types and their deductions to terms. Crudely, a type gives rise to the proposition that the type has an element, and a proposition to the type whose elements are its proofs.

Indeed, as soon as we take some care over it, we have no alternative but to treat the hypothesis for $\Rightarrow$ alongside the generic value of $\forall$ and the bound variable of λ. Similarly Sections 4.3 and 5.3 show that midconditions go with program-variables. Other analogies with types *versus* terms are games *versus* strategies, problems *versus* solutions and specifications *versus* implementations of programs.

Calling the analogy an *isomorphism* overstates the case. Terms may or may not satisfy equations, but proofs are **anonymous**: we do not usually bother either to equate or to distinguish between them. The *difference* between propositions and types is that the former are much simpler; we exploit this by treating them first, in Chapters I and III. Posets and induction concern propositions, categories and recursion are about types. (Some authors say *proof-irrelevance* instead of *anonymity*.)

The propositions as types analogy ought not to be confused with the earlier but superficial one between predicates and classes, which merely states the axiom of comprehension (Definition 2.2.3) in a fixed domain of discourse, and has no algorithmic content. Far more striking is Jan Łukasiewicz's re-evaluation of the history of logic in 1934 (in which he condemned earlier historians for their ignorance of the modern study of the subject): he attributed the identity on propositions ($\phi \vdash \phi$) to the Stoics and that on terms ($x : X \vdash x : X$) to Aristotle.

Numerals and combinators.

EXAMPLE 2·4·1:   The proofs of the proposition $\alpha \rightarrow (\alpha \rightarrow \alpha) \rightarrow \alpha$ may be characterised up to $\beta\eta$-equivalence by a number, *viz.* the number of times modus ponens ($\rightarrow\mathcal{E}$) is used.

These proofs correspond to the λ-terms $0 = \lambda xf.\, x$, $1 = \lambda xf.\, fx$ and $2 = \lambda xf.\, f(fx)$, called the **Church numerals** (Exercise 2.44ff).

1 α		hyp
2 $\alpha \to \alpha$		hyp
3 α		(1)
$\alpha \to (\alpha \to \alpha) \to \alpha$		$\to\mathcal{I}$

(0)

1 α		hyp
2 $\alpha \to \alpha$		hyp
3 α		$\to\mathcal{E}(2,1)$
$\alpha \to (\alpha \to \alpha) \to \alpha$		$\to\mathcal{I}$

(1)

1 α		hyp
2 $\alpha \to \alpha$		hyp
3 α		$\to\mathcal{E}(2,1)$
4 α		$\to\mathcal{E}(2,3)$
$\alpha \to (\alpha \to \alpha) \to \alpha$		$\to\mathcal{I}$

(2)

EXAMPLE 2·4·2: The **Schönfinkel combinators**[2] are

$$\mathsf{I} = \lambda x.\, x \qquad : \alpha \to \alpha$$
$$\mathsf{K} = \lambda x, y.\, x \qquad : \alpha \to \beta \to \alpha$$
$$\mathsf{S} = \lambda x, y, z.\, xz(yz) \; : (\alpha \to \beta \to \gamma) \to (\alpha \to \beta) \to \alpha \to \gamma$$
$$\mathsf{T} = \lambda x, y, z.\, xzy \qquad : (\alpha \to \beta \to \gamma) \to \beta \to \alpha \to \gamma$$
$$\mathsf{Z} = \lambda x, y, z.\, x(yz) \qquad : (\beta \to \gamma) \to (\alpha \to \beta) \to \alpha \to \gamma$$

These types are the axioms of implicational logic (Remark 1.6.10): I is the identity, K corresponds to weakening, S to contraction, T to exchange and Z to cut. Exercise 2.26 shows how to express any λ-term using S and K alone; for example $\mathsf{I} = \mathsf{SKK}$ and $\mathsf{Z} = \mathsf{S(KS)K}$. Proposition 2.8.6 gives further examples of the relationship between proofs and terms.

The correspondence. What is the content of a proof?

REMARK 2·4·3: Arend Heyting and Andrei Kolmogorov independently gave this interpretation of intuitionistic logic in 1934. To prove

(a) **truth**: there is a trivial proof $\star$ of T;

(b) an **atomic predicate**: we must go out into the world and find out whether it's true (*cf.* Remarks 1.6.12 and 1.7.2);

(c) **conjunction**: a proof of $\phi \wedge \psi$ is a pair $\langle a, b \rangle$ consisting of a proof a of ϕ and a proof b of ψ;

(d) **disjunction**: a proof of $\phi \vee \psi$ is a pair $\langle i, b \rangle$, where either $i = 0$ and b is a proof of ϕ, or $i = 1$ and b is a proof of ψ;

---

[2]The letters stand for the German words Identitätsfunktion, Konstanzfunktion, Verschmelzungsfunktion (alloying or amalgamation function), Vertauschungsfunktion (exchange function) and Zusammensetzungsfunktion (composition function). The last two were called C and B respectively by Haskell Curry [CF58], who introduced several others: see [Bar81, p. 163]. In 1921, three years before Moses Schönfinkel's paper, Alfred Tarski showed that K and S suffice for implicational logic, but Gottlob Frege had known this in 1879.

(e) **implication**: a proof of $\phi \Rightarrow \psi$ is a function f which takes a proof x of ϕ as argument and returns a proof $f(x)$ of ψ as result;

(f) **universal quantification**: a proof of $\forall x. \phi[x]$ is a function f taking an argument x in the range of quantification (or maybe a proof that x lies in this range) to a proof $f(x)$ of $\phi[x]$;

(g) **existential quantification**: a proof of $\exists x. \phi[x]$ is a pair $\langle a, b \rangle$, where a is a value in the range, and b is a proof of $\phi[a]$.

These may be read off from the rules of Sections 1.4, 1.5 and 2.3.

REMARK 2·4·4: In particular, the direct logical rules correspond to the operation-symbols of simple type theory:

$$\begin{array}{ccccccc} \top\mathcal{I} & \wedge\mathcal{I}(a,b) & \wedge\mathcal{E}0(f) & \wedge\mathcal{E}1(f) & \vee\mathcal{I}0(a) & \vee\mathcal{I}1(b) & \Rightarrow\mathcal{E}(f,a) \\ \star & \langle a,b \rangle & \pi_0(f) & \pi_1(f) & \nu_0(a) & \nu_1(b) & \mathrm{ev}(f,a). \end{array}$$

Direct deductions are, then, sequences of declarations, *i.e.* assignments to intermediate variables of expressions which involve previously declared variables. The utility of the proof box method as a way of composing proofs lies in its similarity, under this correspondence, to the *declarative* style of programming, which we shall discuss in Section 4.3.

REMARK 2·4·5: The indirect rules correspond to λ-abstraction, in which the bound variable is

(a) a hypothesis in the case of $(\Rightarrow\mathcal{I})$,

(b) an object-language variable for first order $(\forall\mathcal{I})$ and $(\exists\mathcal{E})$,

(c) a proposition- or type-variable for second order $(\forall\mathcal{I})$ and $(\exists\mathcal{E})$, and

(d) a tag (of type $2 = \{0,1\}$: see Exercise 2.13) in the case of $(\vee\mathcal{E})$.

In particular, the box lying above the $(\Rightarrow\mathcal{I})$-rule is a program which transforms a given proof x of ϕ into a proof $p(x)$ of ψ, the proof of $\phi \Rightarrow \psi$ being the *abstraction* $\lambda x. p(x)$. Conversely, the $(\Rightarrow\mathcal{E})$-rule *applies* the proof f of $\phi \Rightarrow \psi$ to the proof a of ϕ to yield a proof fa of ψ. If f was in fact $\lambda x. p$ then there is a β-*reduction* of $(\lambda x. p)a$ into $[a/x]^* p$, which is obtained by removing the box around p and replacing its hypothesis x with the given proof a (Remark 1.5.10).

WARNING 2·4·6: The λ-calculus variables introduced in this translation of indirect rules denote *proofs*, not propositions. They do not occur in terms or predicates and can only be bound by λ, not by $\forall$ or $\exists$. They should not be confused with the variables which denote elements of the object-language, occur free in predicates and are bound by first order quantifiers.

Programs out of proofs. It is sometimes claimed that, as a corollary of the analogy, programs may be *extracted* from proofs. Indeed [GLT89, chapter 15] shows that any function which can be proved to be total in second order logic is definable in Girard's System F. Unfortunately, to achieve this in practice it currently appears to be necessary to add so much detail to a proof that it would be easier to write the program in the first place. Maybe in the future it will be possible to automate the process of filling in this detail.

The constructive existential quantifier. The logical symbol for which the type-theoretic version involves the most extra detail compared to its propositional form is the existential quantifier or dependent sum. Here, then, is where the *isomorphism* has its strongest consequences: in particular Choice is a *theorem*. So conversely this is where it is perhaps the most overstated: the existential quantifier, as understood by ordinary mathematicians, *does not* provide a particular witness, so is weaker than a dependent sum. So judicious consideration leads us to ask where the principle is to be followed, and where it is essentially inapplicable.

REMARK 2·4·7: We observed that the extended use of the notation for set-comprehension (Remark 2.2.7) hides an existential quantifier. Some of the footnotes in the next chapter point out other disguised quantifiers resulting from our blindness to *why* one thing is "less than" another; this is the abstract version of only considering prov*ability*. Inequalities, structures "generated" by a set, and any words which end in "-able" or "-ible" probably all conceal similar secrets. Chapter IV provides ways of accounting for the unstated reasons.

EXAMPLE 2·4·8: Square roots, on the other hand, can manifestly be exhibited, but the extreme constructivist would seem, like Buridan's ass, to be unable to make a choice of them.

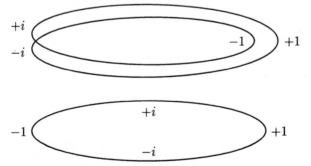

The diagram illustrates the squaring function on the unit circle in the complex plane. It is well known that there is no *continuous* choice of square roots which can be made all the way around the circle, but there are, according to Brouwer, no real- (or complex-) valued functions apart from continuous ones (Remark 1.8.6). We seem to be unable to say that squaring is a *surjective* function, $\forall x.\ \exists y.\ x = y^2$.

One answer which a constructivist type-theorist might give falls back on the Cauchy sequences (Definition 2.1.2) used to define the real and complex numbers. Using only the first approximation $a_0 \in \mathbb{Q}[i]$ to x, we make an arbitrary selection of b_0 such that $b_0^2 \approx a_0$; this is admissible since $\mathbb{Q}$ has decidable equality. Subsequent approximants b_n are chosen to be nearer to b_0 than to $-b_0$. The resulting square root "function" does not preserve the equality relation on its nominal source S^1. So *tokens* for mathematical objects (the actual terms of Cauchy sequences in this case), rather than the objects themselves, must be used to give the witnesses of existential quantifiers. A similar idea will emerge from constructions of algebras using generators and relations in Section 7.4.

I have to say that I am not convinced, and feel that the Example shows that the *unwitnessed* existential quantifier is important in mathematics. This and the dependent sum should be seen as two cases (in some sense the extremes) of a single more general concept, and we shall treat them as such in Section 9.3. To rely on coding is certainly not *conceptual* mathematics, even if Per Martin-Löf regards this as constructive.

Another reason for wanting to ignore the witnesses is that, in the study of *systems*, we need to be able to discard as much information as we can about the lower levels in order to comprehend the higher ones.

Example 6.6.7 considers *how many* square roots there are.

The existence and disjunction properties. Classical logic asserts (and thereby gives a trivial proof of the fact) that $\phi \vee \neg\phi$ is true, without necessarily giving a proof of either ϕ or $\neg\phi$.

REMARK 2·4·9: The normalisation theorem for the sum type (which we shall prove in Section 7.7) says that any proof $\vdash c : \phi \vee \psi$ is either of the form $\nu_0(a)$ with $\vdash a : \phi$ or of the form $\nu_1(b)$ with $\vdash b : \psi$. So the identification of $\vee$ with $+$ means that *the only way* to prove $\phi \vee \psi$ is by first proving one of the disjuncts. This is known as the *disjunction property* of intuitionistic logic.

Similarly, if $\exists x.\ \phi[x]$ is provable then there is some a — which may be found by examination of the given proof — for which $\phi[a]$ is also provable (the *existence property*). This is what justifies the *uniform proofs* in Fact 1.7.1. The classical inter-definability of $\forall$ and $\exists$ *via* negation breaks down because $\forall$ has no analogous property. For example, let $\phi[n]$ be the statement that n is *not* the code of a valid proof in Peano arithmetic whose conclusion is that $0 = 1$. Then $\phi[n]$ has a proof for each n, but

Kurt Gödel (1931) showed that $\forall x.\ \phi[x]$ has no proof in the same system (Theorem 9.6.2).

Classical logic. The failure of the disjunction property in classical logic seems to mean that it has no constructive interpretation.

REMARK 2·4·10: Peirce's law (which, with $\psi = \bot$, is excluded middle),

$$\big((\vartheta \Rightarrow \psi) \Rightarrow \vartheta\big) \Rightarrow \vartheta,$$

may be expressed in sequent form as

$$\frac{\Gamma, \phi \vdash \vartheta \qquad \Gamma, \phi \Rightarrow \psi \vdash \vartheta}{\Gamma \vdash \vartheta}$$

and provides a useful proof box idiom for classical logic. If something in the predicate calculus is provable intuitionistically, the proof can often be found very easily using the methods of Section 1.7. When proof by contradiction is needed, the negation of the goal often has to be introduced repeatedly and un-intuitively as a temporary hypothesis.

The sequent above has been called the **restart rule** because it is invoked when we have failed to prove the local goal: it is sufficient to prove *any pending goal* from an enclosing box. Note that this is a valid classical proof of the main goal ϑ, but not necessarily of the sub-goal $\phi \Rightarrow \psi$.

1 ϕ	hypothesis for $\Rightarrow\mathcal{I}(5)$	
2 $\vdots$		
3 ϑ	by some ordinary proof	
4 ψ	unproved, but closed by restart$(3, 7)$	
5 $\phi \Rightarrow \psi$	$\Rightarrow\mathcal{I}$	
6 $\vdots$		
7 ϑ		

This idiom seems a little less bizarre if we imagine, not a proof, but a program such as a compiler with a recursive structure tied closely to that of the user's input. The easiest way of dealing with potential *errors* in the input is to abandon the structure, *i.e.* to jump directly out of scope. Recently, methods have been developed for compiling higher order programs into very efficient continuation-passing object code. Nevertheless, such idioms are notoriously difficult to understand: the verification must take account of the fact that there are two ways of getting to line 7.

Returning to logic, suppose we have a classical proof which doesn't use $(\bot\mathcal{E})$. Harvey Friedman (1978) observed that the symbol $\bot$ (and its implicit use for negation) may validly be replaced throughout such a proof by *any* formula — such as the one to be proved (Remark 1.4.4).

Peirce's law serves for excluded middle in this translation; this explains why box proofs that involve negation or classical logic are very difficult to find. Hence any proposition of the form $\forall n.\exists m.\phi[n,m]$ with a classical proof also has an intuitionistic one, so long as ϕ is primitive recursive, and in particular decidable. This result does not extend to more complex formulae, for example $\exists m.\,\forall n.\,f(m)\leqslant f(n)$ is provable in classical Peano arithmetic, for any $f:\mathbb{N}\to\mathbb{N}$, but (using Gödel again) $\forall n.\,f(m_0)\leqslant f(n)$ need not be provable in the same system for any m_0.

See [Coq97] for a survey of the constructive interpretations which can be given to classical logic.

Although we do not exploit such interpretations in this book, continuations do feature throughout, in the background. The open-ended $(\exists \mathcal{E})$-box and the commuting conversion rule for sums (Remarks 1.6.5 and 2.3.13ff) were our first encounter with them, and we shall always write them as the symbol ʒ. They will arise in more interesting ways in our treatments of lists and **while** programs.

Term and type assignment. Another difference between propositions and types apparent in the simpler examples of symbolic reasoning is that algebra, the λ-calculus and programming typically involve many terms but few distinct types, whereas in logic we mention many propositions, but the identity of their proofs is unimportant and usually left implicit.

REMARK 2·4·11: The uninteresting information therefore gets elided, and we might expect development tools to recover it automatically:

(a) The type discipline is important, but it is a nuisance for the programmer to have to write this information in longhand. So one of the jobs of a compiler is to interpolate it, *i.e.* to verify that (base types can be chosen such that) each sub-expression obeys the type discipline. This process is known as *type assignment* or *reconstruction*. For algebraic expressions this is very easy, but it becomes more useful (and difficult) in polymorphic languages such as ML

(b) Conversely, filling in the proof of a proposition from given hypotheses is called *term assignment* (Section 1.7).

The Cn notation. We shall exploit the analogy between propositions as types, deliberately confusing algebra, logic and type theory.

NOTATION 2·4·12:   We write

$$Cn(\Gamma, X)$$

for the set of ($\alpha\beta\eta\delta$-equivalence classes of) terms of type X in context Γ. In universal algebra this is known as the _clone_, whilst in model theory it is useful to consider the set $\mathsf{Cn}(\Gamma)$ of all <u>consequences</u> of a set of hypotheses. In the last case, $\mathsf{Cn}(\Gamma, \vartheta)$ is the set of proofs of ϑ, so the consequences are just those ϑ for which this is non-empty. We shall develop this in Chapters IV and VIII.

2.5 INDUCTION AND RECURSION

After the connectives $\times$, $\to$ and $+$, the next thing we want to salvage from set theory is the system of natural numbers, equipped not only with arithmetic but also with induction and recursion. The latter have a long history in number theory, but the idioms are readily transferred to lists, trees and recursive programs. Set theory itself, with the axiom of foundation, admits a more general notion of induction on the $\in$ relation, of which transfinite iteration is a useful special case. Another similar, also rather disorderly, form of induction is that needed to prove properties of self-calling programs. These will be our starting point.

Induction and recursion are often confused in symbolic logic, but the difference between them is another one between propositions and types: we prove theorems by induction, but construct programs by recursion. The two are linked by the General Recursion[3] Theorem 6.3.13 (it is not enough "to prove by induction that $p(x)$ is defined", as this wrongly treats existence as a predicate).

DEFINITION 2·5·1: A self-calling program is said to obey the **recursive paradigm** if it defines a function $p(x)$ of exactly one argument and it proceeds as follows:

(a) from the given argument t, by means of code known to terminate, it derives zero or more other **sub-arguments** $\underset{\sim}{u}$ of the same type,

(b) then it applies p to each sub-argument u_j in parallel, *i.e.* with no interaction among the sub-computations,

(c) finally, from the **sub-results** returned by the recursive calls together with the original argument, by means of another piece of code which is known to terminate, it computes its own result to be output.

We shall call phases (a) and (c) **parsing** and **evaluation** respectively, and write $u \prec t$ whenever u is a sub-argument of t. This *may* mean that u is symbolically a sub-expression of t, in which case we're doing structural recursion, but we shall use the notation and terminology for

---

[3]The name "General Recursion Theorem" is traditional: it is not connected with the notion of general recursion in recursion theory, *i.e.* the strengthening of primitive recursion by the minimalisation operation (Exercise 6.29).

any recursive problem which fits the paradigm, whatever the means by which the *us* are obtained from *t*.

Recursive definitions of functions with many arguments, and systems of functions which call each other (**mutual recursion**), are no more general than the paradigm, because the arguments or results may be packaged into a "tuple". The complication which we have not allowed is that a sub-*result* might be fed back to help generate another sub-*argument*. The effect of the paradigm is to allow the parsing of hereditary sub-arguments to be exhausted before any evaluation is done; this will be expressed more formally in Definition 6.3.7. A special case, tail recursion, is discussed in Section 6.4, where it is shown to be equivalent to the imperative **while** construct.

EXAMPLES 2·5·2:

(a) The familiar example of the factorial function

$$\text{fact } n \quad = \quad \textbf{if } n = 0 \textbf{ then } 1 \textbf{ else } n * \text{fact}(n-1) \textbf{ fi}$$

fits the paradigm because, for $n \geqslant 1$, it first prepares the argument $n-1$, calls itself with this argument, and finally multiplies the result of the recursive call by the original argument to give its own result.

(b) A compiler for a programming language takes the source text of an entire program and extracts the immediate sub-expressions, which it feeds back to itself. It puts the digested code-fragments together as the translation of the whole program.

(c) Carl Friedrich Gauss (1815) showed how to transform a polynomial of degree $2n$ into one of degree $n(2n-1)$, in such a way that each root of the latter may be used to obtain two roots of the former by means of a quadratic equation. Any polynomial of odd degree has a real root by the intermediate value theorem. Here the sub-argument is the auxiliary polynomial and the sub-result one of its roots.

How do we prove that such programs are correct? There is nothing to say in the case of the factorial function, because it is evaluated directly from its definition. The correctness of the compiler, *i.e.* of its output (object code) relative to the input (source), is given case by case for the connectives of the language: for example the Floyd rules in Sections 4.3, 5.3 and 6.4 for imperative languages say that if the sub-expressions satisfy certain conditions then so does the whole program(-fragment). Finally, Gauss's paper gave a method of deriving the auxiliary polynomial and showed that if it has a root then this provides a root of the original one.

Induction. In all of these arguments, and whenever we want to prove consequences of recursive definitions, it is necessary to demonstrate

> a typical instance, $\vartheta[t]$, of the required property,
> from $\forall u.\, u \prec t \Rightarrow \vartheta[u]$, the **induction hypothesis**,
> that all of the sub-arguments have the property,

by methods which are peculiar to the application. Common usage often forgets that the $\vartheta[t]$ so proved is not that of the ultimate conclusion, $\forall x.\, \vartheta[x]$, but only a stepping stone. In order to justify *a single* $\vartheta[a]$, the induction step must be proved for *every* t (or at least for all descendants of a). Such sloppiness in the presentation can give the impression that induction is circular.

To obtain the desired result, we have to use *an additional rule*, of which the proof "by induction" is the *premise*.

DEFINITION 2·5·3: The **induction scheme** is the rule

$$\frac{\forall t.\ \big(\forall u.\, u \prec t \Rightarrow \vartheta[u]\big) \implies \vartheta[t]}{\forall x.\ \vartheta[x]} \quad \prec\text{-induction}$$

where $\prec$ is a binary relation on a set X. We say that $\prec$ is **well founded** if the induction scheme is valid. It is a *scheme* because it is asserted for each predicate ϑ (which may involve parameters) on X; quantification over predicates introduces second order logic (Section 2.8).

The important point formally is that the variable t in the premise and the induction hypothesis which it satisfies are bound in an $(\forall\Rightarrow\mathcal{I})$ proof box, so nothing is to be assumed about t apart from this.

$\forall t$ ₁ $\forall u.\, u \prec t \Rightarrow \vartheta[u]$	induction hypothesis
₂ $\vdots$	
₃ $u_i \prec t$	various terms u_i
₄ $\vartheta[u_i]$	$\forall \mathcal{E}(1,3)$
₅ $\vdots$	
₆ $\vartheta[t]$	the property

₇ $\forall t.\ \big(\forall u.\, u \prec t \Rightarrow \vartheta[u]\big) \implies \vartheta[t]$ $\forall\Rightarrow\mathcal{I}$

₈ $\forall x.\, \vartheta[x]$ $\prec$-induction for ϑ

Both the property ϑ and the variable t must be given *formal* names, and the box clearly delimited. Idiomatically, this is done by saying "we shall prove by $\prec$-induction on t that ϑ" to introduce what (line 7 being omitted) is apparently a proof of $\forall x.\, \vartheta[x]$.

We have followed tradition in using a symbol $\prec$ suggesting order for well founded relations, but this is extremely misleading. The motivating examples of *immediate* sub-processes and sub-expressions above are not transitive, and Corollary 2.5.11 shows that $\prec$ is *ir*reflexive.[4]

Genealogical analogies can be useful, but there is a conflict of intuition about the direction. In set theory it is traditionally synthetic: "on the first day, God made $\varnothing$" [Knu74]. Our recursive paradigm is analytic, and sub-processes are usually called *children*, so $\varnothing$ is the ultimate *descendant*.

DEFINITION 2·5·4:   If $u \prec t$ or $u \prec\!\!\!\prec t$ then we refer to u as a **child** or a **descendant** of t, respectively, where $\prec\!\!\!\prec$ is the transitive closure of $\prec$. For any predicate ϕ on X, we say that t is **hereditarily** ϕ if all of its descendants satisfy ϕ, *i.e.*

$$\forall u.\, u \prec\!\!\!\prec t \Rightarrow \phi[u].$$

Theorem 3.8.11 uses this to show that the **strict induction scheme**

$$\frac{\forall t.\, (\forall u.\, u \prec t \Rightarrow \psi[u]) \Leftrightarrow \psi[t]}{\forall x.\, \psi[x]}$$

(with $\Leftrightarrow$) is equivalent to the one in Definition 2.5.3.

EXAMPLES 2·5·5:

(a) If no instances of the induction hypothesis are actually used in the body of the proof, it is simply an $(\forall \mathcal{I})$ (Definition 1.5.1).

(b) Let $n \prec n+1$ on $\mathbb{N}$. The rule reduces to the familiar **Peano scheme**,

$$\frac{\vartheta[0] \qquad \forall n.\, \vartheta[n] \Rightarrow \vartheta[n+1]}{\forall m.\, \vartheta[m]}$$

which is also known as **simple**, **primitive**, or just **mathematical induction**. $\sum_{k=1}^{n} k^2 = \frac{1}{6}n(n+1)(2n+1)$ and similar results may be proved like this.

(c) Let $\prec$ be the strict arithmetical order $<$ on $\mathbb{N}$. Then we have **course-of-values induction** (sometimes called **complete induction**):

$$\frac{\forall n.\, \big(\forall i < n.\, \vartheta[i]\big) \Rightarrow \vartheta[n]}{\forall m.\, \vartheta[m]}$$

This relation $\prec$ is the transitive closure of the previous one. Notice that in the base case $(n = 0)$ the hypothesis $\forall i < n.\, \vartheta[i]$ is vacuous. Whereas simple induction deals with orderly, step-by-step reductions (like nuclear decay by α- or β-radiation), numerous algorithms split up graphs into parts whose size may be anything between zero and

---

[4]Exercise 3.3 gives an alternative formulation which is reflexive. This only disguises the fact that irreflexivity results from induction over a *higher order* predicate. Recognising this is the first step towards the development of a subtler theory of induction which would, for example, be capable of handling fixed points. See Proposition 3.7.11, Remarks 6.3.10 and 6.7.14 and [Tay96a].

the original size (like fission). For such problems we need course-of-values induction. But the problem for a graph with n nodes does not usually reduce to sub-problems involving graphs of *every* size $< n$ (the products of fission of ^{235}U by neutrons typically have masses about 90 and 145), so $\prec$ is usually sparser than $<$.

(d) For a recursive program in a functional language, the invariant or induction hypothesis behaves just as in the mathematical setting: the induction premise is that the result on an argument t is correct as long as the recursive calls (u_j) have been computed correctly.

(e) Since the order of any subgroup or quotient divides that of a finite group, many properties of groups (notably the Sylow theorems, Fact 6.6.8, and their applications) are shown by induction on the *divisibility* orders on $\mathbb{N}$. This may be seen as just another example of course-of-values induction (c). Alternatively, prime factorisation expresses the positive integers as (part of) an infinite product of copies of $\mathbb{N}$, which may be given either the product order or, as with Gauss's proof, the lexicographic one.

Section 2.7 considers induction and recursion for numbers and lists.

Minimal counterexamples. If some property is not universally true of the natural numbers, there is a least number for which it fails. (Georg Cantor generalised this idea to define the ordinals, Section 6.7.) This *least number principle* depends on excluded middle, but with a little care such proofs can be made intuitionistic.

PROPOSITION 2·5·6: (Richard Montague, 1955) A relation $\prec$ is well founded iff every non-empty subset has a $\prec$-minimal element.

PROOF: Let $\varnothing \neq V \subset X$, and put $\vartheta[x] = (x \notin V)$. This means that the conclusion of the induction scheme $(\forall x.\, \vartheta[x])$ fails. The scheme is therefore valid iff the premise also fails, *i.e.*

$$\exists t.\ [\forall u.\, u \prec t \Rightarrow u \notin V] \wedge (t \in V).$$

This t is what we mean by a minimal element of V. □

REMARK 2·5·7: The classical idiom of induction is to use the minimal counterexample to show that $\forall x.\, \vartheta[x]$ by reductio ad absurdum. Like Example 1.8.2, this is very often gratuitous. The hypothesis "let t be a minimal counterexample" breaks into two parts:

(a) t is a counterexample, $\neg\vartheta[t]$, and

(b) anything less, $u \prec t$, is not a counterexample, *i.e.* $\forall u \prec t.\, \vartheta[u]$.

The second part is the induction hypothesis as given in Definition 2.5.3. Commonly, $\vartheta[t]$ can be deduced *without using* the first part, so this may be eliminated to give a sound intuitionistic argument.

Imagine a proof by induction as the tide climbing up a beach: in the end the whole beach gets submerged. The induction scheme requires us to show that any part which is above *only* water soon gets wet. The classical induction hypothesis is that there is somewhere which is on the water-line but still *completely dry* — a rather implausible supposition!

COROLLARY 2·5·8:   Any (properly) recursive definition of a function which terminates on all values includes an analysis (**parsing**) into

(a) *base cases* (leaves or ≺-minimal elements) at which the value of the function is immediately given in the code (in Examples 2.5.2 these are respectively $0! = 1$, constants of the language and odd-degree polynomials), and

(b) *induction steps* (branches) at which the value of the function is given by a k-ary operation applied to the values at recursive arguments; $k = 1$ in three of the examples, but for a compiler the arities are exactly those of the connectives of the language.   □

This feature of well founded structures and free algebras will recur many times, in the rest of this chapter, in Sections 3.7–3.8, and in Chapter VI.

Descending chains. Dependent Choice (Definition 1.8.9) gives another condition which is non-constructively equivalent to well-foundedness.

PROPOSITION 2·5·9: $(X, \prec)$ is well founded iff there is no sequence of the form $\cdots \prec u_3 \prec u_2 \prec u_1 \prec u_0$.

PROOF:

[⇒] If there is a sequence u_n with $\forall n. u_{n+1} \prec u_n$ then $U = \{u_n \mid n \in \mathbb{N}\}$ is a non-empty set with no ≺-minimal element; alternatively the predicate $\vartheta[x] \equiv (\forall n. x \neq u_n)$ does not satisfy the induction scheme.

[⇐] If $(X, \prec)$ is not well founded, it has a non-empty subset $U \subset X$ with no minimal element. In other words, $\forall t. t \in U \Rightarrow \exists u. u \in U \wedge u \prec t$ and $u_0 \in U$, so the axiom of dependent choice applies.   □

The geometrical form of Euclid's algorithm (Example 6.4.3) was perhaps the first statement of induction: an infinite sequence of numbers is found, each less than the one before, which, as *Elements* VII 31 says quite clearly, is impossible amongst whole numbers. When this algorithm is applied to the side and diagonal of a square or pentagon, however, the successive quotients form a periodic sequence. David Fowler [Fow87] has made an appealing reconstruction of classical Greek mathematics, in which he claims that Euclid's book was motivated by problems like this. (The story that the discovery of the irrationality of $\sqrt{2}$ led to the downfall of the Pythagorean sect seems to have been an invention of Iamblichus nearly 1000 years later.)

In the investigation of the triangle which bears his name, Blaise Pascal (1654) stated lemmas for one case and for any "following" one, and concluded that the result holds for all numbers. John Wallis (1655) may also have been aware of the logical principle, but Pierre de Fermat (1658) was the first to make non-trivial use of the method of **infinite descent** to obtain positive results in number theory (Exercise 2.33).

A variant of this result is (Dénes) **König's lemma** (1928). A **tree** is an oriented graph with a distinguished node (its root) from which to each node there is a *unique* oriented path. Since in most applications the branches (outgoing arrows) from any node are labelled in some fashion, the Choice of branch may usually be made canonically. To any programmer, the procedure in König's lemma is **depth-first search**.

COROLLARY 2·5·10: If, in a tree, every node has finitely many branches and there is no infinite oriented path, then the tree is finite.

PROOF: As there is no ω^{op}-sequence, the branch relation is well founded, so induction on it is valid. Using this, every sub-tree is finite. □

Using this formulation of induction is not so easy as it may appear: somehow we have to find that forbidden infinite sequence, which usually requires König's lemma.

Notice that we never said that the terms in the sequence had to be distinct; of course if only finitely many values may occur then any infinite sequence of them must contain repetitions (*loops*).

COROLLARY 2·5·11: If the relation has a cycle, $u \prec u$, or $u \prec v \prec u$, or $u \prec v \prec w \prec u$, *etc.*, then $\prec$ is not well founded. □

Proof trees. The graphical presentation, as a tree without cycles, shows the difference between the induction step which is to be proved and the induction hypothesis which can be assumed.

$$\frac{\vartheta[0] \quad \dfrac{\forall n.\ \vartheta[n] \Rightarrow \vartheta[n+1]}{\vartheta[0] \Rightarrow \vartheta[1]}}{\vartheta[1]}$$

$$\frac{\dfrac{\vartheta[1] \quad \dfrac{\forall n.\ \vartheta[n] \Rightarrow \vartheta[n+1]}{\vartheta[1] \Rightarrow \vartheta[2]}}{\vartheta[2]} \quad \dfrac{\forall n.\ \vartheta[n] \Rightarrow \vartheta[n+1]}{\vartheta[2] \Rightarrow \vartheta[3]}}{\vartheta[3]}$$

REMARK 2·5·12: The ingredients for an inductive proof of $\vartheta[n]$ by the Peano primitive induction scheme are the axiom/zero $\vartheta[0]$ and the rule/successor $\forall n.\ \vartheta[n] \Rightarrow \vartheta[n+1]$. If we need to prove $\vartheta[3]$, then these may be composed, showing the different roles of the occurrences of ϑ. Compare this with the proofs we called 0, 1 and 2 in Example 2.4.1, and notice how *the structure of the term directly gives that of the proof.*

Termination. Infinite descending sequences are very familiar to anyone who has tried to execute an ill founded recursive program: the sequence is stored on the machine stack, which overflows. (Non-terminating loop programs are usually silent.)

REMARK 2·5·13:   (Robert Floyd, 1967) In what circumstances can a recursive program p which obeys the paradigm (Definition 2.5.1) fail to terminate on a particular argument u_0? If, and only if, the execution of $p(u_0)$ involves the computation of some $p(u_1)$, which in turn requires some $p(u_2)$ and so on *ad infinitum.*

For any two values t and u of the domain of definition, write $u \prec_p t$ if u is one of the sub-arguments generated in the first step of the computation of $p(t)$. Then the program p terminates on all arguments iff $\prec_p$ is well founded. Let $\vartheta[x]$ denote that p terminates on the argument x. (This is a higher order predicate; indeed its mathematical content is really that $\prec_p$ restricted to the descendants of x is well founded.) The premise of the induction scheme holds by assumption on the form of p, that the first and third stages of the program terminate. The conclusion is universal termination, so this is the case iff the induction scheme is valid.   □

A good definition in mathematics is one which is the meeting point of examples and theorems, *cf.* introduction and elimination rules in logic. Definition 2.5.3 is such a meeting point. Minimal counterexamples and descending sequences may perhaps help to give you a picture of how induction works, but they do not *prove* it. Without excluded middle, nor can they even prove the results for which induction is needed.

2.6   CONSTRUCTIONS WITH WELL FOUNDED RELATIONS

This section collects a number of methods of building new well founded structures from old. In each case, we must prove the validity of the new induction scheme by means of induction on the given structure. Since the induction scheme involves embedded $(\forall\Rightarrow)$-formulae, these proofs must use nested $(\forall\Rightarrow\mathcal{I})$-rules. This makes them good exercises in the proof box method which we introduced in Section 1.5 (see [Tay96a] for a bigger example). The heuristics of Section 1.7 are not guaranteed to succeed, because the cut-elimination procedure does not provide the induction predicate. However, it is not as difficult to find proofs such as these by this method as an adherent to the classical ways of Proposition 2.5.6ff might suppose. Giving the quantifiers explicitly also has the advantage that we can see how much higher order logic is really needed.

Complexity measures. A recursive program terminates iff its sub-argument relation is well founded, but usually the only way to calculate the number of iterations needed is to execute the program itself. This doesn't matter because, as the first result shows, if we can show that *something* is reduced at each iteration then the loop terminates. This quantity is called a *loop measure*.

DEFINITION 2·6·1: A function $f : (X, <) \to (Y, \prec)$ between sets with binary relations is said to be *strictly monotone* if

$$\forall x, x'.\, x' < x \implies f(x') \prec f(x).$$

Beware that constant functions are *not* strictly monotone, because of irreflexivity of $\prec$.

PROPOSITION 2·6·2: If $(Y, \prec)$ is well founded then so is $(X, <)$.

PROOF:

	$_1$ $\forall x.\, (\forall x'.\, x' < x \Rightarrow \vartheta[x']) \Rightarrow \vartheta[x]$		
	$_2$ $\psi[y] \stackrel{\text{def}}{=\!=} \forall x.\, (f(x) = y) \Rightarrow \vartheta[x]$	def	
$\forall y$	$_3$ $\forall y'.\, y' \prec y \Rightarrow \psi[y']$	hyp	
$\forall x$	$_4$ $f(x) = y$	hyp	
$\forall x'$	$_5$ $x' < x$	hyp	
	$_6$ $f(x') \prec f(x) = y$	strict monotonicity	
	$_7$ $\psi[f(x')]$	$\forall \mathcal{E}(3, f(x')/y')$	
	$_8$ $\vartheta[x']$	$\forall \mathcal{E}(\text{def}(2, f(x')/y))$	
	$_9$ $\forall x'.\, x' < x \Rightarrow \vartheta[x']$	$\forall \mathcal{I}$	
	$_{10}$ $\vartheta[x]$	$\forall \mathcal{E}(1)$	
	$_{11}$ $\forall x.\, (f(x) = y) \Rightarrow \vartheta[x]$	$\forall \mathcal{I}$	
	$_{12}$ $\psi[y]$	def(2)	
	$_{13}$ $\forall y.\, (\forall y'.\, y' \prec y \Rightarrow \psi[y']) \Rightarrow \psi[y]$	$\forall \mathcal{I}$	
	$_{14}$ $\forall y.\, \psi[y]$	induction for ψ on y	
	$_{15}$ $\forall x.\, \vartheta[x]$	(2)	

See Exercise 3.54 for an alternative proof. ☐

EXAMPLES 2·6·3: For the majority of applications, including the first two of Examples 2.5.2 (factorial and compiler), termination is very easy, because we can see directly that the sub-argument is a smaller number, a shorter string of symbols or a shallower tree. Gauss's proof is shown to terminate with a little more effort: the sub-arguments are polynomials of degree $2^k o_0$, $2^{k-1} o_1$, ..., $2 o_{k-1}$, o_k, where the o_i are odd numbers.

EXAMPLE 2·6·4: Normalisation for the λ-calculus (Fact 2.3.3) is a qualitatively more difficult problem. For weak normalisation, we must define a reduction strategy, *i.e.* a way of *choosing* a redex of any term t, such that the relation $u \prec t$ is well founded, where u is the result of the chosen reduction. This is a tail recursion because there is just one reduced term, which we regard as a sub-argument (in the sense of Definition 2.5.1(a)), and whose normal form is that of the original term. The entire normalisation process happens in the parsing phase of the recursive paradigm, with trivial evaluation, *i.e.* it is tail-recursive.

The strictly monotone function used to show termination takes the term t to the set of *types* of its subterms, and so the proof depends on subject reduction (Definition 1.2.12), and fails for the untyped λ-calculus.

$$\text{raw } \lambda\text{-term } t \;\mapsto\; \text{its redexes} \;\mapsto\; \text{their types} \;\mapsto\; \langle n, m \rangle$$

Suppose a typed λ-term has $m \geqslant 1$ redexes whose type involves n-fold nested exponentials (Convention 2.3.2), but none more complex than this. The reduction strategy is to reduce the innermost of these, so there is no duplication of redexes of complexity n. Although this may create or duplicate arbitrarily many redexes of complexity $< n$, the pair $\langle n, m \rangle$ is reduced in lexicographic order (Proposition 2.6.8).   □

Unions. Well-foundedness is a local property.

DEFINITION 2·6·5: A subset $U \subset X$ is called an ***initial segment*** if
$$\forall t, u{:}X . \; t \prec u \in U \implies t \in U.$$

PROPOSITION 2·6·6: Let $(X, \prec)$ be a set with a binary relation. Suppose for every $x \in X$ there is a well founded initial segment $X' \subset X$ with $x \in X'$. Then X is itself well founded.

PROOF: (Essentially Remark 2.5.13.) Let ϑ be a predicate on X for which the premise of the induction scheme holds, and $x \in X$: we have to show $\vartheta[x]$. Let $X' \subset X$ be a well founded initial segment with $x \in X'$. The premise still holds when restricted to X', and by well-foundedness the conclusion, $\forall x'. \, x' \in X' \Rightarrow \vartheta[x']$, follows. In particular, $\vartheta[x]$.   □

It follows, for example, that the disjoint union of sets with well founded relations is well founded.

Products. There are three ways of putting a well founded relation on the product of two sets, depending on how strict we are about descending on the two sides.

PROPOSITION 2·6·7: The cartesian product of a well founded relation $(Y, \prec)$ with an arbitrary relation $(X, <)$ is well founded:

$$\langle x', y' \rangle \prec \langle x, y \rangle \quad \text{if} \quad x' < x \wedge y' \prec y.$$

In particular the product of two or more well founded relations is well founded, so we may do induction on several variables simultaneously:

$$\frac{\forall x, y. \left(\forall x', y'.\, x' < x \wedge y' \prec y \Rightarrow \vartheta[x', y'] \right) \Rightarrow \vartheta[x, y]}{\forall x, y.\, \vartheta[x, y]}$$

PROOF: The projection $\pi_1 : X \times Y \to Y$ is strictly monotone. □

The next construction was popular in classical mathematics because it preserves trichotomy; in particular this is how to multiply ordinals. See Exercise 2.40 for a more general result.

PROPOSITION 2·6·8: The *lexicographic product*, defined by

$$\langle x', y' \rangle \prec \langle x, y \rangle \quad \text{if} \quad (x' \prec^X x) \vee (x' = x \wedge y' \prec^Y y),$$

is well founded if both $\prec^X$ and $\prec^Y$ are.

PROOF: The induction hypothesis is of the form $(\alpha \vee \beta) \Rightarrow \vartheta$; this is equivalent to $(\alpha \Rightarrow \vartheta) \wedge (\beta \Rightarrow \vartheta)$, which is more convenient.

$$\begin{array}{lll}
_1 \ \forall x, y. \left(\begin{array}{l} \forall x', y'.\, x' \prec x \Rightarrow \vartheta[x', y'] \\ \wedge \ \forall y'.\, y' \prec y \Rightarrow \vartheta[x, y'] \end{array} \right) \Rightarrow \vartheta[x, y] & \\
_2 \ \psi[x] \overset{\text{def}}{=\!=\!=} \forall y.\, \vartheta[x, y] & \text{def} \\
\end{array}$$

| $\forall x$ | $_3$ $\forall x'.\, x' \prec x \Rightarrow \psi[x']$ | hyp |
| | $_4$ $\forall x', y'.\, x' \prec x \Rightarrow \vartheta[x', y']$ | def(2) |

| | $\forall y$ | $_5$ $\forall y'.\, y' \prec y \Rightarrow \vartheta[x, y']$ | hyp |
| | | $_6$ $\vartheta[x, y]$ | $\forall \mathcal{E}(1, 4, 5)$ |

	$_7$ $\forall y. \left(\forall y'.\, y' \prec y \Rightarrow \vartheta[x, y'] \right) \Rightarrow \vartheta[x, y]$	$\forall \mathcal{I}$
	$_8$ $\forall y.\, \vartheta[x, y]$	induction for ϑ on y
	$_9$ $\psi[x]$	def(2)

$$\begin{array}{lll}
_{10} \ \forall x. \left(\forall x'.\, x' \prec x \Rightarrow \psi[x'] \right) \Rightarrow \psi[x] & \forall \mathcal{I} \\
_{11} \ \forall x.\, \psi[x] & \text{induction for } \psi \text{ on } x \\
_{12} \ \forall x.\, \forall y.\, \vartheta[x, y] & \text{def(2)} \qquad \square \\
\end{array}$$

PROPOSITION 2·6·9: The *interleaved product* relation is defined by

$$\langle x', y' \rangle \prec \langle x, y \rangle \quad \text{if}$$
$$(x' \prec x \wedge y' \prec y) \vee (x' \prec x \wedge y' = y) \vee (x' = x \wedge y' \prec y),$$

in which descent is necessary on both sides, at least one of them being strict. This is well founded, assuming that both $\prec^X$ and $\prec^Y$ are.

PROOF: It is sparser than the lexicographic order. □

Although box proofs such as these are not difficult to find, they are nevertheless quite complicated. To express them in the vernacular would require an *ad hoc* lemma for each box (Remark 1.6.2), but this would only take us backwards conceptually. Intuitionism has forced us to devise auxiliary predicates (ψ), and it is by investigating their role that we make progress in Chapter VI and [Tay96b].

The predicate calculus has a better claim to being the "machine code" of mathematics than set theory or the Sheffer stroke does, but machine code is always rather clumsy in handling higher level idioms. Monotonicity,

$$\text{if } x : X \vdash \phi[x] \Rightarrow \psi[x] \text{ then } \vdash \left(\forall x{:}X.\ \phi[x]\right) \Rightarrow \left(\forall x{:}X.\ \psi[x]\right),$$

which was elided from the end of the first proof, is the first of many concepts which need to be coded *on top* of the predicate calculus, but which have foundational status themselves. Monotonicity is the subject of the next chapter. In fact the modal operator (Definition 3.8.2)

$$[\succ]\vartheta[t] \equiv \left(\forall u.\ u \prec t \Rightarrow \vartheta[u]\right) \qquad (\boldsymbol{necessarily}\ \vartheta)$$

is the key to reducing these proofs to easy calculations; Definition 2.6.1, for example, says that $[\succ][f]\vartheta \Rightarrow [f][>]\vartheta$. Although well-foundedness of the transitive closure and the strict induction scheme can be studied using the methods which we have described, this will be much simpler with modal logic in Theorem 3.8.11 and Exercise 3.54. Arithmetic for the ordinals (Section 6.7) packages the techniques of this section.

2.7   Lists and Structural Induction

Now we discuss the idioms for list recursion from functional programming, and then set out the type-theoretic rules for $\mathsf{List}(X)$ analogous to those for $+$ in Section 2.3. We treat the natural numbers in parallel with lists because each sheds light on the other.

Structural recursion over free algebras is crucial for foundations because, from the outside, the mathematical world is just a string of symbols: to handle it at the most basic level we need concatenation and parsing operations. On the other hand, van Kampen's Theorem 5.4.8 illustrates that lists pervade mathematics way beyond foundational considerations. Finite sets arise even more frequently, but to count a set means to form an exhaustive, non-repeating *list* of its elements (Section 6.6).

Induction for numbers and lists.   Richard Dedekind (1888) studied the natural numbers as well as the reals, and gave the following axiomatisation, but it is usually attributed to Giuseppe Peano (1889).

DEFINITION 2·7·1:

(a) $0 : \mathbb{N}$;

(b) if $n : \mathbb{N}$ then $\operatorname{succ} n : \mathbb{N}$;

(c) if $n : \mathbb{N}$ then $0 \neq \operatorname{succ} n$;

(d) if $n, m : \mathbb{N}$ and $\operatorname{succ} n = \operatorname{succ} m$ then $n = m$;

(e) if $U \subset \mathbb{N}$ is a subalgebra, *i.e.* $0 \in U$ and $\forall n{:}\mathbb{N}. \ n \in U \Rightarrow \operatorname{succ} n \in U$, then $U = \mathbb{N}$.

The last is the induction scheme, which we have already mentioned as an example of well founded induction in Example 2.5.5(b).

The set of **lists** of elements of a set X may be defined in a similar way. The set X is sometimes called an **alphabet**, and lists are **words**. Other names for lists are strings, paths and texts.

DEFINITION 2·7·2:

(a) The empty list, written nil, $\star$ or $[\,]$, is in $\operatorname{List}(X)$;

(b) if $h : X$ and $t : \operatorname{List}(X)$ then $\operatorname{cons}(h, t) : \operatorname{List}(X)$, the list <u>cons</u>tructed from the **head** h and **tail** t; some authors write $h :: t$ for this;

(c) $\star \neq \operatorname{cons}(h, t)$;

(d) if $\operatorname{cons}(h, u) = \operatorname{cons}(k, v)$ then $h = k$ and $u = v$;

(e) if $U \subset \operatorname{List}(X)$ is a subalgebra, *i.e.* $\star \in U$ and
$\forall h{:}X. \ \forall t{:}\operatorname{List}(X). \ t \in U \Rightarrow \operatorname{cons}(h, t) \in U$, then $U = \operatorname{List}(X)$.

Stephen Kleene used the notations $X^* = \operatorname{List}(X)$ and $X^+ = X^* \setminus \{[\,]\}$ in the theory of regular grammars, *cf.* $\leadsto^*$ for the transitive reduction relation (Definition 1.2.3), but we shall not use them. It is usual to write

$$\operatorname{cons}(x_1, \operatorname{cons}(x_2, \cdots \operatorname{cons}(x_n, \star) \cdots)) \quad \text{as} \quad [x_1, x_2, \ldots, x_n].$$

The head and tail operations are also known as car and cdr. These are fossils of John McCarthy's original implementation of LISP in 1956 on an IBM 704. This machine had a 36-bit <u>r</u>egister with two readily accessible 15-bit parts called <u>a</u>ddress and <u>d</u>ecrement, of which car and cdr extracted the <u>c</u>ontents. Lists also feature in a dynamic or imperative context as **stacks**, where cons is **push** and (head, tail) together correspond to **pop**.

DEFINITION 2·7·3: The last axiom says that the relation $t \prec \operatorname{cons}(h, t)$ is well founded and gives rise to **list induction**:

$$\frac{\vartheta[\star] \qquad \forall h. \forall t. \ \vartheta[t] \Rightarrow \vartheta[\operatorname{cons}(h, t)]}{\forall \ell. \ \vartheta[\ell]}$$

For the one-letter alphabet $X = \{s\}$, this is just Peano induction on $\operatorname{List}(\{s\}) \cong \mathbb{N}$, where cons is the successor.

Concatenation. Many of the uses of lists can be seen as simply "adding them up", where the notion of *addition* is some associative operation with a unit, *i.e.* it defines a *monoid*. Concatenation of lists is the generic such operation. In the following examples, notice that functions of two lists may often be defined by recursion on only one of them.

DEFINITION 2·7·4:

(a) The *concatenation* of two lists is defined by structural recursion on the first of them (this operation also called append):
$$\star \,;\ell = \ell \qquad \mathsf{cons}(h,t)\,;\ell = \mathsf{cons}\big(h,(t\,;\ell)\big).$$

Section 4.2 explains why we use a semicolon for this as well as for relational composition (Definition 1.3.7).

(b) Let M be a set with an element e and a binary operation m, and $f : X \to M$ any function. Define a function $\mathsf{List}(X) \to M$ by

$$\mathsf{fold}\,(e,m,f,\star) \qquad\qquad = \quad e$$
$$\mathsf{fold}\,\big(e,m,f,\mathsf{cons}(h,t)\big) \quad = \quad m\big(f(h),\mathsf{fold}(e,m,f,t)\big),$$

so $f(x) = \mathsf{fold}\,\big(e,m,f,[x]\big)$.

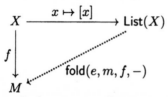

For practical purposes, these recursive definitions are rather inefficient; Exercise 6.28 gives extensionally equivalent versions of fold and append which are *tail-recursive*, *i.e.* they are essentially **while** programs. But we need to know much more about lists and monoids to understand how to transform programs in this way.

PROPOSITION 2·7·5:

(a) Lists form a monoid with $\star$ as unit and (;) as composition:
$$\ell\,;\star = \ell = \star\,;\ell \qquad (\ell\,;\ell_1)\,;\ell_2 = \ell\,;(\ell_1\,;\ell_2).$$

(b) If (M,e,m) is a monoid then $\ell \mapsto \mathsf{fold}(e,m,f,\ell)$ is a homomorphism for the binary operations:
$$\mathsf{fold}\,\big(e,m,f,(\ell\,;\ell_1)\big) \;=\; m\big(\,\mathsf{fold}(e,m,f,\ell),\mathsf{fold}(e,m,f,\ell_1)\big).$$

PROOF: To use list induction in each case we have to identify the variable ℓ and the predicate $\vartheta[\ell]$ (namely the displayed equation), then prove the *base case* $\vartheta[\star]$ and the *induction step* $\forall h.\,\forall t.\,\vartheta[t] \Rightarrow \vartheta[\mathsf{cons}(h,t)]$. We are given one unit law, $\star\,;\ell = \ell$, and so the base case, $\star\,;\star = \star$, for the other.

[right unit, step]   Since $(t \,; \star) = t$ by the induction hypothesis,

$$\mathsf{cons}(h, t) \,; \star = \mathsf{cons}(h, (t \,; \star)) = \mathsf{cons}(h, t).$$

[associativity, base] $\star \,; (\ell_1 \,; \ell_2) = \ell_1 \,; \ell_2 = (\star \,; \ell_1) \,; \ell_2.$

[step] $\begin{aligned}[t] (\mathsf{cons}(h, t) \,; \ell_1) \,; \ell_2 \ &= \ \mathsf{cons}\big(h, (t \,; \ell_1)\big) \,; \ell_2 \\ &= \ \mathsf{cons}\big(h, ((t \,; \ell_1) \,; \ell_2)\big) \\ &= \ \mathsf{cons}\big(h, (t \,; (\ell_1 \,; \ell_2))\big) \quad \text{by hypothesis} \\ &= \ \mathsf{cons}\,(h, t) \,; (\ell_1 \,; \ell_2) \end{aligned}$

Write $[\![\ell]\!] = \mathsf{fold}(e, m, f, \ell)$, so

[base] $m\big([\![\star]\!], [\![\ell_1]\!]\big) = m\big(e, [\![\ell_1]\!]\big) = [\![\ell_1]\!] = [\![\star \,; \ell_1]\!]$

[step] $\begin{aligned}[t] m\big([\![\mathsf{cons}(h, t)]\!], [\![\ell_1]\!]\big) \ &= \ m\big(m(f(h), [\![t]\!]), [\![\ell_1]\!]\big) \\ &= \ m\big(f(h), m([\![t]\!], [\![\ell_1]\!])\big) \quad \text{associativity} \\ &= \ m\big(f(h), [\![t \,; \ell_1]\!]\big) \quad\quad \text{by hypothesis} \\ &= \ [\![\mathsf{cons}\,(h, (t \,; \ell_1))]\!] \\ &= \ [\![\mathsf{cons}\,(h :: t) \,; \ell_1]\!] \qquad\qquad \square \end{aligned}$

The operation fold has countless programming applications.
We freely mix (functional) programs and mathematical notation.

EXAMPLES 2·7·6: Let $\ell \in \mathsf{List}(X)$ and $y \in X$.

(a) $\mathsf{length}(\ell) = \mathsf{fold}(0, +, (\lambda x.\, 1), \ell) \in \mathbb{N}.$

(b) $\mathsf{reverse}(\ell) = \mathsf{fold}(\star, (\lambda \ell_1, \ell_2.\ \ell_2 \,; \ell_1), (\lambda x.\ \mathsf{cons}(x, \star)), \ell) \in \mathsf{List}(X).$

(c) $\mathsf{fold}(0, +, \mathsf{id}, \ell)$ is the sum of the elements.

(d) $\mathsf{map}(f, \ell) = \mathsf{fold}(\star, (;), f, \ell)$ applies a function $f : X \to Y$ to each of
     the elements of a list "in parallel", so

$$\mathsf{map}\, f \,(\mathsf{cons}(h :: t)) \ = \ \mathsf{cons}\big(f(h), \mathsf{map}(f, t)\big)$$

(it is the effect of the *functor* $\mathsf{List} : Set \to Set$, Section 4.4).

(e) $\mathsf{fold}(\bot, \vee, (\lambda x.\, y = x), \ell)$ is a proposition, namely whether the value
     y occurs as an element of ℓ.

(f) $\mathsf{fold}(\varnothing, \cup, (\lambda x.\, \{x\}), \ell) \in \mathcal{P}(X)$ provides the set of elements of ℓ.

(g) If equality in X is decidable, $\mathsf{fold}(0, +, (\lambda x.\, x = y), \ell) \in \mathbb{N}$ counts the
     occurrences of y in ℓ, where "$x = y$" is the function which returns 1
     if they're equal and 0 otherwise.

(h) $\ell \mapsto \mathsf{fold}(\star, (;), \mathsf{id}, \ell) : \mathsf{List}(\mathsf{List}(X)) \to \mathsf{List}(X)$ flattens a list of lists.[5]

Useful though they are, fold and append are not in fact the *fundamental*
operations of recursion over lists.

----

[5]This is the "multiplication" μ of the List monad (Section 7.5), the unit being
$x \mapsto [x] \equiv \mathsf{cons}(x, \star)$.

Primitive recursion for the natural numbers.

REMARK 2·7·7:   Zero and successor are the introduction rules for $\mathbb{N}$:

$$\frac{}{0 : \mathbb{N}}\ \mathbb{N}\mathcal{I}0 \qquad\qquad \frac{n : \mathbb{N}}{\mathsf{succ}(n) : \mathbb{N}}\ \mathbb{N}\mathcal{I}\ \mathsf{succ}$$

The elimination rule resembles that for the sum type (case analysis, Remark 2.3.10) in that it has to provide for these two cases,

$$\frac{\Gamma \vdash z_\Theta : \Theta \qquad\qquad \Gamma \vdash s_\Theta : \Theta \to \Theta}{\Gamma, n : \mathbb{N} \vdash \mathsf{rec}(z_\Theta, s_\Theta, n) : \Theta}\ \mathbb{N}\mathcal{E}$$

where z_Θ and s_Θ are arbitrary terms of these types, possibly involving parameters from the context Γ. The two cases are each *matched* by the β-rules, which say how to compute with rec,

$$\mathsf{rec}(z_\Theta, s_\Theta, 0) \qquad \leadsto \qquad z_\Theta \qquad\qquad\qquad (\mathbb{N}\beta 0)$$

$$\mathsf{rec}(z_\Theta, s_\Theta, \mathsf{succ}\, n) \qquad \leadsto \qquad s_\Theta\big(\mathsf{rec}(z_\Theta, s_\Theta, n)\big). \qquad (\mathbb{N}\beta\ \mathsf{succ})$$

The η-rule recovers the number n from a trivial rec program.

$$\mathsf{rec}(0, \mathsf{succ}, n) \qquad = \qquad n \qquad\qquad\qquad\qquad (\mathbb{N}\eta)$$

EXAMPLE 2·7·8:   Addition and multiplication are defined by

$$m + n = \mathsf{rec}(m, \mathsf{succ}, n) \quad \text{and} \quad m \times n = \mathsf{rec}\big(0, (m + (-)), n\big).$$

For factorial, we need the argument n in the evaluation phase (Definition 2.5.1(c)). Exercise 2.46 uses pairs to reduce this to the basic case.

Predecessor and pattern matching.   Successor and cons would be bijective functions, but for the omission of one element (zero or the empty list) from their images. It is tempting to overlook this and define "inverses", $\mathsf{pred}(n + 1) = n$, $\mathsf{tail}(\mathsf{cons}(h, t)) = t$ and $\mathsf{head}(\mathsf{cons}(h, t)) = h$.

REMARK 2·7·9:   The new operations are only partially defined, but we may extend them as we like to total functions, since the support is complemented. For example $\mathsf{pred}(0)$ may be taken to be zero, "error" or "exception". (The last is a non-judgemental word for error, such as the exit from a loop.)

Although the operations can be forced to be total in this fashion, the rules, such as $\mathsf{succ}(\mathsf{pred}(x)) = x$, are only conditionally valid. A similar situation arises with division by zero, and we shall discuss how algebraic methods may be extended to handle it in Examples 4.6.4 and 5.5.9.

A more flexible approach is to say that each case offers a ***pattern*** $r(x)$ against which the terms may be matched. Then we may define functions as we like by case analysis, so long as the patterns are mutually exclusive (non-overlapping). The function so defined is total iff the patterns are also exhaustive.

In particular, definitions of "well formed" formulae (wffs) in complex type theories (Section 6.2) may be given by side-conditions involving the outermost operation-symbol(s). Sometimes we have to parse more than once, the general procedure being unification (Section 6.5).

Type-theoretic rules for lists. We shall now give the Gentzen-style presentation for lists, but if you are not yet familiar with the rules for the sum type you should skip the remainder of this section.

REMARK 2·7·10: Empty list and cons are introduction rules for $\mathsf{List}(X)$:

$$\frac{}{\star : \mathsf{List}(X)}\;\mathsf{List}\mathcal{I}\star \qquad\qquad \frac{h : X \qquad t : \mathsf{List}(X)}{\mathsf{cons}(h,t) : \mathsf{List}(X)}\;\mathsf{List}\mathcal{I}\mathsf{cons}$$

Again the elimination rule has to provide for these two cases,

$$\frac{\Gamma \vdash z_\Theta : \Theta \qquad\qquad \Gamma \vdash s_\Theta : X \times \Theta \to \Theta}{\Gamma, \ell : \mathsf{List}(X) \vdash \mathsf{listrec}(z_\Theta, s_\Theta, \ell) : \Theta}\;\mathsf{List}\mathcal{E}$$

which are *matched* by the β-rules, saying how to compute with $\mathsf{listrec}$,

$$\mathsf{listrec}(z_\Theta, s_\Theta, \star) \qquad\qquad \rightsquigarrow \quad z_\Theta \qquad\qquad\qquad (\mathsf{List}\,\beta\star)$$

$$\mathsf{listrec}(z_\Theta, s_\Theta, \mathsf{cons}(h,t)) \quad \rightsquigarrow \quad s_\Theta(h, \mathsf{listrec}(z_\Theta, s_\Theta, t)).$$

$$(\mathsf{List}\,\beta\,\mathsf{cons})$$

The η-rule recovers the list ℓ from a trivial $\mathsf{listrec}$ program.

$$\mathsf{listrec}(\star, \mathsf{cons}, \ell) \qquad = \quad \ell \qquad\qquad\qquad (\mathsf{List}\,\eta)$$

The operator $\mathsf{listrec}$ must also be invariant under substitution. By using a function s_Θ instead of a term with free variables of types X and Θ we have avoided introducing yet another variable binder together with its α-equivalence (*cf.* Remark 2.3.11).

Let us compare cons and $\mathsf{listrec}$ with append and fold.

COROLLARY 2·7·11: $\mathsf{List}(X)$ — in particular $\mathbb{N} \cong \mathsf{List}(1)$ — is the free algebra for two different theories:

(a) One constant (variously called 0, $\star$, $[\,]$, empty or nil), together with one *unary* operation (succ or $\lambda n.\, n+1$), or in general ($\lambda t.\, \mathsf{cons}(h,t)$) for each element $h \in X$, *and no laws*. For any other such structure $(\Theta, z_\Theta, s_\Theta)$, $\mathsf{listrec}$ defines the unique mediating homomorphism.

(b) A single *associative binary* operation ($+$ or $;$) with a unit (0, *etc.*) and a generator 1, or h for each element $h \in X$. The mediating homomorphism is given by fold. $\qquad\qquad \square$

No analogue of Proposition 2.7.5 is needed in the unary case. In fact

$$\mathsf{fold}(e, m, f, \ell) = \mathsf{listrec}(e, s_\Theta, \ell), \quad \text{where } s_\Theta(x, u) = m(f(x), u),$$

defines fold in terms of listrec. However, to make the elimination rule into a bijective adjoint correspondence we must use fold and not listrec, as the term below the line is a monoid homomorphism (Example 7.2.8).

REMARK 2·7·12: The terms $z_\Theta : \Theta$ and $s_\Theta : X \times \Theta \to \Theta$ in the premises of (List$\mathcal{E}$) are called the *seed* and *action* of X on Θ. The continuation rule says that if $\jmath : \Theta \to \Theta'$ is a *homomorphism* for this structure,

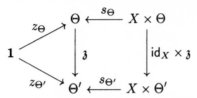

then it is a *recursion invariant* (*cf.* Theorem 6.4.17 for **while**).

$$\frac{\jmath(z_\Theta) = z_{\Theta'} \qquad \forall x{:}X.\ \forall u{:}\Theta.\ \jmath(s_\Theta(x,u)) = s_{\Theta'}(x, \jmath(u))}{\jmath\big(\mathsf{listrec}(z_\Theta, s_\Theta, \ell)\big) = \mathsf{listrec}(z_{\Theta'}, s_{\Theta'}, \ell)}$$

(There is no connection between $\jmath$ and z_Θ at the moment: this notation will make more sense in Section 6.4.)

Although we have written PROOF after these assertions, it will take us until Section 6.3 to show that $\mathsf{List}(X)$, the type of lists with elements from a given set X, actually exists in Zermelo type theory and has the required properties. On the other hand, in functional programming it is more appropriate to treat List as a *new* type constructor like $+$, $\times$ and $\to$, together with the rules in Remark 2.7.10.

List recursion is a type-theoretic phenomenon for which we haven't yet given the propositional analogue. This is induction for the reflexive–transitive closure of any binary relation, and is discussed in Sections 3.8 and 6.4. This will be based on another general induction idiom (on closure conditions), given in Definition 3.7.8. Similar methods are used for finite subsets in Section 6.6, which we shall also discuss using lists. In Chapter VI we shall introduce a new approach to induction based on a categorical analysis of free algebras. Theorem 6.2.8(a) constructs the free category on an oriented graph using the list idea.

2.8   HIGHER ORDER LOGIC

Well-foundedness is a *second order* property because it is defined by *quantifying* the induction scheme over *all predicates* $\vartheta[x]$. In higher order logic, predicates (or, by comprehension, subsets) are first class citizens, allowing quantification over predicates on predicates.

First order schemes. Before ascending to the second order, let us note first that there is a tradition (with almost a strangle-hold over twentieth century logic [Sha91]) of reading any quantification over predicates or types as a *scheme* to be instantiated by each of the formulae which can be defined in the first order part. This has a profound qualitative effect.

REMARK 2·8·1: The *completeness* of first order model theory — the fact that the syntax and semantics exactly match (Remark 1.6.13) — has strange corollaries for the cardinality of its models. If a theory has arbitrarily large finite models then it has an infinite one (the *compactness theorem*), and in this case there are models of any infinite cardinality (the *Löwenheim–Skolem Theorems*). Second order logic has no such property.

In particular, first order logic is unable to characterise $\mathbb{N}$ or $\mathbb{R}$ up to isomorphism. Abraham Robinson [Rob66] exploited this fact to develop *non-standard analysis*: it is possible to construct models which obey exactly the same *first order* statements as the standard $\mathbb{R}$, but which contain infinite or infinitesimal magnitudes. If we can demonstrate some *first order* property in the non-standard structure, possibly employing infinitesimals, then it also holds in the standard one. Since results in the differential calculus can be obtained very rapidly using this technique, it is sometimes advocated as a way of teaching this subject, but it is easy to forget that it depends on a rather sophisticated logical phenomenon.

The first order theory of $\mathbb{R}$ is of algebraic interest in itself. It suffices to say that -1 is not a sum of squares, and every odd-degree polynomial equation has a root, *cf.* Example 2.5.2(c).

Addition and multiplication are definable by recursion from successor in second order Peano arithmetic (Example 2.7.8). Mojżesz Presburger (1930) showed that, in stark contrast, first order arithmetic with addition is decidable, as it can only express (in)congruence *modulo* fixed numbers and so-called linear programming. Arithmetic with addition and multiplication allows Diophantine equations, which are undecidable.

The type of propositions. Even though set theory can be presented in a first order *meta-language*, the subject itself is plainly intended to handle higher order logic. By the definition of powerset (Definition 2.2.5ff), the type of propositions is isomorphic to $\mathcal{P}(\mathbf{1})$.

NOTATION 2·8·2: We shall use the symbol Ω for the type of propositions. It is a hybrid of the Greek letter Ω (omega) used for this purpose in topos theory, and the digit 2, since classically $\Omega = \{\bot, \top\}$.

As Johann Lambert observed, any subset $U \subset X$ gives rise to a function by $x \mapsto \top$ if $x \in U$ and $\perp$ otherwise. Intuitionistically, the correspondence can be made precise simply by a change of notation.

PROPOSITION 2·8·3: $\mathcal{P}(X) \cong \mathcal{Q}^X$.

PROOF: Write $\lambda x{:}X.\ \phi$ instead of $\{x : X \mid \phi[x]\}$ for the term of type $\mathcal{P}(X)$ corresponding to the predicate $\phi[x]$. The membership relation $(a \in U)$ is application (Ua) and the β-, η- and equality rules for the powerset are special cases of those for the λ-calculus.   □

REMARK 2·8·4:   The η-rule for the powerset says that

inter-provable propositions $(\phi \Leftrightarrow \psi)$ are equal $(\phi = \psi)$,

so bi-implication $(\Leftrightarrow)$ is a congruence (Definition 1.2.12). This special case of the extensionality axiom (Remark 2.2.9(a)) is rather mysterious (Exercise 2.54); Remark 9.5.10 considers what need there is for it.

In particular ϑ is equivalent to $(\vartheta = \top)$, *cf.* Exercise 2.21.

Definability of the connectives. In Remarks 2.3.11 and 2.7.10 we made use of the λ-calculus to avoid the need to introduce new variable-binding operations for the $(+\mathcal{E})$- and (List $\mathcal{E}$)-rules.

REMARK 2·8·5:   We may also introduce constants

$\top, \perp : \mathcal{Q}$ equal $: X \to (X \to \mathcal{Q})$
and, or, implies $: \mathcal{Q} \to (\mathcal{Q} \to \mathcal{Q})$ some, all $: (X \to \mathcal{Q}) \to \mathcal{Q}$
member $=$ ev $: X \to ((X \to \mathcal{Q}) \to \mathcal{Q})$,

and then treat $\exists x{:}X.\ \phi$ as an abbreviation for some$(\lambda x{:}X.\ \phi)$.

This is little more than a shift of notation, but there is a further reduction based on a more profound idea. In the following formulae (due to Russell) we use ϑ for the variable bound by $\forall$, because we shall find that it always plays the same special role, namely that of the arbitrary conclusion or result type in the elimination rules for $\perp$, $\vee$, $+$, $\exists$ and List.

PROPOSITION 2·8·6:   In the second order predicate calculus,

$$\perp \quad \dashv\vdash \quad \forall\vartheta.\ \vartheta$$
$$\top \quad \dashv\vdash \quad \forall\vartheta.\ \vartheta \Rightarrow \vartheta$$
$$\phi \quad \dashv\vdash \quad \forall\vartheta.\ (\phi \Rightarrow \vartheta) \Rightarrow \vartheta$$
$$\phi \wedge \psi \quad \dashv\vdash \quad \forall\vartheta.\ (\phi \Rightarrow (\psi \Rightarrow \vartheta)) \Rightarrow \vartheta$$
$$\phi \vee \psi \quad \dashv\vdash \quad \forall\vartheta.\ (\phi \Rightarrow \vartheta) \Rightarrow ((\psi \Rightarrow \vartheta) \Rightarrow \vartheta)$$
$$\exists x.\ \upsilon[x] \quad \dashv\vdash \quad \forall\vartheta.\ (\forall x.\ \upsilon[x] \Rightarrow \vartheta) \Rightarrow \vartheta,$$

where x has any type X (which may be $\mathcal{Q}$), $\upsilon : X \to \mathcal{Q}$ and the other variables are of type $\mathcal{Q}$. In fact the formulae on the right satisfy the introduction and elimination rules for the connectives, even when the latter are not in the language.

PROOF: Consider the third: $\phi \vdash \forall \vartheta. (\phi \Rightarrow \vartheta) \Rightarrow \vartheta$ amounts to ($\Rightarrow \mathcal{E}$); for the converse put $\vartheta = \phi$. The other results are obtained by substituting a first order equivalence into this one.   □

We shall give a detailed proof for the conjunction in Example 2.8.12, from which we see that these equivalences do not depend on the extensionality axiom (Remark 2.8.4). There is a similar result for equality, *cf.* congruence, Definition 1.2.12.

PROPOSITION 2·8·7: (***Leibniz' principle***, 1666)

$$x = y \dashv\vdash \forall \vartheta. \vartheta[x] \Leftrightarrow \vartheta[y],$$

where $x, y : X$ and $\vartheta : X \to 2$. In particular the same holds in the case $X = 2$, where equality is replaced by bi-implication.   □

Cantor's diagonalisation theorem.

PROPOSITION 2·8·8:   There is no injective function $m : \mathcal{P}(X) \hookrightarrow X$.

PROOF: Suppose first that we also have $p : X \twoheadrightarrow \mathcal{P}(X)$ splitting m, *i.e.* $p \circ m = \mathrm{id}_{\mathcal{P}(X)}$. Put $R = \{y \mid \neg(y \in p(y))\}$ and $z = m(R)$, so

$$
\begin{aligned}
z \in R &\Leftrightarrow \neg(z \in p(z)) &&\text{definition of } R \\
&\Leftrightarrow \neg(z \in p(m(R))) &&\text{definition of } z \\
&\Leftrightarrow \neg(z \in R) &&p(m(R)) = R
\end{aligned}
$$

Using a guarded quantifier (Remark 1.5.2), we may in fact define

$$\alpha[x, y] \quad \text{as} \quad \forall U{:}\mathcal{P}(X). (m(U) = x) \Rightarrow y \in U,$$

whence $p : x \mapsto \{y : X \mid \alpha[x, y]\}$ satisfies the argument above because

$$
\begin{aligned}
y \in V &\Leftrightarrow \forall U{:}\mathcal{P}(X). (U = V) \Rightarrow y \in U \\
&\Leftrightarrow \forall U{:}\mathcal{P}(X). (m(U) = m(V)) \Rightarrow y \in U \\
&\Leftrightarrow \alpha[m(V), y] \Leftrightarrow y \in p(m(V))
\end{aligned}
$$

since $(m(U) = m(V)) \Leftrightarrow (U = V)$, as m is an injective function.   □

According to Cantor (1891), $\mathcal{P}(X)$ is "bigger than" X, since there is an injective function $x \mapsto \{x\}$ in one direction but none the other way. He was ignoring Galileo's 1638 warning about "how gravely one errs in trying to reason about infinities by using the same attributes that we apply to finites", in response to the observation that the squares form a proper but equinumerous subset of $\mathbb{N}$, from which he concluded that "equal, greater and less have no place in the infinite". Cantor's interpretation has prevailed (so far), even though it is well known that his motivations were religious at least as much as they were mathematical [Dau79]. Much has also been made of the self-referential nature of this and similar results such as Gödel's Incompleteness Theorem [Hof79]. We shall return to these matters at the end of the book.

Second order types. Quantification over *predicates* was needed for induction, so (applying the ideas of Section 2.4) quantification over *types* should tell us something about recursion. We leave discussion of the *type* of types to the final section of the book.

EXAMPLE 2·8·9: If we delete the dependency on n, Peano induction for $\mathbb{N}$ becomes the second order formula

$$\forall \vartheta. \ \vartheta \to (\vartheta \to \vartheta) \to \vartheta,$$

whose proofs correspond to the Church numerals (Example 2.4.1).

DEFINITION 2·8·10: The **second order polymorphic λ-calculus** was introduced independently in proof theory by Jean-Yves Girard (1972) and in programming by John Reynolds (1974). To the base types of the λ-calculus (Section 2.3) it adds type variables, and an operation of "quantification over types" (α),

$$\frac{\Gamma, \alpha \vdash U \ \mathsf{type}}{\Gamma \vdash \Pi\alpha. \, U \ \mathsf{type}}$$

where we list the free type variables $(\alpha, \beta, ...)$ together with the ordinary variables $(x, y, ...)$ on the left of the turnstile, subject to the condition that for each ordinary variable $x : V$ in Γ, any free type variables in V must occur beforehand (Definition 2.2.8). In particular, when we write "Γ, α" as a context, we presuppose that Γ was already a valid context, so α *must not be free in the type of any variable* $x \in \Gamma$. This is important because, if $x \in \mathsf{FV}(p)$, then x *remains* free in $\lambda\alpha. \, p$ in the $(\Pi\mathcal{I})$-rule.

$$\frac{\Gamma, \alpha \vdash p : U}{\Gamma \vdash (\lambda\alpha. \, p) : (\Pi\alpha. \, U)} \, \Pi\mathcal{I} \qquad \frac{\Gamma \vdash f : \Pi\alpha. \, U \quad \Gamma \vdash V \ \mathsf{type}}{\Gamma \vdash fV : [V/\alpha]^* U} \, \Pi\mathcal{E}$$

Chapter 11 of [GLT89] provides an accessible introduction to this very powerful calculus. In particular it can define many interesting pure types, whereas predicative type theory cannot get off the ground without some user-supplied base types. Girard argues that it has the advantage of eliminating a lot of the *ad hoc* type declarations in programming languages. The (strong) normalisation theorem holds (*op. cit.*, Chapter 15), from which it follows that the closed normal forms of the type called $\mathbb{N}$ below are numerals.

Examples 9.6.10 briefly describe some models of this calculus.

REMARK 2·8·11: The type analogue of Proposition 2.8.6 would be

$$\varnothing \cong \Pi\vartheta.\,\vartheta$$
$$\mathbf{1} \cong \Pi\vartheta.\,\vartheta \to \vartheta$$
$$\{\top,\bot\} \cong \Pi\vartheta.\,\vartheta \to \vartheta \to \vartheta$$
$$U + V \cong \Pi\vartheta.\,(U \to \vartheta) \to ((V \to \vartheta) \to \vartheta)$$
$$U \times V \cong \Pi\vartheta.\,(U \to (V \to \vartheta)) \to \vartheta$$
$$\Sigma x.\,U[x] \cong \Pi\vartheta.\,(\Pi x.\,U[x] \to \vartheta) \to \vartheta$$
$$\mathbb{N} \cong \Pi\vartheta.\,\vartheta \to ((\vartheta \to \vartheta) \to \vartheta)$$
$$\mathsf{List}(U) \cong \Pi\vartheta.\,\vartheta \to ((U \to \vartheta \to \vartheta) \to \vartheta),$$

but John Reynolds showed that the type $\Pi\vartheta.\,((\mathcal{P}(\mathcal{P}(\vartheta)) \to \vartheta) \to \vartheta)$ violates Cantor's theorem.

EXAMPLE 2·8·12: Consider the fifth of these in detail.

	1 $\phi \wedge \psi$		$p_1 : U \times V$
$\forall\vartheta$	2 $\phi \Rightarrow (\psi \Rightarrow \vartheta)$	hyp	$p_2 : U \to (V \to \vartheta)$
	3 ϕ	$\wedge\mathcal{E}0(1)$	$p_3 = \pi_0(p_1) : U$
	4 $\psi \Rightarrow \vartheta$	$\Rightarrow\mathcal{E}(2,3)$	$p_4 = p_2(p_3) : V \to \vartheta$
	5 ψ	$\wedge\mathcal{E}1(1)$	$p_5 = \pi_1(p_1) : V$
	6 ϑ	$\Rightarrow\mathcal{E}(4,5)$	$p_6 = p_4(p_5) : \vartheta$
	7 $\forall\vartheta.\,\big(\phi \Rightarrow (\psi\Rightarrow\vartheta)\big) \Rightarrow \vartheta$	$\forall\Rightarrow\mathcal{I}$	$p_7 = \lambda\vartheta.\,\lambda p_2.\,p_6$

In full, $\quad p_7 = \lambda\vartheta.\,\lambda p_2.\,p_2(\pi_0(p_1))(\pi_1(p_1)) : \Pi\vartheta.\,(U \to (V \to \vartheta)) \to \vartheta.$

Conversely,

1 $\forall\vartheta.\,(\phi \Rightarrow (\psi \Rightarrow \vartheta)) \Rightarrow \vartheta$	hyp	q_1
2 ϕ	hyp	$q_2 : U$
3 ψ	hyp	$q_3 : V$
4 $\phi \wedge \psi$	$\wedge\mathcal{I}(2,3)$	$q_4 = \langle q_2, q_3 \rangle : U \times V$
5 $\psi \Rightarrow \phi \wedge \psi$	$\Rightarrow\mathcal{I}$	$q_5 = \lambda q_3.\,q_4$
6 $\phi \Rightarrow (\psi \Rightarrow (\phi \wedge \psi))$	$\Rightarrow\mathcal{I}$	$q_6 = \lambda q_2.\,q_5$
7 $\phi \wedge \psi$	$\forall\mathcal{E}(1,6)$	$q_7 = q_1[U \times V](q_6)$

Substituting p_7 for q_1, we find that $U \times V$ is a *retract* of the polymorphic type, but they are not necessarily isomorphic. Proposition 2.8.6 relied on the anonymity of proofs, *i.e.* the principle that any two proofs of the same proposition are considered equal. Alternatively, if we use the polymorphic types as *definitions* of the other connectives, the β-rules are satisfied, but the η-rules fail. $\qquad\square$

Impredicativity. Notice that all of the examples above are of the form

$$\Pi\vartheta.\,(T[\vartheta] \to \vartheta) \to \vartheta,$$

where T is a type-expression in which ϑ only occurs positively in the sense of Remark 1.5.9, *i.e.* a covariant functor (Definition 4.4.1). Indeed $T[\vartheta]$ is either the type on the left of the isomorphism (and doesn't depend on ϑ) or a sum of powers of ϑ such as $1 + \vartheta$ in the case of $\mathbb{N}$ and $\mathsf{List}(X)$. We shall study functors of this kind in Section 6.1.

The fact that so many constructions may be characterised in just this *one* idiom is nevertheless very remarkable, and we shall exploit the idea, albeit slightly altered (as universal properties), in the rest of the book.

This form is said to be *impredicative* because it involves *quantification* over all types, *including the type being characterised* (in a private rather than official capacity). Henri Poincaré (1912) and Hermann Weyl (1919) put an objection to the circularity of this mode of definition. As we explore the mathematical world, we discover or postulate new objects (individuals, types and predicates), so at any moment the world remains *open*. The meaning of any definition referring to *all* objects is therefore liable to change as a result of subsequent work. For example, the *least* upper bound of a set can be undermined by the later discovery of an element tucked in between the set and its previously known bounds. For Poincaré and Weyl this was a metaphysical possibility, but Example 3.2.8 shows how it happens concretely.

But induction, which is obviously important and has not *so far* been shown to be inconsistent, is necessarily second order. Those who reject impredicativity therefore have to treat as *axioms* induction principles in which the meta-language either uses completed infinities or gets mixed up with the object-language. Moreover there are many such principles, whereas one impredicative idea unifies them all. It will also serve as our common formulation for all type-theoretic constructions.

The issues of well-foundedness raised by Poincaré, Weyl and in the "inductive conception of sets" are, in modern type theory, transferred from the ontology to the syntax. They will be addressed in Chapter VI and used in Chapters VIII and IX.

The objection to least upper bounds is taken on board as part of our mathematical development. We don't think of a *single, growing* world which is never complete, but a succession of worlds, each complete as far as it goes. (Compare the expansion of the number system, our starting point in this chapter: nowadays number theory considers fields in which *particular* polynomials split, rather than the *absolute* universe $\mathbb{C}$.) Arbitrary expansions of the structure indeed do not preserve impredicatively defined objects, so if these are important *we only allow certain types of expansion*, the homomorphisms for the relevant structure.

In answering the objection to impredicativity, we must distinguish two notions of **definition**, namely **specification** and **implementation**, which Reductionism confuses. Example 2.1.7 showed *how* to construct the disjoint union $X + Y$ of two sets, but *why* we want such a construction is explained by its relationship $X + Y \to \Theta$ to all other sets, which is an impredicative property (Remark 2.3.10). This difference between recipes and concepts is the heart of the modern foundations of mathematics.

EXERCISES II

1. Give a construction of the integers ($\mathbb{Z}$) from the natural numbers such that $z = \{\langle m, n \rangle \mid m - n \leqslant z\}$. Define addition and subtraction for both this coding and the one in Example 1.2.1.

2. Show how to add and multiply complex numbers as pairs of reals, verifying the commutative, associative and distributive laws and the restriction of the operations to the reals.

3. The volume-flow (in $\mathrm{m}^3\,\mathrm{s}^{-1}$) down a pipe of radius r of a liquid under pressure p is $c\eta^n r^m p^k$ for some dimensionless c, where η is the *dynamic viscosity*, in units of $\mathrm{kg\,m}^{-1}\,\mathrm{s}^{-1}$. Find n, m and k.

4. Show how to add Dedekind cuts and multiply them *by rationals*, justifying the case analysis of the latter into positive, zero and negative. What do your definitions say when the cuts represent rationals? Verify the associative, commutative and distributive laws.

5. Express $\sqrt{2}$, $\sqrt{3}$ and $\sqrt{6}$ as Dedekind cuts, and hence show that $\sqrt{2} \cdot \sqrt{3} = \sqrt{6}$.

6. Let $x = (L, U)$ and $y = (M, V)$ be Dedekind cuts of $\mathbb{Q}$ and put

$$
\begin{aligned}
N_1 &= \{x'y + xy' - x'y' \mid x' \in L, y' \in M\} \\
N_2 &= \{x'y + xy' - x'y' \mid x' \in U, y' \in V\} \\
W_1 &= \{x'y + xy' - x'y' \mid x' \in L, y' \in V\} \\
W_2 &= \{x'y + xy' - x'y' \mid x' \in U, y' \in M\}
\end{aligned}
$$

Show that (the lower closure of) $N_1 \cup N_2$ and (the upper closure of) $W_1 \cup W_2$ define a Dedekind cut of $\mathbb{R}$. Calling it xy, verify the usual laws for multiplication, without using case analysis [Con76].

7. For any Cauchy sequence (a_n), show that there is an equivalent sequence (b_m) which satisfies $\forall n, m. \ |b_n - b_m| < \frac{1}{m} + \frac{1}{n}$ [BB85].

8. Show how to add Cauchy sequences and to multiply them by rational numbers.

9. Show how to reduce a Cauchy sequence of reals (*i.e.* of Cauchy sequences) to a Cauchy sequence of rationals [Hint: diagonalise].

10. Define multiplication of Cauchy sequences, without using a case analysis according to sign.

11. Let (a_n) be a Cauchy sequence and (L, U) a Dedekind cut. Formulate the predicate $\phi[(a_n), (L, U)]$, that they denote the same real value. Show that for every Cauchy sequence there is a cut, and that

$$\phi[(a_n), (L, U)] \wedge \phi[(b_n), (L, U)] \iff (a_n) \sim (b_n).$$

Hence ϕ is an injective functional relation $\mathbb{R}_C \hookrightarrow \mathbb{R}_D$. Show that it respects addition and multiplication. Show, using excluded middle, that ϕ is bijective (see also Exercise 2.30).

12. Explain the difference between $p(U)$ in Example 2.1.5 and $f_!(U)$ in Remark 2.2.7.

13. Show that each element $\langle U, V \rangle$ of $X + Y$, defined in Example 2.1.7, is either of the form $\langle \{x\}, \varnothing \rangle$ with $x \in X$, or else of the form $\langle \varnothing, \{y\} \rangle$ with $y \in Y$, but not both. [Hint: use $(\vee \mathcal{E})$ on the first clause.] Deduce that for any functions $f : X \to \Theta$ and $g : Y \to \Theta$, the subset $f_!(U) \cup g_!(V) \subset \Theta$ has exactly one element. So there is a unique function $[f, g] : X + Y \to \Theta$ such that $[f, g] \circ \nu_0 = f$ and $[f, g] \circ \nu_1 = g$.

14. Using the isomorphism $\mathcal{P}(X + Y) \cong \mathcal{P}(X) \times \mathcal{P}(Y)$, construct the product $X \times Y$ from powersets and disjoint unions. [Hint: consider those $U \subset X + Y$ with exactly one element in each component.]

15. For any set X, show that there is a function $c : \mathcal{P}(X) \setminus \{\varnothing\} \to X$ such that $\forall U. \, c(U) \in U$, iff the axiom of choice holds (Definition 1.8.8). (This was Zermelo's formulation of the axiom of choice.)

16. Any predicate $\phi[x]$ gives rise to an equivalence relation $\sim$ on $\{0, 1\} \times X$ so that $\forall x. \, (\phi[x] \Leftrightarrow \langle 0, x \rangle \sim \langle 1, x \rangle)$. Let $Y = (\{0, 1\} \times X)/\sim$ and $p : Y \twoheadrightarrow X$ by $p[i, x] = x$. Prove that $\forall y. \, \exists j. \, y = [j, p(y)]$, making your use of $(\exists \mathcal{E})$ explicit. Show that if this has a Choice function then ϕ is decidable. Hence the axiom of choice implies excluded middle.

17. Show that

$$
\begin{aligned}
\{x \mid \phi[x]\} \times \{y \mid \psi[y]\} &\cong \{\langle x, y \rangle \mid \phi[x] \wedge \psi[y]\} \\
\mathcal{P}(\{x : X \mid \phi[x]\}) &\cong \{U : \mathcal{P}(X) \mid \forall x. \, x \in U \Rightarrow \phi[x]\} \\
\forall x : \{y : Y \mid \phi[y]\}. \, \psi[x] &\Leftrightarrow \forall y : Y. \, \phi[y] \Rightarrow \psi[y] \\
\exists x : \{y : Y \mid \phi[y]\}. \, \psi[x] &\Leftrightarrow \exists y : Y. \, \phi[y] \wedge \psi[y],
\end{aligned}
$$

giving the functional relations involved. Use these to show that the axiom of comprehension may be eliminated from Zermelo type theory, in the sense that every type is of the form $\{x : U \mid \phi[x]\}$ where U is a type-expression built from $\mathbf{1}$, $\mathbb{N}$, $\times$ and $\mathcal{P}$ alone. Hence develop a formalism for Zermelo type theory with unordered contexts.

18. Let $a, b, c, d \in X$, not necessarily distinct. Explain carefully what the left hand side means (*cf.* Example 2.1.6(c)) and prove that

$$\{a, b\} = \{c, d\} \iff (a = c \wedge b = d) \vee (a = d \wedge b = c)$$

by simplifying a disjunction of 16 cases, 14 of which give $a = b = c = d$.

19. For $x, y, x', y' \in Z$, write

$$\mathsf{WKP}(x, y) = \{\{x\}, \{x, y\}\} \in \mathcal{P}^2(Z)$$

for the **Wiener–Kuratowski pair**. Using Exercise 2.18, show that

$$\mathsf{WKP}(x, y) = \mathsf{WKP}(x', y') \iff x = x' \wedge y = y',$$

without using excluded middle on $x = y$ *versus* $x \neq y$. Give the functional relations $\mathcal{P}^2(Z) \rightharpoonup Z$ defining the product projections π_0 and π_1. Hence for $X, Y \subset Z$ show that $\{\mathsf{WKP}(x, y) \mid x \in X \wedge y \in Y\} \subset \mathcal{P}^2(Z)$ satisfies the rules for the product type (Remark 2.2.2). This formula is the one surviving instance of an early application of the representation of the formal poset $x \leqslant y$ by subsets (Definition 3.1.7).

20. Explain how $\mathcal{P}^n(\mathbb{N}) \subset \mathcal{P}^m(\mathbb{N})$ for $n \leqslant m$. Using Exercises 2.17 and 2.19, show that the types U with $U \subset \mathcal{P}^n(\mathbb{N})$ for some $n \in \mathbb{N}$ form a model of pure Zermelo type theory. It is known as the **von Neumann hierarchy** $V_{\omega 2}$.

21. Give the λ-terms which define the isomorphism between X and $X^1 \times 1^X$, and verify that they are mutually inverse.

22. Show how to orient the β- and η-rules for pairs (Remark 2.2.2) to make them confluent and strongly normalising (Definitions 1.2.5 and 1.2.8). The normal form says how the types are bracketed.

23. A (raw) λ-term is in **head normal form** if it looks like

$$\lambda y_1. \, \lambda y_2. \, \ldots \lambda y_n. \, x u_1 u_2 \ldots u_m,$$

where x is either a constant or a variable, possibly one of the y_i, and no sub-sequence $x u_1 \ldots u_k$ is a δ-redex. Show that a term is in normal form iff it is *hereditarily* head normal.

24. The **de Bruijn index** of a bound variable is the number of λs which separate its use from its declaration in the tree structure of the term. For example $\lambda x. \, (\lambda y. \, xy)x$ becomes $\lambda.(\lambda.2\,1)1$. (*Cf.* the way in which a compiled procedure accesses local variables relative to the current stack pointer.) Give the formal translation of raw λ-terms from variables to de Bruijn indices and *vice versa*. Show that, using indices, substitution is performed textually, except that in $[b/y]^*(\lambda x. \, a)$ the *free* indices within b are incremented.

25. Show that α-conversion is unavoidable in the reduction of the untyped term $(\lambda x.\ xx)(\lambda x, y.\ xy)$.

26. **Combinatory algebra** is given by one binary operation called application and two constants S and K with laws $Sxyz = xz(yz)$ and $Kxy = x$ (Example 2.4.2). Show **functional completeness**, that any term p in variables $x_1, \ldots, x_n$ is equivalent to some $fx_1x_2\cdots x_n$, where f is a term with no (free) variables. [Hint: by structural recursion, use

$$u = (Ku)x \quad \text{and} \quad (ux)(vx) = (Suv)x$$

to eliminate x_n first and similarly the other variables.] Translate the term $T = \lambda x, y, z.\ xzy$.

27. Recall from Remark 1.6.10 that the two axioms of **implicational logic** are the types of S and K. Use Exercise 2.26 to prove the **deduction theorem**, that if ϕ is provable from hypotheses $\alpha_1, \ldots, \alpha_n$ then $\alpha_1 \Rightarrow \cdots \Rightarrow \alpha_n \Rightarrow \phi$ is provable under no hypothesis. This says that each instance of the $(\Rightarrow\mathcal{I})$-rule is derivable in this calculus.

28. Suppose that $X^R \lhd R$, i.e. $j : X^R \hookrightarrow R$, $q : R \twoheadrightarrow X^R$ with $j\ ;q = \mathrm{id}_{X^R}$. Show that any $f : X \to X$ has a fixed point.

29. Use the elimination rule for $\varnothing$ to show $\exists x{:}\varnothing.\ \phi[x] \dashv\vdash \bot$ and $\forall x{:}\varnothing.\ \phi[x] \dashv\vdash \top$. Formulate the substitution rule and use it to show that any $X \to \varnothing$ is invertible. Show also that $\exists x{:}\mathbf{1}.\ \phi \dashv\vdash [\star/x]^*\phi \dashv\vdash \forall x{:}\mathbf{1}.\ \phi$.

30. Consider the last clause in the definition of a Dedekind cut (Remark 2.1.1). Show that an assignment $(x_\varepsilon, y_\varepsilon)$ of witnesses (as in Remark 2.4.3(g)) provides a pair of Cauchy sequences which define the same real number in the sense of Exercise 2.11.

31. Show that **Peirce's law**, $((\vartheta \Rightarrow \phi) \Rightarrow \vartheta) \Rightarrow \vartheta$ (Remark 2.4.10), excluded middle and the restart rule are intuitionistically equivalent.

32. What, in a word, is a well founded functional relation?

33. (Pierre de Fermat) Show that a prime number can be expressed as the sum of two squares iff it is (either 2 or) of the form $4n + 1$. [Hint: define $q \prec p$ on primes of this form if $q < p$ and there are numbers k and $1 \leqslant m < p$ such that $kpq = m^2 + 1$.]

34. Prove the results of Section 2.6 using minimal counterexamples, and also using descending sequences. Explain where excluded middle and König's Lemma are needed when using these forms of induction.

35. Suppose that finitely many types of polygonal tile are given, with the property that a disc of arbitrary radius may be covered with copies of the tiles, without overlaps. Using König's Lemma (Corollary 2.5.10), show that the whole plane ($\mathbb{R}^2$) may be covered.

36. Let $\prec$ be a decidable binary relation with no cycles on a finite set. Show that it is well founded (*cf.* Corollary 2.5.11). [Hint: consider the number of descendants of each element and use Proposition 2.6.2.]

37. Show directly that the interleaved product relation (Proposition 2.6.9) is well founded. Investigate whether any of the three product relations have well founded analogues for infinite families of sets. Is the order of potential words in a dictionary well founded?

38. Show that the lexicographic product of two transitive relations is transitive, and similarly for the trichotomy law.

39. Find an example of a union of well founded relations which is *not* well founded, to show that *lower* is necessary in Proposition 2.6.6.

40. Let $f : (X, <) \to (Y, \prec)$ be a function between sets with binary relations which is (non-strictly) monotone in the sense that

$$x' < x \implies fx' \prec fx \ \vee \ fx' = fx$$

where $(Y, \prec)$ is well founded. Suppose that every set $f^{-1}[y] \subset (X, <)$ is well founded. Show that X is also well founded.

41. Using proof boxes, show that the transitive closure of a well founded relation is also well founded (Theorem 3.8.11)

42. Write a program to check for repeats in a list.

43. Let $f : X \twoheadrightarrow Y$ be any surjective function between sets. Show that $\mathsf{map}\, f : \mathsf{List}(X) \twoheadrightarrow \mathsf{List}(Y)$ (Example 2.7.6(d)) and deduce that a finite set of witnesses may be chosen without invoking the axiom of choice (Definition 1.8.8).

44. Show that for the Church numerals (Example 2.4.1), the terms

$$\mathsf{succ} = \lambda n.\, \lambda x.\, \lambda f.\, f(nxf) \quad \mathsf{rec} = \lambda z.\, \lambda s.\, \lambda n.\, nz(\lambda x.\, sxn)$$

define the successor and ***primitive recursion operator***:

$$\mathsf{rec}\, z\, s\, 0 \rightsquigarrow z \qquad \mathsf{rec}\, z\, s\, (\mathsf{succ}\, n) \rightsquigarrow s(\mathsf{rec}\, z\, s\, n).$$

[For any type α, put $N = \alpha \to (\alpha \to \alpha) \to \alpha$; then $n : N$, $x, z : \alpha$, $f, z : \alpha \to \alpha$, $\mathsf{succ} : N \to N$ and $\mathsf{rec} : \alpha \to (\alpha \to \alpha) \to N \to \alpha$.] Show that $(m + n) = \mathsf{rec}\, m\, \mathsf{succ}$ and similarly define multiplication and other arithmetical functions, *cf.* Remark 2.7.7 and Example 2.7.8.

45. Show that if α is the n-element set then the Church numerals of type $\alpha \to (\alpha \to \alpha) \to \alpha$ up to $M + n - 1$ are distinct, but that afterwards they repeat with period M, where M is the least common multiple of $1, 2, \ldots, n$. Estimate M.

46. Show how recursive programs such as factorial of the form

$$p(0) = z \qquad p(n + 1) = s(p(n), n)$$

can be written without using the argument n in the evaluation phase. [Hint: what is the recursive definition of $q(n) = \langle p(n), n\rangle$?]

47. A set X is **Dedekind-infinite** if there are an element $z \in X$, a subset $U \subset X$ with $z \notin U$ and an injection $s : X \hookrightarrow U$. Show that (X, z, s) is exactly a model of the first four Peano axioms (Definition 2.7.1). Using powerset and comprehension but not the existence of $\mathbb{N}$, show that X contains a definable model of all five axioms.

48. Prove by list induction that

$$\text{read}(\text{write}(\ell, m, a), n) = \begin{cases} a & \text{if } n = m \leqslant \text{length}(\ell) \\ \text{read}(\ell, n) & \text{if } n \neq m \leqslant \text{length}(\ell) \end{cases}$$

where
$$
\begin{aligned}
\text{read}\,(\text{cons}(h, t), 0) &= h \\
\text{read}\,(\text{cons}(h, t), n + 1) &= \text{read}(t, n) \\
\text{write}(\star, 0, a) &= \text{cons}(a, \star) \\
\text{write}(\text{cons}(h, t), 0, a) &= \text{cons}(a, t) \\
\text{write}(\text{cons}(h, t), m + 1, a) &= \text{cons}(h, \text{write}(t, m, a))
\end{aligned}
$$

49. Show that $\forall \ell.\ \text{rotate}(\text{length}(\ell), \ell) = \ell$, where
$$
\begin{aligned}
\text{rotate}(0, \ell) &= \ell \\
\text{rotate}(n + 1, \star) &= \star \\
\text{rotate}(n + 1, \text{cons}(h, t)) &= \text{rotate}(n, (t\,;\,(\text{cons}(h, \star))))
\end{aligned}
$$

50. Justify the following binary list induction scheme:
$$\frac{\vartheta[\star] \qquad \forall h.\ \vartheta[\text{cons}(h, \star)] \qquad \forall \ell_1, \ell_2.\ \vartheta[\ell_1] \wedge \vartheta[\ell_2] \Rightarrow \vartheta[\ell_1\,;\,\ell_2]}{\forall \ell.\ \vartheta[\ell]}$$

51. Consider the language of higher order predicate calculus with only the connectives $\forall$ and $\Rightarrow$. Show that the formulae on the right hand side of Proposition 2.8.6 satisfy the introduction and elimination rules for the other connectives.

52. Prove Proposition 2.8.7, that $x = y \dashv\vdash \forall \vartheta.\ \vartheta[x] \Leftrightarrow \vartheta[y]$, making clear where the congruence rules are used. Express the proof in such a way as to make equality a derived connective. By substituting the two-way formula for ϑ, show that the result remains true with one-way implication: $\forall \vartheta.\ \vartheta[x] \Rightarrow \vartheta[y]$.

53. Show that 2 satisfies Exercise 1.10(d) but that (c) gives excluded middle. If you know some sheaf theory, show that (b) fails in $\text{Sh}(\mathbb{R})$.

54. Let $i : 2 \hookrightarrow 2$. Show that $\forall x.\ i(x) \Rightarrow (x = i(\top))$. On the assumption that $i(i(\top)) = \top$, show that $i^2 = \text{id}$ and $i(x) = (x = i(\top))$. Without this assumption, deduce that $i(x) = [i(i(\top)) \wedge (x = i(\top))]$. By computing $i(\top)$ and $i^3(\top)$, show that the assumption is redundant.

III

Posets and Lattices

ORDER STRUCTURES provide some simple tools for investigating semantics. For us, they serve a double purpose, describing *systems* of propositions, and also as the substance of *individual* types.

The *system* of propositions is an order structure with respect to the provability relation, in which the logical connectives are characterised as algebraic operations (meet, join, *etc.*) satisfying certain laws. In the later chapters we shall discuss the analogous operations for types. For example implication is discussed here briefly using Heyting algebras, and the function-type in Chapter IV with cartesian closed categories.

Similar operations, sometimes with weaker laws or in infinitary form, arise in many mathematical situations beyond logic. For example, the lattice of sub*algebras* often throws much light on the structure of an algebra, topological spaces are described by their *open* sets and programs in terms of partial evaluation. This ubiquity has led some authors to try to force other concepts such as well founded relations (Section 2.5) into the same mould, a tendency which we aim to reverse.

Our other use of order structures is as *individual* types. As we remarked in Section 2.2, we need something subtler than sets (as described by Zermelo type theory) to illustrate many of the phenomena of reasoning, especially about non-terminating computation. We do this using posets that have directed joins in Sections 3.3–3.5. Later in the book we shall give examples where types are interpreted as topological spaces.

Implication, the quantifiers and infinitary meets and joins are just a few examples of adjunctions or universal properties. We study them in detail here because many of the features of this central concept of category theory can already be seen in the simpler order-theoretic context. In practice, if a function preserves all meets then its left adjoint tends to be used formulaically, without appreciating the important *theorem* which is involved. Here and in subsequent chapters we shall indicate some of the huge number of mathematical results which can be obtained from simple observations about adjunctions.

The last three sections are devoted to adjunctions between powersets. To fulfil a promise that everything which is later done for categories is treated first for posets, some material has been included that is disproportionately more difficult than the rest of the chapter, so you should feel free to skip it on first reading. The intermediate status of posets, ambiguously individual types or systems of them, is unfortunately also reflected in some schizophrenic notation.

Sections 1.4, 1.5 and 2.3 defined the logical connectives in terms of their *introduction* and *elimination* rules. Algebraic operation-symbols can also be seen as introduction rules; at the propositional level, these are known as *closure conditions*. They arise as the conditions for subalgebras and congruences, including reflexivity, symmetry, transitivity and convexity, and also as logic programs. The corresponding elimination rules are familiar or novel induction schemes, which Section 3.7 also describes.

The final section introduces the construction of semantics from syntax which will be developed throughout the book.

3.1   POSETS AND MONOTONE FUNCTIONS

DEFINITION 3·1·1:   A **poset** is a set $\mathcal{X}$ together with a binary "order" relation $\leqslant$ which is **reflexive**, **transitive** and **antisymmetric**:

$$\frac{}{x \leqslant x}\ \text{refl} \qquad \frac{x \leqslant y \quad y \leqslant z}{x \leqslant z}\ \text{trans} \qquad \frac{x \leqslant y \quad y \leqslant x}{x = y}\ \text{antisymm}$$

If $\leqslant$ is only known to be reflexive and transitive then we call it or the set $\mathcal{X}$ a **preorder**. We write $y \geqslant x$ as a synonym for $x \leqslant y$.

EXAMPLES 3·1·2:   The following are posets:

(a) any set with the **discrete order**, $x \leqslant y$ iff $x = y$;

(b) $\mathbb{N}$, $\mathbb{Z}$, $\mathbb{Q}$ and $\mathbb{R}$ with the usual arithmetical order;

(c) $\mathbb{N}$ with the divisibility order, $n \mid m$ iff $\exists k.\, nk = m$;

(d) the two-element set $\{\bot, \top\}$ with $\bot \leqslant \top$ but $\top \not\leqslant \bot$;

(e) $\mathcal{P}(\mathbb{X})$ with the inclusion order, $\subset$, for any set $\mathbb{X}$;

(f) in particular $2 = \mathcal{P}(\bullet)$, the type of propositions or truth values under implication, which is reflexive and transitive; antisymmetry in this case is the η-rule for the powerset, which says that inter-provable propositions are equal (Remark 2.8.4);

(g) the set of open subsets of a topological space under inclusion;

(h) the set of subgroups of a group, and so on.

(i) The **specialisation order** between points in a topological space,

$$x \leqslant y \quad \text{if} \quad \forall U \subset \mathcal{X} \text{ open. } x \in U \Rightarrow y \in U,$$

is in general a preorder; the space is called T_0 if the specialisation order is antisymmetric, and T_1 if it is a discrete order (*cf.* Leibniz' Principle, Propositions 2.8.7 and 3.8.14).

(j) The "bracket nesting" order on sub-expressions may be regarded as a poset, but its purpose is structural recursion, for which a well founded (and in particular irreflexive) relation is needed.

(k) Formulae form a preorder under prov*ability* ($\vdash$, Definition 1.4.7), though there may be many different proofs.

(l) Expressions form a preorder under reducibility ($\rightsquigarrow$, Definition 1.2.3), though there may be many different reduction paths.

Remember that the transitivity of a relation has significant mathematical force: don't assume it without checking it (*cf.* Example 9.4.9(d))!

The word poset is a corruption of "partially ordered set", where a **total** or **linear order** is one satisfying

DEFINITION 3·1·3: The earliest use of order relations was arithmetical, where the reflexive order ($\leqslant$) is accompanied by an irreflexive relation ($<$) of equal importance. $\mathbb{N}$, $\mathbb{Z}$ and $\mathbb{Q}$ have the **trichotomy** property,

$$\forall x, y : \mathcal{X}. \quad x < y \ \lor \ x = y \ \lor \ x > y,$$

so that $x \leqslant y \iff x < y \lor x = y \iff x \not> y$, where again we write $x > y$ for $y < x$. For $\mathbb{R}$ trichotomy relies on classical logic.

Posets are *partial* in the sense that there may be pairs of elements which are **incomparable**, *i.e.* fail to stand in the order relation either way round, *cf.* Galileo's remarks after Proposition 2.8.8. (This is what people usually mean by "equality" in politics.) As there is no connection between total orders and total functions, or between linear orders and preserving sums, it is best to forget these terms and treat "poset" as a *bona fide* English word for the fundamental notion that it is.

Moving away from arithmetical examples, trichotomy usually fails for posets arising in logic, and is destroyed by products and function-spaces. Imposing trichotomy can be a nuisance in technical situations (*cf.* Corollary 3.5.13, but see the ordinals in Section 6.7). It has also given rise to a great deal of misleading terminology. For example, without trichotomy, $x \not> y$ ("no more than") is no longer a synonym for $x \leqslant y$ ("at most"). Some authors, whilst trying to be careful about non-strict inequalities, fall into a greater error by saying "non-decreasing" for monotone.

The symbol $\leqslant$, like subset inclusion (Remark 2.2.6), is irreducible, but unfortunately too well established to replace; an arrow would be better, both graphically and theoretically.

WARNING 3·1·4:   Any irreflexive relation $<$ can in fact be recovered from $(<) \cup (=)$, but we need excluded middle to recover the reflexive relation $\leqslant$ from $(\leqslant) \cap (\neq)$ (Exercise 3.3). Sometimes we use $<$ and $\leqslant$ together in the same passage: beware that they are *not* assumed to be related in this classical fashion.

Monotone functions. Whenever we define predicates such as $\leqslant$ and operations such as $\wedge$, $\vee$, $\times$ or $+$ we are always interested in the functions which preserve them.

DEFINITION 3·1·5:   Let $(\mathcal{X}, \leqslant^{\mathcal{X}})$ and $(\mathcal{Y}, \leqslant^{\mathcal{Y}})$ be posets (or preorders) and $f : \mathcal{X} \to \mathcal{Y}$ a function. Then we say f is **monotone, covariant** or **order-preserving** if

$$\forall x_1, x_2. \quad x_1 \leqslant^{\mathcal{X}} x_2 \implies f(x_1) \leqslant^{\mathcal{Y}} f(x_2).$$

A function for which the converse implication holds is said to **reflect order**. It is **full** if both directions hold; if $\mathcal{X}$ is a poset this requires f to be injective, identifying $\mathcal{X}$ with a subset of $\mathcal{Y}$, where this subset is equipped with (the restriction of) the same order relation $\leqslant^{\mathcal{Y}}$. (The property of functors which most naturally corresponds to fullness is that they be full *and faithful*, Definition 4.4.8.)

A function which, in the same sense, preserves an *irreflexive* order is called **strictly monotone**, *cf.* Definition 2.6.1 for well founded relations.

If we want to say that

$$\forall x_1, x_2. \quad x_1 \leqslant^{\mathcal{X}} x_2 \implies f(x_1) \geqslant^{\mathcal{Y}} f(x_2),$$

then we call f **antitone, contravariant** or **order-reversing**. There's nothing new in this because antitone functions $\mathcal{X} \to \mathcal{Y}$ are just monotone functions $\mathcal{X} \to \mathcal{Y}^{\mathrm{op}}$ or $\mathcal{X}^{\mathrm{op}} \to \mathcal{Y}$. Here $\mathcal{X}^{\mathrm{op}}$ is the **opposite poset**, $(\mathcal{X}, \geqslant)$, whose order relation is the converse of $\leqslant$ (Definition 1.3.9).

EXAMPLES 3·1·6:

(a) The formulae $\alpha \wedge \phi$, $\phi \vee \alpha$, $\phi \Rightarrow \alpha$, $\forall x.\, \alpha$ and $\exists x.\, \alpha$ are monotone functions of the propositional variable α.

(b) The formulae $\neg \alpha$ and $\alpha \Rightarrow \phi$ are antitone functions of α.

(c) The composite of two monotone functions is monotone, as is the identity function.

(d) A preorder in which $\leqslant$ is symmetric is an equivalence relation. In this case monotone functions are functional in the sense of Remark 1.3.2.

(e) A bijective monotone function has a (monotone) inverse iff it is full.

(f) Recall that any function $f : X \to 2$ is a predicate and so defines a subset $U = \{x \mid f(x) = \top\} \subset X$ by comprehension, and conversely by $f(x) = (x \in U)$ (*cf.* Definition 2.2.5 and Notation 2.8.2).

For a poset $\mathcal{X}$, the function f is monotone iff U is an **upper set**:

$$\forall \vartheta, x : \mathcal{X}. \; \vartheta \geqslant x \in U \implies \vartheta \in U.$$

We write $x \downarrow \mathcal{X} = \{\vartheta \mid x \leqslant \vartheta\}$ for the **up-closure** of the singleton x, *i.e.* the upper set which it generates.[1]

Representation of orders by subset-inclusion. The fundamental example of an order relation is *provability*, to which implication and the containment of subsets are immediately related. Example 3.2.5(f) shows that this also accounts for the arithmetical orders.

DEFINITION 3·1·7: An antitone predicate on $\mathcal{X}$, *i.e.* a monotone function $\mathcal{X} \to 2^{\mathrm{op}}$, defines a **lower set** $A \subset \mathcal{X}$,

$$\forall \gamma, x : \mathcal{X}. \; \gamma \leqslant x \in A \implies \gamma \in A.$$

We write $\mathrm{sh}(\mathcal{X})$ for the collection of lower subsets, ordered by inclusion.[2] The **down-closure** of $x \in \mathcal{X}$ is written $\mathcal{X} \downarrow x = \{\gamma \mid \gamma \leqslant x\}$, and any subset of this form we call a **representable lower set**. We have used γ and ϑ as a reminder of the roles of Γ as hypotheses and ϑ as a conclusion or result type.

Using lower subsets, every partial order may be seen as an inclusion order. This is called the **covariant regular representation**. We introduce this terminology here, while the technology remains simple, to prepare for the Yoneda Lemma for categories in Sections 4.2 and 4.8.

---

[1] Many authors write $\uparrow x$ for our $x \downarrow \mathcal{X}$. Sometimes this is extended to $\uparrow \mathcal{C}$ for the upper set generated by a subset $\mathcal{C}$, with the singleton $\uparrow\{x\}$ as a special case. Similarly $\downarrow x$, $\downarrow \mathcal{C}$ and $\downarrow\{x\}$ are used for the down-closure.

Our notation is a special case of that for comma categories (Definition 7.3.8), and $\mathcal{X} \downarrow x$ is a slice category (Definition 5.1.8). In general, given two monotone functions $F : \mathcal{C} \to \mathcal{S}$ and $U : \mathcal{A} \to \mathcal{S}$, the **comma poset** is $F \downarrow U = \{\langle x, a \rangle \mid F(x) \leqslant^{\mathcal{S}} U(a)\}$. In particular, with $F : \mathcal{C} \hookrightarrow \mathcal{X}$ and $U = \mathrm{id}_{\mathcal{X}}$, there is a surjection $\mathcal{C} \downarrow \mathcal{X} \twoheadrightarrow \uparrow\mathcal{C}$. This is an example of one of the many hidden existential quantifiers in lattice theory, *cf.* Remark 2.2.7.

See Exercise 3.55 for the order relation on subsets induced by inclusion of the lower subsets that they generate.

[2] We write sh for the order-theoretic notion of sheaves (Definition 3.9.6), and Sh for the categorical one; they take values in 2 and *Set* respectively. If some coverage or Grothendieck topology is present, we let it say for itself that the sheaves are to respect it; if there is none, by "sheaf" we mean what is usually called a presheaf.

PROPOSITION 3·1·8:   Let $\mathcal{X}$ be a poset and $x, y \in \mathcal{X}$. Then

(a) $(\mathcal{X} \downarrow x) \subset (\mathcal{X} \downarrow y) \iff x \leqslant y$, *i.e.* the function $\mathcal{X} \downarrow (-) : \mathcal{X} \to \mathsf{sh}(\mathcal{X})$
is monotone and full;

(b) for any lower set $A \subset \mathcal{X}$, we have $\mathcal{X} \downarrow x \subset A \iff x \in A$. $\square$

Upper sets provide a similar *contravariant* representation. There are two
different senses of "representation" here: a subset of the form $\mathcal{X} \downarrow x$ is
represented by the element x, whereas $\mathsf{sh}(\mathcal{X})$ represents the poset $\mathcal{X}$.

EXAMPLE 3·1·9:   Consider a collection of propositions (a **Lindenbaum
algebra**, Remark 1.4.6), ordered by provability. Then

(a) the covariant regular representation represents a proposition ϕ by
its **reasons**, *i.e.* the set $\{\gamma \mid \gamma \vdash \phi\}$ (a "reason" for something may
be some other assumption, from which the thing follows), and

(b) the contravariant representation (using upper sets) represents ϕ by
its **consequences**, $\mathsf{Cn}(\phi) = \{\vartheta \mid \phi \vdash \vartheta\}$ (Notation 2.4.12). $\square$

The regular representation of a preorder identifies x and y iff both $x \preccurlyeq y$
and $y \preccurlyeq x$, since then $\mathcal{X} \downarrow x = \mathcal{X} \downarrow y$. We may cut down the representation
from $\mathcal{X} \downarrow x = \{y \mid y \preccurlyeq x\}$ to $\{y \mid y \sim x\}$.

PROPOSITION 3·1·10:   Let $(\mathcal{X}, \preccurlyeq)$ be a preorder. Then

$$x \sim y \iff x \preccurlyeq y \wedge y \preccurlyeq x$$

is an equivalence relation. The quotient $\mathcal{X}/\sim$ carries an *antisymmetric*
order $\leqslant$ such that the function $\eta : \mathcal{X} \to \mathcal{X}/\sim$ is monotone and full.

Moreover if Θ is any poset and $f : \mathcal{X} \to \Theta$ is a monotone function,
then $x \sim y \Rightarrow f(x) = f(y)$ and there is a unique monotone function
$p : \mathcal{X}/\sim \to \Theta$ such that $f = \eta \,;\, p$.

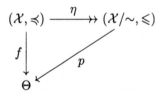

PROOF: First, $\sim$ is reflexive and transitive because $\preccurlyeq$ is, and symmetric
by construction. So it is an equivalence relation, of which Example 2.1.5
gave the quotient $\mathcal{X}/\sim$ as the set of equivalence classes, together with
the mediating function p. We want $[x] \leqslant [y]$ if $x \preccurlyeq y$; this is well defined
because, if $x \sim x'$, $y \sim y'$ and $x \preccurlyeq y$, then $x' \preccurlyeq y'$ by transitivity. Then
$\leqslant$ inherits reflexivity and transitivity from $\preccurlyeq$, and is antisymmetric *on
equivalence classes* by construction of $\sim$. Finally, p is monotone. $\square$

3.2 MEETS, JOINS AND LATTICES

DEFINITION 3·2·1: Let $(\mathcal{X}, \leqslant)$ be a preorder and $u \in \mathcal{X}$. Then u is

(a) a *least element* or *bottom* if $\forall \vartheta. \, u \leqslant \vartheta$,

(b) a *locally least element* if $\forall x, y. \, x \leqslant y \geqslant u \Rightarrow u \leqslant x$ (Exercise 3.5),

(c) a *minimal element* if $\forall x. \, x \leqslant u \Rightarrow u \leqslant x$.

Likewise by reversing the order we say that u is

(d) a *greatest element* or *top* if $\forall \gamma. \, u \geqslant \gamma$,

(e) a *maximal element* if $\forall x. \, x \geqslant u \Rightarrow u \geqslant x$.

Bottom and top, if they exist, are written $\bot$ and $\top$ respectively; if $\leqslant$ is antisymmetric then they are unique. (They are in any case unique up to $\sim$ of Proposition 3.1.10.) The terms maximum and minimum (which are *nouns*) are also used to mean greatest and least elements, but should be avoided as a source of confusion. Local maxima and minima in the sense of elementary calculus are not formally related to any of these concepts.

EXAMPLES 3·2·2:

(a) False and true are least and greatest formulae under provability ($\vdash$).

(b) $\varnothing$ and X are the least and greatest subsets of any set X under $\subset$.

(c) If $(\leqslant) = (<) \cup (=)$ with $<$ irreflexive, then this definition agrees with the notion of minimality used in Proposition 2.5.6 (Exercise 3.3).

(d) A lower subset is representable (Definition 3.1.7) iff it has a greatest element.

Meets and joins.

NOTATION 3·2·3: By abuse of notation, for $\gamma, \vartheta \in \mathcal{X}$ and $\mathcal{I} \subset \mathcal{X}$,

$$\gamma \leqslant \mathcal{I} \quad \text{means} \quad \forall x \in \mathcal{I}. \, \gamma \leqslant x$$
$$\mathcal{I} \leqslant \vartheta \quad \text{means} \quad \forall x \in \mathcal{I}. \, x \leqslant \vartheta.$$

We do not extend the symbol $\leqslant$ *to* $U \leqslant V$, but see Exercise 3.55 for three such orders on the set of subsets.

DEFINITION 3·2·4: Let $(\mathcal{X}, \leqslant)$ be a preorder and $\mathcal{I} \subset \mathcal{X}$.

(a) If $\gamma \leqslant \mathcal{I}$ then we call γ a *lower bound* for $\mathcal{I}$.

(b) A *greatest* lower bound, *i.e.* a greatest element of $\{\gamma \mid \gamma \leqslant \mathcal{I}\} \subset \mathcal{X}$, the set of lower bounds, if such exists, is called a *meet* or *infimum*, and is denoted by $\bigwedge \mathcal{I}$; then the set of lower bounds is *representable* (Definition 3.1.7), namely by the meet.

(c) Similarly, if $\mathcal{I} \leqslant \vartheta$ then we call ϑ an **upper bound** for $\mathcal{I}$. A *least* upper bound, if any, is called a *join* or *supremum*, $\bigvee \mathcal{I}$.

If $\leqslant$ is antisymmetric then meets and joins, where they exist, are *unique*. Otherwise they are unique up to $\sim$ from Proposition 3.1.10. See Exercises 3.5, 3.21 and 3.33ff for minimal and locally least upper bounds.

EXAMPLES 3·2·5:

(a) $\bigvee \varnothing = \bot = \bigwedge \mathcal{X}$ and $\bigwedge \varnothing = \top = \bigvee \mathcal{X}$, if these exist.

(b) If $\mathcal{I} = \{\phi, \psi\}$ then we write $\bigvee \mathcal{I} = \phi \vee \psi$ and $\bigwedge \mathcal{I} = \phi \wedge \psi$.

(c) Meets and joins of sets of subsets in the inclusion order are called intersections ($\bigcap$) and unions ($\bigcup$) respectively.

(d) The union or intersection of a family of lower subsets is again lower.

(e) The Dedekind reals, $\mathbb{R}_D$ (Remark 2.1.1), have meets and joins with respect to the arithmetical order for all bounded inhabited subsets; these are usually written as inf and sup respectively.

(f) For the divisibility order on $\mathbb{N}$, the meet and join of two numbers are called their **greatest common divisor** and **least common multiple** respectively. The extremal elements are $\bot = 1$ and $\top = 0$. This conflict with the conventions of logic is resolved by considering ideals (Example 2.1.3(b)), for which $I \mid J \iff J \subset I$; this is the contravariant regular representation (Example 3.1.6(f)ff).

(g) Arbitrary meets and joins in the type 2 of propositions are found using the guarded quantifiers (Remark 1.5.2):

$$\bigvee \mathcal{I} = \exists \phi. (\phi \in \mathcal{I}) \wedge \phi \qquad \bigwedge \mathcal{I} = \forall \phi. (\phi \in \mathcal{I}) \Rightarrow \phi.$$

(h) If $\mathcal{I} \subset \{\bot, \top\}$ is a *complemented* subset (*i.e.* the predicate ($x \in \mathcal{I}$) is decidable), then $\mathcal{I}$ has a meet and a join. Indeed there are only four cases to consider, depending on whether each of $\bot$ and $\top$ does or does not belong to $\mathcal{I}$. Intuitionism has nothing to do with the finiteness of $\{\bot, \top\}$ in this case, but is concerned with the nature of the subset $\mathcal{I}$ (or the predicate characterising it): we need excluded middle to say that $\{\bot, \top\}$ has meets and joins of *all* subsets.

DEFINITION 3·2·6:   Let $f : \mathcal{X} \to \mathcal{Y}$ be a monotone function between preorders. If u is a meet of $\mathcal{I} \subset \mathcal{X}$ then fu is a lower bound of $f_!(\mathcal{I}) \subset \mathcal{Y}$ by monotonicity of f. (Recall the notation $f_!$ from Remark 2.2.7.) Then *f* **preserves the meet** if fu is a *greatest* lower bound.

(If $\mathcal{X}$ and $\mathcal{Y}$ are posets then of course these meets are unique; the point of stressing preorders and being *a* meet is that no *choice* of meet is needed to define preservation, by an argument similar to Lemma 1.2.11. This will be important for limits and colimits in categories, Definition 4.5.10.)

We also say that f **creates the meet** if (a) $y = \bigwedge f_!(\mathcal{I})$ exists, (b) there is a unique $x \in \mathcal{X}$ (up to $\sim$) such that $y = f(x)$ and $x \leqslant \mathcal{I}$, and, having fixed x, (c) in fact $x = \bigwedge \mathcal{I}$.

Meets and joins of lower sets. The covariant regular representation (by lower sets, Definition 3.1.7) has both meets and joins, but it behaves quite differently with respect to them. Both cases are important.

PROPOSITION 3·2·7: The embedding $\mathcal{X} \downarrow (-) : \mathcal{X} \hookrightarrow \mathsf{sh}(\mathcal{X})$

(a) is full and preserves any meets which exist, *i.e.*

$$\mathcal{X} \downarrow \left(\bigwedge \mathcal{I} \right) = \bigcap_{x \in \mathcal{I}} (\mathcal{X} \downarrow x),$$

since an element γ belongs to either side iff $\gamma \leqslant \mathcal{I}$;

(b) but freely adds joins.

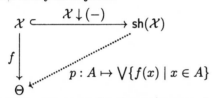

This means that, for any monotone function $f : \mathcal{X} \to \Theta$ to a poset which has joins of all subsets, there is a unique join-preserving function $p : \mathsf{sh}(\mathcal{X}) \to \Theta$ such that $p(\mathcal{X} \downarrow x) = f(x)$. □

EXAMPLE 3·2·8: In the poset $\mathcal{X}$ on the left, u is the join of a and b.

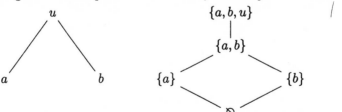

In $\mathsf{sh}(\mathcal{X})$, on the right, $\{a, b\}$ is a *new* "formal" join of this pair. □

Theorem 3.9.7 shows how to add new joins, but *keep* specified old ones.

Diagrams. It is often more convenient[3] to define meets and joins with respect to arbitrary functions $\mathcal{I} \to \mathcal{X}$ instead of just subsets $\mathcal{I} \subset \mathcal{X}$.

---

[3]Then when we consider the effect of a function $f : \mathcal{X} \to \mathcal{Y}$ on meets and joins in $\mathcal{X}$, we can use the composite function $\mathcal{I} \to \mathcal{X} \to \mathcal{Y}$ as the diagram, keeping $\mathcal{I}$ as the shape, instead of having first to find its image in $\mathcal{Y}$. As we noted in Remark 2.2.7, forming this image uses an existential quantifier, which is itself an example of a join, so using diagrams avoids "contamination" of the meet or join under study.

The shape $\mathcal{I}$ need only carry a relation R such that $iRj \Rightarrow x_i \leqslant x_j$, or may be an oriented graph (sketch) for which $x_i \leqslant x_j$ whenever there is an edge $i \to j$.

DEFINITION 3·2·9:   A **diagram** in a poset $\mathcal{X}$ is a function to $\mathcal{X}$ from a set $\mathcal{I}$, or a monotone function from a poset or preorder. We call $\mathcal{I}$ the **shape** of the diagram, and write x_i for the value of the function at $i \in \mathcal{I}$ and $x_{(-)} : \mathcal{I} \to \mathcal{X}$ for the function as a whole.

Then $\vartheta \in \mathcal{X}$ is an **upper bound for the diagram** if $\forall i.\ x_i \leqslant \vartheta$, i.e. ϑ bounds the image, $\{x_i \mid i : \mathcal{I}\} \subset \mathcal{X}$. We write $\bigvee_i x_i$ for the join (least upper bound) of the diagram, and similarly $\bigwedge_i x_i$ for the meet.

The notations $x \vee y$ and $x \wedge y$ are, strictly speaking, examples of a join and a meet of a *diagram* of shape $\mathbf{2} = \{\bullet\ \bullet\}$ rather than of a subset. We shall study certain non-discrete diagrams in the next section.

Any diagram has the same meets and joins as its image. Indeed it has the same joins as the down-closure of its image, and more generally we can say exactly when a comparison between diagrams induces the same joins. This result will be used in Corollary 3.5.13.

PROPOSITION 3·2·10:   Let $U : \mathcal{J} \to \mathcal{I}$ be monotone. Then the following are equivalent:

(a) U is **cofinal**:[4] $\forall i.\ \exists j.\ i \leqslant U(j)$;

(b) for every preorder $\mathcal{X}$, diagram $x_{(-)} : \mathcal{I} \to \mathcal{X}$ and element $\vartheta \in \mathcal{X}$, ϑ is an upper bound for $\{x_i \mid i : \mathcal{I}\}$ iff it is an upper bound for $\{x_{U(j)} \mid j : \mathcal{J}\}$;

(c) for all such diagrams the *minimal* upper bounds coincide;

(d) for all such diagrams the *locally least* upper bounds coincide;

(e) for all such diagrams the *least* upper bounds coincide.

PROOF: The proof is easy except for (e)$\Rightarrow$(a). For this case we use the representation by lower sets, so let $\mathcal{X} = \mathsf{sh}(\mathcal{I})$ and $x_i = \mathcal{I} \downarrow i$. Then ϑ is an upper bound of $\{x_i \mid i : \mathcal{I}\}$ iff $\vartheta = \mathcal{I}$, and of the subset $\{x_{U(j)} \mid j : \mathcal{J}\}$ iff $U_!(\mathcal{J}) \subset \vartheta$, so $\mathcal{I}$ must be the lower closure of $U_!(\mathcal{I})$.   □

Lattices. Now we shall give an algebraic characterisation of *finite* meets and joins, motivated by the logical connectives $\wedge$ and $\vee$ (Section 1.4). Gottfried Leibniz defined order in terms of joins, and Johann Lambert recognised the idempotent law. Ernst Schröder was the first to see that the distributive law was not automatic. These laws, plus associativity, commutativity and the units for lattices, express the structural rules of logic (Definition 1.4.8), as do the reflexivity and transitivity axioms for a preorder. Antisymmetry, on the other hand, is a side-effect of the algebraic formulation of logic in terms of lattices.

---

[4]This is $\mathcal{I} \leqslant^b U_!(\mathcal{J})$ in the notation of Exercise 3.55. This definition was first used in the theory of ordinals (Section 6.7), and is due to Felix Hausdorff (1907).

PROPOSITION 3·2·11: Let $(\mathcal{X}, \leqslant)$ be a poset with a greatest element ($\top$) and meets of pairs $(x \wedge y)$. Then the operation $\wedge : \mathcal{X} \times \mathcal{X} \to \mathcal{X}$ is

associative: $x \wedge (y \wedge z) = (x \wedge y) \wedge z$,
commutative: $x \wedge y = y \wedge x$,
idempotent: $x \wedge x = x$, and
$\top$ is a *unit*: $\top \wedge x = x = x \wedge \top$.

Moreover $x \leqslant y$ iff $x \wedge y = x$. Conversely if $(\mathcal{X}, \top, \wedge)$ satisfies these laws then this condition defines a partial order for which $\wedge$ is the binary meet and $\top$ the greatest element. $\qquad \square$

DEFINITION 3·2·12: Such a structure is called a *semilattice*; a function which preserves $\wedge$ and $\top$ is called a semilattice homomorphism, and is therefore monotone. Notice that *the algebraic laws do not force the direction of the order relation*: $\vee$ and $\bot$ satisfy them as well. When we speak of an algebraic structure as a *join-* or *meet*-semilattice we are therefore imposing the direction by convention.

Since finitary meets and joins are characterised by the same laws and each alone can uniquely determine the order, there must be an additional law forcing the two orders to coincide. It suffices to say that $x, y \leqslant x \vee y$, where $\leqslant$ is the relation defined by $\wedge$, or *vice versa*. Eliminating the inequality, either of these conditions is known as the *absorptive law*:

$$x \wedge (x \vee y) = x \quad \text{and} \quad x \vee (x \wedge y) = x.$$

The nullary cases, $y \wedge \bot = \bot$ and $y \vee \top = \top$, follow from these (with $x = \bot, \top$) and the unit laws.

PROPOSITION 3·2·13: Let $(\mathcal{X}, \leqslant)$ be a poset with finite meets and joins. Then $(\mathcal{X}, \bot, \vee)$ and $(\mathcal{X}, \top, \wedge)$ are both semilattices, and the absorptive laws hold. Conversely, if $(\mathcal{X}, \bot, \vee, \top, \wedge)$ obeys both sets of semilattice laws and either of the absorptive laws then the orders agree and the meets and joins are as given. Then $\mathcal{X}$ is called a *lattice*; lattice homomorphisms by definition preserve $\bot$, $\vee$, $\top$, and $\wedge$. $\qquad \square$

The logical connectives obey another law:

DEFINITION 3·2·14: A lattice is *distributive* if

$$(x \wedge y) \vee (x \wedge z) = x \wedge (y \vee z)$$
$$x \vee (y \wedge z) = (x \vee y) \wedge (x \vee z),$$

where LHS $\leqslant$ RHS hold in an arbitrary lattice. In fact one can be derived from the other in the context of the lattice laws, and we have already stated the nullary versions ($x \wedge \bot = \bot$ and $x \vee \top = \top$). Whereas expressions in the theory of lattices may need many nested brackets, the distributive laws give rise to the **disjunctive normal form**, *e.g.*

$$(a_{11} \land a_{12} \land \cdots) \lor (a_{21} \land a_{22} \land a_{23} \land \cdots) \lor \cdots$$

and the **conjunctive normal form**, *e.g.*

$$(b_{11} \lor b_{12} \lor \cdots) \land (b_{21} \land b_{22} \lor b_{23} \lor \cdots) \land \cdots$$

respectively. We consider implication in Proposition 3.6.14ff.

There are also major applications of lattices to the structure theory of algebra. The lattices of congruences of certain familiar theories, notably groups, rings and vector spaces, obey the **modular law** (Exercise 3.24), which is weaker than distributivity and was first identified by Richard Dedekind (1900). This gives a sense in which these algebras are made of "building blocks", of which the dimension of a vector space and the Jordan–Hölder theorem for groups are examples.

We shall return to lattices with arbitrary meets and joins in Section 3.6. Semilattices will be used in Theorem 3.9.1 to describe Horn theories (Remark 1.7.2ff). Now we shall turn from finitary meets and joins to those which are in a sense "purely infinitary".

3.3   FIXED POINTS AND PARTIAL FUNCTIONS

Definition 2.5.1 described the paradigm for recursive programs, which we studied assuming that they *terminate* and so define *total functions*. Because of the Halting Problem there is no way to guarantee this by imposing conditions on programming languages without destroying their expressive power.

Taking a different attitude, we may accept non-termination as a first class value. By extending the domain of mathematical values which the two sides may be understood to have, we may then treat recursive definitions of programs as equations.

EXAMPLE 3·3·1:   Consider the program

$$\text{fact } n \;=\; \textbf{if } n = 0 \textbf{ then } 1 \textbf{ else } n * \text{fact}(n-1) \textbf{ fi}$$

To make this equation into a definition we must read

$$\text{fact} \;=\; \imath p. \left(p = T(p)\right),$$

where T is the right hand side of the program with p instead of fact. For this to be meaningful the equation must be a *description*, *i.e.* have a unique solution (Definition 1.2.10). The recursive paradigm always yields

equations of this form,[5] namely that the partial function so defined is a
fixed point of a functional T which is an operator on programs.

REMARK 3·3·2: Any recursive program nevertheless has a well defined,
albeit partial, *de facto* meaning. When the program is given a particular
argument, say 2, what the machine actually executes is

$$\text{fact } 2 \ = \ \textbf{if } 2 = 0 \textbf{ then } 1 \textbf{ else } 2 * (\textbf{if } 1 = 0 \textbf{ then } 1$$
$$\textbf{else } 1 * (\textbf{if } 0 = 0 \textbf{ then } 1 \textbf{ else } u \textbf{ fi}) \textbf{ fi}) \textbf{ fi}$$

where u never gets called. So it may be anything — for example the
program **while** yes **do skip od** which goes straight into a tight unending
loop. The latter is interpreted as the empty (totally undefined) partial
function $\bot$, and the version of the factorial function as executed for the
argument 2 is $T^3(\bot) \equiv T(T(T(\bot)))$.

In general, $T^{n+1}(\bot)$ suffices to give fact(n). Intuitively, the programs

$$\bot = T^0(\bot) \sqsubseteq T(\bot) \sqsubseteq T^2(\bot) \sqsubseteq \cdots$$

are progressively better approximations, as we shall now justify.

The poset of partial functions. Agreement of total functions is all or
nothing, but two partial functions (Definition 1.3.1(b)) may agree on the
intersection of their supports, whilst one offers more information than
the other on some other values.

DEFINITION 3·3·3: Let $f, g : X \rightharpoonup Y$ be partial functions between sets.
We write $f \sqsubseteq g$ and say that g **extends** f if

$$\forall x{:}X. \ \forall y{:}Y. \ x \overset{f}{\mapsto} y \implies x \overset{g}{\mapsto} y.$$

In other words, $g(x)$ is defined whenever $f(x)$ is, and when they are both
defined they are *equal*, but $g(x)$ may be defined when $f(x)$ is not.

Partial functions may be expressed by *spans* of total ones:

DEFINITION 3·3·4: The subset supp $f = \{x \mid \exists y. \ x \overset{f}{\mapsto} y\} \subset X$ is called
the **support** of f.

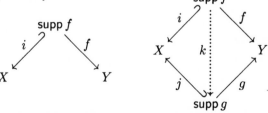

---

[5]The equation $p = T(p)$ is only a crude approximation to the paradigm. We should
consider the action of T on *types* as well as programs, making it a *functor*. Then
the paradigm is the equation $p = $ **parse** ; Tp ; **ev** (Definition 6.3.7). However, what
we then do to the functor is the categorical analogue of the present order-theoretic
discussion, so this does remain of pedagogical value.

Then $f \sqsubseteq g$ iff $\operatorname{supp} f \subset \operatorname{supp} g$ and f is the restriction of g to $\operatorname{supp} f$. So the inclusion $k : \operatorname{supp} f \hookrightarrow \operatorname{supp} g$ satisfies $i = k \,;\, j$ and $f = k \,;\, g$.

LEMMA 3·3·5: Partial functions $X \rightharpoonup Y$ between sets form a poset under the extension relation. Moreover

(a) The empty relation is the least partial function, called $\bot$.

(b) Any total function is maximal (Definition 3.2.1(e)), but not greatest (unless $X = \emptyset$ or $Y = 1$). Total functions at higher types, as in Convention 2.3.2, are not, however, characterised by maximality.

(c) Any inhabited set of partial functions has a meet; its support is the set of elements on which all the partial functions are defined and agree, and it takes the agreed values there.

(d) Any set of partial functions which pairwise agree on their common support has a join in the extension order; its support is the union of the supports and it takes the same value at an argument as any (and hence all) of the partial functions which are defined there. See Exercise 3.21 for a notion of domain based on this fact.   □

Instead of modifying the source of a function we can change its target. The trick we employ to do this is applicable to any relation:

LEMMA 3·3·6:   There is a bijection between relations $R : X \nrightarrow Y$ and functions $\tilde{R} : X \to \mathcal{P}(Y)$, defined by $\tilde{R}(x) = \{y \mid xRy\}$. If the relation R is functional, then each subset $\tilde{R}(x)$ has at most one element.   □

DEFINITION 3·3·7:   The **lift** or **partial function classifier** of a set Y is the set of subsets with at most one element:

$$Y_\bot = \{V \subset Y \mid \forall y_1, y_2 \in V.\, y_1 = y_2\},$$

where we also define $\mathsf{lift}_Y : Y \hookrightarrow Y_\bot$ by $y \mapsto \{y\}$ and write $\bot = \emptyset \in Y_\bot$.

We tend to identify $y \in Y$ with $\mathsf{lift}\, y \in Y_\bot$. (Classically, $Y_\bot = Y \cup \{\bot\}$, and we put $f(x) = \bot$ for $x \notin \operatorname{supp} f$.) The lemma restricts to a bijection between partial functions $X \rightharpoonup Y$ and total functions $X \to Y_\bot$.

$Y_\bot$ is the set of partial functions $\{\star\} \rightharpoonup Y$, and the extension order for these agrees with the inclusion order on subsingleton subsets of Y; we call it the **information order**. It is rather sparse, being discrete when restricted to Y itself, with $\bot \sqsubseteq y$. Example 3.9.8(c) extends the definition of the lift to the case where there is already an order on Y, Exercise 3.71 shows how to construct it for topological spaces and locales, and Example 9.4.11(a) explains its significance in type theory. Function-types in Section 3.5 are typically less "flat" than $Y_\bot$ is.

LEMMA 3·3·8: Partial functions $X \rightharpoonup Y$ extend to functions $X_\perp \to Y_\perp$ by

$$U \mapsto f_!(U) = \{y \mid \exists x.\ x \in U \wedge x \overset{f}{\mapsto} y\},$$

which are monotone with respect to the information order and are **strict**, *i.e.* $\perp \mapsto \perp$. □

We expect computable functions to be monotone with respect to the information order $x_1 \sqsubseteq x_2$. Otherwise *providing the information* that x_2 does beyond that provided by x_1 would result in *retracting the guarantees* which $f(x_1)$ has already given about the output.

On the other hand, strictness of a program means that it tries to *use* the input (*cf.* Remark 2.3.4 and Example 6.1.10). The input of T is the sub-program u in Remark 3.3.2, which, as we saw, need not be called (used), so in general $T(\perp) \neq \perp$.

The fixed point theorem. We may define fact as the join $\bigsqcup_n T^n(\perp)$ in the poset of partial functions $\mathbb{N} \rightharpoonup \mathbb{N}$. It is important to appreciate that the order implicit here is the extension or information order, not the arithmetical one.

Now the poset $\mathbb{N} \rightharpoonup \mathbb{N}$ does not have all joins, so how can we be sure that this particular join exists? We have $\perp \sqsubseteq T(\perp)$ since $\perp$ is the least element, and then since T^n is monotone it follows that

$$\perp = T^0(\perp) \sqsubseteq T(\perp) \sqsubseteq T^2(\perp) \sqsubseteq \cdots \sqsubseteq T^n(\perp) \sqsubseteq T^{n+1}(\perp) \sqsubseteq \cdots$$

DEFINITION 3·3·9: A poset $\mathcal{X}$ is ω-**complete** if any ω-sequence, *i.e.* any diagram $x_{(-)} : \omega \to \mathcal{X}$ where ω is $\mathbb{N}$ with the arithmetical order, has a join. A monotone function $f : \mathcal{X} \to \mathcal{Y}$ between ω-complete posets is ω-**continuous** if it preserves all such joins.

EXAMPLE 3·3·10: The poset $[\mathbb{N} \to \mathbb{N}_\perp]$ of partial endofunctions of $\mathbb{N}$ is ω-complete, by Lemma 3.3.5(d). Any functional $T : [\mathbb{N} \to \mathbb{N}_\perp] \to [\mathbb{N} \to \mathbb{N}_\perp]$ which codes a recursive program is, in fact, ω-continuous.

PROPOSITION 3·3·11: Let $\mathcal{X}$ be an ω-complete poset which has a least element $\perp$ and let $T : \mathcal{X} \to \mathcal{X}$ be an ω-continuous function. Then the join $\bigvee_n T^n(\perp)$ exists, is a fixed point of T and is indeed the least such.

PROOF:

(a) We have already observed that $n \mapsto T^n(\perp)$ is an ω-sequence, so the join exists.

(b) The sequence $n \mapsto T(T^n(\perp)) = T^{n+1}(\perp)$ is the same as $n \mapsto T^n(\perp)$, apart from the missing $T^0(\perp) = \perp$, but this does not affect the join.

(c) Since T preserves joins of ω-sequences, $T(\bigvee_n T^n(\perp)) = \bigvee_n T^{n+1}(\perp)$, so this is a fixed point.

(d) If $T(\vartheta) = \vartheta$ then $T^n(\bot) \leqslant \vartheta$ by induction on n (since if $T^n(\bot) \leqslant \vartheta$ then $T^{n+1}(\bot) \leqslant T(\vartheta) = \vartheta$), so $\bigvee_n T^n(\bot) \leqslant \vartheta$.   □

This is often (inaccurately) called **Tarski's theorem**: the result that Alfred Tarski actually proved (1955, Exercise 3.39) is that any *monotone* endofunction of a complete *lattice* has a least fixed point, and indeed a complete lattice of fixed points (see also Proposition 3.7.11ff).

REMARK 3·3·12: *Algebraic* topology gives other, quite different, reasons why continuous endofunctions of certain spaces must have fixed points: the closed interval $[0, 1] \subset \mathbb{R}$ and disc $\overline{B}^2 = \{(x, y) \mid x^2 + y^2 \leqslant 1\} \subset \mathbb{R}^2$ are the simplest examples (the former is essentially the **intermediate value theorem**). The latter is due to Jan Brouwer, which is ironic because such results rely on excluded middle, and the fixed points do not depend continuously on the given endofunction.

Tarski's theorem does assign fixed points continuously — a fact which is crucial to denotational semantics. A traditional fixed point theorem which is more closely related to what we require is that for *contraction mappings* on a complete metric space (X, d), *i.e.* functions $f : X \to X$ such that $\forall x, y.\, d(f(x), f(y)) \leqslant k\, d(x, y)$ for some constant $0 \leqslant k < 1$. This analogy is closer if the symmetry law, $d(x, y) = d(y, x)$, for metric spaces is dropped, since certain concrete domains can be equipped with such a (pseudo)metric.

We shall return to partial functions in Sections 5.3 and 6.3.

3.4   DOMAINS

Domains abstract what is needed to prove Tarski's fixed point theorem. For the interpretation of programming languages, joins of ω-sequences suffice, so many authors just consider ω-complete posets with $\bot$, though there are very many more specialised notions in the literature.

Directed diagrams. Aside from the fixed point theorem, carrying the completeness property for *sequences* verbatim through our working gets us caught in pedantic and ultimately confusing notation. For example, if we take such a join over two indices n and m we no longer have a sequence. The following definition is equivalent (in the sense of Proposition 3.2.10), but it is invariant under operations such as products.

It may be important that the diagram be *computable* (and *a fortiori* countable), but this is an issue which is best studied separately.

DEFINITION 3·4·1:   A poset $\mathcal{I}$ is **directed** if every finite subset $F \subset \mathcal{I}$ has an upper bound $F \leqslant i \in \mathcal{I}$. Specialising this to the cases $F = \varnothing$ and $F = \{i_1, i_2\}$, directedness is equivalent to

$\mathcal{I}$ is non-empty, *i.e.* it has an element, and
$\forall i_1, i_2. \, \exists i. \, i_1 \leqslant i \wedge i_2 \leqslant i.$

Two similar but weaker notions turn out to be useful.

(a) If the binary form (alone) holds then we say that $\mathcal{I}$ is **semidirected**.
    Classically, a poset is semidirected iff it is either directed or empty.

(b) We say that $\mathcal{I}$ is **confluent** (*cf.* Definition 1.2.5) if

$$\forall i_0, i_1, i_2. \quad i_0 \leqslant i_1 \wedge i_0 \leqslant i_2 \Rightarrow \exists i. \, i_1 \leqslant i \wedge i_2 \leqslant i.$$

A directed *lower* subset of a poset $\mathcal{X}$ is called an **ideal**, and we write $\mathsf{Idl} \, \mathcal{X}$ for the poset of them, ordered by inclusion. The analogy between lattices and rings which lies behind the name ideal (*cf.* Example 2.1.3(b)) was first noticed by Marshall Stone in 1935 (Exercises 3.10 and 3.11).

EXAMPLES 3·4·2:

(a) Any poset which has a greatest element is directed.

(b) $\mathbb{N}$ with the arithmetical order is directed (ω, Definition 3.3.9).

(c) Any join-semilattice (*cf.* Definition 3.2.12) is directed.

(d) A lower subset of a join-semilattice is directed (an ideal) iff it is a
    (lower) subsemilattice.

(e) For any set X, the powerset $\mathcal{P}(X)$ and, more usefully, the **finite
    powerset** $\mathcal{P}_{\mathsf{fg}}(X)$ consisting of the finite subsets of X, are directed
    posets under inclusion; see Lemma 6.6.10(e).

(f) Raw λ-terms form a confluent preorder under (reverse) $\beta\eta$-reduction
    (the Church–Rosser Theorem, Fact 2.3.3).

(g) A poset is confluent iff every connected component (Lemma 1.2.4)
    is directed.

Lemma 3.5.12 and its corollary, about functions of two arguments, show how much easier it is to use directedness than sequences.

Posets with directed joins.

NOTATION 3·4·3: If $\mathcal{I}$ is directed then we indicate this fact by an arrow when writing its **directed join**:

$$\bigvee_{i \in \mathcal{I}} x_i.$$

Often this arrow is used instead of saying in words that the relevant set is assumed or has been shown to be directed.

DEFINITION 3·4·4: A poset which has all directed joins is called **directed complete** or a **dcpo** for short. If it also has a least element — from which it follows (without using excluded middle) that it has joins of all semidirected sets — then we call it an **inductive poset** or **ipo**.

The term **complete partial order** or **cpo** is commonly found instead of ipo in the literature, but we avoid it on the grounds that it conflicts with complete *categories* (*cf.* complete semilattices), which *also* have (the analogue of) finite joins. This confusion has been made worse by authors who use "cpo" for dcpo, *i.e.* not necessarily with bottom.

A (monotone) function between dcpos which preserves directed joins is said to be **Scott-continuous**. These are the most useful morphisms between ipos because we wish to allow programs to ignore inputs and so terminate even if these are not specified. This is peculiar from the point of view of universal algebra because not all of the structure is preserved. The fixed point theorem is essentially the only use of $\perp$.

REMARK 3·4·5: Peter Freyd [Fre91] observed that the three notions of

(a) domain (having, but with functions not necessarily preserving, $\perp$),

(b) **predomain** (not necessarily having $\perp$ at all) and

(c) **lift-algebra** (both having and preserving $\perp$)

ought to be formulated in tandem (Example 7.5.5(c)). Since we only intend to investigate function-spaces, where $\perp$ gets in the way, we shall concentrate on dcpos (predomains). Notice that *when the morphisms change*, in this case whether or not they have to preserve $\perp$, *we change the names of the objects too*, from domain or ipo to lift-algebra.

LEMMA 3·4·6: The composite of two Scott-continuous functions is Scott-continuous, as is the identity function. $\square$

The Scott topology. Continuity may be expressed in terms of open or closed sets, but the correspondence partly depends on excluded middle.

DEFINITION 3·4·7: A subset $A \subset \mathcal{X}$ of a dcpo is said to be **Scott-closed** if it is a lower subset and also closed ("upwards") under directed joins. In particular any representable lower set $\mathcal{X} \downarrow x$ is Scott-closed.

LEMMA 3·4·8: A function $f : \mathcal{X} \to \mathcal{Y}$ between dcpos is Scott-continuous iff the inverse image $f^{-1}(B)$ of every Scott-closed subset $B \subset \mathcal{Y}$ is again Scott-closed.

PROOF: For monotonicity, let $x' \leqslant x \in \mathcal{X}$ and consider $B = \mathcal{Y} \downarrow f(x)$, so $x \in f^{-1}(B)$. Then $x' \in f^{-1}(B) \Leftrightarrow f(x') \leqslant f(x)$.

Suppose that $f^{-1}(B)$ is Scott-closed, where $x = \bigvee x_i$, $y = \bigvee f(x_i)$ and $B = \mathcal{Y} \downarrow y$. Then $x_i \in f^{-1}(B)$ by monotonicity, so $x \in f^{-1}(B)$ by the order-theoretic definition of Scott-closure, *i.e.* $f(x) \in B$ and $f(x) = y$ by monotonicity.

Conversely, let $B \subset \mathcal{Y}$ be any Scott-closed subset and $x_i \in f^{-1}(B) \subset \mathcal{X}$ be a directed set with $\bigvee f(x_i) = f(\bigvee x_i) \in B$. Then $\bigvee x_i \in f^{-1}(B)$ as required. $\qquad\qquad\square$

Classically there is a bijective correspondence between closed subsets and their complementary open subsets, but we shall define them separately.

PROPOSITION 3·4·9: Let $(\mathcal{X}, \leqslant)$ be a dcpo. Then for a subset $U \subset \mathcal{X}$, the characteristic function $\chi_U : \mathcal{X} \to 2$ is Scott-continuous iff

(a) U is an upper set and

(b) it is ***inaccessible by directed joins***, *i.e.* for any directed diagram $x_{(-)} : \mathcal{I} \to \mathcal{X}$, if $\bigvee_i x_i \in U$ then $\exists i.\, x_i \in U$.

Such subsets are said to be ***Scott-open***, and they form a topology. For any Scott-continuous function $f : \mathcal{X} \to \mathcal{Y}$ between dcpos, the inverse image of any open subset $V \subset \mathcal{Y}$ is open. The converse holds so long as there are *enough* Scott-open sets to make the specialisation order (Example 3.1.2(i)) coincide with the given order.

PROOF: Directedness is needed to show that the whole set $\mathcal{X}$ is Scott-open, and that intersections of open subsets are open. Closure under unions is easy, and we leave the rest as an exercise, adapting the previous argument for closed sets. $\qquad\qquad\square$

DEFINITION 3·4·10: The type of truth values is playing a topological role here, in which the point $\top$ is open and $\bot$ is closed. This is called the ***Sierpiński space***, S. Intuitionistically, it is intermediate between 2 and 2, though the considerations bearing on what its definition ought to be lie outside the scope of this book (see the footnote on page 502), so as with $\mathbb{R}$ we shall avoid questions that rely on the distinctions.

Classically, subsets of the form $\{x \mid x \nleqslant y\}$ are Scott-open, but the Scott topology need not be sober [Joh82, p. 46]. For an intuitionistic result, we restrict attention to a smaller class of domains, drawing upon another source of upper (so possibly open) subsets: those of the form $x \downarrow \mathcal{X}$.

DEFINITION 3·4·11: Let $x \in \mathcal{X}$ in a dcpo. If $x \downarrow \mathcal{X}$ is Scott-open, *i.e.*

$$x \leqslant \bigvee y_i \implies \exists i.\, x \leqslant y_i$$

for all directed sets, then x is said to be ***compact*** or ***finitely generable***. We write $\mathcal{X}_{\mathrm{fg}} \subset \mathcal{X}$ for the subset of compact elements. If, for each $x \in \mathcal{X}$,

$$\mathcal{X}_{\mathrm{fg}} \downarrow x \equiv \{y \mid x \geqslant y \in \mathcal{X}_{\mathrm{fg}}\} \text{ is directed with join } x$$

then $\mathcal{X}$ is called an **algebraic dcpo**. The name arose by extension from algebraic *lattices* (Theorem 3.9.4), but since their algebraic aspect has really been lost, **finitary** or **locally finitely generable** would be better names. Algebraic dcpos satisfy the qualification of the previous result, as do the more general continuous dcpos [GHK⁺80, Joh82].

PROPOSITION 3·4·12:   A dcpo $\mathcal{X}$ is algebraic iff it is isomorphic to $\mathsf{Idl}(\mathcal{Y})$ for some poset $\mathcal{Y}$. Then for any dcpo Θ there is a bijection

$$\frac{\text{Scott-continuous functions}\quad p : \mathcal{X} \equiv \mathsf{Idl}(\mathcal{Y}) \to \Theta}{\text{monotone functions}\quad f : \mathcal{Y} \equiv \mathcal{X}_{\mathsf{fg}} \to \Theta}$$

In particular the topology is the lattice of *monotone* functions $[\mathcal{Y} \to 2]$.

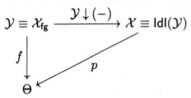

PROOF: [⇒] For $\mathcal{X}$ algebraic, put $\mathcal{Y} = \mathcal{X}_{\mathsf{fg}}$. Then $\mathcal{I} = \{y \mid x \geqslant y \in \mathcal{Y}\}$ and $x = \bigvee \mathcal{I}$ give the isomorphism. [⇐] For any ideal $\mathcal{I} \in \mathsf{Idl}(\mathcal{Y})$, $\mathcal{I} = \bigcup\{\mathcal{Y}\downarrow y \mid y \in \mathcal{I}\}$. $\mathcal{I}$ is compact iff $\mathcal{I} = \mathcal{Y}\downarrow y$ for some $y \in \mathcal{Y}$. Finally, any continuous function $f : \mathcal{X} \to \Theta$ is determined by its values at compact elements.   □

3.5   PRODUCTS AND FUNCTION-SPACES

Examples 2.1.4 and 2.1.7 gave recipes for the type constructors $\to$ and $+$, for sets and functions. In Section 2.3 we gave type-theoretic rules for these, and promised that we would find interpretations of them in other mathematical structures. We shall now do this with posets and domains. The sum of posets or domains is formed in the obvious way, as a disjoint union, *i.e.* the components are incomparable, so we concentrate on products and function-spaces.

Products of posets, domains, lattices and diagrams.

PROPOSITION 3·5·1:   Let $(\mathcal{X}, \leqslant^{\mathcal{X}})$ and $(\mathcal{Y}, \leqslant^{\mathcal{Y}})$ be preorders. We equip the cartesian product $\mathcal{X} \times \mathcal{Y}$ with the **componentwise order**:

$$\langle x_1, y_1 \rangle \leqslant^{\mathcal{X} \times \mathcal{Y}} \langle x_2, y_2 \rangle \quad \text{if} \quad x_1 \leqslant^{\mathcal{X}} x_2 \text{ and } y_1 \leqslant^{\mathcal{Y}} y_2.$$

Then this makes $\mathcal{X} \times \mathcal{Y}$ a preorder too. Also

(a) the projections $\pi_0 : \langle x, y \rangle \mapsto x$ and $\pi_1 : \langle x, y \rangle \mapsto y$ are monotone;

(b) if $a : \Gamma \to \mathcal{X}$ and $b : \Gamma \to \mathcal{Y}$ are monotone functions then so is the function $\langle a, b \rangle : \Gamma \to \mathcal{X} \times \mathcal{Y}$ defined by $\langle a(-), b(-) \rangle$;

(c) a function $p : \mathcal{X} \times \mathcal{Y} \to \mathcal{Z}$ is (jointly) monotone iff it is monotone in each argument, for each constant value of the other;

(d) if $\mathcal{X}$ and $\mathcal{Y}$ are posets, *i.e.* their order relations $\leqslant^{\mathcal{X}}$ and $\leqslant^{\mathcal{Y}}$ are antisymmetric, then $\mathcal{X} \times \mathcal{Y}$ is also a poset. □

The corresponding result for domains applies equally well to semilattices, lattices, Heyting (semi)lattices and complete (semi)lattices, so we state it for individual diagram shapes. We shall consider part (c) later.

PROPOSITION 3·5·2: Let $\mathcal{X}$ and $\mathcal{Y}$ be preorders and $\mathcal{I}$ a diagram shape.

(a) Let $\langle x_{(-)}, y_{(-)} \rangle : \mathcal{I} \to \mathcal{X} \times \mathcal{Y}$ be a diagram. Then the joins

$$\bigvee_{i \in \mathcal{I}}^{(\mathcal{X} \times \mathcal{Y})} \langle x_i, y_i \rangle = \left\langle \bigvee_{i \in \mathcal{I}}^{\mathcal{X}} x_i, \bigvee_{i \in \mathcal{I}}^{\mathcal{Y}} y_i \right\rangle$$

coincide in the sense that if one exists then so does the other, and then they are equal.

(b) If $\mathcal{X}$ and $\mathcal{Y}$ have all joins of shape $\mathcal{I}$ then so does $\mathcal{X} \times \mathcal{Y}$.

(c) The projections $\pi_0 : \langle x, y \rangle \mapsto x$ and $\pi_1 : \langle x, y \rangle \mapsto y$ preserve joins.

(d) If $a : \Gamma \to \mathcal{X}$ and $b : \Gamma \to \mathcal{Y}$ preserve joins of shape $\mathcal{I}$ then so does the pair $\langle a, b \rangle : \Gamma \to \mathcal{X} \times \mathcal{Y}$.

PROOF: In (a), $\langle \vartheta, \phi \rangle$ is an upper bound for the set of pairs iff ϑ bounds the first components and ϕ the second. Hence $\langle \vartheta, \phi \rangle$ is least (or locally least or mimimal) iff both ϑ and ϕ are. The other parts follow. □

COROLLARY 3·5·3: If $\mathcal{X}$ and $\mathcal{Y}$ are dcpos or ipos then so is $\mathcal{X} \times \mathcal{Y}$, the projections are Scott-continuous and pairing preserves continuity. □

Products of *diagrams* are used to handle multiple suffixes. Doubly indexed joins can be rearranged, as may doubly indexed meets, but Example 3.5.14 shows that meets cannot be interchanged with joins.

LEMMA 3·5·4: Let $x_{\langle (-), (=) \rangle} : \mathcal{I} \times \mathcal{J} \to \mathcal{X}$ be a diagram. If either the expression on the left or that on the right is defined in

$$\bigvee_{i \in \mathcal{I}} \bigvee_{j \in \mathcal{J}} x_{\langle i, j \rangle} = \bigvee_{\langle i, j \rangle \in \mathcal{I} \times \mathcal{J}} x_{\langle i, j \rangle} = \bigvee_{j \in \mathcal{J}} \bigvee_{i \in \mathcal{I}} x_{\langle i, j \rangle},$$

then so is the one in the middle and then they are equal.

PROOF: Assuming that the intermediate joins exist, for $\vartheta \in \mathcal{X}$,

$$\left(\forall i. \ \vartheta \geqslant \bigvee_{j \in \mathcal{J}} x_{\langle i, j \rangle} \right) \Leftrightarrow \left(\forall i, j. \ \vartheta \geqslant x_{\langle i, j \rangle} \right) \Leftrightarrow \left(\forall j. \ \vartheta \geqslant \bigvee_{i \in \mathcal{I}} x_{\langle i, j \rangle} \right)$$

and each of these is equivalent to ϑ lying above the corresponding join. Substituting each of the three joins for ϑ, they must be equal. □

Pointwise meets and joins.

PROPOSITION 3·5·5: Let $\mathcal{Y}$ be a preorder and $f, g : \mathcal{X} \to \mathcal{Y}$ be functions (the order, if any, on $\mathcal{X}$ is unimportant). Then the **pointwise order** between f and g is given by

$$f \leqslant^{[\mathcal{X} \to \mathcal{Y}]} g \quad \text{if} \quad \forall x{:}\mathcal{X}.\ f(x) \leqslant^{\mathcal{Y}} g(x).$$

The preorder of monotone functions from $\mathcal{X}$ to $\mathcal{Y}$ is written $[\mathcal{X} \to \mathcal{Y}]$ or $\mathcal{Y}^{\mathcal{X}}$ and is called the **function-space**. Then

(a) $\mathsf{ev} : [\mathcal{X} \to \mathcal{Y}] \times \mathcal{X} \to \mathcal{Y}$, given by $\langle f, x \rangle \mapsto f(x)$, is monotone;

(b) if $p : \mathcal{T} \times \mathcal{X} \to \mathcal{Y}$ is monotone then so is $\tilde{p} : \mathcal{T} \to [\mathcal{X} \to \mathcal{Y}]$, defined by $\lambda x.\, p(-, x)$;

(c) $[\mathcal{X} \to \mathcal{Y}]$ is a poset, *i.e.* the pointwise order is antisymmetric, if $\mathcal{Y}$ is.

PROOF:

[a] If $\langle f, x_1 \rangle \leqslant \langle g, x_2 \rangle$ then $f \leqslant g$ and $x_1 \leqslant x_2$, so using the pointwise order, monotonicity and transitivity, we have a square

$$x_1 \ \leqslant \ x_2$$

$$f \quad fx_1 \ \leqslant \ fx_2$$

$$/\!\!\wedge \ \ /\!\!\wedge \qquad\quad /\!\!\wedge$$

$$g \quad gx_1 \ \leqslant \ gx_2$$

[b] Joint monotonicity implies separate.

[c] $f \leqslant g \leqslant f$ iff $\forall x{:}\mathcal{X}.\ f(x) \leqslant^{\mathcal{Y}} g(x) \leqslant^{\mathcal{Y}} f(x)$ whilst $f = g$ iff $\forall x{:}\mathcal{X}.\ f(x) = g(x)$. $\square$

EXAMPLES 3·5·6:

(a) $[X \to \Omega]$, where X is a set with the discrete order and Ω is the type of propositions, is isomorphic to the powerset $\mathcal{P}(X)$ under inclusion.

(b) $[\mathcal{X}^{\mathrm{op}} \to \Omega]$, where $\mathcal{X}$ is a poset, is isomorphic to $\mathsf{sh}(\mathcal{X})$, the lattice of lower sets (Definition 3.1.7).

(c) Similarly $[\mathcal{X} \to \Omega]^{\mathrm{op}}$ is the lattice of upper sets (Example 3.1.6(f)).

(d) $[X \to Y_\perp]$, where X and Y are (discrete) sets and $Y_\perp$ carries the information order (Definition 3.3.7), is isomorphic to the poset of partial functions $X \rightharpoonup Y$ with the extension order (Definition 3.3.3).

LEMMA 3·5·7: Let $f_{(-)} : \mathcal{I} \to [\mathcal{X} \to \mathcal{Y}]$. If the joins on the right of

$$\bigvee_{i \in \mathcal{I}}^{[\mathcal{X} \to \mathcal{Y}]} f_i = \lambda x. \bigvee_{i \in \mathcal{I}}^{\mathcal{Y}} f_i(x)$$

exist *then* so does the join on the left, and the equation holds.

PROOF:

(a) If we have $\{f_i \mid i : \mathcal{I}\} \leqslant \vartheta$ (in particular if the function ϑ is the join) then each set $\{f_i(x) \mid i : \mathcal{I}\}$ has an upper bound, namely $\vartheta(x)$:

$$\text{if } \exists \vartheta. \forall x. \forall i. \, f_i(x) \leqslant \vartheta(x) \text{ then } \forall x. \exists u. \forall i. \, f_i(x) \leqslant u.$$

(b) The function $p : x \mapsto \bigvee^{\mathcal{X}}_{i \in \mathcal{I}}(f_i(x))$, if it exists, *is* monotone, because if $x' \leqslant x$ then $\forall i. \, f_i(x') \leqslant f_i(x) \leqslant p(x)$ and so since $p(x')$ is the least upper bound we have $p(x') \leqslant p(x)$. It follows that p is the join of the functions. $\qquad\square$

REMARK 3·5·8:

(a) However, knowing that there is some $\{f_i(x) \mid i : \mathcal{I}\} \leqslant u_x$ for each $x \in \mathcal{X}$ is not sufficient to give a bound for the set of functions, because $x \mapsto u_x$ need not be monotone. Indeed we need the axiom of choice (Definition 1.8.8) even to make this exist as a function.

That part (b) above works and this doesn't illustrates the value of uniqueness, *cf.* Lemma 1.2.11; "locally least" may be good enough, but inserting "minimal" doesn't help.

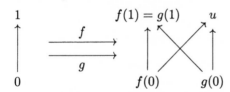

(b) Let ϑ be a monotone upper bound of the two functions illustrated, so $\vartheta(1) = f(1) = g(1)$, but $\vartheta(0)$ must also take this value. This constant function is therefore the least upper bound in the function-space, but as $\vartheta(0) \not\leqslant u$ it is not the pointwise least upper bound. $\square$

COROLLARY 3·5·9: Let $\mathcal{Y}$ be a poset with all joins of shape $\mathcal{I}$.

(a) Then $[\mathcal{X} \to \mathcal{Y}]$ also has all joins of shape $\mathcal{I}$, computed pointwise;

(b) $\mathrm{ev}(-, x)$ preserves them for each x;

(c) of course $\mathrm{ev}(f, -) \equiv f$ preserves them iff f does;

(d) for $p : \Gamma \times \mathcal{X} \to \mathcal{Y}$, if $p(-, x)$ preserves such joins for each $x \in \mathcal{X}$ then so does $\tilde{p} : \Gamma \to [\mathcal{X} \to \mathcal{Y}]$ by $\lambda x. \, p(-, x)$. $\qquad\square$

So far we have been discussing the poset of *all monotone functions*, which is not what we want for domains. Abusing notation, we also write $[\mathcal{X} \to \mathcal{Y}]$ for the poset of *Scott-continuous* functions from $\mathcal{X}$ to $\mathcal{Y}$, with the pointwise order, $\mathcal{X}$ and $\mathcal{Y}$ now being dcpos. Recall that pointwise joins gave joins of functions.

PROPOSITION 3·5·10:   The same holds with $[\mathcal{X} \to \mathcal{Y}]$ reinterpreted to consist only of continuous functions. In particular, $[\mathcal{X} \to \mathcal{Y}]$ is a dcpo, and an ipo with $\perp_{[\mathcal{X} \to \mathcal{Y}]} = \lambda x. \perp_{\mathcal{Y}}$ if $\perp_{\mathcal{Y}}$ exists.

PROOF: From Lemma 3.5.4, if each f_j preserves $\mathcal{I}$-indexed joins (for each $j \in \mathcal{J}$), and $\bigvee_j f_j$ exists pointwise, then it too preserves $\mathcal{I}$-indexed joins. Moreover it is the join amongst join-preserving functions.   □

Beware that we have *not* said that ev : $[\mathcal{X} \to \mathcal{Y}] \times \mathcal{X} \to \mathcal{Y}$ preserves joins.

Joint continuity. Proposition 3.5.2 stated the properties of products of domains analogous to the result preceding it, apart from part 3.5.1(c). This deserves special consideration.

EXAMPLE 3·5·11: The function $f : \mathbb{R} \times \mathbb{R} \to \mathbb{R}$ defined by

$$f(x,y) = \frac{2xy}{x^2 + y^2} \qquad \text{except for } f(0,0) = 0$$

is continuous in x for each fixed value of y and *vice versa*, but it is not continuous at $(0,0)$ as a function of two variables.   □

There are several ways of reacting to this; in particular Exercise 3.20 shows what is missing in the case of binary meets or joins. For directed joins it turns out that the difficulty does not arise, but perhaps this just shows the poverty of order-theoretic representations of semantics.

LEMMA 3·5·12:   The product of two directed, semidirected or confluent (Definition 3.4.1) posets has the same property. Moreover the diagonal function $\mathcal{I} \to \mathcal{I} \times \mathcal{I}$ is cofinal (Proposition 3.2.10) iff $\mathcal{I}$ is semidirected, whilst the function $\mathcal{I} \to \{\star\}$ is cofinal iff $\mathcal{I}$ is inhabited.   □

COROLLARY 3·5·13:   A function $f : \mathcal{X} \times \mathcal{Y} \to \mathcal{Z}$ of two arguments between dcpos is (jointly) continuous iff it is separately continuous. In particular ev : $[\mathcal{X} \to \mathcal{Y}] \times \mathcal{X} \to \mathcal{Y}$ is jointly continuous.

PROOF: Let $\mathcal{I}$ be a directed diagram. Since $\mathcal{I} \to \mathcal{I} \times \mathcal{I}$ is cofinal by Lemma 3.5.4 we have

$$\bigvee_i \langle x_i, y_i \rangle = \bigvee_{\langle i,j \rangle} \langle x_i, y_j \rangle = \left\langle \bigvee_i x_i, \bigvee_j y_j \right\rangle$$

and separate continuity of $f(-, \bigvee y_j)$ and $f(x_i, -)$ gives

$$f\left(\bigvee_i x_i, \bigvee_j y_j \right) = \bigvee_i f\left(x_i, \bigvee_j y_j \right) = \bigvee_i \bigvee_j f(x_i, y_j),$$

whence joint continuity follows by cofinality again.   □

We may compute binary *meets* pointwise if they commute with directed joins; a dcpo with this property is called a **preframe**. Infinite meets in $[\mathcal{X} \to \mathcal{Y}]$ may exist but need not be computed pointwise.

EXAMPLE 3·5·14: Consider $\mathcal{X} = \mathcal{Y} = [0, 1] \subset \mathbb{R}$ (the unit interval) and (with $\mathcal{J} = \mathbb{N}$) let $f_n : \mathcal{X} \to \mathcal{Y}$ by $f_n(x) = x^n$. The pointwise meet is discontinuous at 1 in both the Cauchy and Scott senses; the constantly 0 function is the meet in the function-space. □

It is common for function-spaces to inherit the properties of the target domain, irrespective of the source, because the function-space is often a subalgebra of a product. We have shown that this is the case for the existence of various classes of joins (see also Exercise 3.21), but there are simple counterexamples to this behaviour for the trichotomy property.

Scott's thesis. The origin of the work in these three sections was Dana Scott's unexpected discovery that non-syntactic models of the untyped λ-calculus could be constructed from certain topological spaces (1969). These very un-geometrical spaces were algebraic lattices endowed with Scott's topology. Their retracts, known as ***continuous lattices***, also arose from abstract work in topological lattice theory [GHK+80].

Scott proposed that topological continuity be used as an approximation to computability. There were precedents for this idea in recursion theory around 1955. Yuri Ershov observed the analogy between the lattice of recursively enumerable sets and a topology. The Rice–Shapiro Theorem says that any recursively enumerable set of partial recursive functions (*i.e.* sets of codes such that if one code for a function $\mathbb{N} \rightharpoonup \mathbb{N}$ belongs to the set then so do all others) is Scott-open. The Myhill–Shepherdson Theorem says that any recursive $f : [\mathbb{N} \rightharpoonup \mathbb{N}] \rightharpoonup [\mathbb{N} \rightharpoonup \mathbb{N}]$, as we would write it domain-theoretically, is Scott-continuous.

Christopher Strachey and others applied Scott's work to ***denotational semantics*** of programming languages, where the lattice element ⊤ was inappropriate. Scott and his followers repeatedly simplified the theory for this new audience, with the result that order theory replaced topology in the formal development. In particular, the term ***Scott domain*** came to be applied to any boundedly complete algebraic dcpo $\mathcal{X}$ for which $\mathcal{X}_{fg}$ is countable (Exercise 3.21).

Domain theory can solve, not only fixed *point* equations (Example 3.3.1), but also *type*-equations, such as $X \cong [X \to X]$ for the untyped λ-calculus. The right hand side may involve any of the type-constructors on ipos in an arbitrarily complicated way, giving a domain of mathematical meanings for objects with functions, case analysis and non-determinism.

Here and in Section 4.7 we interpret the λ-calculus, primitive recursion on $\mathbb{N}$ and the fixed point operator Y. Gordon Plotkin [Plo77] considered these as a programming language (PCF), with call-by-name evaluation

(Remark 2.3.4). He showed that any *program* (closed term of type $\mathbb{N}$) whose denotation in $\mathcal{IPO}$ is a numeral in $\mathbb{N}_\perp$ (*i.e.* not $\perp$) *terminates* with that value, so there is a link back from the semantics to the syntax. This can now be proved by methods like those in Section 7.7.

However, **parallel or** ($\mathsf{por}(\mathsf{yes}, \perp) = \mathsf{yes} = \mathsf{por}(\perp, \mathsf{yes})$) is also interpreted in $\mathcal{IPO}$, but is not definable in PCF, whose programs execute "sequentially". The tight link is broken for higher order terms, as there exist such terms that can "recognise" por as an argument. By adding por and a similar "existential quantifier" to PCF, Plotkin was able to extend the correspondence to higher types. Gérard Berry eliminated por with a different notion of "domain", *cf.* Example 4.51, but more complicated examples recur. The sequentiality and "full abstraction" problems remained open for two decades; they were solved for PCF in 1994 by Abramsky, Hyland, Jagadeesan and Ong, using games.

Without bounded completeness, function-spaces of algebraic dcpos need not be algebraic. Achim Jung showed that his own L-domains (Exercise 3.34) and Plotkin's SFP domains are the two maximal cartesian closed categories (*i.e.* closed under function-spaces) of algebraic dcpos [Jun90].

The search for cartesian closed or "convenient" categories in topology is much older, and equally inconclusive. The function-space S^X (where S is the Sierpiński space, Definition 3.4.10) only exists with the properties it should have when X is locally compact, and even when Y^X exists it need not be locally compact. A famous cartesian closed full subcategory of compact Hausdorff spaces was found by John Kelley [Kel55], and there are other approaches to topology with different notions of function-space. Example 9.4.11(f) shows how a certain generalised function-space first arose geometrically.

Because of the generality of the infinitary joins required in topology, a function may be topologically continuous without being computable. One can add the word "effective" throughout the theory, but it seems to me to be very clumsy to bolt together two subjects like this. There ought to be a common axiomatisation, of which the free model would be equivalent to recursion theory, but with another model consisting of certain spaces. Synthetic domain theory abolishes non-computable functions between sets themselves, by refining the underlying logic.

This cannot be done in classical logic, because the extra axioms (such as the Church–Turing thesis) conflict with excluded middle. However, the use of excluded middle so infests existing accounts of mathematical foundations that it was necessary to start from the beginning, although synthetic domain theory is beyond the scope of this book.

3.6   ADJUNCTIONS

Adjunctions unify the treatment of the logical connectives. Generalised from propositions to types, they not only handle the operations $\times$, $\rightarrow$, $+$, *etc.* but also account for the "universal properties" that we have been meeting, such as Propositions 3.1.10, 3.2.7(b) and 3.4.12. Adjunctions themselves also arise as a common method of construction involving powersets. In view of this and the shift in the next chapter to categories, we now call the posets $\mathcal{S}$ and $\mathcal{A}$, and their elements $X \in \mathcal{S}$ and $A \in \mathcal{A}$.

DEFINITION 3·6·1:   Let $F : \mathcal{S} \rightarrow \mathcal{A}$ and $U : \mathcal{A} \rightarrow \mathcal{S}$ be functions (not *a priori* monotone) between preorders. We say that F and U are

(a) ***adjoint***, written $F \dashv U$, if

$$\forall X, A. \quad \frac{F(X) \;\leqslant^{\mathcal{A}}\; A}{X \;\leqslant^{\mathcal{S}}\; U(A)}$$

we also say that F is the ***lower*** or ***left adjoint*** of U, and U is the ***upper*** or ***right adjoint*** of F (also, "F is left adjoint *to* U" *etc.*);

(b) a ***Galois connection*** if $\forall X, A. \ A \leqslant^{\mathcal{A}} F(X) \Leftrightarrow X \leqslant^{\mathcal{S}} U(A)$;

(c) a ***co-Galois connection*** if $\forall X, A. \ F(X) \leqslant^{\mathcal{A}} A \Leftrightarrow U(A) \leqslant^{\mathcal{S}} X$.

Galois and co-Galois connections are also known as ***contravariant*** or ***symmetric adjunctions***, on the right and left respectively. They are the same as ***covariant adjunctions*** (*i.e.* the first kind) between $\mathcal{S}$ and $\mathcal{A}^{\text{op}}$, but Section 3.8 shows how (a) and (b) arise between full powerset lattices in two idiomatic but different ways. Exercise 4.27 uses all three as interesting type-connectives between complete semilattices.

LEMMA 3·6·2:   The above definitions are respectively equivalent to

(a) F and U being monotone with $\mathsf{id}_{\mathcal{S}} \leqslant U \circ F$ and $F \circ U \leqslant \mathsf{id}_{\mathcal{A}}$;

(b) F and U being antitone with $\mathsf{id}_{\mathcal{S}} \leqslant U \circ F$ and $\mathsf{id}_{\mathcal{A}} \leqslant F \circ U$;

(c) F and U being antitone with $U \circ F \leqslant \mathsf{id}_{\mathcal{S}}$ and $F \circ U \leqslant \mathsf{id}_{\mathcal{A}}$;

where $\leqslant$ is the pointwise order (Proposition 3.5.5). Moreover if $\mathcal{S}$ and $\mathcal{A}$ are posets, then in all three cases $F = F \circ U \circ F$ and $U = U \circ F \circ U$.

PROOF: Suppose $F \dashv U$. Let $X \leqslant Y$ and put $A = F(Y)$; then $F(Y) \leqslant A$ so $X \leqslant Y \leqslant U(A)$, whence $F(X) \leqslant A$. Next, $F(X) \leqslant F(X)$ so $X \leqslant U(F(X))$. Monotonicity of U and the other inequality are the same. Conversely if $F(X) \leqslant A$ then $X \leqslant U(F(X)) \leqslant U(A)$, and similarly $F(X) \leqslant F(U(A)) \leqslant A$. Finally $F(X) \leqslant F(U(F(X)))$ since $X \leqslant U(F(X))$, and $F(U(F(X))) \leqslant F(X)$ since $F(U(A)) \leqslant A$ so $F = F \circ U \circ F$ by antisymmetry. The results for Galois connections are obtained by reversing some of the inequalities.   □

REMARK 3·6·3:   A third way of expressing adjunctions, that

$$F(X) \text{ is the } least \text{ } A \text{ such that } X \leqslant U(A),$$

is the one which we shall prefer for categories:

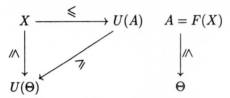

we call this the **universal property** of $F(X)$.

Order, composition and equivalence.

LEMMA 3·6·4:   Let $F \dashv U$ and $G \dashv V$. Then $F \leqslant G$ iff $U \geqslant V$. $\square$

COROLLARY 3·6·5:   If $\mathcal{S}$ is a poset and $F : \mathcal{S} \to \mathcal{A}$ has a right adjoint U then it is unique. Similarly for left adjoints. $\square$

LEMMA 3·6·6:   If $F \dashv U$ and $G \dashv V$ then $F \, ; G \dashv U \circ V$.

$$\mathcal{X} \xrightarrow[\underset{U}{\perp}]{F} \mathcal{Y} \xrightarrow[\underset{V}{\perp}]{G} \mathcal{Z}$$

This is one of the circumstances in which it is convenient to use both the left-handed ($\circ$) and right-handed (;) notations together. $\square$

REMARK 3·6·7:

(a) Any isomorphism between preorders is an adjunction, in two ways, with $\leqslant$ replaced by equality. Likewise any *duality* (isomorphism $\mathcal{S} \cong \mathcal{A}^{\mathrm{op}}$) is both a Galois connection and a co-Galois connection.

(b) Replacing $\leqslant$ in Definition 3.6.1 by the equivalence relation $\sim$ (the conjunction of $\leqslant$ and $\geqslant$, Proposition 3.1.10), we obtain the notion of a **strong equivalence** of preorders: $\mathrm{id}_{\mathcal{S}} \sim U \circ F$ and $F \circ U \sim \mathrm{id}_{\mathcal{A}}$.

(c) A full monotone function $F : \mathcal{S} \to \mathcal{A}$ for which

$$\forall A{:}\mathcal{A}. \; \exists X{:}\mathcal{S}. \; F(X) \sim A$$

is called an **equivalence function**; for example Proposition 3.1.10 showed that any preorder is equivalent to a poset. Using the axiom of choice, every equivalence function is part of a strong equivalence (see Exercise 3.26 for why Choice is necessary).

The last does not, as it stands, give an equivalence relation on the *class* of preorders: it is reflexive and transitive but not symmetric. But as it is also confluent (Exercise 3.27) it is easy to find the equivalence closure:

(d) Two preorders S and T are said to be **weakly equivalent** if a third preorder A and equivalence functions $S \to A \leftarrow T$ are given.

Although Proposition 3.1.10 makes these notions somewhat redundant for preorders, we shall need them for categories (Definition 4.8.9).

The adjoint function theorem. Theorem 3.6.9 is the most important result about infinitary meets and joins. However, it is rarely stated: since the formula for the adjoint is notationally simple, it is used, embedded in more complicated calculations, without appreciating the significance of the theorem.

PROPOSITION 3·6·8:   Let $F \dashv U$. Then

> F preserves any (locally) least upper bounds which exist,
> and U preserves and greatest lower bounds which exist.

Functions which are symmetrically adjoint on the left preserve least upper bounds and functions which are symmetrically adjoint on the right preserve greatest lower bounds.

PROOF: Let C be an upper bound of $\mathcal{I} \subset S$ and ϑ likewise of the image, $F_!(\mathcal{I}) = \{F(X) \mid X \in \mathcal{I}\}$. Then

(a) $F(C)$ is an upper bound of $F_!(\mathcal{I})$ by monotonicity.
(b) Since $\forall X. F(X) \leqslant \vartheta$, we have $\forall X. X \leqslant U(\vartheta)$ by adjointness, so $U(\vartheta)$ is an upper bound of $\mathcal{I}$.
(c) If C is a least upper bound of $\mathcal{I}$ then $C \leqslant U(\vartheta)$, so $F(C) \leqslant \vartheta$.
(d) If C is locally least and $F(C) \leqslant A \geqslant \vartheta$ then $C \leqslant U(A) \geqslant U(\vartheta)$, so $C \leqslant U(\vartheta)$ and $F(C) \leqslant \vartheta$.

The other cases are the same, switching arguments and inequalities. □

The adjoint function theorem can be proved in a similar way for meets and joins of *diagrams* (Definition 3.2.9) rather than subsets. Exercises 3.33ff generalise to *minimal* and *locally* least upper bounds.

THEOREM 3·6·9:   Let S be a poset with all joins, A a preorder and $F : S \to A$ a function. Then F has a right adjoint $U : A \to S$ iff it preserves all joins. Similarly a function from a poset with all meets has a left adjoint iff it preserves all meets.

PROOF: As S has arbitrary joins and F preserves them, the lower subset $\{X : S \mid F(X) \leqslant A\}$ is representable, *i.e.* of the form $S \downarrow X_0$ for some unique $X_0 \in S$ (Definition 3.1.7). Indeed

$$U(A) \stackrel{\text{def}}{=\!=} X_0 = \bigvee \{X \mid F(X) \leqslant A\}.$$

Similarly $F(X) = \bigwedge \{A : S \mid X \leqslant U(A)\}$.   □

EXAMPLES 3·6·10:   Meets and joins are themselves adjoints.

(a) The unique function $F : S \to \{\star\}$ has a right adjoint iff S has a greatest element. The right adjoint U is $\star \mapsto \top$, and $\top$ is the join of the whole poset (considered as a diagram $\mathrm{id} : S \to S$).

(b) The diagonal function $F : S \to S \times S$ has a right adjoint iff S has meets of pairs, and then the right adjoint is $U(X, Y) = X \wedge Y$.

(c) More generally, let $\mathcal{I}$ be any diagram shape, and put $\mathcal{A} = S^{\mathcal{I}}$ and $K(X) = \lambda i.\, X \in \mathcal{A}$. Then $\gamma \in S$ is a lower bound for a diagram $d : \mathcal{I} \to S$ iff $K(\gamma) \leqslant^{\mathcal{A}} d$. The right adjoint, $K \dashv U$, provides the meet, $U(d) = \bigwedge d_!(\mathcal{I})$; it is given by $\bigvee\{\gamma : S \mid K(\gamma) \leqslant^{\mathcal{A}} d\}$. (Notice that this repeats the same calculation at each co-ordinate $i \in \mathcal{I}$.)

Similar results hold for $\bot$ and joins, which are given by *left* adjoints to the diagonal or constant functions. In particular,

(d) the regular representation $\mathcal{X} \downarrow (-) : \mathcal{A} \to \mathsf{sh}(\mathcal{A})$ (Definition 3.1.7), which preserves meets by Proposition 3.2.7(a), has a left adjoint iff $\mathcal{A}$ has all joins, and then the left adjoint is $\bigvee$.

REMARK 3·6·11:   These observations, that $\bigvee \dashv K \dashv \bigwedge$ and meets or joins commute with adjoints, are summed up by the diagram

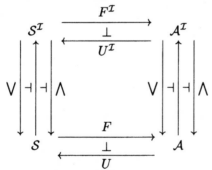

where $K : X \mapsto \lambda i.\, X : S \to S^{\mathcal{I}}$. Using the composition (Lemma 3.6.6) and uniqueness (Corollary 3.6.5) of adjoints, $U \,;\, K_S = K_{\mathcal{A}} \,;\, U^{\mathcal{I}}$ implies $F \circ \bigvee_S = \bigvee_{\mathcal{A}} \circ F^{\mathcal{I}}$, and $F \,;\, K_{\mathcal{A}} = K_S \,;\, F^{\mathcal{I}}$ implies $U \circ \bigwedge_{\mathcal{A}} = \bigwedge_S \circ U^{\mathcal{I}}$.

So long as there are enough meets and joins, the converse holds.

DEFINITION 3·6·12:   A poset is a ***complete join-semilattice*** if every subset has a join, and similarly a ***complete meet-semilattice*** if every subset has a meet; if both of these hold we call it a ***complete lattice***.

Any complete join-semilattice is a complete meet-semilattice and *vice versa*, but we make the distinction between complete meet- and join-semilattices and complete lattices because, although the existence of one structure forces that of the other, a monotone function preserving all meets need not preserve all joins, *cf.* Remark 3.4.5.

By Corollary 3.6.5, a function can have at most one adjoint on each side, but it can have both of them, and there are strings of any finite length of successively adjoint monotone functions. Remark 3.8.9 characterises complete sublattices of powersets, *i.e.* inclusions with both adjoints.

EXAMPLES 3·6·13: Certain weaker conditions are of interest:

(a) The inverse image operation $F = g^*$ on open sets for a continuous function g between topological spaces preserves *finite intersections*.

(b) In domain theory, if $\mathrm{id}_{\mathcal{S}} = U \circ F$ and $F \circ U \leqslant \mathrm{id}_{\mathcal{A}}$ with U Scott-continuous (*i.e.* preserving *directed joins*) then $F : \mathcal{S} \hookrightarrow \mathcal{A}$ is called an **embedding** and $U : \mathcal{A} \twoheadrightarrow \mathcal{S}$ a **projection**. Scott-continuous adjunctions with $\mathrm{id}_{\mathcal{S}} \leqslant U \circ F$ also arise in this subject.

Frames and Heyting algebras. Let us consider when $F : 2 \to 2$ by $\alpha \mapsto \alpha \wedge \phi$ has a right adjoint.

PROPOSITION 3·6·14:

(a) Let $U : \beta \mapsto (\phi \Rightarrow \beta)$, so $F \dashv U$; then these operations preserve joins and meets respectively,
$$(\bot \wedge \phi) = \bot \qquad (\phi \Rightarrow \top) = \top$$
$$\phi \Rightarrow (\alpha \wedge \beta) = (\phi \Rightarrow \alpha) \wedge (\phi \Rightarrow \beta),$$
but the most important case is the **distributive law**,
$$(\alpha \vee \beta) \wedge \phi = (\alpha \wedge \phi) \vee (\beta \wedge \phi).$$

(b) Conversely, by the adjoint function theorem, in a *complete* lattice, F has a right adjoint iff binary meet distributes over arbitrary joins,
$$\phi \wedge \bigvee_i \alpha_i = \bigvee_i (\phi \wedge \alpha_i) \qquad (\phi \Rightarrow \beta) = \bigvee \{\alpha \mid \phi \wedge \alpha \leqslant \beta\},$$
cf. the Frobenius law, $\phi \wedge \exists x. \, \alpha[x] = \exists x. \, \phi \wedge \alpha[x]$ (Lemma 1.6.3). □

DEFINITION 3·6·15: A **Heyting semilattice** $(\mathcal{X}, \top, \wedge, \Rightarrow)$ is a meet-semilattice with an additional binary operation $\Rightarrow$ satisfying
$$\frac{z \wedge x \leqslant y}{z \leqslant (x \Rightarrow y)}$$
cf. the introduction and elimination rules for implication (Remark 1.4.3 and Definition 1.5.1). Exercise 3.28 gives a purely equational version. *Heyting* semilattice homomorphisms by definition preserve $\Rightarrow$, whereas *semilattice* homomorphisms need not.

A **Heyting lattice** also has $\bot$ and $\vee$, and these are to be preserved by Heyting lattice homomorphisms. A **Boolean algebra** is a Heyting lattice in which $x = \neg\neg x$, where $\neg x$ is $x \Rightarrow \bot$, for which the truth table (Remark 1.8.4) gives a normal form. In the language of Heyting semilattices, by contrast, expressions may require nested implications, *cf.* Convention 2.3.2 for bracketing function-types. Powersets in classical and intuitionistic logic provide examples of complete Boolean and Heyting lattices respectively. See [Joh82, p. 35] for a picture of the free Heyting lattice on one generator.

We may instead concentrate on the infinitary joins.

DEFINITION 3·6·16:   A complete join-semilattice in which $\wedge$ distributes over joins is called a *frame*. The open sets of any topological space form a frame, and the inverse image operation of any continuous function preserves finite meets and arbitrary joins. Implication and infinite meets exist, but need not be preserved by *frame* homomorphisms: once again, the name of the objects changes when the morphisms change. Frames will be discussed in Theorem 3.9.9ff.

3.7   CLOSURE CONDITIONS AND INDUCTION

For any adjunction $F \dashv U$, it is an easy corollary of Lemma 3.6.2 that the composite $M = U \circ F$ is *idempotent*, and in this section we concentrate on its fixed points. In the common situation where the ambient poset S is a powerset $\mathcal{P}(\mathbb{X})$, the fixed points of M are often called *closed subsets* of $\mathbb{X}$. They feature in analysis as those subsets which contain the limits of their convergent sequences. Maurice Fréchet's approach to general topology (1906) was to consider the operations which take an arbitrary subset X to the smallest closed set containing it (often written $\overline{X}$, but called MX here), and to the largest open set inside it (its *interior*).

DEFINITION 3·7·1:   A monotone endofunction $M : S \to S$ of a poset such that $\mathsf{id}_S \leqslant M = M \circ M$ is called a *closure operation*.

LEMMA 3·7·2:   For any adjunction, U is injective iff F is surjective iff $F \circ U = \mathsf{id}_A$. Every closure operation M arises as $U \circ F$ like this, where

$$\mathcal{A} \overset{F}{\underset{U}{\rightleftarrows}} \mathcal{S}$$

split the idempotent M (Definition 1.3.12); F takes an element X to the least A which is a fixed point of M above X (Remark 3.6.3).   □

PROPOSITION 3·7·3:   Let $\mathcal{A} \subset S$ be the image of a closure operation, so meets in $\mathcal{A}$ agree with those in S by Proposition 3.6.8. To calculate joins in $\mathcal{A}$, we first find them in $\mathcal{S}$ and then apply the closure operation. $\mathcal{A}$ is already closed under any joins the closure operation preserves.   □

In particular, topological closures preserve finite joins, and the finitary closure operations which arise in algebra are Scott-continuous.

Closure conditions. Behind this static definition lies a *dynamic* point of view, which leads us into a notion of induction. Unfortunately the established terms "operation" and "condition" convey exactly the wrong idea. This kind of induction is actually more general than that treated in Section 2.5, as it allows for the *non-deterministic* recursive paradigm which lies behind resolution with backtracking in logic programming.

Given any subset X of some space, let TX be the set of limits t (in the sense of analysis) of sequences $(u_i) \subset X$. Now TX need not be closed, so we form the set T^2X of limits of sequences $(u_i) \subset TX$, and so on, but, unlike Proposition 3.3.11, even $\bigcup_n T^n X$ need not be closed. Nevertheless, MX exists by the Adjoint Function Theorem 3.6.9.

DEFINITION 3·7·4:   A **system of closure conditions** on a set Σ is given by any relation $\rhd \colon \mathcal{P}(\Sigma) \rightharpoonup \Sigma$, without restriction. Then a subset $A \subset \Sigma$ is said to be $\rhd$-**closed** if

$$\forall K \subset \Sigma. \, \forall t{:}\Sigma. \quad \frac{K \subset A \qquad K \rhd t}{t \in A}$$

so for example in analysis we would write "sequence $\rhd$ limit". We shall also write $\mathcal{L} = (\Sigma, \rhd)$ and call A a **model** of (or **algebra** for) $\mathcal{L}$.

Let $\mathcal{A} \equiv \mathcal{M}od(\mathcal{L})$ be the set of $\rhd$-closed subsets, ordered by inclusion, and $U \colon \mathcal{M}od(\mathcal{L}) \hookrightarrow \mathcal{P}(\Sigma)$. Since this is closed under intersections, the Adjoint Function Theorem 3.6.9 gives $F \dashv U$, and we call $F(X)$ the **closure** of $X \subset \Sigma$. (We shall write $FX \in \mathcal{A}$ for the closure considered as a model or closed subset, and $MX \in \mathcal{S}$ for its underlying set.)

The left hand side of $K \rhd t$ is called the **arity** of this instance of the closure condition: the choice of letter reflects our use of k for the arity of operation-symbols. If this is always a finite set, we sometimes write it as a list, and say that $\rhd$ is **finitary**.

EXAMPLES 3·7·5:   There are many familiar **binary** closure conditions.

(a) If $\rhd$ is a functional relation such as

$$\{u, v\} \rhd u + v, \qquad i.e. \; \forall u, v{:}\Sigma. \; u, v \in A \Rightarrow u + v \in A,$$

then $\rhd$-closed subsets are subalgebras with respect to $+$.

(b) Convexity in an affine space: $\{u, v\} \rhd \lambda u + (1 - \lambda)v$ for $0 \leqslant \lambda \leqslant 1$.

(c) Convexity for an order: $\{u, v\} \rhd t$ whenever $u \leqslant t \leqslant v$.

(d) Transitivity is a closure condition on pairs (instances of $\leqslant$),

$$\{(x, y), (y, z)\} \triangleright (x, z),$$

rather than on elements. Reflexivity and symmetry are nullary and unary conditions on pairs: $\oslash \triangleright (x, x)$ and $\{(x, y)\} \triangleright (y, x)$.

(e) Ideals and normal subgroups are subgroups satisfying an extra unary closure condition: $\{u\} \triangleright ru$ or $\{u\} \triangleright a^{-1}ua$ respectively.

REMARK 3·7·6: A **nullary** closure condition $\oslash \triangleright t$ says that t must be in *every* closed set. If $\triangleright$ consists only of such conditions, for $t \in G$, then $F(X) = G \cup X$ and $\mathcal{A}$ is the upper set $G \downarrow \mathcal{S}$ (Example 3.1.6(f)).

For any system of closure conditions, in order to find the closure $F(G)$ of a set G (of **generators**), we may instead consider the extended system $\triangleright'$, with an extra nullary closure condition $\oslash \triangleright' x$ for each $x \in G$. We write $\triangleright +G$ for $\triangleright'$. Without loss of generality we therefore need only consider $F(\oslash)$, which is the *smallest* $\triangleright$-closed set, so long as we study closure conditions *in general*, for example in Corollary 3.9.2.

Unary closure conditions are considered in detail in the next section.

REMARK 3·7·7: In any situation $A \subset \Sigma$, we may always regard Σ as a set of propositions: each element $x : \Sigma$ names the proposition "$x \in A$". Then we can read any finitary closure condition

$$\{\phi_1, \ldots, \phi_k\} \triangleright \vartheta \qquad \text{as} \qquad \phi_1 \wedge \cdots \wedge \phi_k \Rightarrow \vartheta,$$

i.e. a propositional Horn clause (Remark 1.7.4). A nullary condition is called an axiom. Conversely, any (single, predicate) Horn clause is a *scheme* of closure conditions given by instantiating terms for its free variables. A system of finitary closure conditions is therefore also called a **Horn theory** (a system of Horn clauses involving *predicates* is the dependent type analogue, Chapter VIII).

In logic programming (Remark 1.7.2), the collection of terms generated by the constants and operation-symbols (of the underlying term calculus) is known as the **Herbrand base**, and the set Σ of all instances of the predicates is the **Herbrand universe**. A program is then a Horn theory.

Induction on closures. Each instance $K \triangleright t$ of a closure condition may be seen as an *introduction rule* for the (term t of) type $A = F(\oslash)$. Conversely there is an *elimination rule*, with a premise corresponding to each case of introduction. This manifests itself as an induction scheme, which is to propositions what **structural recursion** is to types.

DEFINITION 3·7·8: The **induction scheme** for a system of closure conditions ▷ is

$$\frac{\text{for all } K \vartriangleright t \quad \big(\forall u.\, u \in K \Rightarrow \vartheta[u]\big) \Rightarrow \vartheta[t]}{\forall a.\, a \in A \Rightarrow \vartheta[a]} \vartriangleright\text{-induction}$$

where $A = F(\aleph)$ is the smallest closed subset. This is valid because the premise says that $\Theta = \{x : \Sigma \mid \vartheta[x]\}$ is itself a ▷-closed subset.

EXAMPLES 3·7·9:

(a) By Remark 3.7.7, the induction scheme says that if the Horn clauses are sound then all of the propositions in $A = F(\aleph)$ are true.

(b) Structural induction is of this form, *e.g.* to prove associativity of concatenation of lists (Proposition 2.7.5), put

$$t \vartriangleright \mathsf{cons}(h, t) \quad \text{and} \quad \vartheta[x] = \big(\forall \ell_1, \ell_2.\, (x\,;\ell_1)\,;\ell_2 = x\,;(\ell_1\,;\ell_2)\big).$$

(c) Course-of-values induction on $\mathbb{N}$ is given by $\{0, 1, \ldots, n-1\} \vartriangleright n$.

(d) The correctness of a logic program is established by induction on the system of closure conditions which it codes. Conversely, any Horn theory has a procedural reading, the goal being to prove $x \in A$.

(e) For any binary relation $\prec$ on a set Σ, let $\{u \mid u \prec t\} \vartriangleright t$ for each $t \in \Sigma$. Then induction on ▷ as a closure condition is the same as well founded induction for $\prec$ (Definition 2.5.3).

(f) To recover $\prec$ from ▷, for each $t \in \Sigma$ there must be a *unique* set K with $K \vartriangleright t$. (It's called $\mathsf{parse}(t)$ in Example 6.3.3.) This typically fails for subalgebras, for example $\{u, v\} \vartriangleright u + v$ in a vector space does not allow the recovery of u and v. It is also the reason why backtracking is needed in logic programs (Remark 1.7.3).

Using results from earlier in this chapter, we can derive stronger idioms of induction for the set $A = F(\aleph)$. In particular, there is a "systematic" way of generating it.

LEMMA 3·7·10: Define $T : \mathcal{P}(\Sigma) \to \mathcal{P}(\Sigma)$ by

$$TX = \{t : \Sigma \mid \exists K.\, K \subset X \wedge K \vartriangleright t\},$$

i.e. the application of the conditions *once*. Then A is ▷-closed iff $TA \subset A$, and the smallest ▷-closed subset of Σ is the least fixed point of T. Every monotone endofunction T arises in this way, with $K \vartriangleright' t$ if $t \in TK$. Then any subset is ▷'-closed iff it is ▷-closed, though the original ▷ itself cannot be recovered from T. □

Exercises 3.40 and 3.42 give the induction scheme in terms of T. We can use induction on closure conditions to investigate the least fixed point.

PROPOSITION 3·7·11:   Let S be a complete (join-semi)lattice and T a monotone endofunction of it. Then T has a least fixed point $A \in S$, whose properties may be deduced from the following induction scheme:

$$\frac{\forall \mathcal{I}. \ \left(\forall U. U \in \mathcal{I} \Rightarrow \vartheta[U]\right) \Rightarrow \vartheta\left[\bigvee \mathcal{I}\right] \qquad \vartheta[U] \Rightarrow \vartheta\left[T(U)\right]}{\vartheta[A]}$$

PROOF: Define a closure condition on $\Sigma = S$ by $\mathcal{I} \triangleright \bigvee \mathcal{I}$ for all $\mathcal{I} \subset \Sigma$, and also $\{U\} \triangleright T(U)$. (Notice that we have *not* said $U \geqslant V \Rightarrow U \triangleright V$: see Exercise 3.41. Nor is this the closure condition derived from T in the lemma.) In particular $\oslash \triangleright \bot$, and any closed subset $\mathcal{A}$ has a greatest element $A = \bigvee \mathcal{A} \in \mathcal{A}$. By $\triangleright$-induction, $\forall X. X \in \mathcal{A} \Rightarrow X \leqslant T(X)$, whilst $\forall X. X \in \mathcal{A} \Rightarrow X \leqslant \vartheta$ for any fixed point $\vartheta = T(\vartheta)$. But $T(A) \in \mathcal{A}$ so $A = T(A)$ is the least fixed point of T. Other properties of A may be proved by instantiation of $\triangleright$-induction at $A \in \mathcal{A} = F(\oslash)$.   □

REMARK 3·7·12:   For *finitary* closure conditions, the function T defined in Lemma 3.7.10 is Scott-continuous, and $F(\oslash) = \bigcup_{n \in \mathbb{N}} T^n(\oslash)$ is the least fixed point by Proposition 3.3.11. Then in the induction scheme it is enough to consider countable directed subsets $\mathcal{I}$, or just ω-sequences. This idiom of induction is due to David Park (1976). When $S = \mathcal{P}(\Sigma)$, as in Lemma 3.7.10, this construction says that if $X \in F(\oslash) \subset \Sigma$ then $X \in T^n(\oslash)$ for some n. Classically, the least such n is called the **rank** of X, but in general this is not well defined intuitionistically.

It is a common but very clumsy idiom to prove something over a closed set by induction on the first time n when each element X gets in. For example, the rank of a goal in a logic program is the depth of its proof tree, which gives a lower bound on the execution time. But it is both difficult to calculate, and extremely crude, as the corresponding upper bound is k^n, where k is the maximum arity. As in Remark 2.5.7, the assumption that n is *least* is unnecessary: it would be clearer to use induction on the *closure* condition directly. By Lemma 3.7.10 these two idioms of induction are equally expressive.

We have seen that Tarski's Theorem can be proved without the Scott-continuity assumption, and so can properties of the least fixed point:

THEOREM 3·7·13:   Let $\triangleright$ be a (possibly infinitary) system of closure conditions on a set Σ and $\vartheta[X]$ any predicate on $S = \mathcal{P}(\Sigma)$ such that

(a) $\vartheta[\oslash]$ holds;

(b) if $\vartheta[X]$ holds and $X \supset K \triangleright t$ then $\vartheta[X \cup \{t\}]$ also holds;

(c) if $X = \bigcup X_i$ such that each $\vartheta[X_i]$ holds then $\vartheta[X]$ holds.

Then $\vartheta[A]$ holds, where A is the smallest $\triangleright$-closed subset.

PROOF: (Thierry Coquand) Let ϑ be the least such predicate. Consider
$$\phi[X] \;\equiv\; \forall Y.\, \vartheta[Y] \Rightarrow \vartheta[X \cup Y],$$
and check that it satisfies (a)–(c), so $\vartheta[X] \Rightarrow \phi[X]$, since ϑ is least. Then $\vartheta[X] \wedge \vartheta[Y] \Rightarrow \vartheta[X \cup Y]$, so ϑ preserves arbitrary unions by (a) and (c), and there is a greatest X_0 with $\vartheta[X_0]$ by the adjoint function theorem. By (b), X_0 is ▷-closed, so $A \subset X_0$. The predicate $\psi[X] \equiv (X \subset A)$ also satisfies the conditions, so $\vartheta[X] \Rightarrow \psi[X]$ and $A = X_0$. ☐

Classically, the least fixed point can be approached by a *transfinite* union, which we investigate in Section 6.7, but Exercise 3.40 defines an intrinsic notion of ordinal for a particular problem.

For the obvious reasons of constructivity, we are primarily interested in finitary algebra, and we shall find in Section 5.6 that there is a significant obstacle to the extension of equational reasoning to infinitary operations. Why, then, should we be interested in infinitary algebra at all? Since there is a duality between finiteness and (infinitary) directed joins, the more we concentrate on the finitary, the more we need to know about infinitary operations and Scott-continuous functions between domains.

Section 3.9 considers the lattice of models of any Horn theory, and the semilattice which "classifies" models, making use of induction on closure conditions. We shall study algebraic theories in Section 4.6, using closure conditions in Sections 5.6 and 7.4 to impose relations. Recursion for *free* algebraic theories is the subject of Chapter VI. Closure operations are generalised to reflective subcategories in Examples 7.1.6 and further to monads in Section 7.5.

3.8 MODALITIES AND GALOIS CONNECTIONS

This section characterises the covariant and contravariant adjunctions between full powerset lattices. In each case, the adjunction is fully determined by a binary relation between the sets, and gives rise to useful idioms in many areas of mathematics. We can say more about these idioms when the relation is transitive, functional, *etc.*

Modal logic. Consider the special case of a system of *unary* closure conditions, where we abbreviate $\{u\} \vartriangleright t$ to $u < t$. A subset is closed under such a condition iff it is an upper subset with respect to the reflexive–transitive closure $\leqslant$ of $<$. (Recall Warning 3.1.4, in particular $<$ need not be transitive.) The "one-way" closure of a point under a unary closure condition is called its ***trajectory***, and the "two-way" closure under a symmetric relation is known as the ***orbit***.

EXAMPLES 3·8·1:   The orbits of

(a) an equivalence relation are equivalence classes;

(b) $u < u^n$ or $u < nu$ in a monoid are cycles;

(c) $u < g^{-1} ; u ; g$ in a group are conjugacy classes.

These examples illustrate that the relation $<$ may arise naturally as a *function*, and we do not always replace it in our considerations by its transitive closure. For this reason we shall allow the relation $<: \mathbb{X} \rightharpoonup \mathbb{Y}$ to have different sets for its source and target.

Recall that, for finitary closure conditions, the functions U, M and T preserve directed joins, *i.e.* "all but" finitary joins (Exercise 3.14). In the unary case they preserve *all* joins, so have right adjoints, which we write as $\langle < \rangle \equiv T \dashv [>]$ and $\langle \leqslant \rangle \equiv M \dashv [\geqslant]$. It may help to think of $x < y$ as meaning that x is the present and y is a potential *future* world. Modal logic has medieval and even ancient roots, but its modern study was begun by Clarence Lewis (1918) and models based on order relations were first given by Saul Kripke (1963).

DEFINITION 3·8·2:   The **modal operators** are defined as follows:

$$\langle < \rangle V \;=\; \{x : \mathbb{X} \mid \exists y{:}\mathbb{Y}.\ x < y \in V\} = TV \qquad \text{``\textbf{\textit{possibly} V}''}$$
$$[<]V \;=\; \{x : \mathbb{X} \mid \forall y{:}\mathbb{Y}.\ x < y \Rightarrow y \in V\} \qquad \text{``\textbf{\textit{necessarily} V}''}$$

for $V \subset \mathbb{Y}$. If the relation is unambiguous, we just write $\Diamond$ and $\Box$. With the opposite relation, $\langle > \rangle U$ and $[>]U \subset \mathbb{Y}$ are similarly defined for $U \subset \mathbb{X}$. Other adverbs used for $\Box$ include **hereditarily** and **stably**.

EXAMPLE 3·8·3:   In Sections 4.3, 5.3 and 6.4 we shall show how to prove correctness of (imperative) programs by means of statements of the form $U \subset \langle < \rangle V$ and $U \subset [<]V$, *i.e.* if the initial state belongs to U then

(a) *some* execution which terminates does so with a final state in V, or

(b) *every* execution which terminates does so with a final state in V.

For deterministic programs, the first is simply that

(a′) the program *does* terminate, *and* the final state is in V,

which is called *total correctness*. For terminating deterministic programs, the statements are the same, but *partial correctness*, that

(b′) *if* the program terminates, *then* the final state is in V,

is more useful for **while** programs (Remark 6.4.16).

EXAMPLE 3·8·4: Let $H \subset G$ be a subgroup of a group, and $u < g^{-1} ; u ; g$ the conjugacy relation (Example 3.8.1(c)). Then the **core** of H,

$$\Box H = \{x : G \mid \forall g{:}G.\ g^{-1} ; x ; g \in H\},$$

is the kernel of the regular action of G on the cosets of H. □

Modal logic is the fragment of predicate calculus consisting of formulae with just one free variable. Since the predicate quantifiers $\exists$ and $\forall$ bind a variable, in order to stop the calculus from degenerating altogether, we allow the quantifiers $\Diamond$ and $\Box$ to introduce a new free variable for each bound one, by means of a binary relation $<$.

PROPOSITION 3·8·5: There are adjunctions

$$
\begin{array}{ccc}
\mathbb{X} & \mathcal{P}(\mathbb{X}) & \mathcal{P}(\mathbb{X}) \\
<\ \Big\uparrow & \langle<\rangle \Big\uparrow \dashv \Big\downarrow [>] & \langle>\rangle \Big\uparrow \dashv \Big\downarrow [<] \\
\mathbb{Y} & \mathcal{P}(\mathbb{Y}) & \mathcal{P}(\mathbb{Y})
\end{array}
$$

so $\Diamond$ preserves disjunction and $\Box$ preserves conjunction, and they obey the intuitionistic $(\neg\Diamond = \Box\neg)$ de Morgan law. They also satisfy

$$\Box U \cap \Diamond V \subset \Diamond(U \cap V),$$

which is the Frobenius law, *cf.* the *IAI* syllogism (Exercise 1.25). The relation $<$ can be recovered from any of the modalities (Exercise 3.65), and every (covariant) adjunction between powersets arises like this.

PROOF: $\langle<\rangle V \subset U$ and $V \subset [>]U$ say respectively that

$$\forall x.\ (\exists y.\ x < y \in V) \Rightarrow x \in U \quad \text{and} \quad \forall y.\ y \in V \Rightarrow (\forall x.\ x < y \Rightarrow x \in U),$$

which are equivalent. The Frobenius law is

$$\left(\forall y.\ x < y \Rightarrow y \in U\right) \land \left(\exists y.\ x < y \in V\right) \ \Rightarrow \ \left(\exists y.\ x < y \in U \cap V\right),$$

and $x < y \iff x \in \langle<\rangle\{y\} \iff y \in \langle>\rangle\{x\}$. $\Box$

The usual features of binary relations translate directly into properties of modal operators (Lemmas 3.8.8 and 3.8.12). Since we're interested in $<$-closed subsets, we consider preorders first, and functions later.

The transitive closure. The following account is due to Gottlob Frege (1879), and is the propositional analogue of the unary treatment of lists in Remark 2.7.10. It will be used for **while** programs in Section 6.4.

DEFINITION 3·8·6: Let $(<), \Theta : X \rightharpoonup X$ be binary relations, which need not be transitive (Warning 3.1.4). Instead of using the binary closure condition on pairs (x, y) that was used to axiomatise transitivity in Example 3.7.5(d), consider the nullary and unary ones

$$\varnothing \rhd (x, x) \text{ for all } x \quad \text{and} \quad \{(y, z)\} \rhd (x, z) \text{ whenever } x < y,$$

so Θ is $\rhd$-closed iff it is reflexive $(\Delta \subset \Theta)$ and $\forall x, y, z.\ x < y\, \Theta\, z \Rightarrow x \Theta z$.

We write $\leqslant$ for the smallest such relation (set $\Theta \subset \Sigma = X \times X$ of pairs); by Definition 3.7.8 it satisfies an induction scheme of the form

$$\frac{\Delta \subset \Theta \qquad (<)\,;\Theta \subset \Theta}{(\leqslant) \subset \Theta}$$

This corresponds to unary Peano induction for cons and listrec (Remark 2.7.10). We now show that $\leqslant$ *with this definition* is the smallest reflexive–transitive relation which contains $<$. The binary closure condition which states transitivity and its associated induction scheme correspond to **append** and **fold** for lists in Definition 2.7.4ff.

PROPOSITION 3·8·7:

(a) $\leqslant$ is transitive,

(b) it satisfies a base/step parsing rule (*cf.* empty/head+tail),

$$x \leqslant z \iff x = z \ \vee \ \exists y.\, x < y \leqslant z,$$

(c) and a binary induction scheme (*i.e.* it is the transitive *closure*),

$$\frac{\Delta \subset \Theta \qquad (<) \subset \Theta \qquad \Theta\,;\Theta \subset \Theta}{(\leqslant) \subset \Theta}$$

PROOF: These follow from the unary induction scheme for various Θ.

[a] Consider $x\Theta y \equiv (\forall z.\, y \leqslant z \Rightarrow x \leqslant z)$.

[b] Put Θ for the right hand side, so $\Delta \subset \Theta$ and

$$w < x \Theta z \iff w < x = z \ \vee \ \exists y.\, w < x < y \leqslant z,$$

which implies $w < x \leqslant z$ and so $w\Theta z$.

[c] From the premise of the binary rule we have $(<)\,;\Theta \subset \Theta\,;\Theta \subset \Theta$, so the unary rule applies. $\square$

Modal logic for preorders. This insight into the relationship between $[<]$ and $[\leqslant]$ will enable us to complete some unfinished business from Section 2.6, where we promised that modal logic would greatly facilitate the study of well-foundedness.

LEMMA 3·8·8: A binary (endo)relation $<$ on a set Σ is

(a) reflexive iff $>$ is reflexive iff $\mathrm{id} \leqslant \langle < \rangle$ iff $\mathrm{id} \geqslant [<]$, and

(b) transitive iff $>$ is transitive iff $\langle < \rangle^2 \leqslant \langle < \rangle$ iff $[<]^2 \geqslant [<]$.

So for a preorder $\leqslant$, $\langle \leqslant \rangle = M$ is a closure and $[\leqslant]$ a coclosure.

The propositional calculus extended with these two modalities arising from a preorder is known as the *modal logic S4*. Both $\square$ and $\Diamond$ split into adjunctions, which combine with those which we identified earlier.

REMARK 3·8·9: For *any* subset $U \subset \Sigma$, $\langle\leqslant\rangle U$ is the down-closure of U and $[\geqslant]U$ is the largest lower subset contained in U, so these closure and coclosure operations have the same image.

$$
\begin{array}{cc}
\mathcal{P}(\Sigma) & \mathcal{P}(\Sigma) \\
\langle\leqslant\rangle \left|\dashv\right|\dashv\left|[\geqslant]\right. & \langle\geqslant\rangle\left|\dashv\right|\dashv\left|[\leqslant]\right. \\
\mathcal{A} = \mathsf{sh}(\Sigma, \leqslant) & \mathsf{sh}(\Sigma, \geqslant)
\end{array}
$$

Moreover every complete sublattice $\mathcal{A} \subset \mathcal{P}(\Sigma)$ arises in this way. □

An informal way of putting this is that, whereas any closure operation rounds *up*, in the unary case we can also round *down* to a closed subset.

The lattices $\mathsf{sh}(\Sigma, \leqslant)$ and $\mathsf{sh}(\Sigma, \geqslant)$ are the covariant and contravariant regular representations (Example 3.1.6(f)ff) of $\leqslant$, or of any relation $<$ of which it is the reflexive–transitive closure.

Now we turn to well founded relations, for which we need

LEMMA 3·8·10: The parsing rule for the transitive (irreflexive) closure is

$$v \twoheadleftarrow t \iff v \prec t \vee \exists u.\, v \prec u \twoheadleftarrow t.$$

So the transitive $\twoheadleftarrow$ and reflexive–transitive $\leqslant$ closures of any binary relation $\prec$ satisfy the parsing formulae

$$[\twoheadrightarrow] = [\twoheadrightarrow] \circ [\succ] \wedge [\succ] \quad \text{and} \quad [\geqslant] = [\geqslant] \circ [\succ] \wedge \mathsf{id}.$$

Moreover $[\twoheadrightarrow] \circ [\succ] = [\succ] \circ [\twoheadrightarrow]$ and $[\geqslant] \circ [\succ] = [\succ] \circ [\geqslant]$. □

THEOREM 3·8·11:

(a) The well founded **induction scheme** (Definition 2.5.3) is equivalent to its **strict** ($\iff$) form.

$$\frac{\forall t.\, \bigl(\forall u.\, u \prec t \Rightarrow \vartheta[u]\bigr) \iff \vartheta[t]}{\forall x.\, \vartheta[x]} \qquad\qquad \frac{[\succ]\vartheta \iff \vartheta}{\vartheta}$$

(b) The transitive closure of a well founded relation is also well founded.

PROOF: Put $\psi = [\twoheadrightarrow]\vartheta$ (*hereditarily* ϑ, Definition 2.5.4) in both cases.

[a] Suppose that ϑ satisfies the lax $\prec$-induction premise, $[\succ]\vartheta \Rightarrow \vartheta$, and the strict $\prec$-induction scheme is valid: if $[\succ]\psi \iff \psi$ then $\forall x.\, \psi[x]$. So $[\succ][\twoheadrightarrow]\vartheta = [\twoheadrightarrow][\succ]\vartheta \Rightarrow [\twoheadrightarrow]\vartheta$ by the lax $\prec$-premise, but by parsing $[\twoheadrightarrow]\vartheta = [\succ][\twoheadrightarrow]\vartheta \wedge [\succ]\vartheta \Rightarrow [\succ][\twoheadrightarrow]\vartheta$, so $\psi \equiv [\twoheadrightarrow]\vartheta \iff [\succ]\psi$. Hence ψ holds by strict $\prec$-induction, and *a fortiori* so does $[\succ]\vartheta$, whence ϑ follows by the lax $\prec$-premise again. (The other way is trivial.)

[b] Suppose that ϑ satisfies the $\twoheadleftarrow$-induction premise, $[\twoheadrightarrow]\vartheta \Rightarrow \vartheta$, and the $\prec$-induction scheme is valid, *i.e.* if $[\succ]\psi \Rightarrow \psi$ then $\forall x.\, \psi[x]$.

Parsing says $[\succ\!\!\!\!-]\vartheta = [\succ][\succ\!\!\!\!-]\vartheta \wedge [\succ]\vartheta$, but this is just $[\succ][\succ\!\!\!\!-]\vartheta$, since the first term implies the second by the $\prec\!\!\!\!-$-premise. So $\psi \equiv [\succ\!\!\!\!-]\vartheta$ holds by $\prec$-induction, so ϑ follows by the $\prec\!\!\!\!-$-premise again.   □

Functions and quantifiers. Since modal logic is its unary fragment, in order to recover the predicate calculus we must make use of pairs to encode many-place predicates. Consider $(\gamma, x) < \gamma$, the product projection $\pi_0 = \hat{x} : \Gamma \times X \longrightarrow \Gamma$. For a context Γ consisting of typed variables, recall that $\mathsf{Cn}(\Gamma)$ is the set of formulae in these variables. For an extra variable x, there is an inclusion $\hat{x}^* : \mathsf{Cn}(\Gamma) \subset \mathsf{Cn}\big([\Gamma, x : X]\big)$ called *weakening* (Remark 2.3.8). Similar results hold for any functional relation in place of π_0.

LEMMA 3·8·12:   From Definition 1.3.1, $<: X \rightharpoonup Y$ is

(a) functional (single valued) iff $\langle<\rangle \leqslant [<]$, iff $\langle<\rangle$ preserves binary $\wedge$, iff $[<]$ preserves binary $\vee$, iff $< \circ > \ \leqslant \Delta$, iff $\mathsf{id} \leqslant [>] \circ [<]$, iff $\langle>\rangle\circ\langle<\rangle \leqslant \mathsf{id}$, where Δ is the identity relation on Y, *cf.* Lemma 1.2.11 and Exercise 1.16, and

(b) total (entire) iff $[<] \leqslant \langle<\rangle$, iff $\langle<\rangle$ preserves $\top$, iff $[<]$ preserves $\bot$, iff $\Delta_X \leqslant \ > \circ <$, iff $[<] \circ [>] \leqslant \mathsf{id}$, iff $\mathsf{id} \leqslant \langle<\rangle \circ \langle>\rangle$.

Injectivity and surjectivity are characterised by similar conditions on the opposite relation.   □

REMARK 3·8·13: $\exists x \equiv \langle \pi_0^{\mathsf{op}} \rangle$ and $\forall x \equiv [\pi_0^{\mathsf{op}}]$.

(a) The adjoints to weakening are the quantifiers, *cf.* Remark 1.5.5.

$$\begin{array}{ccc} \Gamma \times X & \quad & \mathsf{Cn}\big([\Gamma, x : X]\big) \\[4pt] \hat{x} \Big\downarrow \pi_0 & \quad & \exists x \Big\vert \dashv \hat{x}^* \dashv \Big\vert \forall x \\[4pt] \Gamma & \quad & \mathsf{Cn}(\Gamma) \end{array}$$

Consequently $\exists x.\ -$ and $\forall x.\ -$ preserve joins and meets respectively.

(b) For any function, $\langle f \rangle = [f] = f^*$ is the inverse image map, and this has adjoints on both sides, the *guarded quantifiers*, *cf.* Remark 1.5.2.

$$\langle f^{\mathsf{op}} \rangle = f_! = \exists_f \ \dashv \ [f] = f^* = \langle f \rangle \ \dashv \ [f^{\mathsf{op}}] = f_* = \forall_f$$

Such adjunctions give a deeper analysis of the quantifiers in Chapter IX.

Galois connections. Evariste Galois's name was given to Definition 3.6.1(b), by Øystein Ore, not because he spent his short life (1811–32) considering such definitional minutiae, but because the correspondence between intermediate fields and subgroups of the Galois group of a field extension (Example 3.8.15(j)) was the first such situation known.

But the basic properties of the correspondence do not at all depend on groups and fields, so they are repeatedly re-proved in the literature. In fact *any* binary relation $\perp: \mathbb{S} \rightharpoonup \mathbb{A}$ gives rise to a Galois connection. It must not be confused with bottom or falsity, although it often has a negative connotation: for $x \perp a$ we read "x is **orthogonal** to a".

PROPOSITION 3·8·14: For $X \subset \mathbb{S}$ and $A \subset \mathbb{A}$,
$$X^{\perp} = F(X) = \{a \mid \forall x \in X.\, x \perp a\}$$
$$^{\perp}A = U(A) = \{x \mid \forall a \in A.\, x \perp a\}$$
form a Galois connection $\mathcal{S} \rightleftarrows \mathcal{A}$, where $\mathcal{S} = \mathcal{P}(\mathbb{S})$ and $\mathcal{A} = \mathcal{P}(\mathbb{A})$. (This is what suggested the notation $\rightharpoonup$ for relations in Section 1.3.) Notice that the implication is in the opposite direction from that defining $[<]$. By Lemma 3.6.2,
$$X^{\perp} = \left(^{\perp}(X^{\perp})\right)^{\perp} \quad \text{and} \quad {}^{\perp}A = {}^{\perp}((^{\perp}A)^{\perp}).$$
So the composites $X \mapsto {}^{\perp}(X^{\perp})$ and $A \mapsto (^{\perp}A)^{\perp}$ are closure operations, and there is an order-reversing bijection
$$\{X \in \mathcal{S} \mid X = {}^{\perp}(X^{\perp})\} \underset{^{\perp}(-)}{\overset{(-)^{\perp}}{\underset{\cong}{\rightleftarrows}}} \{A \in \mathcal{A} \mid A = (^{\perp}A)^{\perp}\}$$

Conversely, any Galois connection between powersets is of this form:
$$a \in F(\{x\}) \iff x \perp a \iff x \in U(\{a\}).$$
A Galois connection is often presented as the lower set
$$\{\langle X, A\rangle \mid (X \leqslant {}^{\perp}A) \equiv (A \leqslant X^{\perp}) \equiv (\forall x \in X.\, \forall a \in A.\, x \perp a)\},$$
which is closed under unions in each component (separately).

The $\perp$-closed subsets on either side are also closed under any algebraic operations or other closure conditions that $\perp$ respects: for example they are *automatically* subgroups, subfields, *etc.*

We may also define a **specialisation order** on $\mathbb{S}$, *cf.* Example 3.1.2(i),
$$x \leqslant y \quad \text{if} \quad \forall a : \mathbb{A}.\ x \perp a \Rightarrow y \perp a,$$
which is reflexive and transitive. If this is antisymmetric (a poset) we say that there are **enough** $\mathbb{A}$s (*viz.* to distinguish the $\mathbb{S}$s). $\qquad\square$

EXAMPLES 3·8·15:

(a) (Abraham Lincoln, 1858) "You can fool all the people some of the time, and some of the people all the time, but you cannot fool all the people all the time."

(b) Let $\mathbb{S} = \mathbb{A}$ be any set and $x \perp a$ the inequality relation, $x \neq a$, on $\mathbb{S}$. Then $X^{\perp} = \mathbb{S} \setminus X$, and there are enough $\mathbb{A}$s iff $\neg\neg(x = y) \Rightarrow x = y$ (a presheaf $\mathbb{S}$ with this property is **separable** in sheaf theory).

(c) Let $\mathbb{S} = \mathbb{A}$ be a preorder and $(\perp) \equiv (\leqslant)$ its order relation. Then $x \in {}^{\perp}A$ iff x is a lower bound for $A \subset \mathbb{S}$, and similarly $a \in X^{\perp}$ iff a is an upper bound. Then each $X^{\perp}$ is an upper set and each ${}^{\perp}A$ is a lower set. There are enough $\mathbb{A}$s iff $\leqslant$ is antisymmetric, and $\mathbb{S}$ is a complete lattice iff every closed set is of the form ${}^{\perp}\{a\}$ or $\{x\}^{\perp}$. For $\mathbb{S} = \mathbb{Q} \cap [0, 1]$, the closed subsets are one-sided Dedekind cuts (*cf.* Remark 2.1.1).

(d) Let $\mathbb{S}$ be a collection of "individuals" and $\mathbb{A}$ be some "properties" with $(\perp) \equiv (\vDash)$ the satisfaction relation. The set $\{x\}^{\perp}$ of properties of an individual is closed with respect to the logical connectives. If there are *enough properties* then $\{x\}^{\perp}$ is a *description* and **Leibniz' principle** holds, *cf.* Proposition 2.8.7. The specialisation order on properties $(\vDash)$ is *semantic entailment*, and it coincides with $\vdash$ iff there are *enough models*, *i.e.* the system is *complete* (Remark 1.6.13).

(e) Let $\mathbb{S}$ be the set of points and $\mathbb{A}$ the set of lines in the plane, with $\perp$ the incidence relation. Then ${}^{\perp}\{a\}$ is the set of points which lie on a line a and $\{x\}^{\perp}$ the pencil of lines passing through a point x.

(f) As a special case of (d), let $\mathbb{A}$ be a topology on $\mathbb{S}$ and $x \perp a$ the relation that a point belongs to an open set; this even-handed view is the one taken in [Vic88]. Then $X^{\perp}$ consists of the neighbourhoods of the subset X and is closed under arbitrary unions and finite intersections. ${}^{\perp}\{a\}$ is the extent (set of points) of an open set. A spatial locale is one with enough points, a T_0 space has enough open sets (*cf.* Proposition 3.4.9).

(g) In topology *à la* Fréchet, let $x \perp a$ be instead the relation $x \notin a$. This respects convergence of sequences (accumulation points) in $\mathbb{S}$ and arbitrary unions in $\mathbb{A}$.

(h) Let $\mathbb{S}$ be a vector space and $\mathbb{A} = (\mathbb{S} \multimap K)$ the dual space, with $x \perp a$ if $a(x) = 0$. Then $X^{\perp}$ and ${}^{\perp}A$ are both called *annihilators*; they are closed under addition and scalar multiplication.

(i) Let $\mathbb{S} = \mathbb{A}$ be a group and $x \perp a$ the property that they commute. Then $X^{\perp}$ is the *centraliser subgroup* for a subset X.

(j) Let $\mathbb{S}$ be a field of numbers, $\mathbb{A}$ its group of automorphisms and $\perp$ the relation that the automorphism $a \in \mathbb{A}$ fixes the number $x \in \mathbb{S}$, *i.e.* $a(x) = x$. The pair (F, U) is the original Galois connection. Each $X^{\perp}$ is a subgroup which is closed in a certain topology, whilst ${}^{\perp}A$ is a subfield such that if $x^p \in {}^{\perp}A$ then $x \in {}^{\perp}A$, where $p \neq 0$ is the characteristic of the field.

For us, the most important example of a Galois connection will be that defining the factorisation of functions into epis and monos in Section 5.7, which we shall use to study the existential quantifier in Sections 5.8 and 9.3. Unary theories are considered further in Sections 4.2 and 6.4, but the unary closure condition that most concerns us is invariance under substitution or pullback in Chapter VIII.

3.9 CONSTRUCTIONS WITH CLOSURE CONDITIONS

We shall now use closure conditions to fashion the first of our "exotic" worlds containing a generic model, and use it to prove completeness of Horn logic. Actually, it's not very exotic, but really rather common: we show that *every* semilattice arises in this way, for which we have to set out the so-called "internal language". The aim is to make syntax and semantics *equivalent*. We prefer to call the language *canonical*,[6] because the construction first arose in sheaf theory; in particular, the most important closure condition is that under arbitrary unions, called the **canonical coverage** of a (complete) semilattice. Example 3.8.15(g) showed that this is the Galois dual of the convergence of sequences in analysis with which we opened Section 3.7.

As sheaf theory is outside the scope of this book, we just develop it far enough for lattices to illustrate the connection. The notion of stable saturated (Exercise 3.36) coverage (Definition 3.9.6) is the analogue for posets of a Grothendieck topology on a category, which is itself an intuitionistic version of the *forcing* technique invented by Paul Cohen (1963) to show independence of Cantor's continuum hypothesis. Saul Kripke used similar ideas to model intuitionistic logic (1965). For a full account, see [MLM92]; the propositional analogue of this subject, the theory of locales, is expertly described in [Joh82].

The Lindenbaum algebra for conjunction.

THEOREM 3·9·1: For any Horn theory $\mathcal{L} = (\Sigma, \triangleright)$ there is a semilattice $\mathsf{Cn}^\wedge_{\mathcal{L}}$ which **classifies** its models in the sense that there is a bijection

$$\frac{\text{semilattice homomorphisms } \mathsf{Cn}^\wedge_{\mathcal{L}} \to \mathsf{2}}{\text{models ($\triangleright$-closed subsets) } A \subset \Sigma}$$

$\mathsf{Cn}^\wedge_{\mathcal{L}}$ is the *free* semilattice on Σ subject to the relations which $\triangleright$ codes.

---

[6] Although it is written in an abstract "alphabet", the language is not *internal* in the formal sense of Example 4.6.3(f). See Sections 4.2, 4.6, 6.2 and 7.6 for more discussion of this point. The word *vocabulary* would perhaps be better, as by the "language" $\mathcal{L}$ we mean the collection of symbols, axioms, *etc.* peculiar to a particular theory. The "grammar" which generates all well formed sentences from the vocabulary by means of connectives, quantifiers, *etc.* (generically written □) is called a **fragment** or **doctrine** of logic.

PROOF: The *elements* of $\text{Cn}_{\mathcal{L}}^{\wedge}$ are contexts Γ, *i.e.* lists or finite subsets of elements of Σ considered as conjunctions of propositions. The preorder relation $\vdash$ is generated by (*i.e.* is the reflexive–transitive closure of)

(a) weakening: $\Gamma, \phi \vdash \Gamma$, so $\Delta \vdash \Gamma$ whenever Γ is a subset of Δ, and

(b) the axioms: $K \vdash \phi$ if $K \rhd \phi$.

Recall that, for us, lists on *both* sides of the turnstile mean conjunction, not disjunction. So $\Gamma \vdash \Delta$ iff each proposition ϑ in Δ has a proof tree with root ϑ whose leaves are certain subsets of Γ and whose nodes are instances of the $\rhd$ relation. Remark 1.7.2 showed how to find such a proof. The union of two contexts defines the meet in this preorder: we can obtain a semilattice in the sense of Definition 3.2.12 as a quotient using Proposition 3.1.10.

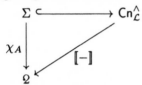

In the presence of a subset $A \subset \Sigma$, Remark 3.7.7 gave a propositional meaning, which we now write as $[\![\phi]\!]$, to each element $\phi \in \Sigma$. We extend this to contexts by conjunction, so that $[\![-]\!] : \text{Cn}_{\mathcal{L}}^{\wedge} \to \mathcal{Q}$ is a semilattice homomorphism. At least it is so long as we ensure that it is monotone. We have to check the two generating cases of $\vdash$, but clearly $[\![\Gamma, \phi]\!] \Rightarrow [\![\Gamma]\!]$. For the other case, A obeys the closure condition $K \rhd \phi$ iff $[\![K]\!] \Rightarrow [\![\phi]\!]$, so A has to be a model. The restriction to single-proposition contexts recovers (the characteristic function χ_A of) the model from $[\![-]\!]$.   □

COROLLARY 3·9·2: Horn logic is complete in the sense of Remark 1.6.13.

PROOF: The smallest $\rhd$-closed subset $A = F(\varnothing) \subset \Sigma$ consists of the $\rhd$-provable propositions, *i.e.* $\phi \in A \iff (\vdash_{\mathcal{L}} \phi)$. All such ϕ are identified with $\top \in \text{Cn}_{\mathcal{L}}^{\wedge}$ when the order on $\text{Cn}_{\mathcal{L}}^{\wedge}$ is made antisymmetric. But A is also a model, satisfying exactly those propositions that are true in all models, since it *is* the set of propositions which are *in* all models.

We deduce completeness by varying $\rhd$. Suppose that $\Gamma \vDash_{\mathcal{L}} \vartheta$, *i.e.* for every model $\Gamma \subset A \subset \Sigma$, also $\vartheta \in A$. In particular $\vartheta \in A = F(\Gamma)$. Consider the theory $\rhd +\Gamma$ in which the propositions in Γ are added to $\rhd$ as axioms (Remark 3.7.6). Now A is the smallest closed subset (model) of $\rhd +\Gamma$, so ϑ is provable in this theory, *i.e.* from the hypotheses Γ in the theory $\rhd$. (Notice the change of demarcation of Γ, *cf.* page 3.)   □

By the completeness theorem, $\vdash$ coincides with the semantic entailment $\vDash$ in Example 3.8.15(d); then since there are enough models, to prove ϑ

we need only show (classically) that there is no **counterexample**, *i.e.* a model which distinguishes ϑ from $\top$.

This way of constructing the relation $\vdash$ in $\mathsf{Cn}^{\wedge}_{\mathcal{L}}$, as the reflexive–transitive closure of two generating cases, will be developed in Sections 4.3 and 8.2.

As promised, we have a stronger completeness result:

PROPOSITION 3·9·3: Every semilattice $\mathcal{C}$ classifies some Horn theory.

PROOF: In the **canonical language** $\mathcal{L} = \mathsf{L}^{\wedge}(\mathcal{C}) = (\Sigma, \triangleright)$, Σ is the set of elements of the semilattice $\mathcal{C}$, with the closure conditions $\varnothing \triangleright \top$ and $\{u, v\} \triangleright u \wedge v$. The elements of the *preorder* $\mathsf{Cn}^{\wedge}_{\mathcal{L}}$ are finite subsets of Σ, with $\Gamma \vdash_{\mathcal{L}} \Delta$ iff $\bigwedge \Gamma \leqslant_{\mathcal{C}} \bigwedge \Delta$, so the quotient poset is the given $\mathcal{C}$. □

Beware that if $\mathcal{C} = \mathsf{Cn}^{\wedge}_{\mathcal{L}_0}$ was the classifying semilattice for some Horn theory $(\Sigma_0, \triangleright_0)$, then the new Σ is bigger than Σ_0 and the systems of closure conditions are also different.

Algebraic lattices. Now we consider the lattice $\mathcal{M}od(\mathcal{L})$ of models of a finitary Horn theory. The classifying semilattice is static, and loses the dynamic information in the original theory; as the lattice of models is also static, there is no harm in using the classifying semilattice $\mathcal{C}$ to represent the theory in the next result.

THEOREM 3·9·4: The models of a finitary Horn theory form an algebraic lattice. Every algebraic lattice arises uniquely in this way, in the sense that the classifying semilattice is unique up to unique isomorphism.

PROOF: By Proposition 3.7.3, any directed union of models ($\triangleright$-closed subsets) is a model, and a model is finitely generable in the sense of Definition 3.4.11ff iff it is the closure of some finite subset. By Theorem 3.9.1, models of $\mathcal{C}$ correspond to semilattice homomorphisms $\mathcal{C} \to \mathbf{2}$, and so to upper subsets containing $\top$ and closed under $\wedge$. Since such subsets are *ideals* of $\mathcal{C}^{\mathrm{op}}$ (Example 3.4.2(d)),

$$\mathcal{SL}at(\mathcal{C}, \mathbf{2}) \cong \mathsf{Idl}(\mathcal{C}^{\mathrm{op}}) \cong \mathcal{M}od(\mathcal{L}) \equiv \mathcal{A} \cong \mathsf{Idl}(\mathcal{A}_{\mathrm{fg}}),$$

using Proposition 3.4.12. Hence we recover $\mathcal{C}$ as $(\mathcal{A}_{\mathrm{fg}})^{\mathrm{op}}$. □

COROLLARY 3·9·5: There is an order-reversing bijection between finitely generated models $A \in \mathcal{A}_{\mathrm{fg}}$ and contexts $\Gamma \in \mathsf{Cn}^{\wedge}_{\mathcal{L}}$. □

This justifies the name **algebraic lattice**: recall that subalgebras and congruences were described by Horn theories. In fact any algebraic lattice arises as the lattice of subalgebras of some algebra for some theory.

Any system of (possibly infinitary) closure conditions has a complete lattice of closed subsets. This lattice is algebraic — characterised in terms of *directed joins* — iff every instance $K \triangleright t$ of the closure condition contains a *finite* condition $K' \subset K$ with $K' \triangleright t$. Given that directedness can be defined by a nullary and a binary condition, this result sheds some light on the notion of finiteness, but we shall defer a full discussion (using closure conditions) to Section 6.6.

Adding and preserving joins. Now we turn our study of closure conditions back onto the order theory from which it came, repeating for arbitrary joins the treatment which we have just given to finite meets. *Beware that we have dropped stability under meets from the way in which the following ideas are usually presented.* Recall that $\mathcal{P}(\Sigma)$ has arbitrary joins, which in fact it freely adds to the set Σ. Similarly, $\mathsf{sh}(\Sigma, \leqslant)$, which consists of the $\leqslant$-lower subsets of a poset $(\Sigma, \leqslant)$, freely adds joins respecting the order $\leqslant$ (Proposition 3.2.7(b)).

Now we want to *force* some of the joins to have particular values, and in the extreme case retain all joins which already exist in Σ. This can be done with the closure condition $K \triangleright \bigvee K$ which we have already met in Proposition 3.7.11, Example 3.8.15(g) and (for $\wedge$) Proposition 3.9.3.

The order relation $x \leqslant y$ can be coded using joins ($x \vee y = y$), and so by a closure condition, as in the previous section. However, we prefer to take the lattice $\mathsf{sh}(\Sigma, \leqslant)$ of $\leqslant$-lower subsets as our raw material.

DEFINITION 3·9·6: Let $(\Sigma, \leqslant)$ be a poset. A closure condition $\triangleright$ on Σ

(a) is a *coverage* if whenever $K \triangleright t$ we already have $K \leqslant t$,

(b) and is *subcanonical* if whenever $K \triangleright t$ and $K \leqslant x$ then $t \leqslant x$; in particular, if $x = \bigvee K$ then $t = x$, so the coverage is *only* used to nominate actual joins for preservation by the embedding η below.

(c) The *canonical coverage* of a complete join-semilattice Σ has $K \triangleright \bigvee K$ for all $K \subset \Sigma$.

(d) A $\triangleright$-closed $\leqslant$-lower subset $A \subset \Sigma$ is called a $\triangleright$-*sheaf* or $\triangleright$-*ideal*:
$$x \leqslant a \in A \Rightarrow x \in A \qquad K \triangleright t \,\&\, K \subset A \Rightarrow t \in A.$$

In particular a coverage $\triangleright$ is subcanonical iff every representable lower subset $\Sigma \downarrow x$ (Definition 3.1.7) is a $\triangleright$-sheaf.

THEOREM 3·9·7: Let $\triangleright$ be a subcanonical coverage for a poset $(\Sigma, \leqslant)$. Write $\mathcal{A} = \mathsf{sh}(\Sigma, \leqslant, \triangleright)$ for the lattice of sheaves; these are the elements of $\mathsf{sh}(\Sigma, \leqslant)$ which "think" that t is the join of K.

For each $x \in \Sigma$, the set $\eta(x) \equiv \Sigma \downarrow x \equiv \{a : \Sigma \mid a \leqslant x\}$ belongs to $\mathcal{A}$, and $\eta : \Sigma \to \mathcal{A}$ is monotone and obeys $K \triangleright t \Rightarrow \eta(t) = \bigvee_{\mathcal{A}} \{\eta(u) \mid u \in K\}$.

Moreover it is universal: let Θ be another complete join-semilattice and $f : \Sigma \to \Theta$ a monotone function such that $f(t) = \bigvee_{\Theta} \{f(u) \mid u \in K\}$

whenever $K \rhd t$. Then there is a unique function $p : \mathcal{A} \to \Theta$ which preserves all joins and satisfies $\eta \,; p = f$.

The map η is also full and preserves arbitrary meets.

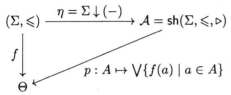

PROOF: For a subcanonical coverage, each $\Sigma \downarrow x$ is $\rhd$-closed, whence η is full and preserves arbitrary meets by Proposition 3.2.7(a). Similarly $p(\Sigma \downarrow x) = \bigvee \{ f(a) \mid a \leqslant x \} = f(x)$ by monotonicity of f. If the mediator p exists then it must be given by the formula in the diagram, since $A = \bigvee_{\mathcal{A}} \{ \Sigma \downarrow a \mid a \in A \}$. There is a right adjoint to p, which is given by $\vartheta \mapsto \{ a \mid f(a) \leqslant \vartheta \}$, so p preserves joins.

We can also show that p preserves joins by $\rhd$-induction. Let $A_i \in \mathcal{A}$ for $i \in \mathcal{I}$ such that $\forall i.\, p(A_i) \leqslant \vartheta \in \Theta$. We have to show that $f(a) \leqslant \vartheta$ for all $a \in A = \bigvee_{\mathcal{A}} A_i$, given that it holds for all $a \in A_i$. To satisfy Definition 3.7.8 we need $\big(\forall u.\, u \in K \Rightarrow f(u) \leqslant \vartheta \big) \Rightarrow f(t) \leqslant \vartheta$ whenever $K \rhd t$, but this was the hypothesis on f. □

EXAMPLES 3·9·8:

(a) Let $\diameter \rhd \perp$ and $\{u, v\} \rhd u \vee v$ for each $u, v \in \Sigma$ (a join-semilattice). Then A is in $\mathcal{A}$ iff it is a directed lower set, so $\mathcal{A} = \mathsf{Idl}(\Sigma)$, which is an algebraic lattice (Definitions 3.4.1 and 3.4.11). The map $\Sigma \hookrightarrow \mathsf{Idl}(\Sigma)$ preserves finite joins by construction, but if Σ also has meets (*i.e.* it is a lattice) then these are preserved too, *cf.* Exercise 3.17. If Σ is a distributive lattice then $\mathsf{Idl}(\Sigma)$ is a frame.

(b) Let $K \rhd t$ if K is directed with join t. Then A is $\rhd$-closed iff it is Scott-closed (Definition 3.4.7). In fact $\mathcal{A} = \mathcal{A}^+ \cup \{ \diameter \}$, where $\mathcal{A}^+$ is defined in the same way but with K *semi*directed (so $\diameter \rhd \perp$ and $\perp \in A$ for all $A \in \mathcal{A}^+$). $\mathcal{A}^+$ is called the **Smyth powerdomain**.

(c) Let $K \rhd t$ if K is inhabited with join t. Then any inhabited subset A is $\rhd$-closed iff it has a greatest element, but $\diameter$ is also $\rhd$-closed. Then $\mathcal{A}$ is the *lifting* $\Sigma_\perp$ for complete semilattices, *cf.* Definition 3.3.7.

Joins which are stable under meet. In conclusion, we restore the distributivity condition which was stripped from Theorem 3.9.7.

THEOREM 3·9·9: Let $\rhd$ be a subcanonical coverage on a semilattice $(\Sigma, \leqslant)$. Suppose that it is **stable**, *i.e.*

whenever $K \rhd t$ and $t' \leqslant t$ then $\{ u \wedge t' \mid u \in K \} \rhd t'$.

Then $\mathcal{A} \equiv \mathsf{sh}(\Sigma, \leqslant, \rhd)$ is a frame (Definition 3.6.16) and the left adjoint F of its inclusion in $\mathsf{sh}(\Sigma, \leqslant)$ preserves finite meets. If Θ is also a frame and $f : \Sigma \to \Theta$ preserves finite meets then so does $p : \mathcal{A} \to \Theta$.

PROOF: We always have $F(X \cap Y) \subset FX \cap FY$. A typical element of the right hand side is $a \wedge b$ with $a \in FX$ and $b \in FY$, so consider

$$X \times Y \subset C = \{(a,b) : \Sigma^2 \mid (a \wedge b) \in F(X \cap Y)\}.$$

We use a double induction to show that $C = FX \times FY$. Suppose $J \rhd a$, $K \rhd b$ with $J \times K \subset C$. Then $J_k = \{j \wedge k \mid j \in J\} \subset F(X \cap Y)$, but $J_k \rhd a \wedge k$ by stability, so $a \wedge k \in F(X \cap Y)$, since this is closed. Therefore $K_a = \{a \wedge k \mid k \in K\} \subset F(X \cap Y)$, but again $K_a \rhd a \wedge b$ by stability, so $a \wedge b \in F(X \cap Y)$, *i.e.* $(a,b) \in C$ as required.

Since $\mathsf{sh}(\Sigma, \leqslant)$, whose elements we call *presheaves*, is a frame, and p preserves joins, algebraic manipulation easily shows that $\mathcal{A}$ is also a frame, and p preserves finite meets if f does and Θ is a frame.   □

EXAMPLES 3·9·10:

(a) Let $\mathbb{A} \subset \mathbb{X}$ be any subset of a topological space, with the subspace topology, and $F : \mathcal{S} \twoheadrightarrow \mathcal{A}$ the inverse image induced by the inclusion. This preserves meets and has a right adjoint U. Indeed $\mathcal{A}$ is the image of a **nucleus** M on $\mathcal{S}$, *i.e.* a meet-preserving closure operation.

(b) In particular, if $\mathbb{A} \subset \mathbb{X}$ is open, then $M = (\mathbb{A} \Rightarrow (-))$. In this case, M has a left adjoint $\mathbb{A} \wedge (=)$, and $\mathcal{A} \cong \mathcal{S} \downarrow \mathbb{A}$.

(c) If $\mathbb{A}$ is the complementary closed subset to $\mathbb{B}$ then $M = (\mathbb{B} \vee (-))$.

(d) Double negation $\neg\neg$ is a nucleus, for which $\mathcal{A}$ is a Boolean algebra. The open subsets of $\mathbb{X}$ that are fixed by $\neg\neg$ are called *regular*; in $\mathbb{R}$ these are unions of non-touching intervals, so $(0,1) \cup (2,3)$ is regular open but $(0,1) \cup (1,2)$ isn't.

(e) Let $\mathcal{X}$ and $\mathcal{Y}$ be the frames of open subsets of topological spaces $\mathbb{X}$ and $\mathbb{Y}$. Put $\Sigma = \mathcal{X} \times \mathcal{Y}$ with the componentwise order. Where $a \in \mathcal{X}$ and $b \in \mathcal{Y}$ represent open subsets ("intervals") in $\mathbb{X}$ and $\mathbb{Y}$, (a,b) will be the open "rectangle" in $\mathbb{X} \times \mathbb{Y}$. Whenever $a = \bigvee K \in \mathcal{X}$ and $b = \bigvee J \in \mathcal{Y}$, let $\{(a,j) \mid j \in J\} \rhd (a,b)$ and $\{(k,b) \mid k \in K\} \rhd (a,b)$. This is a stable coverage and $\mathsf{sh}(\Sigma, \leqslant, \rhd)$ is the **Tychonov product**, which is the topology on $\mathbb{X} \times \mathbb{Y}$ [Joh82, pp. 59–62].

Generalising from propositions to types. The remainder of the book will develop the analogues for categories and types of the poset and propositional ideas in this chapter. In the propositional terminology, Chapter IV studies posets, monotone functions, the transitive closure, universal properties, Horn theories, Heyting semilattices and the pointwise order. The phenomena in Chapter V are largely new to the level of types, but it does discuss distributive lattices (and use closure conditions to construct quotient algebras). Chapter VI is about induction on infinitary closure conditions. Most of the categorical developments are to be found in Chapter VII: adjunctions, closure operations, adding meets and joins, the adjoint function theorem and the canonical language; we only touch on Galois connections, modal logic and sheaf theory. Chapter VIII restores the predicates to Horn theories, and in the final chapter we see how types and propositions interact in the behaviour of the quantifiers.

EXERCISES III

1. Give an example of a strictly monotone function which is not injective.

2. Let $(\mathcal{X}, \preccurlyeq)$ be a preorder. Show that the quotient poset $\mathcal{X}/\sim$ constructed in Proposition 3.1.10 is isomorphic to the image of $\mathcal{X}$ in its regular representation.

3. A binary relation $<$ is ***irreflexive*** if $\forall x.\ x \not< x$. Show that

(a) any irreflexive relation can be recovered from its reflexive closure, characterising (in terms of decidability of equality) those reflexive binary relations that arise in this way from irreflexive ones;

(b) Proposition 2.5.6 and Definition 3.2.1(c) agree on minimality;

(c) the interleaved product agrees with the componentwise order, as defined in Propositions 2.6.9 and 3.5.1;

(d) any function which is injective and monotone (*i.e.* it preserves the reflexive relation) is strictly monotone (it preserves the irreflexive one), but not conversely.

Reformulate the induction scheme (Definition 2.5.3) with respect to the reflexive relation.

4. Show that $(\mathcal{X}, \leqslant)$ satisfies $\forall x, y.\ x \leqslant y \lor y \leqslant x$ iff it has binary joins and *every* monotone function $f : \mathcal{X} \to \mathcal{Y}$ preserves them. [Hint: use the representation by lower subsets.]

5. Show that an element of a poset is locally least iff it is least in its connected component (Lemma 1.2.4), and that if the poset has ***pullbacks*** (meets of pairs which are bounded above) then it suffices that the element be minimal.

6. For $\mathcal{I} \subset \mathbf{2}$, show that $\bigvee \mathcal{I} \Leftrightarrow (\{\top\} \subset \mathcal{I})$ and $\bigwedge \mathcal{I} \Leftrightarrow (\mathcal{I} \subset \{\top\})$.

7. Describe the greatest common divisor and least common multiple of a pair of ideals in a commutative ring.

8. Show that the various forms of the absorptive law mentioned before Proposition 3.2.13 agree.

9. Show that any lattice homomorphism between Boolean algebras also preserves $\neg$ and $\Rightarrow$. Find a sublattice of a Boolean algebra which is a Heyting lattice but for which the implication operations are different.

10. Let R be a ring such that $\forall x{:}R.\ x^2 = x$. Show that $\forall x.\ x{+}x = 0$ and $\forall x, y.\ x * y = y * x$, and that $x \vee y \equiv x + y - x * y$, $x \wedge y = x * y$ make R a Boolean algebra. Conversely, define the ring operations on any Boolean algebra, and show that a function between such structures is a ring homomorphism iff it is a homomorphism for the logical operations.

11. Let $\mathcal{X}$ be a distributive lattice, but write 0, $+$ and $\times$ instead of $\bot$, $\vee$ and $\wedge$. Show that any subset $\mathcal{I} \subset \mathcal{X}$ is an ideal in the sense of ring theory (Example 2.1.3(b)) *via* this notation iff it is a directed lower set.

12. Write $\mathsf{Y}_{\mathcal{X}} : [\mathcal{X} \to \mathcal{X}] \to \mathcal{X}$ for the operation which yields the least fixed point of a continuous function on an ipo $\mathcal{X}$. Express $\mathsf{Y}_{\mathcal{X}}$ as $\mathsf{Y}_{\mathcal{Y}}(F)$ for some $F : \mathcal{Y} \to \mathcal{Y}$ and hence deduce (without any further order-theoretic calculation) that $\mathsf{Y}_{\mathcal{X}}$ is itself continuous.

13. (Hans Bekič) Let $h : \mathcal{D} \times \mathcal{E} \to \mathcal{D} \times \mathcal{E}$ in $\mathcal{IPO}$. For any $e \in \mathcal{E}$ devise some $f_e : \mathcal{D} \to \mathcal{D}$ and hence define $p(e) \in \mathcal{D}$ as its fixed point. Similarly $q : \mathcal{D} \to \mathcal{E}$. Using $\mathsf{Y}_{\mathcal{D}}$ again, obtain a fixed point of h.

14. Show that if a poset $\mathcal{X}$ has, or a function $\mathcal{X} \to \mathcal{Y}$ preserves, both finite $(\bot, \vee)$ and directed $(\bigvee)$ joins, then it has or preserves all joins.

15. Let $U : \mathcal{J} \to \mathcal{I}$ be a cofinal function between posets. Show that if $\mathcal{J}$ is directed then so is $\mathcal{I}$, and conversely if U is also full. Show that any countable directed poset has a cofinal sequence, and hence that the corresponding notions of countable Scott continuity coincide.

16. Suppose that every ipo has a maximal element (this assertion is known as **Zorn's Lemma**), and assume excluded middle. Deduce the axiom of choice (Definition 1.8.8). [Hint: consider the ipo of partial functions contained in the given entire relation.]

17. Let $\mathcal{X}$ be a *meet*-semilattice. Show that $\mathsf{Idl}\,\mathcal{X}$ is a preframe, *i.e.* it has meets and they distribute over directed join in each argument. Also show that $\mathcal{X} \to \mathsf{Idl}\,\mathcal{X}$ preserves $\top$ and meets.

18. Let $\mathcal{Y}$ be a preframe and $\mathcal{X}$ a dcpo. Show that $[\mathcal{X} \to \mathcal{Y}]$, the dcpo of Scott-continuous functions, is also a preframe, binary meets being computed pointwise.

19. Let $\mathcal{X}$ and $\mathcal{Y}$ be posets or domains. Show that their disjoint union $\mathcal{X} + \mathcal{Y}$ (with no instances of the order relation linking the two summands) obeys the rules for sums (Remark 2.3.10).

20. Let $\mathcal{X}$, $\mathcal{Y}$ and $\mathcal{Z}$ be meet-semilattices and $f : \mathcal{X} \times \mathcal{Y} \to \mathcal{Z}$ a monotone function such that $f(x, -)$ and $f(-, y)$ preserve binary meets, for each $x \in \mathcal{X}$ or $y \in \mathcal{Y}$. Find a necessary and sufficient condition

(in terms of the order relations but not the meets) which makes f a semilattice homomorphism (*cf.* Corollary 3.5.13).

21. A poset $\mathcal{X}$ is **boundedly complete** if any subset $\mathcal{I} \subset \mathcal{X}$ which has some upper bound $\mathcal{I} \leqslant \vartheta$ actually has a join. Show that the product and function-space of two such posets are boundedly complete. Do the same for boundedly complete domains (in which directed subsets also have joins), with the Scott-continuous function-space.

22. Construct the meet and join of a bounded inhabited set of real numbers, considered as Dedekind cuts. Assume excluded middle. Under what (simple) condition does a monotone function $\mathbb{R} \to \mathbb{R}$ have adjoints? Express lim sup and lim inf as extrema.

23. Show that $\mathsf{FV} \dashv \mathsf{Cn}$ (Notation 2.3.6).

24. Show that for *any* lattice $\mathcal{X}$ there is an adjunction

$$(y \wedge z \downarrow \mathcal{X} \downarrow z) \; \underset{z \wedge x \,\leftarrowtail\, x}{\overset{x \,\mapsto\, x \vee y}{\underset{\perp}{\rightleftarrows}}} \; (y \downarrow \mathcal{X} \downarrow y \vee z).$$

If this is an isomorphism then $\mathcal{X}$ is called a **modular lattice**.

25. Suppose $F \dashv U$ and $U \dashv F$ between posets. Show that they are mutually inverse, or form a strong equivalence in the case of preorders.

26. Let $U : A \twoheadrightarrow X$ be a surjective function between sets. Define a preorder $\leqslant^A$ such that U is an *equivalence function* (Remark 3.6.7), where $\leqslant^X$ is discrete. Show that U is part of a strong equivalence iff it is split epi. In other words, every weak equivalence is strong iff the axiom of choice holds.

27. Let $F_1 : S \to A_1$ and $F_2 : S \to A_2$ be equivalence functions between preorders. Find equivalence functions $G_i : A_i \to B$ with $G_1 \circ F_1 = G_2 \circ F_2$. [Hint: let the underlying set of B be the union of those of A_1 and A_2.]

28. Show that a semilattice equipped with an additional binary operation $\Rightarrow$ is a Heyting semilattice (Definition 3.6.15) iff it satisfies

$$(x \Rightarrow x) = \top \qquad (x \wedge (x \Rightarrow y)) = (x \wedge y)$$
$$y \wedge (x \Rightarrow y) = y \qquad (x \Rightarrow (y \wedge z)) = ((x \Rightarrow y) \wedge (x \Rightarrow z)).$$

29. Describe implication and infinitary meet in the frame of open sets of any topological space. [Hint: do negation first.] Let $f : X \to Y$ be a continuous function. Describe the adjunctions $f^* \dashv f_*$ between the frames of open subsets of X and Y.

30. For any poset $\mathcal{X}$, show that $\mathsf{sh}(\mathcal{X})$ is a complete Heyting lattice with $[A \Rightarrow B] = \{x \mid \forall y.\, x \geqslant y \in A \Rightarrow y \in B\}$. [Hint: use Proposition 3.1.8(b) and $(-) \cap A \dashv (A \Rightarrow (=))$.] Show that $x \mapsto \mathcal{X} \downarrow x$ preserves $\Rightarrow$.

This interpretation of intuitionistic implication is due to Saul Kripke and was inspired by Definition 3.8.2 for modal logic.

31. Show that a poset is boundedly complete (Exercise 3.21) iff it has meets of all inhabited subsets.

32. Let $\mathcal{X}$ be a poset with $\top$ and $U : \mathcal{X} \to \mathcal{Y}$ a monotone function to any poset. Show that it is cofinal iff $U(\top)$ is the top element of $\mathcal{Y}$.

33. Let $F \dashv U$ with F injective. Show that F preserves minimal upper bounds, but U need not preserve maximal elements. Conversely, find a criterion involving minimal upper bounds for $F : \mathcal{X} \subset \mathcal{A}$ to have a right adjoint.

34. Show that the following are equivalent for a poset $\mathcal{X}$:

(a) $\mathcal{X} \downarrow \vartheta$ is a complete lattice for each $\vartheta \in \mathcal{X}$;

(b) $\mathcal{X}$ has meets of all subsets which are bounded above (*wide pullbacks*, Example 7.3.2(h)), but not necessarily $\top$;

(c) for each diagram $\mathcal{I} \subset \mathcal{X}$ with an upper bound $\mathcal{I} \leqslant \vartheta$ there is a unique minimal upper bound for $\mathcal{I}$ in $\mathcal{X}$ below ϑ.

$\mathcal{X}$ is then called an *L-poset*, and an *L-domain* if it also has all directed joins. Formulate and prove an adjoint function theorem for L-posets. Show that if $U : \mathcal{X} \to \mathcal{Y}$ preserves wide pullbacks and is cofinal then it has a left adjoint. (Notice how introducing a degree of uniqueness improves the result by allowing us to drop the injectivity assumption.)

35. Describe the closure and coclosure operations arising from the examples of adjunctions in Section 3.6, and characterise the (co)closed subsets or elements.

36. Let $M : \mathcal{P}(\Sigma) \to \mathcal{P}(\Sigma)$ be any closure operation. Show that it arises from the system of closure conditions

$$K \rhd t \quad \text{if} \quad t \in M(K),$$

and that this is a *saturated system of closure conditions*, i.e.

$$\{t\} \rhd t \quad \text{and} \quad \frac{\{u_i \mid i \in K\} \rhd t \qquad \forall i{:}K.\ J_i \rhd u_i}{\left(\bigcup_{i \in K} J_i\right) \rhd t}$$

but induction for $\rhd$ is essentially trivial. Show that the saturation of a unary closure condition $<$ is its reflexive–transitive closure $\leqslant$.

37. Show that if M is Scott-continuous in the previous exercise then it suffices to use finite K.

38. Let M be a closure operation on a poset or preorder $(\mathcal{S}, \leqslant)$. Let $\mathcal{A}$ be the set $\mathcal{S}$ equipped with the relation that $A \preceq B$ if $M(A) \leqslant B$. Show that $(\mathcal{A}, \preceq)$ is a preorder which is equivalent to the set of fixed points of M on $\mathcal{S}$. Exercise 3.56 is an example of this construction.

39. (Tarski) Let S be a complete (*meet*-semi)lattice and $T : S \to S$ a *monotone* function. Put $A = \bigwedge\{X : S \mid TX \leqslant X\}$ and show that $TA \leqslant A$. Using $T(TA) \leqslant TA$, deduce that A is the least fixed point.

40. With the same data, call an element $X \in S$ *well founded* if $X \leqslant TX$ and $\forall U. TU \wedge X \leqslant U \leqslant X \Rightarrow U = X$. Show that if X is well founded then so is TX, and that any join of well founded elements is again well founded, so there is a greatest well founded element, and it is fixed by T. Show that if X is well founded and $T\vartheta \leqslant \vartheta$ then $X \leqslant \vartheta$. Compare this with the proof of Proposition 3.7.11. Show that the simpler condition that $X \leqslant TX$ and $\forall U. TU \wedge X \leqslant U \Rightarrow U = X$ is equivalent to well-foundedness.

41. Let $T : S \to S$ be a $\wedge$-preserving endofunction of a frame, $X \in S$ a well founded element and $Y \leqslant X$. Show that if also $Y \leqslant TY$ then Y is well founded. [Hint: given $TV \wedge Y \leqslant V \leqslant Y$ consider $U = (Y \Rightarrow V) \wedge X$.] Show that the property $Y \leqslant TY$ is not automatic. [Hint: $S = \mathcal{P}(2)$.]

42. Let $T : \mathcal{P}(\Sigma) \to \mathcal{P}(\Sigma)$ be defined as in Lemma 3.7.10 from a closure condition $\vartriangleright$. Show that X is a well founded element of $\mathcal{P}(\Sigma)$ in the sense of Exercise 3.40 iff X has no non-trivial *relatively* $\vartriangleright$-closed subset (*cf.* Definition 3.7.8). Show also that, when $\vartriangleright$ is itself defined from a binary relation $\prec$ by Example 3.7.9(e), this is equivalent to well-foundedness of $\prec$ on X (Definition 2.5.3).

43. Proposition 2.5.6ff described classical approaches to induction. Discuss induction on closure conditions (Definition 3.7.8) in this fashion, giving a condition for an element *not* to be in the smallest closed subset.

44. Use Proposition 3.7.11 to show that the set of closure operations on a complete lattice S is itself the image of a closure operation on $[S \to S]$. Now let S be a dcpo. Show that the set of Scott-continuous closure operations on S is the image of a closure operation defined on $\mathrm{id} \downarrow [S \to S] \equiv \{T : S \to S \mid \mathrm{id}_S \leqslant T\}$.

45. (Dmitri Pataraia) By Lemma 3.5.7, the set of inflationary monotone functions $\mathrm{id} \leqslant f : A \to A$ on any dcpo A forms an ipo. Show that there is a greatest such function, g. [Hint: use composition to show that the set is directed.] If A has bottom, show that $g(\bot)$ is a fixed point for any inflationary function f.

46. Let $s : \mathcal{X} \to \mathcal{X}$ be a monotone endofunction of an ipo. Consider the smallest subset $A \subset \mathcal{X}$ with $\bot \in A$ which is closed under s and $\bigvee$ (*cf.* Exercise 6.53). Using Definition 3.7.8, show that the restriction of s is inflationary on A, and so has a least fixed point. Show that this is in fact the only fixed point in A, and is its greatest element, and also that it is the least fixed point in $\mathcal{X}$. Applying this to $\mathcal{X} = \{X \subset \Sigma \mid \vartheta[X]\}$,

prove Theorem 3.7.13. If $\mathcal{X}$ has binary meets, show that its subset of well founded elements (Exercise 3.40) has similar properties to $\mathcal{A}$.

47. Use infinitary conjunction to interpret $\forall \vartheta.\,(T\vartheta \Rightarrow \vartheta) \Rightarrow \vartheta$ in a complete Heyting lattice and verify Proposition 2.8.6.

48. For any monotone function $T : 2 \to 2$, show that

$$m = \mu\alpha.\,T(\alpha) = \forall\alpha.\,(T(\alpha) \Rightarrow \alpha) \Rightarrow \alpha$$

is the least fixed point of $T : 2 \to 2$. [Hint: show $(T\alpha \Rightarrow \alpha) \Rightarrow (m \Rightarrow \alpha)$ first, and deduce $m \Rightarrow (T\alpha \Rightarrow \alpha) \Rightarrow \alpha$ for all α, so $Tm \Rightarrow m$ and $T^2 m \Rightarrow Tm$, so $m \Rightarrow Tm$.]

49. Let H and K be congruences on a lattice L. Write down the closure conditions on $\Sigma = L \times L$ which define the congruence $H \vee K$ generated by H and K. Use induction on these conditions, and the formula $(x \vee y) \wedge (y \vee z) \wedge (z \vee x)$, to show that the lattice of congruences of L is distributive. Since the theory is finitary, congruences form a Heyting algebra: is there a formula for $H \Rightarrow K$?

50. Show that the function $\alpha \mapsto (\alpha \Rightarrow \phi)$ is symmetrically adjoint to itself on the right (a Galois connection) for any ϕ. Express this in the notation of Proposition 3.8.14. Hence it sends joins to meets:

$$(\bot \Rightarrow \phi) \Leftrightarrow \top \quad \text{and} \quad ((\alpha \vee \beta) \Rightarrow \phi) \Leftrightarrow ((\alpha \Rightarrow \phi) \wedge (\beta \Rightarrow \phi)).$$

For $\phi = \bot$, deduce the intuitionistic de Morgan law:

$$\neg(\alpha \vee \beta) \Leftrightarrow ((\neg\alpha) \wedge (\neg\beta)) \qquad (\text{also } \neg\bot \Leftrightarrow \top).$$

51. Assuming excluded middle, show that $\neg(-)$ is also symmetrically adjoint to itself on the left (a co-Galois connection), so the classical **de Morgan law** holds (Theorem 1.8.3). In fact this is weaker than excluded middle, giving only $\vdash \neg\alpha \vee \neg\neg\alpha$.

52. Show that $[(\exists x.\,\phi[x]) \Rightarrow \vartheta] \Leftrightarrow \forall x.\,(\phi[x] \Rightarrow \vartheta)$, where x is not free in ϑ, by identifying a Galois connection.

53. Let $R : \mathbb{X} \nrightarrow \mathbb{Y}$ be a relation between sets which themselves carry binary relations $<$ and $\prec$ respectively. Then R is called a **simulation** (of $(\mathbb{X}, <)$ by $(\mathbb{Y}, \prec)$) if the first implication is satisfied,

$$(\prec)\,;R^{\mathrm{op}} \Rightarrow R^{\mathrm{op}}\,;(<) \qquad (<)\,;R \Rightarrow R\,;(\prec),$$

and a **bisimulation** if both are. (Beware that bisimulation *does not* mean that R and R^{op} are both simulations.) Show how modal formulae can be transferred from $\mathbb{X}$ to $\mathbb{Y}$ and *vice versa*.

54. Show that $f : \mathbb{X} \to \mathbb{Y}$ is strictly monotone iff it satisfies Lemma 3.8.12 and $[\succ] \circ [f] \leqslant [f] \circ [\succ]$. Hence prove Proposition 2.6.2, that f reflects well-foundedness, using modal operators (*cf.* Theorem 3.8.11).

Similarly, show that f *preserves* well-foundedness if it is a surjective simulation. Can you prove Proposition 2.6.8 in the same way?

55. Let $U, V \subset \mathcal{X}$ and write $\downarrow U$ for the down-closure of (*i.e.* the lower set generated by) U. Show $\downarrow U \subset \downarrow V$ iff $\forall u \in U.\, \exists v \in V.\, u \leqslant v$. Carl Gunter has suggested the musical notation $U \leqslant^\flat V$ for this, the *lower order*, on subsets, together with $U \leqslant^\sharp V$ if $\forall v \in V.\, \exists u \in U.\, u \leqslant v$ corresponding to $\uparrow U \supset \uparrow V$. (Beware that the converse interchanges them: $U \leqslant^\flat V \iff V \geqslant^\sharp U$.) The conjunction, $U \leqslant^\flat V \land U \leqslant^\sharp V$, is written $U \leqslant^\natural V$ and is called the *Egli–Milner order*. These order relations arise in the study of non-deterministic programs, for which they are used to construct *powerdomains*.

56. Show that $\uparrow U$ is the $\subset$-largest subset V with $U \leqslant^\flat V \land V \leqslant^\flat U$ (*cf.* Proposition 3.1.10), and that the corresponding property for the Egli–Milner order $\leqslant^\natural$ gives the convex hull of U (Example 3.7.5(c)), $\{y \mid \exists u_1, u_2 \in U.\, u_1 \leqslant y \leqslant u_2\}$.

57. Show that $(R\,;S)^\flat = R^\flat\,;S^\flat$ for binary relations $R : X \rightsquigarrow Y$ and $S : Y \rightsquigarrow Z$. [Hint: to show that $U\,(R\,;S)^\flat\,W \Rightarrow U(R^\flat\,;S^\flat)W$, consider the set $V = \{y \mid \exists z \in W.\, ySz\}$.] What is $\mathrm{id}^\flat$?

58. Show that, for symmetry of $<$, any inequality between modal operators of the same kind, *e.g.* $\langle > \rangle \leqslant \langle < \rangle$ or $[>] \leqslant [<]$, suffices.

59. Show that $<$ is confluent iff $\langle > \rangle \circ [>] \leqslant [>] \circ \langle > \rangle$.

60. Write down the predicate on the set $\mathsf{List}(<)$ which says that a list of instances of any binary relation $<$ is composable (a *path*). Defining the reflexive–transitive closure $\leqslant$ as a subquotient of $\mathsf{List}(<)$ in this way, develop the unary and binary properties in Definition 3.8.6ff.

61. Suppose that $F \dashv U$ in Remark 3.8.13(b). What is the relationship amongst $F_!$, F^*, F_* and $U_!$, U^*, U_*? Use this to give examples of arbitrarily long chains of adjunctions between posets.

62. Show that $f_!$ preserves $\top$ iff f is surjective iff $f_* \circ f^* = \mathrm{id}$, and that $f_!$ preserves $\cap$ iff f is injective iff $f^* \circ f_* = \mathrm{id}$.

63. Let $f : \mathbb{X} \hookrightarrow \mathbb{Y}$ and $g : \mathbb{Y} \hookrightarrow \mathbb{X}$ be two injective functions. Regarding them as functional relations, interpret the subsets

$$U = \mathsf{Y}\big(\langle g^{\mathrm{op}}\rangle \circ [f^{\mathrm{op}}]\big) \subset \mathbb{X} \quad \text{and} \quad V = \mathsf{Y}\big(\langle f^{\mathrm{op}}\rangle \circ [g^{\mathrm{op}}]\big) \subset \mathbb{Y},$$

where Y is the least fixed point operator. [Hint: consider chains of alternating relations, and *cf.* Exercise 2.47.] Using excluded middle, show that $g^{-1} : U \hookrightarrow \mathbb{Y}$ and $f : \mathbb{X} \setminus U \hookrightarrow \mathbb{Y}$ with complementary images. Hence prove the *Schröder–Bernstein Theorem*, that there is a bijection between $\mathbb{X}$ and $\mathbb{Y}$. (This conjecture of Cantor was actually first proved by Richard Dedekind in 1887.)

64. What are the specialisation orders (Proposition 3.8.14) between elements of groups or fields in Galois theory (in characteristic zero)?

65. For any binary relation $(<) : \mathbb{X} \rightsquigarrow \mathbb{Y}$, show that $x < y$ iff $\forall V \subset \mathbb{Y}. (x \in [<]V \Rightarrow y \in V)$.

66. Let $(\Sigma, \leqslant)$ be a complete semilattice and $\triangleright$ its canonical coverage (Definition 3.9.6). Show that $\eta : \Sigma \cong \mathsf{sh}(\Sigma, \leqslant, \triangleright)$, and also that $\triangleright$ is stable iff Σ is a frame.

67. Show that $\neg\neg$ on a Heyting lattice is the greatest closure operation which preserves $(\top,)$ $\wedge$ and $\perp$ (but not $\vee$).

68. Let $F \dashv U$ be an adjunction between algebraic lattices. Show that F is Scott-continuous iff U preserves finite generability.

69. Let $M : \mathcal{S} \to \mathcal{S}$ be a closure operation on a Heyting semilattice and $\mathcal{A} \subset \mathcal{S}$ its set of fixed points. Show that $X \Rightarrow^{\mathcal{S}} A$ is the implication in $\mathcal{A}$ for all $X \in \mathcal{S}$, $A \in \mathcal{A}$ iff M preserves $\wedge$ (*cf.* Theorem 3.9.9ff).

70. Let $U : \mathcal{A} \to \mathcal{S}$ be a meet-semilattice homomorphism, where $\mathcal{S}$ is a frame, and consider $\mathcal{S} \downarrow U = \{(I, \Gamma) \mid I \in \mathcal{S}, \Gamma \in \mathcal{A}, I \leqslant F\Gamma\}$ with the componentwise order. Show that $\pi_1 : \mathcal{S} \downarrow U \to \mathcal{A}$ has adjoints on both sides, given by $\Gamma \mapsto (\perp, \Gamma)$ and $(F\Gamma, \Gamma)$. Investigate π_0 too.

71. Let $\mathcal{X}$ be the frame of open subsets of a topological space $\mathbb{X}$ and $\mathbb{A} \subset \mathbb{X}$ a particular open subspace. Show that, from the frames $\mathcal{A}$ and $\mathcal{S}$ of open subsets of $\mathbb{A}$ and its complementary closed subspace, $\mathcal{X}$ may be recovered as $\mathcal{S} \downarrow U$ for some $U : \mathcal{A} \to \mathcal{S}$, and that $\pi_0 : \mathcal{X} \to \mathcal{S}$ and $\pi_1 : \mathcal{X} \to \mathcal{A}$ are the inverse image functions. Show in particular how to construct the lift $\mathbb{A}_\perp$, $\mathcal{S}$ being the topology on the new closed point $\perp$.

72. Let $\mathcal{A}$ in Exercise 3.70 be a Heyting semilattice. Show that $\mathcal{S} \downarrow U$ is also a Heyting semilattice, in which $(I, \Gamma) \Rightarrow (J, \Delta)$ is $(H, [\Gamma \Rightarrow \Delta])$, where $H = [I \Rightarrow J] \wedge U[\Gamma \Rightarrow \Delta]$, and that the inclusion of a poset in its free Heyting semilattice is full.

IV

Cartesian Closed Categories

CATEGORY THEORY unifies the symbolic (Formalist) and model-based (Platonist) views of mathematics. In particular it offers an agnostic solution to the question that we raised in Section 1.3 of whether a function is an algorithm or an input–output relation.

Traditionally, categories were *congregations*, each object being a *set with structure*: a topological space, an algebra or a model of some theory. The morphisms are functions that preserve this structure (homomorphisms), so the notion of composition is ultimately that for relations (Definition 1.3.7). As an approach to logic, this went round "three sides of a square" (*cf.* Remark 1.6.13 for model theory) and so ran into some foundational problems over the category of all categories.

In informatics, the principal examples are constructed from λ-calculi and programming languages; being syntactic, they are typically recursively enumerable. Composition is by substitution of terms (Definition 1.1.10), or by the cut rule (Definition 1.4.8), which uses old conclusions as new hypotheses. A category $\mathsf{Cn}_{\mathcal{L}}^{\square}$ of this kind encodes a certain theory $\mathcal{L}$ itself, instead of collecting its models; we call it the *category of contexts and substitutions* by analogy with categories of *objects and homomorphisms* in semantics.

We shall give a novel construction of $\mathsf{Cn}_{\mathcal{L}}^{\square}$ that embodies well-established techniques for proving correctness of programs and works uniformly for any fragment of logic. At the unary level, the ideas come from geometry and physics (groups), automata and topology; we also carry it out for algebraic theories ($\mathsf{Cn}_{\mathcal{L}}^{\times}$) before turning to the λ-calculus ($\mathsf{Cn}_{\mathcal{L}}^{\rightarrow}$).

The fragment of logic in question, $\square$, corresponds to certain categorical structure defined by universal properties: products and exponentials in this chapter, coproducts and factorisation systems in the next. The recursively defined interpretation functor $[\![-]\!] : \mathsf{Cn}_{\mathcal{L}}^{\square} \to \mathcal{S}$ preserves this structure, so the semantic universe $\mathcal{S}$ must also have it. $\mathsf{Cn}_{\mathcal{L}}^{\square}$ is also called the *classifying category* for (models of) the theory, because there is a correspondence between such functors and models in $\mathcal{S}$.

4.1 CATEGORIES

DEFINITION 4·1·1: A *category* C consists of

(a) a class of *objects*, ob C,

(b) for each pair of objects $X, Y \in \text{ob}\,C$, a set of *morphisms*, $C(X,Y)$ (for $f \in C(X,Y)$ we write $f : X \to Y$, calling X the *source* and Y the *target*; the words *domain* and *codomain* are synonyms for source and target, and *map* and *arrow* are synonyms for morphism[1]; as a throwback to the time when categories were only used for algebras and *homo*morphisms, $C(X,Y)$ is called the *hom-set*),

(c) for each object $X \in \text{ob}\,C$, an *identity*, $\text{id}_X \in C(X,X)$, and

(d) for each triplet of objects, $X, Y, Z \in \text{ob}\,C$, a *composition*,

$$C(X,Y) \times C(Y,Z) \to C(X,Z),$$

which we write *synonymously* as $f;g$ or $g \circ f$ (composition is a *function* between hom-*sets*, whereas $f : X \to Y$ is an abstract arrow),

such that $f ; \text{id} = f = \text{id} ; f$ and $f ; (g ; h) = (f ; g) ; h$, so the composition (;) is *associative* and id is a *unit*:

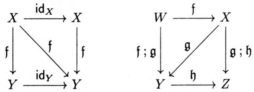

The *opposite category*, C^{op}, is obtained by "reversing the arrows" — it has the same object-class but $C^{\text{op}}(X,Y) = C(Y,X)$ and the order of composition is the other way round.

CONVENTION 4·1·2: The order of composition is a contentious issue. The left-handed notation ($\circ$) is older, both in category theory and in mathematics as a whole, and comes from the custom of writing function-*application* on the left — apart from the factorial function! We shall not challenge this custom itself: juxtaposition will always mean application on the left; moreover we shall adopt the convention from the λ-calculus that $\Phi F X$ means $(\Phi F)X$. If functional composition arises by abstraction of application, it is clearer to use the left-handed notation.

For those literate in Arabic or Hebrew, diagram-chasing in the right-to-left notation may present no problem, but for the rest of us it can be rather confusing. There are even situations, such as the composition of adjunctions (Lemma 3.6.6), where it helps to use both conventions

---

[1] Frequently I have chosen whichever gives the best line-break!

together. Of course we distinguish them by using two different symbols (; and ∘), and we will always use one or other of them, without relying on juxtaposition. In practice there is no need to decorate these symbols with the objects to which they apply, though it is useful to annotate id_X.

Categories as theories. In this book we shall be most interested in how categories can describe type-theoretic phenomena. The objects are contexts (lists of typed variables and predicates) and the morphisms are assignments (substitutions) of terms or proofs to these variables.

Categories as congregations. First, however, we give the usual list of mathematical structures and their homomorphisms. These categories are traditionally named by their objects, but in some cases such as matrices and programs the arrows are more prominent: following Peter Freyd and Andre Scedrov [FS90], we speak of the category *composed of* named maps, here relations.

LEMMA 4·1·3: Sets and binary relations form a category $\mathcal{Rel}$, where the identity on X is the equality relation $(=_X)$, and the composition is the relational one given in Definition 1.3.7.

PROOF: Let $R : X \nrightarrow Y$. Then
$$x((=_X)\,;R)y \iff \exists x':X.\ x = x' \land x'Ry \iff xRy.$$
Showing that $R\,;(=_Y) = R$ is similar. For associativity,

$$
\begin{aligned}
&w\big(Q\,;(R\,;S)\big)z \\
&\iff\ \exists x.\ wQx \land x(R\,;S)z \\
&\iff\ \exists x.\ wQx \land (\exists y.\ xRy \land ySz) \\
&\iff\ \exists x.\ \exists y.\ wQxRySz \qquad \text{Frobenius law (Lemma 1.6.3)} \\
&\iff\ \exists y.\ (\exists x.\ wQx \land xRy) \land ySz \\
&\iff\ \exists y.\ w(Q\,;R)y \land ySz \\
&\iff\ w\big((Q\,;R)\,;S\big)z
\end{aligned}
$$

where $Q : W \nrightarrow X$ and $S : Y \nrightarrow Z$. □

PROPOSITION 4·1·4: The following form categories:

	objects	morphisms	Section
Set	sets	(total) functions	1.3
Pfn	sets	partial functions	
Bin	binary endorelations	functions preserving the relation	
Wfr	well founded relations	strictly monotone functions	2.5
Preord	preorders	monotone functions	3.1
Pos	posets	monotone functions	

$\mathcal{P}os^{\dashv}$	posets	(left) adjoints	3.6
$\mathcal{CSLat}$	complete semilattices	$\bigvee$–preserving functions	
$\mathcal{SLat}$	meet-semilattices	$(\top, \wedge)$-homomorphisms	3.2
$\mathcal{Lat}$	lattices	$(\top, \bot, \wedge, \vee)$-homomorphisms	
$\mathcal{DLat}$	distributive lattices	$(\top, \bot, \wedge, \vee)$-homomorphisms	
$\mathcal{BA}$	Boolean algebras	$(\top, \bot, \wedge, \vee)$-homomorphisms	1.8
$\mathcal{HSL}$	Heyting semilattices	$(\top, \wedge, \Rightarrow)$-homomorphisms	3.6
$\mathcal{Heyt}$	Heyting lattices	$(\top, \wedge, \bot, \vee, \Rightarrow)$-homomorphisms	
$\mathcal{Frm}$	frames	$(\top, \wedge, \bigvee)$-homomorphisms	
$[\mathcal{Loc}$	locales	continuous maps; $\mathcal{Loc} = \mathcal{Frm}^{\text{op}}]$	
$\mathcal{Dcpo}$	directed complete posets	(Scott-)continuous functions	3.4
$\mathcal{IPO}$	inductive partial orders	(Scott-)continuous functions	
$\mathcal{Sp}$	topological spaces	continuous functions	
$\mathcal{Mon}$	monoids	$(1, *)$-homomorphisms	2.7
$\mathcal{Gp}$	groups	$(1, * [, (-)^{-1}])$-homomorphisms	
$\mathcal{CMon}$	commutative monoids	homomorphisms	5.4
$\mathcal{Ab}$	Abelian groups	$(0, + [, -])$-homomorphisms	
$\mathcal{Fld}$	fields of numbers	$(0, 1, +, *)$-homomorphisms	
$\mathcal{Vsp}$	vector spaces	linear maps	
$\mathcal{Rng}$	rings	$(0, 1, +, *)$-homomorphisms	
$\mathcal{CRng}$	commutative rings	$(0, 1, +, *)$-homomorphisms	

PROOF: In all cases the morphisms are at least (functional) relations between sets, the diagonal relation is the identity and composition is the relational one. The identity and associativity laws are therefore already known to hold: it is only necessary to check that *composition preserves the defining properties of the morphisms.* For total and partial functions this follows from Lemma 1.6.6 and for Scott-continuity from Lemma 3.4.6. Adjunctions compose by Lemma 3.6.6, and hence by the Adjoint Function Theorem (3.6.9) so do join-preserving functions between complete semilattices. The finitary algebraic examples may be verified equationally.   □

Categories as structures. Many mathematical *structures* may themselves be seen as special cases of the definition of a category. Moreover Examples 4.4.2 show that functors are the appropriate homomorphisms.

PROPOSITION 4·1·5: Every preorder X (Definition 3.1.1) may be viewed as a category: the objects are the elements of X, and for each pair of elements with $x \leqslant y$ there is a single (anonymous) map $x \to y$. Identities are given by reflexivity and composition by transitivity. Conversely any category arises (uniquely) in this way iff there is at most one map in each hom-set $\mathcal{C}(x, y)$, which we may now call the **hom-predicate**.   □

This result recalls our reservations about calling the propositions as types analogy of Section 2.4 an *isomorphism*: propositions form a poset, whereas types form a category. In a category we have to specify whether *parallel maps* (*i.e.* with the same source and target) are equal or not. This is how categories uncover the existential quantifiers hidden behind lattice theory. Nevertheless the analogy is an important one, because many properties of posets contain the germs of categorical ideas. The former are generally much simpler — as a rule of thumb they take a tenth of the space to describe — so wherever possible we shall give pointers to them in Chapter III. You would do well to be sure that you understand each concept for posets before moving on to the categorical form. (Little remains of Chapter V when restricted to posets, however.)

The analogy is particularly relevant in domain theory, where much of the published work fails to mention the corresponding — and often prior — results in categorical algebra. Accounts of the poset version often amount to a list of isolated facts: like the shadows on the wall of Plato's cave, they lack the depth and colour of the real phenomena they represent.

Preorders restrict the hom-sets: we may restrict the object class instead.

DEFINITION 4·1·6: A category which

(a) has only one object is a *monoid*, whose elements are the maps;

(b) has every morphism invertible is a *groupoid*;

(c) satisfies both of these is a *group*;

(d) is both a preorder and a groupoid is an equivalence relation on its object class (Definition 1.2.3).

So a category is a *preorder with proofs* or a *monoid with types*.

The notion of group has wide and important applications, which will give us valuable insight into the meaning of categories. Indeed when Sammy Eilenberg and Saunders Mac Lane first introduced categories in 1940, one of the motivations was to extend the Erlanger Programm. Indeed, our category composed of substitutions echoes the early development of groups, which were first called "systems of substitutions". We build on these intuitions in the next section, and subsequently use them to construct categories for type theory and programming. Ronald Brown's survey article [Bro87] explains why many of the familiar applications of groups make more sense in terms of groupoids.

EXAMPLES 4·1·7: Monoids, groups and categories.

(a) $\mathbb{N}$ is a monoid and $\mathbb{Z}$, $\mathbb{Q}$, *etc.* are groups under addition ($\circ = +$).

(b) The polynomial; primitive, total and partial recursive; polynomial-time *etc.*, functions on $\mathbb{N}$ form monoids with function-composition.

(c) $\mathsf{List}(X)$, the set of finite lists of elements of a set X, forms a monoid with concatenation for composition and the empty list as identity (Definition 2.7.4).

(d) Square matrices of a fixed size form a monoid under multiplication of matrices. The invertible ones alone form a group, and collecting those of all sizes gives a groupoid. They all lie within a *category* composed of all matrices, whose objects are numbers: the source and target of a matrix say how many columns and rows it has.

(e) For any object $X \in \mathsf{ob}\,\mathcal{C}$ of any category, $\mathsf{End}_{\mathcal{C}}(X) = \mathcal{C}(X, X)$ is the monoid of **endomorphisms** and its invertible elements form the group $\mathsf{Aut}_{\mathcal{C}}(X)$ of **automorphisms** (*cf.* Section 1.3).

(f) In particular, Galois groups are fragments of the abstract category of fields (Example 3.8.15(j)).

(g) The symmetries of any geometrical structure form a group. In the geometry of Euclid, figures are *congruent* if one may be transformed into the other by rotation and translation, and *similar* if scaling is also allowed. Reflection may or may not be permitted. In affine geometry, shears (different x- and y-scaling) are added, with the effect that a square can be transformed into a parallelogram. Finally, projective geometry also admits inversion, which can turn a straight line into a circle. Felix Klein's *Erlanger Programm* axiomatised these various forms of geometry in terms of the *groups* involved and the features left *invariant* by them.

(h) This also works in physics. A dynamical system whose laws are invariant under translation, rotation and temporal translation obeys respectively conservation of momentum, angular momentum and energy (as proved by Emmy Noether, who was born in Erlangen). The fact that the laws of mechanics and electromagnetism due to Isaac Newton and James Clerk Maxwell are invariant under *different* groups (called after Galileo Galilei and Hendrik Lorentz) led Albert Einstein to discover special relativity. In quantum mechanics, the irreducible matrix representations of the symmetry group of the system correspond to particles.

(i) The points of a topological space are the objects of a category whose morphisms are the paths between pairs of points (Exercise 4.43). The paths may be concatenated, but must be "reparametrised" if they are defined as continuous functions from the unit interval $[0, 1] \subset \mathbb{R}$. Reversal provides the inverse, the maps now being homotopy classes, giving the first homotopy or **fundamental groupoid**.

This may be reduced to a group by choosing a base point; in a path-connected space the result is unique, but up to non-unique isomorphism: see Exercise 4.37.

(j) François Viète showed that the coefficients of a polynomial can be expressed as polynomials in its roots; the coefficients are invariant under permutations of the roots. The solution of quadratic, cubic (Example 4.3.4) and quartic equations uses other expressions, known as **discriminants**, which are invariant under *proper* subgroups of the permutations. Evariste Galois showed that the discriminants may be successively expressed as radicals iff each corresponding subgroup is normal in the next with cyclic quotient (Example 5.6.2(d)). He also showed that the group S_n of all $n! = 120$ permutations of five roots has no such chain of subgroups. So the *generic* quintic has no solution by radicals, as Niels Abel had already shown directly. (Both Abel and Galois died at an early age in tragic circumstances.)

Size issues. The word *class* has reared its ugly head in the definition of category, not just as a synonym of set (like family or collection) but as something which may be "too big" to be a set. This is to allow us to speak of the category of *all* sets, the category of *all* posets, and so on.

REMARK 4·1·8: The familiar categories listed in Proposition 4.1.4 are populated by structures that can easily be coded as a finite system of sets. (Some authors insist that, for example, a monoid $\langle M, e, m \rangle$ is *in substance* a Wiener–Kuratowski triple, Exercise 2.19.) The problem in these cases is essentially no more difficult than for sets themselves, where a notion of class due to Frege, von Neumann and Bernays has traditionally been used: it is able to collect *all* such codings but cannot itself be a member of a set or class. We only need to be able to test whether a given lump of set-theoretic data is a well formed code for a structure of the appropriate kind.

Such a set theory is not enough to let us treat a functor (Definition 4.4.1) as a *function* on objects (together with its effect on maps) defined in the extensional way as an input–output relation. This can be done using a calculus of "big" sets, with the same behaviour as the usual ones; these are called Grothendieck universes, of which a modern definition is a full internal topos (Proposition 9.6.4). What this amounts to is that, just as $\infty = 2^{64}$ rarely causes any harm in programming, so $\mathbf{ob}\,Set = \mathcal{P}^{64}(\mathbb{N})$ suffices for most mathematical purposes.

REMARK 4·1·9: The *intensional* view better describes actual practice. We argue about *what it is to be* a ring, or whatever, not the *totality* of them (*cf.* Remark 1.6.13). Similarly, a functor is a *scheme* of constructions (an *algorithm*), which says how to transform the generic ring into its spectral space, for example. To formalise this we must consider dependent type theory in detail, as we do in Chapters VIII and IX. Even then, recursively defined functors such as $[\![-]\!]$, like recursive algorithms, require a new principle, in the form of the axiom-scheme of replacement.

REMARK 4·1·10:   Some authors collect all of the morphisms into a single class, mor $\mathcal{C}$, together with functions src, tgt : mor$\mathcal{C} \rightrightarrows$ ob $\mathcal{C}$. Composition then becomes a *partial* operation, defined only when the target of the first morphism agrees with the source of the second. This point of view is discussed briefly in Remark 5.2.9, but the one we prefer, using dependent types, is explained in Example 8.1.12.

Besides this, we know that there really is a *set* (and not just a class) of morphisms between two sets, posets, algebras, *etc.* A category for which this is the case is called ***locally small*** in the literature. A ***small category*** is one for which ob $\mathcal{C}$ and mor $\mathcal{C}$ are both known to be sets.

4.2   ACTIONS AND SKETCHES

Although in the examples from geometry and physics the operations are untyped and always reversible, they illustrate that groups, monoids and categories not only are syntactic abstract algebraic structures, but also have a semantic influence. In the latter, the identity is "do nothing" and the composition is "do this, then do that", which is clearly associative.

Actions with a single sort.

DEFINITION 4·2·1:   A ***covariant action*** of a group or monoid $(M, \mathrm{id}, \circ)$ on a set A is a binary operation $(-)_*(=) : M \times A \to A$ such that

$$\mathrm{id}_* a \quad = \quad a$$
$$(\mathfrak{g} \circ \mathfrak{f})_* a \quad = \quad \mathfrak{g}_*(\mathfrak{f}_* a)$$

for all $a \in A$. Similarly a ***contravariant action*** is a binary operation $(-)^*(=) : M \times A \to A$ such that $\mathrm{id}^* a = a$ and

$$(\mathfrak{g} \circ \mathfrak{f})^* a \quad = \quad \mathfrak{f}^*(\mathfrak{g}^* a) \quad = \quad (\mathfrak{f} ; \mathfrak{g})^* a.$$

Usually we treat the star as a unary operation on its first argument, so the abstract arrow $\mathfrak{f}$ is represented by a concrete function $\mathfrak{f}_* : A \to A$ or $\mathfrak{f}^* : A \to A$ such that id_* and id^* are the identity on A and

$$(\mathfrak{g} \circ^M \mathfrak{f})_* = \mathfrak{g}_* \circ^{Set} \mathfrak{f}_* \qquad (\mathfrak{g} \circ^M \mathfrak{f})^* = \mathfrak{f}^* \circ^{Set} \mathfrak{g}^* \equiv \mathfrak{g}^* ; \mathfrak{f}^*.$$

Notice that these laws link the composition in M to that in Set.

A contravariant action of M is the same as a covariant action of M^{op}, the opposite monoid, *cf.* Definitions 1.3.9, 3.1.5 and 4.1.1.

A *faithful action* is one for which things are semantically equal only when they are syntactically the same:

$$(\forall a{:}A.\ \mathfrak{f}_* a = \mathfrak{g}_* a) \Longrightarrow \mathfrak{f} = \mathfrak{g}.$$

When the structure A is considered to be variable, we shall write $\mathfrak{f}_*$ as whichever of $\mathfrak{f}_A$ or $A_{\mathfrak{f}}$ better expresses the emphasis intended at the time, and similarly $\mathfrak{f}^*$ as $\mathfrak{f}^A$ or $A^{\mathfrak{f}}$.

EXAMPLE 4·2·2: Let $M = (\mathbb{R}, 1, \times)$ and A be a vector space. Then multiplication of a vector $a \in A$ by a scalar $\mathfrak{f} \in \mathbb{R}$ is an action. Notice that it preserves the (additive) structure of A as well as the multiplicative (and additive) structure of $\mathbb{R}$.

Actions may be used to give the effective definition of a monoid or group.

EXAMPLE 4·2·3: **Rubik's cube** consists of 27 pieces jointed in such a way that any of the six faces (each with nine pieces) can be rotated through a quarter turn. From the home position, in which each face is uniformly coloured, $2^{12} \cdot 12! \cdot 3^8 \cdot 8!/12 \approx 4.3 \cdot 10^{19}$ positions can be reached. The quarter turns generate a group of this order which acts on the set of components of the cube.

To solve this puzzle, *i.e.* to restore a jumbled cube to its home position, you need to know the complicated laws which the generators satisfy. However, if I presented a group with six generators by just giving these laws, without telling you that it acts on this structure, the problem would be *more* difficult: there is no general algorithm to decide whether the group defined by an arbitrary presentation is finite, or on the other hand whether it is non-trivial. Syntax — the expression of elements of the group as strings of generators subject to laws ("relations") — gives us very little help in understanding the group. Indeed *the structure on which it acts is the only thing we can use to give an explicit description.* The semantics gives a kind of tally of the syntax, and then this group may be characterised quite straightforwardly in the language of group theory. The imperative interpretation (Remark 4.3.3) develops the "tally" idea.

It was originally thought that the following result, **Cayley's Theorem**, would eliminate the need for the *abstract* study of groups.

THEOREM 4·2·4: Every group has a faithful action of each variance.

PROOF: Let $(M, \text{id}, \circ)$ be a group or monoid, and put $A = M$, *i.e.* the underlying set. Then $\mathfrak{f}_* a = \mathfrak{f} \circ a$ and $\mathfrak{f}^* a = \mathfrak{f}; a = a \circ \mathfrak{f}$ define the covariant and contravariant **regular actions**, respectively. They may be seen to be faithful by considering the effect on $\text{id} \in A$. $\qquad\square$

The importance of the result is undisputed, as is shown by the fact that it will shortly turn into a Lemma[2] named after someone else, but the reductionist motive was mistaken. Both the abstract and concrete approaches are needed to complement one another. Indeed, the same group, *e.g.* $A_5 \cong PSL_2(5)$ (page 80), may have two intuitively unrelated concrete representations. Likewise, the beautifully simple but powerful theory of matrix representations of finite groups relies precisely upon considering *all* such representations. Again this supports the thesis that *what mathematical objects do is more important than what they are.*

The Rubik cube group was specified by means of its six **generators** and their action on the cube, from which we could see the **laws** or **relations**. The converse process — deriving the concrete form from the laws — is, as we have said, notoriously difficult. As for expressions in Lemma 1.2.4ff, we just have to accept that the equivalence classes can be constructed *somehow*. In Sections 5.6 and 7.4 we shall examine such quotienting (for arbitrary laws) categorically.

Sketches. Let us transfer these ideas from groups to categories, starting with generators and laws.

DEFINITION 4·2·5:   A **unary theory**, $\mathcal{L}$, consists of

(a) named *base* types or sorts, X, Y, ... (there are no constructors);

(b) one variable $x : X$ for each occurrence of each sort;

(c) unary operation-symbols or constructors $\mathfrak{r}$ (in preparation for our treatment of type theory we write $x : X \vdash \mathfrak{r}(x) : Y$ or just $X \vdash \mathfrak{r} : Y$, using the turnstile $\vdash$ instead of an arrow to emphasise that there is no function-type);

(d) laws, $\mathfrak{r}_n(\mathfrak{r}_{n-1}(\cdots \mathfrak{r}_2(\mathfrak{r}_1(x))\cdots)) = \mathfrak{s}_m(\mathfrak{s}_{m-1}(\cdots \mathfrak{s}_2(\mathfrak{s}_1(x))\cdots))$.

We write Σ for the set of sorts. A **free unary theory** has no laws.

EXAMPLES 4·2·6:   This basic tool has many different names, and it would be instructive to read through the remainder of this section several times, substituting each of the following points of view in turn.

(a) If there is only one sort, the operation-symbols generate a monoid (whose elements are the terms) subject to the given laws.

(b) If, further, there are no laws then these terms are simply *lists* of operation-symbols (Definition 2.7.2).

---

[2]Lemmas do the work in mathematics: Theorems, like management, just take the credit. A good lemma also survives a philosophical or technological revolution.

(c) Instead of specific laws it may be understood that *all* parallel pairs of maps are equal (Proposition 4.1.5). The sorts are just individuals (from a set Σ) without internal structure, and the operation-symbols are the instances of a binary relation $<$ on Σ. Terms are instances of the reflexive–transitive closure $\leqslant$ (Section 3.8 and Exercise 3.60).

(d) In particular, if the types are *propositions*, the operation-symbols are deduction steps and the terms are proofs.

(e) A free unary theory is just a (labelled) **oriented graph**. A graph in this sense may have loops and multiple edges, unlike in combinatorics (*cf.* Example 5.1.5(e)), and we use the word *oriented* to avoid confusion with directed diagrams (Definition 3.4.1). The types are called vertices or nodes, and the operation-symbols are (oriented) edges. Terms are paths. This is the many-sorted version of (b).

(f) A free unary theory may be seen as a **deterministic automaton**. The types are called states and the operation-symbols are actions. It is deterministic because the action labelled $\mathfrak{r}$ has a unique target. Now terms are *acceptable words* or *behaviour traces*, and may be described by a **regular grammar**.

(g) Unary theories are also called **linear** or **elementary sketches**. The types are objects, the operation-symbols are arrows or generating morphisms, and the laws are called commutative polygons.

(h) A unary theory also describes a **polyhedron** in which the types are vertices, the operation-symbols are oriented edges and the laws are faces, though it need not be embeddable in $\mathbb{R}^3$. Each raw term is given by an oriented path, but the faces generate an equivalence relation, as follows. A path which follows (all of) one half of a face may be "dragged across it" and is equal to the path taking the alternative route. The terms are homotopy classes, and *in*equations are holes (⊹).

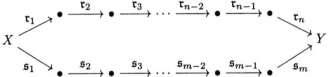

A law is a (so-called **commutative**) $(n+m)$-sided **polygon** which has exactly one source and exactly one target (or **sink**): at every other node there must be just one incoming and one outgoing edge. The target is the type of the term and the source is (the type of) the free variable. Variables are redundant, as is the notion of application: only associative graphical *composition* of arrows remains. Indeed commutative diagrams are the best way to illustrate unary first order equational reasoning.

When we draw sketches, as for instance in Example 4.6.3(f), we may name several nodes with the same type X. This is to avoid appearing to state unintended laws, where it is understood informally that the polygons which we draw are meant to commute. (This convention is made formal in [FS90], using the **puncture** symbol ∻ to indicate that an apparent law is not required, though this does not mean that the equation is forbidden in any interpretation.) All occurrences of the same named type X must nevertheless be interpreted by the same set A_X.

DEFINITION 4·2·7: A **model** (also called an **algebra, interpretation, representation** or **covariant action**) of an elementary sketch is

(a) an assignment of a set A_X to each type name X and

(b) a function $\mathfrak{r}_* : A_X \to A_Y$ to each operation-symbol $\mathfrak{r}$, such that

(c) each law $\mathfrak{r}_n(\cdots \mathfrak{r}_2(\mathfrak{r}_1(x))\cdots) = \mathfrak{s}_m(\cdots \mathfrak{s}_2(\mathfrak{s}_1(x))\cdots)$ holds, in the sense that $\mathfrak{r}_{n*}(\cdots \mathfrak{r}_{2*}(\mathfrak{r}_{1*}(a))\cdots) = \mathfrak{s}_{m*}(\cdots \mathfrak{s}_{2*}(\mathfrak{s}_{1*}(a))\cdots) \in Y$ for all $a \in A_X$, where $\mathsf{src}\,\mathfrak{r}_1 = X = \mathsf{src}\,\mathfrak{s}_1$ and $\mathsf{tgt}\,\mathfrak{r}_n = Y = \mathsf{tgt}\,\mathfrak{s}_m$.

EXAMPLES 4·2·8: An action of

(a) $\mathsf{List}(X)$ is given by $z_A : A$ and $s_A : X \times A \to A$ in Remark 2.7.10;

(b) a unary closure condition ▷ in 2 is a ▷-closed subset or trajectory in the set Σ of sorts (Examples 3.8.1);

(c) a polyhedron is a geometric realisation;

(d) an automaton is the regular language which it recognises.

PROPOSITION 4·2·9: Every elementary sketch has a faithful covariant action on its clones $\mathcal{H}_X = \bigcup_\Gamma \mathsf{Cn}_{\mathcal{L}}(\Gamma, X)$. Substitution for the variable defines a *contravariant* action on $\mathcal{H}^Y = \bigcup_\Theta \mathsf{Cn}_{\mathcal{L}}(Y, \Theta)$, also faithful.

PROOF: Recall from Notation 2.4.12 that the **clone**, $\mathsf{Cn}_{\mathcal{L}}(\Gamma, X)$, is the set of terms of type X in the context Γ, subject to the laws. In our case Γ consists of a single typed variable, so we abuse notation by writing Γ for the type too. A term of type X is a composable string of unary operation-symbols applied to a variable $\sigma : \Gamma$ (σ for "state"). The actions of $\mathfrak{r} : X \to Y$ on $\mathsf{Cn}(\Gamma, X) \subset \mathcal{H}_X$ and $\mathsf{Cn}(Y, \Theta) \subset \mathcal{H}^Y$ are

$$\mathfrak{r}_* \mathfrak{a}_n(\cdots \mathfrak{a}_2(\mathfrak{a}_1(\sigma))\cdots) = \mathfrak{r}(\mathfrak{a}_n(\cdots \mathfrak{a}_2(\mathfrak{a}_1(\sigma))\cdots)) \in \mathsf{Cn}_{\mathcal{L}}(\Gamma, Y)$$
$$\mathfrak{r}^* \mathfrak{z}_m(\cdots \mathfrak{z}_2(\mathfrak{z}_1(y))\cdots) = \mathfrak{z}_m(\cdots \mathfrak{z}_2(\mathfrak{z}_1(\mathfrak{r}(x)))\cdots) \in \mathsf{Cn}_{\mathcal{L}}(X, \Theta)$$

with $\sigma : \Gamma$, $x : X$ and $y : Y$; they are faithful by considering the empty strings ($n = m = 0$). $\qquad\square$

Analogously, Definitions 3.1.7 and 3.9.6ff represented each element x of a poset X covariantly as the lower subset $X \downarrow x$ and contravariantly as the upper set $x \downarrow X$. Lists were used to form the transitive closure of a binary relation in Exercise 3.60. We shall postpone the analogue of Proposition 3.1.8 (called the Yoneda Lemma) to Theorem 4.8.12.

EXAMPLE 4·2·10:   The Lindenbaum algebra (Example 3.1.9) gives the regular representation of *propositions*. For *types* this is the term model.

(a) The covariant representation of a term is the effect of substituting values for its free variable.

(b) The contravariant representation is the result of substituting the term itself for a free variable in other terms (***continuations***).   □

Convention 4.1.2, that juxtaposition means application on the left, is clearly not very appropriate for unary theories. Many algebraists, in group theory in particular, apply functions on the right, and abstract this to composition without any sign; this is unambiguous in that subject because the language is first order and strongly typed. As we shall soon be passing on to many-argument and higher order languages, we shall put up with the earlier convention.

This interpretation is **sound**: the semantics obeys the rules specified by the syntax. It is also **complete**: any two terms which are semantically equal may be proved to be so using the given rules, because we only made them equal when the rules said so; this is what *faithful* means.

Generating a category. A faithful action can be used to represent a category concretely: the object X *is* the set A_X and maps $f : X \to Y$ *are* those functions ("homomorphisms") $A_X \to A_Y$ which arise as actions. In the case of the Cayley–Yoneda action, we shall characterise them intrinsically in terms of naturality in Example 4.8.2(f).

The category obtained from an elementary sketch *via* its action has *the same objects* as the sketch. The maps of the category are composites of those of the sketch. In the case of a *free* unary theory, the maps are lists (Definition 2.7.2) of generators, though we postpone the proof to Theorem 6.2.8(a). With laws, they are equivalence classes of lists.

Since the action is faithful, the only equations amongst maps are those provable from those given in the sketch plus the axioms for a category. Lemma 1.2.4 gave the preorder version of this construction; recall that it made objects *interchangeable* (isomorphic), but didn't pretend to make them *equal*.

LEMMA 4·2·11: The hom-set $\mathsf{Cn}_{\mathcal{L}}(X, Y)$ in the category is the clone of the same name. The identity id_X is the variable $x : X$ *quâ* term, and composition is substitution, which is associative.   □

The category *saturates* the sketch by adding into $\mathsf{Cn}_\mathcal{L}(X,Y)$ all **derived operations** of type Y (these are just composites in the unary case, *cf.* the reflexive–transitive closure of a relation), and making all provable identifications amongst them. To sum up the precise relationship,

THEOREM 4·2·12:   Every elementary sketch presents a category, and conversely, any small category $\mathcal{C}$ is presented by some sketch $\mathcal{L}$ in the sense that $\mathcal{C} \cong \mathsf{Cn}_\mathcal{L}$. We write $\ulcorner - \urcorner$ for this isomorphism and call the sketch $\mathcal{L} = \mathsf{L}(\mathcal{C})$ the **canonical elementary language** of $\mathcal{C}$ (*cf.* Proposition 3.9.3). It is defined as follows:

(a) the sorts $\ulcorner X \urcorner$ of $\mathsf{L}(\mathcal{C})$ are the objects X of $\mathcal{C}$,

(b) the operation-symbols $\ulcorner \mathfrak{f} \urcorner$ are its morphisms $\mathfrak{f}$, and

(c) the laws are $\ulcorner \mathsf{id} \urcorner(x) = x$ and $\ulcorner \mathfrak{g} \urcorner(\ulcorner \mathfrak{f} \urcorner(x)) = \ulcorner \mathfrak{g} \circ \mathfrak{f} \urcorner(x)$,

using $\ulcorner X \urcorner$ and $\ulcorner \mathfrak{f} \urcorner$ to distinguish the linguistic sorts and operation-symbols from the objects and morphisms of the original category.   □

We are accustomed to writing languages in alphabets of 26 or maybe 128 enumerated symbols, but for this construction we need a "letter" for each object and morphism of $\mathcal{C}$. Questions such as whether we can distinguish between letters or form a dictionary of the words now arise, to which one answer might be a severe restriction on the applicability of this result. This is not the line which we follow, but the issues deserve separate consideration, which we defer to Section 6.2.

The notion of sketch interpolates between those of oriented graph and category: it mentions *some* composites, where the extremes require none or all of them. For a category, the representations of an object X given by Proposition 4.2.9 reduce simply to $\bigcup_\Gamma \mathcal{C}(\Gamma, X)$ and $\bigcup_\Theta \mathcal{C}(X, \Theta)$, where we no longer have to form explicit composites.

The geometrical interpretation shows that unary theories have an input–output symmetry. This is broken by the *term* calculi of the richer type theories, but the closer analogy is retained if we consider programs or substitutions instead. By examining the covariant and contravariant actions of the latter, we shall next give a construction of the category of contexts and substitutions which is directly applicable to a very wide class of formal languages. The use of sketches to do this is new.

This section has shown how categories can present combinatorial data in algebra, topology and logic. The proofs are not complicated, but nor are the ideas trivial or immediately grasped: you should go back and use them to express any familiar examples as a category. This is important because our account of substitutions and hence the semantics of type theory depend on it.

4.3   CATEGORIES FOR FORMAL LANGUAGES

The algebraic notation which is standard today only became so in the seventeenth century. By this time general methods for solving quadratic (in ancient Babylonia), cubic and quartic equations (Renaissance Italy) had already been found and expressed in an algorithmic or declarative style. Indeed if we put the solution of the cubic into closed or normal form it would become incomprehensible. This presentation is convenient for us too, because it is *already* the category we require.

DEFINITION 4·3·1:   The **direct declarative language** has the following syntax for ⟨program⟩s (sometimes called commands):

> **skip**
> **put** ⟨variable⟩ = ⟨term⟩
> **discard** ⟨variables⟩
> ⟨program⟩ ; ⟨program⟩

subject to the **weak variable convention** that no declaration may be made of a variable (with the same name as one) which is already in scope. A variable is said to be **in scope** from its **put** to its **discard** command; of course only the variables in scope may occur freely in ⟨term⟩s. In the construction of this section, the notion of ⟨term⟩ is meant to be an indeterminate one, which will be taken to be algebra in Section 4.6, and the λ-calculus in Section 4.7. Conditionals, loops and other constructs will be added later. The **put** command is commonly called **let**, but we wish to maintain the distinction made in Remark 1.6.2 between definite and indeterminate values.

The direct declarative language alone is rather mundane, so we shall borrow an extension from Section 5.3 to present a famous example.

Operational interpretation. Treating the language as *imperative*, the **state** at any point of the execution of a program like this is determined by the tuple of current values of the variables in scope. The type of states is the cartesian product of the sets over which the variables range. Notice that the **put** and **discard** commands *change not only the value of the state σ but also its type*, so we really need a category and not just a monoid (or semigroup, *i.e.* monoid without identity) to interpret the language.

REMARK 4·3·2:   As in Example 4.2.10, programs are represented

(a) contravariantly by their effect on continuations, and

(b) covariantly by changing the **initialisations** of program-variables,

where composition (;) performs substitution of values for variables.

REMARK 4·3·3:   Our language does not seem to have an imperative flavour, but in fact the weak variable convention allows **assignment** to be defined in it:

$$x := a \quad \text{is the same as} \quad \begin{cases} \textbf{put } x' = a; \\ \textbf{discard } x; \\ \textbf{put } x = x'; \\ \textbf{discard } x' \end{cases}$$

where x' is a new variable, *i.e.* not in scope. This is a sleight of hand: from outside this fragment of code it appears that the value of x has been changed, *cf.* "without loss of generality" (Remark 1.6.9).

EXAMPLE 4·3·4:   This program[3] finds the one real (x_0) and two possibly complex roots ($x_\pm$) of the cubic equation $x^3 + ax^2 + bx + c = 0$. The index n takes each of the values 0, +1 and −1, and $\omega = -\frac{1}{2} + \frac{1}{2}\sqrt{-3}$ denotes a complex cube root of unity.

$$\begin{aligned}
&\textbf{put } p = \tfrac{1}{9}a^2 - \tfrac{1}{3}b; && \{b = \tfrac{1}{3}a^2 - 3p\} \\
&\textbf{put } q = -\tfrac{1}{27}a^3 + \tfrac{1}{6}ab - \tfrac{1}{2}c; && \{c = \tfrac{1}{27}a^3 - 2q - ap\} \\
&\textbf{discard } b, c; \\
&\textbf{if } p^3 \leqslant q^2 \\
&\textbf{then put } s = \sqrt{q^2 - p^3}; \\
&\qquad \textbf{discard } p; \\
&\qquad \textbf{put } u, v = \sqrt[3]{q \pm s}; && \{p = \sqrt[3]{q^2 - s^2} = uv\} \\
&\qquad \textbf{discard } q, s; && \{2q = u^3 + v^3\} \\
&\qquad \textbf{put } y_n = \omega^n u + \omega^{-n} v; && \{y_0 + y_+ + y_- = 0\} \\
&\qquad \textbf{discard } u, v && \{y_0 y_+ y_- = u^3 + v^3\} \\
& && \{y_0 y_+ + y_0 y_- + y_+ y_- = -3uv\} \\
&\textbf{else put } r = \sqrt{p}; && \{p = r^2\} \\
&\qquad \textbf{put } \vartheta = \cos^{-1}(q/r^3); && \{q = r^3 \cos\vartheta\} \\
&\qquad \textbf{discard } p, q; \\
&\qquad \textbf{put } y_n = 2r \cos \tfrac{1}{3}(\vartheta + 2\pi n); && \{y_0 + y_+ + y_- = 0\} \\
&\qquad \textbf{discard } r, \vartheta && \{y_0 y_+ y_- = 2r^3 \cos\vartheta\} \\
& && \{y_0 y_+ + y_0 y_- + y_+ y_- = 6r^2 \cos\tfrac{2\pi}{3}\} \\
&\textbf{fi}; && \{y_n^3 - 3p y_n - 2q = 0\} \\
&\textbf{put } x_n = y_n - \tfrac{1}{3}a; && \{x_n^3 + ax_n^2 + bx_n + c = 0\} \\
&\textbf{discard } a, y_n
\end{aligned}$$

---

[3]The **then** part of the conditional was discovered by Scipione del Ferro about 1500. Niccolo Tartaglia found it independently and, after years of intense pressure, told it in verse to Gerolamo Cardano in 1539, on condition that it remain secret. Cardano, however, published it (with due credit) in his *Ars Magna*, in 1545, together with the similar method for quartics discovered by his student Ludovico Ferrari. The **else** part, known throughout the sixteenth century as the *casus irreducibilis*, was found by François Viète in 1591. See [Str69].

Logical interpretation. What the covariant imperative action means for more complex languages becomes less clear. It also has the weakness that it acts on terms which may be substituted for the free variables, which must therefore belong to the same calculus, here a programming language. The contravariant action, by substitution *into* expressions, is not restricted in this way: these expressions may belong to a much richer language, such as higher order logic. Recall that we also used predicates (or subsets) to represent posets in Definition 3.1.7ff.

This method of proving partial correctness of programs is due to Robert Floyd [Flo67], though he presented it for flow charts. The notation $\gamma\{u\}\phi$ is due to Tony Hoare. Floyd also gave the criterion for termination (Remark 2.5.13).

REMARK 4·3·5: For any program u, we write

$$\{\gamma\}u\{\vartheta\}$$

to mean "if γ was true before executing u then ϑ will be true afterwards", γ and ϑ being known as *pre-* and *postconditions* respectively.

These satisfy the sequent-style rules

$$\big\{[a/x]^*\vartheta\big\} \textbf{ put } x = a \ \{\vartheta\}$$

$$\{\vartheta\} \textbf{ skip } \{\vartheta\}$$

$$\{\hat{x}^*\vartheta\} \textbf{ discard } x \ \{\vartheta\}, \text{ where } x \notin \mathsf{FV}(\vartheta)$$

$$\frac{\{\gamma\} \, u \, \{\phi\} \qquad \{\phi\} \, v \, \{\vartheta\}}{\{\gamma\} \, u \, ; v \, \{\vartheta\}}$$

$$\frac{\gamma' \Rightarrow \gamma \qquad \{\gamma\} \, u \, \{\vartheta\} \qquad \vartheta \Rightarrow \vartheta'}{\{\gamma'\} \, u \, \{\vartheta'\}}$$

Or, in the informal notation (without $\hat{x}^*$ from Notation 1.1.11),

$$\{\vartheta[a]\} \textbf{ put } x = a \ \{\vartheta[x]\} \qquad \{\vartheta\} \textbf{ discard } x \ \{\vartheta\}.$$

In the case of the whole program, γ and ϑ constitute the *specification*, for instance in the Example it is that the program produces the solution to the cubic equation. Of course we may always take $\gamma = \bot$ and $\vartheta = \top$, but this vacuous specification says that the program is good for nothing.

It is natural to insert *midconditions* between the lines of the program, instead of this repetitive sequent-style notation. A fully proved program consists of phrases of proof interrupted by single commands. The latter, together with the proof lines either side, must obey the Floyd rules.

The midconditions need not be computable. Even when they are, it would often be more difficult to compute them than the program itself. Sometimes they involve universal quantification over infinite sets, so they are of strictly greater logical (quantifier) complexity. The proofs may be arbitrarily complicated: it is true that given integers $x, y, z \geqslant 1$ and $n \geqslant 3$ the program $x^n + y^n - z^n$ will always produce a non-zero answer, but the proof-phrase took 357 years!

REMARK 4·3·6: We may also read **put** as an *abbreviation, local definition* or (as we called it in Definition 1.6.8) a ***declaration*** within the proof. We showed there that a declaration may itself be treated as an $(\exists \mathcal{E})$-proof box — the scope of the variable. The box is open-ended: it extends until the end of the argument, or of the enclosing conditional branch.

As we saw, the weak variable convention allows us to define assignment, but the proof rules then become much more complicated. The original (strong) convention gives ***referential transparency***, the free ability to import formulae (*cf.* Lemma 1.6.3). But if we have shown that x is odd and then do $x := 2$ we may no longer use the previous knowledge. For this reason assignment is to be avoided in programming.

The logical interpretation gives a *contravariant* action:

REMARK 4·3·7:   For any u and ϑ there is a ***weakest precondition***, obtained by letting u act on ϑ by substitution:

$$\{\gamma\} u \{\vartheta\} \iff \gamma \vdash u^* \vartheta.$$

Normal forms. The operational interpretation — the execution of the program on numerical values — is not the only notion of computation. As in the λ-calculus, there are rules for rewriting programs with the aim of putting them in normal (or, as it is called in algebra, closed) form.

REMARK 4·3·8:   The operational and logical interpretations satisfy

$$\begin{aligned}
\textbf{put } x = a \, ; \textbf{discard } x &\equiv \textbf{skip} \\
\textbf{put } x = a \, ; \textbf{put } y = b &\equiv \textbf{put } y = [a/x]^* b \, ; \textbf{put } x = a \\
\textbf{put } x = a \, ; \textbf{discard } y &\equiv \textbf{discard } y \, ; \textbf{put } x = a \\
\textbf{discard } x \, ; \textbf{discard } y &\equiv \textbf{discard } y \, ; \textbf{discard } x \\
\textbf{put } y = x \, ; \textbf{discard } x \, ; \textbf{put } x = y \, ; \textbf{discard } y &\equiv \textbf{skip}
\end{aligned}$$

where x and y are distinct variables with $x, y \notin \mathsf{FV}(a)$ and $y \notin \mathsf{FV}(b)$. Of course **skip** is the identity and ; is composition. The last law, which is not redundant, says that $x := x$ does nothing. Assignment is the simplest case of conflict between the names of input and output variables, which must be resolved by renaming. We shall discuss the orientation of these laws as reduction rules in Remark 8.2.7.

THEOREM 4·3·9: Every program of the direct declarative language is equivalent (with respect to the above laws) to one in **normal form**:

put $z = q$; **discard** x ; **put** $y = z$; **discard** z

where $\mathsf{FV}(q) \subset x$ are the inputs and y the outputs. This means

(a) first the (renamed) outputs are declared by $\langle \text{term} \rangle$s in which only the input variables occur, not the output ones;

(b) then the input variables are **discard**ed;

(c) finally the output variables are renamed. □

The proof (by induction on the length of the program) is left as a valuable exercise. The point is to show that the laws suffice to capture the familiar process of eliminating intermediate variables (p, q, *etc.* in the Example). Beware that we are only normalising the connectives defined in this section: the theorem says nothing about any normal forms of the $\langle \text{term} \rangle$s themselves (*e.g.* from the λ-calculus). The normal form is unique (up to order, which may be canonised, and the choice of new names), so it may be used to compare programs and make deductions about commutative diagrams (Remark 7.6.12). But it is not a good way to define the category, as (the proof of) the theorem is needed every time we compose two morphisms, *cf.* Example 1.2.1.

REMARKS 4·3·10:

(a) Normalisation uses the second law from left to right, but (in the opposite sense) the derived law

put $x = [u/z]^* a \rightsquigarrow$ **put** $z = u$; **put** $x = a$; **discard** z

decomposes terms in algebra into operation-symbols and sub-terms, and hence to operation-symbols alone. On a machine the latter are calls to library routines, so this process (**compilation**) reduces any program to a sequence of such calls. The second law then becomes redundant, being replaced by those of the language itself.

(b) The weak variable convention is necessary to define the category composed of programs.

The object at each semicolon
is the set of typed variables in scope there.
In particular the source and target of the program are the lists of types of the input and output variables respectively. It is convenient to assume that these have no local variable names in common.

(c) From the example and theorem, we see that the **put** and **discard** commands do not form a nested system like proof boxes; indeed if they did the program would just throw away its results.

Remark 1.6.5 already gives some freedom to choose when to close $(\exists\mathcal{E})$-boxes and definitions. The laws for **discard** extend this *conservatively*, *i.e.* they give no new equivalences between properly nested boxes. In fact the proof of Lemma 1.6.6 illustrates that the *natural* scope of $(\exists\mathcal{E})$-boxes is not necessarily nested.

(d) The **discard** commands (which Floyd called **undefine**) have been added for an exact match with the categorical concepts which are used to interpret the language, but experience shows that a compiler can do a more accurate and efficient job if it is given better type and resource information about the intentions of the programmer. If the variables obey the strong Convention 1.1.8 (*i.e.* they cannot be reused, so assignment is not allowed) the laws allow us to move all of these commands to the end of the program. As compilers do this automatically, **discard** is redundant in programming.

The language defines a sketch whose objects are sets of typed variables; the maps are programs whose source and target objects are the variables which are in scope before and after. We have given familiar covariant and contravariant actions, and five laws which are *sound* for them. The normal form theorem shows that the substitution action is *faithful*, and indeed that we need only consider its effect on variables; in other words the five laws provide a *complete* axiomatisation.

The category of contexts and substitutions. These programs are not exactly the notation which we introduced in Section 1.1, but the difference is "syntactic sugar".

DEFINITION 4·3·11:   The *category of contexts and substitutions*, which is called $\mathsf{Cn}_{\mathcal{L}}^{\times}$, is presented by the following elementary[4] sketch:

(a) The objects of $\mathsf{Cn}_{\mathcal{L}}^{\times}$ are the contexts of $\mathcal{L}$ (Notation 2.3.6), *i.e.* finite lists of distinct variables, $[x_1 : X_1, x_2 : X_2, \ldots, x_k : X_k]$, together with their types. In shorthand, $[\underline{x} : \underline{X}]$, $[\underline{X}]$, $\underline{x}$ or Γ, with len $\underline{x} = k$. As far as Section 4.6, the X_i will just be base types (sorts), then in Sections 4.7 and 5.3 we shall begin to allow *expressions* such as

---

[4]The category which is presented has finite products, given by concatenation of contexts. Since all of the objects of the intended category have already been mentioned in the sketch, we do not want to add any more by regarding it as a finite product sketch. In fact it *can* be treated as a finite product sketch, by specifying that each context (list of types) is the vertex of a product cone consisting of those types. The reason for not taking this approach is that it no longer works for generalised algebraic theories, where the contexts have to be generated recursively *as a whole* (Definition 8.2.2). We do not treat product sketches *as such* in this book, though Section 7.6 constructs the canonical language of a semantic category using essentially sketch-theoretic ideas.

$X \to Y$ and $X + Y$ for the types of the terms. Lists of types on both sides of the turnstile mean products.

(b) The generating morphisms are **put** (declaration) and **discard**, *i.e.*
- single substitutions $[a/x] : \Gamma \to [\Gamma, x : X]$ for each term $a : X$ in the context Γ, where x is a new variable of the same type X;
- single omissions $\hat{x} : [\Gamma, x : X] \longrightarrow \Gamma$ for each variable $x : X$; a morphism of this special form is called a ***display map***, for reasons which will emerge in Chapter VIII.

The extra variable x may be inserted *anywhere* in the list.

(c) The Extended Substitution Lemma (1.1.12) gives the laws:

$$[a/x] \, ; \hat{x} \qquad\qquad = \quad \mathsf{id}$$
$$[a/x] \, ; [b/y] \qquad\quad = \quad [[a/x]^* b/y] \, ; [a/x]$$
$$[a/x] \, ; \hat{y} \qquad\qquad = \quad \hat{y} \, ; [a/x]$$
$$\hat{x} \, ; \hat{y} \qquad\qquad\quad = \quad \hat{y} \, ; \hat{x}$$
$$[x/y] \, ; \hat{x} \, ; [y/x] \, ; \hat{y} \quad = \quad \mathsf{id}$$

where $x \not\equiv y$, $x, y \notin \mathsf{FV}(a)$ and $y \notin \mathsf{FV}(b)$. They are shown as commutative diagrams in Section 8.2.

For a Horn theory, Theorem 3.9.1 generated the analogous preorder $\mathsf{Cn}_{\mathcal{L}}^{\wedge}$ of contexts under provability, by the two cases of omitting a proposition from a context and using a single instance of the closure condition.

There is nothing in this construction which is peculiar to either algebra, programming or the λ-calculus: it may be applied to any typed calculus of substitutions. The maps have been written with a substitution notation, because *this* is the notion of composition. The λ-calculus defines *another* composition operation *via* abstraction and application, but it is only associative after the β- and η-rules (Definition 4.7.6) have been imposed, and then the two forms of composition agree. Substitution is a primitive of symbolic manipulation, the λ-calculus is not.

In Section 4.7 we shall begin to add type constructors such as $\to$ to the logic. Then $[f : (X \to Y)]$ will be a valid context, and so will be added to the category as a new object. We shall write $\mathsf{Cn}_{\mathcal{L}}^{\to}$ for the larger category, in which the morphisms are formed by λ-abstraction and application. But it turns out (and this is an important *theorem*) that there will be no additional maps between the *old* objects, nor do maps which were previously distinguished become equal: we say that $\mathsf{Cn}_{\mathcal{L}}^{\to}$ is a ***conservative extension*** of $\mathsf{Cn}_{\mathcal{L}}^{\times}$. It may provide more powerful methods of reasoning, without doing anything which we couldn't have done before. In other words, it gives short proofs of facts which were already true in the simpler system, but which would have taken much (maybe hyper-exponentially) longer to prove. Other type constructors extend the category further, and we write $\mathsf{Cn}_{\mathcal{L}}^{\square}$ for the generic situation; in fact we have already made such an extension from the unary case in Section 4.2. We shall discuss conservativity in Sections 7.6 and 7.7.

Terms as sections. Theorem 4.3.9 substantiates the remarks about simultaneous substitution which we made after we first introduced the Substitution Lemma 1.1.5.

NOTATION 4·3·12: Any map may be written uniquely as a **multiple** or **simultaneous substitution**,

$$[\underline{x} : \underline{X}] \xrightarrow{[\underline{a}/\underline{y}]} [\underline{y} : \underline{Y}],$$

i.e. a sequence of bindings of terms to variables, where y_j and a_j have type Y_j, and $\mathsf{FV}(a_j) \subset \{\underline{x}\}$. Composition is by simultaneous substitution and the identity is $[\underline{x}/\underline{x}]$.

The sources and targets of the maps are ambiguous in this notation. There may be more variables in the source context than are mentioned in the substituted terms, and it is not clear whether they should survive or be forgotten in the target. Indeed we took advantage of the ambiguity by saying that the substitutions for different variables *commute*. The (strict) notion of commutativity in monoids does not extend to categories because the sources must agree with the targets, but it becomes meaningful in situations like this, where maps with different endpoints have "essentially" the same effect, differing only in their passive contexts. This is what led to the "commutative" diagram terminology.

COROLLARY 4·3·13:

(a) A map $\mathfrak{a} : \Gamma \to [\Gamma, x : X]$ is a single substitution iff it is a **section of the display** $\hat{x} : [\Gamma, x : X] \twoheadrightarrow \Gamma$, *i.e.* $\mathfrak{a} \,;\, \hat{x} = \mathrm{id}_\Gamma$. Then $\mathfrak{a} = [a/x]$, where the term $\Gamma \vdash a : X$ is determined by $a = \mathfrak{a}^* x$.

(b) Sections $\mathfrak{a} : \Gamma \to [\Gamma, x : X] \cong \Gamma \times X$ correspond bijectively to **total functional relations** $\Gamma \rightharpoonup X$, *cf.* Exercise 1.14.

(c) The *clone* $\mathsf{Cn}_{\mathcal{L}}(\Gamma, X)$ is isomorphic to the *hom-set*[5] $\mathsf{Cn}_{\mathcal{L}}^{\times}(\Gamma, [x : X])$, whence the deliberate use of similar notation.

(d) More generally, $\mathsf{Cn}_{\mathcal{L}}^{\times}(\Gamma, [\underline{x} : \underline{X}]) \cong \prod_{i=1}^{\mathrm{len}\,\underline{X}} \mathsf{Cn}_{\mathcal{L}}(\Gamma, X_i)$. □

We shall use display maps and their sections to recover a generalised algebraic from a category in Chapter VIII.

---

[5]The square brackets around a context serve to disambiguate the use of the comma — it separates both the source and target in a hom-set and items in lists — we write $[\Gamma, x : X]$ instead of $\Gamma, [x : X]$ for augmented contexts.

Use of variables. In the last section terms of a *unary* language were called (oriented) *strings* of operations.

REMARK 4·3·14: Let's say *wires* instead; then a morphism in the many-argument version is to a term in the unary one as a multi-core *cable* is to a single wire. We now need a way to distinguish the wires, where we did not before: this is what variables do. We use 𝔊erman and *italic* letters for the cables and *single wires* respectively.

In the foregoing account we have chosen to *colour* the wires with variable-names. Many authors prefer to *number* the pins in the plugs and sockets instead. That is, they specify that $x_1, \ldots, x_n$ are the *actual* names rather than meta-variables (*cf.* Definition 1.1.9), and have to *re*number them at every stage. According to this convention, a morphism is a list of *equivalence classes* of assignments of the form $[x_1, \ldots, x_n \mapsto a]$, where

$$[x_1, \ldots, x_n \mapsto a] = [y_1, \ldots, y_n \mapsto [\underline{y}/\underline{x}]^* a].$$

This ubiquitous renaming needlessly complicates the formal discussion of the notation. It is useful as an *informal* way of naming a single map, to avoid choosing names for the target variables. However, in a diagram, where such maps are to be composed, it is often more convenient to attach distinct variables to the objects (*i.e.* to adopt our convention) in order to use them in a global symbolic argument about commutativity.

In this book we shall keep an explicit distinction between *free* variables, so that $[x : X]$ and $[y : X]$ are isomorphic but unequal contexts. The difference is that our category has many isomorphic duplicates of each object X, but the leaner one may be obtained from it by a straight-forward construction from abstract category theory (Exercise 4.7). We have chosen this convention in order to take best advantage of variables as they are normally used in mathematics, namely to relate quantities defined in one part of any argument to their use in another.

A change of *free* variables we call **open α-equivalence**. When we come to λ-abstraction, terms differing only by the names of corresponding *bound* variables will be considered to be the same. The context says which free variables are allowed, but the bound ones are unlimited.

A map is in fact not just a cable but a device with inputs and outputs. Corollary 4.3.13(d) says that if the output is a tuple then it may be split into several (multiple-input,) single-output devices. Yet another common metaphor is to think of each term as a *tree* (Remark 1.1.1) and the maps are *forests* (collections of trees); composition is by substitution of the roots of one forest for the leaves of another! Theorem 3.9.1 also described $\mathsf{Cn}_{\mathcal{L}}^{\wedge}$ using proof trees.

4.4 FUNCTORS

Since an action takes composition in the syntax to composition in the semantics, it is an example of a *homomorphism* of categories.

DEFINITION 4·4·1: A *functor* $F : C \to \mathcal{D}$ between categories is

(a) a (class) function $F_o : \mathrm{ob}\, C \to \mathrm{ob}\, \mathcal{D}$, together with

(b) a function $F_{X,Y} : C(X, Y) \to \mathcal{D}(F_o X, F_o Y)$ for each pair of objects $X, Y \in \mathrm{ob}\, C$

which preserves the structure (identity and composition), *i.e.*

$$F_{X,X}(\mathrm{id}_X^C) = \mathrm{id}_{F_o(X)}^{\mathcal{D}} \qquad F_{X,Z}(\mathfrak{f} \,;^C \mathfrak{g}) = F_{X,Y}(\mathfrak{f}) \,;^{\mathcal{D}} F_{Y,Z}(\mathfrak{g}).$$

A functor $F : C \to \mathcal{D}^{\mathrm{op}}$ or $F : C^{\mathrm{op}} \to \mathcal{D}$ may be called a *contravariant* functor from C to $\mathcal{D}$, the usual case being styled *covariant* if emphasis is needed. To avoid the confusion caused by discussing morphisms of an opposite category explicitly when describing contravariant functors, it is usual simply to define $F_{X,Y} : C(X, Y) \to \mathcal{D}(F_o Y, F_o X)$.

Since the essence of a functor is that it is defined in a "coherent" fashion for all objects and morphisms together, the subscripts and superscripts are omitted: we write FX and $F\mathfrak{f}$ for the application of the functor to an object or morphism. If it is defined on objects by built-in notation such as $C(X, -)$ or $Y^{(-)}$ this can look a bit strange when applied to maps.

Of course given another functor $G : \mathcal{D} \to \mathcal{E}$ we can apply this too, writing the result as $G(FX)$ or $G(F\mathfrak{f})$, with the brackets. The abstract theory of functors is a good example of a unary language (Definition 4.2.5), and would be clearer in the left-to-right notation without operators or brackets. For the sake of conformity with other notations and concepts, we shall, however, always write composition of functors from right to left as $G \cdot F$, and not using juxtaposition.

EXAMPLES 4·4·2: Following 4.1.5 and 4.1.6, a functor

(a) between preorders considered as categories is exactly a monotone function (Definition 3.1.5), and a contravariant functor is antitone;

(b) between monoids, groups or groupoids is exactly a homomorphism;

(c) between equivalence relations is a function between their quotients (Remark 1.3.2, Examples 2.1.5 and 3.1.6(d));

(d) from a group to the category of vector spaces is a linear or matrix representation of the group;

(e) from a poset $\mathcal{X}$ to $\mathfrak{Q}$ is an upper subset of $\mathcal{X}$ (Example 3.1.6(f)), and $\mathcal{X}^{\mathrm{op}} \to \mathfrak{Q}$ is a lower subset (Definition 3.1.7);

(f) from $\mathcal{C}$ to *Set* is a covariant action of $\mathcal{C}$ (Definition 4.2.7);

(g) from $\mathcal{C}^{\text{op}}$ to *Set* is a contravariant action of $\mathcal{C}$ on sets; it is also called a **presheaf** on $\mathcal{C}$, *cf.* Definitions 3.1.7 and 3.9.6. □

Constructions as functors.

EXAMPLES 4·4·3: The following are often known as **forgetful functors** or **underlying set functors**. This terminology should only ever be used when the meaning, *i.e.* just what is being forgotten, is completely clear from the presentation of the category. Notice that we may forget (a) properties of objects, (b) properties of morphisms or (c) structure on an object together with the property of morphisms that they preserve the structure. The last is the commonest situation. In all cases composition is preserved because it is defined in the same way on both sides.

(a) $\mathcal{P}os \to \mathcal{P}reord$, $\mathcal{D}\mathcal{L}at \to \mathcal{L}at$, $\mathcal{IPO} \to \mathcal{D}cpo$, $\mathcal{A}b \to \mathcal{G}p$, $\mathcal{CM}on \to \mathcal{M}on$ *etc.* which forget the significance of laws and the descriptions of special elements such as $\bot$;

(b) $\mathcal{P}os^{\dashv} \to \mathcal{P}os$, $\mathcal{S}et \to \mathcal{P}fn$ and $\mathcal{P}fn \to \mathcal{R}el$ which forget that all joins exist, and totality and functionality of relations;

(c) $\mathcal{H}eyt \to \mathcal{L}at \to \mathcal{SL}at \to \mathcal{P}os \to \mathcal{S}et$, $\mathcal{D}cpo \to \mathcal{P}os$, $\mathcal{M}on \to \mathcal{S}et$ and $\mathcal{R}ng \to \mathcal{A}b$ which forget operations and their preservation.

Besides forgetting things, functors also arise from constructions, where now one may need to check preservation of (identities and) composites.

EXAMPLES 4·4·4: The following are functors:

(a) $\mathcal{P}fn \to \mathcal{P}os$, which takes a set X to its lift, $X_\bot$, with the information order (Definition 3.3.7), and a partial function $f : X \rightharpoonup Y$ to the monotone function $(U \in X_\bot) \mapsto \{y \mid \exists x \in U. \, x \overset{f}{\mapsto} y\}$;

(b) $\mathcal{R}el \to \mathcal{CSL}at$ by $X \mapsto \mathcal{P}(X)$ and $R \mapsto (U \mapsto \{y \mid \exists x \in U. \, xRy\})$;

(c) $\mathcal{CSL}at \to \mathcal{P}os^{\dashv}$, which equips a function that preserves all joins with its unique right adjoint (Theorem 3.6.9);

(d) $\mathcal{D}cpo \to \mathcal{S}p$ by the Scott topology (Proposition 3.4.9);

(e) $\mathcal{S}p \to \mathcal{L}oc \equiv \mathcal{F}rm^{\text{op}}$ by the frame of open sets;

(f) $\mathcal{S}p \to \mathcal{P}reord$ by the specialisation order (Example 3.1.2(i)).

PROOF:

[a] First check that the result of the functor applied to a map is a map of the right kind, in this case $\{y \mid \exists x. \, x \in U \wedge \langle x,y \rangle \in f\} \in Y_\bot$. This

and preservation of composition are technically the same as the fact
that relational composition preserves functionality (Lemma 1.6.6).

[b] Powersets have arbitrary joins, given by unions. These are preserved
by the formula shown for morphisms, and in fact any join-preserving
function $\mathcal{P}(X) \to \mathcal{P}(Y)$ arises uniquely in this way.

The other examples rely on composition of adjunctions (Lemma 3.6.6)
and of continuous functions.   □

A classifying category. We saw that a category is what is required to
express a unary algebraic theory. An interpretation of such a theory is
similarly given by a functor. Any category may play the role of *Set*: we
restrict to the special case simply because we did in Section 4.2.

THEOREM 4·4·5:   Let $\mathcal{L}$ be a unary language and $\mathsf{Cn}_{\mathcal{L}}$ the category it
presents by Theorem 4.2.12. Then *interpretations of $\mathcal{L}$ correspond to
functors* $\mathsf{Cn}_{\mathcal{L}} \to Set$.

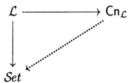

PROOF: Let the interpretation be A_X on sorts and $A_{\mathsf{r}} : A_X \to A_Y$ on
operations. These are already part of the required data for a functor
$A_{(-)} : \mathsf{Cn}_{\mathcal{L}} \to Set$, but it remains to define its effect on strings. *This is
uniquely determined by preservation of (the identity and) composition.*
Using list recursion, the identity is the base case and composition (cons)
the recursion step. Proposition 2.7.5, which showed that append is associative, guarantees that this too is preserved.

Where a law is given to hold in the interpretation of $\mathcal{L}$, this means
exactly that the functor takes equal values on the corresponding strings
of operation-symbols. Conversely any functor $\mathsf{Cn}_{\mathcal{L}} \to Set$ restricts to
the sorts and generating arrows in a way which satisfies the laws.   □

How interpretations and functors correspond is what matters here, not
just the *fact* that they do. (Category theory is in a real sense constructive
logic, since the proofs are usually needed to give an accurate statement
of the theorems.) Theorem 4.6.7 extends the result to $\mathsf{Cn}_{\mathcal{L}}^{\times}$ and, since
it discusses algebra, has a more type-theoretic flavour; later we shall do
the same for larger fragments of logic. Example 4.8.2(d) shows how the
correspondence deals with homomorphisms.

The propositional analogue at the unary level is simply that a function
$f : (\Sigma, <) \to (\Theta, \leqslant_{\Theta})$ obeys $x < y \Rightarrow f(x) \leqslant_{\Theta} f(y)$ iff the same function

$p : \mathsf{Cn}_{\mathcal{L}} = (\Sigma, \leqslant) \to \Theta$ is monotone when the source is considered to carry the reflexive–transitive closure (Section 3.8). A model of a unary Horn theory is a $<$-upper subset. Similarly any function $f : \Sigma \to \Theta$ from a set to a monoid extends uniquely to a monoid homomorphism $\mathsf{Cn}_{\mathcal{L}} = \mathsf{List}(\Sigma) \to \Theta$ (Section 2.7).

Theorem 4.2.12 generated a category *freely* from a sketch in the sense (of universal algebra) that it satisfies only those laws which are forced. By Theorem 4.4.5, it is the free category in the categorical sense of satisfying a universal property (next section). Classifying categories for algebra and the λ-calculus will be given in Theorem 4.6.7 and Remark 4.7.4.

The force of functoriality. It is easy to get into the (bad) habit of only defining the effect of a functor on objects, since we usually write them in this way. The force of functoriality, however, lies in the definition on morphisms and the preservation of composition.

EXAMPLES 4·4·6: The following are *not* functors.

(a) The map $\mathcal{C} \nrightarrow M$ from any category which takes isomorphisms to id and everything else to $\mathfrak{e}$, where M is the monoid $\{\mathsf{id}, \mathfrak{e}\}$ with $\mathfrak{e}^2 = \mathfrak{e}$.

(b) The **centre of a group**, $Z(G) = \{x : G \mid \forall g.\, xg = gx\}$. The result of applying a homomorphism $f : G \to H$ to a central element of G need not be central in H, so $Z(-)$ is *not defined* on maps. In my experience this is the commonest fallacy: not checking that the "expected" action on morphisms is well defined.

(c) *Set* $\nrightarrow$ *Set* by $X \mapsto X^X$ is also not defined on morphisms, because it is the "restriction to the diagonal" of a functor of **mixed variance** *Set*$^{\mathsf{op}} \times$ *Set* $\to$ *Set*.

(d) Operations satisfying the definition of a functor apart from the preservation of identities are called **semifunctors**; they were first studied by Susumi Hayashi. For an example $\mathcal{Rel} \to \mathcal{Rel}$, take the powerset on sets and the lower order, $(-)^{\flat}$, on relations (Exercises 3.55 and 3.57). Any semifunctor gives rise to a functor by splitting idempotents in the categories (Definition 1.3.12, Exercise 4.16).[6]

Category theory was first used in algebraic topology, which aims to assign an (easily calculable) algebraic structure to each topological space in

---

[6]In the example given we obtain an endofunctor of the category of continuous lattices and functions which preserve arbitrary joins. This functor is part of a comonad (Definition 7.5.1) whose co-Kleisli category has (the same objects,) Scott-continuous functions as maps, and is cartesian closed. Raymond Hoofman (1992) observed that some of the early models of the untyped λ-calculus (in particular those for which the η-rule fails) are more naturally seen as semifunctors; he thereby gave some very simple models of linear logic (*cf.* Exercises 4.27ff and 7.41).

order to distinguish between spaces. For example (only) the nth reduced homology group is non-trivial for the sphere S^n which embeds in $(n+1)$-dimensional space. The *homeomorphisms* (*i.e.* topological isomorphisms, and even the continuous functions) between the spaces also give rise to isomorphisms (respectively homomorphisms) between the corresponding groups. It is this property which enables algebraic structures to distinguish the spaces.

REMARK 4·4·7: Suppose $X_1 \cong X_2$ in $\mathcal{C}$, *i.e.* there is a pair of morphisms $\mathfrak{u} : X_1 \to X_2$ and $\mathfrak{v} : X_2 \to X_1$ with $\mathfrak{u}\,;\mathfrak{v} = \mathrm{id}_{X_1}$ and $\mathfrak{v}\,;\mathfrak{u} = \mathrm{id}_{X_2}$; we say that the two objects are *isomorphic* (*cf.* Lemma 1.3.11).

(a) Any structure carried by X_1 may be transferred to X_2, because any morphism $\Gamma \to X_1$ or $X_1 \to \Theta$ may be turned into a morphism $\Gamma \to X_2$ or $X_2 \to \Theta$ by composition with either $\mathfrak{u}$ or $\mathfrak{v}$, and the process is reversible. Hence $\cong$ is a *congruence* (Definition 1.2.12) with respect to categorically definable properties.

(b) Any functor $F : \mathcal{C} \to \mathcal{D}$ preserves this property: $FX_1 \cong FX_2$.

Hence if $X_1, X_2 \in \mathcal{C}$ are two objects and $F : \mathcal{C} \to \mathcal{D}$ is a functor (such as homology) for which FX_1 and FX_2 are *not* isomorphic in $\mathcal{D}$, then X_1 and X_2 are not isomorphic in $\mathcal{C}$.   □

Full and faithful. Since objects are only defined up to isomorphism, it is harmless (and often useful) to make isomorphic duplicates of them (for example in our use of variables, Remark 4.3.14). For this reason injectivity and surjectivity on objects are not particularly important for functors. The force of functoriality is, as we have said, on morphisms.

DEFINITION 4·4·8: Let $F : \mathcal{C} \to \mathcal{D}$ be a functor.

(a) F is *faithful*[7] if the functions $F_{X,Y} : \mathcal{C}(X,Y) \to \mathcal{D}(FX, FY)$ are injective, *i.e.* given $\mathfrak{f}, \mathfrak{g} : X \rightrightarrows Y$ in $\mathcal{C}$, if $F\mathfrak{f} = F\mathfrak{g}$ then $\mathfrak{f} = \mathfrak{g}$.

(b) F is *full* if each function $F_{X,Y}$ is surjective, *i.e.* given $X, Y \in \mathrm{ob}\,\mathcal{C}$ and $\mathfrak{h} : FX \to FY$ in $\mathcal{D}$ there is some $\mathfrak{f} : X \to Y$ with $F\mathfrak{f} = \mathfrak{h}$. Notice that (unlike surjectivity of functions) $\mathcal{C}$ *gives* as well as *takes* in this definition; that is, the objects of $\mathcal{C}$ must be specified, not just the morphism of $\mathcal{D}$. In particular $\varnothing \to \mathcal{C}$ is full. Fullness is often accompanied by faithfulness, just as uniqueness is more important than existence (see the remarks after the proof of Lemma 1.2.11).

(c) F is *essentially surjective (on objects)* or has *representative image* if for every object $A \in \mathrm{ob}\,\mathcal{D}$ there are some object $X \in \mathrm{ob}\,\mathcal{C}$ and an isomorphism $FX \cong A$ in $\mathcal{D}$.

---

[7][BW85] and [ML71] agree with this usage, but [FS90] also requires that invertibility be reflected, for reasons which we illustrate in Examples 4.4.9 and 4.5.5(b).

(d) F is **replete**[8] if for every $X \in \text{ob}\,\mathcal{C}$ and isomorphism $\mathfrak{v} : A \cong FX$ in $\mathcal{D}$ there is a (not necessarily unique) isomorphism $\mathfrak{u} : Y \cong X$ in $\mathcal{C}$ such that $FY = A$ and $F\mathfrak{u} = \mathfrak{v}$. A forgetful functor which is replete *reflects* the means of exchange in the sense that the underlying object may be exchanged for an isomorphic copy and the structure will follow. This is a feature of the presentation, but in their usual form most of the functors we describe are replete.

(e) F **reflects invertibility** if every morphism $\mathfrak{u} : X \to Y$ in $\mathcal{C}$ for which $F\mathfrak{u} : FX \cong FY$ in $\mathcal{D}$ is already itself invertible in $\mathcal{C}$.

(f) F **reflects** the existence of **isomorphisms** if every $X, Y \in \text{ob}\,\mathcal{C}$ such that $FX \cong FY$ in $\mathcal{D}$ are already themselves isomorphic in $\mathcal{C}$.

(g) F is an **equivalence functor** if it is full, faithful and also essentially surjective (see also Definition 4.8.9(c)).

Similarly, a **full subcategory** is one whose inclusion functor is full, so it shares the same hom-sets and is determined by its objects. Conversely, a **wide** or **lluf subcategory** is one with the same objects, but perhaps fewer morphisms. A **replete subcategory** $\mathcal{U} \subset \mathcal{C}$ is one which is full with respect to isomorphisms and is such that if $X \in \text{ob}\,\mathcal{U}$ and $X \cong Y$ in $\mathcal{C}$ then $Y \in \text{ob}\,\mathcal{U}$; this happens, for example, when $\mathcal{U}$ is defined by a universal property.

EXAMPLES 4·4·9:

(a) Every monotone function between posets is faithful. It is injective iff it reflects the existence of isomorphisms, and surjective iff it is essentially surjective. The notions of fullness and reflecting invertibility are relevant to posets, and repleteness to preorders.

(b) Monoid and group homomorphisms are faithful and full iff they are respectively injective and surjective. The monoid inclusions $\mathbb{N} \hookrightarrow \mathbb{Z}$ (under addition) and $\mathbb{Z} \setminus \{0\} \hookrightarrow \mathbb{Q} \setminus \{0\}$ (under multiplication) do not reflect invertibility. A group homomorphism is replete iff it is surjective.

(c) The functor $\mathcal{S}p \to \mathcal{L}oc$ (giving the open-set lattice) is not faithful, but becomes so exactly when restricted to T_0-spaces.

(d) An action of a category is faithful *quâ* functor iff the action is faithful, *i.e.* maps which have identical effect are equal (Definition 4.2.7).

(e) Examples 4.4.3(a) are full and faithful, as are the Scott topology $\mathcal{D}cpo \to \mathcal{S}p$, the powerset functor $\mathcal{R}el \to \mathcal{CSL}at$ and the forgetful

---

[8]This term is standardly applied only to full subcategories. Our definition is chosen to be necessary and sufficient for the pseudo-pullback (Proposition 7.3.9) of F along arbitrary functors to be (weakly) equivalent to the strict pullback.

functor $CSLat \to Pos^{\dashv}$. The topology functor $Sp \to Loc$ is full and faithful exactly when restricted to sober spaces.

(f) Forgetful functors from categories of algebras, such as $Gp \to Set$ and $Lat \to Set$, also reflect invertibility: any bijective homomorphism is an isomorphism.

(g) The forgetful functors $Pos \to Set$ and $Sp \to Set$ are faithful in the sense given, but *do not reflect invertibility* (Remark 3.1.6(e)).

(h) For a forgetful functor to reflect the *existence* of isomorphisms, each carrier set must support at most one algebraic structure.

(i) Examples 4.4.3(b) are wide subcategories.

(j) Let L and L' be unary languages (elementary sketches) with the same sorts and operation-symbols, but such that L' has additional laws. Then $Cn_L \to Cn_{L'}$ is full but not faithful.

There is a moral to this: the full subcategory of $CSLat$ consisting of powerset lattices has forgetful functors successively to $CSLat$, $Pos^{\dashv}$, Pos, Set and finally Rel, but is itself equivalent to Rel. This shows that *it is misleading to regard forgetful functors as providing a hierarchy of simplicity amongst categories*: the notion is entirely dependent upon presentation, and indeed some of the functors in Examples 4.4.4 would be regarded as forgetful by certain authors.

The "true form" (in Plato's sense) of a mathematical object is the *totality* of constructions from it — its presentations are only *images* (in both the Platonist and functorial senses). This is in the modern spirit of object-oriented programming, in which data-objects are available only *via* constructions and not as their substance (machine representation).

4.5   A Universal Property: Products

At first sight, the definition of a category is not sufficiently general to interpret λ-abstraction or many-argument operations: it seems that it is only suitable for unary ones. In Section 4.3 we effectively extended the definition of category by admitting a *list* of objects as the source of each morphism; such structures, called **multi-categories**, were formalised by Joachim Lambek [Lam89]. But this is the thin end of a wedge: for every new constructor we would need to modify the definition of a category again to make the semantics follow slavishly after the syntax. For example allowing function-types in the source too would be useful to simplify the definition of sums and lists (Remarks 2.3.11 and 2.7.10).

Our strategy will be to keep faith with categories as they are, uncovering their latent structure — what is there already. We shall show that this matches the type-theoretic phenomena of interest. Even graph theory does this in a primitive fashion, by studying features such as bridges, *i.e.* edges whose removal would disconnect the graph. Particular nodes and edges may be characterised *à la* Leibniz (Proposition 2.8.7) by such features — by the way in which the others see them.

But a graph may have many (different) bridges: we need *descriptions* (Definition 1.2.10) — properties for which the definite article (*the*) can be used. We use *superlatives* such as *greatest*. These must be justified by *comparatives* (*greater*) with all other objects. Indeed French and Italian (*il più bel prodotto*) use the definite article to make superlatives out of comparatives. In a category the comparatives are the morphisms.

The Leibnizian method of description is based on *quantification* over the whole category, and we shall use either[9] Γ or Θ for the bound variable. This quantification means that universal properties are *impredicative* definitions; they are (loosely) related to those in Remark 2.8.11.

Variation over the ambient category may instead be expressed as an axiom *scheme* with an instance for each comparative object (Γ or Θ). The maps $a : \Gamma \to X$ used to test a universal property may be seen as terms or elements of (type) X, so quantifying over the objects Γ in the category says "for all terms $a : X$" without specifying the parameters Γ occurring in them. But *the definitions must be preserved by substitution for these parameters, cf.* Lemma 1.6.3.

The terminal object. We shall begin by looking closely at the very simplest case. Category theory attracts from mathematicians derisive names such as "abstract nonsense" and "empty-set theory" because of definitions like this one and the remarks which follow it. Informaticians will be better aware of the need to get base cases right: although they appear trivial, they are where the actual work of a construction gets done, and are also where most program bugs lie. In category theory, *any* universal construction is the terminal object of some category.

DEFINITION 4·5·1: $1 \in \mathrm{ob}\,\mathcal{C}$ is a *terminator, terminal object* or *final object* if for each object $\Gamma \in \mathrm{ob}\,\mathcal{C}$ there is a *unique* morphism $!_\Gamma : \Gamma \to 1$.

EXAMPLES 4·5·2:

(a) A terminal object of a preorder *quâ* category is a greatest element (Definition 3.2.1(d)); truth is the terminal formula under provability.

(b) $\{\star\}$ is the terminal object of *Set*, and also of *Preord*, *Pos*, *CSLat*, *SLat*, *Lat*, *DLat*, *HSL*, *Heyt*, *Frm*, *Dcpo*, *IPO*, *Sp*, *Mon*, *Gp*, *CMon*

---

[9]The Greek words Γένεσις and Θάνατος mean birth and death respectively.

and *Ab* with the unique structure. Lemma 2.1.9(e) showed that there is exactly one *total* function, $x \mapsto \star$.

(c) ⊘ is the terminal object of *Rel* and *Pfn*; ! is now the empty relation.

(d) The empty sequence of types is the terminal object of the category of contexts and substitutions; ! is the empty sequence of terms.

(e) The terminal type in C and JAVA is called void.

REMARK 4·5·3:   In *Set*, for each element $a \in X$ there is a function $\{\star\} \to X$ selecting a (Lemma 2.1.9(d)), so maps from the terminal object are called **global elements**. Although this usage came from topology *via* sheaf theory (Example 8.3.7(a)) its meaning is exactly what might be expected from programming: a term in the global (empty) context. By Corollary 4.3.13, a general map $\mathfrak{a} : \Gamma \to X$ is a term $\Gamma \vdash a : X$ in the arbitrary or local context of the parameters Γ, so it is called a **generalised** or **local element**.

The distinction is well known and important in recursion theory: we say that two functions $f, g : \mathbb{N} \to \mathbb{N}$ are **numeralwise equal** if $f(0) = g(0)$, $f(1) = g(1)$, $f(2) = g(2)$, ..., *i.e.* the two composites

$$1 \xrightarrow{\ n\ } \mathbb{N} \underset{g}{\overset{f}{\rightrightarrows}} \mathbb{N}$$

agree (*cf.* a predicate $\phi[n]$ provable for each n separately), but this is not enough to prove $f = g$ or $\forall x.\ \phi[x]$ (Remark 2.4.9, Theorem 9.6.2).

Proposition 4.2.9 showed that every category has a faithful covariant action $\mathcal{H}_X$ on its generalised elements (see also Theorem 4.8.12).

DEFINITION 4·5·4:   If $\mathcal{G} \subset \mathrm{ob}\,\mathcal{C}$ is such that the restriction of the action to $(\mathcal{G}_X = \bigcup_{\Gamma \in \mathcal{G}} \mathcal{C}(\Gamma, X) : X \in \mathrm{ob}\,\mathcal{C})$ remains faithful then it is a **class of generators**. Classically this is often stated as follows: if $\mathfrak{f} \neq \mathfrak{g} : X \rightrightarrows Y$ in $\mathcal{C}$ then there are some $\Gamma \in \mathcal{G}$ and $\mathfrak{a} : \Gamma \to X$ with $\mathfrak{f} \circ \mathfrak{a} \neq \mathfrak{g} \circ \mathfrak{a}$. To say that there is some *set* $\mathcal{G}$ of generators is a widely applicable way of restricting the size of a category (*cf.* locally small, Remark 4.1.10).

EXAMPLES 4·5·5:

(a) In *Set* the singleton $\mathcal{G} = \{\mathbf{1}\}$ suffices; a category for which $\{\mathbf{1}\}$ is a set of generators is sometimes called **well pointed**.

(b) *Pos* is well pointed in the weak sense that $Pos(\mathbf{1}, -) : Pos \to Set$ is faithful, but this functor doesn't reflect invertibility. Theorem 4.7.13 needs this stronger property, which $\mathcal{G} = \{\mathbf{1}, \{\bot < \top\}\}$ does satisfy.

(c) A well pointed semantics for the λ-calculus is called a *model*, where the word *algebra* is used for the general case.

Some writers have specialised the term **concrete category** to this case, where the action on global elements is faithful. The standard usage of this term is that there is *some* faithful action. For example $\mathcal{CSLat}$ is manifestly described "concretely", but $\mathcal{CSLat}(1, X)$ always has exactly one element, so $\{1\}$ cannot generate. There are other categories which we would like to call concrete but in which there are few maps $1 \to X$, or which don't have a terminal object at all. What "concrete" really means is that "the action you first thought of" is faithful, so neither definition is particularly satisfactory. However, there is no real need for any of this terminology. See also Exercise 4.12.

Unique up to unique isomorphism. The account of the theory of descriptions which we gave in Section 1.2 was more liberal than Russell's, allowing interchangeable objects to share the description. This is the nature of universal properties.

THEOREM 4·5·6: The terminal object, if it exists, is unique up to unique isomorphism. Conversely, any object which is isomorphic to a terminal object is itself terminal.

PROOF: Let T_1 and T_2 be terminal objects. Then putting T_1 for $\mathbf{1}$ and T_2 for Γ, there is a unique map $\mathfrak{u} : T_2 \to T_1$, and with the opposite assignment a unique map $\mathfrak{v} : T_1 \to T_2$.

To show that these are mutually inverse, consider id_{T_1} and $\mathfrak{v}\,;\mathfrak{u}$. These are both maps $T_1 \to T_1$, but casting T_1 in both roles (as $\mathbf{1}$ and as Γ), there is only one such map and so *the two candidates for it must be the same, cf.* Lemma 1.2.11. So $\mathfrak{v}\,;\mathfrak{u} = \mathrm{id}_{T_1}$ and similarly $\mathfrak{u}\,;\mathfrak{v} = \mathrm{id}_{T_2}$.

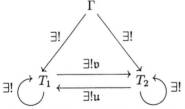

Finally, let T_1 be terminal and suppose $\mathfrak{u}$ and $\mathfrak{v}$ form an isomorphism with T_2. Then for any object Γ, the unique map $\Gamma \to T_1$ extends by $\mathfrak{v}$ to T_2, but if $\mathfrak{a}, \mathfrak{b} : \Gamma \rightrightarrows T_2$ then $\mathfrak{a}\,;\mathfrak{u} = \mathfrak{b}\,;\mathfrak{u}$ since T_1 is terminal, and then $\mathfrak{a} = \mathfrak{a}\,;\mathfrak{u}\,;\mathfrak{v} = \mathfrak{b}\,;\mathfrak{u}\,;\mathfrak{v} = \mathfrak{b}$ since $\mathfrak{u}\,;\mathfrak{v} = \mathrm{id}_{T_2}$. $\qquad\square$

Notice how the *uniqueness* of maps to $\mathbf{1}$ enabled us to deduce the equality of id and $\mathfrak{u}\,;\mathfrak{v}$ and hence transfer known properties from one to the other. Mere existence of morphisms between objects is nothing like as useful.

Products. The cartesian product can also be described in this way. It is what we use in type theory to describe functions of two variables.

DEFINITION 4·5·7:   Let P, X and Y be objects of a category C. Then $(\pi_0 : P \to X, \pi_1 : P \to Y)$ is a **product** if for every object $\Gamma \in C$ and pair of morphisms $(a : \Gamma \to X, b : \Gamma \to Y)$ in C there is a unique mediating map $f : \Gamma \to P$ such that $\pi_0 \circ f = a$ and $\pi_1 \circ f = b$. The product is written $X \times Y$, and the mediator f, which is called the **pair**, is written $\langle a, b \rangle$.

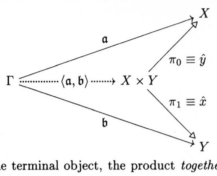

Like the terminal object, the product *together with the projections*, π_0 *and* π_1, is unique up to unique isomorphism. Conversely, any isomorphic object $u : Q \cong X \times Y$ may be given the structure of the product, with projections $u ; \pi_0$ and $u ; \pi_1$ and pairs $u^{-1} ; \langle a, b \rangle$.

The π notation signifies *selection* (4.5.8(f)), but we often want to specify the *omission* of a component (4.5.8(i)); for this we use the hat notation ($\hat{x}$) and the triangle arrowheads. Beware that the latter indicate a type-theoretic interpretation and not an intrinsic property such as surjectivity.

EXAMPLES 4·5·8:

(a) A product in a poset or preorder is a meet (Definition 3.2.4(b), Proposition 3.2.11) and in particular is conjunction for provability of formulae (Definition 1.4.2).

(b) The product in *Set* is the cartesian product (Remark 2.2.2) with its usual projection functions. Given $a : \Gamma \to X$ and $b : \Gamma \to Y$, the mediator is $\langle a, b \rangle : z \mapsto \langle a(z), b(z) \rangle$, where $\Gamma = [z : Z]$.

(c) The componentwise operations on the cartesian product give the categorical product in *SLat*, *Lat*, *DLat*, *HSL*, *Heyt*, *BA*, *Frm* and *Gp*.

(d) The componentwise order gives the product in *Pos*, *Preord*, *SLat*, *CSLat*, *Dcpo* and *IPO* by Propositions 3.5.1 and 3.5.2.

(e) This is also the componentwise order in *Lat*, *DLat*, *HSL*, *Heyt* and *Frm* by Proposition 3.2.11.

(f) A **record** consists of data assignments $t(i) \in X_i$ to each **field** $i \in I$. The type of records whose fields are of given types is the product of those types. A **selector** is a component projection π_i.

(g) Let $\mathcal{C}$ be the monoid of primitive recursive functions, *quâ* category with one object called $\mathbb{N}$. Then $\mathbb{N} \times \mathbb{N} \cong \mathbb{N}$ in $\mathcal{C}$.

(h) Remark 2.2.2 gave the type-theoretic rules for pairing,

$$\frac{\Gamma \vdash a = \pi_0(f) : X \qquad \Gamma \vdash b = \pi_1(f) : Y}{\Gamma \vdash f = \langle a, b \rangle : X \times Y}$$

though this only defines the product for two *types*, not for objects of $\mathsf{Cn}_{\mathcal{L}}^{\times}$, which are contexts.

(i) For two disjoint[10] contexts $[\underset{\sim}{x} : \underset{\sim}{X}] \times [\underset{\sim}{y} : \underset{\sim}{Y}]$, the product is their concatenation $[\underset{\sim}{x} : \underset{\sim}{X}, \underset{\sim}{y} : \underset{\sim}{Y}]$, and the projections π_0 and π_1 are the sub-sequences of variables $[\underset{\sim}{x}/\underset{\sim}{x}]$ and $[\underset{\sim}{y}/\underset{\sim}{y}]$. The pair $\langle [\underset{\sim}{a}/\underset{\sim}{x}], [\underset{\sim}{b}/\underset{\sim}{y}] \rangle$ is also the concatenation $[\underset{\sim}{a}/\underset{\sim}{x}, \underset{\sim}{b}/\underset{\sim}{y}]$ (Corollary 4.3.13(d)).

See also Exercises 4.17, 4.21, 4.22, 5.22 and 5.30.

REMARK 4·5·9: The product $X \times_{\mathcal{C}} Y$ in an abstract category and the cartesian product of hom-sets in *Set* are related by the equation

$$\mathcal{C}(\Gamma, X) \times_{Set} \mathcal{C}(\Gamma, Y) \cong \mathcal{C}(\Gamma, X \times_{\mathcal{C}} Y).$$

This is a bijection $\langle a, b \rangle \leftrightarrow f$ because (*cf.* Corollary 4.3.13(d))

$$\pi_0 \langle a, b \rangle = a \qquad \pi_1 \langle a, b \rangle = b \qquad \langle \pi_0(f), \pi_1(f) \rangle = f.$$

Notice that 1 has just one generalised element $\Gamma \to 1$, whilst the set of elements of $X \times Y$ is the cartesian product of those of X and of Y.

Preservation and creation of products. There is no need to make a *choice* of products to define what it means for a functor to preserve them: if one product P is preserved then so is any other $Q \cong P$.

DEFINITION 4·5·10: Let $X, Y \in \mathrm{ob}\,\mathcal{C}$. A functor $U : \mathcal{C} \to \mathcal{D}$

(a) **preserves the product** of X and Y if, *whenever* $X \leftarrow P \to Y$ obeys the universal property defining a product cone in $\mathcal{C}$, then so does $UX \leftarrow UP \to UY$, but in $\mathcal{D}$ (*cf.* Definition 3.2.6 for posets);

(b) preserves this product **on the nose** if *choices* of product cones have been made in both categories and U takes one choice to the other;

---

[10]The square (product with itself) of a context therefore poses a problem: in general repeated variable names must be renamed using α-equivalence. Exercise 6.50 does this. However, we shall find in Chapter VIII that taking the product of two contexts is not a natural operation anyway: in practice they are augmented by variables and hypotheses *one at a time*. *Cf.* the footnote to Proposition 9.4.6.

(c) **creates the product** if

▸ for each product cone $UX \xleftarrow{\pi_0} Q \xrightarrow{\pi_1} UY$ which exists in $\mathcal{D}$,

▸ there is a cone $X \xleftarrow{a} P \xrightarrow{b} Y$ in $\mathcal{C}$ with an isomorphism

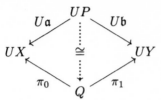

in $\mathcal{D}$, such that this diagram is unique up to unique isomorphism (as such, *without* the next condition),

▸ and moreover $X \xleftarrow{a} P \xrightarrow{b} Y$ is a product cone in $\mathcal{C}$.

So if U is a forgetful functor there is a unique structure that can be put on the product of the underlying sets and is consistent with the structure of the given objects.

EXAMPLES 4·5·11:   Recall Examples 4.4.9(f) and (g).

(a) The forgetful functor $\mathcal{L}at \to \mathcal{S}et$ (or $\mathcal{M}od(\mathcal{L}) \to \mathcal{S}et$ for any single-sorted algebraic theory $\mathcal{L}$) *creates* products: once we have found the product of the carriers in a diagram of lattices, *there is only one structure* such that the projections are homomorphisms.

(b) A functor creates *unary* products (*sic*) iff it reflects invertibility, *cf.* Definition 4.4.8(e).

(c) The forgetful functor $\mathcal{S}p \to \mathcal{S}et$ *preserves* products, but it does not create them: apart from the Tychonov topology (which gives the categorical product), the projections are also continuous when the product of the underlying sets is given the discrete topology.   □

Using the existence of products. To say that a category $\mathcal{C}$ "has products" is a statement of the form

$$\forall X, Y : \text{ob}\,\mathcal{C}. \ \exists P. \ \exists \pi_0, \pi_1. \ X \xleftarrow{\pi_0} P \xrightarrow{\pi_1} Y \text{ is a product cone,}$$

which is proved and used by the rules for $\forall$ and $\exists$ in Section 1.5.

REMARK 4·5·12:   In particular, whenever we have two particular objects A and B, we may (instantiate the universal quantifiers and) suppose that some product cone $A \leftarrow P \rightarrow B$ is at our disposal. The way that we do this is just the same as the idiom for $(\exists \mathcal{E})$ in any other circumstances (Remark 1.6.5): it is a formal property of the existential quantifier that we may invent a name for such a cone and continue to use it for as long as the given A and B remain in scope. *Nothing in this procedure assumes a global assignment of products to all pairs of objects in the category.*

Suppose, on the other hand, that an object P has already been introduced by some other means, and that we can show that it possesses the universal property needed to be a product. Suppose also that no other name has so far been given to any product of X and Y. Then, since being a product is a description up to isomorphism (Lemma 1.2.11, Theorem 4.5.6), we may write $P = X \times Y$, electing P as the specified product, instead of $P \cong X \times Y$.

Usually in category theory, as throughout mathematics in action, the quantifiers do their entrances and exits unannounced. The process of inventing names for witnesses for existential statements may be repeated any finite number of times (and that number may be unspecified: we may use products of pairs of objects from a list of indeterminate length, Exercise 2.43). In a sense it may even be done infinitely: if, by methods such as in Chapter VIII, we have some way of collecting an "indexed family" as one object then the uniqueness feature of universal properties means that one application of a product construction suffices to provide it uniformly for all members of the family.

The point at which such methods break down is where we want to make some construction (usually a functor) *globally* throughout the category. There are three ways to proceed. Classically, the axiom of choice selects products once and for all. Logically, such constructions may be regarded as *schemes*: to be instantiated to each situation as required. The third way is to replace the category itself with an equivalent (interchangeable) one on which products are indeed defined globally. We shall show how to do this in Section 7.6, *without* using the axiom of choice.

Universal properties give functors. We see why it is important for mediators to be unique when we extend universal constructions to maps.

PROPOSITION 4·5·13: Let U be an object of a category $\mathcal{C}$ for which all products $X \times U$ for $X \in \mathrm{ob}\,\mathcal{C}$ exist and are specified. Then there is a unique functor $\mathcal{C} \to \mathcal{C}$ whose effect on objects is $X \mapsto X \times U$ and such that $(\mathfrak{f} \times U)\,;\pi_0 = \pi_0\,;\mathfrak{f}$ and $(\mathfrak{f} \times U)\,;\pi_1 = \pi_1$. It is called $(-) \times U$.

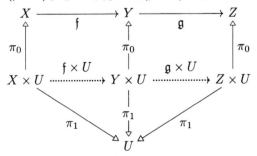

PROOF: The equations amount to one side of the commutative diagram. Clearly the universal property of products provides a unique fill-in, but it still remains for us to check that this is a functor. In the case $f = id_X$, the identity $id_{X \times U}$ is such a fill-in, and so by uniqueness this must be it: $(id_X \times U) = id_{X \times U}$. Similarly, ignoring $Y \times U$, there is a unique fill-in $(f ; g) \times U : X \times U \to Z \times U$, but $(f \times U) ; (g \times U)$ also makes the diagram commute, and so by uniqueness they are equal.   □

There is similarly a functor $\times : \mathcal{C} \times \mathcal{C} \to \mathcal{C}$ of two arguments which yields the product of an arbitrary pair of objects or morphisms.

The terminal object and binary product are the nullary and binary cases of an *n*-ary connective on objects, but as usual they suffice.

PROPOSITION 4·5·14: If a category has a terminal object and a product for every pair of objects then it has a product for every finite list of objects.   □

REMARK 4·5·15: The cartesian products $X \times (Y \times Z)$ and $(X \times Y) \times Z$ of sets are not equal: their typical elements are $\langle x, \langle y, z \rangle \rangle$ and $\langle \langle x, y \rangle, z \rangle$ respectively. Although what these products *are* is different, what they *do* is the same: they both satisfy the universal property of a product of three objects, and so they must be *uniquely* isomorphic. That is why definition up to isomorphism is the (accepted) nórm, and definition up to equality is meaningless.

Set has a canonical way of assigning binary products, but we have just seen that this is not associative. By convention, we shall take the **left-associated product** $(\cdots (((\mathbf{1} \times X_1) \times X_2) \times X_3) \times \cdots \times X_{n-1}) \times X_n$, but notice that this depends on a notion of atomic type. We use it in Remark 4.6.5 to define $[\![\Gamma, x : X]\!] = [\![\Gamma]\!] \times X$ inductively and hence to interpret expressions in algebras and programming languages.

All universal properties are terminal objects.

LEMMA 4·5·16: Let $X, Y \in \mathrm{ob}\,\mathcal{C}$. Then there is a category, which we call $\mathcal{C} \downarrow \langle X, Y \rangle$, whose objects are **spans**, *i.e.* pairs of $\mathcal{C}$-maps $X \xleftarrow{a} U \xrightarrow{b} Y$, and whose morphisms from this are $\mathcal{C}$-maps $f : U \to U'$ making the two triangles below commute.

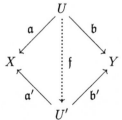

The terminal object (if it exists) of this category is the product cone for X and Y in $\mathcal{C}$.

PROOF: As with the "concrete" categories of Proposition 4.1.4, to show that $\mathcal{C} \downarrow \langle X, Y \rangle$ is a category we just have to verify that (the identity id_U is a morphism and) composition preserves the defining property. This is easy, but note that it uses the associativity law in $\mathcal{C}$:

$$\mathfrak{a} = \mathfrak{f} \, ; \mathfrak{a}' = \mathfrak{f} \, ; (\mathfrak{g} \, ; \mathfrak{a}'') = (\mathfrak{f} \, ; \mathfrak{g}) \, ; \mathfrak{a}''.$$

Comparing the definitions of $\mathcal{C} \downarrow \langle X, Y \rangle$ and $X \times Y$ shows that the terminal object in $\mathcal{C} \downarrow \langle X, Y \rangle$ is the product cone in $\mathcal{C}$. □

Mainstream mathematicians tend to view this as a rather abstruse way of saying something quite simple, but the method is like programming with abstract data types. The complexity of the constructions (in this case the commutativity of the triangles in the above diagram) is hidden from the user and later applications work in an apparently effortless way, but at a certain price. The components (modules) must be fully equipped with their ancillary operations and specifications: in this case the product projections and the associativity law of the given category, which are needed to make the definitions of respectively the objects and maps meaningful. What seems to be secondary, even optional, structure in the data turns out to be primary and necessary structure in the derived constructions; conversely "trivial" features of the derived construction unravel into something non-trivial in the given presentation.

In order to see this ancillary structure in action, you should demonstrate in detail that the product of two objects (if it exists) in a category is unique up to unique isomorphism, both directly in terms of the definition and indirectly *via* the category $\mathcal{C} \downarrow \langle X, Y \rangle$.

The type disciplines of category theory and the abstract data type idiom in programming often ensure that the ancillary structure is complete in the following sense: When the Real Mathematician — and this applies equally to the Real Programmer — is up to his elbows in the grime of a difficult procedure, he frequently needs to prove (or to leave to the reader) lemmas and sub-lemmas (respectively to program sub-routines or perform in-line hacks) which import the notation of the immediate application but in fact only reproduce in a multitude of special cases that ancillary structure which he had disregarded as trivial. By contrast, the categorist *begins* by applying the parsimonious categorical tools which she has learned to trust. She then approaches applications with the calm self-confidence of one who knows that she is prepared for the occasion.

We shall collect more examples of universal properties in Chapter VII.

4.6 ALGEBRAIC THEORIES

Section 4.2 began the semantics of formal languages with the unary case, for which plain categories were needed. The constructors of more complex languages will be interpreted using appropriate structure in categories, in particular products for multiary operations in this section and function-spaces for λ-abstraction in the next. In general, *appropriate structure* is defined by universal properties, *interpretation* by a structure-preserving functor from the category of contexts to the semantics.

$$\begin{array}{ccc} \text{formal} & \xrightarrow{\quad\text{interpretation}\quad} & \text{semantics with} \\ \text{languages} & \text{functor } [\text{--}] & \text{universal properties} \end{array}$$

DEFINITION 4·6·1: A (finitary many-sorted) **algebraic theory** $\mathcal{L}$ has

(a) base types or sorts, X (there is as yet no need for a product type constructor, as many-variable contexts can handle products for us); we write Σ for the set of sorts;

(b) an inexhaustible collection of variables $x_i : X$ of each sort;

(c) **operation-symbols**, $X_1, \ldots, X_k \vdash r : Y$, each of which has an **arity**, *i.e.* a list of input sorts X_i, and an output sort Y; and

(d) **laws** between terms (it is a **free theory** if there are none).

DEFINITION 4·6·2: An $\mathcal{L}$-**algebra** A in a category $\mathcal{C}$ with products is

(a) an object A_X of $\mathcal{C}$ for each sort X, together with

(b) a **multiplication table** for each operation-symbol r, *i.e.* a map $r_A : A_{X_1} \times \cdots \times A_{X_k} \to A_Y$ in $\mathcal{C}$ from an appropriate product,[11]

such that the polygons (such as those in Example 4.6.3(f)) which express the laws commute.

A **homomorphism** $A \to B$ of $\mathcal{L}$-algebras is an assignment to each sort X of a $\mathcal{C}$-morphism $\phi_X : A_X \to B_X$ between the corresponding objects which preserves each operation r in the sense that the squares of the following form commute.

$$\begin{array}{ccc} A_{X_0} & \xleftarrow{\quad r_A \quad} & A_{X_1} \times \cdots \times A_{X_k} \\ {\scriptstyle \phi_{X_0}}\downarrow & & \downarrow {\scriptstyle \phi_{X_1} \times \cdots \times \phi_{X_k}} \\ B_{X_0} & \xleftarrow{\quad r_B \quad} & B_{X_1} \times \cdots \times B_{X_k} \end{array}$$

---

[11]There are many isomorphic copies of the product, but it doesn't matter which is chosen. Remark 4.6.5, where the map r_A is actually used, *supplies its own choice* of product, which depends not only on the arity but also on the other variables in scope at the time. The unique mediator between these choices of product (each equipped with its projections) converts r_A as given into the required form.

Note that we have already used the product *functor* (Proposition 4.5.13) in this definition, so we would now be stuck if we hadn't insisted on uniqueness of the mediator in the definition of product.

Algebras and homomorphisms in C form a category, called[12] $Mod_C(\mathcal{L})$; the subscript is omitted if $C = Set$ is understood.

For each sort there is a forgetful functor $Mod_C(\mathcal{L}) \to C$, but it is usually not faithful: if it were, this one sort would suffice. There is a faithful functor to C^Σ reflecting invertibility, where Σ is the set of sorts.

Examples. It is clearer to fix the meaning of these widely applicable definitions by means of familiar examples than by formality. But note that, at present, we intend each of the sorts, operation-symbols and laws to be named concretely. Typical examples of theories in mathematics have one or two sorts, half a dozen operation-symbols (of arity zero, one, two, or occasionally three) and a dozen laws.

In Chapter VI we shall develop *internal* theories, whose linguistic classes may themselves be objects of the model, and there may be arbitrarily many sorts and operation-symbols. The arities will be allowed to be "infinite", *i.e.* again objects of the model, but there will be no laws. As we shall see in Section 5.6, this is because, for laws to behave as intended, each operation must have a finitely enumerated family of arguments. Nevertheless, when there are laws, there may be as many as we please.

EXAMPLES 4·6·3:

(a) An internal algebra in *Set* is an algebra in the ordinary sense.

(b) The propositional form of an algebraic theory is a **Horn theory**, *i.e.* a system of finitary closure conditions ▷ on the set Σ of sorts (Sections 1.7, 3.7 and 3.9). A model in $\mathbf{2}$ is an assignment of a truth-value to each element of Σ, *i.e.* a subset $A \subset \Sigma$, which is ▷-closed. A homomorphism is simply a containment $A \subset B$.

(c) Natural numbers ($\mathbb{N}$), lists and trees are described by free theories (Sections 2.7 and 6.1). Zero, the empty list and the leaves of trees are nullary operations or constants. Successor and adding an item to a list are unary operations, and the node-types of a tree are operations

---

[12]The reason why we don't write $Mod^\times(\mathcal{L})$, and more generally $Mod^\square(\mathcal{L})$, is that the language $\mathcal{L}$ is understood to say for itself what structure is meant to be preserved (*cf.* the use of sh for the *presheaves* on a poset for which no coverage has been specified in the footnote to Definition 3.1.7). *How* it does this is discussed in Section 7.6. Indeed, if $\mathcal{L}$ is, say, a unary theory then its models *quâ* unary theory in a category C which happens to have products are exactly the models of $\mathcal{L}$ *quâ* algebraic theory. In Sections 6.1 and 7.5 we shall write $Mod(T)$ and $Mod(M, \eta, \mu)$ for the algebras for a functor or monad respectively.

of various arities. Bourbaki used the word **magma** for an algebra with one binary operation and no laws, *i.e.* the theory whose free algebra consists of binary trees. Remark 2.7.12 gave the continuation rule for lists in terms of homomorphisms.

(d) The **abstract syntax** of a **context-free language** is the same thing as a free theory. In the design of programming languages the sorts are unfortunately called *syntactic categories*, *e.g.* ⟨program⟩, ⟨term⟩, ⟨number⟩. The keywords of the language are the operation-symbols, but these needn't be represented graphically. For example, in

$$⟨\text{integer}⟩ \quad ::= \quad ⟨\text{number}⟩ \mid -⟨\text{number}⟩ \mid +⟨\text{number}⟩$$

$$⟨\text{number}⟩ \quad ::= \quad ⟨\text{digit}⟩ \mid ⟨\text{number}⟩⟨\text{digit}⟩$$

$$⟨\text{digit}⟩ \quad ::= \quad 0 \mid 1 \mid 2 \mid 3 \mid 4 \mid 5 \mid 6 \mid 7 \mid 8 \mid 9$$

$0, \dots, 9$ are constants and $+, -$ are unary operations, but there are an invisible binary operation ⟨number⟩ × ⟨digit⟩ → ⟨number⟩ and two unary ones ⟨digit⟩ ↪ ⟨number⟩ ↪ ⟨integer⟩. The variables for each sort are called **meta-variables**, since ⟨variable⟩ may itself be one of the sorts. We shall consider the meta-language of type theory (variables, terms, types, contexts, substitutions) in Section 6.2.

(e) The specification for a **program module** consists of data-sorts, operations and laws; its **implementations** are algebras.

(f) An **internal monoid** in a category $\mathcal{C}$ with finite products is an object $M \in \text{ob}\,\mathcal{C}$ together with unit and multiplication maps,

$$e : 1 \to M \quad \text{and} \quad m : M \times M \to M$$

such that the following diagrams commute:

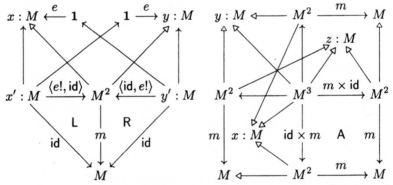

The square and triangles marked A, L and R express the substance of the associative and unit laws; the rest of the diagram specifies that M^2 and M^3 are products (Example 7.6.3). Similarly an **internal group** also has an inversion operation and corresponding laws.

(g) Internal groups in the categories of topological spaces, manifolds and algebraic varieties are topological, Lie and algebraic groups.

(h) A ring together with a module is an example of a two-sorted algebra.

(i) The morphisms of a category with a fixed set O of objects form an algebra for a theory with $O \times O$ sorts, namely the hom-sets for each source–target pair. There are O constants (identity on each object), O^3 binary operations (composition) and $O^2 + O^2 + O^4$ (unit and associative) laws. When $O = \{\star\}$ this is just a monoid.

(j) A particular model of a theory may be specified by **generators and relations**, *i.e.* as the free algebra for the theory augmented by constants and laws. List$(\{a, b, c\})/(ab \sim ba)$, for example, specifies the monoid with three generators, two of which commute (Section 7.4).

EXAMPLES 4·6·4: The following are *not* algebraic theories, as there are exceptional values at which the operations are not defined, but it is often profitable to apply algebraic ideas to them.

(a) Lists, with head and tail, because of the empty list (Section 2.7).

(b) Number fields, because of division by zero.

(c) An abstract projective plane consists of two sets (of "points" and "lines") and an ("incidence") relation between them. Through any two *distinct* points a unique line passes, and any two *distinct* lines intersect in a unique point (Example 3.8.15(e)).

These theories involve *conditional* properties ("either ... or ..."), which we shall discuss briefly in Examples 5.5.9.

Semantics of expressions. Now we shall extend Theorem 4.4.5 to algebraic theories. We have already seen several notations for the effect of a functor, and now we shall introduce yet another: Dana Scott's **semantic brackets**. These are convenient because the construction is applied to lengthy expressions; the brackets also draw attention to the difference between syntax and semantics. When more than one model A is under discussion, we write $A[\![X]\!]$, $A[\![\Gamma]\!]$, $A[\![r]\!]$, $A[\![u]\!]$, etc.

REMARK 4·6·5: An algebra A in a category $\mathcal{C}$ gives the meaning of each sort X as an object A_X. This extends canonically to the contexts by

$$[\![\,]\!] \quad = \quad 1$$
$$[\![\Gamma, x : X]\!] \quad = \quad [\![\Gamma]\!] \times A_X,$$

using the left-associated product (Remark 4.5.15). A also gives the meaning of each operation-symbol $Y_1, \ldots, Y_k \vdash r : Z$ and each constant $c : Z$ as a morphism $r_A : A_{Y_1} \times \cdots \times A_{Y_k} \to A_Z$ or $c_A : 1_{\mathcal{C}} \to A_Z$. These extend uniquely to the ⟨term⟩s in the context $\Gamma \equiv [x_1 : X_1, \ldots, x_n : X_n]$ by structural recursion, as follows,

$$[\![x_i]\!] \qquad\qquad : [\![\Gamma]\!] \cong A_{X_1} \times \cdots \times A_{X_n} \xrightarrow{\ \pi_i\ } A_{X_i}$$

$$[\![c]\!] \qquad\qquad : [\![\Gamma]\!] \xrightarrow{\ !\ } 1 \xrightarrow{\ c_A\ } A_Z$$

$$[\![r(u_1, \ldots, u_k)]\!] : [\![\Gamma]\!] \xrightarrow{\ \langle[\![u_1]\!], \ldots, [\![u_k]\!]\rangle\ } A_{Y_1} \times \cdots \times A_{Y_k} \xrightarrow{\ r_A\ } A_Z$$

where $[\![u_i]\!] : [\![\Gamma]\!] \to A_{Y_i}$ are the interpretations of the sub-expressions.

To illustrate this, consider the first line of Example 4.3.4,

$$\tfrac{1}{9}a^2 - \tfrac{1}{3}b,$$

in the ring $(\mathbb{R}, +, *)$ in the category *Set*. The operation-symbols have been marked in the diagram, along with the product projections that show which variables and sub-expressions are the arguments. Using the universal property of each product, there is a unique way of filling in other maps which define the sub-expressions as (polynomial) functions of $\langle a, b, c\rangle \in \mathbb{R}^3$. The last is the whole expression as a map $\mathbb{R}^3 \to \mathbb{R}$.

In general, the ⟨program⟩s are interpreted as follows:

$$[\![\textbf{skip}]\!] \qquad\quad = \ \ \text{id}$$

$$[\![\mathtt{u}\,;\mathtt{v}]\!] \qquad\quad = \ \ [\![\mathtt{u}]\!]\,;[\![\mathtt{v}]\!]$$

$$[\![\textbf{discard } x]\!] \ = \ \ \pi_0 : [\![\Gamma, x : X]\!] \equiv [\![\Gamma]\!] \times A_X \longrightarrow [\![\Gamma]\!]$$

$$[\![\textbf{put } x = a]\!] \ = \ \ \langle\text{id}, [\![a]\!]\rangle : [\![\Gamma]\!] \to [\![\Gamma]\!] \times A_X$$

To show that this is well defined we must show that the laws of the sketch (Remark 4.3.8) and of the theory $\mathcal{L}$ are respected. Those involving **discard** are easy. The Substitution Lemma itself holds because we used Remark 4.3.10(a) to give the interpretation of terms. Terms a and b which are shown to be equal by a single application of a law $a' = b'$ in $\mathcal{L}$ correspond to programs

$$\mathtt{u}\,;\textbf{put } x = a'\,;\mathtt{3} \quad \text{and} \quad \mathtt{u}\,;\textbf{put } x = b'\,;\mathtt{3},$$

but these have the same interpretation because A is an algebra, *i.e.* it is given to satisfy $[\![a']\!] = [\![b']\!]$. □

EXAMPLE 4·6·6: YACC[13] generates a parser in C from a context-free grammar (Example 4.6.3(d)). The rule

```
expr:   expr '+' expr    { $$ = addition ($1, $2); }
```

states the syntax of the token + in the program text, and defines its semantics to be the C subroutine `addition`.

Notice that the variables disappear in passing from syntax to semantics, so when we go the other way we must make an arbitrary choice of them (*cf.* Remark 4.3.14).

The classifying category. The notions of functor and product exactly capture algebraic theories and their models. The single-sorted version is due to Bill Lawvere (1963, Exercise 4.29).

THEOREM 4·6·7: Let $\mathcal{L}$ be an algebraic theory. Then

(a) $\mathsf{Cn}_{\mathcal{L}}^{\times}$ has a choice of finite products, and a ("generic") model of $\mathcal{L}$.

(b) Let $\mathcal{C}$ be another category with a choice of finite products and a model of $\mathcal{L}$. Then the functor $[\![-]\!] : \mathsf{Cn}_{\mathcal{L}}^{\times} \to \mathcal{C}$ preserves (the choice of)[14] finite products and the model, and is the unique such functor.

(c) Any functor $\mathsf{Cn}_{\mathcal{L}}^{\times} \to \mathcal{C}$ which preserves finite products also preserves the $\mathcal{L}$-model.

PROOF: Recall that the objects of $\mathsf{Cn}_{\mathcal{L}}^{\times}$ are contexts and the morphisms are programs or substitutions.

[a] Products are given by concatenation of contexts (Example 4.5.8(i) and its footnote). The model is as follows:

- ▶ The base type X of $\mathcal{L}$ is interpreted as a single-variable context $[x : X] \in \mathrm{ob}\, \mathsf{Cn}_{\mathcal{L}}^{\times}$, in which the name x is chosen arbitrarily.
- ▶ The operation-symbol $\underset{\sim}{X} \vdash r : Y$ of $\mathcal{L}$ is interpreted as a (single) substitution $[r(\underset{\sim}{x})/y] : [\underset{\sim}{x} : \underset{\sim}{X}] \to [y : Y]$.

[b] The interpretation is given by Remark 4.6.5.

[c] The sorts and operation-symbols of the model are given by the effect of the functor, essentially as in part (a). □

Theorem 3.9.1 found the classifying semilattice $\mathsf{Cn}_{\flat}^{\wedge}$ for a Horn theory, using induction on closure conditions to show that $[\![-]\!]$ is monotone, and

---

[13] "Yet another compiler-compiler", S. C. Johnson, 1975.

[14] By construction of the interpretation functor, the *choices* of products used in Remark 4.6.5 (namely certain left-associated ones) are in fact preserved on the nose. This is an issue which attracts far more attention than it deserves, so we shall leave those very few readers who are worried about it to work out for themselves from that Remark precisely what is meant by this.

A more serious issue is that $\mathcal{C}$ must either be a small category or contain some kind of universe to enclose the recursion. On this occasion $\prod_{X \in \Sigma} (1 + [\![X]\!])^{\mathbb{N}}$ is enough, but for larger fragments of logic we shall need the axiom-scheme of replacement.

Exercise 3.36 gave the saturated form of any closure condition. Now $\mathsf{Cn}_{\mathcal{L}}^{\times}$ is the saturation as an algebraic theory, with all possible derived operations of every finite arity, *modulo* all provable equalities between them (*cf.* regarding expressions as operation-symbols in Remark 1.1.2). Hence the term **clone**, which was introduced by Philip Hall in 1963.

Example 4.8.2(e) extends the Theorem to homomorphisms.

The model is generic in the category $\mathsf{Cn}_{\mathcal{L}}^{\times}$ in the same way that the value x_0 was generic in a proof box in Remark 1.5.6: inside this world it may be treated as an ordinary object, and there are special rules for importing and exporting it. Indeed, just as in Remark 1.5.10, given any such object in the outside world, there is a β-reduction of the generic object which reproduces it, namely the functor $[\![-]\!]$.

Hence algebra is **complete** (Remark 1.6.13, Corollary 3.9.2).

COROLLARY 4·6·8:   Let ϕ be a formula involving equations between terms of $\mathcal{L}$, conjunction and universal quantification, and let Γ be some possibly infinite set of such formulae. We write $\Gamma \vDash_{\mathcal{L}} \phi$ if every $\mathcal{L}$-algebra satisfying all of the formulae in Γ also satisfies ϕ. Then $\Gamma \vDash_{\mathcal{L}} \phi$ iff $\Gamma \vdash_{\mathcal{L}} \phi$.

PROOF: Without loss of generality ϕ is a single equation and we may strip conjunctions and universal quantification from Γ. Let $\mathcal{L}'$ be the theory $\mathcal{L}$ together with the equations in Γ as additional laws. Then an "$\mathcal{L}$-algebra satisfying all of the formulae in Γ" is just an $\mathcal{L}'$-algebra. If $\Gamma \vDash_{\mathcal{L}} \phi$ then in particular $\mathsf{Cn}_{\mathcal{L}'}^{\times}$ satisfies ϕ, but this model satisfies exactly those equations that are provable from $\mathcal{L}'$, so $\Gamma \vdash_{\mathcal{L}} \phi$. The other direction is soundness.   □

We shall give a generalisation of the universal property of products in Section 5.1. In the corresponding wider notion of algebraic theory, which includes categories themselves as an example, operations may be defined or laws imposed only when certain conditions hold. Remark 5.2.9 and Chapter VIII set out two ways of formulating such theories.

4.7   INTERPRETATION OF THE LAMBDA CALCULUS

The business of this section is to establish the connection between the *symbolism* of the λ-calculus (λ-abstraction, free variables, β-reduction and so on) and its *semantics*, involving sets, Scott domains and cartesian closed categories. On the face of it, these are very different beasts. Usually, some kind of *translation* is defined, which takes one system to the other. Such an approach tends to exaggerate the separation in order to make a greater triumph of the reconciliation.

In this chapter, we already have a technique which brings both parties together on the same categorical platform. Then the ways in which they each express the same essential features can be compared directly.

The syntactic category was constructed in Section 4.3. Its objects are lists of ⟨type⟩d variables and its morphisms are lists of ⟨term⟩s, where we left the notions of ⟨type⟩ and ⟨term⟩ undefined. In Section 4.6, these meant sorts (base types) and algebraic expressions respectively. Now we shall allow the types to be expressions built up from some given sorts using the binary connective →, and the terms to be λ-expressions.

We already have the notion of composition: it is given by substitution, as before, and is not something new involving λ-abstraction, which we do not yet understand. The Normal Form Theorem 4.3.9 for substitutions still holds — not to be confused with that for the λ-calculus, Fact 2.3.3. The category has specified products, given by concatenation.

As the *raw* syntax is now a category, we can already ask about any universal properties it might have. We begin by defining the ⟨term⟩s to be $\alpha\delta$-equivalence classes, *i.e.* taking account of any algebraic laws between operation-symbols in the language, but not the $\beta\eta$-rules. These take the form of equations between morphisms of the category, and we shall argue *from* them *towards* cartesian closure. The technique is a generic one, and will be applied to binary sums, dependent sums and dependent products in Sections 5.3ff, 9.3 and 9.4 respectively.

It is easy to be fooled by syntactic treatments into thinking that for a type to be *called* $[X \to Y]$ is necessary and sufficient for it to *behave* as a function-type. Our development (here and in Chapter IX) is based on how this type is *used* (application, abstraction and the $\beta\eta$-rules): *any* type-expression or semantic object is *a priori* eligible for the role.

REMARK 4·7·1: Application is a binary operation-symbol: every instance of it is obtained uniquely by substitution for f and x into

$$\mathsf{ev}_{X,Y} \stackrel{\mathrm{def}}{=\!=} [fx/y] : [f : [X \to Y], x : X] \to [y : Y].$$

The raw calculus. Abstraction, which is what the function-type is about, is much more interesting. In Sections 1.5 and 2.3 sequent rules were needed: λ is not an operation on terms but a *meta*-operation on terms-*in-context*. It defines a bijection,

$$\frac{[\Gamma, x : X] \xrightarrow{\;[p/y]\;} [y : Y]}{\Gamma \xrightarrow{\;[\lambda x.\, p/f]\;} [f : [X \to Y]]} \lambda_{\Gamma,X,Y}$$

as in Definition 2.3.7, so it does the same to their interpretations.

Categorically, the effect of an operation-symbol is by postcomposition, affecting only the target of a morphism; invariance under substitution is automatic as that works by precomposition, at the source. Abstraction, however, affects the source as well as the target, so it must be defined as a *meta*-operation on hom-sets, *cf.* the indirect transformations of trees on page 5. An extra condition must be added to handle substitution; this requires a new concept, naturality, which we shall study in the next section (Example 4.8.2(h)).

So first we shall concentrate on the meaning of the new operation λ in the *raw* calculus, and in particular on its invariance under substitution,[15] adding the β- and η-rules later. With or without these rules, substitution remains the notion of composition, and we shall refer to the **categories composed of λ-terms** and of **raw λ-terms** respectively.

DEFINITION 4·7·2: Let C be a category with a specified terminal object and for which each pair of objects has a *specified* product together with its projections. Then a **raw cartesian closed structure** assigns to each pair of objects X and Y

(a) an object of C, called the **function-type** or **-space**, $[X \to Y]$,

(b) a C-morphism, called **application**, $\text{ev}_{X,Y} : [X \to Y] \times X \to Y$, and

(c) for each object Γ, a function $\lambda_{\Gamma,X,Y} : C(\Gamma \times X, Y) \to C(\Gamma, [X \to Y])$ between hom-sets called **abstraction**, obeying the **naturality law**

$$\lambda_{\Gamma,X,Y}(\mathfrak{p} \circ (\mathfrak{u} \times \text{id}_X)) = (\lambda_{\Delta,X,Y}(\mathfrak{p})) \circ \mathfrak{u}$$

for each $\mathfrak{u} : \Gamma \to \Delta$ and $\mathfrak{p} : \Delta \times X \to Y$.

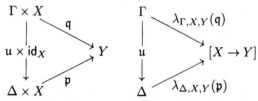

In defining products we didn't reserve any special treatment for base (atomic) types — nor do we here. In the semantic case they are not special anyway, but in the syntactic category an object is a *list* of types. We use **Currying** (Convention 2.3.2) to exponentiate by a list.

EXAMPLES 4·7·3: The following have raw cartesian closed structures:

(a) In *Set*, the specified products are cartesian ones, the function-space Y^X (Example 2.1.4) serves as $[X \to Y]$ and application provides

---

[15] Another reason for this approach is to illustrate that category theory can deal with dynamic as well as static phenomena. Exercise 4.34 shows how 2-categories handle reduction and even predict ideas which have only arisen syntactically.

$\text{ev}_{X,Y}$. The abstraction $\lambda_{\Gamma,X,Y}$ takes the function $p : \Gamma \times X \to Y$ of two variables to the function (of one variable $\sigma : \Gamma$) whose value is the function $x \mapsto p(\sigma, x)$. The naturality law is clearly valid.

(b) Any Heyting semilattice (Definition 3.6.15); the interpretation of the language is that of propositions as types (Remark 2.4.3). The rules $(\Rightarrow\mathcal{I})$ and $(\Rightarrow\mathcal{E})$ correspond to abstraction and evaluation, and nothing need be said about naturality.

(c) We write $\text{Cn}^{\times}_{\mathcal{L}+\lambda}$ for the category composed of raw λ-terms, given by Definition 4.3.11, in which a $\langle\text{term}\rangle$ is an $\alpha\delta$-equivalence class. The function-type $[[\underline{x} : \underline{X}] \to [\underline{y} : \underline{Y}]]$ in $\text{Cn}^{\times}_{\mathcal{L}+\lambda}$ is $[\underline{f} : \underline{F}]$, where $\text{len}\,\underline{F} = \text{len}\,\underline{Y}$, the $\underline{f}$ are new variables and

$$
\begin{aligned}
F_j &= X_1 \to [X_2 \to \cdots [X_k \to Y_j]\cdots] \\
\text{ev}_{\underline{x},\underline{y}} &= [(\cdots((\underline{f}x_1)x_2)\ldots x_k)/\underline{y}] \\
\lambda_{\Gamma,\underline{x},\underline{y}}[\underline{p}/\underline{y}] &= [(\lambda x_1.\,(\lambda x_2.\,\ldots(\lambda x_k.\,\underline{p})\cdots))/\underline{f}].
\end{aligned}
$$

The expressions in $\underline{f}$, $\underline{p}$ and $\underline{y}$ must be read as "for each j, ..." (which explains the bracket conventions). Naturality with respect to $\mathsf{u} = \hat{z}$ and $[c/z]$ is the substitution rule in Definition 2.3.7. By the Normal Form Theorem 4.3.9, any substitution u may be expressed as a composite of these two special cases, from which the naturality law above follows.

Interpretation. The raw λ-calculus extends the language of algebra by some new types $[X \to Y]$ and operation-symbols $\text{ev}_{X,Y}$. As yet these have no special significance, and they can be handled as if they were algebra: hence the notation $\text{Cn}^{\times}_{\mathcal{L}+\lambda}$ (in contrast to $\text{Cn}^{\to}_{\mathcal{L}}$ below).

REMARK 4·7·4: Let $\mathcal{C}$ be a category together with a raw cartesian closed structure, in which the base types, constants and operation-symbols of $\mathcal{L}$ have an assigned meaning. Then the language $\mathcal{L} + \lambda$ has a unique interpretation, and this defines a functor $[\![-]\!] : \text{Cn}^{\times}_{\mathcal{L}+\lambda} \to \mathcal{C}$.

PROOF: Building on Remark 4.6.5, by structural recursion,[16]

(a) the *base types* are given to be certain objects;

(b) the *function-types* are those of the raw cartesian closed structure;

(c) from these the *contexts* are (specified) products;

(d) the *variables* and *operation-symbols* (including *evaluation*, ev) are treated as in algebra, and the *laws* are satisfied;

(e) the last clause of the raw cartesian closed structure says how to perform λ-*abstraction*;

---

[16]$\mathcal{C}$ must either be a small category, or admit a universe such as a model of the untyped λ-calculus, or satisfy the axiom-scheme of replacement.

(f) the *morphisms* are treated as in the direct declarative language.

By Theorem 4.6.7 this is a product-preserving functor, and by the present construction it also preserves the new structure. □

The β- and η-rules. We have constructed $\mathsf{Cn}^{\times}_{\mathcal{L}+\lambda}$ from the syntax, so it has names for its objects and morphisms. But it is also a category, so we may compare these two modes of expression, from which we shall derive a universal property.

LEMMA 4·7·5:   The interpretation satisfies the rules iff

$$\mathsf{ev}_{X,Y} \circ \big(\lambda_{\Gamma,X,Y}(\mathfrak{p}) \times \mathsf{id}_X\big) \quad = \quad \mathfrak{p} \tag{β}$$

$$\lambda_{[X\to Y],X,Y}(\mathsf{ev}_{X,Y}) \quad = \quad \mathsf{id}_{[X\to Y]} \tag{η}$$

PROOF: These are the interpretations of

$$[fx/y] \circ [\lambda x.\, p/f, x/x] = [(\lambda x.\, p)x/y] \rightsquigarrow [p/y]$$

and $[\lambda x.\, fx/f]\rightsquigarrow[f/f]$, where x can be free in p, and f is a variable.   □

DEFINITION 4·7·6:   A **cartesian closed structure** on a category is a raw cartesian closed structure which satisfies the β- and η-rules.

EXAMPLES 4·7·7:   The following are cartesian closed structures:

(a) *Set* and any Heyting semilattice.
(b) The category of contexts and λ-terms, $\mathsf{Cn}^{\to}_{\mathcal{L}}$. This is again given by Definition 4.3.11, but now a ⟨term⟩ is an $\alpha\beta\delta\eta$-equivalence class.
(c) Using domain-theoretic techniques developed by Dana Scott, it is possible to construct a space X such that $X \cong X \times X \cong [X \to X]$. Then X is a model of the untyped λ-calculus with surjective pairing and $\mathsf{End}(X)$ is called a **C-monoid** [Koy82]. We only need to add a terminal object to get a two-object cartesian closed category $\{X, \mathbf{1}\}$. (But splitting idempotents (Definition 1.3.12, Exercise 4.16) gives a richer category.)

REMARK 4·7·8:   Currying or λ-abstraction gives an alternative way of handling (many-argument) operation-symbols as *constants*, in the same way that Hilbert's implicational logic (Remark 1.6.10) eliminated all of the rules apart from $(\Rightarrow\mathcal{E})$ in favour of axioms.

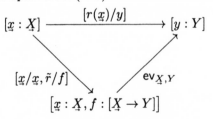

Let $X_1, \ldots, X_k \vdash r : Y$ be an operation-symbol (of type Y and arity k). Then $\tilde{r} \equiv \lambda \underset{\sim}{x} : \underset{\sim}{X}.\ r(\underset{\sim}{x}) : [\underset{\sim}{X} \to Y]$ is a constant (of exponential type), from which the original operation is recovered by the β-rule.

In the unary case, $r(a)$ is an operation applied to a value. In a cartesian closed category this equals $\text{ev}(\tilde{r}, a) = \tilde{r}a$ in the sense of λ-application. The former uses composition by substitution (which we treat as the standard notion), whilst the λ-calculus provides $g \circ f = \lambda x.\ g(fx)$. These coincide iff the β- and η-rules hold.

The universal property.

DEFINITION 4·7·9: An object F, with a morphism $\text{ev} : F \times X \to Y$, is an **exponential** or **function-space** if for every object $\Gamma \in \text{ob}\,\mathcal{C}$ and morphism $\mathfrak{p} : \Gamma \times X \to Y$ there is a unique morphism $\mathfrak{f} : \Gamma \to F$ such that $\mathfrak{p} = \text{ev} \circ (\mathfrak{f} \times \text{id}_X)$. The exponential F is written $[X \to Y]$ or Y^X.

The notation $\tilde{\mathfrak{p}}$ for the **exponential transposition** is commonly found in category theory texts, but it is clearly inadequate to name all of the morphisms of a cartesian closed category. So they frequently go without a name. All too often, proofs are left to rely on verbal transformations of unlabelled diagrams, without regard to the categorical precept that morphisms are at least as important as objects. The λ-calculus gives the general notation we need.

PROPOSITION 4·7·10: A cartesian closed structure on a category is given exactly by the choice of a product and a function-space for each pair of objects, together with the projections and evaluation.

PROOF: Suppose we have the structure of Definition 4.7.2 and Lemma 4.7.5. The two definitions of function-space share the same data and also the β-rule, so that $\lambda_{\Gamma,X,Y}(\mathfrak{p})$ serves for $\tilde{\mathfrak{p}}$.

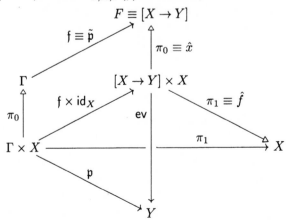

If $\lambda_{\Gamma,X,Y}$ satisfies η and the naturality law, whilst $\tilde{p}$ obeys β, then

$$
\begin{aligned}
\lambda_{\Gamma,X,Y}(p) &= \lambda_{\Gamma,X,Y}(\text{ev}_{X,Y} \circ (\tilde{p} \times \text{id}_X)) && \text{by } \beta \\
&= \lambda_{[X \to Y],X,Y}(\text{ev}_{X,Y}) \circ \tilde{p} && \text{naturality} \\
&= \text{id}_{[X \to Y]} \circ \tilde{p} = \tilde{p} && \text{by } \eta
\end{aligned}
$$

so $\tilde{p}$ is unique, *i.e.* the universal property holds. Conversely, naturality and the β-rule follow from the universal property by uniqueness (as in Proposition 4.5.13). The η-rule holds as id serves for $\lambda(\text{ev})$, and the naturality law holds because its right hand side serves for the left. □

Example 7.2.7 gives an alternative proof.

COROLLARY 4·7·11: The exponential object $[X \to Y]$ is unique up to unique isomorphism. It defines a functor which is contravariant in the first or raised argument and covariant in the other.

PROOF: Loosely speaking, $[u \to \mathfrak{z}](f) = u \,;\, f \,;\, \mathfrak{z}$. □

THEOREM 4·7·12: The category $\text{Cn}_{\mathcal{L}}^{\rightarrow}$ has a cartesian closed structure and a model of the λ-calculus with base types and constants from $\mathcal{L}$. Any other such interpretation in a category $\mathcal{C}$ is given by a unique functor $[\![-]\!] : \text{Cn}_{\mathcal{L}}^{\rightarrow} \to \mathcal{C}$ that preserves the cartesian closed structure and the model. Conversely, any such functor is an interpretation.

PROOF: As in Theorem 4.4.5 and Remark 4.6.5. Remark 4.7.4 extends the interpretation to the λ-calculus; in particular the function-types have to be preserved. By Proposition 4.7.10, these must be exponentials. □

Cartesian closed categories of domains. The category of sets and total functions is the fundamental interpretation of the typed λ-calculus, but it does not have the fixed point property (Proposition 3.3.11) needed for denotational semantics. During the 1970s and 1980s a veritable cottage-industry arose, manufacturing all kinds of domains with Scott-continuous maps, each with its own peculiar proof of cartesian closure. In fact these categories (necessarily) share the same function-space as in *Dcpo*: what is needed in each case is not a repetition of general theory, but the verification that the special semantic property is inherited by the function-space.

THEOREM 4·7·13: *Pos* has a cartesian closed structure.

PROOF: The universal property tells us what the exponential $[X \to Y]$ must be. Taking $\Gamma = \{\star\}$, it is the set of monotone functions, whilst for ev to be monotone we must have $f \leqslant g \Rightarrow \forall x.\ f(x) \leqslant g(x)$. Now consider $\Gamma = \{\bot < \top\}$, *cf.* Example 4.5.5(b). If $f \leqslant g$ pointwise then there is a monotone function $\Gamma \times X \to Y$ by $\langle \bot, x \rangle \mapsto f(x)$ and $\langle \top, x \rangle \mapsto g(x)$.

The exponential transpose is $\bot \mapsto f$ and $\top \mapsto g$, so $f \leqslant g$ as elements of the function-space.

For this to give a cartesian closed structure we must verify that

(a) $X \times Y$ and $[X \to Y]$ are well defined objects;

(b) π_0, π_1 and ev are well defined morphisms;

(c) $\langle -, - \rangle$ and λ take morphisms to morphisms;

(d) pairing, naturality and the β- and η-rules are satisfied.

The first three parts were proved in Propositions 3.5.1 and 3.5.5, but it is the notion of cartesian closed category which makes sense of the collection of facts in Section 3.5. The laws in part (d) are inherited from the underlying sets and functions. □

The result for Scott-continuous functions (redefining $[X \to Y]$) is proved in the same way.

THEOREM 4·7·14: *Dcpo* has a cartesian closed structure.

PROOF: For similar reasons, $[X \to Y]$ must be the set of Scott-continuous functions with the pointwise order. Propositions 3.5.2 and 3.5.10 gave the details, based on a discussion of pointwise joins, and in particular Corollary 3.5.13 about joint continuity of ev. □

Algebraic lattices, boundedly complete posets, L-domains and numerous other structures form cartesian closed categories with Scott-continuous functions as their morphisms. The issue of making ev preserve structure *jointly* in its two arguments may be resolved in a different way, as Exercise 4.51 shows.

At the end of the next section we shall show that categories themselves may be considered as domains and form a cartesian closed category. First we need to introduce the things which will be the morphisms of the exponential category; this turns out to be the abstract notion which we needed for substitution-invariance of λ. Section 7.6 returns to the relationship between syntax and semantics, bringing the term model into the picture. Function-spaces for dependent types are the subject of Section 9.4.

4.8 NATURAL TRANSFORMATIONS

The first task of category theory is an organisational one: after various kinds of objects (types, sets, posets, complete semilattices and dcpos) and maps (terms, relations; partial, total, monotone, continuous and structure-preserving functions; and adjunctions) have been introduced, we were able to put them in a common framework as categories. For categories as objects we provided functors as morphisms, so why have we not yet discussed the category of categories and functors?

One reason is the problem of "size" mentioned in Remark 4.1.8, but there is also an algebraic one. As we have said, mathematical constructions generally define objects only up to isomorphism, because frequently there is a different but equally useful representation which can be substituted. For example there are two versions of the three-fold cartesian product. But once the representation is chosen, (the elements and more generally) the morphisms have a unique construction.

Such constructions of objects are, with a few rare exceptions, always functors, albeit frequently contravariant or even of "mixed" variance (Example 4.4.6(c)). In particular, *an algebra* (the interpretation of a theory $\mathcal{L}$) *is a functor* $\mathrm{Cn}_{\mathcal{L}}^{\times} \to Set$ (Theorem 4.6.7). Thus functors are often *parametric objects* and so, like objects, are intrinsically defined only up to isomorphism. Whereas morphisms of a category are in some sense isolated from one another, functors (like the objects which are their values) have a kind of fluidity between them, given by the morphisms of the target category, which we haven't taken into account.

DEFINITION 4·8·1:   Let $F, G : \mathcal{C} \rightrightarrows \mathcal{D}$ be two parallel functors. Then a *natural transformation* $\phi : F \to G$ consists of, for each object $X \in \mathrm{ob}\,\mathcal{C}$, a morphism $\phi X : FX \to GX$ in $\mathcal{D}$ such that, for each map $\mathfrak{f} : X \to Y$ of $\mathcal{C}$, the following square commutes:

$$
\begin{array}{ccc}
X & \xrightarrow{\;\mathfrak{f}\;} & Y \\[2mm]
\end{array}
$$

$$
\begin{array}{cccc}
F & FX & \xrightarrow{\;F\mathfrak{f}\;} & FY \\[2mm]
\phi\downarrow & \phi X\downarrow & \text{N} & \downarrow\phi Y \\[2mm]
G & GX & \xrightarrow[G\mathfrak{f}]{} & GY
\end{array}
$$

Often the object X is put as a subscript (ϕ_X), but here we have written ϕX as an application to an object of $\mathcal{C}$ yielding a morphism of $\mathcal{D}$. This is the counterpart of using the same notation for the result of a functor applied to a morphism as to an object (Definition 4.4.1). Indeed the naturality square is the application of $\phi : F \to G$ to $\mathfrak{f} : X \to Y$. The square is not symmetrical, and occasionally we shall indicate whether the vertical is natural with respect to the horizontal (as above) or *vice versa* by an "N" or "Z" in the middle.

EXAMPLES 4·8·2:

(a) If C and D are preorders then a natural transformation $F \to G$ is (an instance of) the pointwise order $F \leqslant G$ (Proposition 3.5.5).

(b) Proposition 4.5.13 defined the effect of the product functor $(-) \times U$ by making $\pi_0 : (- \times U) \to \text{id}$ natural between functors $C \to C$.

(c) Application $\text{ev}_X : (-)^X \times X \to (-)$ is natural for each $X \in \text{ob}\,C$, by Definition 4.7.2(c).

(d) Theorem 4.4.5 showed that functors $A : \text{Cn}_{\mathcal{L}} \to Set$ correspond to algebras for a unary theory $\mathcal{L}$. Similarly natural transformations $\phi : A \to B$ correspond to their homomorphisms. It suffices to test naturality with respect to the operation-symbols (generating maps) since by composition of squares it follows automatically for general maps of $\text{Cn}_{\mathcal{L}}$, as they are composites of operation-symbols.

(e) Natural transformations between the (product-preserving) functors $A, B : \text{Cn}_{\mathcal{L}}^{\times} \rightrightarrows Set$ which interpret algebraic theories correspond to homomorphisms $\phi : A \to B$. In this case the definition of ϕ must be extended from base types to contexts in the same way as in the remarks after Definition 4.6.2.

(f) Consider $\mathcal{H}_X = C(-, X) : C^{\text{op}} \to Set$ as a *functor* (instead of the union $\bigcup_\Gamma \text{Cn}_{\mathcal{L}}(\Gamma, X)$ in Proposition 4.2.9). Then postcomposition with $u : X \to Y$ gives a natural transformation $\mathcal{H}_X \to \mathcal{H}_Y$.

(g) The pairing operation $\langle\ ,\ \rangle : C(-, X) \times C(-, Y) \to C(-, X \times Y)$ is also a natural transformation between functors $C^{\text{op}} \to Set$.

(h) The abstraction operation $\lambda_{-,X,Y} : C(- \times X, Y) \to C(-, [X \to Y])$ of a (raw) cartesian closed structure (Definition 4.7.2(c)) is another natural transformation between functors $C^{\text{op}} \to Set$.

Natural transformations show up even when we only set out to consider categories and functors.

PROPOSITION 4·8·3: Let C and Γ be categories.

(a) Their product $\Gamma \times C$ in the category *Cat* of categories and functors has object class $(\text{ob}\,\Gamma) \times (\text{ob}\,C)$ and hom-sets

$$(\Gamma \times C)(\langle E', X'\rangle, \langle E, X\rangle) = \Gamma(E', E) \times C(X', X)$$

with componentwise composition, *cf.* Proposition 3.5.1. The pairing operation and projection functions are as expected.

(b) In particular for groups (one-object categories), the factors C and Γ are isomorphic to subgroups of the product $\Gamma \times C$ by $\mathfrak{f} \mapsto \langle \text{id}, \mathfrak{f}\rangle$ and $\mathfrak{h} \mapsto \langle \mathfrak{h}, \text{id}\rangle$, and these commute.

(c) Let $P : \mathcal{T} \times \mathcal{C} \to \mathcal{D}$ be a functor ("of two variables") and $\mathfrak{h} : E' \to E$ a morphism of $\mathcal{T}$. Then $P(\mathfrak{h}, -) : P(E', -) \to P(E, -)$ is a natural transformation between functors $\mathcal{C} \to \mathcal{D}$.

$$
\begin{array}{ccc}
P(E', X') & \xrightarrow{\ P(E', \mathfrak{f})\ } & P(E', X) \\
\Big\downarrow{\scriptstyle P(\mathfrak{h}, X')} & & \Big\downarrow{\scriptstyle P(\mathfrak{h}, X)} \\
P(E, X') & \xrightarrow{\ P(E, \mathfrak{f})\ } & P(E, X)
\end{array}
$$

(d) Suppose that the value $P_o(E, X)$ of the functor on objects is given, together with $P(E, \mathfrak{f})$ and $P(\mathfrak{h}, X)$, functorially in each argument separately, *such that the square above commutes*. Then the ("joint") functor $P : \mathcal{T} \times \mathcal{C} \to \mathcal{D}$ can be defined, *cf.* 3.5.1(c).   □

It is often the case that "naturally defined" constructions are functorial or natural in the formal sense by completely routine calculation. However normality, like functoriality (Examples 4.4.6ff), does carry mathematical force, since it provides an equation, and *may be the point at issue*, as it is in Theorem 7.6.9.

Composition. We shall use the following scheme to discuss composition of natural transformations; it also explains the geometrical terminology.

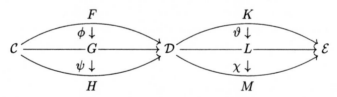

DEFINITION 4·8·4: The *vertical composite* $\phi\,;\psi : F \to H$ is defined by $(\phi\,;\psi)X = (\phi X)\,;(\psi X)$, as in the following diagram:

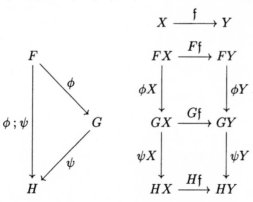

The identity for this composition is defined by $(\mathsf{id}_F)X = \mathsf{id}_{FX} = F\mathsf{id}_X$, but it is often called just F. We never use $(;)$ to compose functors. For posets, vertical composition is the transitivity of the pointwise order.

On the other hand, functors themselves apply to morphisms and hence to natural transformations, giving $K \cdot \phi$ and $L \cdot \phi$. These are natural because the functors K and L preserve commutativity of the naturality square for ϕ. Natural transformations also apply to the results of the functors on the objects, to give $\vartheta \cdot F$ and $\vartheta \cdot G$, which are natural by instantiation. These are related by the square,

$$
\begin{array}{ccc}
K(FX) & \xrightarrow{\;K(\phi X)\;} & K(GX) \\[2pt]
\Big\downarrow{\scriptstyle \vartheta(FX)} & \searrow{\scriptstyle \vartheta(\phi X)} & \Big\downarrow{\scriptstyle \vartheta(GX)} \\[2pt]
L(FX) & \xrightarrow{\;L(\phi X)\;} & L(GX)
\end{array}
$$

which commutes by naturality of ϑ. Note that the object X is completely passive; in fact if we replace it by a morphism $\mathsf{f} : X \to Y$ we obtain the commutative cube which shows naturality of $\vartheta \cdot \phi$. For posets, we are simply applying a monotone function to the pointwise order.

DEFINITION 4·8·5: The common composite is a natural transformation

$$\vartheta \cdot \phi = (K \cdot \phi)\,;\,(\vartheta \cdot G) = (\vartheta \cdot F)\,;\,(L \cdot \phi) : (K \cdot F) \to (L \cdot G)$$

which is called the **horizontal composite**. The notation is inherited from composition of functors, and is consistent with writing the name of a functor for its vertical identity natural transformation. The identity for this composition is the (componentwise) identity natural transformation *on the identity functor*, $\mathsf{id} = \mathsf{id}_{\mathsf{id}_{\mathcal{C}}} : \mathsf{id}_{\mathcal{C}} \to \mathsf{id}_{\mathcal{C}}$.

LEMMA 4·8·6: The two composition operations are related by the **middle four interchange law**,

$$(\vartheta\,;\chi) \cdot (\phi\,;\psi) = (\vartheta \cdot \phi)\,;\,(\chi \cdot \psi)$$

as suggested by the diagram opposite. $\qquad\square$

LEMMA 4·8·7: A natural transformation ϕ is vertically invertible iff every component ϕX is invertible, and is then called a **natural isomorphism**. It is horizontally invertible iff *also* the functors which it relates are themselves invertible. $\qquad\square$

EXAMPLES 4·8·8: The following are natural isomorphisms:

(a) Pairing and λ-abstraction (Examples 4.8.2(g) and (h)) make

$$\mathcal{C}(-, X) \times \mathcal{C}(-, Y) \cong \mathcal{H}_{X \times Y} \quad \text{and} \quad \mathcal{C}(- \times X, Y) \cong \mathcal{H}_{[X \to Y]}.$$

(b) The double transpose $(-)^{\perp\perp} : V \to V^{**}$ for a finite-dimensional vector space; this was the original example considered by Saunders Mac Lane and Sammy Eilenberg (1945), in order to distinguish this from the "unnatural" (basis-dependent) isomorphism $V \cong V^*$.

(c) Any natural transformation between functors $F, G : \mathcal{C} \rightrightarrows \mathcal{D}$ for which either $\mathcal{C}$ or $\mathcal{D}$ is a group, groupoid or equivalence relation.

(d) In particular between permutation or matrix representations of a group.

Equivalences. Functors are the means of exchange between categories, so since functors are only defined up to isomorphism, exchange between categories is a notion of isomorphism that is further weakened by putting isomorphism for equality. In Section 7.6 we exploit the difference between strong and weak equivalences to resolve the issue of whether products, exponentials, *etc.* in a category are structure or properties.

DEFINITION 4·8·9:   A functor $F : \mathcal{S} \to \mathcal{A}$ is

(a) an ***isomorphism of categories*** if there is a functor $U : \mathcal{A} \to \mathcal{S}$ such that $\mathrm{id}_{\mathcal{S}} = U \cdot F$ and $F \cdot U = \mathrm{id}_{\mathcal{A}}$ *as functors*, *i.e.* on both objects and morphisms;

(b) a ***strong equivalence of categories*** if the following are given together with F:

  ▸ a functor $U : \mathcal{A} \to \mathcal{S}$, called its ***pseudo-inverse***,
  ▸ natural isomorphisms $\eta : \mathrm{id}_{\mathcal{S}} \cong U \cdot F$ and $\varepsilon : F \cdot U \cong \mathrm{id}_{\mathcal{A}}$, and
  ▸ the laws $F(\eta X); \varepsilon(FX) = \mathrm{id}_{FX}$ and $\eta(UY); U(\varepsilon Y) = \mathrm{id}_{UY}$ hold;

(c) an ***equivalence functor*** if it is full, faithful and also essentially surjective (Definition 4.4.8(g)).

(d) As in Remark 3.6.7(d) we also say that two categories $\mathcal{S}$ and $\mathcal{T}$ are ***weakly equivalent*** if a third category $\mathcal{A}$ and equivalence functors $F : \mathcal{S} \to \mathcal{A}$ and $G : \mathcal{T} \to \mathcal{A}$ are given.[17]

We write $\mathcal{S} \simeq \mathcal{A}$ to indicate that categories are equivalent, making it clear in each context whether we mean strongly or weakly. See Definition 3.6.7 for the preorder version and Exercise 3.26 for the need for Choice.

We shall show in Corollary 7.2.10(c) that, with Choice, strong and weak equivalence coincide, *i.e.* any equivalence functor F has a pseudo-inverse, but this is determined only up to unique isomorphism, not equality. For given $Y \in \mathrm{ob}\,\mathcal{A}$ there may be many objects $X \in \mathrm{ob}\,\mathcal{S}$ with $FX \cong Y$, and any such object may itself have many automorphisms. (The reason for postponing the proof is simply to avoid repetition, since it is technically

---

[17]Theorem 8.4.10 shows that the useful notion in type theory is actually that of a weak equivalence *on slices*.

the same as the relationship between universal properties and categorical adjunctions.) Exercises 4.36ff explore equivalences for monoids.

Functor categories. As we observed in Proposition 4.1.5ff, categories may arise as structures as well as congregations. In particular, some of the more exotic "domains" in the literature [HP89, Tay89] are categories rather than posets.

THEOREM 4·8·10: The category *Cat* of small categories and functors (Remark 4.1.8) has a cartesian closed structure.

PROOF: We shall follow the four-point plan set out in Theorem 4.7.13, but Proposition 4.8.3 has already discussed the product. To compare with Proposition 3.5.5, think of a category as a "preorder with proofs" (Definition 4.1.6). The generalisation forces us to give notation explicitly for them: $\phi : F \to G$ and $\mathfrak{f} : X_1 \to X_2$ are "the reasons why $f \leqslant g$ and $x_1 \leqslant x_2$" and monotonicity becomes the idea of a functor.

(a) We know that functors $\mathcal{X} \to \mathcal{Y}$ and natural transformations between them form a category with vertical composition.

(b) Next we show that ev : $[\mathcal{X} \to \mathcal{Y}] \times \mathcal{X} \to \mathcal{Y}$ is a functor. A map in its source category consists of a natural transformation $\phi : F \to G$ and a morphism $\mathfrak{f} : X_1 \to X_2$. The naturality square (Definition 4.8.1) for ϕ at $\mathfrak{f}$ corresponds to that in Proposition 3.5.5. Its diagonal defines evaluation on maps, by $\mathsf{ev}(\phi, \mathfrak{f}) = \phi\mathfrak{f}$. A similar diagram of nine objects in four squares shows that composites are preserved.

(c) To show that λ preserves functoriality, let $P : \Gamma \times \mathcal{X} \to \mathcal{Y}$ be a functor. Then so are $\tilde{P}(U') = P(U', -)$ and $\tilde{P}(U) = P(U, -)$; moreover $\mathfrak{h} : U' \to U$ gives rise to a natural transformation between them, $\tilde{P}(\mathfrak{h}) = P(\mathfrak{h}, -)$ (Example 4.8.3(c)).

(d) Finally, naturality and the β- and η-rules must be tested on maps as well as on objects. □

REMARK 4·8·11: As *Set* is not a small category, the size problems have to be handled differently in the next result. In practice, the easiest way is to continue to treat functors as schemes of constructions, but the objects of $\mathcal{S}et^{\mathcal{C}^{op}}$ also have an alternative representation by the Grothendieck construction (Proposition 9.2.7), so long as at least $\mathcal{C}$ remains small. For the (large) category-domains mentioned above, it is still possible to control the size of the functor categories, because the functors in question are Scott-continuous and are therefore determined by their values on "finite" objects as in Proposition 3.4.12. We restrict attention to those locally finitely presentable categories (Definition 6.6.14(c)) which have a *set* of generators in the sense of Definition 4.5.4.

The Yoneda Lemma. The following theorem is the abstract result which underlies the regular (Cayley) representation studied in Section 4.2 (and, for posets, in Sections 3.1 and 3.2). Unfortunately, the *abstract* version is often all that is presented, leaving students unenlightened and, more seriously, depriving them of a powerful technique. Section 4.3 used it to construct the category of contexts and substitutions of a formal language, which we shall develop in Chapter VIII. Proposition 3.1.8 gave the poset analogues of parts (b) and (c).

THEOREM 4·8·12: Let C be a category and $X, Y \in \mathrm{ob}\, C$.

(a) Let $\phi_\Gamma : C(\Gamma, X) \to C(\Gamma, Y)$ be any system of functions. Then $\phi_{(-)}$ is natural iff it arises by postcomposition with some map $f : X \to Y$ as in Example 4.8.2(f), and then f is unique.

(b) Hence $\mathcal{H}_{(-)} : C \hookrightarrow Set^{C^{op}}$ (where $\mathcal{H}_X \equiv C(-, X) : C^{op} \to Set$) is a full and faithful functor, known as the **Yoneda embedding**.

(c) For any functor $F : C \to Set$ and object $X \in \mathrm{ob}\, C$,

$$Set^{C^{op}}(\mathcal{H}_X, F) \cong FX$$

naturally. (This part is called the **Yoneda Lemma** itself, 1954.)

$$
\begin{array}{ccc}
\Gamma = X & \mathrm{id}_X \in C(\Gamma, X) \xrightarrow{\ \phi_\Gamma\ } C(\Gamma, Y) \ni f = \phi_X(\mathrm{id}_X) \\
\Big\downarrow{\scriptstyle u} & \Big\downarrow{\scriptstyle \mathrm{pre}(u)} \qquad\qquad\qquad \Big\downarrow{\scriptstyle \mathrm{pre}(u)} \\
\Delta & u \in C(\Delta, X) \xrightarrow{\ \phi_\Delta\ } C(\Delta, Y) \ni u\,;f = \phi_\Delta(u)
\end{array}
$$

PROOF:

[a] Put $f = \phi_X(\mathrm{id}_X)$, so this is uniquely determined by ϕ. Then by naturality with respect to $u : \Gamma = X \to \Delta$, we have $\phi_\Delta(u) = u\,;f$.

[b] Verify that $\mathrm{post}(f)$ is functorial in f; it is full and faithful by the previous part.

[c] Let $\phi_{(-)} : \mathcal{H}_X \to F$ be natural. Put $a = \phi_X(\mathrm{id}_X) \in FX$. Then by naturality with respect to $u : \Gamma \to X$, $\phi_\Gamma(u) = Fu(a)$. Conversely, verify that this defines a natural transformation for any $a \in FX$. □

By Exercises 4.40 and 4.41, the Yoneda embedding preserves products (indeed all limits) and exponentials. Section 7.7 is a powerful application of the Yoneda lemma to the equivalence between semantics and syntax.

DEFINITION 4·8·13: A **representable functor** is one which is naturally isomorphic to some $\mathcal{H}_X = C(-, X)$ (*cf.* Definition 3.1.7).

EXAMPLES 4·8·14: By Examples 4.8.8(a),

(a) $X \times Y$ represents $\mathcal{C}(-, X) \times \mathcal{C}(-, Y) \cong \mathcal{C}(-, X \times Y)$;

(b) X^Y represents $\mathcal{C}(- \times Y, X)$ as $\mathcal{C}(-, X^Y)$;

(c) $\mathcal{P}(Y)$ represents $\mathcal{Rel}(-, Y)$ in Lemma 3.3.6 as $\mathcal{Set}(-, \mathcal{P}(Y))$;
    for $Y = \mathbf{1}$, $\mathbf{2} = \mathcal{P}(\mathbf{1})$ represents Sub (Proposition 5.2.6);

(d) $Y_\perp$ represents $\mathcal{Pfn}(-, Y)$ in Definition 3.3.7 as $\mathcal{Set}(-, Y_\perp)$.

We shall relate representable functors to universal properties in general in Corollary 7.2.10(a).

2-Categories. Since it is equipped with natural transformations as well as functors, the class of categories is an example of a two-dimensional generalisation of categories themselves.

DEFINITION 4·8·15: A **2-category** has

(a) a class of *0-cells*;

(b) for each pair of 0-cells, a *category*, whose objects and morphisms we call *1-cells* and *2-cells* respectively; the given 0-cells are called the *left* and *right hand ends* of these 1- and 2-cells; the source and target of the 2-cells are called their *top* and *bottom sides* or *edges* and the composition is styled *vertical*;

(c) for each pair of 1-cells or 2-cells of which the right side of the first is the left side of the second, a *horizontal* composite 1- or 2-cell;

such that the vertical and horizontal associativity and identity laws and the middle four interchange law hold. There is a corresponding notion of 2-functor. Beware that 2-cells are not square but lens-shaped, with two ends and two sides, *cf.* the diagram before Definition 4.8.4.

EXAMPLES 4·8·16: The following each define the 0-, 1- and 2-cells of 2-categories:

(a) $\mathfrak{Cat}$: categories, functors and natural transformations;

(b) $\mathfrak{Pos}$: posets, monotone functions and (instances of) the pointwise order; horizontal composition is defined by the monotonicity of functional composition (*cf.* the proof of Proposition 3.5.5);

(c) $\mathfrak{Rel}$: sets, binary relations and inclusion;

(d) $\mathfrak{Sp}$: topological spaces, continuous functions and the specialisation order (Example 3.1.2(i)) pointwise;

(e) the objects and morphisms of any category, with only identity 2-cells (the discrete 2-category on a category);

(f) $\mathfrak{Cn}^\rightarrow$: the types, raw terms and standard reduction paths of the λ-calculus (Exercise 4.34).

Sometimes composition is only defined up to isomorphism, satisfying certain coherence equations that were identified by Saunders Mac Lane and Max Kelly (1963); structure of this kind is called a **bi-category**.

(g) The points, paths and homotopies in a topological space form a bi-category (Exercise 4.43); a 2-functor $\pi_1(X) \to \mathcal{C}$ takes a homotopy, *i.e.* a continuous function $I \times I \to X$, to a commutative square in $\mathcal{C}$, a fact which we use to prove van Kampen's Theorem 5.4.8.

(h) Conversely, Exercise 4.49 about λ-abstraction of natural transformations may be seen as homotopy of functors.

Since there are two directions of motion, there are two independent ways of forming **opposite 2-categories**, and a third by doing both of them. Hence there are three notions of **contravariant 2-functor**. We say that a functor is "contravariant at the 1- and/or 2-level".

EXERCISES IV

1. Formulate and prove the fact that any polygon commutes iff it can be decomposed into commuting polygons.

2. Using the Cayley–Yoneda action, show that every monoid arises as a submonoid of H^H for some set H. For any set X, characterise the constant functions $\lambda x.\, a$ solely in terms of composition in X^X. Describe the Cayley–Yoneda action of any monoid M on its constants.

3. By choosing some combinatorial structure, such as the cosets of subgroups, show that every group arises up to *isomorphism* as the group of all automorphisms of some algebraic structure.

4. Restate Proposition 4.2.9 for categories. Show that $\mathcal{C}$ has two faithful regular actions: on $\mathcal{H}^X = \bigcup_\Theta \mathcal{C}(X, \Theta)$ (contravariantly by precomposition) and $\mathcal{H}_X = \bigcup_\Gamma \mathcal{C}(\Gamma, X)$ (covariantly by postcomposition). Reformulate these as functors $\mathcal{H}^X = \mathcal{C}(X, -)$ and $\mathcal{H}_X = \mathcal{C}(-, X)$.

5. Prove the Normal Form Theorem 4.3.9. Note that **skip**, **put** and **discard** each require special treatment. Say where each of the five laws is used, explaining why the last is not redundant.

6. Let $\mathcal{L}$ be a free theory (with no laws). Show that the isomorphisms of $\mathsf{Cn}_{\mathcal{L}}^\times$ are just the variable renamings (open α-equivalences).

7. Let $\mathcal{C}$ be any category and $\mathcal{K} \subset \mathcal{C}$ a wide subcategory which is an equivalence relation, *i.e.* it contains all objects and identities, and there is at most one morphism between any two objects, whose inverse exists and is in $\mathcal{K}$. Construct the category $\mathcal{C}/\mathcal{K}$ whose objects are $\mathcal{K}$-equivalence classes of $\mathcal{C}$-objects and show that $\mathcal{C} \twoheadrightarrow \mathcal{C}/\mathcal{K}$ is an equivalence functor.

Show also that for any functor $F : C \to \Theta$ which takes all $\mathcal{K}$-maps to identities there is a unique mediating functor $P : C/\mathcal{K} \to \Theta$.

8. Show that Example 4.3.4 correctly solves the cubic equation; explain where the Floyd rules (Remark 4.3.5) are used.

9. Formulate the side-conditions on the use of variables which are needed to adapt the Floyd rules to assignment.

10. Explain Examples 4.4.2, in particular that functors $C \to \mathcal{S}et$ and $C^{op} \to \mathcal{S}et$ are actions.

11. Prove Example 2.7.6(d), that List $(-)$ is a functor.

12. Why are forgetful functors between concrete categories usually faithful but not full?

13. Show that the functor $\mathcal{R}el \to \mathcal{CSL}at$ of Example 4.4.4(b) is full and faithful.

14. For a locale (frame) A, define $\mathsf{pts}(A) = \mathcal{F}rm(A, 2)$. Extend this definition to a functor $\mathsf{pts} : \mathcal{L}oc \to \mathcal{S}p$. When is it faithful? Full?

15. Let F_0, F_1 and F_2 be the sets of vertices, edges and faces of some polyhedron with triangular faces, for example these sets have respectively 12, 30 and 20 elements in the case of an icosahedron. Using the functions $\partial_n^0, \partial_n^1, \ldots, \partial_n^{n-1} : F_n \to F_{n-1}$, describe the **boundary** of the n-dimensional faces and explain how the data (F_n, ∂_n^i) fit together into a functor $\Delta \to \mathcal{S}et$, where Δ is the opposite of the category of finite sets and injective functors. Such a structure is called a **simplicial complex**.

16. Verify that there is a category $\mathcal{K}(C)$, known as the **Karoubi completion**, whose objects are idempotents $(\mathfrak{e} ; \mathfrak{e} = \mathfrak{e}$, Definition 1.3.12) and whose morphisms $\mathfrak{e} \to \mathfrak{d}$ are C-maps $\mathfrak{f}$ with $\mathfrak{e} ; \mathfrak{f} = \mathfrak{f} = \mathfrak{f} ; \mathfrak{d}$. Show that every idempotent in $\mathcal{K}(C)$ is **split**, *i.e.* expressible as $\mathfrak{p} ; \mathfrak{i}$ where $\mathfrak{i} ; \mathfrak{p} = \mathrm{id}$. Let $F : C \to \Theta$ be any functor, where Θ has given splittings of idempotents. Show that there is a unique functor $\mathcal{K}(C) \to \Theta$ extending it and formulate this as a universal property. Finally, show that functors $P : \mathcal{K}(C) \to \Theta$ correspond to semifunctors $F : C \to \Theta$ (Example 4.4.6(d)).

17. Show that groups and groupoids, considered as categories, have no non-trivial products (*cf.* Exercise 5.1).

18. Prove Proposition 4.5.14, that the terminal object and products of pairs suffice to give all finite products.

19. Justify Example 4.5.8(g), *i.e.* that $\mathbb{N} \cong \mathbb{N} \times \mathbb{N}$ in the category composed of primitive recursive functions.

20. Propositions 2.6.7–2.6.9 defined three constructions with well founded relations. Are they products? If so, identify the categories.

21. Characterise products in $\mathcal{Pfn}$, assuming excluded middle.

22. Show that the Tychonov product (Example 3.9.10(e)) gives the (categorical) product in $\mathcal{Sp}$ and also in $\mathcal{Dcpo}$ and $\mathcal{IPO}$.

23. Let $\mathcal{L}$ be the two-sorted algebraic theory of a commutative ring together with a module. Characterise the objects of $\mathsf{Cn}_{\mathcal{L}}^{\times}$ by a pair of natural numbers, and the morphisms by a list of polynomials together with a matrix (*cf.* Remark 4.6.5).

24. Let $\mathcal{L}$ be a single-sorted algebraic theory. Show that the forget-ful functor $|-| : \mathcal{Mod}(\mathcal{L}) \to \mathcal{Set}$ creates products (Definition 4.5.10(c)). What is the analogous result for many-sorted theories?

25. Let $\mathcal{L}$ be an algebraic theory all of whose operation-symbols happen to be unary. Consider its classifying categories $\mathsf{Cn}_{\mathcal{L}}$ and $\mathsf{Cn}_{\mathcal{L}}^{\times}$ *quâ* unary and algebraic theories respectively. Show that $\mathsf{Cn}_{\mathcal{L}}$ is strongly equivalent to the full subcategory of $\mathsf{Cn}_{\mathcal{L}}^{\times}$ with of one-variable contexts, *i.e.* the algebraic theory is a conservative extension of the unary one.

26. Show that any internal monoid in the category $\mathcal{Mon}$ of monoids is commutative. [Hint: consider $m(\langle x, e \rangle \langle e, y \rangle)$.]

27. For complete join-semilattices A and B let $A \otimes B$, $A \,\mathcal{B}\, B$ and $A \multimap B$ be respectively the semilattices of Galois connections, co-Galois connections and left adjoints from A to B (Definition 3.6.1). Show that

(a) $A \otimes B$ satisfies the Mac Lane–Kelly laws with unit 2;

(b) $\mathcal{CSLat}(A, B \multimap C) \cong \mathcal{CSLat}(A \otimes B, C)$;

(c) $A \otimes 2 \cong A \cong 2 \otimes A$;

(d) $(A \multimap 2^{\mathrm{op}}) \cong A^{\mathrm{op}}$;

(e) $A \,\mathcal{B}\, B \cong A^{\mathrm{op}} \multimap B \cong (A^{\mathrm{op}} \otimes B^{\mathrm{op}})^{\mathrm{op}}$ also satisfies the Mac Lane-Kelly laws, with unit 2^{op};

(f) negation defines $\mathcal{CSLat}$-maps $2 \to 2^{\mathrm{op}}$ and $A \otimes B \to A \,\mathcal{B}\, B$.

Such a structure is called a *-**autonomous category** [Bar79].

28. Let $!A$ be the lattice of Scott-closed sets (Proposition 3.4.9) of a complete lattice A. Show that $!1 = 2$, $!(A \times B) \cong (!A) \otimes (!B)$ and $[A \to B] \cong (!A) \multimap B$, the latter being the lattice of Scott-continuous functions. Hence show that the category of complete lattices and Scott-continuous functions is cartesian closed.

29. (Bill Lawvere) Let $\mathcal{L}$ be a single-sorted algebraic theory. Show that $\mathsf{Cn}_{\mathcal{L}}^{\times} \simeq \mathcal{C}$ for a category with $\mathrm{ob}\,\mathcal{C} = \mathbb{N}$ such that addition is the categorical product. Conversely show that every such category $\mathcal{C}$ arises in this way.

30. Show that any group in which $\forall g.\ g^2 = \mathrm{id}$ is Abelian. Express this argument in the various notations we have used.

31. Show that if $f(x, y)$ is a derived operation in the theory of groups such that $f(x, f(y, z)) = f(f(x, y), z)$ then $f(x, y)$ is one of the forms xy, yx, x, y or id. [Hint: it is enough to consider $\mathbb{Z}^2$.]

32. Describe the category (analogous to that in Lemma 4.5.16) of which the function-space Y^X is the terminal object.

33. Let $\mathcal{C}$ be a cartesian closed category. Show that $\mathcal{K}(\mathcal{C})$ is too (Exercise 4.16).

34. Consider the category $\mathcal{C}$ whose *objects* are raw λ-terms of type X in a context Γ, so $\mathrm{ob}\,\mathcal{C} = \mathsf{Cn}^{\times}_{\mathcal{L}+\lambda}(\Gamma, X)$, where the maps of $\mathcal{C}$ are reduction paths. What equivalence relation on paths is needed in order to make the diamond in the Church–Rosser Theorem (Lemma 1.2.4) commute in $\mathcal{C}$? A canonical choice amongst equivalent paths is given by reducing these redexes from left to right; this is called a **standard reduction**: see [Bar81, Definition 11.4.1].

35. By considering the action of substitution in the λ-calculus, show that the equivalence relation of the previous exercise is needed to make horizontal composition in the 2-category $\mathfrak{Cn}^{\rightarrow}$ of contexts, raw λ-terms and $\beta\eta\delta$-reduction well defined. Use it to characterise normal λ-terms, distinguishing them from terms which reduce to themselves.

36. Show that there is a natural transformation between parallel group homomorphisms iff they are **conjugate**. Give an example of a strong equivalence (Definition 4.8.9(b)) between groups for which η and ε are not determined by the isomorphisms F and U involved.

37. A category is **skeletal** if $A \cong B \Rightarrow A = B$; in particular a preorder is skeletal iff it is antisymmetric, *i.e.* a poset. Show that any preorder is weakly equivalent to a poset in a unique way, up to unique isomorphism (Proposition 3.1.10). For any category $\mathcal{C}$ show (using Choice) that there are equivalence functors $\mathcal{C} \to \mathcal{D}$ and $\mathcal{D} \to \mathcal{C}$ for some skeletal category $\mathcal{D}$. By considering the groupoid with exactly two objects and exactly two morphisms in each hom-set, show that these are not unique. Discuss this in the case of the fundamental groupoid of a topological space (Example 4.1.7(i)).

38. Show how to compose and invert strong equivalences, and how to compose equivalence functors. Show also that equivalence functors are confluent (*cf.* Exercise 3.27).

39. Which parts of Definition 4.4.8 are invariant under weak equivalence? What are the invariant versions of those which are not, and also of injectivity and surjectivity on objects?

40. Let $\mathcal{C}$ be a category with finite products. Show that the Yoneda embedding $\mathcal{H}_{(-)} : \mathcal{C} \hookrightarrow \mathcal{S}et^{\mathcal{C}^{op}}$ (Theorem 4.8.12(b)) preserves products.

[Hint: see Proposition 3.2.7(a) and Remark 4.5.9.]

41. For any small category $\mathcal{C}$, show that $\mathcal{S} \equiv \mathcal{S}et^{\mathcal{C}^{op}}$ is cartesian closed. [Hint: $[F \to G](X) = \mathcal{S}(\mathcal{H}_X, [F \to G]) = \mathcal{S}(\mathcal{H}_X \times F, G)$ by the Yoneda Lemma, Theorem 4.8.12(c).] Deduce that the Yoneda embedding preserves function-spaces. (See Exercise 3.30 for the poset version.)

42. Find a category $\mathcal{C}$ (which need only have two objects) in which the Schröder–Bernstein Theorem (Exercise 3.63) fails. Using the Yoneda embedding, show that it also fails in $\mathcal{S}et^{\mathcal{C}^{op}}$. Since this is a topos (model of Zermelo type theory), this theorem relies on excluded middle.

43. Let X be a topological space. Show how to define the **path category** $\mathcal{C}$, whose objects are the points of X and whose morphisms $x \to y$ are continuous functions $f : [0, n] \to X$ from real intervals of length n with $f(0) = x$ and $f(n) = y$. [Hint: composition of paths adds lengths.]

A **homotopy** $f \to g$ between paths of the same length n is a function $h : [0, n] \times [0, 1]$ with $h(t, 0) = f(t)$, $h(t, 1) = g(t)$, $h(0, s) = x$ and $h(n, s) = y$. Modify this definition to allow f and g to have different lengths, and in particular so that $f \,;\, f^-$ and $f^- \,;\, f$ are homotopic to identities in $\mathcal{C}$, where $f^-(t) = f(n - t)$. Define a bi-category of points, paths and homotopies.

Now show that the existence of a homotopy $f \to g$ is an equivalence relation, and that its quotient is a groupoid, the **fundamental groupoid** $\pi_1(X)$ of X.

44. Show that evaluation has the **dinaturality** property on the left.

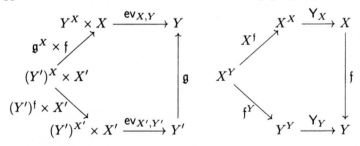

45. Show that the fixed point operator $Y_X : X^X \to X$ (Proposition 3.3.11) satisfies the dinaturality property on the right, and conversely that any transformation with this property yields fixed points. [Hint: put $Y = X$ and consider $\mathrm{id} \in X^X$.]

46. Formulate the general notion of dinaturality.

47. Let $\mathcal{C}$ be any category. Taking $\mathcal{I}$ to be $\varnothing$, $\{\bullet\}$, $\{\bullet \to \bullet\}$, $\{\bullet \rightrightarrows \bullet\}$ and $\{\bullet \to \bullet \to \bullet\}$, describe $[\mathcal{I} \to \mathcal{C}]$.

48. Taking Γ to be successively $\{\bullet\}$, $\{\bullet \to \bullet\}$ and $\{\bullet \to \bullet \to \bullet\}$ use the universal property to show that the exponential $[\mathcal{X} \to \mathcal{Y}]$ in the cartesian closed 1-category Cat must have as objects the functors $\mathcal{X} \to \mathcal{Y}$, as morphisms the natural transformations between such functors and as composition vertical composition of natural transformations. Finally, use the diagram with four objects to show that this is associative.

49. The proof of Theorem 4.8.10 defined the λ-abstraction $\widetilde{P}$ of any *functor* $P : \Gamma \times \mathcal{X} \to \mathcal{Y}$. Show how to define $\widetilde{\psi}$ for any natural transformation $\psi : P \to Q$ between such functors. [Hint: replace Γ by $\Gamma \times \{\bullet \to \bullet\}$]. Show that this is natural in the sense of Definition 4.7.2(c), preserves vertical composition, and defines a bijection between the natural transformations $P \to Q$ and $\widetilde{P} \to \widetilde{Q}$.

50. Write down the definition of a 3-category, together with its notions of composition and the laws which hold between them. Give some geometrical and categorical examples. [Hint: a "middle four" law between any two levels suffices: there is no "middle eight".]

51. By a **pullback** in a poset is meant the meet of two elements *which have a common upper bound* (Exercises 3.5, 3.20 and 3.34). Let $\mathcal{C}$ be the category of posets with pullbacks and pullback-preserving monotone functions. For $X, Y \in \mathrm{ob}\,\mathcal{C}$, write $[X \to Y]$ for $\mathcal{C}(X, Y)$ with an order relation which is to be determined; show that for $\mathrm{ev} : [X \to Y] \times X \to Y$ to preserve pullbacks $(f \leqslant g) \Rightarrow \forall x, x'.\, x' \leqslant^X x \Rightarrow fx' = fx \wedge gx'$ is needed. Show that this does in fact make $\mathcal{C}$ cartesian closed.

Let V be the three-point poset $\{\mathsf{no} \geqslant \perp \leqslant \mathsf{yes}\}$. Show that there is no pullback-preserving function $\mathsf{por} : V \times V \to V$ with $\mathsf{por}(\mathsf{no}, \mathsf{no}) = \mathsf{no}$ and $\mathsf{por}(\mathsf{yes}, \perp) = \mathsf{yes} = \mathsf{por}(\perp, \mathsf{yes})$.

V

Limits and Colimits

PRODUCTS may easily be defined for infinitely many factors, and also considered in the opposite category. However, unlike meets and joins in posets, this does not exhaust the possible types of limits and colimits in categories. Since most of the interesting phenomena may be observed more clearly in the concrete cases of pullback, equaliser, pushout and coequaliser, we postpone the abstract definition to Section 7.3.

This chapter is an account of first order logic, originally motivated by the needs of homological algebra — the understanding of sets, functions and relations came later. As we said in Chapter II, products, equalisers, sums and quotients of equivalence relations provide the real foundation for algebra, rather than the powerset. The diversity of the behaviour of finite limits and colimits is striking: the basic features of groups, rings, vector spaces and topology may often be discovered just by looking for the coproducts in these categories.

Besides these traditional applications in mathematics, we also show how stable disjoint coproducts (as in *Set*) interpret the conditional declarative language, with **if then else fi**. The extension to **while** in Section 6.4 shows that general coequalisers in *Set* are much more complicated than equalisers or quotients.

As in the λ-calculus, the need for the operations of logic to respect substitution must be expressed in category theory. In Chapters VIII and IX we shall show in terms of syntax that pullbacks effect substitution, but in this chapter relational algebra exhibits the same behaviour.

Limits and colimits also interact in that we can try to construct one from the other as in Theorem 3.6.9. However, size issues arise in the case of categories, where they did not for lattices, so we postpone the general result until Theorem 7.3.12. By way of preparation we consider factorisation systems, which abstract the *image* of a function. This useful technique is frequently relegated to an exercise, and the lemmas repeatedly re-proved in papers which use it. Here we treat it in full as we shall need it for the study of the existential quantifier in Section 9.3.

5.1 PULLBACKS AND EQUALISERS

Pullbacks are everywhere, as Serge Lang observed in the 1950s. Indeed they occur spontaneously, as it were: commutative squares arising for other reasons often happen to satisfy the universal property.

DEFINITION 5·1·1:

(a) Let $\mathfrak{f}, \mathfrak{g} : X \rightrightarrows Y$ be a **parallel pair** of maps in a category $\mathcal{S}$. Then an object E together with a morphism $\mathfrak{m} : E \to X$ is an **equaliser** of $\mathfrak{f}$ and $\mathfrak{g}$ if $\mathfrak{m} \,;\, \mathfrak{f} = \mathfrak{m} \,;\, \mathfrak{g}$, and whenever $\mathfrak{a} : \Gamma \to X$ is another morphism such that $\mathfrak{a} \,;\, \mathfrak{f} = \mathfrak{a} \,;\, \mathfrak{g}$ there is a unique map $\mathfrak{h} : \Gamma \to E$ with $\mathfrak{a} = \mathfrak{h} \,;\, \mathfrak{m}$.

(b) Let $\mathfrak{f} : X \to Z$ and $\mathfrak{g} : Y \to Z$ be two maps in a category. Then an object P together with a pair of maps $\pi_0 : P \to X$ and $\pi_1 : P \to Y$ is a **pullback** if $\pi_0 \,;\, \mathfrak{f} = \pi_1 \,;\, \mathfrak{g}$ and whenever $\mathfrak{a} : \Gamma \to X$ and $\mathfrak{b} : \Gamma \to Y$ also satisfy $\mathfrak{a} \,;\, \mathfrak{f} = \mathfrak{b} \,;\, \mathfrak{g}$ there is a unique map $\mathfrak{h} : \Gamma \to P$ such that $\mathfrak{h} \,;\, \pi_0 = \mathfrak{a}$ and $\mathfrak{h} \,;\, \pi_1 = \mathfrak{b}$.

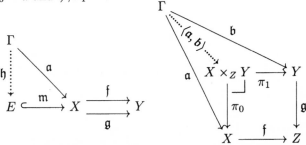

Pullbacks and equalisers are unique up to unique isomorphism where they exist, *cf.* Theorem 4.5.6.

The pullback and mediator are written $X \times_Z Y$ and $\langle \mathfrak{a}, \mathfrak{b} \rangle$. However, one should remember that this notation hides not only the projection maps (which were already absent from the product and exponential notations we have used) but also the maps $\mathfrak{f}$ and $\mathfrak{g}$ which are part of the data. The object $X \times_Z Y$ or the map $\pi_1 : X \times_Z Y \to Y$ is also called the pullback of $\mathfrak{f}$ **along** or **against** $\mathfrak{g}$. In Proposition 5.1.9, where we write $\pi_0 = \mathfrak{f}^* \mathfrak{g}$, we shall see that this asymmetrical language is more typical of the way pullbacks arise than the diagram above suggests.

Pullbacks are often indicated with the right angle symbol, which was suggested by William Butler in 1974 and popularised by Peter Freyd. In Section 9.4 we will no longer be able to afford the space for it, and will instead adopt the convention that the ubiquitous pullbacks are drawn as *parallelograms*. This emphasises how pullbacks do "parallel translation" of structure, in particular by substitution in syntax.

There is no widely used notation for an equaliser, but Freyd and Scedrov [FS90] write $\xrightarrow[g]{f} \bullet$ or just $\xrightarrow{\quad} \bullet$ for the map $\mathfrak{m}$. The hook notation will be discussed in the next section.

The following easy result will turn out to be *extremely* useful:

LEMMA 5·1·2: Suppose that the two squares commute below, and that the one on the right is a pullback.

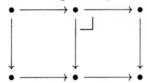

Then the rectangle is a pullback iff the left hand square is.   □

Applications.

EXAMPLES 5·1·3:

(a) Pullbacks in a poset are just meets, but of pairs with a common upper bound (Exercises 3.5, 3.20, 3.34 and 4.51). Since all squares commute, what the bound *is* doesn't matter, only that it *exists*. Again, since any two parallel maps are equal, they trivially have an equaliser: the identity.

(b) A parallel pair $\mathfrak{f}, \mathfrak{g} : [\underline{x} : \underline{X}] \rightrightarrows [\underline{y} : \underline{Y}]$ in the category of contexts and substitutions is given by a list of pairs of terms $f_j(\underline{x}), g_j(\underline{x}) : Y_j$. A substitution $\mathfrak{a} = [\underline{a}/\underline{x}]$ has equal composites with them iff it is a **unifier**, *i.e.* $[\underline{a}/\underline{x}]^* f_j = [\underline{a}/\underline{x}]^* g_j$, or, in the informal notation, $f_j(\underline{a}) = g_j(\underline{a})$. The equaliser, if any, is the **most general unifier** (Remark 1.7.8), since any other unifier is a substitute. Similarly pullbacks solve $f_j(\underline{x}) = g_j(\underline{y})$. In Section 6.5 we shall construct the most general unifier in the simplest case, a free algebraic theory.

(c) A pullback rooted at the terminal object $Z = 1$ is a product. In any category with binary products, pullbacks may be constructed from equalisers and *vice versa* (*cf.* Remark 5.2.3).

EXAMPLES 5·1·4: Pullbacks of sets and functions have several names.

(a) The equaliser of two parallel functions $f, g : X \rightrightarrows Y$ is (the inclusion of) the subset $E = \{x \mid f(x) = g(x)\}$. If $a : \Gamma \to X$ with $f \circ a = g \circ a$ then the results of a at all elements of Γ lie in $E \subset X$, so a restricts to $h : \Gamma \to E \subset X$.

(b) The pullback of any two functions $f : X \to Z$ and $g : Y \to Z$ is the subset $\{\langle x, y \rangle \mid f(x) = g(y)\}$ with the usual projections. A pair of maps from Γ gives rise to a map $\langle a, b \rangle : \Gamma \to X \times Y$ to the product, which restricts to the pullback in the same way as for the equaliser.

(c) If $g : Y \subset Z$ is a subset inclusion then the pullback is the inverse image, $f^{-1}[Y] \subset X$.

(d) If f and g are both subset inclusions then the pullback, $X \cap Y \subset Z$, is their intersection (Example 2.1.6(d)).

EXAMPLES 5·1·5: Some other familiar semantic categories.

(a) The restriction of the order on the source gives the equaliser and on the product gives the pullback in $\mathcal{P}reord$, $\mathcal{P}os$, $\mathcal{CSLat}$, $\mathcal{SLat}$, $\mathcal{Lat}$, $\mathcal{DLat}$, $\mathcal{BA}$, $\mathcal{F}rm$, $\mathcal{HSL}$, $\mathcal{H}eyt$ and $\mathcal{D}cpo$. This fails, however, in many popular categories of domains.

(b) Equalisers and pullbacks in $\mathcal{S}et^{\mathcal{C}^{op}}$ are constructed pointwise and the Yoneda embedding (Theorem 4.8.12(b)) $\mathcal{H}_{(-)} : \mathcal{C} \hookrightarrow \mathcal{S}et^{\mathcal{C}^{op}}$ preserves whatever limits exist.

(c) The category $\mathcal{M}od(\mathcal{L})$ of models of any algebraic theory $\mathcal{L}$ has pullbacks and equalisers. Indeed the forgetful functor $\mathcal{M}od(\mathcal{L}) \to \mathcal{S}et^{\Sigma}$ creates limits (*cf.* Definition 4.5.10(c)), Σ being the set of sorts.

(d) The category of trichotomous orders (Definition 3.1.3) and *strictly monotone* functions does not have products, but it has got pullbacks and equalisers, constructed in $\mathcal{S}et$. This is because its maps are all injective, so pullbacks are intersections.

(e) A **coherence space**[1] is a set X with two *symmetric* relations $\frown$ (joined or coherent) and $\smile$ (un-joined or incoherent) satisfying a trichotomy law: exactly one of $x \frown y$, $x = y$ and $x \smile y$ holds. An **embedding** is a function which preserves all three relations, whence it also reflects them and is injective, so pullbacks are intersections.

(f) The category of fields also has pullbacks and equalisers but not binary products or a terminal object. We can say that two elements $x, y \in K$ of a field are *apart* in a *positive* sense which is preserved by homomorphisms, if $\exists z. (x - y)z = 1$, which we write as $x \# y$. This obeys the *di*chotomy law that exactly one of $x = y$ and $x \# y$ holds, whence all homomorphisms are injective. The full subcategory consisting of those fields in which a particular polynomial has a root (say $x^2 + 1 = 0$) has pullbacks but not equalisers.

EXAMPLES 5·1·6: Some squares which are "spontaneous" pullbacks.

(a) Exercise 4.34 showed that the Church–Rosser Theorem (Fact 2.3.3)

---

[1] This term is used in [GLT89]. It is the traditional notion of **graph** in combinatorics: unoriented and without loops or multiple edges. What is usually required in category theory and informatics is Example 4.2.6(e).

may be expressed as a commutative square in a certain category. This square is in fact a pushout (a pullback in the opposite category) [Bar81, Exercise 12.4.4].

(b) The effect of the product functor on morphisms (Proposition 4.5.13) gives rise to pullback squares. These also occurred in the diagrams Example 4.6.3(f) and Definition 4.7.9.

(c) Let $\mathfrak{f} : X' \to X$ in $\mathcal{C}$ and $\mathfrak{g} : E' \to E$ in Γ be maps in any two categories. Then this square (Proposition 4.8.3(c)) is a pullback:

$$
\begin{array}{ccc}
\langle E', X' \rangle & \xrightarrow{\langle \mathsf{id}_{E'}, \mathfrak{f} \rangle} & \langle E', X \rangle \\
{\scriptstyle \langle \mathfrak{g}, \mathsf{id}_{X'} \rangle} \Big\downarrow & & \Big\downarrow {\scriptstyle \langle \mathfrak{g}, \mathsf{id}_X \rangle} \\
\langle E, X' \rangle & \xrightarrow{\langle \mathsf{id}_E, \mathfrak{f} \rangle} & \langle E, X \rangle
\end{array}
$$

This means that the result of applying a pullback-preserving functor $P : \Gamma \times \mathcal{C} \to \mathcal{D}$ to this square yields another pullback: *cf.* Exercises 3.20 and 5.9.

Slices. Pullbacks arise in three important ways in type theory:

(a) they generalise products (so are also called **fibred products**),

(b) they have something to do with equality, and

(c) in Chapter VIII their primary role is substitution, *cf.* Exercise 5.6.

In the first two cases there is a symmetry between the two legs, but this is not true of substitution, where one is definitely acting on the other and not *vice versa*. Let's look at (b) and then (c).

REMARK 5·1·7:   The pullbacks below are in a sense trivial, in that the pullback of an identity must be an isomorphism. But if we think of the pullback as a subset $X \times_Z Y \subset X \times Y$, on the left we obtain the **graph of a function**, *cf.* Definition 1.3.1, Exercise 1.14 and Corollary 4.3.13(a).

$$
\begin{array}{ccccccc}
\mathsf{graph}(f) & \longrightarrow & Y & \qquad & (=_X) & \longrightarrow & X \\
{\scriptstyle \cong} \Big\downarrow & & \Big\downarrow {\scriptstyle \mathsf{id}} & & \Big\downarrow & & \Big\downarrow {\scriptstyle \mathsf{id}} \\
X & \xrightarrow{\ f\ } & Y & & X & \xrightarrow{\ \mathsf{id}\ } & X
\end{array}
$$

The even simpler case on the right gives the binary **equality relation**, $(=_X) \subset X \times X$, to which we return in Remark 8.3.5.   □

To see pullbacks as products exactly, we need to formalise the idea used in Lemma 4.5.16. For the terminal object, see Exercise 5.4.

DEFINITION 5·1·8: Let X be any object of any category S. The **slice category** $S \downarrow X$ has

(a) as *objects* the S-morphisms $\mathfrak{d} : Y \to X$,

(b) as *morphisms* the commuting triangles in S,

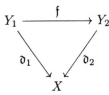

(c) and as *identity* and *composition* those for S.

PROPOSITION 5·1·9: If S has chosen pullbacks against $u : X' \to X$ then $u^* : S \downarrow X \to S \downarrow X'$ is a functor. This is a contravariant action (Section 4.2), except that id^* need only be *isomorphic* to the identity and $(u \,;\, \mathfrak{v})^*$ to $u^* \cdot \mathfrak{v}^*$.

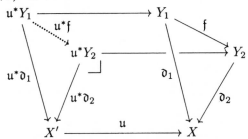

PROOF: The effect of u^* on the morphism $\mathfrak{f}$ is the mediator between the pullback parallelograms. Identities and composites are preserved by the same argument by uniqueness of mediators as in Proposition 4.5.13. □

The analogue of the slice for posets is just the lower set generated by an element (Definition 3.1.7). In the next section we shall use the same construction, but with the $\mathfrak{d}$s restricted to be monos. It will appear in a more general form in Definition 8.3.8, where the $\mathfrak{d}$s belong to a specified class but u and $\mathfrak{f}$ will be arbitrary. An important common generalisation of pullbacks and slices (comma categories) is given in Definition 7.3.8.

5.2   SUBOBJECTS

This section and the next study the categorical notions of subsets and injective functions, *cf.* Exercise 1.16; Remark 5.8.4 discusses surjective functions for their own sake, rather than by duality. In *Set* a function which is both mono and epi is an isomorphism, but this fails in most other categories of interest: Proposition 5.8.10 shows what is needed.

DEFINITION 5·2·1:   A map $m : U \to X$ in a category is called

(a) a ***monomorphism*** or ***mono*** if the post*cancellation property* holds: given any $a, b : \Gamma \rightrightarrows U$, if $a \,;\, m = b \,;\, m$ then $a = b$ (reading maps from Γ as generalised elements in the sense of Remark 4.5.3, this says exactly that m is injective Definition 1.3.10(a)),

(b) a ***regular mono*** if it is the equaliser (Definition 5.1.1(a)) of some parallel pair (*cf.* Lemma 5.6.6(a) for a canonical such pair), and

(c) a ***split mono*** if there is a postinverse $e : X \twoheadrightarrow U$ with $m \,;\, e = \mathrm{id}_U$.

Dually we have (regular, split) ***epimorphisms*** or ***epis*** — please, *not* "epics". Monos are often indicated by a hook on the arrow ($\hookrightarrow$ or $\rightarrowtail$) and similarly epis by $\twoheadrightarrow$ (or, *but not in this book*, by $\longrightarrow\!\!\!>$); notice that the hook is at the end where the cancellation properties hold.

See Examples 5.7.5 for the familiar cases.

PROPOSITION 5·2·2:   Let $m : U \to X$ be any map of any category.

(a) Then m is mono iff the square is a pullback:

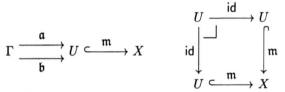

(b) In particular this happens if m is the structure map of an equaliser.

Let $F : \mathcal{C} \to \mathcal{D}$ be a functor. Then

(c) if m is split mono then so is Fm;

(d) if F preserves pullbacks and m is mono then so is Fm;

(e) if F preserves equalisers and m is regular mono then so is Fm.

For any two maps $m : X \to Y$ and $n : Y \to Z$,

(f) if m and n are mono then so is $m \,;\, n$, and similarly for split monos, but Example 5.7.5(d) shows that this may fail for regular monos;

(g) conversely if $m \,;\, n$ is (split) mono then so is m;

(h) if $m \,;\, n$ is regular mono and n is mono then m is regular mono. $\square$

REMARK 5·2·3:   The class of split monos is not closed under pullback, but instead generates the class of regular monos (so a regular mono is "possibly split" in the sense of Definition 3.8.2). For if m is the equaliser of $f, g : X \rightrightarrows Z$ then it is the pullback along $\langle f, g \rangle$ of the diagonal

$Z \to Z \times Z$ (assuming that this exists). The pullback $u^*m : V \to Y$ of m along any map $u : Y \to X$ is also regular mono, being the equaliser of $(u \,; f)$ and $(u \,; g)$.

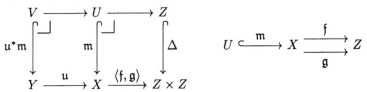

The class of plain monos is also stable under pullback. See Example 5.4.6(e) and Lemma 5.5.7 regarding coproduct inclusions $X \to X + Y$, and Exercise 5.37 for pullbacks of split epis. ☐

The lattice of subobjects.

DEFINITION 5·2·4: A *subobject* of an object X is an equivalence class of monos $m : U \hookrightarrow X$, where this is identified with $n : V \hookrightarrow X$ if there is an isomorphism $f : U \cong V$ such that $m = f \,; n$.

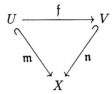

If there is *any* map f (necessarily a unique mono, by the cancellation properties) making this triangle commute, we write $m \leqslant n$ or $U \subset_X V$.

REMARK 5·2·5: We write $\mathsf{Sub}_S(X)$ for the class of subobjects. It is the full subcategory $(S \! \downarrow \! X)$ of the slice $S \downarrow X$ (Definition 5.1.8) consisting of those objects $\partial : U \to X$ for which ∂ is mono. There is no *a priori* restriction on the morphisms, but *in the case of monos* (and not in the similar but more general situation, $S \! \downarrow \! X$, of Definition 8.3.8) f is forced to be mono and unique. The latter makes $\mathsf{Sub}_S(X)$ a preorder, so we take the poset reflection (Proposition 3.1.10). If this is small (a set) for all X then we say that S is *well powered*.

We said that pullbacks arise in type theory as equality types, products and substitution. When restricted to subobjects, the equality types become trivial (Proposition 5.2.2(a)), but the product and substitution are known as *intersection* and *inverse image* (u^{-1}). The pullback u^* considered as a *functor* says that if $U \subset V$ then $u^{-1}(U) \subset u^{-1}(V)$.

If the category S is well powered and has inverse images then there is a functor $\mathsf{Sub} : S^{\mathrm{op}} \to S\mathcal{L}at$, with $\mathsf{Sub}(u) = u^{-1}$.

PROPOSITION 5·2·6:   In *Set*, the type 2 of propositions is a **subobject classifier**. That is, for any mono $m : U \hookrightarrow X$, there is a *unique* function $\chi_m : X \to 2$ which makes the square a pullback:

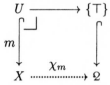

In other words, the natural transformation $\mathcal{S}(-, 2) \to \mathsf{Sub}(-)$ defined by the square is an isomorphism, so the functor Sub is **representable** (Definition 4.8.13) by 2 (we introduced the symbol 2 in Notation 2.8.2).

PROOF: For $x : X$, define $\chi_m(x) \equiv (x \in U) : 2$.

Although the pullback is given only up to isomorphism, the *subobject* is unique, being defined as an isomorphism class. To show that the isomorphism $\mathcal{S}(-, 2) \to \mathsf{Sub}(-)$ is *natural*, use Lemma 5.1.2.   □

Using cartesian closure, the functor $\mathsf{Sub}_{\mathcal{S}}(- \times X)$ is also representable, by the powerset $\mathcal{P}(X) = 2^X$ (Definition 2.2.5). These properties, including the existence of pullbacks, are precisely what we need to model Zermelo type theory, *i.e.* to axiomatise the category of sets and functions. Such a category is called an **elementary topos**. These properties extend to the functor category $Set^{\mathcal{C}^{op}}$ for any small category $\mathcal{C}$, by Exercises 4.41, 5.8 and 5.13, so *presheaves* form a topos.

COROLLARY 5·2·7:   Every mono m in *Set* is regular, because it is the equaliser of χ_m and $x \mapsto \top$.   □

We are running way ahead of ourselves in discussing higher order logic at this point: the purpose of this chapter is to present the traditional categorical account of the *first* order logic of sets. Section 9.5 returns to the comprehension and powerset axioms in type theory.

Sets of solutions of equations. There is no recognisable axiom of comprehension in toposes, because they use monos instead of actually considering the predicate calculus (*cf.* Remark 2.2.4). We do, however, have a theory of equations, and can reconstruct a certain amount of logic around them. Recall the context notation $[\underline{x} : \underline{X}]$ from Definition 4.3.11.

REMARK 5·2·8:   A subobject defined as the equaliser of a pair of maps,

$$E \xhookrightarrow{\quad\quad} [\underline{x} : \underline{X}] \xrightarrow[b]{a} A,$$

is the set of solutions to the equation $a(\underline{x}) = b(\underline{x})$.

For example, this could be the set of roots of a polynomial, or some geometrical figure[2] such as the circle $x^2 + y^2 = r^2$.

Either by replacing A with a context, or by using pullbacks (intersection of subobjects), this treatment of a single equation extends easily to the simultaneous solution of a family of them, $\mathfrak{a} = \mathfrak{b}$, *i.e.* to conjunction.

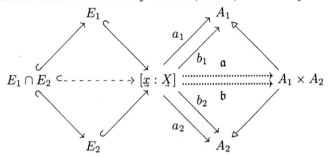

Given the interpretations of equations $\mathfrak{a} = \mathfrak{b}$ and $\mathfrak{c} = \mathfrak{d}$, we may express the (semantic) entailment (*cf.* Corollary 4.6.8),

$$\underset{\sim}{x} : \underset{\sim}{X} \vDash \mathfrak{a}(\underset{\sim}{x}) = \mathfrak{b}(\underset{\sim}{x}) \implies \mathfrak{c}(\underset{\sim}{x}) = \mathfrak{d}(\underset{\sim}{x}),$$

i.e. that the set of solutions to one is contained in the other, by saying that there is a fill-in as shown, making this square commute:

$$
\begin{array}{ccc}
\{\underset{\sim}{x} \mid \mathfrak{a}(\underset{\sim}{x}) = \mathfrak{b}(\underset{\sim}{x})\} & \lhook\joinrel\longrightarrow & [\underset{\sim}{x} : \underset{\sim}{X}] \underset{\mathfrak{b}}{\overset{\mathfrak{a}}{\rightrightarrows}} [A] \\
\vdots & & \| \\
\downarrow & & \| \\
\{\underset{\sim}{x} \mid \mathfrak{c}(\underset{\sim}{x}) = \mathfrak{d}(\underset{\sim}{x})\} & \lhook\joinrel\longrightarrow & [\underset{\sim}{x} : \underset{\sim}{X}] \underset{\mathfrak{d}}{\overset{\mathfrak{c}}{\rightrightarrows}} [C]
\end{array}
$$

We may consider a general map (substitution) $\mathfrak{u} : [\underset{\sim}{x} : \underset{\sim}{X}] \to [\underset{\sim}{y} : \underset{\sim}{Y}]$ instead of repeating the object $[\underset{\sim}{x} : \underset{\sim}{X}]$ in this diagram, but this only substitutes $\mathfrak{u}(\underset{\sim}{x})$ for $\underset{\sim}{y}$ in $\mathfrak{c}$ and $\mathfrak{d}$.

If the fill-in is invertible then $\mathfrak{a} = \mathfrak{b} \Leftrightarrow \mathfrak{c} = \mathfrak{d}$.

---

[2]Peter Freyd argues [FS90, §1.43], on the basis that this was René Descartes' work, that categories with finite products and equalisers (and hence all finite limits) should be called **cartesian categories**. Other authors (wrongly, in Freyd's opinion) have applied the word cartesian to categories which merely have products. I dissent from his emphasis since, in Chapter VIII, we shall require *some but not all* finite limits; the justification of Freyd's usage depends on claiming that Descartes meant to consider equalisers of *all* pairs of functions. We shall use the term **lex category** for one with all finite limits; this word is a corruption of left exact, which (as Freyd also says) is an inappropriate left-over from the theory of Abelian categories, so we shall treat it as an autonomous English word in the same way as we did "poset" (page 127).

Besides equality and conjunction, there is another connective which can be expressed using finite limits.

REMARK 5·2·9: Recall that we wrote $\hat{y} : X \times Y \longrightarrow X$ for the product projection in Definition 4.5.7. In this case, if there is a fill-in

$$\{\langle \underline{x}, y\rangle \mid \mathfrak{a}(\underline{x}, y) = \mathfrak{b}(\underline{x}, y)\} \lhook\joinrel\longrightarrow [\underline{x} : \underline{X}, y : Y]$$

$$\cong \qquad\qquad \hat{y}$$

$$\{\underline{x} \mid \mathfrak{c}(\underline{x}) = \mathfrak{d}(\underline{x})\} \lhook\joinrel\longrightarrow [\underline{x} : \underline{X}]$$

then

$$\forall \underline{x}.\, (\exists y.\, \mathfrak{a}(\underline{x}, y) = \mathfrak{b}(\underline{x}, y)) \implies \mathfrak{c}(\underline{x}) = \mathfrak{d}(\underline{x}).$$

But if the fill-in is an isomorphism we also have the converse,

$$\forall \underline{x}.\, \mathfrak{c}(\underline{x}) = \mathfrak{d}(\underline{x}) \implies \exists! y.\, \mathfrak{a}(\underline{x}, y) = \mathfrak{b}(\underline{x}, y),$$

where the existential quantifier is *unique.*

The inverse map, which we have already supposed to be present in the category, is an operation of arity $\underline{X} \vdash Y$, defined *conditionally* on the equations $\mathfrak{c}(\underline{x}) = \mathfrak{d}(\underline{x})$. We may give it a name, *i.e.* conservatively extend the language, using the $\imath$-calculus (Lemma 1.2.11).

Entailments of this form may be used as axioms for a *generalised* algebraic theory. For example, categories may be defined using the condition

$$\mathsf{tgt}\,\mathfrak{f} = \mathsf{src}\,\mathfrak{g}\ \wedge\ \mathsf{tgt}\,\mathfrak{g} = \mathsf{src}\,\mathfrak{h} \implies (\mathfrak{f}\,;\mathfrak{g})\,;\mathfrak{h} = \mathfrak{f}\,;(\mathfrak{g}\,;\mathfrak{h}).$$

Peter Freyd styled such theories ***essentially algebraic*** [Fre72], since products, coproducts and equalisers (but not quotients) of models are found in the same way as for the algebraic theories in Section 4.6. Michel Coste set out the fragment of predicate calculus consisting of equality, conjunction and unique existence [Cos79]; he used the term $\underrightarrow{\lim}$*-theory*, since to define a classifying category analogous to that in Theorem 4.6.7 we must use (at least some) pullbacks.

Peter Gabriel and Friedrich Ulmer [GU71] proved the duality linking small lex categories *quâ* theories to their categories of models, which are locally finitely presentable categories (Definition 6.6.14(c)). Chapters 2 and 3 of [MR77] provide a textbook account of this approach (there's no need to read Chapter 1 first), as does [BW85], which uses sketches.

Proposition 5.6.4 describes equivalence relations categorically in this way. Remark 5.8.5 discusses $\exists$ without the uniqueness condition. In my view it is more natural to describe the theory of categories using dependent types (Example 8.1.12).

Special subobjects. Algebraic equations typically describe a smaller class of subobjects than general predicates do. We may similarly wish to restrict attention to recursively enumerable sets or those of some other low degree of logical complexity. In categories other than *Set*, perhaps not all monos are regular or "well behaved" in some other sense. For such a class to be useful it must at least be closed under inverse image, *i.e.* be invariant under substitution.

DEFINITION 5·2·10:   A class $M \subset \text{mor}S$ of monos such that

(a) all isomorphisms are in M,

(b) if $m : X \to Y$ and $n : Y \to Z$ are in M then so is $m \,;\, n$, and

(c) if $m : X \to Y$ is in M and $u : Y' \to Y$ is arbitrary then the pullback $u^*m : X' \to Y'$ exists and is in M

is called a ***class of supports*** or a ***dominion***. Of particular interest is the case where there is an object Σ equipped with a global element $\top : 1 \to \Sigma$ having the same property for M that 2 has for *all* monos in *Set*. This is called a ***support classifier*** or ***dominance***. If both Σ and 2 exist then there is a $\wedge$-semilattice inclusion $\Sigma \hookrightarrow 2$ (Exercise 5.12).

EXAMPLES 5·2·11: The following are dominions:

(a) all monos in *Set*, classified by 2 (Proposition 5.2.6);

(b) upper subsets in *Pos*, also classified by 2 (Example 3.1.6(f));

(c) algebraic equations in $CRng^{\text{op}}$, the category of affine varieties (not classified);

(d) open subsets in *Dcpo*, *IPO*, *Sp* or *Loc*, classified by the Sierpiński space S (Definition 3.4.10);

(e) recursively enumerable subsets in a certain category composed of total recursive functions; Exercise 5.10 describes the classifier.

Classes of monos will be used as the supports of partial functions in the next section. Pullback-stable classes of maps which are not necessarily monos will be needed to interpret dependent types in Chapters VIII and IX. We study support classifiers and the powerset in Section 9.5.

5.3   PARTIAL AND CONDITIONAL PROGRAMS

The direct declarative language, with **put**, **discard** and composition, was presented as a category in Section 4.3. Besides its obvious covariant imperative action on states, it has a contravariant action on predicates. We saw that these *midconditions* $\phi[\underline{x}]$ belong naturally at the semicolons between commands, at each of which there is also an associated list $\underline{x}$ of program-variables, namely those which are in scope at that point. But we only defined the contexts (objects of the category) to consist of the typed variables, so now we shall also incorporate the midconditions.

We aim to identify the essential categorical features of the imperative action, where the interpretation of such a context is the set of states of $\underline{x}$ which satisfy $\phi[\underline{x}]$, *i.e.* the comprehension $\{\underline{x} \mid \phi[\underline{x}]\}$. If $\phi[\underline{x}]$ is a family of equations, we already know how to express this subobject directly in category theory, as an equaliser; this will be extended to full first order logic in Section 5.8.

Midconditions were introduced with a view to proving correctness of programs, but they are also used to control (non-)termination. Although we cannot *cause* non-termination in the programs we *build* until we introduce recursion or **while** loops in Section 6.4, the building *blocks* may themselves be non-terminating if applied indiscriminately. We use conditionals to ensure that they only get called in circumstances where we know that they are defined and produce correct results.

EXAMPLE 5·3·1:   In the program for the cubic equation (Example 4.3.4), some of the operations involved are only defined on part of $\mathbb{R}$, or have ambiguous results. We must be clear, for example, that $\sqrt{}$ and $\cos^{-1}$ are the inverses of maps $(-)^2 : P \to P$ and $\cos : H \to I$, where the subsets $P = [0, \infty)$, $H = [0, \pi]$ and $I = [-1, 1]$ of $\mathbb{R}$ must also be objects of the category.

If, as in Section 4.3, the meaning of a program is to be a morphism of a category whose objects are simply lists of program-variables, then this morphism must be allowed to be a partial one. We can restore totality by introducing **virtual objects**, defined by program-variables together with midconditions. In particular, the troublesome primitive operations in the cubic equation program are treated as partial functions (Section 3.3) whose supports are given by known predicates, *i.e.* the preconditions which guarantee termination and correctness.

For this approach to be adequate for recursive or **while** programs, the logic used to define the virtual objects must be strong enough to express the termination or otherwise of the program-fragments which are about to be executed, and in particular *stronger than the program itself could verify*. This is the case if category S incorporates the predicate calculus, with quantification over $\mathbb{N}$, or is simply assumed to be (like) *Set*. This chapter adopts the point of view of traditional categorical logic, which aims to describe the category of sets and functions. Our account therefore falls short of a purely syntax-driven construction.

REMARK 5·3·2: Let $u : \Gamma \to \Delta$ be a terminating (everywhere defined) program, where $\Gamma = [x : X]$. The interpretation of its restriction to the subset on which $\phi[x]$ holds is of course the composite

$$[\Gamma, \phi] \equiv \{x \mid \phi[x]\} \lhook\joinrel\longrightarrow \Gamma \xrightarrow{\;u\;} \Delta,$$

where different programs $u, v : \Gamma \rightrightarrows \Delta$ may have the same effect when restricted to $[\Gamma, \phi]$. The Floyd assertion (Remark 4.3.5)

$$\{\phi\} \; u \; \{\vartheta\}$$

says that the *target* may also be restricted to the subset $[\Delta, \vartheta] \hookrightarrow \Delta$, *i.e.* that (in the interpretation as sets of states) there is a function which completes the square

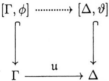

The top map is unique because the right hand one is an inclusion. This square is a pullback iff ϕ is the **weakest precondition** $u^*\vartheta$ for which the property is valid (Remark 4.3.7).

In the terminology of Example 3.8.3, we still aim for **total correctness**; partial correctness is more useful for **while** programs (Remark 6.4.16).

REMARK 5·3·3: Notice that we have returned to the more general notion of context developed for the predicate calculus (Definition 1.5.4), which involves both typed variables and logical formulae. Given that the Floyd rules are anyway not fully specified without a statement of the variables involved, with their types, we now see them as a minor variant of our standard notation $u : [\Gamma, \phi] \to [\Delta, \vartheta]$.

Computationally, Γ is the physical range of the program-variables. The state is confined to $[\Gamma, \phi]$ only by a "gentlemen's agreement", namely the proof of the preceding program-fragment: the program could be started (such as in a debugging environment) at this intermediate point in *any* state $a \in [\Gamma]$. Consider, for example, the type of primes (with the usual binary representation of numbers), or the type of codes for programs that terminate. These acquire their meaning by *intension*, not extension.

The **discard** commands divide Example 4.3.4 into seven phrases. Each one is interpreted as a bijection, whose inverse is essentially given by the comments. When a set is defined using comprehension in Zermelo type theory, it is commonly understood to be interchangeable with any isomorphic set, defined by some other set-theoretic formula. However, in the case of a program such as this, to dismiss these isomorphisms as mere changes of representation of the same object would trivialise this historic achievement in algebra. So, in a computational intuition this virtual object is more closely related to the ambient Γ than to any isomorphic $[\Delta, \vartheta]$. There is a rigid **division of the context** between typed variables and predicates. The axiom of comprehension makes the division permeable, emancipating the subset (Sections 9.2 and 9.5).

Partial morphisms. Now we drop the assumption that $u : \Gamma \to \Delta$ is total on the physical range of the program-variables. We begin by defining partial functions in terms of a given class of total functions in a category S such as *Set*. Afterwards, we consider how the total functions and virtual objects can be recovered when the possibly non-terminating programs are the raw material.

DEFINITION 5·3·4: A **partial function** with **source** X, **target** Y and **support** U in a category S is an isomorphism class of diagrams like

The $m : U \hookrightarrow X$ may be required to belong to a class $\mathcal{M}$ of supports (Definition 5.2.10). When m is an isomorphism, the function is total.

Definition 3.3.4 defined equality and the extension order ($f \sqsubseteq g$) on partial functions, exactly analogously to the definitions for subobjects in Section 5.2. They also bear the same relation to the category of spans used in Lemma 4.5.16 as $S \updownarrow X \simeq \mathrm{Sub}_S(X)$ did to the slice $S \downarrow X$.

PROPOSITION 5·3·5: Composition of partial functions, $U \,;V$, defined by pullback, is associative, *cf.* Lemmas 1.6.6 and 4.1.3. (Note that $U \,; V$ is defined as a *subobject*, *i.e.* as an *isomorphism class* of such pullbacks, so it is legitimate to say that associativity holds up to equality.)

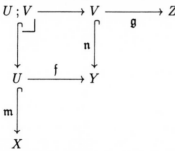

We write $\mathcal{P} = \mathsf{P}(S, \mathcal{M})$ for the category composed of partial functions.

PROOF: Use Lemma 5.1.2.

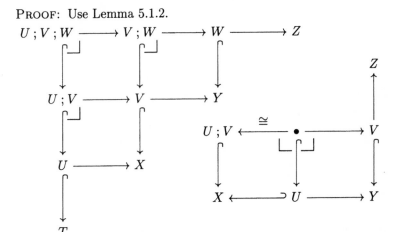

The diagram on the right shows that a second pullback also arises; it is used in Exercise 5.53 and Section 6.4. □

REMARK 5·3·6: An alternative view of partial functions starts from the category $\mathcal{P}$ and tries to recover $\mathcal{S}$ and $\mathcal{M}$:

$$\mathsf{supp}(a) = [\![\mathbf{put}\ x = a\ ;\mathbf{discard}\ x]\!]$$

is a partial endofunction $\langle \mathfrak{m}, \mathfrak{m} \rangle : [\![\Gamma]\!] \rightharpoonup [\![\Gamma]\!]$ where $\langle \mathfrak{m}, \mathfrak{f} \rangle = [\![a]\!]$. (In the total case Remark 4.3.8 made this the identity.) This may be regarded as the inclusion of a (virtual) subobject of $[\![\Gamma]\!]$. Exercise 5.15 uses this idea to characterise $\mathcal{P}$, and shows that it may be fully *embedded* in some $\mathsf{P}(\mathcal{S}, \mathcal{M})$. The latter is, in general, somewhat bigger, because $\mathcal{S}$ contains the virtual objects definable from but not necessarily present in $\mathcal{P}$. For example, if $\mathcal{P}$ is the monoid composed of partial recursive functions on $\mathbb{N}$, then all recursively enumerable subsets of $\mathbb{N}^k$ occur as objects of $\mathcal{S}$.

Proposition 5.8.7 considers an approach to relations analogous to this one for partial functions. A similar construction using virtual objects applies. Section 6.4 shows how **while** programs can be interpreted using coequalisers; in this case the virtual objects form a far larger structure than is needed to prove the soundness result.

Conditionals. Even when no proof of correctness has been supplied for a program, it is still natural to think of the branches of a conditional as defined only on the virtual subobjects described by the condition or its negation. The word "if" is misleading: the test is not a proposition as in implication but a computable function. Failure is not enough to cause execution of the **else** branch: that only happens when the test has *succeeded* in producing the second value. This all too common confusion with logic may be avoided by regarding the condition as a *question*, to which yes and no are possible responses: there may be no answer at all. The two parts are put together in the same way as those of a disjunction $(\vee\mathcal{E})$-box or a coproduct.

DEFINITION 5·3·7:   The **conditional declarative language** extends the ⟨program⟩s of Definition 4.3.1 by

> **if** ⟨Boolean expression⟩ **then** ⟨program⟩ **else** ⟨program⟩ **fi**

For correct scoping, the two branches must finish up with the same variables; in particular any that only one declares must be discarded.

ASSUMPTION 5·3·8:   Without loss of generality, the test c terminates without side-effect. This can be ensured by inserting **put** $z = c$ before the conditional. The test is then just that of a *variable* of type **2** (Boolean), *i.e.* of a single **bit** (<u>bi</u>nary dig<u>it</u>). We shall need this in the following account because it only uses the test to select one of the branches, and does not incorporate it into the flow of control.

REMARK 5·3·9:   The interpretation of a conditional with test c is defined in terms of the restrictions of the two branches to the virtual objects

$$Y = [\Gamma_0, \phi \wedge c], \qquad N = [\Gamma_0, \phi \wedge \neg c] \quad \text{and} \quad \Theta = [\Theta_0, \vartheta],$$

where Γ_0 and Θ_0 are the lists of variables in scope before and after the conditional, and ϕ and ϑ are the pre- and postconditions. Then the two squares on the left are pullbacks:

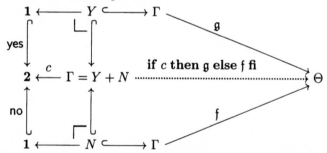

The conditional is then the mediator $[\mathfrak{g}, \mathfrak{f}] : \Gamma \to \Theta$, which is a bijective correspondence so long as the triangles $Y \to \Gamma \to \Theta$ and $N \to \Gamma \to \Theta$ commute. This requires the β- and η-rules

$$\mathfrak{g} \quad = \quad \textbf{if yes then } \mathfrak{g} \textbf{ else } \mathfrak{f} \textbf{ fi}$$

$$\mathfrak{f} \quad = \quad \textbf{if no then } \mathfrak{g} \textbf{ else } \mathfrak{f} \textbf{ fi}$$

$$\mathfrak{p} \quad = \quad \textbf{if } c \textbf{ then } \mathfrak{p} \textbf{ else } \mathfrak{p} \textbf{ fi},$$

and it is also possible to recover the value of the test,

$$c \quad = \quad \textbf{if } c \textbf{ then yes else no fi.}$$

The β- and η-rules make Γ the **coproduct** $Y + N$ of the virtual objects for the two branches. However, this is not the typical coproduct situation, as ours also has to agree with a pair of pullbacks.

EXAMPLE 5·3·10: In Sections 4.3 and 4.6 we saw how to interpret the individual lines of the cubic equation program (Example 4.3.4). The whole program is the composite along the top of

$$[a,b,c:\mathbb{R}] \xrightarrow{\ 1\ } [a,p,q:\mathbb{R}] \dashrightarrow [a,y_0:\mathbb{R},y_\pm:\mathbb{C}] \xrightarrow{\ 7\ } [x_0:\mathbb{R},x_\pm:\mathbb{C}]$$

$$\hat{a}\downarrow \qquad\qquad\qquad \hat{a}\downarrow$$

$$[p,q:\mathbb{R}] \longrightarrow [y_0:\mathbb{R},y_\pm:\mathbb{C}]$$

where the dotted line is a product mediator, and we still have to define the lower map in terms of the conditional. (The numbers refer to the seven phrases into which the **discard** commands break the program.)

Writing $Y, N \subset \mathbb{R}^2$ for the complementary[3] subsets of $\mathbb{R}^2$ on which the condition succeeds and fails, the two maps are then as shown.

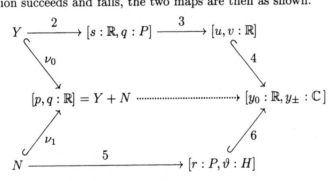

REMARK 5·3·11: The Floyd rule (Remark 4.3.5) for conditionals is

$$\frac{\{\phi \wedge c\}\,\mathfrak{g}\,\{\vartheta\} \qquad\qquad \{\phi \wedge \neg c\}\,\mathfrak{f}\,\{\vartheta\}}{\{\phi\}\ \text{if } c \text{ then } \mathfrak{g} \text{ else } \mathfrak{f} \text{ fi}\ \{\vartheta\}}$$

---

[3]At this point we are using excluded middle in $\mathbb{R}$. Indeed we are well aware of the peril of trying to compute numerically the square root of (or divide by) a number near zero, or the inverse cosine of a number near ± 1. From the point of view of complex analysis, in which the distinction between the cases is artificial anyway, the program works by performing several conformal transformations. Ignoring the boundaries, these are $Y \cong \mathbb{R} \times P \cong \mathbb{R}^2$ and $N \cong P \times H$, in which the critical case ($p^3 = q^2$) is a singularity. It is also an algebraic singularity, because there two solutions coincide, and all three may be obtained as rational functions of the coefficients. Notice how the disparate considerations of logic, numerical computation, analysis and algebra all have something to say about this simple problem! However, the aim of this Example is to illustrate the use of the coproduct. Charles Sturm (1829) showed how to estimate the roots of polynomials [Coh77, Chapter 7]; hence there is a cubic-closed subfield with a decidable computational representation that can be used instead.

which is particularly natural in the proof box idiom:

$\{\phi\}$

if c

then $\{c\}$	**else** $\{\neg c\}$
$\mathfrak{g}$	$\mathfrak{f}$
$\{\vartheta\}$	$\{\vartheta\}$

fi $\{\vartheta\}$

The next two sections treat coproducts in detail. If you do not have a background in pure mathematics, you may prefer to go directly to Section 5.5, as the next section is concerned with coproducts in categories which are *unsuitable* for interpreting conditional programs. But it would be a pity not to look at it at all, as it describes the phenomena which gave rise to category theory in the first place.

5.4   COPRODUCTS AND PUSHOUTS

The universal property of products and coproducts was formulated by Saunders Mac Lane in his study of categories of modules in homological algebra (1950). The resulting description of Abelian categories put too much emphasis on the duality of the axioms [ML88, p. 338]: although the definitions are precisely opposite, colimits behave very differently from limits in most other categories of interest. Distributivity for lattices treats meets and joins symmetrically, but this too fails for the category of sets. Example 7.3.4(c), calculating colimits, is perhaps the one useful application of the literal analogy between *Set* and *Set*^{op}.

We aim to redress the balance in our survey, which illustrates several other important themes in mathematics (more particularly in topology, but this may be a historical accident). Coproducts are very simple in *Set* — they are called disjoint unions and interpret conditionals — but get more complicated as algebraic operations are added. In universal algebra coproducts and pushouts were called free products and free compositions respectively, because of the way they are constructed for groups.

DEFINITION 5·4·1:

(a) An object **0** is an *initial object* if for each $\Theta \in \text{ob}\,\mathcal{S}$ there is exactly one morphism $\mathbf{0} \to \Theta$.

(b) Let $N, Y \in \text{ob}\,\mathcal{S}$. An object C together with maps $\nu_0 : N \to C$ and $\nu_1 : Y \to C$ is a *coproduct* if for each object Θ and pair of maps $N \xrightarrow{\mathfrak{f}} \Theta \xleftarrow{\mathfrak{g}} Y$ there is a unique map $\mathfrak{p} : C \to \Theta$ such that $\nu_0\,;\mathfrak{p} = \mathfrak{f}$ and $\nu_1\,;\mathfrak{p} = \mathfrak{g}$. We usually write $N + Y$ for C and $[\mathfrak{f}, \mathfrak{g}]$ for $\mathfrak{p}$. In the case $Y = N = \Theta = X$ and $\mathfrak{f} = \mathfrak{g} = \mathrm{id}$, we write $\nabla_X : X + X \to X$ for the **codiagonal**.

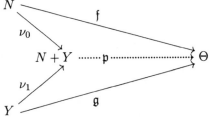

In a poset, the initial object is the least element and the coproducts are joins (Definition 3.2.4); in particular they are falsity and disjunction for formulae under the provability order.

Disjoint unions. These are discussed more fully in the next section.

EXAMPLES 5·4·2:

(a) The empty set, $\varnothing$, is the initial object of *Set*, *Pos*, *Preord*, *Dcpo*, *Sp*, *Cat* and *Gpd*, by Proposition 2.1.9(a).

(b) The coproduct in *Set* was constructed in Example 2.1.7 and is called the **disjoint union**. Exercise 2.13 showed how to find the mediator $p = [f, g] : N + Y \to \Theta$, and that it is unique.

(c) Coproducts in *Pos*, *Preord*, *Dcpo*, *Pfn*, *Sp*, *Gpd* and *Cat* are computed as in *Set*, where in the first three cases $\nu_i(x) \leqslant \nu_j(y)$ iff $i = j$ and $x \leqslant y \in X_i$.

(d) The **Boolean type** is the coproduct $\mathbf{2} = \mathbf{1} + \mathbf{1}$ in *Set*. The inclusions are called yes and no and the coproduct mediator is the conditional.

(e) More generally, a **variant field** in a record consists of a **tag** $i \in I$ and a data assignment $x \in X_i$. See Exercise 5.36 for how **switch** is typically implemented.

(f) JAVA allows constructors with the same name and result type but different arities; the source of such a constructor is effectively the coproduct of the arities.

(g) JAVA and ML provide idioms for exception-handling, which amount to returning a result of coproduct type, *cf.* Remark 2.7.9.

Abelian categories. In categories of vector spaces and modules for rings, finite products coincide with the corresponding coproducts.

DEFINITION 5·4·3: An object of a category which is both terminal (**1**) and initial (**0**) is known as a **zero object**. In *Vsp*, the zero object is the space consisting only of the zero vector, and in general the singleton is the zero algebra for any single-sorted theory which has exactly one definable constant (*i.e.* every expression of the form $r(c, c, \ldots)$ must also be provably equal to c), for example $\mathcal{M}on$, $\mathcal{CM}on$, $\mathcal{SL}at$, $\mathcal{CSL}at$, $\mathcal{HSL}$, $\mathcal{G}p$ and $\mathcal{A}b$. On the other hand, $\varnothing$ is the zero object in $\mathcal{R}el$ and $\mathcal{P}fn$.

The composite $X \to \mathbf{1} = \mathbf{0} \to Y$ is called the **zero map** $0 : X \to Y$.

EXAMPLE 5·4·4:   The coproduct in the category of *commutative* monoids agrees with the product, in which $\nu_0 : N \to N+Y \cong N \times Y$ is $x \mapsto \langle x, 0 \rangle$, $\nu_1 : y \mapsto \langle 0, y \rangle$ and $[f, g] : \langle x, y \rangle \mapsto f(x) + g(y)$. In particular

$$\mathbb{N} \times (\mathbb{N} + \mathbb{N}) \cong \mathbb{N}^3 \quad \text{but} \quad (\mathbb{N} \times \mathbb{N}) + (\mathbb{N} \times \mathbb{N}) \cong \mathbb{N}^4.$$

The common construction is known as the **biproduct** or **direct sum**, written $N \oplus Y$. Since morphisms

$$X_1 \oplus \cdots \oplus X_n \to Y_1 \oplus \cdots \oplus Y_m$$

may be seen as going *from* the coproduct *to* the product, they are determined by an $(n \times m)$ **matrix** of maps $X_i \to Y_j$. Composition is matrix multiplication, in which we take composition of homomorphisms between components as the notion of "scalar multiplication".

Conversely, any category with a zero object and biproducts is $\mathcal{CM}on$-**enriched**, *i.e.* the hom-sets carry a commutative monoid structure for which composition is linear in each argument separately (Exercise 5.20). The categories $\mathcal{A}b$, $\mathcal{V}sp$, $\mathcal{SL}at$, $\mathcal{R}el$ and $\mathcal{CSL}at$ are $\mathcal{CM}on$-enriched (Exercise 5.22); in the last three cases "addition" means join or union.   □

REMARK 5·4·5:   Homological algebra was the progenitor of category theory. Generalising Leonhard Euler's formula $f + v = e + 2$ for the faces, vertices and edges of a convex polyhedron, Enrico Betti defined numerical invariants of spaces by formal addition and subtraction of faces of various dimensions; Henri Poincaré formalised these and introduced homology. Emmy Noether stressed the fact that these calculations go on in Abelian groups, and that the operation ∂_n taking a face of dimension n to the alternating sum of faces of dimension $n - 1$ which form its boundary is a *homomorphism*, and it also satisfies $\partial_n \cdot \partial_{n+1} = 0$. There are many ways of approximating a given space by polyhedra, but the quotient $H_n = \mathsf{Ker}\, \partial_n / \mathsf{Im}\, \partial_{n+1}$ is an invariant, the **homology group**. Since Noether, the groups have been the object of study instead of their dimensions, which are the Betti numbers [Die88].

The categories used for homology are $\mathcal{A}b$-enriched (additive) — but more. It emerged in the 1950s that one could argue in them by chasing diagrams involving kernels and cokernels instead of elements. (Kernels and their quotients are the subject of the later parts of this chapter.) David Buchsbaum axiomatised "Abelian" categories, in his thesis (under Sammy Eilenberg's supervision, but without knowing about Mac Lane's work) and in [CE56, appendix]. Alexander Grothendieck, again independently, showed that *sheaves* of vector spaces and modules also form Abelian categories (1957). We defer to Definition 5.8.1(d) discussion of the extra condition which an *Ab*-enriched category must satisfy in order to be Abelian, since it also applies to sets and other algebraic theories. Abelian categories are covered thoroughly in [Fre64], [ML71], [FS90] and in any modern homology text.

In domain theory, Dana Scott (1970) discovered an analogous infinitary property, that certain filtered colimits coincide with cofiltered limits. This may be used to find domains satisfying equations such as $X \cong X^X$. Michael Smyth (1982) showed that it arises in *Dcpo*-enriched categories. For a treatment of more general diagrams, see [Tay87]. More recently, Peter Freyd [Fre91] has emphasised the coincidence of initial algebras and final coalgebras for certain functors.

Stone duality. For certain algebraic theories — the ones with which the discipline of universal algebra, despite its name, is mainly concerned — $\mathcal{M}od(\mathcal{L})^{op}$ has a *spatial* flavour: the lattices of congruences of groups, rings and modules are modular, and for lattices they are distributive (Exercise 3.49). This suggests viewing their quotients as monos in the opposite category. Marshall Stone (1937) showed how any Boolean algebra arises as the lattice of clopen (*i.e.* both open and closed) subsets of some compact Hausdorff totally disconnected topological space. This was the first real theorem linking logic to the mainstream of mathematics, and dualities like this are explored in [Joh82]. In particular, coproducts of certain algebras signify topological *products*.

EXAMPLES 5·4·6:

(a) The two-element lattice is the initial object of $\mathcal{D}\mathcal{L}at$, $\mathcal{B}\mathcal{A}$ and $\mathcal{H}eyt$.

(b) $\mathbb{Z}$ is the initial ring.

(c) There is no initial field, but $\mathbb{Q}$ is initial in the full subcategory of fields of characteristic zero.

(d) The coproduct of two *commutative* rings $R, S \in \text{ob}\,\mathcal{C}\mathcal{R}ng$ is given by their ***tensor product*** $R \otimes_{\mathbb{Z}} S$ as Abelian groups or $\mathbb{Z}$-modules;

(e) in particular the coproduct of $\mathbb{Z}/(2)$ and $\mathbb{Z}/(3)$ is the trivial ring (with $0 = 1$), so the "inclusions" are not mono.

(f) More generally, any homomorphism $T \to R$ of commutative rings makes R into a T-module; then the pushout of $T \to R$ and $T \to S$ is the tensor product $R \otimes_T S$ of T-modules (Example 7.4.7).

(g) The initial object of $\mathcal{F}rm$ is 2, and its coproducts and pushouts can also be found by a tensor product construction (Example 3.9.10(e)).

Free (co)products and van Kampen's theorem. Coproducts of algebras *in general* tend to be rather chaotic, $\mathcal{M}on$ being typical.

EXAMPLE 5·4·7:   By Corollary 2.7.11, $\mathsf{List}(X)$ is the free monoid on a set X. Then $\mathsf{List}(N +_{Set} Y) \cong \mathsf{List}(N) +_{\mathcal{M}on} \mathsf{List}(Y)$ and in particular $N +_{\mathcal{M}on} N \cong \mathsf{List}(2)$ consists of words in the letters a and b, so

$$N \times (N + N) \quad\cong\quad \mathsf{List}(\{a, c, d\})/\langle ac \sim ca, ad \sim da\rangle$$
$$(N \times N) + (N \times N) \quad\cong\quad \mathsf{List}(\{a, b, c, d\})/\langle ac \sim ca, bd \sim db\rangle.$$

Elements of such monoids have normal forms in which we choose the first representative in lexicographic (dictionary) order. The situation for algebraic theories with more operations gets progressively worse.

Coproducts of monoids are clearly relevant to formal languages, but one might think that the only other value of this construction is in universal algebra. The following famous result shows that, on the contrary, it is also of interest to the geometric tradition. These topological intuitions were already present in Section 4.2.

We only intend to give a sketch of the algebraic idea in its simplest topological form. It is not necessary that both maps be open inclusions, but there are some topological counterexamples which we do not want to consider. The interested reader should see *e.g.* [Bro88, Section 6.7].

Those not familiar with topology may ignore the compactness and open sets, considering finite networks instead. Indeed the edges of the network may be oriented, in which case there is a category but no meaningful group(oid) of paths. For example the paths in an oriented figure of 8 beginning and ending at the cross-over form the monoid $\mathsf{List}(2)$.

THEOREM 5·4·8:   Let X be a topological space and $U, V \subset X$ be open subspaces with $U \cup V = X$. Put $W = U \cap V$, so the diagram shown is (both a pullback and) a pushout in $\mathcal{S}p$, and also in $\mathcal{S}et$.

Then the functor π_1 which assigns the fundamental groupoid (Exercise 4.43) to any space preserves this pushout (it trivially preserves $\otimes$).

PROOF: As $\mathrm{ob}\,\pi_1(X)$ is by definition the underlying set of the space X, it is the pushout of $\mathrm{ob}\,\pi_1(U)$ and $\mathrm{ob}\,\pi_1(V)$ from $\mathrm{ob}\,\pi_1(W)$, so we must consider the morphisms (paths). Let $F : \pi_1(U) \to C$ and $G : \pi_1(V) \to C$ be functors which agree on $\pi_1(W)$.

In order to define $[F, G](s)$ for any path $s : I \to X$ (with $I = [0, 1] \subset \mathbb{R}$), s must be expressed as a composite of paths in U and V. The open sets $s^{-1}(U), s^{-1}(V) \subset I$ are each unions of open intervals; altogether they cover I, but this is compact so finitely many suffice. Hence the path s is $a_1 ; b_1 ; a_2 ; b_2 ; \cdots ; a_n ; b_n$ where each a_i is a path in U and each b_i is a path in V (so the changeover points lie in $W = U \cap V$).

By a similar argument using compactness of $I \times I$, any homotopy between composites s and t of this form may be decomposed into a (rectangular) patchwork whose cells each lie wholly in either U or V (so the boundaries between cells of different kinds are in W). Since F and G are defined on homotopy classes, they map each of these cells to a commutative square in $\mathcal{C}$ (Example 4.8.16(g)). By composing this array of commutative squares, $[F, G](s) = [F, G](t)$. Hence $[F, G] : \pi_1(X) \to \mathcal{C}$ is well defined, and it preserves identities and composition. $\square$

The group of *endo*-paths of a *point* $a \in X$ is known as a fundamental *group* of X and written $\pi_1(X, a)$. When U and V are path-connected and W is contractible (so in particular every path is homotopic to a point, and the fundamental groupoid of W is trivial), the theorem reduces to saying that the fundamental group for X is the coproduct (in $\mathcal{G}p$) of those for U and V. This special case is commonly attributed to Edgar van Kampen (1935), though Herbert Seifert proved an earlier result for simplicial complexes. Van Kampen stated his result using generators and relations (Section 7.4), so the proof is very difficult to follow. Ronald Brown (1967) proved the groupoid form, and Richard Crowell formulated it in terms of a universal property with a modern proof.

Van Kampen wanted to find the fundamental groups of the complements of algebraic curves in $\mathbb{C}^2$. The case where W is not connected is needed even in the simplest example of the fundamental group of a circle (or the complement of a point in $\mathbb{R}^2$). His results may be deduced from the group*oid* form, but not solely from the result for groups.

We did not need to *construct* the pushout of groupoids, because $\pi_1(X)$ already has this property. Yet the fundamental groups of a wide range of spaces of traditional interest in geometric topology (such as a many-handled torus) may be deduced from this theorem, starting from the easy case of contractible spaces. This illustrates the power of categorical methods, both for producing the "right" object for algebraic study, and for manipulating constructions with it.

These examples show that coproducts in *Set*, *Ab*, *Mon* and *CRng* behave very differently from one another. For coproducts in general algebraic theories we must resort to generators and relations (Lemma 7.4.8). The next section considers coproducts in *Set*, *Frm*^{op} and *CRng*^{op}.

5.5 EXTENSIVE CATEGORIES

Now we shall look more closely at disjoint unions as we find them in *Set*, in some categories of "spaces", in type theory and in programming languages. In particular, in the last setting we find *two* constructs which involve branching or case analysis, namely **if then else**, and the tagged sum type. In mathematics, Examples 4.6.4 (including the theory of fields and the characterisation of *free* algebras such as $\mathbb{N}$ and $\mathsf{List}(X)$) have an algebraic flavour but involve exceptions such as zero; coproducts are needed to define classifying categories for them.

The notion of stable (or universal) disjoint coproduct was recognised by the Grothendieck school (*c.* 1960), and is part of Jean Giraud's characterisation of sheaf toposes. However, our categorical account is based on ideas of Steve Schanuel and Bill Lawvere from about 1990. The details are due to Robin Cockett [Coc93], Aurelio Carboni, Steve Lack and Robert Walters [CLW93].

The weaker of the two notions which we consider corresponds exactly to a type theory with products and sums (Section 2.3), but I do not know of a syntactic calculus for the stronger one, other than by restricting full predicate logic. This is unfortunate, as it is this one which we want, but the actual difference between the two is less than it may seem.

Distributivity. Recall the second diagram in Example 5.3.10: this omitted the variable a because it is not used during the conditional. Clearly we do not want the local behaviour of the program to depend on how many (unused) global variables are also present.

DEFINITION 5·5·1: A category with finite products and coproducts is said to be **distributive** if, for all $\Gamma, Y, N \in \mathrm{ob}\,\mathcal{S}$,

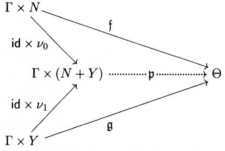

is a coproduct diagram (*cf.* Definition 5.4.1). The analogous property for nullary coproducts is $\Gamma \times \mathbf{0} \cong \mathbf{0}$, which is equivalent to saying that any map $\Gamma \to \mathbf{0}$ is an isomorphism (*cf.* Proposition 2.1.9(b)); in this case we call $\mathbf{0}$ a **strict initial object**.

This restores the parameter-context Γ to the rule in Remark 2.3.10,

$$\frac{\Gamma, x : N \vdash f = p \circ \nu_0 : \Theta \qquad \Gamma, y : Y \vdash g = p \circ \nu_1 : \Theta}{\Gamma, z : N + Y \vdash p = [f, g] : \Theta} +\mathcal{E}$$

Categorically, the functor $\Gamma \times (-)$ preserves coproduct diagrams:

$$
\begin{array}{ccccc}
\Gamma \times N & \longrightarrow & \Gamma \times (Y + N) & \longleftarrow & \Gamma \times Y \\
\Big\downarrow & & \Big\downarrow & & \Big\downarrow \\
N & \xrightarrow{\;\nu_0\;} & Y + N & \xleftarrow{\;\nu_1\;} & Y
\end{array}
$$

(Recall from Definition 4.5.7 the use of $\longrightarrow\!\!\!\!\!\shortmid$ for the product projection π_1.) In this case the maps

$$(\Gamma \times Y) + (\Gamma \times N) \xrightarrow[\langle [\pi_0, \pi_0], [\pi_1, \pi_1]\rangle]{[\langle \pi_0, \pi_1; \nu_0\rangle, \langle \pi_0, \pi_1; \nu_1\rangle]} \Gamma \times (Y + N),$$

which are in any case equal by the universal properties, are invertible. The bottom row is given to be a coproduct, whilst the squares are easily seen to be pullbacks, as they are defined in terms of products; then we conclude that the top row is also a coproduct.

In particular, for total programs $f, g : \Gamma \rightrightarrows \Theta$ and $c : \Gamma \to 2$, we have

$$\textbf{if } a \textbf{ then } f \textbf{ else } g \textbf{ fi} : \Gamma \xrightarrow{\langle \text{id}, c\rangle} \Gamma \times 2 \cong \Gamma + \Gamma \xrightarrow{[f, g]} \Theta.$$

In Examples 5.4.4 and 5.4.7 we observed that $\mathcal{M}on$ and $\mathcal{CM}on$ are not distributive; indeed very few categories of algebras are [Joh85, Joh90].

REMARK 5·5·2: Theorem 4.6.7 showed that each algebraic theory $\mathcal{L}$ is *classified* by a category $\text{Cn}_{\mathcal{L}}^{\times}$ with products. The objects of $\text{Cn}_{\mathcal{L}}^{\times}$ were contexts, consisting of variables of base type. In Section 4.7 we gave a similar construction $(\text{Cn}_{\mathcal{L}}^{\rightarrow})$ for the λ-calculus, where the objects are still contexts, but consisting of type-*expressions* in the constructor $\to$. If $\mathcal{L}$ was just an algebraic theory then $\text{Cn}_{\mathcal{L}}^{\rightarrow}$ is the *free* cartesian closed category on $\text{Cn}_{\mathcal{L}}^{\times}$.

The same can be done for sum types, by allowing the types listed in contexts to be expressions such as $X + Y$ rather than $X \to Y$. The terms may now involve the coproduct injections ν_0 and ν_1 together with the conditional $[\,,\,]$, for which the β-, η-, substitution and continuation rules were given in Section 2.3. The category $\text{Cn}_{\mathcal{L}}^{\times+}$ of such contexts and substitutions is distributive (from the substitution rule), and the models of $\mathcal{L}$ in $\mathcal{S}$ correspond to $(\times, +)$-preserving functors $\text{Cn}_{\mathcal{L}}^{\times+} \to \mathcal{S}$. As these automatically preserve Definition 5.5.1, the semantic category $\mathcal{S}$ must be distributive too. If $\mathcal{L}$ was only an algebraic theory then $\text{Cn}_{\mathcal{L}}^{\times+}$ is the *free* distributive category on $\text{Cn}_{\mathcal{L}}^{\times}$; the inclusion $\text{Cn}_{\mathcal{L}}^{\times} \hookrightarrow \text{Cn}_{\mathcal{L}}^{\times+}$ can be shown to be full and faithful, so the extension is conservative (Sections 7.6–7.7).

Extensive categories. For categories rather than lattices, it is natural to state the distributive law more generally: that coproducts are to be *stable under pullback*. That is, if we replace $\Gamma \times (Y + N)$ above with an arbitrary object C and form the two squares as pullbacks, the top row still has the universal property of a coproduct. (In fact this stronger property does already hold for distributive lattices, but trivially, as pullbacks exist but are no more general than meets, by Example 5.1.3(a).)

What distinguishes the sums we require from joins in lattices is that joins are *idempotent*, but we want components to be *disjoint*. This property can be formulated as the "converse" of stability under pullback. The term *extensive* describes properties such as mass, volume and force which increase with quantity, as opposed to *intensive* properties like density and acceleration which remain the same.

DEFINITION 5.5.3:   An ***extensive category*** is one which has finite coproducts, such that every commutative diagram of the form

$$
\begin{array}{ccccc}
N & \longrightarrow & C & \longleftarrow & Y \\
\downarrow & & \downarrow & & \downarrow \\
A & \xrightarrow{\ \nu_0\ } & A + B & \xleftarrow{\ \nu_1\ } & B
\end{array}
$$

is a pair of pullbacks *iff* the top row is a coproduct diagram.

If the category has **1**, by Lemma 5.1.2 it suffices to consider $A = B = \mathbf{1}$.

In a distributive lattice, the rows can be coproducts (joins) without the squares being pullbacks (meets), as $N \cap B$ and $Y \cap A$ can be non-trivial.

LEMMA 5.5.4: Consider the following commutative diagram, in which the rows are coproducts.

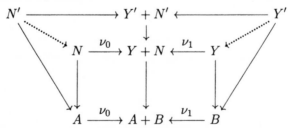

(a) Then the two lower squares are pullbacks iff

(b) for all commutative trapezia there are unique $N' \to N$ and $Y' \to Y$ making the whole diagram commute.

PROOF: To test Y, put $N' = \mathbf{0}$.   □

COROLLARY 5·5·5: A category S with finite coproducts is extensive iff

$$+ : (S \downarrow A) \times (S \downarrow B) \to (S \downarrow (A + B))$$

is an equivalence functor for every $A, B \in \mathrm{ob}\, S$. In particular, the functor category S^2 is equivalent to the slice category $S \downarrow \mathbf{2}$.

PROOF: By the lemma, this functor is full and faithful iff coproducts imply pullbacks. Essential surjectivity says that the source of any map into $A + B$ must be a coproduct $N + Y$ that makes the squares commute, but these squares are pullbacks by the first part. $\square$

EXAMPLES 5·5·6:

(a) *Set* enjoys these properties by Exercise 2.13.

(b) *Pos*, *Dcpo*, *Sp*, *Gpd* and *Cat* are extensive, since the forgetful functor to *Set* creates coproducts and pullbacks.

(c) *Loc* $\equiv$ *Frm*^{op} and *CRng*^{op} are also extensive (Exercise 5.31). $\square$

Exercise 5.35 explains, using partial maps, why the category of virtual objects in Section 5.3 must be extensive rather than just distributive to interpret conditionals.

Stable disjoint sums. This is the traditional formulation.

LEMMA 5·5·7: In an extensive category with finite limits,

(a) coproducts are **stable**, *i.e.* the pullback of any coproduct diagram is another one (in particular the distributive law holds);

(b) the initial object $\mathbf{0}$ is strict;

(c) the components of the coproduct are **disjoint**, *i.e.* the square on the left is always a pullback.

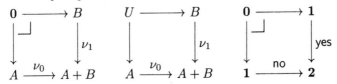

From these three properties, rather than extensivity, it follows that

(d) if the middle square commutes then $U \cong \mathbf{0}$;

(e) for disjointness of general binary coproducts, it suffices that $\mathbf{0}$ be strict and that the square on the right be a pullback, *i.e.* yes $\neq$ no;

(f) the maps $Y \to Y + N \leftarrow N$ are monos (*cf.* Example 5.4.6(e));

(g) in fact they are *regular* monos, assuming extensivity.

PROOF: [a] is part of Definition 5.5.3. For [b,c] consider

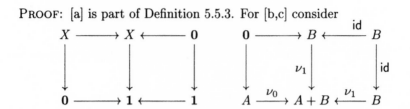

which are clearly coproducts, therefore pairs of pullbacks; so $X \leftarrow 0$ is the pullback of the isomorphism $1 \leftarrow 1$.

[d] The mediator to the pullback is $U \to 0$ by (c), so $U \cong 0$ by (b).

[e] Form a cube from the middle and right hand squares.

[f] Form the pullbacks U and V in the left hand diagram below, then $U \cong 0$ by disjointness; but $B = U + V$ by stability, so $B \cong V$ is the pullback in Proposition 5.2.2(a).

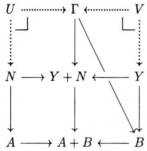

[g] The inclusions $1 \hookrightarrow 2 \hookleftarrow 1$ are *split* mono, so general coproduct inclusions are *regular* mono by Remark 5.2.3.   □

THEOREM 5·5·8:   A category with pullbacks is extensive iff it has stable disjoint sums and a strict initial object.

PROOF: It remains to show that Y is a pullback in the diagram below, so take a commutative square with vertex Γ. Form the pullbacks U and V, so $\Gamma = U + V$ since the coproduct $Y + N$ is stable (*cf.* the proof of Lemma 5.5.7(f)). Now we have a commutative trapezium from U to $A + B$ *via* A and B, so $U \cong 0$ by Lemma 5.5.7(d).

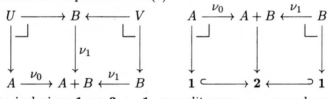

$\Gamma \cong V \to Y$ is the pullback mediator: the triangle $\Gamma \to Y + N$ is actually the top right square, which commutes; that to B commutes because the composites as far as $A + B$ are equal and $B \hookrightarrow A + B$ by Lemma 5.5.7(f). The mediator is unique because $Y \hookrightarrow Y + N$.   □

Interpretation of theories with disjunction. Examples 4.6.4 listed some mathematical structures that are almost algebraic theories, but which have *exceptions* such as (division by) zero and (**pop**ping) the empty stack, *cf.* the predecessor (Remark 2.7.9).

EXAMPLES 5·5·9: The theories of natural numbers, lists (Section 2.7), number fields and projective planes involve axioms of the form

$$x = 0 \quad \vee \quad \exists! y.\, x = \mathsf{succ}(y)$$
$$\ell = [\,] \quad \vee \quad \exists! h.\, \exists! t.\, \ell = \mathsf{cons}(h, t)$$
$$x = 0 \quad \vee \quad \exists! y.\, x \times y = 1$$
$$p = q \quad \vee \quad \exists! \ell.\, p \in \ell \wedge q \in \ell$$
$$\ell = m \quad \vee \quad \exists! p.\, p \in \ell \wedge p \in m,$$

in which there is only one witness to each existential quantifier, and only one term of each disjunction can be satisfied. These things can be proved from the theory because they also have axioms of the form

$$\mathsf{succ}(y) = 0 \vdash \bot \qquad \mathsf{cons}(h, t) = [\,] \vdash \bot \qquad 0 \times y = 1 \vdash \bot.$$

Negation and unique disjunction are the nullary and binary forms of an associative operation, so these are called *disjunctive theories*.

REMARK 5·5·10: Unique existential quantification has already arisen for "essentially" algebraic theories (Remark 5.2.9), but the unique disjunction is a new feature of extensive categories. The property

$$\Gamma \vDash (\phi \wedge \neg\psi) \vee (\neg\phi \wedge \psi)$$

can be expressed by the (stable disjoint) coproduct

$$[\Gamma, \phi] \hookrightarrow \Gamma \longleftarrow [\Gamma, \psi],$$

and similarly the negation $\neg\phi$ says that $[\Gamma, \phi] \cong \mathbf{0}$.

How can we define a classifying category for such a theory, short of interpreting the whole of the predicate calculus? The sum calculus of Remark 2.3.10ff can express $N \cong \mathbf{1} + N$ and $L \cong \mathbf{1} + X \times L$, but, without extending the fragment $\square = (\times, +)$ of logic to include pullbacks, there is nothing to force $[\![-]\!] : \mathsf{Cn}_\mathcal{L}^\square \to \mathcal{S}$ to preserve the pullback which expresses disjointness. (*A priori* we do not even know that this square *is* a pullback in $\mathsf{Cn}_\mathcal{L}^{\times+}$, though it follows from Section 7.7 that in fact it is.) However, the tradition of sketch theory, like that of model theory, has only considered the situation where $\mathcal{S} = Set$, so the coproduct diagram in the semantics is necessarily stable and disjoint.

There is no avoiding other finite limits in the cases of fields and projective planes, as the axioms themselves involve equations.

REMARK 5·5·11: The category $\mathcal{Mod}(\mathcal{L})$ of models and homomorphisms of such a theory need not have products, though it still has pullbacks, equalisers, and indeed limits of all connected diagrams (Examples 5.1.5).

The axioms above say that lists and natural numbers can be *parsed*, an idea which we shall take up again in Chapter VI. Infinitary sums will be treated type-theoretically in Section 9.3. The rest of this chapter is about coequalisers, but we postpone stability for them until Section 5.8, where it is also combined with extensivity to give the notion of pretopos. Virtual objects reappear there and in Section 6.4, which interprets an imperative language with **while**.

5.6   Kernels, Quotients and Coequalisers

Coequalisers in *Set*, like coproducts, behave very differently from their duals. The construction is very complicated in the general case: only quotients of equivalence relations have a straightforward description in Zermelo type theory, although it is much simpler for Abelian categories. The method extends to congruences for finitary algebraic theories, and is used to impose laws (Example 4.6.3(j), Section 7.4).

Kernels. We begin by using pullbacks to code Definition 1.2.3.

DEFINITION 5·6·1:   The *kernel pair* or *level* of a morphism $f : A \to B$ in a category with pullbacks is the pullback square

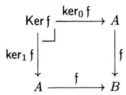

EXAMPLES 5·6·2:

(a) Any map f is a mono iff $\ker_0 f = \ker_1 f = \mathrm{id}_A$ (Proposition 5.2.2(a)).

(b) In *Set*, $\mathsf{Ker}\, f = \{\langle x,y\rangle \mid f(x) = f(y)\} \subset A^2$ is just an equivalence relation.

(c) In $\mathcal{Mod}(\mathcal{L})$, where $\mathcal{L}$ is a single-sorted algebraic theory, $\mathsf{Ker}\, f$ is also a subalgebra of $A \times A$. For example in $\mathcal{Lat}$,
   ► the constants $\bot = \langle \bot, \bot\rangle$ and $\top = \langle \top, \top\rangle$ are in $\mathsf{Ker}\, f$ because it's reflexive, or because f preserves them,
   ► if $\langle x,y\rangle$ and $\langle u,v\rangle$ are in $\mathsf{Ker}\, f$ then so are $\langle x \vee u,\ y \vee v\rangle$ and $\langle x \wedge u,\ y \wedge v\rangle$, because f is a homomorphism,
   ► for theories with unary operations (*e.g.* logical and arithmetical

negation), Ker f is closed under them as well, and likewise under operations of arbitrary arity.

A fortiori, the operations on Ker f also satisfy the laws of $\mathcal{L}$.

(d) In the category of groups, the subset $N = \{x \mid f(x) = 1\} \subset A$ suffices to define the kernel pair, as $K = \{\langle x, y \rangle \mid x \,; y^{-1} \in N\}$. N is the equaliser of f with the trivial (constantly 1) homomorphism. It is characterised as a **normal subgroup**, *i.e.* a subgroup satisfying $\forall x, n : A. \; n \in N \Rightarrow x^{-1} \,; n \,; x \in N$.

(e) Similarly in the category of vector spaces (or modules for a ring) the subspace or submodule $U = \{x \mid f(x) = 0\}$ determines the kernel pair by $K = \{\langle x, y \rangle \mid x - y \in U\}$. Again U is the equaliser with the zero map, but this time every submodule arises in this way.

(f) The kernel pair for rings has yet another representation peculiar to that category, as a (two-sided) **ideal** $I = \{x \mid f(x) = 0\}$, Example 2.1.3(b). The kernel pair can be recovered in the same way as for vector spaces, but I is not an equaliser or subring unless B is the trivial ring; it has the property that $\forall x, i : A. \; i \in I \Rightarrow xi, ix \in I$.

(g) In a many-sorted algebraic theory, kernels are constructed for each sort independently of the others, as the inclusion $\mathcal{M}od(\mathcal{L}) \subset \mathcal{S}et^{\Sigma}$ creates pullbacks (Definition 4.5.10(c)), where Σ is the set of sorts.

(h) Let M be a module for a commutative ring R. Then a kernel consists of an ideal $I \subset R$ and a submodule $U \subset M$, with the *additional* condition that $IM \subset U$. So the sorts do interact, and the general case is more complicated than this example.

The trick in (d)–(f) does not work for other theories such as lattices, but needs division or subtraction to translate everything to the origin. However, the kernel pair of a homomorphism of *complete* semilattices can also be represented as a subset, which is order-isomorphic to its quotient: see Lemma 5.6.14 below and Exercises 5.39 and 5.41.

Congruences. The kernel pair of any map, where it exists in a category, is an equivalence relation (Definition 1.2.3), which we express in the style of Remark 5.2.8. For an algebraic theory $\mathcal{L}$, we can regard the following diagrams as being in $\mathcal{S}et$, $\mathcal{S}et^{\Sigma}$ or $\mathcal{M}od(\mathcal{L})$, since $\mathcal{M}od(\mathcal{L}) \to \mathcal{S}et^{\Sigma}$ creates limits.

DEFINITION 5·6·3: A mono $K \hookrightarrow A \times A$ (or the pair $K \rightrightarrows A$) for which these diagrams exist is called a **congruence**. (This *loosely* agrees with Definition 1.2.12 in that it transfers algebraic properties.)

In $\mathcal{S}et$ a congruence is just an equivalence relation; in $\mathcal{M}od(\mathcal{L})$ it is also a subalgebra of $A \times A$.

PROPOSITION 5·6·4:   In a category with pullbacks, kernel pairs satisfy the reflexive, symmetric and transitive laws, in the sense defined by the following diagrams.

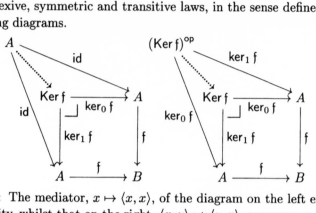

PROOF: The mediator, $x \mapsto \langle x, x \rangle$, of the diagram on the left expresses reflexivity, whilst that on the right, $\langle x, y \rangle \mapsto \langle y, x \rangle$, expresses symmetry. The bottom and right faces of the cube represent the hypotheses for the transitive law: $f(x) = f(y)$ and $f(y) = f(z)$. The back face forms their conjunction, and the front the conclusion; since this is a pullback and the cube commutes, the mediator exists and states the entailment.   □

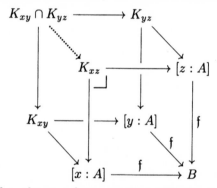

The other two faces are also pullbacks (by Lemma 5.1.2), and in fact the eighth vertex is the ternary pullback of the first three edges, and also the limit of the first three faces. See Exercise 5.40 for an alternative proof.

Quotients.

DEFINITION 5·6·5:   The **coequaliser** of a parallel pair $u, v : R \rightrightarrows A$ is the universal map $q : A \to Q$ with $u \,;\, q = v \,;\, q$.

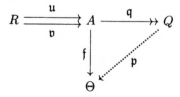

So, for any other map $\mathfrak{f} : A \to \Theta$ with $\mathfrak{u}\,;\mathfrak{f} = \mathfrak{v}\,;\mathfrak{f}$, there is a unique mediator $\mathfrak{p} : Q \to \Theta$ such that $\mathfrak{f} = \mathfrak{q}\,;\mathfrak{p}$. As usual, coequalisers are unique up to unique isomorphism. We shall concentrate on congruences since they are easier to handle than general pairs, but also because of

LEMMA 5·6·6: Let $\mathcal{S}$ be a category with kernel pairs and coequalisers.

(a) If a map $A \to Q$ is the coequaliser of *some* pair $W \rightrightarrows A$ then it is also the coequaliser of its kernel pair. It is called a **regular epi**.

(b) If $K \rightrightarrows A$ is the kernel pair of *some* map $A \to \Theta$ then it is also the kernel pair of its coequaliser.

PROOF: Consider the relation $(W \rightrightarrows A) \perp (A \to \Theta)$ that the composites are equal; this induces a Galois connection (Proposition 3.8.14) between the class of pairs of maps into A and that of single maps out, or between the double slice $\mathcal{S} \downarrow \langle A, A \rangle$ (Lemma 4.5.16) and the coslice $A \downarrow \mathcal{S}$. Then the kernel pair of $A \to Q$ is the terminal object of $^{\perp}(A \to Q)$, whilst the coequaliser of $W \rightrightarrows A$ is initial in $(W \rightrightarrows A)^{\perp}$. The result follows from the idempotence of Galois connections. $\qquad\square$

DEFINITION 5·6·7: We want to use equivalence relations to specify that pairs of elements are to be identified. The coequaliser $\mathfrak{q} : A \twoheadrightarrow Q$ of a *congruence* $K \rightrightarrows A$ is known as its **quotient** and written $Q = A/K$. It is said to be **effective** if K is the kernel pair of $\mathfrak{q}$, *i.e.* the quotient identifies the specified pairs *and no more*.

PROPOSITION 5·6·8: In *Set*, every congruence has an effective quotient.

PROOF: Let $Q \subset \mathcal{P}(A)$ be the set of equivalence classes. Example 2.1.5 showed that Q is the coequaliser (the mediator p is unique since $A \twoheadrightarrow Q$ is surjective). This is effective because $[x] = [y] \iff \langle x, y \rangle \in K$. $\qquad\square$

THEOREM 5·6·9: Let $\mathcal{L}$ be a *finitary* algebraic theory. Then the category $\mathcal{M}od(\mathcal{L})$ also has effective quotients of congruences.

PROOF: We begin with semilattices, as a typical single-sorted finitary algebraic theory. The constant $\top$ in the quotient is the equivalence class $[\top]$. Similarly $[a] \wedge [b] = [a \wedge b]$, which we must show to be well defined with respect to choices of representatives: the equation $[a] = [a']$ means that $\langle a, a' \rangle \in K$; if also $\langle b, b' \rangle \in K$, then $\langle a \wedge b, a' \wedge b' \rangle \in K$ since it's a subalgebra, so $[a \wedge b] = [a' \wedge b']$.

The same applies to any finitary operation-symbol $X_1, \ldots, X_k \vdash r : Y$.

Categorically, the two squares from K_X to A_Y exist as K is a subalgebra; indeed they are pullbacks and $K_{\underset{\sim}{X}}$ is a congruence on $A_{\underset{\sim}{X}}$. But in order to define the dotted map, the top row must be a coequaliser. It is, because the product functors preserve coequalisers.

$$K_{X_1} \times \cdots \times K_{X_k} \rightrightarrows A_{X_1} \times \cdots \times A_{X_k} \twoheadrightarrow \frac{A_{X_1}}{K_{X_1}} \times \cdots \times \frac{A_{X_k}}{K_{X_k}}$$

with vertical maps r_K, r_A, $r_{(A/K)}$ to the bottom row

$$K_Y \rightrightarrows A_Y \xrightarrow{\ q\ } A_Y/K_Y$$

The laws of $\mathcal{L}$ are inherited from A, again by choosing representatives. They may be expressed diagrammatically as pairs of derived operations $u, v : \underset{\sim}{X} \rightrightarrows Y$; if these are equal for A then they are also equal for A/K since q is epi. $\qquad\square$

It is essential here that the functors $(-)^k$ preserve regular epis. In the next chapter we shall extend some of the methods of universal algebra to infinitary theories: the theory may have infinitely many sorts or laws without causing difficulty. But this will be at the cost of the ability to treat laws in general; the problem lies in the above result, though this obstacle can be removed by brute force (the axiom of choice). By *finitary* we really do mean that the arity k has to be finitely *enumerated* (Definition 6.6.2(a)). This is not finitist dogma: the condition may be formulated abstractly, and is called projectivity, Remark 5.8.4(e).

COROLLARY 5·6·10: In a category with pullbacks and effective quotients of congruences, the Galois connection used to prove Lemma 5.6.6 reduces to an order-isomorphism between

(a) quotients (regular epis out) of A, *cf.* Sub(A) in Remark 5.2.5, with id$_A$ as the *least* element and $! : A \to 1$ as the greatest, and

(b) congruences on A, with the diagonal as least element (the *discrete* congruence) and $A \times A$ as greatest (the *indiscriminate* one). $\qquad\square$

General coequalisers. It remains to convert an arbitrary parallel pair into a congruence, using zig-zags (Lemma 1.2.4).

LEMMA 5·6·11: In the following diagram in a category with kernel pairs, suppose that $u = e \,;\, m \,;\, n \,;\, \pi_0$ and $v = e \,;\, m \,;\, n \,;\, \pi_1$, the maps are epi and mono as marked and K is the smallest congruence containing R.

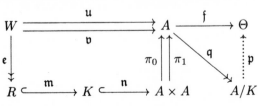

Then the following are equivalent:

(a) $u\,;f = v\,;f : W \to \Theta$

(b) $m\,;n\,;\pi_0\,;f = m\,;n\,;\pi_1\,;f : R \to \Theta$;

(c) $n\,;\pi_0\,;f = n\,;\pi_1\,;f : K \to \Theta$.

If the quotient $q : A \twoheadrightarrow A/K$ exists, it is also equivalent that

(d) f factor through q, *i.e.* there be a unique map p such that $f = q\,;p$.

Hence A/K is the coequaliser of u and v.

PROOF: [a$\Rightarrow$b]: e is epi. [b$\Rightarrow$c]: $R \subset \mathsf{Ker}\,f$, so $K \subset \mathsf{Ker}\,f$ by hypothesis. [c$\Rightarrow$d]: by definition of A/K. The converses hold by composition. $\square$

COROLLARY 5·6·12: *Set* has coequalisers for all parallel pairs, as does the category $\mathcal{M}od(\mathcal{L})$ of algebras for any *finitary* algebraic theory $\mathcal{L}$.

PROOF: Put $R = \{\langle x_0, x_1 \rangle \mid \exists y.\, u(y) = x_0 \wedge u(y) = x_1\}$, and let K be its congruence closure. In *Set* this is the set of zig-zags. $\square$

Note that to find e and m from u and v in Lemma 5.6.11 uses the image factorisation, which we shall discuss in the next section.

REMARK 5·6·13: In the case of the category of algebras for a many-sorted algebraic theory, the congruence $(K_X \subset A_X \times A_X)_{X \in \Sigma}$ has to be generated for all of the sorts and operation-symbols together, because operation-symbols with results of one type may make use of arguments of any of the other types. Treating it as a subset of $\bigcup_{X \in \Sigma}(A_X \times A_X)$, the reflexive, symmetric and transitive laws, and each operation-symbol (with respect to which the inclusion must be a homomorphism), give rise to a closure condition (Example 3.7.5(a)).[4]

Colimits by duality. Although this way of constructing coequalisers is not in general available for infinitary operations, most of those of interest (meets, joins, limits and colimits) are defined by universal properties. We can use our knowledge of adjunctions to make the necessary choices canonically. Indeed in some categories it is much easier to find colimits in this way than it is in *Set* by the combinatorial technique.

---

[4]Notice that the set $\Sigma_1 = \bigcup_{X \in \Sigma} A_X \times A_X$ on which this system of closure conditions is defined is much bigger than the set Σ of sorts for the algebraic theory. Did we make a mistake in Definition 3.7.4 and only give a single-sorted notion of closure condition? No, we didn't: Example 4.6.3(b) showed that this was the propositional analogue of the *many*-sorted algebraic theories in Section 4.6. The set Σ_1 of new sorts is for a more complicated theory which we mention again in Remark 7.4.10 and Example 9.2.4(h).

For example, since the category of complete join-semilattices is self-dual, i.e. $\mathcal{CSLat} \cong \mathcal{CSLat}^{op}$, its colimits follow immediately from its limits. The next result was, of course, discovered by unwinding this corollary, but it is instructive as an example of a construction which does not simply work by symbol-pushing.

LEMMA 5·6·14: $\mathcal{CSLat}$ has coequalisers.

PROOF: We are given $u, v : X \rightrightarrows A$, which preserve all joins. Put

$$Q = \{a \mid \forall x.\, u(x) \leqslant a \Leftrightarrow v(x) \leqslant a\} \subset A$$

and observe that it is closed under all *meets* in A. Using Theorem 3.6.9, Q is a complete join-semilattice, and the inclusion $j : Q \hookrightarrow A$ has a (join-preserving) left adjoint, $e \dashv j$. In fact

$$e(a) = \bigwedge\{q \mid a \leqslant q\} \quad \text{and} \quad \bigvee_i^Q q_i = e\left(\bigvee_i^A q_i\right).$$

Then
$$e(u(x)) = \bigwedge\{q \mid u(x) \leqslant q\} = \bigwedge\{q \mid v(x) \leqslant q\} = e(v(x)).$$

Now given $f : A \to \Theta$ preserving joins and such that $f \circ u = f \circ v$, define $p : Q \to \Theta$ by $p(q) = f(q)$; since $e \circ j = \mathrm{id}_Q$ this is the unique function with $p \circ e = f$. Using the above formula for joins in Q, we see that p preserves them. $\qquad\square$

Although $\mathcal{SLat}$, $\mathcal{Ab}$, $\mathcal{Vsp}$ and categories of modules for rings are not quite self-dual, a similar technique applies. The particular case of the coequaliser of a linear map u with the zero map $v = 0$ (Example 5.4.4) is called its **cokernel**.

Section 6.4 uses coequalisers to interpret loops, and Sections 6.5 and 7.4 consider laws in algebras. The rest of this chapter shows how limits and monos interact with colimits and epis.

5.7   FACTORISATION SYSTEMS

Lattice theory could handle quotients (or the congruences defining them) and subalgebras separately, but things became difficult when both were needed at the same time. Several standard results with non-standardised names like "second isomorphism theorem" were formulated that link subgroups to normal subgroups and subrings to ideals. Some of these hold for general algebras, whilst others need the congruence lattices to be modular (Exercise 3.24).

It was Emmy Noether who shifted the emphasis from subalgebras and congruences to homomorphisms. Including both in the same structure shows us the universal property that distinguishes and re-unites them.

The result also explains the existential quantifier, so often obscured by lattice-theoretic methods, as we shall see in Sections 5.8 and 9.3.

DEFINITION 5·7·1:   We say that two maps $e : X \to A$ and $m : B \to \Theta$ in S are **orthogonal** and write $e \perp m$ if, for any two maps f and $\mathfrak{z}$ such that the square commutes, there is a unique morphism $p : A \to B$ making the two triangles commute:

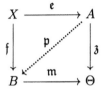

For any classes of maps $\mathcal{E}, \mathcal{M} \subset S$, we write (as in Proposition 3.8.14)

$$\mathcal{E}^{\perp} = \{m \in \mathrm{mor}S \mid \forall e \in \mathcal{E}.\, e \perp m\}$$
$$^{\perp}\mathcal{M} = \{e \in \mathrm{mor}S \mid \forall m \in \mathcal{M}.\, e \perp m\}.$$

If $\mathcal{E} = {}^{\perp}\mathcal{M}$ and $\mathcal{M} = \mathcal{E}^{\perp}$, we call them a **prefactorisation system**.

For technical reasons it is also useful to say that "$e \perp m$ *with respect to* $\mathfrak{z}$" if the fill-in property above holds for all f but just this particular $\mathfrak{z}$.

DEFINITION 5·7·2:   A **factorisation system** [FK72] on a category S is a pair of classes of morphisms $(\mathcal{E}, \mathcal{M})$ of S such that

(a) the classes $\mathcal{E}$ and $\mathcal{M}$ each contain all isomorphisms, and are closed under composition on either side with isomorphisms (we shall find that they are non-full replete subcategories),

(b) every morphism $f : X \to \Theta$ of S can be expressed as $f = e\,;\,m$ with $e \in \mathcal{E}$ and $m \in \mathcal{M}$, and

(c) $e \perp m$ for every $e \in \mathcal{E}$ and $m \in \mathcal{M}$.

If the pullback of any composite $e\,;\,m$ against any map $u : \Gamma \to \Theta$ exists, and the parts lie in $\mathcal{E}$ and $\mathcal{M}$ respectively, then we call $(\mathcal{E}, \mathcal{M})$ a **stable factorisation system**, cf. stable coproducts in Section 5.5.

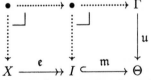

In *Set*, image factorisation is stable: this is necessary in Lemma 5.8.6 to make relational composition associative, and in Theorem 9.3.11 for the existential quantifier to be invariant under substitution.[5] The only

---

[5]For this $\mathcal{M}$ does not have be the class of monos, nor need *all* maps factorise. It is the *factorisation* of (factorisable) maps $f = e\,;m$ which must be stable under pullback, rather than the *class* $\mathcal{E}$.

pullback-stability properties that factorisation systems in general have are Lemmas 5.7.6(f) and 5.7.10. Although the image factorisation is the most familiar and accounts for the notation, there are other important examples in topology, categorical logic and domain theory. Exercise 9.5 describes one that is related to virtual objects (Remark 5.3.2).

Image factorisation. First we shall look at the motivating examples, so let $\mathcal{S}$ be a category that has kernel pairs and their coequalisers.

LEMMA 5·7·3:   If e is regular epi and m mono then $e \perp m$. Conversely, if m satisfies $e \perp m$ for every regular epi e then m is mono.

PROOF: Given a coequaliser and a mono in a commutative square as shown, the composites $K \rightrightarrows X \to B \hookrightarrow \Theta$ are equal; hence so are those $K \to B$ and by the universal property there is a unique fill-in.

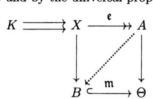

Conversely, apply orthogonality to the square with id $: X \to B$; the diagonal fill-in shows that e is invertible. $\square$

So with $\mathcal{M}$ and $\mathcal{E}$ the classes of monos (inclusions) and regular epis (quotients or surjections), we have $\mathcal{E}^{\perp} = \mathcal{M}$ in any category which has kernels and quotients.

LEMMA 5·7·4:   If the class of regular epis is closed under composition, then together with the class of monos it forms a factorisation system.

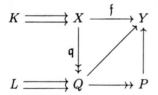

PROOF: To factorise $f : X \to Y$, let $q : X \twoheadrightarrow Q$ be the coequaliser of the kernel pair $K \rightrightarrows X$ of f; by Lemma 5.6.6(b) this is also the kernel pair of q. We must show that $Q \hookrightarrow Y$, so form its kernel pair $L \rightrightarrows Q$ and let P be the coequaliser. The kernel pair of the composite $X \twoheadrightarrow Q \twoheadrightarrow P$ is sandwiched (as a subobject of $X \times X$) between those of $X \to Y$ and $X \to Q$, which are both K. By hypothesis $X \twoheadrightarrow Q \twoheadrightarrow P$ is regular epi, so it is the quotient of its kernel pair $K \rightrightarrows X$. But $X \twoheadrightarrow Q$ was already the quotient of this pair, so $L \cong Q \cong P$. By Proposition 5.2.2(a), $X \twoheadrightarrow Q \hookrightarrow Y$. $\square$

We would like to say that whenever the relevant finite limits and colimits exist, so does the **image factorisation** into regular epis and monos, and also dually the co-image factorisation into epis and regular monos. Unfortunately this is not so in general, but it is when the class of regular epis is closed under pullback (Proposition 5.8.3). In any case we call $^\perp\mathcal{M}$-maps **covers**.

EXAMPLES 5·7·5:

(a) In a preorder all morphisms are both epi and mono, but only the isomorphisms are regular. (Example 5.7.9 nevertheless gives a non-trivial prefactorisation system in a poset.)

(b) If, as in $\mathcal{S}et$ by Corollary 5.2.7, all monos are regular, then the dual of this lemma shows that epis and monos form a factorisation system. From Lemma 5.7.6(a) it follows that all epis are regular.

(c) A homomorphism of algebras for a single-sorted finitary algebraic theory $\mathcal{L}$ is regular epi in the category $\mathcal{M}od(\mathcal{L})$ iff it is surjective on its carriers, and mono iff it is injective (Exercise 5.38). These classes form a factorisation system.

(d) In $\mathcal{CM}on$ regular monos do not compose. Consider the submonoids $U = \langle 3, 5 \rangle$ and $V = \langle 3, 5, 7 \rangle$ of $\mathbb{N}$; the inclusions $U \hookrightarrow V$ and $V \hookrightarrow \mathbb{N}$ are regular monos but their composite is not, because if $f, g : \mathbb{N} \rightrightarrows \Theta$ agree at 3 and 5 then (as $2 + 5 = 2 + 3 + 2 = 5 + 2$) they do at 7 too.

(e) Although any field homomorphism is mono (Example 5.1.5(f)), it is regular iff it is a separable extension. There are non-trivial epis, namely totally inseparable extensions, such as $K = \mathbb{F}_p[x] \hookrightarrow K[\sqrt[p]{x}]$ (*cf.* Example 3.8.15(j), and see [Coh77, Theorem 6.4.4]).

(f) In $\mathcal{S}p$, continuous functions are epi or mono according as they are surjective or injective on points, but are regular iff the topologies are the quotient or subspace ones. Both factorisations arise.

(g) Let $\mathcal{E}$ be the class of full functors which are bijective on objects, and $\mathcal{M}$ the class of faithful functors.

(h) Regular epis in $\mathcal{C}at$ do not compose: $(\bullet \to \bullet) \twoheadrightarrow \mathbb{N} \twoheadrightarrow \mathbb{Z}/3$.

Instead of allowing *all* subsets to be in $\mathcal{M}$, we may restrict to those that are *closed* in some sense (Section 3.7).

(i) In $\mathcal{S}p$, let $\mathcal{M}$ be the inclusions of closed sets in the topological sense. Then $\mathcal{E}$ is composed of the continuous functions with dense image.

(j) In $\mathcal{P}os$, let $\mathcal{M}$ be the inclusions of lower sets and $\mathcal{E}$ be the class of cofinal maps (Proposition 3.2.10).

Properties of factorisation systems. Proposition 5.2.2 showed that the class of regular monos has weaker cancellation properties than that of all monos. Although the diagonal map $\mathfrak{p}$ is determined in both cases by just one triangle, even this property is not typical. The other closure conditions satisfied by epis and monos do hold for arbitrary factorisation systems, but to show this we must study orthogonality more closely.

LEMMA 5·7·6:

(a) $\mathcal{M} \subset \mathcal{E}^{\perp} \iff \mathcal{E} \subset {}^{\perp}\mathcal{M}$, *i.e.* $\mathcal{E} \mapsto \mathcal{E}^{\perp}$ and $\mathcal{M} \mapsto {}^{\perp}\mathcal{M}$ form a Galois connection (Definition 3.6.1(b)) between classes of morphisms.

(b) If i is invertible then $e \perp i$ for any $e \in \mathcal{E}$, so $\mathcal{E}^{\perp}$ contains all of the isomorphisms, as does ${}^{\perp}\mathcal{M}$ (they are replete, Definition 4.4.8(d)).

(c) If $i \perp i$ then i is invertible, so $\mathcal{M} \cap {}^{\perp}\mathcal{M}$ and $\mathcal{E} \cap \mathcal{E}^{\perp}$ contain *only* isomorphisms.

(d) If $e \perp m_1$ and $e \perp m_2$ then $e \perp (m_2 ; m_1)$, so $\mathcal{E}^{\perp}$ is closed under composition, and likewise ${}^{\perp}\mathcal{M}$.

(e) If $e \perp n$ and $e \perp (m ; n)$ then $e \perp m$, *cf.* Proposition 5.2.2(h).

(f) If $e \perp m$, and $n = u^* m$ is a pullback of m against any map u, then $e \perp n$, *cf.* Remark 5.2.3.

(g) The $\perp$ relation is preserved and reflected by full and faithful functors. Similarly ${}^{\perp}\mathcal{M}$ is a subcategory closed under pushouts.

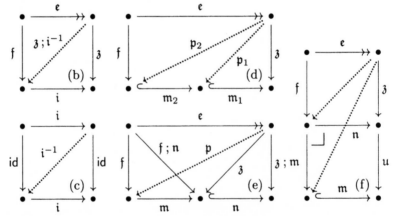

PROOF: The relation $e \perp m$ defines a Galois connection by Proposition 3.8.14; most of the rest is shown in the diagrams.

(d) If $e ; \mathfrak{p} = \mathfrak{f}$ and $\mathfrak{p} ; m_2 ; m_1 = \mathfrak{z}$ then $\mathfrak{p} ; m_2 = \mathfrak{p}_1$ and $\mathfrak{p} = \mathfrak{p}_2$ since the mediators $(\mathfrak{p}, \mathfrak{p}_1, \mathfrak{p}_2)$ are required to be unique.

(e) Any fill-in for the left-hand trapezium serves for the rectangle $(\mathfrak{p})$, but $\mathfrak{z}$ and $\mathfrak{p} ; m$ both serve as the fill-in for the right-hand trapezium.

(f) The fill-in for the rectangle gives one for the upper square by pullback and conversely by composition.   □

PROPOSITION 5·7·7: Any factorisation system is also a prefactorisation system, so its two classes are closed under the above properties.

PROOF: It only remains to show that $\mathcal{E}^\perp \subset \mathcal{M}$. Using the factorisation property, suppose that $(e \,;m) \in \mathcal{E}^\perp$ with $e \in \mathcal{E}$ and $m \in \mathcal{M}$. Then in particular $e \perp (e \,;m)$ and $e \perp m$, so $e \perp e$ is invertible using Lemmas 5.7.6(e) and (c). By repleteness, $(e \,;m) \in \mathcal{M}$. Similarly $^\perp\mathcal{M} \subset \mathcal{E}$. □

Finding factorisations. Given an arbitrary prefactorisation system $(\mathcal{E}, \mathcal{M})$, we can now try to factorise $\mathcal{S}$-morphisms $f : X \to Y$ as $f = e \,;m$ with $e \in \mathcal{E} = {}^\perp\mathcal{M}$ and $m \in \mathcal{M} = \mathcal{E}^\perp$. Any $\mathcal{M}$- or $\mathcal{E}$-morphism we can find that factors appropriately into f will contribute to this.

LEMMA 5·7·8: $\mathcal{M}$ is closed under wide pullbacks (Example 7.3.2(h)), *i.e.* arbitrary intersections in the case of monos. That is, for any wide pullback diagram in $\mathcal{E}^\perp$, if the limit exists in $\mathcal{S}$ then its limiting cone lies in $\mathcal{M}$, as does the mediator for any cone of $\mathcal{M}$-maps. □

Similar results hold for wide pushouts in $\mathcal{E}$. We have to impose algebraic and size conditions to ensure that the limit for $\mathcal{M}$ and colimit for $\mathcal{E}$ exist, but they may still fail to meet in the middle.

EXAMPLE 5·7·9: Here is a prefactorisation system in a poset.

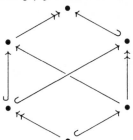

The classes $\mathcal{E}$ and $\mathcal{M}$ consist of the marked arrows, together with all of the identities and a single composite. The wide pullback of the $\mathcal{M}$-maps into any object exists, as does the wide pushout of the $\mathcal{E}$-maps out, but the unmarked broken arrow does not factorise. □

The problem is that the map cannot be in $\mathcal{E}$ because there are distant $\mathcal{M}$-maps to which it is not orthogonal, but parallel translation using pullback which might bring it into $\mathcal{M}$ is not available.

LEMMA 5·7·10: Suppose that the pullback of any $\mathcal{M}$-map against any $\mathcal{S}$-map exists (and so is in $\mathcal{M}$). Then to show $e \in {}^\perp\mathcal{M}$ it suffices to test orthogonality with respect to $\mathfrak{z} = \mathsf{id}$, *i.e.*

if $e = f \,;m$ with $m \in \mathcal{M}$ then $\exists!p.\ p \,;m = \mathsf{id} \wedge e \,;p = f$.

If all $\mathcal{M}$-maps are mono then this condition makes m invertible.

PROOF: Similar to Lemma 5.7.6(f) with $\mathfrak{z} = \mathrm{id}$.   □

We need a solution-set condition such as that for the General Adjoint Functor Theorem 7.3.12 to show that any prefactorisation system with sufficient pullbacks is a factorisation system (Exercise 7.34), so we shall end this section with a special case.

PROPOSITION 5·7·11: Let S be a category such that there is a *functor* $\mathrm{Sub} : S^{\mathrm{op}} \to CSLat$. Explicitly, S is well powered (Remark 5.2.5) and has *arbitrary* intersections of subobjects and inverse images, *i.e.* pullbacks of them along S-maps. For example S may be *Set*, *Sp* or any category of algebras. Then any prefactorisation system $(\mathcal{E}, \mathcal{M})$ which is such that all $\mathcal{M}$-maps are mono is a factorisation system.

PROOF: The factorisation of $\mathfrak{f} : X \to \Theta$ is $\mathfrak{e} \mathbin{;} \mathfrak{m}$ where $\mathfrak{m} : A \hookrightarrow \Theta$ is the intersection of the $\mathcal{M}$-subobjects $B \hookrightarrow \Theta$ through which $\mathfrak{f}$ factors (using Theorem 3.6.9). Then $\mathfrak{e} \in \mathcal{E}$ by Lemma 5.7.10.   □

5.8   REGULAR CATEGORIES

This section sums up the "good behaviour" of coequalisers, and of finite colimits in general, like that of coproducts in Section 5.5. Categories of algebras inherit this good behaviour for quotients of congruences, but not for coproducts or general coequalisers. The following definitions progressively capture the *exactness* properties of *Ab* and *Set*. It is these properties of the sum and quotient, and not the all-powerful operations of Zermelo type theory, which provide the foundation of twentieth century algebra.

DEFINITION 5·8·1: Let S be a category.

(a) If S has finite limits it is called a *lex category*.

(b) If further the image factorisation exists (Lemma 5.7.4) and is stable under pullback then S is called a *regular category*.

(c) If further every congruence is the kernel pair of its coequaliser then S is said to be *effective regular* or *Barr-exact*.

(d) An *Ab*-enriched category (Definition 5.4.3ff) which is also effective regular is known as an *Abelian category*. Equivalent definitions are discussed in [FS90, Section 1.59].

(e) A *prelogos* is a regular category with finite unions of subobjects which are stable under pullback (inverse image).

(f) A prelogos in which the inverse image operation between subobject lattices has a right adjoint is called a *logos*.

(g) By Theorem 3.6.9 any prelogos which has *arbitrary* stable unions of subobjects is a logos; we call it **locally complete**.

(h) A **pretopos** is an effective regular extensive category (Definition 5.5.3); in particular it is a prelogos, but not necessarily a logos.

The original example of a regular category was $\mathcal{A}b$, and in general $\mathcal{M}od(\mathcal{L})$ for any finitary algebraic theory $\mathcal{L}$, by Theorem 5.6.9. From Section 5.6, these categories have *all* coequalisers, so there is some ambiguity in the literature as to whether all coequalisers are required in the definition: after [FS90, §1.52], we say they are not. Indeed in regular categories coequalisers *of their kernel pairs* are stable under pullback, but even when they exist general coequalisers need not be stable.

EXAMPLES 5·8·2:

(a) *Set* is a pretopos by Example 5.5.6(a) and Proposition 5.6.8.

(b) A poset is lex iff it is a semilattice, and then it is trivially effective regular. It is a prelogos iff it is a distributive lattice, and a logos iff it is a Heyting lattice (Definition 3.6.15). Conversely, in any prelogos or logos, every $\mathsf{Sub}_\mathcal{S}(X)$ is a distributive or Heyting lattice respectively. Posets can never be pretoposes, except degenerately.

(c) $\mathcal{M}od(\mathcal{L})$, where $\mathcal{L}$ is a finitary algebraic theory, is effective regular by Theorem 5.6.9. It also has arbitrary colimits; *filtered* colimits (Example 7.3.2(j), in particular *directed* unions of subobjects) are stable, but finite ones are usually not.

(d) In particular $\mathcal{G}p$ is effective regular. It also has all coequalisers, but the following one is not stable under pullback.

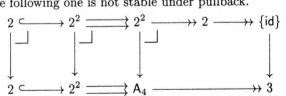

The parallel pairs consist of the inclusion of the normal subgroup

$$\{\text{id}, \ (12)(34), \ (13)(24), \ (14)(23)\} \subset \mathsf{A}_4$$

and the map which interchanges $(13)(24)$ with $(14)(23)$. (Freyd)

(e) **Von Neumann regular rings**, those satisfying $\forall x. \exists y. \, xyx = x$, form a non-effective regular category.

(f) Abelian groups, vector spaces and modules form Abelian categories.

(g) The category of compact Hausdorff spaces and continuous functions is a pretopos, in which directed unions and general coequalisers also exist but are not stable.

(h) Equivalence relations in $\mathcal{P}os$ are not effective: all three points in the quotient of $\{0 < 1 < 2\}$ by $0 \sim 0 \sim 2 \sim 2$, $1 \sim 1$ are identified. The kernel pairs of monotone functions are convex (Example 3.7.5(c)).

Stable image factorisation. To make Lemma 5.7.4 work as intended, we have to show that the class of regular epis is closed under composition.

PROPOSITION 5·8·3:   In a regular category the class of regular epis is closed under pullback and so is the $\mathcal{E}$ class of the image factorisation.

PROOF: Suppose $\mathfrak{e} : X \to Y$ and $\mathfrak{f} : Y \to Z$ coequalise their kernel pairs, $K \rightrightarrows X$ and $L \rightrightarrows Y$ respectively. Let $M \rightrightarrows X$ be the kernel pair of $X \to Y \to Z$; comparing commutative squares with these kernels (*cf.* Lemma 5.6.6(a)), there are mediators $K \to M \to L$.

Let $X \to \Theta$ be a map having equal composites with $M \rightrightarrows X$. Then the composites $K \to M \rightrightarrows X \to \Theta$ are equal and give $Y \to \Theta$. The required mediator $Z \to \Theta$ exists iff the composites $L \rightrightarrows Y \to \Theta$ are equal.

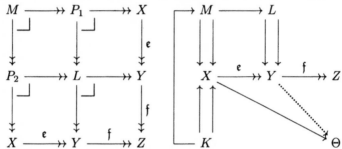

We can achieve this by showing that the maps $M \to P \to L$ are epi. Indeed, both parts are pullbacks of the regular epi $\mathfrak{e} : X \twoheadrightarrow Y$ (against split epis).   □

The last part is sufficient but not necessary. The literature on this subject is confusing, because in certain categories, one may show by slightly different arguments that regular epis compose, without being stable under pullback. There is apparently no convenient necessary and sufficient condition for composability of regular epis.

REMARK 5·8·4:   There are four useful notions of surjectivity $\mathfrak{e} : X \twoheadrightarrow Y$ in lex categories (*cf.* Definition 5.2.1 for monos):

(a) ***split epi***: there is a map $\mathfrak{m} : Y \hookrightarrow X$ with $\mathfrak{m} \,;\, \mathfrak{e} = \mathrm{id}_Y$;

(b) ***regular epi***: $\mathfrak{e}$ is the coequaliser of some (its kernel) pair;

(c) ***cover*** or ***surjective***: $\mathfrak{e} \perp \mathfrak{m}$ for all monos $\mathfrak{m}$;

(d) ***epi***: for any $\mathfrak{f}, \mathfrak{g} : Y \rightrightarrows Z$, if $\mathfrak{e} \,;\, \mathfrak{f} = \mathfrak{e} \,;\, \mathfrak{g}$ then $\mathfrak{f} = \mathfrak{g}$.

By Proposition 5.8.3, the middle two coincide in a regular category. Furthermore, every epi is regular in *Set*, and indeed in any pretopos, but to say that every epi splits would be an internal form of the axiom of choice (Exercise 1.38).

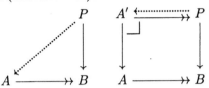

(e) An object P with the lifting property on the left for every cover $A \twoheadrightarrow B$ is called **projective**. In a regular category, the right hand diagram shows that P is projective iff every cover $A' \twoheadrightarrow P$ splits.

Relations. Logically, stability of image factorisation corresponds to the Frobenius law (Lemma 1.6.3). It is also vital for relational calculus.

REMARK 5·8·5: The stable image interprets existential quantification, removing the uniqueness requirement from Remark 5.2.9.

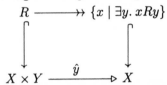

Stable unions of subobjects similarly express disjunction, where once again we have dropped uniqueness from Remark 5.5.10. The right adjoint in a logos gives the universal quantifier. In this way, existential, coherent ($\wedge$, $\vee$, $\exists$) and first order logic may be interpreted in regular categories, prelogoses and logoses respectively. We study these type-theoretically in Chapter IX, and just treat relational algebra here.

LEMMA 5·8·6: Let S be a lex category, in which relations $R : X \rightharpoonup Y$ are interpreted as subobjects $R \hookrightarrow X \times Y$. Then composition of *general* relations, defined by image factorisation, is associative iff S is regular.

PROOF: Let $W \overset{Q}{\rightharpoonup} X \overset{R}{\rightharpoonup} Y \overset{S}{\rightharpoonup} Z$.

(a) As in Proposition 5.3.5, the pullback below gives a subobject, abbreviated as $xRySz \subset X \times Y \times Z$, of the three-fold product, but not necessarily a mono into $X \times Z$. We use image factorisation to get one, as in the bottom row.

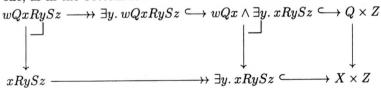

(b) To say that the factorisation is stable means that the mono in the middle of the top row is invertible; this was one of the two crucial steps in the proof of Lemma 4.1.3, the other being similar.

(c) Interchanging existential quantifiers does not cause any problem, because surjections compose (but *cf.* Lemma 5.7.4).

(d) Composition of three particular relations may be associative without stability, as the two subobjects

$$\{w, x, z \mid \exists y.\, wQxRySz\}$$
$$\cup \qquad\qquad \cup$$
$$\{w, z \mid \exists x.\, wQx \wedge (\exists y.\, xRySz)\} \qquad \{w, z \mid \exists y.\, (\exists x.\, wQxRy) \wedge ySz\}$$

may coincide whilst still being proper.

(e) However, if R is functional then the lower epi is invertible, so one subobject is total without hypothesis. Thus associativity in the case where R^{op} is functional and $Z = \mathbf{1}$ implies stability.   □

PROPOSITION 5·8·7:   Relations in a regular category

(a) form another category under relational composition $(\mathrm{id}, R\,;S)$;

(b) with the same source and target admit binary intersection $(R \cap S)$, which is associative, commutative and idempotent; we also define
$$R \subset S \iff R = R \cap S;$$

(c) have monotone composition: $R\,;(S \cap T) \subset (R\,;S) \cap (R\,;T)$;

(d) have opposites, where R^{op} has the source and target interchanged, and satisfies
$$\begin{array}{llll} (R^{\mathrm{op}})^{\mathrm{op}} & = & R & \qquad (R \cap S)^{\mathrm{op}} & = & R^{\mathrm{op}} \cap S^{\mathrm{op}} \\ (R\,;S)^{\mathrm{op}} & = & S^{\mathrm{op}}\,;R^{\mathrm{op}} & \qquad R \subset S & \iff & R^{\mathrm{op}} \subset S^{\mathrm{op}} \end{array}$$

(e) and also obey **modularity**: $(R\,;S) \cap T \subset (R \cap (T\,;S^{\mathrm{op}}))\,;S.$ □

Freyd and Scedrov [FS90] call such a structure an **allegory**. They show how the logical connectives described above may be reformulated in terms of relations rather than functions. For example R is functional iff $R^{\mathrm{op}}\,;R \subset \mathrm{id}$, total iff $\mathrm{id} \subset R\,;R^{\mathrm{op}}$ and confluent iff $R^{\mathrm{op}}\,;R \subset R\,;R^{\mathrm{op}}$.

Using this we can try to recover the category composed of functions from its allegory composed of relations. Any $R \subset \mathrm{id}$ is the support of a partial function (Remark 5.3.6) and must be introduced as a virtual object. Every allegory is thereby embedded in the category of relations of a regular category. The utility of the technique, as we shall see in Theorem 6.4.19, is that allegories can sometimes be constructed from one another more simply than the corresponding categories, and results deduced without the extraneous objects.

Stable unions. To make full use of relational algebra we need unions of possibly overlapping subsets. In a pretopos (effective regular extensive category), unions of subobjects may be defined as quotients of sums and are stable. First we consider the stable unions alone.

LEMMA 5·8·8:

(a) Image factorisation provides the left adjoint to the inverse image map $f^* : \mathsf{Sub}(X) \to \mathsf{Sub}(Y)$, so preserves unions by Proposition 3.6.8.

(b) Relational composition distributes over stable unions.

(c) In a logos there are adjoints $(-) \, ; R \dashv (-)/R$ and $R \, ; (-) \dashv R\backslash(-)$, which both reduce to **Heyting implication** $R \Rightarrow (-)$ if $R \subset \Delta$. These satisfy $R \, ; (R\backslash S) \subset S$ and $(S/R) \, ; R \subset S$. The analogous structure for relations is called a **division allegory**.

The slash notation, suggesting division, is due to Joachim Lambek, who used it first in linguistics [Lam58] and later for modules for rings [Lam89]. (Don't confuse division with $\backslash$ for subset difference, Example 2.1.6(e).) We shall use it to study transitive closures in the proof of Lemma 6.4.9.

PROOF: $R \subset f^*U \Rightarrow I \subset U$ since $(R \twoheadrightarrow I) \perp (U \hookrightarrow X)$,

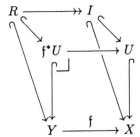

and conversely since f^*U is a pullback. Composition is given by the image of a pullback, and both operations preserve unions or have adjoints as appropriate. □

The elementary properties of *Set* are already beginning to emerge in a prelogos. The next result is known as the Pasting Lemma as it is needed to combine partial functions, for example in Lemma 6.3.12.

LEMMA 5·8·9: Let $U, V \subset X$ and suppose that product distributes over union. Then the diagram on the left, which is a pullback by construction, is also a pushout.

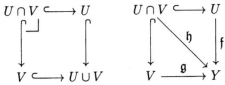

PROOF: Recall that a relation $R \hookrightarrow X \times Y$ is the graph of a function iff $R \hookrightarrow X \times Y \xrightarrow{\pi_0} X$ is an isomorphism. Applying this to $\mathfrak{f}$, $\mathfrak{g}$ and $\mathfrak{h}$ in the diagram on the right, the graph of $\mathfrak{f}$ is isomorphic to U, etc. Then the union of the graphs of $\mathfrak{f}$ and of $\mathfrak{g}$, formed as a subobject of $X \times Y$, is isomorphic to $U \cup V \subset X$ by distributivity. Hence it is the graph of a function $U \cup V \to Y$. □

Pretoposes enjoy even better properties.

PROPOSITION 5·8·10:   Let $Y \xleftarrow{\ m\ } X \xhookrightarrow{\ n\ } Z$ be monos in a pretopos.

(a) Then $(X+Y)+(X+Z) \xrightarrow[{[m;\nu_1,\ \nu_0,\ n;\nu_0,\ \nu_1]}]{\overset{[m;\nu_0,\ \nu_0,\ n;\nu_1,\ \nu_1]}{\rule{0pt}{0pt}}} Y+Z$ is a congruence, whose quotient is the pushout;

(b) this pushout of m and n consists of monos and is a pullback;

(c) every mono is regular (put $m = n$) and

(d) every epi is regular.

PROOF: [a] If $\mathfrak{f} : Y \to \Theta$ and $\mathfrak{g} : Z \to \Theta$ make a commutative square then $[\mathfrak{f}, \mathfrak{g}] : Y + Z \to \Theta$ coequalises the congruence. [b,c] Since the quotient is effective, to verify monos and equalisers, it suffices to inspect the congruence. [d] By Lemma 5.7.3. □

See Exercise 5.27 for the case where the subsets $X \subset Y$ and $X \subset Z$ are decidable, for which the category need only be extensive.

It is very easy to see that *Set* has the properties which we have described. Perhaps it is too easy, as they were only explicitly identified in categorical form long after similar results for Abelian categories (Remark 5.4.5).

The handicap of *Set* is that essentially only one model was known (*cf.* the natural numbers), until the abstract work of the Grothendieck school on the logic of sheaves. Jean Giraud characterised categories of sheaves in terms of their colimits and finite limits; the finitary part of this result is now captured in the notion of pretopos. We shall need the Pasting Lemma, a *categorical* property of sets, to establish the allegedly prior recursive structure of the set-theoretic hierarchy in Section 6.3.

EXERCISES V

   1. Show that any group or groupoid, considered as a category, has pullbacks, *cf.* Exercise 4.17. [Hint: division.]

   2. (Bob Paré) In the diagram on the left, suppose that $n ; p = \mathrm{id}_Y$ and $p ; n ; \mathfrak{v} = \mathfrak{v} ; n ; \mathfrak{v}$. Show that $(n ; \mathfrak{v})$ is idempotent (Definition 1.3.12), so (assuming that idempotents split in the category, *cf.* Exercise 4.16)

we may let $n\,;\,v = q\,;\,m$ with $m\,;\,q = id_Q$. Show that $p\,;\,q = v\,;\,q$ and q is the coequaliser of p and v. It is an **absolute coequaliser** in the sense that (being equationally defined) it is preserved by *any* functor.

Barry Jay observed that $(p, v) : X \rightrightarrows Y$ describes a binary "reduction" relation on Y with a normal form (defined by n).

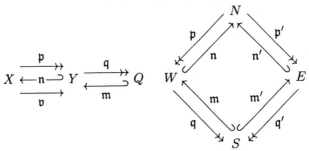

3. Similarly, in the diagram on the right, suppose that the four pairs are split ($n\,;\,p = id_W$, *etc.*) and that the four squares $N \to S$, $S \to N$, $W \to E$ and $E \to W$ commute. Show that S is the pushout of p and p'. In fact S is the absolute coequaliser of p and $v = p'\,;\,n'\,;\,p$.

Some of these conditions are redundant. Let e and e' be idempotents on any object N such that $e\,;\,e'\,;\,e = e'\,;\,e$. Then in the Karoubi completion (Exercise 4.16), the objects $e'\,;\,e$ and $e'\,;\,e\,;\,e'$ are isomorphic quotients (but not isomorphic subobjects) of N and provide the absolute pushout.

4. Show that if S has a terminal object then there is an isomorphism $S{\downarrow}1 \cong S$, and conversely that id_X is the terminal object of $S{\downarrow}X$.

5. Show that products in $S{\downarrow}X$ are exactly pullbacks in S over X.

6. Suppose that $a\,;\,p = id_Z$, $b\,;\,q = id_X$, $a'\,;\,p' = id_Y$ and the two lower squares are pullbacks:

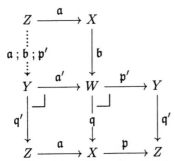

Show that the upper square commutes, and is also a pullback. [Hint: show that $a\,;\,b\,;\,p'\,;\,q' = id$ and use Lemma 5.1.2.]

7. Investigate pullbacks and equalisers in the categories Pfn, IPO, $Pos^{\dashv}$ and Rel.

8. Let $\mathcal{C}$ be a category with pullbacks (equalisers), and $\mathcal{I}$ *any* small category. Show that the functor category $\mathcal{C}^{\mathcal{I}}$ also has pullbacks (equalisers), constructed pointwise. [Hint: *cf.* Lemma 3.5.7.]

9. Let $\mathcal{Pbk}$ be the category whose objects are small categories with binary pullbacks and whose morphisms are pullback-preserving functors. Show that $\mathcal{Pbk}$ is cartesian closed, the morphisms of the functor category being *cartesian transformations*, *i.e.* those natural transformations such that the square in Definition 4.8.1 is a pullback (*cf.* Exercise 4.51).

10. (Giuseppe Rosolini) A **PER** (partial equivalence relation) is a symmetric transitive binary relation on $\mathbb{N}$. A PER R names an object, which we may think of more concretely as the set $\mathbb{N}/R$ of equivalence classes under R; these are disjoint, but their union need not be $\mathbb{N}$, as we have not required R to be reflexive.

(a) Considering $f, g \in \mathbb{N}$ as Gödel numbers for partial recursive programs $\mathbb{N} \rightharpoonup \mathbb{N}$, write $f[R \to S]g$ if f and g induce the same function $\mathbb{N}/R \to \mathbb{N}/S$, and this is total. Formulate this condition directly in terms of the binary relations R and S and show how to define a cartesian closed category with exponential $[R \to S]$.

(b) Let $f \Sigma g$ be the equivalence relation "f terminates iff g does", where these are programs which run without being given any input (they have type $\mathbf{1} \rightharpoonup \mathbf{1}$). Define partial maps by relaxing the totality requirement on PER-maps, and show that Σ is a support classifier.

11. Show that a mono in $\mathcal{Pos}$ or $\mathcal{Dcpo}$ is regular iff it is full. Find a subdcpo of the lazy natural numbers whose Scott topology (Definition 3.4.7ff) is not the subspace topology.

12. Construct the map $\Sigma \to \mathbf{2}$ in Definition 5.2.10 and explain why it is mono and is a semilattice homomorphism.

13. Let $\mathcal{C}$ be any small category and $X \in \mathrm{ob}\,\mathcal{C}$. A class $R \subset \mathrm{mor}\,\mathcal{C}$ of generalised elements of X, *i.e.* maps $\mathfrak{a} : \Gamma \to X$, is called a *sieve* or *crible* on X if it is closed under precomposition with any $u : \Delta \to \Gamma$. Show that a sieve on X is exactly a subobject of $\mathcal{H}_X$ in $Set^{\mathcal{C}^{op}}$. Let $\Omega(X)$ be the set (indeed complete lattice) of sieves on X, and write $\top_X \in \Omega(X)$ for the sieve of *all* generalised elements. Show how to make $\Omega : \mathcal{C}^{op} \to Set$ into a functor and $\top : \mathbf{1} \to \Omega$ a natural transformation, and that this is the subobject classifier of $Set^{\mathcal{C}^{op}}$.

14. Formulate what it is to be a category in $\mathcal{Gpd}$ (the category of groupoids and homomorphisms). Show how *any* small category (in the ordinary sense) can be regarded as such an internal category, in such a way that it is skeletal there (*cf.* Exercise 4.37), in a sense to be defined.

15. (Giuseppe Rosolini, [RR88]) Given partial maps $f : Z \rightharpoonup X$ and $g : Z \rightharpoonup Y$, show how to define $\langle f, g \rangle : Z \rightharpoonup X \times Y$ and hence $\times$ as an endofunctor of $P(\mathcal{S}, \mathcal{M})$. [Hint: take the intersection of their supports.] Although the symbol $\times$ no longer denotes the categorical product, show that the projection $\pi_0 \equiv p_{X,Y} : X \times Y \to X$ is natural in X but that the corresponding square for Y need not commute but involves an inequality. The diagonal, $\partial_X : X \to X \times X$, however, remains natural, and

$$
\begin{array}{lcl}
\partial_X \, ; p_{X,X} & = \; \mathrm{id}_X & = \; \partial_X \, ; q_{X,X} \\
(\mathrm{id}_X \times p_{Y,Z}) \, ; p_{X,Y} & = \; p_{X,Y \times Z} & = \; (\mathrm{id}_X \times q_{Y,Z}) \, ; p_{X,Z} \\
(p_{X,Y} \times \mathrm{id}_Z) \, ; q_{X,Z} & = \; q_{X \times Y, Z} & = \; (q_{X,Y} \times \mathrm{id}_Z) \, ; q_{Y,Z} \\
\partial_{X \times Y} \, ; (p_{X,Y} \times q_{X,Y}) & = \; \mathrm{id}_{X \times Y} &
\end{array}
$$

A category $\mathcal{P}$ equipped with a (tensor) "product" functor $\times : \mathcal{P} \times \mathcal{P} \to \mathcal{P}$ and natural transformations $\partial : \mathrm{id} \to (- \times -)$, $p_{-,X} : - \times X \to \mathrm{id}$ and $q_{X,-} : X \times - \to \mathrm{id}$, satisfying these laws and the Mac Lane–Kelly laws for associativity and commutativity of $\times$, is known as a **p-category**. Construct $\mathrm{supp}_{X,Y} : \mathcal{P}(X, Y) \to \mathcal{P}(X, X)$ and identify $\mathcal{S}$ and $\mathcal{M}$ within $\mathcal{P}$. Finally, show that $(\mathcal{P}, \times)$ has a full embedding $\mathcal{P} \hookrightarrow P(\mathcal{S}, \mathcal{M})$ such that $\times$ restricts to the categorical product in $\mathcal{S}$.

16. Similarly characterise the product structure using the category composed of relations, and, given an allegory satisfying your conditions, show how to embed it in a category of relations (Proposition 5.8.7).

17. Show how the Floyd rules (Remark 4.3.5) define the category of virtual objects (contexts with midconditions) in Remark 5.3.2.

18. By giving the matrices for the (co)projections, (co)diagonals and (co)pairing, show that

(a) $\mathbb{R}^{n+m}$ is the product of $\mathbb{R}^n$ and $\mathbb{R}^m$ in $\mathcal{V}sp$ and $\mathcal{C}Rng$;

(b) it is also their coproduct in $\mathcal{V}sp$;

(c) $\mathbb{R}^{n \times m}$ is their coproduct in $\mathcal{C}Rng$.

19. Show that if $\mathbb{N}^n \cong \mathbb{N}^m$ in $\mathcal{C}Mon$ then $n = m$.

20. Let $\mathcal{C}$ be a $\mathcal{C}Mon$-enriched category, *i.e.* each hom-set $\mathcal{C}(X, Y)$ is a commutative monoid (written using 0 and $+$) and the composition $(;) : \mathcal{C}(X, Y) \times \mathcal{C}(Y, Z) \to \mathcal{C}(X, Z)$ is a monoid homomorphism. Show that the terminal object (if any) is also initial and any product also carries a coproduct structure.

21. Conversely, show that in any category with a zero object (Definition 5.4.3), the zero map is preserved by composition on either side with any map. Suppose further that binary products and coproducts exist and $[\pi_0 \, ; \nu_0, \pi_1 \, ; \nu_1] : X \times Y \cong X + Y$. Show that the category is $\mathcal{C}Mon$-enriched.

22. Show that $\mathcal{R}el \cong \mathcal{R}el^{op}$ and that the product and coproduct are both given by disjoint union. What happens in $\mathcal{CSL}at$ and $\mathcal{SL}at$? Describe the $\mathcal{CM}on$-enriched structures.

23. Use van Kampen's theorem to show that the fundamental group of the circle S^1 is $\mathbb{Z}$. [Hint: U and V are open intervals and W is the disjoint union of two open intervals.]

24. Let M_0 and M_1 be modules for the commutative rings R_0 and R_1 respectively. Show that their coproduct in the category of rings-with-modules is the module $(M_0 \otimes R_1) \oplus (R_0 \otimes M_1)$ for the ring $R_0 \otimes R_1$.

25. (Only if you know some hyperbolic geometry.) Considering the matrices $\left(\begin{smallmatrix} 0 & -1 \\ 1 & 1 \end{smallmatrix} \right)$ and $\left(\begin{smallmatrix} 0 & 1 \\ -1 & 0 \end{smallmatrix} \right)$ as elements of $\mathsf{PSL}_2(\mathbb{Z})$, show that it is the coproduct of the groups $\mathbb{Z}/(3)$ and $\mathbb{Z}/(2)$.

26. Explain how the maps $(\Gamma \times Y) + (\Gamma \times N) \to \Gamma \times (Y + N)$ given in Definition 5.5.1 arise from the universal properties of $+$ and $\times$, and why they are equal. Show that they are invertible in a cartesian closed category. Do the same for $X^{Y+Z} \cong X^Y \times X^Z$ and $(X^Y)^Z \cong X^{Y \times Z}$. Show that these isomorphisms are natural in all three variables, and relate them to Exercise 1.22, Remark 2.3.11 and Proposition 3.6.14(a).

27. Show that this square is a pushout in any category which has coproducts, and that it is also a pullback in an extensive category:

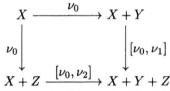

In other words, if two sets $Y' = X + Y$ and $Z' = X + Z$ have a common decidable subset X then, within their union, they intersect *only* in X. Deduce Proposition 5.8.10 for decidable subsets. Although all four maps are monos, the mediator from the pushout to a commuting square of monos need not be mono, since it may identify some $y \in Y$ and $z \in Z$.

28. The following similar property holds in $\mathcal{P}os$. Suppose $X \subset Y'$ and $X \subset Z'$ are *full* subposets (and the underlying sets are decidable subsets). Then Y' and Z' are full subposets of the union, intersecting only in X. Deduce that it holds for embeddings (Example 3.6.13(b)).

29. Let $A, B \subset X$ be ideals of a distributive lattice and $f : A \to \Theta$, $g : B \to \Theta$ be $\vee$-*semilattice* homomorphisms which agree on $A \cap B$. Show that if $a \vee b = a' \vee b'$ then $f(a) \vee g(b) = f(a') \vee g(b')$, where $a, a' \in A$ and $b, b' \in B$. Deduce that, with $C = \{a \vee b \mid a \in A, b \in B\} \subset X$, there is a unique $\vee$-semilattice homomorphism $p : C \to \Theta$ which restricts to f and g.

30. Characterise finite products and coproducts in $\mathcal{L}oc$. [NB: The two-element lattice $\{\bot, \top\}$ is not complete without excluded middle (Example 3.2.5(h)); Example 3.9.10(e) gave the product in $\mathcal{L}oc$.]

31. Show that partitions of $C \in \mathrm{ob}\,\mathcal{L}oc$ (*i.e.* isomorphism classes of coproduct diagrams $A \to C \leftarrow B$) correspond to pairs $\alpha, \beta \in C$ with $\alpha \wedge \beta = \bot$ and $\alpha \vee \beta = \top$. Similarly partitions in $\mathcal{C}Rng^{\mathrm{op}}$ are given by pairs with $\alpha^2 = \alpha$, $\beta^2 = \beta$, $\alpha\beta = 0$, $\alpha + \beta = 1$. Hence show that $\mathcal{L}oc$ and $\mathcal{C}Rng^{\mathrm{op}}$ are extensive.

32. Show that binary coproducts commute with wide pullbacks in an extensive category. Does the analogous property hold for distributive lattices? Let $\mathcal{L}$ be a disjunctive theory; show that $\mathcal{M}od(\mathcal{L})$ has wide pullbacks.

33. Show how to swap the values of two variables using the direct declarative language with assignment. Using this and conditionals, write a program to sort three objects with respect to a given trichotomous order relation. How many comparisons are needed in the worst case?

34. Show that the coproduct in $\mathcal{S}$ extends to a functor on $\mathsf{P}(\mathcal{S}, \mathcal{M})$, where it remains the coproduct. It is not stable disjoint there: explain why we do not want it to be.

35. Let $\mathcal{S}$ be a distributive category with a class $\mathcal{M}$ of supports to which $\mathrm{yes}, \mathrm{no} : \mathbf{1} \rightrightarrows \mathbf{2}$ belong. Suppose that, for each pair of objects, there are partial maps $N \xleftarrow{\ \sigma_0\ } Y + N \xrightarrow{\ \sigma_1\ } Y$ such that $\nu_i \,;\, \sigma_j = \mathrm{id}$ if $i = j$ and $\bot$ otherwise. Show that $\mathcal{S}$ is extensive.

36. Variant records of type $Y + N$ are typically implemented by coding the elements $\nu_0(n)$ and $\nu_1(y)$ as $\langle 0, 0, n \rangle, \langle 1, y, 0 \rangle \in \mathbf{2} \times Y \times N$ respectively, where 0 is a "dummy value" of type Y and N. Formulate the midcondition ϕ which defines this subobject (*cf.* Example 2.1.7). Making clear where extensivity is used, show that the virtual object is indeed the coproduct $Y + N$, and define the **switch**, satisfying the rules of Remark 2.3.10. Modify this construction to remove the assumption that the unused field is *initialised* to the dummy value. Generalise it to the case where the components are themselves (possibly uninhabited) virtual objects; the dummy value now belongs to the underlying real objects but not necessarily to the virtual ones.

37. Show that the pullback of a split epi against an arbitrary map is another split epi. Conversely, show that every split epi is a pullback of its kernel projection, if the kernel pair exists.

38. Let $\mathcal{L}$ be a single-sorted algebraic theory. Show that $\mathfrak{f} : A \to B$ is a mono in $\mathcal{M}od(\mathcal{L})$ iff it is an injective function, and a regular epi iff it is surjective. [Hint: Propositions 5.2.2(a) and 5.6.8.]

39. Characterise the inverse images of $\perp$ and $\top$ under homomorphisms in $\mathcal{DL}at$ and of 0 and 1 in $\mathcal{CR}ng$.

40. Let $K \hookrightarrow A \times A$ be a relation in any category. Define when it is a congruence (Definition 5.6.3) without constructing new pullbacks. [Hint: quantify over commutative diagrams with a variable object Γ in the place of the pullback cube.]

41. Prove the analogues of Theorem 5.6.9 for $\mathcal{CSL}at$ and $\mathcal{F}rm$, and show how to use a closure operator on an object (not its square) to code kernel pairs, analogously to Lemma 5.6.14. [Hint: *cf.* Theorem 3.9.9.]

42. Let $X \times X \to \mathcal{2}$ be the characteristic map of an equivalence relation K on X in *Set*. What is the image factorisation of its exponential transpose, $X \twoheadrightarrow I \hookrightarrow \mathcal{2}^X$?

43. How do you recover the kernel *pair* of $e : A \twoheadrightarrow Q$ from the subset $Q \subset A$ in Lemma 5.6.14?

44. Explain the relationship between factorisation systems and the so-called isomorphism theorems for groups, rings and vector spaces.

45. By Proposition 3.8.14, the orthogonality relation $e \perp m$ in Definition 5.7.1 gives rise to a specialisation order. Investigate its relationship to the category, and what it means to have *enough* epis or monos.

46. Suppose that the rectangle is a pullback and contains a mono and a stable surjection as shown. Show that the two squares are pullbacks.

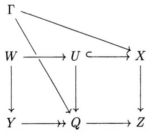

[Hint: form the pullback of $Y \twoheadrightarrow Q$ against $\Gamma \to Q$; this result holds for any stable factorisation, not just the image one.]

47. Let $\mathcal{M}$ be closed under pullback, and $m, n \in \mathcal{M}$. Prove the following tighter form of Lemma 5.7.6(e), that if $e \perp (m\,;n)$ and $e \perp n$ *with respect to* n then $e \perp m$ (*cf.* Exercise 9.27).

48. Let $\mathcal{E}$ be the class of functors which are full and essentially surjective, and $\mathcal{M}$ those that are faithful (Definition 4.4.8). Show that this is a factorisation system in *Cat*.

49. Let U be a relation. Explain what categorical structure is needed

to show that $U \cup \Delta$ and $U \cup U^{op}$ are respectively its reflexive and symmetric closures, and that $U \cup \Delta$ is confluent if U is functional.

50. Show that, in a pretopos, the (stable) union $U \cup V \subset X$ is computed as the image of $U + V \to X$.

51. In a pretopos, show that the pushout of a mono against an arbitrary map is again a mono, and that the square is also a pullback. Deduce that if the opposite category has equalisers then it is regular.

52. Let $(\bullet \to \bullet) \xrightarrow{\ e\ } \mathbb{N} \xrightarrow{\ f\ } \mathbb{Z}/(3)$ in *Cat*. In Proposition 5.8.3, find L and discuss how K, P and M might be defined, noting that e is neither replete nor full. [Hint: ob $K \cong \mathbf{2} + \mathbf{4} \times \mathbb{N}$ in Definition 7.3.8.]

53. In the diagram,

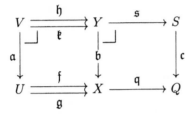

(a) Given $\mathfrak{f}, \mathfrak{g}, \mathfrak{b}, \mathfrak{s}, \mathfrak{q}, \mathfrak{c}$ with the right square a pullback, suppose that $\mathfrak{f}^*\mathfrak{b}$ and $\mathfrak{g}^*\mathfrak{b}$ exist, *a priori* with different vertices V and W. Show that $\mathfrak{a}, \mathfrak{h}, \mathfrak{e}$ may be chosen so that the two left squares are pullbacks.

(b) In a regular category, suppose that the $\mathfrak{q}$ and $\mathfrak{s}$ are coequalisers, $(\mathfrak{h}, \mathfrak{e})$ and $(\mathfrak{f}, \mathfrak{g})$ are kernel pairs, and that the two left squares are pullbacks. Show that the right hand square is also a pullback. [You have to be pretty good at diagram-chasing to do this.]

(c) In *Set*, suppose again that the left squares are pullbacks and the rows are *arbitrary* coequalisers. Assume also that $\mathfrak{c} = \mathrm{id}$, and show by induction on zig-zags that $\mathfrak{b}$ is surjective.

(d) For general $\mathfrak{c}$, deduce that $Y \twoheadrightarrow X \times_Q S$ in the right-hand square.

(e) With $S = Q = \mathbf{1}$, $U = X = \mathbf{2}$, $V = Y = \mathbf{6}$, find maps in *Set* satisfying these conditions, so $\mathfrak{b}$ need not be an isomorphism.

The reason for the difficulty is that the fibres of $\mathfrak{b}$ are carried isomorphically to one another by the transitions. If these have (unoriented) cycles, these isomorphisms may give an automorphism of a single fibre. This can't happen if $\mathfrak{b}$ is mono or $U \rightrightarrows X$ is functional and acyclic.

Structural Recursion

FOUNDATIONS must be built on a sound layer of ballast. The edifice of mathematics rests on calculi of expressions composed of symbols written on paper or coded in a computer. To make a mathematical study of this edifice itself, or of the ways of handling expressions mechanically, we must consider the completed infinities of formulae, such as $\mathbb{N}$. Since these completed infinities have no existence as objects in the real world, but only as characters in the drama of mathematics, this kind of study (*meta*-mathematics) must be made within a pre-existing mathematical context. This is why we are doing it in Chapter VI and not Chapter I.

Mathematically, the algebras of formulae are *free* for certain free theories, whose operations are the connectives and proof rules of logic and type theory. In the category-theoretic tradition, a free algebra is the *initial object* in the category of all algebras. However, the very general nature of universal properties — the fact that this mode of description applies to so many other mathematical phenomena, as we have seen — means that this does not give a "hands on" appreciation of term algebras.

In formal languages, terms cannot be equal unless they have been built in the same way. Since they are *put together* with connectives such as $+$, $\wedge$ and $\exists$, they form an algebra, but, unlike the arithmetical examples which motivated Section 4.6, they can also be *taken apart* — parsed by analysis into cases. This leads to the techniques of structural recursion, which we have been using from the first page of this book, and unification, the subject of Section 6.5. We shall characterise, and thereby construct, the free algebras for free theories, by the property that the parsing operation is well defined and well founded, *i.e.* that it terminates (Section 6.3).

Section 6.4 specialises to tail recursion, which can be implemented more efficiently using imperative **while** programs. The interpretation in this case is described in terms of coequalisers, from the previous chapter.

The properties of term algebras also hold for infinitary operations. More careful consideration, indeed, shows that notions of induction are needed to capture finiteness, and not *vice versa* (Section 6.6).

The mathematisation of symbolism also gives us the means to create new mathematical worlds *inside* existing ones, for example in a model of set theory or in a topos of sheaves. In this book we have been careful to make clear which logical connectives are needed for each application, rather than rely globally on higher order logic: cartesian closed categories are used in functional programming, and pretoposes serve many of the purposes of universal algebra. Certain *other* structure is needed in the *meta-logic* to express or construct these features, so for example we might treat the λ-calculus symbolically using algebraic tools. Category theory, unlike symbolic logic, is well adapted to the parallel treatment of different fragments, because it can both *represent* logical structure very naturally and also *be represented* in a clear combinatorial way.

Set theory tried to deal with both the internal and external structure, though it was particularly ill suited to universal algebra. The inductive hierarchy of sets, on which mathematics was allegedly founded, is defined using the quantifiers, and Lemma 6.3.12 shows that it also relies on the Pasting Lemma 5.8.9. Nevertheless, set theory does throw some light on induction: Section 6.3 uses category theory to generalise some of its ideas and apply them to new inductive situations.

One reason for the difficulty in understanding set-theoretic induction was that the ordinals, as traditionally presented, depend heavily on excluded middle. My original draft of this chapter contained material from set theory which was embarrassingly old fashioned. I tried to develop a simple intuitionistic account of the ordinals; such an account now exists in [JM95], [Tay96a] and [Tay96b], and is summarised at the end of the chapter, though it is hardly simple. But the motivating problem was solved in the mean time in a more elementary way (Exercise 3.45).

The sections of this chapter depend very little on one another.

6.1   Free Algebras for Free Theories

A system of closure conditions describes algebra at the *propositional* level, for example subalgebras and congruences within a fixed domain of discourse. Regarding the closure conditions themselves as introduction rules, Section 3.7 showed that the elimination rule is an *induction* scheme. Similarly, the *type*-theoretic elimination rules provide *recursion*. We had already treated both induction and recursion in our account of natural numbers and lists in Section 2.7, albeit as separate axiomatisations, which we now have to link together.

The algebraic theories for which we are able to study recursion do not have laws. The reason was in Example 3.7.9(f), which showed that a system ▷ of closure conditions arises from a (well founded) relation ≺ iff for each element t there is exactly one set K of preconditions $(K \triangleright t)$. We shall provide free models for infinitary free theories in this section, and also give some applications of them. As for induction on closure conditions, it is a distortion of the theory to restrict the arities to be finite in advance.

We considered algebraic theories in general in Section 4.6, allowing laws and many sorts. But we insisted there that the arities were finite, since Theorem 5.6.9 fails for infinitary operations. The results of this chapter are still needed and useful, as the means whereby the algebras are constructed in Zermelo type theory. We shall develop this technique in Section 7.4, also using it to find colimits of algebras.

Infinitary algebraic theories without laws. What do we mean by *infinite*? In classical mathematics, for any set K, either it can be finitely enumerated, or any attempt to do so is non-terminating and we have $\mathbb{N} \hookrightarrow K$. Cantor and his followers developed *trans*finite numbers (ordinals, Section 6.7) to extend the counting idea. Constructively, this argument does not apply, and the enumeration may fail for many other reasons. For example the set in Exercise 2.16 with two "overlapping" elements is not finite in the strong sense required by Theorem 5.6.9. (We discuss finiteness in Section 6.6.)

Describing the arities and carrier as *in*finite simply means that we *do not restrict them* to being concrete enumerations. (Nor need the arities be ordinals or carry any other special structure.) In this setting we can treat **internal theories**, in which the collection of operation-symbols need no longer be $\{0, 1, +, -, \times, \div\}$ or $\{\bot, \top, \wedge, \vee, \Rightarrow, \neg\}$ but can be a type Ω in the object-language. For example Ω may be a topological space or an abstract data type. The arities ar$[r]$ of the operation-symbols may also be internal objects instead of numbers such as 0 and 2.

For our purpose the arities must nevertheless be specified in advance of the models. The theory of *complete* semilattices — with operations of arbitrarily large arity — is therefore excluded, even though it happens to have free models (Proposition 3.2.7(b)). The apparently similar theory of complete Boolean algebras has no free algebra on $\mathbb{N}$ [Joh82, pp. 33–34].

DEFINITION 6·1·1:   A **free algebraic theory** is given by a set Ω of operation-symbols together with an assignment to each $r \in \Omega$ of a set ar$[r]$, called the **arity** of r. The disjoint union $\kappa = \coprod_{r \in \Omega} \text{ar}[r]$ is known as the **rank**.

The aim is to construct and study the free model, which is also known as an **absolutely free** algebra. It is no loss of generality to consider only the free model without generators, *i.e.* the initial object of the category $\mathcal{M}od(\Omega)$ of algebras and homomorphisms. If we require the algebra generated by a set G, we consider instead the theory $\Omega' = \Omega + G$, where $\mathsf{ar}[x] = \varnothing$ for $x \in G$ (*cf.* Remark 3.7.6 for closure operations).

As in Definition 4.6.2, a **model** or **algebra** of (Ω, ar) is a set A together with a *multiplication table* r_A for each operation-symbol $r \in \Omega$.

$$A^{\mathsf{ar}[r]} \xrightarrow{\ r_A\ } A \qquad TA \overset{\text{def}}{=\!=} \coprod_{r \in \Omega} A^{\mathsf{ar}[r]} \xrightarrow{\ \mathsf{ev}_A\ } A$$

As there is only one sort, these may be combined into one function ev_A, called the **structure map**, as shown on the right. *As there are no laws, any such map defines an algebra* (*cf.* Proposition 7.5.3(b) for monads).

Similarly a **homomorphism** $f : A \to B$ is a function making the square on the left commute for each operation-symbol $r \in \Omega$:

$$
\begin{array}{ccc}
\underline{a} \in A^{\mathsf{ar}[r]} \xrightarrow{\ r_A\ } A & \qquad & TA =\!=\!= \coprod_{r \in \Omega} A^{\mathsf{ar}[r]} \xrightarrow{\ \mathsf{ev}_A\ } A \\
\downarrow{\scriptstyle f^{\mathsf{ar}[r]}} \qquad \downarrow{\scriptstyle f} & & \downarrow{\scriptstyle Tf} \qquad\qquad \downarrow{\scriptstyle f} \\
B^{\mathsf{ar}[r]} \xrightarrow{\ r_B\ } B & & TB =\!=\!= \coprod_{r \in \Omega} B^{\mathsf{ar}[r]} \xrightarrow{\ \mathsf{ev}_B\ } B
\end{array}
$$

These conditions too may be combined, as the single diagram on the right. Compare the role of the functor T here with that of the product functor in the diagram in Definition 4.6.2. Notice that $\Omega \cong T1$, and $\mathsf{ar}[r]$ can also be recovered (Exercise 7.37), but we shall often forget that T has a power series expansion, assuming only that it preserves monos, their inverse images and arbitrary intersections. The lattice analogue of the functor T was given in Lemma 3.7.10, based on closure conditions.

This is quite a different notation from that in which algebraic theories were presented in Definition 4.6.1, so before reading any further you should convince yourself that we have merely presented the relevant data in a more concise form. We shall restore the sorts in Proposition 6.2.6, the generators in Section 6.5 and the laws in Section 7.4.

DEFINITION 6·1·2: Even though the arities are *general types* rather than numbers, it is useful to retain the notation $\underline{a}$ for a typical element of $A^{\mathsf{ar}[r]}$ and a_j for the co-ordinate $\underline{a}(j)$ where $j \in \mathsf{ar}[r]$.

We call an algebra $\mathsf{ev} : TA \hookrightarrow A$ **equationally free** if ev is injective, so

$$r_A(\underline{u}) = s_A(\underline{v}) \implies r \equiv s \wedge \forall j{:}\mathsf{ar}[r].\ u_j = v_j,$$

and **parsable** if ev_A is an isomorphism, so every $u \in A$ is $r(\underline{v})$ for some *unique* $r \in \Omega$ and $\underline{v} \in A^{\mathrm{ar}[r]}$. Compare Example 3.7.9(f), where we needed $\forall t. \exists! K. K \rhd t$. Note that algebraic theories with laws usually have no equationally free models.

EXAMPLE 6·1·3:   Let $\Omega = \{z, s\}$ with $\mathrm{ar}[z] = \varnothing$ and $\mathrm{ar}[s] = \{\star\}$. Any Ω-algebra satisfies the first two of the **Peano axioms** for the natural numbers (Definition 2.7.1). The third and fourth axioms make it equationally free (*cf.* Exercise 2.47). $\mathbb{N} + \mathbb{Z}$ with the obvious structure is parsable but not initial, since it fails Peano's fifth axiom.   □

The following well known properties of initial T-algebras, due to Lambek and to Lehman and Smyth, give a taste of the rest of this chapter.

PROPOSITION 6·1·4: Let $\mathrm{ev} : TA \to A$ be an algebra.

(a) Then $T\mathrm{ev} : T^2A \to TA$ is also an algebra and $\mathrm{ev} : TA \to A$ is a homomorphism.

Let $\mathrm{ev} : TF \to F$ be the initial T-algebra.

(b) Then it is parsable, *i.e.* $\mathrm{ev} : TF \cong F$;

(c) and any T-subalgebra $U \subset F$ is the whole of F. (Out of classical habit, we say that "F has no proper subalgebra".)

Let $\mathrm{ev} : TA \hookrightarrow A$ be an equationally free algebra. Then

(d) if T preserves monos, any subalgebra $U \subset A$ is also equationally free,

(e) but if A has no proper subalgebra then it is parsable.

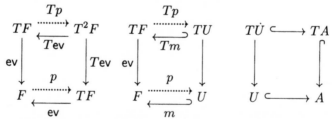

PROOF: [a] Obvious. [b] From (a), since F is initial, there is a unique homomorphism $p : F \to TF$. Then $p \,;\, \mathrm{ev} : F \to F$ is an endomorphism of the initial algebra, so by uniqueness $p \,;\, \mathrm{ev} = \mathrm{id}$. But as p is a homomorphism, $\mathrm{ev} \,;\, p = Tp \,;\, T\mathrm{ev} = \mathrm{id}$, so $p = \mathrm{ev}^{-1}$. [c] Similarly, $p \,;\, m = \mathrm{id}_F$, but m is mono, so $U \cong F$ [LS81, §5.2]. [d] Cancellation of monos, which T preserves. [e] TA is a subalgebra by (a).   □

Given any equationally free algebra, we obtain one which is parsable as the intersection of all subalgebras, by the Adjoint Function Theorem 3.6.9. We shall show in Section 6.3 that this is the initial algebra.

Natural numbers. Being the free algebra for the functor $(-) + \mathbf{1}$ captures the *recursive* properties of $\mathbb{N}$, as set out in Remark 2.7.7.

DEFINITION 6·1·5: The diagram on the left below displays the data for any algebra $(\Theta, z_\Theta, s_\Theta)$ and the homomorphism $p : \mathbb{N} \to \Theta$ for the Peano operations. The second diagram re-expresses this in terms of the functor $T = (-) + \mathbf{1}$.

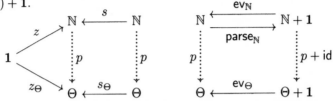

This universal property was identified by Bill Lawvere (1963), and such a structure is called a **natural numbers object** or simply an **NNO**.

Since $\mathbb{N}$ has a universal property, it is unique. But beware that this relies on second order logic (the induction scheme): there are non-standard structures which share all of the first order properties of $\mathbb{N}$ (Remark 2.8.1). Example 6.4.13 gives another characterisation of $\mathbb{N}$.

REMARK 6·1·6: The Lawvere property provides the unique solution for any **primitive recursion** problem of the form

$$p(0) = z_\Theta \qquad p(n+1) = s_\Theta(p(n)),$$

but the elimination rule in Remark 2.7.7,

$$\frac{\Gamma \vdash z_\Theta : \Theta \qquad \Gamma \vdash s_\Theta : \Theta \to \Theta}{\Gamma, n : \mathbb{N} \vdash \mathrm{rec}(z_\Theta, s_\Theta, n) : \Theta} \, \mathcal{NE}$$

allowed z_Θ and s_Θ to be expressions with parameters from a context Γ. Incorporating them, the diagram becomes

$$
\begin{array}{ccc}
& \langle \mathrm{id}_\Gamma, z \rangle \nearrow \Gamma \times \mathbb{N} & \xleftarrow{\mathrm{id}_\Gamma \times \mathrm{succ}} \Gamma \times \mathbb{N} \\
\Gamma & \vdots\, p & \vdots\, (\pi_0, p) \\
& z_\Theta \searrow \Theta & \xleftarrow{\quad s_\Theta \quad} \Gamma \times \Theta
\end{array}
$$

In a cartesian closed category the parametric problem may be reduced to the simple one by putting $\Theta' = \Theta^\Gamma$, $s'_\Theta : g \mapsto \lambda \underline{x}.\, s(\underline{x}, g(\underline{x}))$ and $p' = \tilde{p}$ (Remark 4.7.8). Without these exponentials, Γ is essential to make the definition invariant under substitution. Similarly the target algebra for recursion over a general free theory must be expressed as

$$\Gamma \times T\Theta \equiv \Gamma \times \coprod_r \Theta^{\mathrm{ar}[r]} \cong \coprod_r \left(\Gamma \times \Theta^{\mathrm{ar}[r]} \right) \xrightarrow{\mathrm{ev}} \Theta,$$

where the sums have to be stable (Section 5.5) as shown. For a general abstract functor T to admit parametric recursion, additional structure, known as a **strength**, is needed; see Exercise 6.23. The polynomial functors arising from free theories and the other functors over which we shall consider recursion all admit this structure in an obvious way.

EXAMPLE 6·1·7:   The recursive argument itself may also occur as a parameter, for example in the factorial function (*cf.* Example 2.7.8):

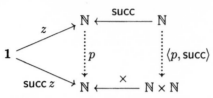

Exercises 2.46, 6.24 and 6.25 show how to handle this case using pairs.

Finally, the theory (Ω, ar) may be parametric, but we shall not consider this possibility in this book. In fact we shall usually omit the parameters Γ and $\mathbb{N}$ as well.

Lists. For any set (alphabet) G there is a free theory of lists, in which the set of operations is $\Omega = G + \{z\}$, with $\mathsf{ar}[x] = 1$ for $x \in G$ and $\mathsf{ar}[z] = \varnothing$, so $TA = \{z\} + G \times A$ (Definition 2.7.2ff). Here G and Ω are general (internal) types, and do not have to be concrete enumerations of symbols. The case $G = \{s\}$ gives the natural numbers.

In a cartesian closed category, using $\mathbb{N}$ we can construct equationally free algebras for the theory of lists, and Exercise 6.10 shows (also using pullbacks) that all finitary free theories have free algebras.

LEMMA 6·1·8:   There is an equationally free algebra of lists on any set.

PROOF: Put $A \equiv (\{\star\} + G)^{\mathbb{N}}$ with

$$\mathsf{ev}(z) \quad = \quad \lambda n. \star$$
$$\mathsf{ev}\langle x, u\rangle \quad = \quad \lambda n. \begin{cases} x & \text{if } n = 0 \\ u(n-1) & \text{if } n \geqslant 1. \end{cases}$$

This is the set of **streams**, in which the symbol $\star$ is being used to indicate the end of a (finite) list. More categorically,

$$\{z\} + G \times A \lhook\joinrel\longrightarrow A + G \times A \cong (\{\star\} + G)^{\{0\}+N} \hookrightarrow (\{\star\} + G)^{N},$$

i.e. $\mathsf{ev} : TA \hookrightarrow A$, for any N such that $N \twoheadrightarrow \{0\} + N$.   □

Infinitary conjunction and disjunction. Using infinitary algebraic theories we can now define $\mathcal{M} \vDash \phi$, the validity of a formula ϕ of the predicate calculus in a model $\mathcal{M}$, in the way sketched in Remark 1.6.12. The meaning of the quantifiers is defined, not by the proof rules, but by infinitary conjunction or disjunction of its instances in $\mathcal{M}$. Note that this model is chosen *before* defining the infinitary theory below.

EXAMPLE 6·1·9: Let $\mathcal{L}$ be a collection of sorts U and relation-symbols ρ. The (closed raw) formulae of first order predicate calculus (Definition 1.4.1) over $\mathcal{L}$ form the free algebra for the free theory with

$$\Omega = \{\top, \bot, \wedge, \vee, \Rightarrow\} \; + \; \{\rho_{\underset{\sim}{u}}\} \; + \; \{\exists_U, \forall_U\}.$$

The last term adds two copies of the *set* Σ of sorts to Ω. The set of symbols of the form $\rho_{\underset{\sim}{u}}$ contributed to Ω depends on the model $\mathcal{M}$ in which we aim to interpret $\mathcal{L}$, specifically the sets $[\![U]\!]$ denoted by the sorts U: $\rho_{\underset{\sim}{u}}$ ranges over the instances of each relation-symbol at each tuple in the interpretation of its arity.

Thus if there are, for example, relation-symbols of arities U^2 and $V \times W$ in the theory then a summand $[\![U]\!]^2 + [\![V]\!] \times [\![W]\!]$ is included in Ω. (The language and model may also have functions, but these do no more than add synonyms for the instances of the relations.)

The arities are given by

$$\begin{aligned}
\mathsf{ar}[\top] \; &= \; \mathsf{ar}[\bot] \; = \; \mathsf{ar}[\rho_{\underset{\sim}{u}}] \; = \; \varnothing \\
\mathsf{ar}[\wedge] \; &= \; \mathsf{ar}[\vee] \; = \; \mathsf{ar}[\Rightarrow] \; = \; \mathbf{2} \\
\mathsf{ar}[\exists_U] \; &= \; \mathsf{ar}[\forall_U] \; = \; [\![U]\!],
\end{aligned}$$

where the arity of the symbols $\exists_U$ and $\forall_U$ also depends on $\mathcal{M}$. The role played before by variable-binding is now taken by the infinite arities. We are interested, not in the *free* Ω-algebra, but in the particular structure $\mathsf{ev} : T2 \to 2$ (Notation 2.8.2) for which

$$\mathsf{ev}(\exists_U, \underset{\sim}{\phi}) \; = \; \bigvee_{u \in [\![U]\!]} \phi_u \qquad \mathsf{ev}(\forall_U, \underset{\sim}{\phi}) \; = \; \bigwedge_{u \in [\![U]\!]} \phi_u$$

and each nullary operation $\rho_{\underset{\sim}{u}}$ is a propositional constant $[\![\rho_{\underset{\sim}{u}}]\!] \in 2$ (which is again prescribed by the model $\mathcal{M}$). The constants $\top$ and $\bot$ and binary operation-symbols $\wedge$, $\vee$ and $\Rightarrow$ have the usual meanings in 2.

Note that ϕ_u is not a single formula with a free variable, but a U-indexed tuple of elements of 2 (a function $U \to 2$). Then a formula ϕ is **valid** in $\mathcal{M}$ if its value in this algebra (calculated, as always with expressions in algebras, by structural recursion) is $\top$. From this we may say what it means for $\mathcal{M}$ to obey certain first order axioms, or to satisfy some other property, as in Remark 1.6.13.

Proofs $\Gamma \vdash \phi$ in the predicate calculus also form a free algebra, whose operation-symbols are named by the proof rules (but the formulation of this algebra is complicated by pattern matching and side-conditions which we shall discuss in the next section). By structural induction on the proof, we may show that if $\mathcal{M} \models \Gamma$ then $\mathcal{M} \models \phi$, *i.e.* the interpretation is sound. For this structural induction, it is only necessary to verify the soundness of each proof rule individually. $\square$

EXAMPLE 6·1·10:   The same *conjunctive interpretation*, in which $r(\underline{\phi}) = \bigwedge_j \phi_j$ in $\underline{2}$ for *every* operation-symbol $r \in \Omega$, is also the basis of *strictness analysis*. Instead of treating the data types in the program as sets or domains and the values as elements, the (base) types are all interpreted as $\underline{2}$ and the constructors as conjunction. The program may then be simplified to a conjunction of some of its inputs, namely those that need to be evaluated in order to execute the program. This subset may be found mechanically by the compiler, which may then detect which arguments actually need to be evaluated.   □

Existence of equationally free models. The following construction is applicable to any free theory; see Example 6.2.7 for the finitary case.

PROPOSITION 6·1·11:   Any free theory has an equationally free algebra.

PROOF: Let $\kappa = \coprod_r \mathsf{ar}[r]$ be the *rank* and $A = \mathcal{P}(\mathsf{List}(\kappa) \times \Omega)$ be the set of sets of lists of odd length. Such lists begin and end with an operation-symbol; each such symbol r (except the last) is followed by a position $j \in \mathsf{ar}[r]$ in its arity. In particular, a nullary operation-symbol can only occur at the end of the list.

Define $\mathsf{ev}(r, \underline{u}) = \{r\} \cup \{[r,j] ; \ell \mid j \in \mathsf{ar}[r], \ell \in u_j\}$.

In list notation, $\ell \in \mathsf{ev}(r, \underline{u})$ iff $\mathsf{head}(\ell) = r$ and

> either $\mathsf{tail}(\ell) = [\,]$,
> or $\mathsf{head}(\mathsf{tail}(\ell)) = j \in \mathsf{ar}[r]$ and $\mathsf{tail}(\mathsf{tail}(\ell)) \in u_j$.

Then r is characterised as the unique singleton list and

$$u_j = \{\ell \mid ([r,j] ; \ell) \in \mathsf{ev}(r, \underline{u})\},$$

so this algebra is equationally free.   □

The idea of this construction is that the terms are (infinitely branching) trees, and are determined by the set of paths through them from the root. Imagine a term being processed by a program; at any moment it is at a certain point in the tree, with the path stored on its stack, *i.e.* as a list. Corresponding to the root there is an operation-symbol, r_0, with a co-ordinate $j_0 \in \mathsf{ar}[r_0]$; the next stage is a similar pair (r_1, j_1) with $j_1 \in \mathsf{ar}[r_1]$ and so on. At the last stage (which is the top of the stack or the head of the list) we have only an operation-symbol r_n without any specified co-ordinate. Otherwise we would not be able to handle the nullary operations, without which the free algebra would be empty.

It still remains to show that the minimal equationally free T-algebra is initial, but we defer this to Section 6.3, devoting the next section to further, more complicated, examples.

6.2 WELL FORMED FORMULAE

Free algebraic theories provide a useful "scaffolding" which can be used during the building of more complicated linguistic structures, such as dependent type theory in Chapters VIII and IX. In this section we shall describe the recursive aspects of arguments about such structures, which are for example used in the construction of the interpretation functor $[\![-]\!] : \mathsf{Cn}_{\mathcal{L}}^{\square} \to \mathcal{S}$ and Gödel's Incompleteness Theorem 9.6.2.

In practice, additional *side-conditions* are required of the terms which are to be admitted to the language. Some of these, such as the *number* of arguments taken by each operation-symbol, can be enforced in advance, but others must be stated by simultaneous recursion together with the expressions themselves. The terms which do satisfy the conditions are traditionally known as *wffs* (*well formed formulae*).

DEFINITION 6·2·1: A *wff-system* is a set X of terms for a free theory (Ω, ar) such that if $r(\underline{u}) \in X$ then $u_j \in X$ for all $j : \mathsf{ar}[r]$. Therefore

$$\mathsf{parse} : X \lhook\joinrel\longrightarrow TX$$

is a total function on X, and is injective (since ev is a partial inverse).

Nonsense results if instead we admit expressions with ill formed subterms, for example the assertion that "*the* unicorn is *the* author of the *Principia*" (page 18). As with $\mathsf{pred}(0)$ (Remark 2.7.9), the cost of trying to make all operations total is a proliferation of exceptions to the rules of inference. These are easily overlooked in complex situations, leading to errors in programs which are extremely hard to track down.

Recursive covers. Having enumerated the raw terms (wffs), we impose laws to equate them. When arguing inductively, the ideal situation is that the laws be oriented, *i.e.* presented as reduction rules, and these be confluent and strongly normalising. The surjection in Definition 6.2.2 then has a (canonical) splitting, the semantic values being identified with the class of normal forms. This class may itself have a recursive characterisation: for example Exercise 2.23 showed that normal λ-terms are hereditarily $\lambda\underline{x}.\,y\underline{u}$ (*i.e.* each sub-term u_j is also normal, and of this form). We can regard this as a (finer) notion of well-formedness, with useful inductive consequences of its own (*e.g.* Remark 7.6.12ff).

Failing this, finitary algebraic theories handle laws using quotients by congruences (Sections 5.6 and 7.4), whilst the properties of adjunctions take care of most of the infinitary theories of interest (*e.g.* Lemma 5.6.14). These methods tend to interact notoriously badly with recursion.

Section 7.6 turns the construction of linguistic structures on its head, in the search for a language which exactly matches a given semantic structure, for example a λ-calculus which is equivalent to a given cartesian closed category. New symbols such as $\ulcorner X \urcorner$ are added to the language for each entity in the semantics, but then terms in the language have equal meanings because this is given in the semantics, rather than because there is a symbolic proof of this fact.

Either way round, the reason for considering raw terms is that they admit structural recursion.

DEFINITION 6·2·2: A ***recursive cover*** of a "semantic" set A is a wff-system X ("syntax", with a well founded "sub-expression" relation $\prec$), together with a *surjective* function $p : X \twoheadrightarrow A$. (Since $X \subset F$, A is a *sub*quotient of the free algebra F if the latter exists.)

EXAMPLES 6·2·3:

(a) To solve Rubik's cube (Example 4.2.3) we must find a list of moves which restores a semantically given position to the home one. We need to split the surjection $F = \mathsf{List}(6) \twoheadrightarrow A$, where A is the group.

(b) The reflexive–transitive closure $\leqslant$ of a relation $<$ is a subquotient of $\mathsf{List}(<)$ (Exercise 3.60).

(c) The finite powerset $\mathcal{P}_{\mathsf{fg}}(X)$ is a quotient of $\mathsf{List}(X)$ (Definition 6.6.9).

(d) Let A be the free algebra for a finitary theory $\mathcal{L}$ and F the free algebra for the free theory which consists of the operation-symbols of $\mathcal{L}$ alone, forgetting the sorts and laws. We construct A in Section 7.4 as the quotient by the laws of the subset $X \subset F$ of terms obeying the type discipline.

(e) Let Ω be the rules of logic. The class Θ of formulae is an Ω-algebra, of which the class $A \subset \Theta$ of formulae true in some class of models is a subalgebra (*cf.* Remark 3.7.7 and Example 6.1.9). The free algebra F consists of proofs and the homomorphism $F \dashrightarrow \Theta$ sends each proof to its conclusion. If $F \twoheadrightarrow A$ then all true formulae are provable.

(f) Let $\mathcal{C}$ be a cartesian closed[1] category (Section 4.7) and $\Theta = \mathsf{mor}\,\mathcal{C}$. Let F be the class of *raw* λ-terms in the canonical language of $\mathcal{C}$ (Section 7.6) and $p : F \dashrightarrow \Theta$ the interpretation function. Its image $A \subset \Theta$ is the class of λ-definable functions.

(g) The same, where F is the class of *normal* λ-terms.

(h) PERs (Exercise 5.10).

---

[1] The interpretation function $[\![-]\!]$ covers the cartesian closed *structure* of the category of λ-terms of $\mathcal{C}$. However, any property which is preserved by weak equivalences will transfer to $\mathcal{C}$ (Proposition 7.6.7(b)).

(i) (Lothar Collatz, 1930s) Let $f : \mathbb{N} \to \mathbb{N}$ by

$$f(n) = \begin{cases} n/2 & \text{if } n \text{ is even} \\ (3n+1)/2 & \text{if } n \text{ is odd;} \end{cases}$$

it is an open problem whether $\forall n.\, \exists m.\, f^m(n) = 1$.

Putting $a(m) = 2m$ and $b(m) = (2m-1)/3$ (defined only when $m \equiv 2 \bmod 3$), there is a partial function $p : \mathsf{List}(\{a,b\}) \rightharpoonup \mathbb{N}$ with $p([\,]) = 1$, $p(\mathsf{cons}(a,\ell)) = a(p(\ell))$ and $p(\mathsf{cons}(b,\ell)) = b(p(\ell))$. The Collatz problem asks whether p is surjective.

So both in the Collatz problem and in logic, the image of one recursive structure in another may be an intractably intricate maze, *cf.* Gödel's Incompleteness Theorem 9.6.2.

The semantics inherits a weak induction principle from the syntax. (For a stronger result see Exercise 3.54.)

PROPOSITION 6·2·4: Let $(X, \prec)$ be well founded and let $p : X \twoheadrightarrow A$ be a surjective partial function. Then for predicates $\phi[a]$,

$$\frac{\forall t{:}X.\ \big(\forall u{:}X.\ u \prec t \Rightarrow \phi[p(u)]\big) \implies \phi[p(t)]}{\forall a{:}A.\ \phi[a]}$$

PROOF: The rule is simply the induction scheme for $\phi{\circ}p$: from $\forall t.\ \phi[p(t)]$ we deduce $\forall a.\ \phi[a]$ by surjectivity. □

Variables. At the lowest level of linguistic analysis, we use variables, operation-symbols and punctuation, and it must be decidable whether any two such things are the same. This is obvious for marks on the page or bits in a computer, but the development of Section 6.1 was intended also to apply to internal structures, where the alphabets such as Ω are themselves types, and to address the issues raised in Theorem 4.2.12.

EXAMPLE 6·2·5: The variable names (or generators, as we call them in this chapter) in a context have to be distinct, so they should come from a population G with decidable equality. Our convention is that they be explicit, but we need some elementary manipulations:

(a) Example 2.7.6(e) tests whether the name x is in the list Γ;

(b) Exercise 6.50 chooses new names from a population G, providing the inexhaustible supply of variables needed for substitution under a λ (Definition 1.1.9ff);

(c) Simple type theory (Section 2.3) is the free algebra Σ for the theory with three operation-symbols ($\times$, $+$ and $\to$) over the base types.

Then $\mathsf{ob}(\mathsf{Cn}_{\mathcal{L}}^{\times}) \subset \mathsf{List}(G \times \Sigma)$, where the variables are not repeated. □

In the same fashion we can define the set $FV(t)$ of free variables of a term in algebra, the λ-calculus and logic and prove the Extended Substitution Lemma (Proposition 1.1.12). The unification algorithm (Section 6.5) also depends on parsing and the ability to distinguish between operation-symbols and generators.

Many-sorted theories. The type discipline is obeyed (globally) iff the (local) side-conditions on the formation of each operation-symbol are satisfied *hereditarily* (Definition 2.5.4). As the type of a term is that of its outermost symbol, the validity of the typing rules may be expressed by a function $F \to 2 \times \Sigma$, where Σ is the class of types. If the operation-symbols have finite arity and are distinguishable from one another, we may instead define $F \to \Sigma + \{\star\}$, the extra element being an error value.

PROPOSITION 6·2·6:   Many-sorted free theories have free models.

PROOF: Let (Ω, ar) be the corresponding untyped theory and F its free algebra. From the type discipline of Notation 1.3.3,

> $r(\underline{u}) \in F$ is well formed and of sort V iff
> the operation r has arity $\prod_{j \in \mathsf{ar}[r]} U_j \vdash V$ and
> for each $j \in \mathsf{ar}[r]$, u_j is well formed and of sort U_j.

Then the carrier F_U of the free typed algebra at sort U is the set of well formed terms of sort U. Also $X = \coprod_U F_U \subset F$ is a wff-system.

If A is any algebra for the typed theory then $\Theta = \coprod_U A_U$ is a partial algebra for the untyped one, and then Theorem 6.3.13 gives the unique solution $p : X \rightharpoonup \Theta$ of the recursion equation. By induction, if $u : F$ is well formed and of type U then $p(u)$ is defined and $p(u) \in \mathcal{A}_U \subset \Theta$.   □

The notion of *type* in this result is just accountancy: it need not be semantic. For example it may simply be the number of terms in a list.

EXAMPLE 6·2·7:   A word $\ell \in \mathsf{List}(\Omega)$ is a well formed sequence of n terms over a finitary theory $\mathsf{ar} : \Omega \to \mathbb{N}$ if

> $\ell = [\,]$ and $n = 0$; or
> $\ell = \mathsf{cons}(h, t)$ where (h is an operation-symbol and)
> t is a well formed sequence of $n + \mathsf{ar}[h] - 1$ terms.

Those words which are *single* terms ($n = 1$) form the free algebra for the free finitary theory. This way of forming expressions, for example $(1+2) \times (3+4)$ is written $[\times, +, 1, 2, +, 3, 4]$, is called **Polish notation**; it is due to Jan Łukasiewicz (1920s). It is used by compilers in reversed form (*i.e.* with operations *after* their arguments) for the evaluation of arithmetic expressions using a **stack** (Exercise 6.8).

THEOREM 6·2·8:

(a) Every directed graph $\text{src}, \text{tgt} : E \rightrightarrows O$ generates a free category.

(b) Let $\mathcal{C}$ be any internal category and $|\mathcal{C}|$ its underlying graph. Then the internal graph homomorphism $\ulcorner-\urcorner$ names objects and arrows of $\mathcal{C}$ as types and unary operation-symbols of $\mathsf{L}(|\mathcal{C}|)$, and hence as objects and arrows of $\mathsf{Cn}_{\mathsf{L}(|\mathcal{C}|)}$. Conversely, the internal functor $[\![-]\!]$ interprets types and terms of $\mathsf{L}(|\mathcal{C}|)$ back in $\mathcal{C}$.

$$\mathsf{Cn}_{\mathsf{L}(|\mathcal{C}|)} \underset{\ulcorner\_\urcorner}{\overset{\mathfrak{p} \equiv [\![-]\!]}{\rightleftarrows}} |\mathcal{C}|$$

These are inverse on objects and form a retraction on morphisms.

(c) $\mathcal{C} \cong \mathsf{Cn}_{\mathsf{L}(\mathcal{C})}$, where $\mathsf{L}(\mathcal{C})$ is the canonical language of Theorem 4.2.12.

PROOF:

[a] Since $\mathsf{List}(E)$ has only one identity (the empty list), we select the set $\mathsf{mor}(\mathcal{C})$ of well formed paths from $O \times \mathsf{List}(E) \times O$ by the rules
$(x, [\,], x)$ is well formed of sort $\mathcal{C}(x, x)$;
$(x, \mathsf{cons}(h, t), z)$ is well formed of sort $\mathcal{C}(x, z)$ if
(y, t, z) is well formed of sort $\mathcal{C}(y, z)$,
where $\text{src}\, h = x$ and $\text{tgt}\, h = y$
(*cf.* Exercise 3.60). By induction on ℓ_1, if (x, ℓ_1, y) and (y, ℓ_2, z) are well formed paths, so is $(x, (\ell_1 \,;\, \ell_2), z)$. Proposition 2.7.5 showed associativity. Theorem 4.4.5 constructed $[\![-]\!] : \mathsf{Cn}_{\mathcal{L}} \to \mathcal{C}$ by list recursion, with base case $[\![\ulcorner\alpha\urcorner]\!] = \alpha$. [b] Follows by construction.

[c] The maps of $\mathsf{Cn}_{\mathcal{L}}$ are lists, so we use list induction to show that $\mathfrak{f} : \ulcorner X \urcorner \to \ulcorner Y \urcorner$ is provably equal to $\ulcorner[\![\mathfrak{f}]\!]\urcorner$. This is immediate for the empty list (identity); for the induction step, if $\mathfrak{f} = \ulcorner[\![\mathfrak{f}]\!]\urcorner$ then

$$\ulcorner\alpha\urcorner \,;\, \mathfrak{f} = \ulcorner[\![\ulcorner\alpha\urcorner]\!]\urcorner \,;\, \ulcorner[\![\mathfrak{f}]\!]\urcorner = \ulcorner[\![\alpha]\!]\urcorner \,;\, \ulcorner[\![\mathfrak{f}]\!]\urcorner = \ulcorner[\![\ulcorner\alpha\urcorner \,;\, \mathfrak{f}]\!]\urcorner.$$

In particular the internal notion of algebraic theory, based on Definition 6.1.1, answers the question after Theorem 4.2.12 of what it is to have an "alphabet" in which the canonical language of a category $\mathcal{C}$ may be written. □

Notice that the laws of $\mathsf{L}(\mathcal{C})$ are confluent and strongly normalising. This makes part (c) easier than the construction of the free category on an *arbitrary* elementary sketch, which we further defer to Example 7.4.5.

Exercise 6.10 uses the finite limit methods of Remark 5.2.8 to give an alternative construction of $\mathsf{mor}(\mathcal{C})$.

Formation rules in type theories. In the last three examples, the class of sorts (Σ) was fixed in advance. Example 6.2.5 gave the types and contexts for the simply typed λ-calculus, again before the terms, so now it remains to generate these (or just their normal forms).

EXAMPLE 6·2·9: Raw λ-terms (even before α-equivalence is imposed) together form the free algebra with one constant and one unary operation (λ-abstraction) for each variable-name, and a single binary operation (application). Such a term is well formed in the simply typed calculus if it obeys the formation rules of Definition 2.3.5, of which we note just

$$t = \lambda x{:}X.\ u \text{ is well formed of type } X \to Y \text{ in context } \Gamma$$
$$\text{if } x \notin \Gamma \text{ and } u \text{ is well formed of type } Y \text{ in } [\Gamma, x : X].$$

Similarly the η-rule $f = \lambda x.fx$ is valid in a context Γ in which f is well formed and $x \notin \Gamma$. ☐

Section 4.3 constructed the classifying category from the contexts and terms. Its morphisms (the multiple substitutions) are given either by a sketch or by normal forms, both of which are certain well formed lists of terms. Interpretations were defined by structural recursion, as the equivalence between syntax and semantics will be in Section 7.6.

For more complex theories, well-formedness is intimately involved with the meaning of the calculus itself (Example 8.2.1). Well formed *histories* justifying the well-formedness of the intended syntactic entities must be used to stand for these entities themselves. As with many-sorted algebras, the distinction between terms, types, contexts, *etc.* is made afterwards. Remark 8.4.1 gives a concise presentation for generalised algebraic theories.

Proof theory. Induction on the structure of a proof may be exploited to derive consequences from an assertion ϕ (or, in particular, to refute it) not by using ϕ as a *hypothesis*, but by asking how, if at all, it might have been *proved* (Example 6.2.3(e)). General models have properties which are not consequences of the given theory, so the semantic application of this idea is to the analysis of *free* models.

The most powerful results from this approach are obtained after showing that if ϕ has some proof then it has one in normal form. Normalisation and cut-elimination themselves make heavy use of structural recursion over the free algebras of proofs or formulae (see for example Chapter 14 of [GLT89]). The disjunction and existence properties of intuitionistic logic follow, although in Section 7.7 we shall prove what we need of them using category theory, without direct symbolic manipulation.

Infinitary operations. The introduction to infinitary free theories in the previous section explained how this treatment was the propositions-to-types generalisation of our account of induction on closure conditions in Section 3.7. It also mentioned the conflict between equations and infinitary operations.

The *induction* scheme in Definition 3.7.8 is an analysis of the prov*ability* of propositions from Horn clauses. *Recursion* for the free algebra for a free theory, by contrast, makes use of the *proof*. Such a proof is a term in the free algebraic theory whose operation-symbols are the instances of the closure conditions (*cf.* resolution, Remark 1.7.6).

Another way of putting this distinction is that closure conditions, being the propositional form, give rise to the order-theoretic notion of *least* subalgebra, which becomes the *initial* algebra in the categorical form. Proposition 6.1.4(c) showed that the initial algebra is minimal, so what is its relationship to other minimal algebras?

REMARK 6·2·10: Consider *each instance $K \triangleright x$ of a closure condition* on a set Θ (Definition 3.7.4) as a partial operation r of arity K, with just one defined instance $r(K) = x$. The formal theory Ω consists of all of these operation-symbols r, and Θ is a *partial* Ω-algebra.

(a) Then any subset $A \subset \Theta$ is $\triangleright$-closed iff it is an Ω-*subalgebra* in the sense that, for each $\underline{a} \in A$, *if $r(\underline{a})$ is defined in Θ then it lies in A.*

(b) Let A be the smallest subalgebra of Θ and F the initial Ω-algebra. Then the unique homomorphism $p : F \to A$ (Theorem 6.3.13) is surjective if Ω is *finitary*, but need not be otherwise.

PROOF: The unique homomorphism $F \to \Theta$ factors through A. Then let $F \twoheadrightarrow U \subset \Theta$ be the image factorisation (in *Set*), so $U \subset A \subset \Theta$. If the subset U is a subalgebra then $U = A$ as required. This is the case for finitary theories by Theorem 5.6.9, but Exercise 6.12 shows that this assertion for arbitrary theories is equivalent to the Axiom of Choice. □

Remark 1.1.1 used trees to represent the elements of the free algebra. The nodes are sub-expressions, which are themselves elements, so the tree is a subset of the algebra (in fact a wff-system). The immediate sub-expression relation, being well founded, is irreflexive, by Corollary 2.5.11, so the tree is highly disconnected *vertically* (in the horizontal direction, which is indexed by the arities, it may be continuous). This means that operations with non-projective arities (Remark 5.8.4(e)) have no opportunity to act on systems of arguments which cut across the ranking of the tree. In a non-free model, a formerly disconnected system of arguments may become connected, allowing the operation-symbol to act on it. In this way, quotient models can be *bigger* than the image of the free algebra.

Beware that much of the literature on infinitary algebraic theories deals with these difficulties by assuming Choice, often without saying so.

Unfortunately, constructive mathematics has not yet faced up to this challenge either: we may have provability without proof, *cf.* our sceptical comments about the constructive existential quantifier in Section 2.4.

6.3   THE GENERAL RECURSION THEOREM

This section makes the link, at last, between the inductive Peano and recursive Lawvere definitions of the natural numbers. For an arbitrary free theory T, the former generalises to a minimal parsable T-algebra, and we have to show that this is initial in the category of all T-algebras.

However, we saw in the previous section that there are many applications in which the required structure is not the initial algebra — consisting of *all* terms — but a subset of "well formed formulae". Wff-systems also have a direct computational meaning: they consist of the hereditary sub-arguments that are actually generated in the course of a particular recursive calculation. So they measure the stack space that it needs, and if an execution goes beyond the largest feasible wff-system then it *overflows*. This approach can also be used when T is a functor such as the powerset which has no initial algebra.

We shall characterise wff-systems as extensional well founded parse-*co*algebras, using a new, categorical, definition of well-foundedness.

The proof of the main theorem — that induction implies recursion — works by pasting together *attempts* (partial solutions); it is similar to the fixed point theorem (Proposition 3.3.11), but without the need for Scott-continuity. In the infinitary case we want to reapply T *after* taking the colimit over infinitely many steps. Transfinite iteration can be done using ordinals (Section 6.7), but the soundness of that technique relies on the result we're about to prove.

Instead of doing this tedious and repetitive job ourselves, imagine (after John Conway) that we have a *class* of servants, who each do what they can before getting tired. Classically, we ask which servant claims to be the most (or maximally) hard-working and, by finding his shortcomings, show how he might have done better. Intuitionistically, the result is obtained by co-operation. It uses second order logic.

Behind this proof is the idea that wff-systems, like attempts, are built up from the empty set by iterating the functor T. The collection of all such coalgebras generalises the **von Neumann hierarchy** in set theory; the general recursion theorem also comes from that tradition, where it was originally stated for the ordinals. As in set theory, the hierarchy exists even when the initial algebra doesn't.

REMARK 6·3·1: In this section, $T : Set \to Set$ will denote any functor that preserves monos and inverse image diagrams, and therefore partial functions and their composition and order (Section 5.3). Extra structure is needed to handle parameters (Exercise 6.23). The functors $\mathcal{P}$, $\mathcal{P}_{fg}$, List and $\sum_{r \in \Omega} (-)^{ar[r]}$ have these properties; in fact they also preserve arbitrary intersections of monos.

Using various such functors, the following development is not absolutely restricted to free theories. Laws of *certain* forms can be handled, such as commutativity (permutation of maybe infinitely many sub-terms) and idempotence (in the sense of the law $r(x, x) = r(x)$, again infinitarily). Adding these to the theory of lists, the list-forming operations such as cons and append make *sets* instead.

One can also generalise to a class of supports $\mathcal{M}$ (Definition 5.2.10) in categories other than *Set* with certain completeness conditions [Tay96b].

Well founded coalgebras. Definition 6.2.1 captures the feature which was common to the examples of that section, using parse rather than ev.

DEFINITION 6·3·2: Any map parse $: X \to TX$ is called a T-*coalgebra*.

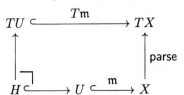

Suppose that whenever the above diagram is a pullback, m is in fact an isomorphism; then we call X a **well founded coalgebra**. Proposition 6.3.9 and Theorem 6.3.13 are examples of arguments which use this as an idiom of induction.

Exercise 3.42 shows how this new notion of well-foundedness, reduced to its lattice form, relates to induction for well founded relations and for closure conditions (Definitions 2.5.3 and 3.7.8).

EXAMPLE 6·3·3: A coalgebra for the covariant powerset functor defines a binary relation $\prec$ by $u \prec t$ if $u \in$ parse(t), so parse$(t) = \{u \mid u \prec t\}$. Well-foundedness agrees with the old sense (Definition 2.5.3) because

$$H \cong \{(t, V) \mid \mathsf{parse}(t) = V \subset U\} \cong \{t \mid \forall u.\, u \prec t \Rightarrow u \in U\}$$

is the induction hypothesis, the inclusion $H \subset U$ is the induction premise (*cf.* $H = U$ in Example 3.8.11(a) and Proposition 6.1.4(c)), and $U = X$ is the conclusion.

The relation is **extensional** in the set-theoretic sense (Remark 2.2.9),

$$(\forall w.\ w \prec u \Leftrightarrow w \prec v) \implies u = v,$$

iff parse is mono. Recall that parse is also mono in a wff-system.

REMARK 6·3·4:   Let $\kappa : T \to P$ be a **cartesian transformation**, *i.e.* a natural transformation whose naturality squares are pullbacks.

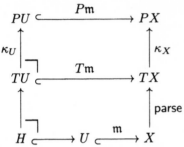

Then parse $: X \to TX$ is a well founded T-coalgebra iff the composite parse $;\kappa_X : X \to PX$ is a well founded P-coalgebra (by Lemma 5.1.2).

There *is* a natural transformation $\kappa : T \to \mathcal{P}$ such that the naturality squares *with respect to monos* are pullbacks iff T preserves arbitrary intersections (as our examples do). Then $\kappa_X : TX \to \mathcal{P}(X)$ by

$$\kappa_X(\tau) = \{v \mid \forall U \subset X.\ \tau \in TU \Rightarrow v \in U\},$$

and we call the relation $v \prec t \iff v \in \kappa_X(\text{parse}(t))$ the **immediate sub-expression** relation. For a free algebraic theory, $\tau = (r, \underline{y})$ and $t = r(\underline{y})$ for some unique $r : \Omega$ and $\underline{y} : X^{\text{ar}[r]}$. Then v is an immediate sub-expression of t iff $v = u_j$ for some $j : \text{ar}[r]$ (consider $U = \{\underline{y}\}$).

From the point of view of induction (though not algebra) this allows us to identify each expression with its *set* of immediate sub-expressions, and to do so hereditarily. Certain ideas from set theory now become useful, where for the purposes of universal algebra in the previous chapter they were a nuisance.

REMARK 6·3·5: (Gerhard Osius [Osi74]) Inclusion $Y \subset X$ between wff-systems, or between sets in the set-theoretic sense, is characterised by an (injective) **coalgebra homomorphism**, *i.e.* a function $f : Y \to X$ making the square on the left commute:

First notice that the inequality $\text{parse}_Y \,;\, \mathcal{P}(f) \subset f \,;\, \text{parse}_X$ holds iff f is strictly monotone (Definition 2.6.1) with respect to the associated $\prec_X$ and $\prec_Y$. Then the square commutes iff the lifting property

$$\forall t{:}Y.\ \forall u{:}X.\ u \prec_X f(t) \implies \exists v{:}Y.\ u = f(v) \wedge v \prec_Y t$$

holds. An *inclusion* $Y \subset X$ has this property iff it is closed downwards with respect to $\prec$; we call it an **initial segment**, *cf.* Definition 2.6.5. As in Remark 6.3.4, if there is a cartesian transformation $\kappa : T \to P$ then initial segments for T and P agree. The unit of the extensional reflection (Example 7.1.6(g)), *i.e.* the map linking a well founded structure to its Mostowski collapse, is a *surjective* coalgebra homomorphism.

$\mathcal{P}$-coalgebra homomorphisms are **simulations** (Exercise 3.53).

We write $\mathcal{C}oalg(T)$ for the category of T-coalgebras and homomorphisms.

COROLLARY 6·3·6: If there is a coalgebra homomorphism $f : Y \to X$ (or an initial segment) with X well founded then Y is well founded too.

PROOF: By Proposition 2.6.2 or Exercise 3.54. There is also a direct categorical proof [Tay96b], in which $f_* \equiv [f^{\mathrm{op}}]$ (Remark 3.8.13(b)) is applied to the subset $V \subset Y$ testing well-foundedness of Y. □

See Exercise 3.41 for the lattice version.

The recursion scheme. Recall from Definition 2.5.1 that the three phases of the recursive paradigm say that $p = \text{parse}\,;\, Tp\,;\, \text{ev}$.

We use partial functions with the extension order $\sqsubseteq$ (Definition 3.3.3).

DEFINITION 6·3·7: (Osius) Let $\text{parse}_X : X \to TX$ be a coalgebra and $\text{ev}_\Theta : T\Theta \rightharpoonup \Theta$ a (partial) algebra. Then $p : X \rightharpoonup \Theta$ is an **attempt** if $p \sqsubseteq \text{parse}_X\,;\, Tp\,;\, \text{ev}_\Theta$, and satisfies the **recursion equation** if these are equal as partial functions, *i.e.* have the same support.

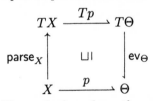

The coalgebra obeys the **recursion scheme** if for every such $(\Theta, \text{ev}_\Theta)$ there is a unique $p : X \rightharpoonup \Theta$ with $p = \text{parse}\,;\, Tp\,;\, \text{ev}_\Theta$. In particular, if $\text{parse}_X = \text{ev}_X^{-1}$ and ev_Θ is total, this diagram says that $\text{ev}_X : TX \to X$ is the *initial T-algebra*.

This is a very convenient diagrammatic form in which to present recursive programs, as we illustrate in Example 6.4.7 and Exercises 6.27ff.

LEMMA 6·3·8:   Any partial attempt $p : X \rightharpoonup \Theta$ is given by a total one $p = (p' \,;\, \mathsf{ev}) : Y \to \Theta$, the support $i : Y \hookrightarrow X$ being an initial segment.

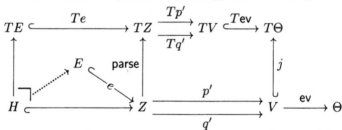

If $f : Z \to Y$ is another coalgebra homomorphism then $f \,;\, p : Z \to \Theta$ (restriction along f) also satisfies the recursion equation.

PROOF: Expand and rearrange the diagram of partial functions.   □

PROPOSITION 6·3·9:   If $\mathsf{parse}_Z : Z \to TZ$ is well founded then there is at most one total function $p : Z \to \Theta$ satisfying the recursion equation.

PROOF: Suppose $p, q : Z \rightrightarrows \Theta$ both satisfy it. Then the two parallel rectangles on the right commute since p and q are total attempts. Let $e : E \hookrightarrow Z \rightrightarrows V$ be the equaliser of p and q, and form the pullback H.

$$
\begin{array}{ccccccc}
TE & \xrightarrow{\;\;Te\;\;} & TZ & \overset{Tp'}{\underset{Tq'}{\rightrightarrows}} TV & \xrightarrow{\;Tev\;} & T\Theta \\[2mm]
\uparrow & & \uparrow & & & \uparrow \\[1mm]
 & E & \;\mathsf{parse}\; & & & j \\[2mm]
H & \longrightarrow & Z & \overset{p'}{\underset{q'}{\rightrightarrows}} V & \xrightarrow{\;ev\;} & \Theta
\end{array}
$$

The composites $H \rightrightarrows T\Theta$ are equal by construction, and j is mono by hypothesis. Hence $H \hookrightarrow Z \rightrightarrows V$ are equal, and $H \hookrightarrow Z$ factors through the equaliser, so $e : E \cong Z$ by well-foundedness of Z, whence $p = q$.   □

Notice that once again we have uniqueness before existence (page 21).

REMARK 6·3·10:   Using the conjunctive interpretation (Example 6.1.10), well-foundedness is also *necessary* for uniqueness. For $T = \mathcal{P}$, $\Theta = \Omega$ and $\mathsf{ev} = \bigwedge = \forall$, the recursion equation reduces to

$$p[t] \iff (\forall u.\, u \prec t \Rightarrow p[u]),$$

which is the strict induction premise (Definition 2.5.4). But the constant function $p : t \mapsto \top$ also satisfies this equation, so *uniqueness* of p is equivalent to the induction scheme. See also Exercise 6.14.   □

This result should be treated with circumspection: taking $\Theta = \Omega$ means that we are using *higher order logic* (a point which is obscured classically, where $\Omega = \mathbf{2}$). Induction for the second order predicate $\phi[x] \equiv (x \not\prec x)$ shows that well founded relations in this sense are irreflexive, and so are too clumsy to analyse fixed points of iteration. By closer examination of the carrier and structure of the *intended* target of recursion, maybe we can restrict the class of subsets (predicates) to those that *need* to be considered, and thereby get a weaker notion of well-foundedness which admits more source structures X but remains sufficient for recursion.

The general recursion theorem. It remains to show that well-foundedness is sufficient for recursion. There is a zero attempt, with $\varnothing$ as support, and we now describe the successor.

LEMMA 6·3·11: Let parse : $X \to TX$ be a coalgebra and $p : X \rightharpoonup \Theta$ be an attempt. Then Tparse : $TX \to T^2X$ is also a coalgebra and $(Tp\,; \mathsf{ev}) : TX \rightharpoonup \Theta$ and $p \sqsubseteq q \equiv (\mathsf{parse}\,; Tp\,; \mathsf{ev}) : X \rightharpoonup \Theta$ are attempts.

$$
\begin{array}{ccccc}
X & \xrightarrow{\ \mathsf{parse}\ } & TX & \xrightarrow{\ T\mathsf{parse}\ } & T^2X \\[2pt]
\downarrow{\scriptstyle p} & \sqsubseteq & \downarrow{\scriptstyle Tp} & \sqsubseteq & \downarrow{\scriptstyle T^2p} \\[2pt]
\Theta & \xleftarrow{\ \mathsf{ev}\ } & T\Theta & \xleftarrow{\ T\mathsf{ev}\ } & T^2\Theta
\end{array}
$$

Note that this is a diagram of partial functions: Remark 6.3.1 says that T acts on such diagrams. □

Let's pause to look at this symbolically. In set theory, given an attempt

$$p(t) \sqsubseteq \mathsf{ev}(\{p(u) \mid u \prec t\}) \equiv q(t) \equiv a,$$

consider also

$$\mathsf{ev}(\{q(u) \mid u \prec t\}) = \mathsf{ev}(\{\mathsf{ev}(\{p(v) \mid v \prec u\}) \mid u \prec t\}) \equiv b.$$

If a is defined then so is each $p(u)$, for $u \prec t$, so $\mathsf{ev}(\{p(v) \mid v \prec u\})$ is also defined and equal to $p(u)$. Then a and b are equal, *i.e.* $a \sqsubseteq b$ since definedness of a was a hypothesis. In algebra we have similarly

$$p(r(s_i(v_{ij}))) \sqsubseteq r(p(s_i(v_{ij}))) \sqsubseteq r(s_i(p(v_{ij}))),$$

where i ranges over $\mathsf{ar}[r]$ and j over $\mathsf{ar}[s_i]$. Sammy Eilenberg, one of the founders of category theory, but whose main work was in subscript-ridden homological algebra (*cf.* Exercise 4.15) commented on a seminar in 1962 that "If you define it right, you won't need a subscript."

LEMMA 6·3·12: Let X be a well founded coalgebra and $p_0, p_1 : X \rightharpoonup \Theta$ be attempts. Then there is an attempt p with $p_0, p_1 \sqsubseteq p$.

PROOF: Let $Y_0, Y_1 \subset X$ be the supports of p_0 and p_1, so Y_0 and Y_1 are initial segments of X (Lemma 6.3.8). By Lemma 5.8.9, the union $Y = Y_0 \cup Y_1 \subset X$ is the pushout over the intersection $Z = Y_0 \cap Y_1$. These are also initial segments: the structure map of Z mediates to $TY_0 \cap TY_1 = T(Y_0 \cap Y_1)$, and that of Y from the pushout. Then Y and

Z are well founded by Corollary 6.3.6. The restrictions of p_0 and p_1 to Z satisfy the recursion equation (Lemma 6.3.8), so agree by Proposition 6.3.9. Hence we may form the union $p : Y \to \Theta$ of the partial functions as a pushout mediator, and $p = \mathsf{parse}_Y \, ; Tp \, ; \mathsf{ev}$ because the right hand side also mediates from this pushout. □

THEOREM 6·3·13: Let X be a well founded coalgebra and Θ a partial algebra. Then there is a greatest attempt $p : X \rightharpoonup \Theta$, and this satisfies the recursion equation, $p = \mathsf{parse}_X \, ; Tp \, ; \mathsf{ev}$. If Θ is total then so is p.

PROOF: Attempts $X \rightharpoonup \Theta$ form an ipo, which we have just shown to be directed as well, so let $p : X \rightharpoonup \Theta$ be the greatest one (by the Adjoint Function Theorem 3.6.9), with support Y. As $\mathsf{parse}_X \, ; Tp \, ; \mathsf{ev}$ is also an attempt (Lemma 6.3.11), by maximality we have $p = \mathsf{parse}_X \, ; Tp \, ; \mathsf{ev}$.

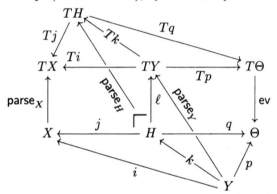

Suppose now that Θ is total. Form the pullback H, with $\mathsf{parse}_H = \ell \, ; Tk$ and $q = \ell \, ; Tp \, ; \mathsf{ev}$. Then $k \, ; q = k \, ; \ell \, ; Tp \, ; \mathsf{ev} = p$ and $\mathsf{parse}_H \, ; Tq \, ; \mathsf{ev} = q$. So $q : H \to \Theta$ is an attempt, but $p : Y \to \Theta$ is the greatest such, so $H = Y$, but then $Y = X$ by well-foundedness. □

Exercise 3.40 gives the lattice analogue.

In applications such as Remark 6.2.10 and Proposition 6.2.6, we need the homomorphism p to be total on the hypothesis that ev_Θ is defined "whenever it needs to be". It is still necessary to use methods of proof by induction, but now the issue is that the *steps* be well defined (partial correctness) rather than that the *whole process* terminate.

Finally, we characterise wff-systems and the term algebra itself.

COROLLARY 6·3·14: Let $F \underset{\mathsf{ev}}{\overset{\mathsf{parse}}{\rightleftarrows}} TF$ be mutually inverse. Then

(a) (F, ev) is initial iff (F, parse) is well founded.

(b) Any equationally free algebra with no proper subalgebra is initial.

(c) If the initial algebra exists, the extensional well founded coalgebras are its initial segments.

PROOF: [a] Well-foundedness is needed by Remark 6.3.10; conversely by Definition 6.3.7 the greatest attempt is the unique homomorphism $p : F \to X$. [b] Minimality is the strict induction scheme, which is equivalent to the lax one (*cf.* Theorem 3.8.11(a)). [c] Regarding parse_X as a partial *algebra* structure, the theorem gives homomorphisms $F \rightharpoonup X$ and $X \to F$, the latter being total since ev_F is. Then the composite $X \to F \rightharpoonup X$ is an attempt (since ev_F ; $\mathsf{parse}_Y = \mathsf{id}_{TF}$), which is the identity by induction, so $X \hookrightarrow F$. $\qquad\qquad\square$

6.4   TAIL RECURSION AND LOOP PROGRAMS

The previous section gave a categorical generalisation of the recursive paradigm (Definition 2.5.1), where the way in which a program calls itself was controlled by a certain functor. Now we shall consider the special case in which the functor is of the form $TX = X + N$, where $N \subset X$ is the set of base cases or exit states.

DEFINITION 6·4·1: In a ***tail-recursive*** program,

(a) there is at most one sub-argument to each recursive call, and

(b) its sub-result is immediately passed out as that of the whole program (in particular the original argument is not needed).

In the functional style this is
$$p(x) = \begin{cases} p(\mathfrak{s}(x)) & \text{if the condition } c(x) \text{ holds (the step)} \\ \mathfrak{z}(x) & \text{otherwise (the base case).} \end{cases}$$
Subordinate clauses, in which the reader (or listener) must remember what's going on in the surrounding text, illustrate recursion in grammar; a tail-recursive version would put them at the end of the sentence.

Operationally, the sub-argument $\mathfrak{s}(x)$ may be *assigned* to the variable x, and the continuation $\mathfrak{z}$ from the (most deeply) nested call is just that from the main program, so the functional idiom of tail recursion translates directly into the imperative **while** program
$$\mathfrak{p} \equiv \textbf{while } c(x) \textbf{ do } x := \mathfrak{s}(x) \textbf{ od } ; \mathfrak{z}.$$

DEFINITION 6·4·2:   The ***simple imperative language*** extends the conditional declarative language of Definition 5.3.7 by the ⟨program⟩

while ⟨Boolean expression⟩ **do** ⟨program⟩ **od**

As in the conditional, the test has type **2** (*not* 2), and the same variables must be in scope at the end of the body of the loop as at the beginning.

EXAMPLE 6·4·3: The following program computes the highest common factor (y) of two integers $a, b \in \mathbb{Z}$ by **Euclid's algorithm**.

put $x = a$;	
put $y = b$;	$\{\psi[x, y]\}$
while $x \neq 0$ **do**	$\{x \neq 0 \wedge \psi[x, y]\}$
put $z = y \bmod x$;	$\{|z| < |x| \wedge \exists n. \ z + nx = y\}$
	$\{\psi[z, x]\}$
$y := x$;	$\{\psi[z, y]\}$
$x := z$;	$\{\psi[x, y]\}$
discard z	
od;	$\{x = 0 \wedge \psi[x, y]\}$
discard a, b, x	

where

$$\psi[u, v] \stackrel{\text{def}}{\Longleftrightarrow} \forall m. \ m \mid a \wedge m \mid b \Leftrightarrow m \mid u \wedge m \mid v$$

is the (strict) loop invariant. From the last comment we can deduce immediately that $y \mid a \wedge y \mid b$, so y is the highest common factor. □

REMARK 6·4·4: Notice that both parts of Definition 6.4.1 are needed for the translation into a **while** program. The definition of the factorial function, read as a program, is not tail-recursive because the argument must be saved for the multiplication with the sub-result. Unary recursion (satisfying just the first condition) can, however, be translated into tail recursion by using an **accumulator** (Exercise 6.26).

It is unlikely that there is such a *uniform* way of reducing the *arity* of a recursion, though it can sometimes be done. For example, the Fibonacci function is defined by

$$\mathsf{fib}(n + 2) = \mathsf{fib}(n) + \mathsf{fib}(n + 1) \qquad \mathsf{fib}(0) = \mathsf{fib}(1) = 1,$$

so is a *binary* recursion, *i.e.* it makes *two* nested calls at each level, and the functor is $T = (-)^2 + \mathbb{N}$. This program therefore has exponential complexity, but there is also a tail-recursive program that calculates the same function in linear or even logarithmic time (Exercise 6.27).

Semantics. Fixed points in order structures (Sections 3.3–3.5) give the semantic treatment of recursion which is perhaps the best known, but they take no notice of the recursive paradigm, let alone the special case of tail recursion. After setting up notation like that in Section 5.3, we shall use various categorical techniques to axiomatise the naïve understanding of iteration. There are several threads in our treatment, describing the data and results in terms of partial functions, relational algebra, recursion for a functor as in the previous section, and the categorical structure of Section 5.8, so you should feel free to skip some parts.

REMARK 6·4·5: We shall consider the program

$$\mathfrak{p} = \mathfrak{w} \, ; \mathfrak{z} \quad \text{where } \mathfrak{w} = \textbf{while } c \textbf{ do } \mathfrak{s} \textbf{ od},$$

subject to Assumption 5.3.8, that the condition c always terminates without side-effect, but now the assignment **put** $y = c$ must be inserted at the end of the body of the loop, as well as before the loop.

$$S \underset{\mathfrak{m}}{\overset{\mathfrak{s}}{\underset{\longrightarrow}{\rightleftarrows}}} X \overset{\mathfrak{n}}{\longleftarrow} N$$

Let $X = Y + N$ be the partition induced by the loop condition and $S \subset Y \subset X$ the support of the partial function which represents the body $\mathfrak{s}$ of the loop, so $S \cap N = \varnothing$.

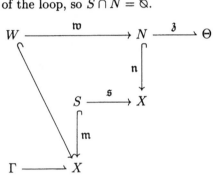

Together with the code $\Gamma \rightharpoonup X$ and $N \rightharpoonup \Theta$ before and after the loop, the whole program is illustrated by the staircase diagram, in which the S step may be repeated any number of times. The letters of course stand for successor and zero, but notice that they count *backwards*: comparing tail recursion with Remark 2.7.7, $\mathfrak{s}$ is the *predecessor*. (This is the reason for using the letter $\mathfrak{z}$ for continuations.) Because of the exit condition, N is the target of the partial function W which represents the whole loop.

REMARK 6·4·6: Rearranging the data, the span

$$X = Y + N \overset{\mathfrak{m} + \text{id}}{\longleftarrow} S + N \overset{\mathfrak{s} + \text{id}}{\longrightarrow} X + N$$

defines a partial map $\textsf{parse} : X \rightharpoonup X + N$, which is a coalgebra for the functor $T = (-) + N$. The algebra is $\textsf{ev} \equiv [\text{id}, \mathfrak{z}] : \Theta + N \to \Theta$, or just the codiagonal $\nabla : N + N \to N$ with no continuation $\mathfrak{z} : N \to \Theta$ after the loop. The two conditions for tail recursion are respectively that the functor and algebra be of these forms; Exercise 6.26 deals with arbitrary algebras $\textsf{ev} \equiv [\mathfrak{a}, \mathfrak{z}] : \Theta \times X + N \to \Theta$. As the functor is always the same, we drop the use of T for it and use this letter for something else.

Recall from Lemma 6.3.8 that a partial attempt $X \rightharpoonup \Theta$ is a total attempt on a subcoalgebra, $X \leftarrow W \rightarrow \Theta$. The rectangle on the right below says that $\mathfrak{p} = $ **if** c **then** $\mathfrak{s}$; $\mathfrak{p}$ **else** $\mathfrak{z}$ **fi**.

$$
\begin{array}{ccccc}
X + N & \longleftarrow & W + N & \xrightarrow{\ \mathfrak{p} + \mathrm{id}\ } & \Theta + N \\
\big\uparrow{\scriptstyle \mathfrak{s} + \mathrm{id}} & & \big\uparrow & & \big\downarrow \\
S + N & \longleftarrow & S \,;W + N & & [\mathrm{id}, \mathfrak{z}] \\
\big\uparrow{\scriptstyle [\mathfrak{m}, \mathfrak{n}]} & & \big\downarrow{\scriptstyle \cong} & & \big\downarrow \\
X & \longrightarrow & W & \xrightarrow{\ \ \mathfrak{p}\ \ } & \Theta
\end{array}
$$

In particular the loop $\mathfrak{w}$ itself (with trivial continuation $\mathfrak{z} = \mathrm{id}$) satisfies

$$\mathfrak{w} = \textbf{if } c \textbf{ then } \mathfrak{s} \textbf{ ; } \mathfrak{w} \textbf{ else skip fi}$$

The semicolons arise from the relational algebra which we shall use.

EXAMPLE 6·4·7: For the Euclidean algorithm $N = \{0\} \times \mathbb{Z} \hookrightarrow X = \mathbb{Z} \times \mathbb{Z}$ and we define parse $: X \to TX = (\mathbb{Z} \times \mathbb{Z}) + \mathbb{Z}$ by

$$
\mathsf{parse}(x, y) = \begin{cases} \nu_0 \langle y \bmod x, x \rangle & \text{if } x \neq 0 \\ \nu_1(y) & \text{if } x = 0 \end{cases}
$$

in the diagram:

$$
\begin{array}{ccccc}
\mathbb{N} + \mathbb{Z} & \longleftarrow & \mathbb{Z} \times \mathbb{Z} + \mathbb{Z} & \xrightarrow{\ \mathfrak{w} + \mathrm{id}\ } & \mathbb{Z} + \mathbb{Z} \\
\big\uparrow & \geqslant & \big\uparrow{\scriptstyle \mathsf{parse}} & & \big\downarrow{\scriptstyle \nabla} \\
\mathbb{N} & \xleftarrow{\ |x|\ } & \mathbb{Z} \times \mathbb{Z} & \xrightarrow{\ \ \mathfrak{w}\ \ } & \mathbb{Z}
\end{array}
$$

The loop terminates since $|x| \in \mathbb{N}$, the **loop measure**, strictly decreases (*cf.* Remark 2.5.13, Proposition 2.6.2 and Corollary 6.3.6).   □

REMARK 6·4·8:   Barry Jay observed that any map $\mathfrak{f} : X \to \Theta$ with $\mathfrak{m} ; \mathfrak{f} = \mathfrak{s} ; \mathfrak{f}$ is *invariant* in the strict sense that its value is restored after each iteration, so it is the same when the loop terminates (if it does) as at the beginning. Such a map factors through the coequaliser Q, which Jay called the *universal loop invariant*.

$$
\begin{array}{ccc}
S + N & \underset{\scriptstyle [\mathfrak{s}, \mathfrak{n}]}{\overset{\scriptstyle [\mathfrak{m}, \mathfrak{n}]}{\rightrightarrows}} & X \xrightarrow{\hspace{2cm}} Q \\
& & \big\downarrow{\scriptstyle \mathfrak{f}} \ \ \ \ \ \\
& & \Theta
\end{array}
$$

The correctness of **while** loops is always shown by finding an appropriate invariant. Indeed every competent programmer writes "the state of the variables at the point of the loop test is ..." or similar. This is vital, as the commonest error is to be out by 1 in an array suffix.

However, the established usage of the term **loop invariant** is for a predicate (so $\Theta = 2$) such that, *if* it holds before execution of the body *then* it holds afterwards. In other words, ∫ may *become* valid when it had not been before, and the converse implication is not relevant. (The heredity operation, Definition 2.5.4, turns such a lax invariant into one in Jay's strict form, *cf.* Theorem 3.8.11(a) and Exercise 6.2.) The coequaliser must also be modified to account for the exit condition N.

In fact we shall show that, in the diagram

$$N \stackrel{\subset}{\longrightarrow} S \,;\, W + N \underset{\subset}{\overset{\subset}{\rightrightarrows}} W \xrightarrow{\;\mathfrak{w}\;} \!\!\!\!\rightarrow N,$$

both composites are id_N and $\mathfrak{w}$ is the coequaliser of the parallel pair. It follows that the recursion equation for any $\mathfrak{z} : N \to \Theta$ has a unique solution, since $\mathfrak{z}$ is the unique mediator from the coequaliser.

We shall use relational algebra to investigate this coequaliser in terms of the associated equivalence relation. The results do not apply to arbitrary categories, but rely on certain exactness properties of *Set*, and our aim, as in the previous chapter, is to find out what these are: we need stable coequalisers not just of congruences but of functional relations. Recall that general coequalisers are computed in several steps (Lemma 5.6.11), not all of which are needed this case, but we still need to consider stable *directed* unions in order to form the transitive closure.

The coequaliser is also peculiar to this situation in another respect, namely that it only works for unary recursion, *cf.* the special properties of unary algebra in Section 3.8. In an arbitrary coalgebra, an element is well founded iff *all* of its children (Remark 6.3.4) are well founded. If there is just one child, the parent is well founded iff the child is, so the well founded elements are exactly those that are related to childless (base) cases by the equivalence generated by the transition relation. It is like König's Lemma, except that there is *no* choice to be made.

Transitive closure. The partial function $\mathfrak{s}$ will now be treated as a binary relation $S : X \rightharpoonup X$ and the subset $N \subset X$ as a subrelation of the diagonal Δ (*cf.* the virtual objects in Proposition 5.8.7). These relations should be thought of as transition graphs on the set X of states; evaluation of the whole program W consists in following the relation S, or rather its transitive closure T, until we arrive inside the subset N, so $W = T \,;\, N$. (The letter T is no longer a functor.)

We take up the story of the transitive closure from Proposition 3.8.7, where it was defined by a *unary* induction scheme. Frege (1879) showed that the transitive closure of any functional relation S is trichotomous (classically). We shall show instead that it is *confluent* (Definition 1.2.5), so the equivalence closure is $K = T\,;T^{\mathrm{op}}$. Working backwards, K is the kernel of a coequaliser, and also gives enough information to allow us to investigate T and W. Exercise 3.60 instead considers the transitive closure as a list of steps, capturing the imperative idiom directly.

Let $R = S \cup \Delta$ be the reflexive closure of S, and $E = (S + N) \cap \Delta$ the equaliser of $(S + N) \rightrightarrows X$, so $N \subset E$. Note that $A + B$ denotes the union $A \cup B$ of two relations, but also signifies that they are disjoint, *i.e.* $A \cap B = \varnothing$ (*cf.* the use of $\bigvee$ in Notation 3.4.3).

LEMMA 6·4·9: Let $S : X \rightarrowtail X$ be a relation in a prelogos with countable stable unions, *or* in a logos, *or* in a division allegory (Lemma 5.8.8(c)).

(a) For any relation $V : X \rightarrowtail \Theta$, if $S\,;V \subset V$ then $T\,;V \subset V$.

(b) For any relation $A : X \rightarrowtail X$, if $A\,;S \subset S\,;A$ then $A\,;T \subset T\,;A$.

(c) $E\,;T = E$.

If $S : X \rightharpoonup X$ is functional then

(d) $R = S \cup \Delta$ is confluent (Definition 1.2.5), *i.e.* $R^{\mathrm{op}}\,;R \subset R\,;R^{\mathrm{op}}$;

(e) T is also confluent (*cf.* Lemma 1.2.4);

(f) $K = T\,;T^{\mathrm{op}}$.

PROOF: In a prelogos with stable countable unions,

$$T = \bigcup_{n=0}^{\infty} S^n = \bigcup R^n \qquad T\,;N = \coprod_{n=0}^{\infty}(S^n\,;N) = \bigcup R^n\,;N$$

from which the results follow by brute force.

They may also be proved in a finitary way using the universal properties in a logos. We show, in the notation of Lemma 5.8.8(c), that

$$\Theta = V\backslash V, \quad A\backslash(T\,;A) \text{ and } E\backslash E$$

satisfy the premises of the unary induction scheme in Definition 3.8.6, and so contain T. (In the present notation, we used T/T there to show that $T\,;T \subset T$.)

[a]

$$\frac{\text{co-unit: } (V/V)\,;V \subset V \qquad S\,;V \subset V}{\cfrac{S\,;(V/V)\,;V \subset V}{\cfrac{S\,;(V/V) \subset (V/V)}{T \subset V/V \iff T\,;V \subset V}}}$$

$$\frac{\text{unit}}{\Delta \subset V/V}$$

[b] Transitivity of $A\backslash(T\,;A)$ is similar; also $A\,;S \subset S\,;A \subset T\,;A$, so $S \subset A\backslash(T\,;A)$ and $T \subset A\backslash(T\,;A)$. Hence $A\,;T \subset T\,;A$.

[c] $E\,;S \subset E$, so by (a^{op}), $E\,;T \subset E = E\,;\Delta \subset E\,;T$.

[d] $S^{op};S \subset \Delta$, so $R^{op};R = \Delta \cup S \cup S^{op} \cup S^{op};S = \Delta \cup S \cup S^{op} \subset R;R^{op}$.

[e] As in (b), $A;R \subset R;A \Rightarrow A;T \subset T;A$; with R^{op} for A, by (d), $R^{op};T \subset T;R^{op}$. By the same principle for the opposite relations, now with T^{op} for A, we have $T^{op};T \subset T;T^{op}$.

[f] $T;T^{op};T;T^{op} \subset T;T;T^{op};T^{op} \subset T;T^{op}$. $\qquad \square$

It is curious that we need *two* forms (a) and (b) of the inductive principle.

The recursion and induction schemes. Now we are equipped to justify Remarks 6.4.6 and 6.4.8 and see how the categorical induction and recursion schemes from Section 6.3 restrict to tail recursion.

THEOREM 6·4·10: The relation $W = K;N$ is functional; it is the least solution (*cf.* Section 3.3) of the recursion equation

$$\mathfrak{w} = \textbf{if } c \textbf{ then } \mathfrak{s};\mathfrak{w} \textbf{ else skip fi},$$

and satisfies $c = \textsf{no}$ on exit.

PROOF: Since $N = N^{op} = E;N = N;T$ by Lemmas 6.4.9(c) and (f),

$$W = K;N = T;T^{op};N = T;N.$$

Then by Lemma 6.4.9(e),

$$W^{op};W = N;T^{op};T;N \subset N;T;T^{op};N = N \subset \Delta,$$

and by Proposition 3.8.7(b) and Lemma 5.8.8(b),

$$W = T;N = (\Delta \cup S;T);N = N \cup S;T;N = N + S;W,$$

which is the recursion equation. This condition is $\textsf{no}$ on exit from $\mathfrak{w}$ because $W = K;N = K;N;N = W;N$.

Finally let $V : X \rightharpoonup X$ be another solution, so $N \subset V \supset S;V$. Then $T;V \subset V$ by Lemma 6.4.9(a), and $W = T;N \subset T;V \subset V$. $\qquad \square$

The discussion of the transitive closure above completes the investigation of the stability of the steps in the construction of general coequalisers in Lemma 5.6.11. The fact that $K;N$ is functional can be restated in terms of coequalisers, and is of interest in itself. It relies on stability of unions: see Example 5.8.2(d) for a counterexample in $\mathcal{G}p$. Without loss of generality (by adding a loop counter), the body always changes the state, *i.e.* $N = E$.

PROPOSITION 6·4·11: Let $U \rightrightarrows X$ be a functional endo-relation in a logos or a pretopos with stable countable unions. Let E and Q be the equaliser and coequaliser. Then the common composite shown is mono:

$$E \hookrightarrow U \underset{\mathfrak{s}}{\overset{\mathfrak{m}}{\rightrightarrows}} X \twoheadrightarrow Q$$

This says that there is at most one fixed point in each component of the transition graph of a functional relation.

PROOF: Two elements $x, y : \Gamma \rightrightarrows E$ of the equaliser become identified in the coequaliser $Q \cong X/K$ iff $\langle x, y \rangle \in K$, since Q is effective. Then

$$\langle x, y \rangle \in E \; ; K \; ; E^{\mathrm{op}} = E \; ; T \; ; T^{\mathrm{op}} \; ; E^{\mathrm{op}} = E \; ; E^{\mathrm{op}} = E \subset \Delta$$

by Lemmas 6.4.9(c) and (f) with $S = U$, so $x = y$ as claimed.   □

REMARK 6·4·12: The construction of W is shown in the diagram below. As a rule, pullbacks of parallel pairs need not have the same vertex, but they do here because, as the composites $S + N \rightrightarrows X \rightarrow Q$ are equal, we may form the pullbacks rooted at Q (Lemma 5.1.2, Exercise 5.53).

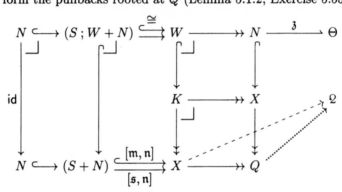

With the same assumptions as before (and in particular in *Set*), $W \twoheadrightarrow N$ is the coequaliser of the pair shown. For suppose that $V : X \rightharpoonup \Theta$ with support W has equal composites; then $N \; ; V \subset V$ and $S \; ; V \subset V$, so

$$W \; ; N \; ; V = T \; ; N \; ; V \subset T \; ; V \subset V : X \rightharpoonup \Theta,$$

but these are functional relations with the same support, so are equal. So $N \; ; V$ is a coequaliser mediator; it is unique as $W \twoheadrightarrow N$ is epi.   □

EXAMPLE 6·4·13: (Peter Freyd [Fre72]) Consider the following special case, with $x : \mathbb{N}$, which always terminates:

while $x > 0$ **do** $x := x - 1$ **od.**

The partition is $N = \{0\}$ and $Y = \{n \mid n \geqslant 1\}$; this and the interpretation of the program are then the coproduct and coequaliser diagrams

$$1 \hookrightarrow \overset{0}{\longrightarrow} \mathbb{N} \overset{\mathrm{succ}}{\longleftarrow} \mathbb{N} \qquad \mathbb{N} \underset{\mathrm{id}}{\overset{\mathrm{succ}}{\rightrightarrows}} \mathbb{N} \longrightarrow\!\!\!\twoheadrightarrow 1,$$

where $\mathbb{N} \rightarrow \mathbb{N} + 1$ is the usual coalgebra structure, which is well founded by Peano induction. (Beware that we have re-indexed this diagram: the above discussion actually gives $\mathrm{id}, \mathrm{pred} : \{n \mid n \geqslant 1\} \rightrightarrows \mathbb{N}$.)

PROPOSITION 6·4·14: In the sense of Definition 6.3.2, W is the largest well founded subcoalgebra of X.

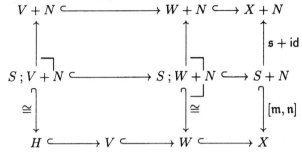

PROOF: Treat $V \subset W$ as a subrelation, so the pullbacks are relational composites as shown. The diagrammatic induction scheme says that

$$\frac{N \subset V \subset W \qquad S\,;V \subset V}{V = W}$$

which we have already deduced from Lemma 6.4.9(a). The way that W was constructed, as a pullback, makes it the largest well founded subcoalgebra by Exercise 6.17. □

COROLLARY 6·4·15: W is the *unique* well founded solution to the fixed point equation in Theorem 6.4.10. □

Partial correctness. Recall the difference between total and partial correctness, expressed in terms of modal logic in Example 3.8.3.

REMARK 6·4·16: For a partial program $u : X \rightharpoonup Y$, the triple

$$\{\phi\}u\{\vartheta\}$$

is redefined to mean that, if u terminates *and* ϕ holds beforehand, then ϑ holds afterwards.

Let $A = \{\langle x, x' \rangle \mid x = x' \wedge \phi[x]\}$ and $B = \{\langle y, y' \rangle \mid y = y' \wedge \vartheta[y]\}$ be subdiagonal relations as before.

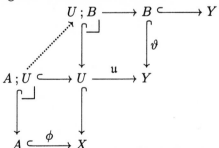

Then $A\,;U \subset A \subset X$ consists of those states for which ϕ holds and u terminates; the effect of u is the composite $A\,;U \hookrightarrow U \to Y$. The partial Floyd triple says that this factors through $B \subset Y$. Hence there is a pullback mediator $A ; U \to U ; B$ as shown, or, in terms of relations, $A ; U \subset U ; B : X \rightharpoonup Y$.

THEOREM 6·4·17: $\mathfrak{w}$ satisfies the partial correctness Floyd rule

$$\{\vartheta\}$$

$$\textbf{while } c \textbf{ do}$$

$$\frac{\{c \wedge \vartheta\} \ \mathfrak{s} \ \{\vartheta\}}{\{\vartheta\} \textbf{ while } c \textbf{ do } \mathfrak{s} \textbf{ od } \{\neg c \wedge \vartheta\}} \qquad \boxed{\begin{array}{l} \{c \wedge \vartheta\} \\ \mathfrak{s} \\ \{\vartheta\} \end{array}}$$

$$\{\neg c \wedge \vartheta\}$$

for any predicate ϑ on X.

PROOF: With $A = \{\langle x, x' \rangle \mid x = x' \wedge \vartheta[x]\}$, the premise says that

$$(Y \cap A) ; S = A ; Y ; S = A ; S \subset S ; A$$

(since S is only defined on Y anyway), so by Lemma 6.4.9(b),

$$A ; W = A ; T ; N \subset T ; A ; N = T ; N ; A = W ; (A \cap N)$$

(since $A, Y, N \subset \Delta$), which is the conclusion. $\square$

By Corollary 6.4.15, the behaviour of a **while** program is captured by proving partial correctness like this, together with termination, *i.e.* well-foundedness of the coalgebra W, which is done by finding a loop measure.

Discussion. Although we have used properties of $\mathcal{S}et$, in particular the transitive closure, to *prove* correctness of the interpretation, it can be *stated* using pullbacks, coproducts and coequalisers alone. This means that any (**exact**) functor which preserves finite limits and finite colimits also preserves the interpretation.

REMARK 6·4·18: Correctness is *reflected* if the functor $F : \mathcal{C} \hookrightarrow \mathcal{S}$ is also full and faithful. For suppose that both categories have the limits and colimits needed to draw the diagrams in this section (so F makes these agree) and that in $\mathcal{S}$ we have shown that $FW \rightrightarrows FX$ is a functional relation satisfying

(a) $FW = FN + FS ; FW = FW ; FN$,

(b) $\forall V. \ FN \subset V \supset FS ; V \Rightarrow FW \subset V$, and

(c) $\forall A. \ A ; FS \subset FS ; A \Rightarrow A ; FW \subset FW ; A$.

Then F preserves composition and intersection of relations, and reflects their containment, so these properties restrict to $\mathcal{C}$. In particular, Freyd's characterisation of $\mathbb{N}$ (Example 6.4.13) says that exact functors between toposes, such as inverse images of geometric morphisms, preserve $\mathbb{N}$. $\square$

When can a category without infinite unions be embedded in one with them, thereby generalising the interpretation?

Stability of the coequaliser is clearly necessary, but unfortunately seems not to be sufficient. But if the coequaliser of $S + N \rightrightarrows X$ and its kernel K exist then the cocone $(\Delta \cup S \cup S^{\mathrm{op}})^n \hookrightarrow K$ indexed by $n \in \mathbb{N}$ is *always* colimiting, even though we have not asked for a general infinitary union *operation*. Then a pretopos $\mathcal{C}$ can be embedded in a topos of sheaves on $\mathcal{C}$ preserving (finite limits, coproducts, quotients of equivalence relations and) the coequaliser iff *this union is stable under pullbacks in* $\mathcal{C}$.

Using this condition there is a simpler way to extend the proof: we only need unions of relations, not the virtual objects in this sheaf topos.

THEOREM 6·4·19: The language is correctly interpreted in any pretopos such that each functional relation has an equivalence closure which is *stably* the union of powers. Any functor which preserves finite limits and colimits also preserves the interpretation.

PROOF: Define an **ideal relation** $A : X \rightharpoonup Y$ to be a *set* of relations $B \subset X \times Y$ which is closed downwards and under whatever *stable* unions already exist (*cf.* Theorem 3.9.9). For example

$$(A) \overset{\mathrm{def}}{=\!=} \{B \mid B \subset A\} \quad \{B \mid B\,;E \subset E\} \quad \{B \mid V\,; B \subset V\}.$$

Ideal relations can be shown to form a **division allegory** (Proposition 5.8.7). The second and third examples are transitive, as is

$$\{B \mid (A\,;B) \subset \mathcal{T}\,;A\}$$

for any transitive ideal relation $\mathcal{T}$, by the same argument as for Lemma 6.4.9. The crucial point is that

$$(K) = \mathcal{T}\,;\mathcal{T}^{\mathrm{op}},$$

where we return from the ideal transitive closure $\mathcal{T}$ to the real equivalence closure K. For this we need that the latter be a *stable* union. $\quad\square$

6.5 UNIFICATION

Now we shall apply the parsing and well-foundedness properties of term algebras to the unification algorithm of Remark 1.7.7. Our approach is a new one, which begins from an idea in universal algebra, and carries this through directly to a very efficient implementation.

The *most general unifier* has a universal property, which may be viewed in two ways: it is the *equaliser* in the category $\mathsf{Cn}_{\mathcal{L}}^{\times}$ of contexts and substitutions (Example 5.1.3(b)), and also the *coequaliser* of *free* algebras.

The connection between these points of view is the fact that the clone $\mathsf{Cn}_{\mathcal{L}}^{\times}([x_1 : X_1, \ldots, x_n : X_n], [y : Y])$ is the free algebra on n generators (Theorem 4.6.7).

The most general unifier is not the coequaliser amongst all algebras.

EXAMPLE 6·5·1: Consider a unary operation s and two variables x and y. Then as a unification problem the equation $s(x) = s(y)$ implies $x = y$, but the coequaliser as an algebra is essentially "$\mathbb{N}$ with two zeros".   □

Recall that the *products* in $\mathsf{Cn}_{\mathcal{L}}^{\times}$ must be preserved in any interpretation of the theory $\mathcal{L}$ in a category. The equalisers are not preserved because terms may have equal value without being provably so in the theory.

We shall discuss coequalisers amongst arbitrary algebras in Section 7.4, and characterise the subcategory of free algebras in Proposition 7.5.3(a). Section 8.2 constructs a version of the classifying category which does have some equalisers (and so whose interpretations preserve them).

For our work in this section we need first to restore the generators which we discarded in Definition 6.1.1. The recursion scheme for them is the universal property. The generators do not behave as extra constants as we said in Definition 6.1.1, because the homomorphism which solves the problem may send them to arbitrary terms in the target algebra Θ.

NOTATION 6·5·2: Write $\eta_G : G \to FG$ for the inclusion of the generators G into the free algebra which they generate.

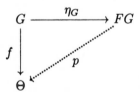

Then any function $f : G \to \Theta$ to another algebra extends uniquely to a homomorphism $p : FG \to \Theta$ with $\eta_G \,; p = f$.   □

Unification. For any homomorphism $f : A \to \Theta$ of algebras, the kernel pair $\{(a, b) \mid f(a) = f(b)\}$ is an equivalence relation on A and is closed under its operation-symbols (Proposition 5.6.4). In a unification problem the kernel pair is also closed under parse:

LEMMA 6·5·3: Let $f : A \to \Theta$ be a homomorphism from any algebra to an equationally free one (so even if A is free, f may send generators to expressions).

If $f(r_A(\underset{\sim}{u})) = f(s_A(\underset{\sim}{v}))$ then $r = s$ and $\forall j{:}\mathsf{ar}[r].\ f(u_j) = f(v_j)$.

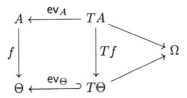

PROOF: Since f is a homomorphism,

$$\mathsf{ev}_\Theta(r, \underrightarrow{f(u)}) = f(r_A(\underset{\sim}{u})) = f(s_A(\underset{\sim}{v})) = \mathsf{ev}_\Theta(s, \underrightarrow{f(v)}),$$

but ev_Θ is mono, so $r = s$ and $\forall j{:}\mathsf{ar}[r].\ f(u_j) = f(v_j)$. □

So two expressions are unifiable iff their outermost operation-symbols agree and the corresponding pairs of sub-expressions are each unifiable, but all by *the same* assignment to the generators. For unification to be possible, we must therefore be able to distinguish the generators and operation-symbols from one another, *cf.* Example 6.2.5.

REMARK 6·5·4: Besides parse, the congruence is also closed under the operations $r \in \Omega$ and the axioms for an equivalence relation.

(a) Transitivity *via* generators gives rise to parsable equations:

$$r(\underset{\sim}{u}) = x \quad x = \cdots = y \quad y = s(\underset{\sim}{v})$$

where $x = \cdots = y$ is a zig-zag of equations *amongst generators*. (This affects the way in which we must prove termination: consider

$$x = u \quad y_1 = v_1 \quad y_2 = v_2 \quad y_3 = v_3 \quad \cdots$$
$$r(x, r(x, r(x, \ldots))) = r(y_1, r(y_2, r(y_3, \ldots)))$$

in which u reappears in a new equation arbitrarily deeply.)

It turns out that the other cases can be avoided:

(b) transitivity of equations between terms — but from $r(\underset{\sim}{u}) = r(\underset{\sim}{v})$ and $r(\underset{\sim}{v}) = r(\underset{\sim}{w})$ we can deduce $\forall j.\ u_j = v_j = w_j$ anyway;

(c) symmetry and reflexivity may be taken as read;

(d) applying operation-symbols: $\{u_j = v_j \mid j : \mathsf{ar}[r]\} \vdash r(\underset{\sim}{u}) = r(\underset{\sim}{v})$;

(e) substitution of $x = u$ into $v = w$, where $x \in \mathsf{FV}(v, w)$. By a similar argument this may also be postponed, unless v or w is the generator x, when transitivity *via* generators applies. □

LEMMA 6·5·5: Let $p : FG \to \Theta$ be a homomorphism from a free algebra to any algebra. Then (*cf.* Lemma 1.3.8)

$$px = pu \wedge py = pv \iff px = pu \wedge py = p([u/x]^* v),$$

PROOF: By structural induction on v, considering x, y, z and $r(\underset{\sim}{w})$. □

REMARK 6·5·6: The previous Remark makes useful optimisations, as *they avoid the need to compare or copy terms*, providing a parallel, *in sitû* algorithm which unifies terms almost as fast as they can be read. The terms may be represented in Polish notation (Example 6.2.7), but it is more usual to code each instance of an operation as a record consisting of the operation-symbol together with pointers to records for each immediate sub-expression (Remark 1.1.1). The pending equations are then held as pairs of pointers to these records. The program needs to store the equivalence relation E on generators, together with the equations R between generators and (pointers to) sub-terms.

The given and subsequently generated equations $u = v$ do not need to be stored, since each may be dealt with as a sub-process:

(a) if it is of the form $r(u') = s(v')$ with $r \not\equiv s$ then we have a clash;

(b) the equation $r(u') = r(v')$ forks into sub-processes for each $u'_j = v'_j$;

(c) $x = u$ or $u = x$ is added to R, unless there is already an equation $x \overset{E}{=} x' \overset{R}{=} v$, in which case $u = v$ is handled as in the previous cases;

(d) $x = y$ is added to E and their equivalence classes amalgamated; if two R-equations become linked by $u \overset{R}{=} x' \overset{E}{=} x = y \overset{E}{=} y' \overset{R}{=} v$, then one of these is deleted and the unification step applied to $u = v$. $\square$

LEMMA 6·5·7:   The unification algorithm (applied to any finite set of equations between terms for a finitary free theory) terminates.

PROOF: Consider the total number of generators and operation-symbols in the outstanding equations, including the terms assigned by R to the generators. The generators listed in E and R are *not* counted. The operation-symbol r in case (b), or the generators x and y in cases (c) and (d), are deducted from this count at each step, so the program terminates by Proposition 2.6.2. (In fact only the outermost operation-symbol of u in case (c) is considered more than once, *cf.* Remark 6.5.4(a).) $\square$

THEOREM 6·5·8:   The algorithm provides the most general unifier.

PROOF: The unification step continues to be applicable as long as there remains any equation relating a term to a term, possibly *via* generators. So when the iteration terminates, the outstanding equations consist of an equivalence relation $E : G \leftrightarrow G$ together with a system of equations $R : G \leftrightarrow FG$ such that, for $x, y \in G$ and $u, v \in FG$,

$$u \overset{R}{=} x \overset{E}{=} y \overset{R}{=} v \;\Rightarrow\; u \equiv v.$$

Then any unifier of $E \rightrightarrows G \to FG$ factors through G/E and so $F(G/E)$.

By construction, the equations R link equivalence classes $[x] \in G/E$ of generators to expressions, so without loss of generality E is $(=_G)$.

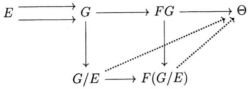

We want to partition the set $G = D + I$ into dependent and independent generators, the former being the support of R.

Define $x < y$ on G if $x \not\equiv y \land x \in \mathsf{FV}(v) \land y \overset{R}{=} v$. If there is any unifier $p : FG \to \Theta$ then $x < y \Rightarrow p(x) \prec p(y)$, by structural induction on v, where $\prec$ is the transitive closure of the sub-expression relation in Θ. Since $\prec$ is well founded (Theorems 6.3.13 and 3.8.11(b)), $<$ must be too (Proposition 2.6.2). In particular the relation $<$ must have no cycles, which, by substitution, would lead to the situation $x = u(x)$.

Since G is finite and $<$ is decidable, to verify well-foundedness it is enough to make the **occurs check**, for such loops (Exercise 2.36).

By $<$-recursion, dependent variables may now be eliminated from R using Lemma 6.5.5. Then we have one equation for each generator, which expresses it as a term in FI, the independent variables standing for themselves, so the left-hand triangle commutes:

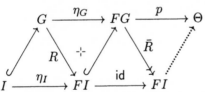

With a little diagram-chasing we see that FI is the required coequaliser.

Each step of the execution has replaced one unification problem by another which is equivalent to it, the last being an assignment of terms to independent variables, which has a coequaliser. Since the latter is unique up to unique isomorphism, the algorithm is confluent, despite being slightly non-deterministic. $\qquad\square$

6.6 FINITENESS

It probably comes as something of a surprise that we have got this far through a book on foundations of mathematics — especially a book which stresses constructivity — without discussing finiteness. Emphasis on finite enumeration was a reaction to the excesses of classical set theory, and clearly any set *which is given by picking its elements* must be finite in order to be represented in a machine.

The reason why we have played down finiteness is that mathematical and computational objects are normally handled *according to their structure*, with no need for explicit enumeration. Besides, enumerative processes are very slow: one does not have to go to a very high order of functions over a two-element base type to exhaust the memory of a computer, or indeed of the Universe. We can nevertheless handle higher order functions very easily with the λ-calculus, and prove certain properties of *all* numbers by induction without examining *each* one.

Arguments in combinatorics usually do make essential use of finite sets. Unfortunately it is for these that the type theory and logic tend to be the most difficult to mechanise (although the combinatorial aspects can be handled directly), whereas Section 1.6 showed that variables, the connectives and the quantifiers can be translated very fluently between the vernacular and formal language. When a proof does need finite sets, considerable bureaucratic manipulation of lists must be added to the level of detail that a (human) mathematician would regard as sound and precise. Even this section elides many such details, since we have already said as much about lists as we intend to say.

When we have spoken of finite sets in this book, notably in the definition of algebraic theories and the crucial result about quotients of algebras, we have usually had in mind particular numbers, such as 3. It is said that "hard cases make bad law": each instance of finiteness is familiar, but it is very difficult to identify the abstract notion.

Adam of Balsham (1132) observed that the difference between finite and infinite sets is that the latter admit proper self-inclusions, such as $n \mapsto 2n$ (*cf.* Exercises 1.1 and 2.47). But, without Choice, only special infinite sets have this property, so the rest are inappropriately called "finite".

As with the exactness properties of *Set* (Section 5.8), progress was made only after similar concepts had been identified in algebra and topology. In these disciplines, many familiar objects, despite having infinitely many elements, are "bounded" in some sense which may be used to investigate their properties. For example $[0, 1] \subset \mathbb{R}$ is **compact** (if it is contained in any directed union of open sets then just one of them suffices), whence any continuous function on it is bounded and attains its bounds.

EXAMPLE 6·6·1: A commutative ring R is said to be **Noetherian** if every directed set of ideals is eventually constant. Assuming excluded middle and Dependent Choice, every element $a \in R$ of such a ring can be expressed as a product of irreducibles (*cf.* Example 2.1.3(b) for *prime factorisation*).

PROOF: If a is reducible then $a = bc$, where the ideals they generate satisfy $(a) \subsetneq (b) \subsetneq R$. Repeating this process, we get either a product as required or, by Dependent Choice (Definition 1.8.9), a properly increasing sequence of principal ideals, which is forbidden by hypothesis. $\square$

Emmy Noether was the pioneer of conceptual mathematics (*begriffliche Mathematik*), but this notion is no mere piece of abstract nonsense. Since it is inherited by quotients, polynomial rings and even formal power series rings, it replaced the "jungle of formulae" (as she described her own doctoral thesis) which had characterised nineteenth century algebra (*cf.* Exercise 6.49).

Three definitions by counting. Of course a set is finite iff there is a listing of its elements, and we know from Section 2.7 what a list is. Classically (more precisely, when equality is decidable), any repetition in such a list may be eliminated, but anyone who has tried to maintain a moderately large database will be aware that this is not a trivial problem in practice. Depending on the amount of control we have over repetition, there are three useful definitions.

DEFINITION 6·6·2: A set X is said to be

(a) ***finitely enumerated*** if $n \in \mathbb{N}$ and $p : \mathbf{n} \cong X$ are given, where $\mathbf{n}$ is the set $\{m : \mathbb{N} \mid m < n\}$;

(b) ***finitely presented*** if $n, k \in \mathbb{N}$ are given together with a coequaliser diagram $\mathbf{k} \rightrightarrows \mathbf{n} \twoheadrightarrow X$; the elements of $\mathbf{k}$ name laws (and also generate a finite equivalence relation);

(c) ***finitely generated*** if $n \in \mathbb{N}$ and a surjection $p : \mathbf{n} \twoheadrightarrow X$ are given;

and finitely enumer*able*, present*able*, or gener*able* if these structures exist but are not specified. The first two are equivalent for sets (Exercise 6.42); the last is usually called ***Kuratowski-finiteness***, but Cesare Burali-Forti was actually the first to formulate it.

EXAMPLES 6·6·3:

(a) Theorem 5.6.9 requires that the positions in the arity of a general finitary operation must be distinguishable, because we want to put semantic values in them separately.

(b) If an object is repeated in a family of which we are forming the coproduct, then we get multiple copies of it, which (in the case of *Set*, Section 5.5) are disjoint.

(c) However, an element of a semilattice may be repeated arbitrarily in the formation of a join (Proposition 3.2.11).

(d) Any closure condition may be repeated without changing its force.

(e) An **alias** in a database specifies that two entries denote the same thing. If we are sure that all such coincidences have been recognised then the database is a finite *presentation* of its subject matter, otherwise it is a finite *generation.*

(f) Any subsingleton $\{\star \mid \phi\} \subset \mathbf{1}$ is finite in all three ("-able") senses iff ϕ is decidable (Exercise 6.41). More generally, a subset of a finite set is finite iff it is complemented.

It does not seem to be an appropriate property to require of a notion of finiteness that *any* subset of a finite set be finite (but *cf.* Remark 9.6.5). As with the connectives of logic and type theory, we shall take natural deduction for finite sets (rather than the prejudices of classical logic) as our guide, and this matches the definitions we have given.

REMARK 6·6·4: There are similar (and in fact prior) notions in algebra. The first corresponds to $F\mathbf{n}$, the free algebra on n generators (such as a finite-dimensional vector space), which replaces the set $\mathbf{n}$ in the others. This analogy is a conceptual one, and is only tenuously related to the size of the carrier sets. The algebraic definitions are also distinguishable even in classical logic.

(a) The free monoid on one generator is infinite.

(b) Addition *modulo m* is finitely presented as a monoid (generated by one element subject to one law), but it is not free; of course the carrier set is finite.

(c) The wreath product $\mathbb{Z} \wr \mathbb{Z}$ is the group generated by symbols a and b subject to the relations that $b^{-n}ab^{n}$ commutes with $b^{-m}ab^{m}$ for all $m, n \in \mathbb{Z}$. It has no finite presentation.

Any algebra with finitely enumerated carrier for a theory with finitely many operation-symbols is finitely presented as an algebra: for laws we just list the instances of the operations (*cf.* the ambiguity in the usage of this word mentioned in Definition 1.2.2). However, there is no other implication. We shall compare and contrast these definitions further in Proposition 6.6.13ff.

The *ability* to count. We have usually avoided discussing the external interpretation of logic, but it is important to understand the meaning of the existential quantifier ("-able") in these definitions.

REMARK 6·6·5: Recall from Remark 5.8.5 that $\Gamma \vdash \exists y{:}Y.\ \phi[y]$ means that $!: \{y : Y \mid \phi[y]\} \to \Gamma$ is epi. Thus an object X is finitely enumer*able* or gener*able* iff

$$! : \Delta \overset{\text{def}}{=\!=} \{\langle n, p\rangle : \mathbb{N} \times \mathcal{P}(\mathbb{N} \times X) \mid \phi[n, p]\} \twoheadrightarrow \Gamma,$$

where $\phi[n, p]$ says that p is a listing of X of length n, *i.e.*

$$[\forall m{:}\mathbb{N}.\ \forall x, y{:}X.\ \langle m, x\rangle, \langle m, y\rangle \in p \Rightarrow x = y]$$
$$\wedge\ [\forall m.\ m < n \Leftrightarrow (\exists x.\ \langle m, x\rangle \in p)]$$
$$\wedge\ [\forall x.\ \exists(!)m.\ \langle m, x\rangle \in p],$$

in which we read $\exists!$ in the case of enumeration and $\exists$ for generation.

The left projection $\Delta \to \mathbb{N}$, an element $n \in \mathbb{N}^\Delta$, may be regarded as a "variable number", *viz.* the generic length of *a* way of listing the set. For $i < n$, *i.e.* $i \in \mathbb{N}^\Delta$ with $\forall \delta.\ i(\delta) < n(\delta)$, we have $x_i(\delta) = p(i(n, p)) \in X$ where $\delta = (n, p) \in \Delta$. In other words, we may write

$$X = \{x_0, \ldots, x_i, \ldots, x_{n-1}\}$$

(either as a set or as a list) so long as the suffixes are understood to be these variable numbers.

Now consider the introduction and elimination rules for the quantifier.

$$\frac{\begin{array}{c} \vdots \\ q : \mathbf{m} \twoheadrightarrow X \end{array}}{\exists n, p.\ p : \mathbf{n} \twoheadrightarrow X}\ \exists \mathcal{I} \qquad \begin{array}{|l} \exists n, p.\ p : \mathbf{n} \twoheadrightarrow X \\ \hline \exists n, x_{(-)} \quad X = \{x_0, \ldots, x_{n-1}\} \\ \vdots \\ \vartheta \end{array}$$

The premise of the introduction rule is just a listing. In the elimination rule, ϑ must not depend in any way on the listing, *i.e.* n, i, or x_i. On the other hand, $(\exists \mathcal{E})$-boxes are always open-ended below (Remark 1.6.5).

COROLLARY 6·6·6: If a set is finite in either sense then we may assume that it has a listing (with or without repetition).   □

This means that we do not need to mention the set Δ when working in the internal logic, nor are we making a *choice* of listing (Remark 1.6.7). But this box, like all others, must be properly nested. So if the set X depends on parameters, the $(\exists \mathcal{E})$-box which encloses the choice of listing must be closed before any surrounding box which quantifies the parameters.

However, there are problems with counting even when we seem to be able to identify a specific number of distinct things.

EXAMPLES 6·6·7: The set X of square roots of a generic number

(a) in $\mathbb{C}$ is *Kuratowski*-finite, since zero has just one;

(b) on the unit circle $S^1 \subset \mathbb{C}$ satisfies $\exists p.\ p : \mathbf{2} \cong X$, where the existential quantifier (*cf.* Example 2.4.8) must be interpreted carefully as above: there is a cover $\Delta \twoheadrightarrow S^1$ by open subsets each equipped with such an isomorphism, but there is no global one. The collection $\{p \mid \mathbf{2} \cong X\}$, as a sheaf on S^1, is called a **torsor**; as an algebraic structure, it retains the Mal'cev operation $(uv^{-1}w)$ of the automorphism group of X, but not the identity element (Example 9.2.12(d)).

To sum up, we are not at liberty to regard the listing of a *dependent* finite set $X[y]$, or even the number of elements needed to cover a Kuratowski-finite set, as a *function* of y.

Counting is unique. *But* if a set can be finitely enumerated then the *length* of such an enumeration is an invariant. Although this must have been known as a fact before any other human intellectual achievement, the realisation that there is (a) a powerful theorem, and (b) something to prove, takes a degree of mathematical sophistication.

Bo Peep's Theorem (Exercise 1.1) amounts to saying that any injection $\mathbf{n} \hookrightarrow \mathbf{n}$ is a bijection. The **pigeonhole principle** is a stronger result: if $f : \mathbf{n} \to \mathbf{m}$ with $m < n$ then $f(i) = f(j)$ for some $i \neq j \in \mathbf{n}$.

These properties fail, of course, for big sets like $\mathbb{N}$, but finiteness is also about granularity. This is the reason for only allowing *decidable* subsets of finite sets to be called finite. For example, if I give you some wool, and some more wool, and a third quantity, and then take some back, and some more, and some more again, you may or may not still be holding some of what I gave you.

Besides its usefulness in agriculture, the power of counting is illustrated by the **Sylow theorems** in group theory.

FACT 6·6·8:   Let G be a group with $p^k m$ elements, where p is a prime not dividing m. Then there is a subgroup $H \subset G$ of order p^k. Moreover the number n of such subgroups (which are conjugate to each other in G) divides m, and $n \equiv 1 \bmod p$. If $n = 1$ then H is normal.   □

By counting Sylow subgroups and elements of particular orders we may easily identify the simple group of order 168 and show that there is no simple group of small non-prime odd order (105 is the first tricky case).

Finite subsets. Now let us look more closely at Kuratowski finiteness, which is apparently an exception to the rule that the poset version of a concept is simpler and older than the categorical analogue. This is the one which is usually relevant for subsets, since the other requires equality of elements to be decidable before we may form the subset $\{a, b\}$.

Although the induction scheme is the main feature to which we want to draw attention, we shall continue the discussion based on lists instead of developing the theory from the closure conditions as we did for the transitive closure in Definition 3.8.6ff (*cf.* Exercises 3.60 and 6.43).

There are rules for adding elements and subsets, corresponding to cons and append for lists. $\mathcal{P}_{\mathrm{fg}}(X)$ is the free semilattice on a set X, just as List(X) is the free monoid (Corollary 2.7.11).

DEFINITION 6·6·9: The *finite powerset* is the image

$$\mathsf{List}(X) \twoheadrightarrow \mathcal{P}_{\mathrm{fg}}(X) \hookrightarrow \mathcal{P}(X)$$

of the "set of elements" function, Example 2.7.6(f). This is a recursive cover (Definition 6.2.2).

LEMMA 6·6·10:

(a) A subset $U \subset X$ belongs to $\mathcal{P}_{\mathrm{fg}}(X)$ iff U is finitely generable; this property is, in particular, independent of the ambient set X.

(b) $\mathcal{P}_{\mathrm{fg}}(X)$ is the smallest set of subsets closed under

$$\varnothing \rhd \varnothing \quad \text{and} \quad \{U\} \rhd U \cup \{x\}$$

(beware that these are *elements* of $\mathcal{P}_{\mathrm{fg}}(X)$ on the right of $\rhd$), so by Definition 3.7.8 it gives rise to the *Kuratowski induction scheme* in its unary form,

$$\frac{\vartheta[\varnothing] \qquad \forall x.\,\forall U.\,\vartheta[U] \Rightarrow \vartheta[U \cup \{x\}]}{\forall U \in \mathcal{P}_{\mathrm{fg}}(X).\ \vartheta[U]} \ \text{K-induction}$$

for predicates ϑ on $\mathcal{P}_{\mathrm{fg}}(X)$, *cf.* Peano induction for $\mathbb{N}$, cons-induction for lists and the unary definition of transitive closure (Definitions 2.7.1, 2.7.3 and 3.8.6). Exercise 6.44 gives the binary induction scheme, corresponding to addition and the append operation.

(c) The image of a finitely generable set is also finitely generable.

(d) The coproduct or union of two finitely generable sets is also finitely generable.

(e) $\mathcal{P}_{\mathrm{fg}}(X)$ is a semilattice, and hence directed (Definition 3.4.1); seen as a diagram in $\mathcal{P}(X)$, the join is X.

PROOF: Many of the parts are immediate corollaries of one another.

[a] "$U \in \mathcal{P}_{\mathrm{fg}}(X)$" and "$U$ is finitely generable" are both existentially quantified statements. The latter is a surjection $p : \mathbf{n} \twoheadrightarrow U$. In the former we have a list $\ell : \mathsf{List}(X)$, with a function $p : \mathbf{n} \to X$ where

$n = \mathsf{length}(\ell)$ and $p(i) = \mathsf{read}(\ell, i)$ (Exercise 2.48). We can show by list induction on ℓ that $p_!(\mathbf{n}) = U$.

[b] The second part is now an example of Proposition 6.2.4.

[c] The third follows from the definition of finite generability.

[d] [ ] lists $\varnothing$; if t lists U then $\mathsf{cons}(h, t)$ lists $\{h\} \cup U$; if ℓ_1 and ℓ_2 list U_1 and U_2 then $\ell_1 \,;\, \ell_2$ lists both $U_1 + U_2$ and $U_1 \cup U_2$.

[e] X is the join (within $\mathcal{P}(X)$) of its singletons, $X \subset \mathcal{P}_{\mathsf{fg}}(X)$.   □

PROPOSITION 6·6·11: $\mathcal{P}_{\mathsf{fg}}(X)$ is the *free* semilattice on X.

PROOF: Let $f : X \to \Theta$ be a function to another semilattice, in which (by Proposition 3.2.11) the operations may be taken to be $\bot$ and $\vee$. By Kuratowski induction, $(\Theta, \leqslant)$ has joins of all finitely generable sets, but if $U \subset X$ is finitely generable then so is its image $p(U) = f_!(U) \subset \Theta$, so it has a join. Hence there is a semilattice homomorphism $\mathcal{P}_{\mathsf{fg}}(X) \to \Theta$ extending f; it is unique because the equaliser of any two such is a subalgebra, but there is no proper such. See also Exercise 6.45(a).   □

The empty set. Corresponding to $\mathsf{List}(X) \cong \{\star\} + X \times \mathsf{List}(X)$, an element of the free semilattice can be *parsed* as empty or inhabited.

COROLLARY 6·6·12: $\mathcal{P}_{\mathsf{fg}}(X)$ is the disjoint union of $\{\varnothing\}$ and $\mathcal{P}_{\mathsf{fg}}^{+}(X)$, the set of inhabited finitely generable subsets; the latter is the free algebra for binary join alone. In other words, it is decidable whether a finitely generable subset is empty or has an element, *cf.* Example 6.6.3(f).

PROOF: $\{\bot, \top\}$ is a semilattice; the unique homomorphism taking all singletons to $\top$ maps everything else there except $\varnothing$.   □

Finiteness and Scott-continuity. Theorem 3.9.4 established a link between the (Kuratowski-)finite arities of Horn theories and preservation of directed joins. We should therefore look for a categorical analogue involving finite presentability.[2] To recap,

PROPOSITION 6·6·13:   The following are equivalent for a set X:

(a) X is finitely generable in the listing sense of Definition 6.6.2(c);

(b) it is finitely generable in the sense of Definition 3.4.11,
     i.e. if $X = \bigcup_{i \in I} U_i$ then $\exists i.\, U_i = X$;

(c) the functor $(-)^X$ preserves directed unions.

Moreover every set is the directed union of its finitely generated subsets.

---

[2]The analogy doesn't work as well as one might like. In a complete Heyting lattice, $X \Rightarrow (-)$ preserves directed unions *whenever* X is complemented, however large it may be (say in the case of a powerset lattice). It seems that we must regard any such X as a subsingleton.

PROOF: [b] is a special case of (c), and the last part is Lemma 6.6.10(e), whence [b⇒a]. [a⇒c] Any function $f : \{x_1, \ldots, x_n\} \to \bigcup_i V_i$ factors through some V_i. □

It can also be shown that a set is finitely *presentable* iff $(-)^X$ preserves *filtered colimits* (Example 7.3.2(j)), and conversely that every set can be expressed as a filtered colimit of finitely presentable sets. This suggests a generalisation, since, for $\mathcal{S}et$, the exponential $(-)^X$ is the same as the hom-set $\mathcal{S}et(X, -)$.

DEFINITION 6·6·14: Let X be an object of a category $\mathcal{S}$ with finite limits and pullback-stable filtered colimits.

(a) If the functor $\mathcal{H}^X \equiv \mathcal{S}(X, -) : \mathcal{S} \to \mathcal{S}et$ preserves filtered colimits then we say that X is **externally finitely presentable**.

(b) If $\mathcal{S}$ is cartesian closed and the functor $(-)^X : \mathcal{S} \to \mathcal{S}$ preserves filtered colimits then X is **internally finitely presentable**.

(c) If every object of $\mathcal{S}$ is a filtered colimit of finitely presentable objects then $\mathcal{S}$ is **locally internally** or **externally finitely presentable**.

The algebraic notions of finiteness (Remark 6.6.4) can be shown to be equivalent to the *external* ones for $\mathcal{S} = \mathcal{M}od(\mathcal{L})$, where $\mathcal{L}$ is a finitary algebraic theory. $\mathcal{M}od(\mathcal{L})$ is locally externally finitely presentable; Peter Gabriel and Friedrich Ulmer [GU71] showed that every locally externally finitely presentable category which *also* has all small colimits is of this form, for some *essentially* algebraic theory in the sense of Remark 5.2.9. The term LFP is traditionally applied only to such categories, in contrast to algebraic dcpos, Definition 3.4.11.

EXAMPLES 6·6·15: To make the distinction clearer, let $\mathcal{S} = \mathcal{S}et^{\mathbb{N}}$ be the category of presheaves on $\mathbb{N}$. Then $X \in \mathrm{ob}\,\mathcal{S}$ is finitely presentable

(a) *internally* iff each of the sets $X(n)$ is finite, although they may grow without bound as a function of n;

(b) *externally* iff $\sum_n X(n)$ is finite, so $\exists m. \forall n > m.\, X(n) = \varnothing$.

The terminal object $\mathbf{1}_{\mathcal{S}} = \lambda n. \{\star\}$ is internally but not externally finite. Nor need finitely enumerable presheaves (finite sets *dependent* on $n \in \mathbb{N}$) be expressible externally as coproducts of copies of the *constant* set $\mathbf{1}$.

This, and the fact that emptiness or habitation of a Kuratowski-finite set is decidable, illustrate that singletons are rather big subsets. In this case how is it that *every* set is the directed union of its finitely generable subsets? If it only consists of partial singletons, the nodes of the diagram (other than $\varnothing$) are themselves partial.

Finiteness, therefore, is not a prerequisite to constructive mathematics, nor is it "obvious" what it means. On the other hand, it is amenable to logical and categorical study.

6.7   THE ORDINALS

The ordinals admit a peculiar kind of arithmetic, into which it is often possible to make a crude translation of syntax. This provides a simple but powerful way of finding loop measures and so justifying termination or induction principles.

It was already known before Georg Cantor that a real-valued function could have a Fourier representation even if it had finitely many points of discontinuity. He saw that this could be extended to functions which were discontinuous on a set X_0 such as $\{0\} \cup \{\frac{1}{n} \mid n \in \mathbb{N}\}$ with an accumulation point $X_1 = \{0\}$. Repeating the argument, X_1 could have a set X_2 of accumulation points and so on. Then in 1870 he realised that the set $\bigcap X_n$ could also be non-empty, and that a "transfinite" sequence of sets X_∞, $X_{\infty+1}$, ... could be defined. We might picture ∞, now called ω, as a diminishing sequence of matchsticks:

$$1 = \big| \qquad 2 = \big|\big| \qquad 3 = \big|\big|\big| \qquad \omega = \big|\big|\big|\big|{\scriptstyle\text{III}\cdots} \qquad \omega + 2 = \big|\big|\big|{\scriptstyle\text{III}\cdots}\big|\big|$$

$$\omega 2 = \big|\big|\big|{\scriptstyle\text{III}\cdots}\big|\big|\big|{\scriptstyle\text{III}\cdots} \qquad \omega 3 = \big|\big|\big|{\scriptstyle\text{III}\cdots}\big|\big|\big|{\scriptstyle\text{III}\cdots}\big|\big|\big|{\scriptstyle\text{III}\cdots}$$

$$\omega^2 + \omega 2 + 3 = \big|\big|\big|{\scriptstyle\text{III}\cdots}\big|\big|{\scriptstyle\text{III}\cdots}\big|{\scriptstyle\text{III}\cdots}\,{\scriptstyle\text{III}\cdots}\,{\scriptstyle\cdots}\big|\big|\big|{\scriptstyle\text{III}\cdots}\big|\big|\big|{\scriptstyle\text{III}\cdots}\big|\big|\big|$$

Although Cantor had developed ordinal arithmetic in 1880, only in 1899 did he formulate a correct definition [Dau79], that an ordinal is an ordered set in which every non-empty subset has a least member: "there's a first time for everything". In other words, a trichotomous (Definition 3.1.3) well founded relation (Proposition 2.5.6).

The easiest way to tell these countable infinities apart is to work backwards, *i.e.* find out what *descending* sub-sequences there are. Every such sub-sequence must eventually stop: that's well-foundedness (Proposition 2.5.9). In the longer ones, there is more opportunity to dawdle down, but from time to time we must leap from the heights of a limit ordinal, landing somewhere that is infinitely far below.

The relation on an ordinal is easily seen to be transitive and extensional (Example 6.3.3), so it is useful to identify each ordinal with its *set* of predecessors, *cf.* $5 = \{0, 1, 2, 3, 4\}$ and Remark 6.3.5.

REMARK 6·7·1: The operations on ordinals are as follows:

(a) Zero (the empty string of matchsticks) is an ordinal.

(b) If α is an ordinal then so is $\mathsf{succ}\,\alpha$, its **successor**, also written α^+. We embed $\alpha \subset \mathsf{succ}\,\alpha$ as an initial segment, and use α to name the extra matchstick on the right. Under the identification of each $\beta \in \alpha$ with $\{\gamma \in \alpha \mid \gamma \prec \beta\}$, the elements β and α of $\mathsf{succ}\,\alpha$ are *subsets* of α, so $\mathsf{succ}\,\alpha$ may alternatively be defined to consist of the initial segments of α. These definitions are equivalent classically, but not intuitionistically, and we shall be ambiguous for the moment as to which is intended.

(c) If α_i for $i \in I$ are ordinals then so is $\bigcup_{i \in I} \alpha_i$, this union being taken with respect to the inclusions of initial segments.

Successor and *directed* union preserve Cantor's definition, but comparing two ordinals (or directed unions) given separately relies on excluded middle: if $\alpha \cap \beta \neq \alpha$ then β is the least element of $\alpha \setminus (\alpha \cap \beta)$.

Between any two ordinals there is a *reflexive* relation $(\subset)$, whether one is an initial segment of the other. On the other hand, the identification between systems and individuals gives rise to an (irreflexive) well founded relation, written as $\in$. They are related by

$$\beta \subset \alpha \quad \Longleftrightarrow \quad \forall \gamma.\, \gamma \in \beta \Rightarrow \gamma \in \alpha$$
$$\beta \in \alpha \quad \Longleftrightarrow \quad \mathsf{succ}\,\beta \subset \alpha.$$

Classically, any ordinal is *either* zero, *or* a successor *or* a **limit** (union). These cases may be distinguished by trying to define the **predecessor**,

$$\mathsf{pred}(\alpha) \overset{\text{def}}{=\!=} \bigcup \alpha = \{\gamma \mid \exists \beta.\, \gamma \in \beta \in \alpha\} \subset \alpha.$$

Then $\mathsf{pred}(\mathsf{succ}(\alpha)) = \alpha$, and either $\mathsf{pred}(\alpha) \in \alpha$ if α is a successor ordinal, or $\mathsf{pred}(\alpha) = \alpha$ if it is a limit.

LEMMA 6·7·2: For any ordinal, $\alpha = \bigcup\{\mathsf{succ}\,\beta \mid \beta \in \alpha\}$. $\qquad\square$

Transfinite recursion. Since ordinals are a special case of well founded relations, for any $\mathsf{ev}_\Theta : \mathcal{P}(\Theta) \to \Theta$ the equation

$$p(\alpha) = \mathsf{ev}_\Theta(\{p(\beta) \mid \beta \in \alpha\})$$

has a unique solution by the General Recursion Theorem 6.3.13.

REMARK 6·7·3: The classical **transfinite recursion theorem** is based on a three-way case analysis. Instead of ev_Θ, it takes an element $z \in \Theta$, and functions $s_\alpha : \Theta \to \Theta$ and $u_\lambda : \Theta^\lambda \to \Theta$. (Since the arity of u depends on the ordinal anyway, we are obliged to treat the case where the argument is used in the evaluation phase of the recursion.) Then there is a unique function p with

$$
\begin{aligned}
p(0) &= z \\
p(\operatorname{succ}\alpha) &= s_\alpha(p(\alpha)) \\
p(\lambda) &= u_\lambda\langle p(\alpha) : \alpha \in \lambda\rangle \qquad \text{if } \lambda \text{ is a limit.}
\end{aligned}
$$

PROOF: Since u may make use of its arguments in an entirely arbitrary way, the general recursion theorem *for free theories* must be used, with operations $r_\alpha : \Theta^{\operatorname{succ}\alpha} \rightharpoonup \Theta$ of *each* arity. For

$$
r_{(\operatorname{succ}\alpha)}\langle x_\beta : \beta \in \operatorname{succ}\alpha\rangle = s_\alpha(x_\alpha),
$$

the operation $r_{(\operatorname{succ}\alpha)}$ must be able to select x_α, its αth argument, by its position. Of course $r_0(\varnothing) = z$ and $r_\lambda = u_\lambda$. $\square$

This arbitrariness is not possible in the intuitionistic version, but recall that infinitary operations tend to be given by universal properties, join being typical. The three conditions are regarded as equations to be solved or "boundary conditions" rather than as a case analysis.

THEOREM 6·7·4: Any system of ordinals obeys the universal property (recursion scheme) for a free partial $(\bigvee, s)$-algebra with s monotone. That is, for any complete join-semilattice Θ equipped with a monotone unary operation $s_\Theta : \Theta \to \Theta$, there is a unique homomorphism p, *i.e.*

$$
\begin{aligned}
p(\operatorname{succ}\alpha) &= s_\Theta(p(\alpha)) \\
p\Big(\bigcup_i \alpha_i\Big) &= \bigvee_i p(\alpha_i).
\end{aligned}
$$

In particular $p(0) = \bot$ (the empty join), p is monotone in the sense that

$$
\beta \subset \alpha \implies p(\beta) \leqslant p(\alpha),
$$

and for any limit ordinal λ we still have $p(\lambda) = \bigvee\{p(\alpha) \mid \alpha \in \lambda\}$.

PROOF: Using Lemma 6.7.2, p must satisfy

$$
p(\alpha) = \bigvee\{s_\Theta(p(\beta)) \mid \beta \in \alpha\},
$$

which has a unique solution by Theorem 6.3.13. Then

$$
\begin{aligned}
p\Big(\bigcup_i \alpha_i\Big) &= \bigvee\{s_\Theta(p(\beta)) \mid \beta \in \bigcup_i \alpha_i\} \\
&= \bigvee_i \bigvee\{s_\Theta(p(\beta)) \mid \beta \in \alpha_i\} = \bigvee_i p(\alpha_i),
\end{aligned}
$$

whence in particular p is monotone. The successor equation,

$$
p(\operatorname{succ}\alpha) = \bigvee\{s_\Theta(p(\beta)) \mid \beta \subset \alpha\} = s_\Theta(p(\alpha)),
$$

is satisfied since, by monotonicity of p and s_Θ, $s_\Theta(p(\alpha))$ is the biggest term in the join. The property for limit ordinals still holds since any such ordinal is closed under successor. $\square$

The seed z need not be $\bot$, as in place of Θ we may consider the upper set $z \downarrow \Theta \equiv \{x \mid z \leqslant x\}$ (Example 3.1.6(f)). The unary operation s may also take an ordinal parameter, in which it must be monotone.

Remark 9.6.16 considers iteration of *functors*, with filtered colimits in place of directed joins, and using the axiom-scheme of replacement.

Rank. It is a striking feature that *systems* of ordinals share many of the properties of the *individuals*. (We have already seen this for wff-systems in Remark 6.2.10.) However, Cesare Burali-Forti (1897) showed that the whole system is a proper class: it cannot be an individual. Dimitry Mirimanoff (1917) showed that this and the Russell paradox don't arise in Zermelo set theory (Remark 2.2.9), by measuring the well-foundedness of each *definable* set. This was later incorporated into Zermelo–Fraenkel set theory as the axiom of foundation.

DEFINITION 6·7·5: For any well founded relation $(X, \prec)$,

$$p(t) = \bigcup \{s(p(u)) \mid u \prec t\}$$

defines the ordinal **rank** function on X, using the axiom of replacement. Remark 3.7.12 gave the finitary version of

PROPOSITION 6·7·6: Let $t \in F$ be a term for a free theory T. Then the rank of t with respect to the immediate sub-expression relation is the first stage at which t occurs, *i.e.* $t \in T^{p(t)}(\varnothing)$. $\square$

Arithmetic. In practice it is not feasible to calculate the rank of a term or proof exactly, *cf.* the number of iterations needed to complete a loop (Remark 2.5.13ff). Instead of the *least* upper bound $\bigcup$ in the previous definition, we assign to $r(u)$ *any* value which is larger than those of the sub-expressions u, as in Examples 2.6.3ff. This can be done using ordinal arithmetic. It was to define ordinal exponentiation that Cantor introduced transfinite recursion.

DEFINITION 6·7·7: The recursive definitions for $\mathbb{N}$ in Example 2.7.8,

$$
\begin{array}{llll}
\alpha + 0 & = & \alpha & \qquad \alpha + (\operatorname{succ} \beta) & = & \operatorname{succ}(\alpha + \beta) \\
\alpha * 0 & = & 0 & \qquad \alpha * (\operatorname{succ} \beta) & = & (\alpha * \beta) + \alpha \\
\alpha^0 & = & 1 & \qquad \alpha^{(\operatorname{succ} \beta)} & = & (\alpha^\beta) * \alpha & (1 \subset \alpha),
\end{array}
$$

are extended transfinitely by making each operation preserve inhabited (directed and binary) joins in the *second* argument β. (Using upper sets, they can instead be considered to preserve *all* joins.)

LEMMA 6·7·8: By induction on γ, one may prove successively that

$$
\begin{array}{lll}
\alpha + (\beta + \gamma) & = & (\alpha + \beta) + \gamma \\
\alpha * (\beta + \gamma) & = & (\alpha * \beta) + (\alpha * \gamma)
\end{array}
$$

$$\alpha * (\beta * \gamma) \quad = \quad (\alpha * \beta) * \gamma$$
$$\alpha^\beta * \alpha^\gamma \quad = \quad \alpha^{\beta+\gamma} \qquad\qquad (1 \subset \alpha)$$
$$(\alpha^\beta)^\gamma \quad = \quad \alpha^{\beta*\gamma} \qquad\qquad (1 \subset \alpha)$$

but

$$1 + \omega = \big|\,\big|\big|\big|\text{\tiny{III}}\text{\tiny{......}} = \omega \qquad 2 * \omega = \big|\big|\,\big|\big|\text{\tiny{III}}\text{\tiny{......}} = \omega$$

$\omega = (2+3) * \omega \neq 2 * \omega + 3 * \omega = \omega * 2$ and $\omega^3 * 2 = (\omega * 2)^3 \neq \omega^3 * 8$. $\square$

It is also important to appreciate that α^β is not an exponential in either a cartesian closed category or a Heyting lattice. One way of seeing this is that the latter have *left* adjoints (product and meet).

REMARK 6·7·9: As these operations preserve unions, they have *right* adjoints (Theorem 3.6.9), called **subtraction**, **division** and **logarithm**,

$$
\alpha \downarrow \mathcal{O}n \qquad\qquad \mathcal{O}n \qquad\qquad 1 \downarrow \mathcal{O}n
$$

$$
\alpha + (-) \ \dashv\ (-) - \alpha \qquad \alpha * (-) \ \dashv\ (-) \div \alpha \qquad \alpha^{(-)} \ \dashv\ \log_\alpha (-)
$$

$$
\mathcal{O}n \qquad\qquad \mathcal{O}n \qquad\qquad \mathcal{O}n
$$

where $\mathcal{O}n$ is the class of ordinals (reflexively) ordered by inclusion. The units and co-units of the adjunctions satisfy

$$\beta = (\alpha + \beta) - \alpha \qquad \alpha + (\gamma - \alpha) = \gamma \qquad\qquad (\alpha \subset \gamma)$$
$$\beta = (\alpha * \beta) \div \alpha \qquad \alpha * (\gamma \div \alpha) \subset \gamma \qquad\qquad (1 \subset \alpha)$$
$$\beta = \log_\alpha(\alpha^\beta) \qquad \alpha^{\log_\alpha \gamma} \subset \gamma \qquad\qquad (2 \subset \alpha,\ 1 \subset \gamma)$$

subject to the conditions on the right. Notice that we subtract matchsticks from the left, so $(-) - 1$ is not pred (for example at $\omega + 1$). The second equation, however, depends on excluded middle: $1 - \phi = \neg\phi$ and $\phi + (1 - \phi) = \phi \vee \neg\phi$ for $\phi \subset 1$. Since for $2 \subset \alpha$ and $1 \subset \gamma$ we also have

$$\gamma \div \alpha^{\log_\alpha \gamma} \prec \alpha \qquad \gamma - (\alpha^\beta * (\gamma \div \alpha^\beta)) \prec \alpha^\beta,$$

it is possible (classically) to write any ordinal γ uniquely as

$$\alpha^{\beta_n} * \delta_n + \alpha^{\beta_{n-1}} * \delta_{n-1} + \cdots + \alpha^{\beta_0} * \delta_0 + \alpha^{\beta_1} * \delta_1,$$

where $\beta_n \succ \beta_{n-1} \succ \cdots \succ \beta_2 \succ \beta_1$ and $\delta_i \prec \alpha$. The β_i may then also be decomposed in the same way. In the case $\alpha = \omega$, this hereditary expression is known as the **Cantor normal form**. $\square$

REMARK 6·7·10: Arithmetic on matchstick pictures may be represented by the lexicographic product (Proposition 2.6.8), and in fact any finite tower $\omega, \omega^\omega, \omega^{\omega^\omega}, \ldots$ can be encoded as an order on $\mathbb{N}$ (or a subset of $\mathbb{Q}$) using primitive recursion. Lemma 6.7.8 may be extended to a system of **reduction rules** (there are some messy extra cases for α^β), and such expressions do have a (Cantor) normal form. Moreover, by comparing successive exponents of ω and their coefficients, the trichotomy law holds, just as it does in $\mathbb{N}$.

The union (equivalently, the collection) of all ordinals which may be expressed using finite towers of ωs is called ε_0. By induction over ε_0 one can prove consistency of Peano arithmetic (with primitive recursion), so by Gödel's Incompleteness Theorem 9.6.2 *general* recursion must be needed to show that ε_0 is well founded. Proof theory, which is typically concerned with extensions of first order logic by specific induction schemes, uses the representability of ordinals to measure the strength of formal systems. (The final section looks at other ways of doing this.)

Classical applications. Refining Burali-Forti's argument, Friedrich Hartogs (1917) actually provided arbitrarily large ordinals.

LEMMA 6·7·11: For any set X, the set $\mathfrak{H}(X)$ of ordinal structures on subsets (or subquotients) of X forms an ordinal, and there is no injective function $\mathfrak{H}(X) \hookrightarrow X$. In particular $\omega_1 = \mathfrak{H}(\mathbb{N})$ is the first uncountable ordinal and others are defined by iteration (Example 9.6.15(b)). □

EXAMPLE 6·7·12: Let $s : \Theta \to \Theta$ be a monotone endofunction of an ipo (Definition 3.4.4). With $\alpha = \mathfrak{H}(\Theta)$, define $p : \mathrm{succ}\,\alpha \to \Theta$ by Theorem 6.7.4. Then $x = p(\alpha)$ satisfies $\neg\neg(x = sx)$. If there is a fixed point, $y \in \Theta$ such that $(y = sy)$, then $x \leqslant y$. □

Assuming excluded middle, x is the least fixed point, and conversely Exercise 6.54 derives ordinal-indexed joins from least fixed points.

Besides transfinite recursion, ordinals also provide a convenient way of performing constructions which require the axiom of choice.

PROPOSITION 6·7·13: The following are classically equivalent:

(a) the axiom of choice as Ernst Zermelo gave it: for any set Θ, there is a function $c : \mathcal{P}(\Theta) \setminus \{\varnothing\} \to \Theta$ with $\forall U.\, c(U) \in U$;

(b) the **well-ordering principle**: any set carries an ordinal structure;

(c) **Zorn's lemma**: any ipo (Definition 3.4.4) has a maximal element;

(d) the axiom of choice as given in Definition 1.8.8.

Recall that the axiom of choice implies excluded middle (Exercise 2.16).

PROOF: [c⇒d] and [d⇔a] were Exercises 3.16 and 2.15.

[a⇒b] Define $p(\alpha) = c\big(\Theta \setminus \{p(\beta) \mid \beta \in \alpha\}\big)$ by the General Recursion Theorem 6.3.13. Hartogs' Lemma bounds the ordinal needed.

[b⇒a] It is the typical use of the well-ordering principle to let $c(U)$ be the *first* element of U.

[a⇒c] Apply Example 6.7.12 to $s(x) = c(\{y \mid x \leqslant y \land x \neq y\})$; as there is no fixed point, the construction must cease to be possible at a certain stage, which is a maximal element.   □

Axiomatisation. It is quite usual for intuitionistic logic to cause a bifurcation of definitions and results. The Cantor Normal Form fails *as a theorem*, but proof theory has found ordinals in this form (as synthetic expressions) valuable as a measure of complexity. On the other hand, what should the analytic second order definition of ordinal be? One objective of such a theory would be to replace the set theory which still litters infinitary algebra, for example linking the "cardinal" rank in Definition 6.1.1 to the ordinal one in Proposition 6.7.6, and to preservation of κ-filtered colimits.

Cantor's definition is equivalent, classically, to requiring the relation $\prec$ to be transitive, extensional and well founded. The successor is just $\alpha \cup \{\alpha\}$ and the condition $\forall x.\, x \leqslant s_\Theta(x)$ suffices for the Transfinite Recursion Theorem 6.7.4. However, intuitionistically, we only have

$$\mathsf{succ}\,\beta \prec \mathsf{succ}\,\alpha \quad \Rightarrow \quad \mathsf{succ}\,\beta \subset \alpha \quad \Leftrightarrow \quad \beta \prec \alpha$$
$$\mathsf{succ}\,\beta \subset \mathsf{succ}\,\alpha \quad \Leftrightarrow \quad \beta \prec \mathsf{succ}\,\alpha \quad \Rightarrow \quad \beta \subset \alpha.$$

There are several intuitionistically inequivalent notions of ordinal, (some of) which André Joyal and Ieke Moerdijk [JM95] characterised as the free structures[3] in the sense of Theorem 6.7.4 for which s (a) satisfies no condition, (b) obeys $x \leqslant s(x)$, (c) is monotone or (d) preserves binary joins. These correspond to sets, transitive, plump and directed ordinals.

The problem lies in the confusion of $\in$ with $\subset$ in the classical theory (where $\beta \subset \alpha \Leftrightarrow \beta \in \alpha \vee \beta = \alpha$). In fact there are *three* relations: the partial order should be treated *a priori* separately even from the inclusion defined in terms of the (irreflexive) well founded relation.

REMARK 6·7·14:   The results of Section 6.3 may be developed for the functor sh : $\mathcal{P}os \to \mathcal{P}os$ (Definition 3.1.7) in the place of the covariant powerset $\mathcal{P} : \mathcal{S}et \to \mathcal{S}et$. A sh-coalgebra is well founded in the sense of Definition 6.3.2 iff $\prec$ is a well founded relation (Definition 2.5.3). The difference between sets and ordinals lies in the notion of "mono" used to define extensionality, *viz.* inclusions of lower sets. The structure map

$$(X, \leqslant) \xhookrightarrow{\alpha \,\mapsto\, \{x \mid x \prec \alpha\}} \{U \subset X \mid \forall u, x.\, x \leqslant u \in U \Rightarrow x \in U\}$$

of a well founded coalgebra is the inclusion of a lower set iff

$$\forall \alpha{:}X.\ \forall U{:}\mathsf{sh}(X).\ U \subset \{x \mid x \prec \alpha\} \Rightarrow \exists \beta{:}X.\ U = \{x \mid x \prec \beta\}.$$

---

[3]The main point of their work was the application of topos-theoretic methods to the control of infinite arities, so that the free structure is a legitimate object, and not a proper class, *cf.* Proposition 9.6.4. Like Hartogs' Lemma, this curtails the natural on-going system of ordinals, but not quite so crudely.

Like Cantor's definition, this property implies that $\prec$ is extensional and transitive in the usual sense, and $\leqslant$ coincides with the inclusion order. However, this definition of **plump ordinal** is much stronger: U is an initial segment of X, and so is also a plump ordinal. Therefore

$$U \subset \alpha \in X \implies U \in X,$$

which, classically, reduces to transitivity as $U \subset \alpha \iff U \in \alpha \lor U = \alpha$.

The successor operation must be adjusted accordingly: now $\operatorname{succ} \alpha$ does indeed consist of all initial segments of α, and all four implications above become reversible. But, as their name suggests, plump ordinals consist of rather more than the matchstick pictures: 2 is 2, 3 is $\operatorname{sh}(2)$ and the axiom-scheme of replacement seems to be needed to construct plump ω.

Unlike the classical ones, neither the transitive nor the plump ordinals need be directed (intuitionistically): this rather useful property has to be imposed separately (and hereditarily). In fact the same strategy, now in the category of binary semilattices (with $\lor$ but not necessarily $\bot$), provides directed plump ordinals. These are better in that only *directed* joins are needed in the Transfinite Recursion Theorem 6.7.4.

Written in set-theoretic language, these intuitionistic notions of ordinal are very complicated [Tay96a]. Category theory has come to the rescue, restoring the unity of the theory by abstract analysis of the old notions of extensionality and well-foundedness. The constructions may then be applied to other functors which do not have the ontological pretensions of the powerset, but the resulting ordinals have quite different properties.

For example, we use least fixed points in informatics, and expect to find a system which, unlike Cantor's, *naturally* stops at ω, this being its own successor, rather than being curtailed perfunctorily *à la* Hartogs [Tay91]. For this, TX in $\mathcal{D}cpo$ is the lattice of Scott-closed subsets of X, where we used lower subsets in $\mathcal{P}os$ above. Domains of this kind have been generalised by Roy Crole and Andrew Pitts [CP92] to "FIX objects" in more complicated recursive structures. The induction scheme must be restricted to a smaller class of predicates, such as the Scott-continuous ones in Theorem 3.7.13.

The proof of Proposition 6.7.13 above was not Zermelo's, which instead constructed the well founded relation on Θ using the substance of Θ itself, rather than the monolithic structure of all ordinals. This monolith has failed quite spectacularly to provide an intuitionistic proof of the fixed point theorem in Example 6.7.12: such a proof was found in the centenary of Burali-Forti, but depends on no such machinery. We have already given it in Exercise 3.46, and in fact it is very similar to Zermelo's 1908 argument (Exercise 6.53).

The remaining chapters build and study $\mathsf{Cn}_{\mathcal{L}}^{\square}$, often using induction, recursion and parsing, but only considering finitary connectives in the fragment $\square$.

EXERCISES VI

1. Construct $\mathsf{FV}(t)$, the set of *free variables* of a term in a free algebra for a free theory, as an example of the conjunctive (or, rather, *disjunctive*) interpretation.

2. Any subset $U \subset A$ has a characteristic function $\chi_U : A \to 2$ (Notation 2.8.2). Suppose that U is a T-subalgebra of A and endow 2 with the conjunctive interpretation. Is χ_U a homomorphism? What is the effect of applying the heredity operator (Definition 2.5.4) to it?

3. Define the *additive* interpretation in $\mathbb{N}$ of any finitary free theory in an analogous way to the conjunctive interpretation in 2. Describe the free algebra $F\mathbf{1}$ on one generator and the interpretation $[\![-]\!] : F\mathbf{1} \to \mathbb{N}$ in terms of trees. Generalise to an interpretation of infinitary free theories with the class of sets as its carrier.

4. Explain how the set of streams in an alphabet G is the final coalgebra for $T = (-) \times G$. Derive the final coalgebra for an infinitary free theory from Proposition 6.1.11.

5. Show that the functor List preserves pullbacks, and $\mathcal{P}_{\mathsf{fg}}$ preserves monos.

6. Show that the functor T of Definition 6.1.1 preserves directed unions iff the arities of all of the operations are finitely generable.

7. Prove Remark 1.6.12, that each of the rules of natural deduction for the predicate calculus preserves validity of formulae in any interpretation, in the sense defined by Example 6.1.9.

8. Define a function $\mathsf{List}(\mathbb{R} + \{+, -, \times, \div\}) \to \mathsf{List}(\mathbb{R}) + \{\star\}$ which evaluates arithmetic expressions written in Polish notation.

9. Formulate and prove the Substitution Lemmas 1.1.5 and 1.1.12.

10. Use the List functor, head, tail, their analogues for reversed lists, and equalisers to construct the set of morphisms of the free category on a graph (Theorem 6.2.8(a)). Use append and a pullback to define composition.

11. Show that if the set Ω of operation-symbols of a finitary free theory has decidable equality then so does the initial algebra.

12. Let $e : X \twoheadrightarrow Y$. Consider the free theory with $\Omega = X + \{r\}$, where $\mathsf{ar}[x] = \emptyset$ and $\mathsf{ar}[r] = Y$. Let F be its free algebra. Consider also

the algebra $A = Y + (Y+1)^Y$ with $x_A = \nu_0(e(x))$ and $r_A(g) = \nu_1(g\,;k)$ for $g \in A^Y$, where $k = (\text{id} + !) : A \to Y + 1$. Show that each $\nu_0(y)$ is in the smallest Ω-subalgebra of A, and hence so is $\nu_1(\bar{\nu}_0)$. But if this lies in the (*Set*-)image of the unique homomorphism $p : F \to A$ then e is split.

13. Give a direct categorical proof of Corollary 6.3.6.

14. Let $\text{ev} : T2 \to 2$ be the characteristic map of the subset $T\{\top\} \subset T2$. Use this to show that the recursion scheme implies the induction scheme for any functor T (*cf.* Remark 6.3.10).

15. Show that if $\text{parse}_X : X \to TX$ is well founded or extensional then so is $T\text{parse}_X : TX \to T^2X$ (*cf.* Exercise 3.40 and Lemma 6.3.11).

16. Show that any colimit of well founded coalgebras and homomorphisms is well founded. Discuss extensionality for filtered colimits and pushouts.

17. Let $W \subset X$ be a subcoalgebra and suppose that W (but not X) is well founded. Show that it is the *largest* well founded subcoalgebra iff the square is a pullback:

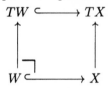

18. Show that for every well founded relation $(X, \prec)$ there is a wff-system (for a free theory) of which it is the immediate sub-expression relation. [Hint: take $\text{tgt} : (\prec) \to X$ as the arity display of the theory.]

19. Show how to apply the General Recursion Theorem 6.3.13 to Remark 6.2.10 and Proposition 6.2.6.

20. Give the symbolic proofs in set theory and algebra corresponding to Proposition 6.3.9, Lemma 6.3.12 and the last part of Theorem 6.3.13, *cf.* the comments after Lemma 6.3.11.

21. Follow the particular cases of the predecessor function and the Euclidean algorithm through Section 6.4.

22. A natural transformation $\sigma_{\Gamma,A} : \Gamma \times TA \to T(\Gamma \times A)$ is called a *strength* for the functor T if it satisfies

$$\sigma_{1,A} = \text{id}_{TA} \qquad \sigma_{\Gamma \times \Delta, A} = (\text{id}_\Delta \times \sigma_{\Gamma,A})\,;\sigma_{\Delta, \Gamma \times A}.$$

Find σ for the covariant powerset and for the functor which codes a free algebraic theory.

23. The notions of T-coalgebra and algebra *with parameters in* Γ are given by $\text{parse} : \Gamma \times X \to TX$ and $\text{ev} : \Gamma \times T\Theta \to \Theta$, in the presence of a strength σ. Explain how the diagram on the left describes the parametric

recursion scheme, in particular in the cases of the powerset and a free algebraic theory:

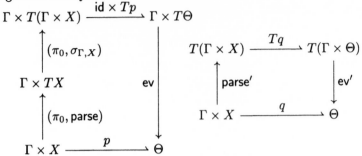

If the category is cartesian closed show how a solution of the non-parametric recursion on the right solves the given one, *cf.* Remark 6.1.6. Using Corollary 6.3.6, show that if $X \to TX$ (doesn't depend on Γ and) is well founded then so is $\Gamma \times X \to T(\Gamma \times X)$. Without using cartesian closure, formulate and prove directly the parametric version of Proposition 6.3.9, and hence of the General Recursion Theorem 6.3.13.

24. Recursion may be parametric in another sense, namely that the argument is used in the evaluation phase. This issue itself divides into two parts, depending on whether the argument is used in parsed or unparsed form. The second case is given by the diagram on the left:

$$
\begin{array}{ccc}
X \times TX \xrightarrow{\text{id} \times Tp} X \times T\Theta & \qquad & TX \xrightarrow{\quad Tq \quad} T(\Theta^X) \\
\uparrow^{(\text{id},\,\text{parse})} \qquad \downarrow^{\text{ev}} & & \uparrow^{\text{parse}'} \qquad \downarrow^{\text{ev}'} \\
X \xrightarrow{\quad p \quad} \Theta & & X \xrightarrow{\quad q \quad} \Theta^X
\end{array}
$$

By a similar method to the previous exercise, show that this is equivalent to the form on the right in a cartesian closed category, where T has a strength σ. Again, formulate and prove the parametric General Recursion Theorem directly without using cartesian closure.

25. If the original argument is used in parsed form, the recursion scheme simplifies to

$$
\begin{array}{ccc}
TX \xrightarrow{\;T\langle p,\,\text{id}\rangle\;} T(\Phi \times X) \\
\uparrow^{\text{parse}} \qquad \downarrow^{\text{ev}} \\
X \xrightarrow{\quad p \quad} \Phi
\end{array}
$$

Derive this from the previous case by putting $\Theta = \Phi \times X$. Conversely, reduce that one to this using $\text{ev}_X = \text{parse}_X^{-1}$, in the case where X is the initial T-algebra (an extensional well founded coalgebra is enough, ev_X being partial).

26. A *unary* recursion problem is one which makes at most one recursive call at each level, but where the argument may be used again after the return from the nested call. Explain how this is coded by a functor of the form $T = K + C \times (-)$ and show how to reduce this to the diagram on the left (allowing the argument as a parameter as in Exercise 6.25):

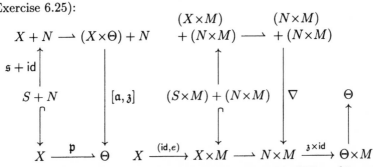

The map $\mathfrak{a} : X \times \Theta \to \Theta$ gives a homomorphism $\mathsf{List}(X) \to [\Theta \to \Theta]$ of monoids. Suppose that this factors through another monoid M, which is called an **accumulator**. Show how to reformulate the problem as a *tail* recursion with functor $T' = (-) + (N \times M)$, as on the right.
Explain how the case $M = \mathsf{List}(X)$ corresponds to a stack, and how $M = \mathbb{N}$ suffices if $\mathfrak{a}$ does not depend on X. The last map $\Theta \times M \to \Theta$ is itself defined by recursion over M: to what part of the execution of a recursive program does it correspond?

27. Express the factorial program in this form and apply the transformation, using the monoid $M = (\mathbb{N}, \times)$. Do the same for the Fibonacci function, which is calculated using the unary recursion on $\mathbb{N}^2$

$$p(0) = (0, 1) \qquad p(n + 1) = (v, u + v) \text{ where } (u, v) = p(n),$$

using the monoid of (2×2) matrices under multiplication. Hence show that $p(n) = (0, 1)\left(\begin{smallmatrix} 0 & 1 \\ 1 & 1 \end{smallmatrix}\right)^n$, so by calculating $\left(\begin{smallmatrix} 0 & 1 \\ 1 & 1 \end{smallmatrix}\right)^{2^k}$ the problem can be solved in logarithmic time. What features of a recursive program must a compiler recognise in order to make use of optimisations of this kind?

28. Express fold and append for lists (Definition 2.7.4) using Exercise 6.26, with $M = \mathsf{List}(X)$, and translate these back into functional programs. What advantage, if any, do the tail-recursive programs have over the original non-tail recursion using a stack?

29. **General recursion** extends primitive recursion by use of the **search** or **minimalisation operator** μ. For any partial recursive function $f : \Gamma \times \mathbb{N} \rightharpoonup 2$, $\mu n.\, f(x, n)$ is the n (if any) for which $f(x, n) = 1$

but $f(x,m) = 0$ for all $m < n$ (if any of these values is undefined then so is $\mu n. f(x,n)$). Express this as a **while** program, and hence as recursion over a certain coalgebra structure $(\Gamma \times \mathbb{N}) \rightharpoonup (\Gamma \times \mathbb{N}) + \mathbb{N}$.

30. Show that $(L, z, s) \cong (\mathsf{List}(X), [\,], \mathsf{cons})$ iff the diagrams

$$1 \xrightarrow{\ z\ } L \xleftarrow{\ s\ } X \times L \qquad X \times L \underset{\pi_1}{\overset{s}{\rightrightarrows}} L \twoheadrightarrow 1$$

are respectively a coproduct and a coequaliser, *cf.* Example 6.4.13.

31. Modify the Floyd rules of Remarks 4.3.5, 5.3.2 and 5.3.9 for partial correctness, as in Remark 6.4.16ff.

32. Express the unification algorithm in the style of Sections 6.3–6.4.

33. Investigate unification in equationally free algebras that are not necessarily well founded, so that the equation $x = r(x)$ can be solved.

34. Show that the unifier is the quotient by the parsing congruence, and that the congruence is the kernel of the unifier, *cf.* Section 5.6. Allow infinitely many equations, and infinite arities.

35. Investigate the non-determinacy in the unification algorithm.

36. Apply the unification algorithm as given to Remark 6.5.4(a).

37. Explain why, in Definition 6.6.2(b), any coequaliser diagram $\mathbf{k} \rightrightarrows \mathbf{n}$ generates a finite equivalence relation on $\mathbf{n}$.

38. Prove the **pigeonhole principle**, which is that, for any function $f : \mathbf{n+1} \to \mathbf{n}$, there are some $i < j \leqslant n$ with $f(i) = f(j)$. Deduce Bo Peep's theorem (Exercise 1.1).

39. Develop a Kuratowski-style unary induction scheme for finitely *enumerable* sets, and use it to prove the pigeonhole principle in an extensive category (Section 5.5) without assuming the existence of $\mathbb{N}$, or using numbers at all.

40. By considering the orders of elements of its Sylow subgroups, show that there is no simple group of order 105.

41. Show that $\{\star \mid \phi\}$ (Remark 2.2.7) is finitely enumerated iff ϕ is decidable. [Hint: *exactly* how many elements does it have?] Show that this also holds for finite generability.

42. Show that every finitely presented set is finitely enumerable.

43. Develop the results about Kuratowski finiteness and the finite powerset starting from the unary induction scheme in Lemma 6.6.10(b), in the style of Definition 3.8.6.

44. Show that the following binary Kuratowski induction scheme is equivalent to the unary one:

$$\frac{\vartheta[\varnothing] \qquad \forall x.\, \vartheta[\{x\}] \qquad \forall U, V.\, \vartheta[U] \wedge \vartheta[V] \Rightarrow \vartheta[U \cup V]}{\forall U.\, \vartheta[U]}$$

45. Let $R : X \rightharpoonup Y$ be any binary relation, and suppose that $U \subset X$, $V \subset Y$ are finitely generable sets which *match* in the sense that

$$\big(\forall u.\, \exists v.\, uRv\big) \quad \wedge \quad \big(\forall v.\, \exists u.\, uRv\big).$$

Show that U and V have matching listings, *i.e.* $U = \{u_0, \ldots, u_{n-1}\}$ and $V = \{v_0, \ldots, v_{n-1}\}$ with $\forall i.\, u_i R v_i$. Deduce the following results by taking R to be equality or an order relation.

(a) The kernel pair of the set of elements function $\mathsf{List}(X) \to \mathcal{P}(X)$ is generated by the laws for a semilattice; use Section 5.6 to deduce that $\mathcal{P}_{\mathrm{fg}}(X)$ is the free semilattice on X.

(b) Let $(X, \leqslant)$ be a preorder. Suppose $\mathcal{P}_{\mathrm{fg}}(X)$ also carries a preorder $\sqsubseteq$ (*not inclusion*) with respect to which $\cup : \mathcal{P}_{\mathrm{fg}}(X) \times \mathcal{P}_{\mathrm{fg}}(X) \to \mathcal{P}_{\mathrm{fg}}(X)$ and $\{-\} : X \to \mathcal{P}_{\mathrm{fg}}(X)$ are monotone. Then if $U, V \subset X$ are finitely generable and match with respect to $\leqslant$, *i.e.* $U \leqslant^{\natural} V$ in the notation of Exercise 3.55, then $U \sqsubseteq V$.

46. Let $\mathcal{L}$ be a single-sorted algebraic theory with only finitely many operation-symbols and laws, and let A be an $\mathcal{L}$-algebra. Show that if the carrier of A is a finite set then A is finitely presentable as an algebra.

47. Let $\mathcal{L}$ be a single-sorted algebraic theory which is finitary in the stronger sense that it has finitely many (*i.e.* a finitely enumerated set of) operation-symbols, each of finite arity, and finitely many laws (where, for example, the associative law counts as one law). Let A be an algebra whose carrier is Kuratowski-finite, so there is a surjection $\mathbf{n} \twoheadrightarrow A$ for some n. Show that there are a finitely enumerated *algebra* B and a surjective homomorphism $B \twoheadrightarrow A$. [Hint: show that $\mathbf{n}$ is an algebra for the operation-symbols of $\mathcal{L}$ but not necessarily the laws, and that the laws of $\mathcal{L}$ define a finite equivalence relation $\mathbf{k} \rightrightarrows \mathbf{n}$.]

48. Find a finitely generated monoid with no finite presentation.

49. (Only for those who have studied ring theory.) Using the fact that polynomial rings are Noetherian, show that every finitely generable commutative ring is finitely presentable.

50. Let G be a set with decidable equality equipped with a map $G \to G$ which has no cycles (*cf.* Exercise 2.47). Construct functions $\nu : G \times \mathcal{P}_{\mathrm{fg}}(G) \to G$ and $\sigma : \mathcal{P}_{\mathrm{fg}}(G) \times \mathcal{P}_{\mathrm{fg}}(G) \to \mathcal{P}_{\mathrm{fg}}(G)$ such that $\nu(x, U) \notin U$ and $\sigma(U, V) \cong U + V$, where the coproduct inclusions are

also to be found. The functions must not depend on the order in which the elements of the finite sets are given.

51. Using the results of Section 2.6, show how to code ω^ω, ω^{ω^ω}, etc. either as special orders on $\mathbb{N}$ or with the arithmetical order on $\mathbb{Q}$.

52. Describe the three product relations on $\alpha \times \alpha$ defined in Propositions 2.6.7–2.6.9, for each of $\alpha = \omega$, $\omega 2$ and ω^2. If the result is not extensional, describe the extensional quotient as a binary operation $\alpha \times \alpha \to \beta$. [Hint: one of them gives ordinal multiplication and another intersection.]

53. Given a Choice function $c : \mathcal{P}(\Theta) \smallsetminus \{\varnothing\} \to \Theta$ (Exercise 2.15), define $T : \mathcal{P}(\Theta) \smallsetminus \{\varnothing\} \to \mathcal{P}(\Theta)$ by $T(U) = U \cup \{c(\Theta \smallsetminus U)\}$ for $U \neq \Theta$ and $T(\Theta) = \Theta$. Let $\alpha \subset \mathcal{P}(\Theta)$ be the smallest set closed under T and union (*cf.* Proposition 3.7.11 and Exercise 3.46). Show that α is in bijection with Θ, and that every non-empty subset of α has a first ($\subset$-greatest) element (without using Hartogs' Lemma; this was Zermelo's proof).

54. Show, using excluded middle, that if a poset $\mathcal{X}$ has a least fixed point for every monotone endofunction $s : \mathcal{X} \to \mathcal{X}$ then $\mathcal{X}$ has a least element and all ordinal-indexed joins. [Hint: for $\bot$, take $s = \mathrm{id}$; for $y \in \mathcal{X}$ and a diagram $x_{(-)} : \alpha \to \mathcal{X}$ consider whether $\{\beta \mid x_\beta \leqslant y\}$ is α or an element of α, and in the second case use the successor.]

55. Show that the relation $x \prec y \equiv (\neg x \wedge y)$ on 2 is extensional and well founded, but that the reflexive closure of this relation is sparser than $\Rightarrow$ unless excluded middle holds. Show that this structure has no non-trivial automorphism, and that (up to isomorphism) this is the only extensional well founded relation on 2.

56. Let X be a set with a bijection $X + X \cong X$. Construct a bijection $\mathfrak{H}(X) \times \mathfrak{H}(X) \cong \mathfrak{H}(X)$. An ordinal κ is said to be a **cardinal** (or an **initial ordinal**) if it is least amongst the ordinal structures on the same underlying set. Show that $\varnothing$, $\mathfrak{H}(-)$ and $\bigcup$ generate the class of cardinals, and hence deduce that $\kappa \times \kappa \cong \kappa$ for any cardinal κ. $\mathfrak{H}(\kappa) \equiv \kappa^+$ is called its successor cardinal.

57. Show that any transitive extensional well founded relation is trichotomous, using proof boxes, making explicit the use of excluded middle in the form $\neg\neg\phi \vdash \phi$. [This is not easy: try Remark 2.4.10.]

Adjunctions

U NIVERSAL PROPERTIES galore have arisen in the earlier chapters, and it is high time we gave a unified framework for them. In 1948 Pierre Samuel identified universal properties as a common formulation of several constructions in topology, and the Bourbaki school used them in their comprehensive account of mathematics. Independently, Daniel Kan introduced *adjoint* pairs of *functors*, with the tensor product and internal hom for vector spaces as his main example (1958). The name was suggested by Sammy Eilenberg, by analogy with $\langle Ta, b \rangle = \langle a, T^*b \rangle$ for operators in a Hilbert space, which notation had itself been proposed by Marshall Stone.

Nowadays, every user of category theory agrees that this is the concept which justifies the fundamental position of the subject in mathematics.

There are several other formulations (such as ends and Kan extensions), and which of them to use is a matter of personal taste. The symmetrical presentation of a pair of adjoint functors between two categories will be given in Section 7.2, but this raises logical questions because of the *choice* of a *particular* product or whatever within its isomorphism class. We prefer *diagrammatic* reasoning, which opens the calculations out to view, especially in complicated situations, and avoids the Choice. We shall also show that the naturality conditions on adjoint functors — all too easily dismissed as bureaucracy — are directly related to substitution- and continuation-invariance of the rules of type theory.

The most commonly used universal constructions are limits and colimits. We devote Sections 7.3–7.5 to them and to how they relate to other adjunctions, using the fact that left adjoints preserve colimits to fashion each from the other. Limits and colimits of topological spaces have an almost completely "soft" construction, and we also investigate free algebras, leading into the theory of monads (Section 7.5).

Often the thing which is required is obtained as a composite of two universal constructions: recognising it as such makes the development more modular. Sometimes the construction can only be done in a few simple cases, others being too complicated to be judged reliable. When it becomes apparent that it is an adjoint — frequently to something completely trivial, for example pullback between slices is right adjoint to composition — the general case quickly falls into line (Exercise 7.42).

Finally we return to the task of showing the equivalence between syntax and semantics. Adjunctions not only describe the logical connectives themselves, but also characterise the category $\mathsf{Cn}_{\mathcal{L}}^{\square}$ of contexts and substitutions. Equivalences, in their strong form themselves examples of adjunctions, settle the issue of the *choice* of the structure (Section 7.6). Finally, Section 7.7 proves some deep results about syntax using just adjunctions and pullbacks.

7.1   EXAMPLES OF UNIVERSAL CONSTRUCTIONS

Because of the importance of universal properties, we devote this section to collecting examples of them. As usual, the purpose of this is to give the *flavour* of the many ways in which this concept can behave, rather than to advance or rely on any particular mathematical specialisation. The most directly useful formulation of universal properties is one of the oldest; it is also the one which can most readily be generalised, *idiomatically* expressing factorisation systems, function-spaces and many other notions. Compare, for example, the next diagram and that in Definition 7.1.7 with the one for orthogonality in Definition 5.7.1.

DEFINITION 7·1·1:   We say that $\eta : X \to UA$ is a **universal map from the object** $X \in \mathrm{ob}\,\mathcal{S}$ **to the functor** $U : \mathcal{A} \to \mathcal{S}$ if, whenever $\mathfrak{f} : X \to U\Theta$ is another morphism with $\Theta \in \mathrm{ob}\,\mathcal{A}$, there is a unique $\mathcal{A}$-map $\mathfrak{p} : A \to \Theta$ such that $\eta\,;U\mathfrak{p} = \mathfrak{f}$ (*cf.* Remark 3.6.3 for posets).

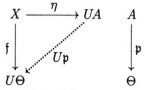

If every object X has a universal map to U then the latter is called an **adjunctible functor**.

As we showed for the terminal object in Theorem 4.5.6, the universal property determines A up to unique isomorphism. So it is a *description*, but in the *interchangeable* sense which we discussed at the end of Section 1.2. Of course this is why we generalised Russell's theory. Although any description allows us to introduce a function (indeed a func*tor*, the left adjoint) there is a loss of logical clarity here because of the choices; we prefer to shift the balance of the formulation back from algebra to logic.

EXAMPLES 7·1·2: Adjoint monotone functions.

(a) Let $U : \mathcal{A} \subset \mathcal{S} = \mathcal{P}(\Sigma)$ be the inclusion of the lattice of closed subsets for a system of closure conditions on a set Σ. Then $\eta : X \to UA$ is the inclusion of an arbitrary subset $X \subset \Sigma$ in its closure (Section 3.7).

(b) The possibility modal operator $\Diamond$, existential quantifier $\exists$ and direct image $f_!$ in Section 3.8. These will be developed in Section 9.3.

(c) In general, let $U : \mathcal{A} \to \mathcal{S}$ be a monotone function with $F \dashv U$. Then $\eta : X \to U(FX)$ is one of the inequalities in Lemma 3.6.2.

Colimits. Examples 3.6.10 gave meets as adjoint monotone functions.

EXAMPLES 7·1·3:

(a) Let $U : \mathcal{A} \to 1$ be the unique functor to the category with one object $(\star)$ and only its identity morphism. Then $\eta = \mathrm{id}_\star : \star \to U0$ is the universal map iff $0 \in \mathrm{ob}\,\mathcal{A}$ is the initial object. The mediating homomorphism $\mathfrak{p} : 0 \to \Theta$ is the unique map, *cf.* Example 3.6.10(a), Definition 4.5.1ff and Definition 5.4.1(a).

(b) Let $U : \mathcal{A} \to \mathcal{A} \times \mathcal{A}$ be the diagonal functor $A \mapsto \langle A, A \rangle$, and N and Y two objects of $\mathcal{A}$. Then the pair $\eta = \langle \nu_0, \nu_1 \rangle : \langle N, Y \rangle \to \langle A, A \rangle$ is universal iff it is a coproduct diagram. Given maps $\mathfrak{f} : N \to \Theta$ and $\mathfrak{g} : Y \to \Theta$, the mediator is $\mathfrak{p} = [\mathfrak{f}, \mathfrak{g}]$, *cf.* Example 3.6.10(b), Definitions 4.5.7ff and 5.4.1(b).

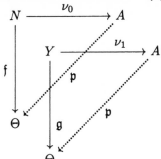
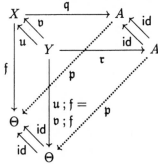

(c) Let $U : \mathcal{A} \to \mathcal{A}^{\rightrightarrows}$ be the diagonal functor to the category whose objects are parallel pairs; $\eta = (\mathfrak{q}, \mathfrak{r})$ is universal iff $\mathfrak{r} = \mathfrak{u} \, ; \mathfrak{q} = \mathfrak{v} \, ; \mathfrak{q}$ and $\mathfrak{q}$ is the coequaliser (Definition 5.1.1(a) and Proposition 5.6.8ff).

(d) Let $\mathcal{I}$ be any diagram-shape and $U : \mathcal{A} \to \mathcal{A}^{\mathcal{I}}$ the diagonal or constant functor $(X \mapsto \lambda i. X)$ into the functor category $\mathcal{A}^{\mathcal{I}}$ (Theorem 4.8.10). Then for $X_{(-)} : \mathcal{I} \to \mathcal{A}$ and $A \in \mathrm{ob}\,\mathcal{A}$, $\eta : X_{(-)} \to UA$ is universal iff it is a colimiting cocone. For any cocone $\mathfrak{f}$, the mediator $\mathfrak{p}$ in Definition 7.1.1 is that from the colimit, *cf.* Example 3.6.10(c).

We shall treat general limits and colimits fully in Section 7.3.

Free algebras. Besides coequalisers, the new feature which arises when we move from posets to categories is the free algebra (Chapter VI).

EXAMPLES 7·1·4:

(a) Let $U : \mathcal{A} \to \mathcal{S}$ be the forgetful functor $\mathcal{C}Rng \to \mathcal{S}et$. Then A is $\mathbb{Z}[X]$, the ring of **polynomials** in a set X of indeterminates, and p evaluates polynomials using the assignment $f : X \to \Theta$.

(b) With $\mathcal{A} = \mathcal{V}sp$, let $A \subset \mathbb{R}^X$ consist of those functions $a : X \to \mathbb{R}$ with non-zero values at only finitely many elements of X. The unit $\eta(x)$ is the xth basis vector and $p(\sum_i a_i x_i) = \sum_i a_i f(x_i)$.

(c) Let $U : Rng \to \mathcal{M}on$, forgetting addition. The free ring A consists of linear combinations of elements of a monoid. When $X = G$ is a group this is called the **group ring** $\mathbb{Z}G$. Similarly, with K any ring, for the coslice $U : K \downarrow Rng \to \mathcal{M}on$ (the objects of $K \downarrow Rng$ are called **K-algebras**), we get the group ring KG.

(d) Let $\mathcal{A} = \mathcal{R}el$, $\mathcal{P}os$, $\mathcal{S}p$, $\mathcal{G}raph$ or $\mathcal{C}at$ and $\mathcal{S} = \mathcal{S}et$, with U the forgetful functor. Then A is X with the **discrete** structure (the empty relation, equality, all subsets open, no arrows, and only identity maps).

Section 7.4 discusses techniques for constructing free algebras.

Completions. An important special case of universal maps arises from "completing" an object, so that when it is complete (re-)applying the construction has no (further) effect. This is most the direct analogue of a closure operation (Section 3.7, but see also Section 7.5).

LEMMA 7·1·5:   Following Definition 4.4.8, any functor $U : \mathcal{A} \to \mathcal{S}$ is

(a) full and faithful iff id : $UA \to UA$ is a universal map from UA to U for each $A \in \text{ob} \mathcal{A}$, and

(b) an equivalence functor iff for every $X \in \text{ob} \mathcal{S}$ there is an invertible universal map $\eta : X \cong UA$.

A (full, and often by convention replete) inclusion $\mathcal{A} \subset \mathcal{S}$ which is adjunctible is called a **reflective subcategory** (Corollary 7.2.10(b)).

PROOF:

[a]   Both conditions say that $\forall \mathfrak{f} : UA \to U\Theta$. $\exists! \mathfrak{p} : A \to \Theta$. $\mathfrak{f} = \text{id} \, ; U\mathfrak{p}$.

[b]   For equivalence we also need, for each $X \in \text{ob} \mathcal{S}$, some isomorphism $\eta : X \cong UA$. The issue is therefore to show that any such η is a

universal map. For any $\mathfrak{f} : X \to U\Theta$ there is a unique $\mathfrak{p} : A \to \Theta$ with $U\mathfrak{p} = \eta^{-1} ; \mathfrak{f} : UA \to U\Theta$, since U is full and faithful. □

EXAMPLES 7·1·6: Reflections are sometimes completions (the universal map is injective) and sometimes quotients (where it is surjective).

(a) $\mathcal{P}os \subset \mathcal{P}reord$ is reflective, the quotient $\eta : X \twoheadrightarrow X/\sim$ described in Proposition 3.1.10 being the universal map.

(b) Let $\mathcal{S}$ be the category of metric spaces and isometries (functions that preserve distance) and $\mathcal{A} \subset \mathcal{S}$ the full subcategory of spaces in which every Cauchy sequence (Definition 2.1.2) converges. This is reflective, and the universal map $\eta : X \hookrightarrow A$ is an inclusion.

(c) The functor $U : \mathcal{S}et \hookrightarrow \mathcal{G}raph$ which treats a set as a discrete graph (Example 7.1.4(d)) is the inclusion of a reflective subcategory, where $\eta : X \twoheadrightarrow X/\sim$ is given by the set of (zig-zag) components (Lemma 1.2.4). Similarly we have the components of a preorder, groupoid or category.

(d) Imposing laws such as distributivity or commutativity on algebras, for example the reflections of $\mathcal{D}\mathcal{L}at \hookrightarrow \mathcal{L}at$, $\mathcal{C}\mathcal{M}on \hookrightarrow \mathcal{M}on$ and $\mathcal{A}b \hookrightarrow \mathcal{G}p$, results in a quotient.

(e) The reflection of $\mathcal{A}b$ in $\mathcal{C}\mathcal{M}on$ adjoins negatives, but unless the monoid has the cancellation property $\forall x, y, z. \ x + z = y + z \Rightarrow x = y$ (as with $\mathbb{N} \mapsto \mathbb{Z}$) there is also identification.

(f) Let $\mathcal{S}$ be the category of integral domains and monomorphisms and $\mathcal{F}ld \subset \mathcal{S}$ be the full subcategory of fields. Then $\eta : X \hookrightarrow A$ is the inclusion of X in its field of fractions, for example $\mathbb{Z} \hookrightarrow \mathbb{Q}$.

(g) Let $\mathcal{S}$ be the category of well founded relations and simulations, so $\mathcal{S} \cong \mathcal{C}oalg(\mathcal{P})$ by Remark 6.3.5. Let $\mathcal{A}$ be the full subcategory which consists of the extensional relations; set-theorists call its objects *transitive sets*, and its maps are set-theoretic inclusions [Osi74]. Then $\mathcal{A} \subset \mathcal{S}$ is reflective; Andrzej Mostowski (1955) used Definition 6.7.5 to find the extensional quotient set-theoretically. See Exercise 9.62 and [Tay96a] regarding the axiom of replacement.

Co-universal properties.

DEFINITION 7·1·7: A map $\varepsilon : FX \to A$ is said to be **co-universal from the functor** $F : \mathcal{S} \to \mathcal{A}$ **to the object** A if for every map $\mathfrak{p} : F\Gamma \to A$ in $\mathcal{A}$ there is a unique map $\mathfrak{f} : \Gamma \to X$ in $\mathcal{S}$ such that $F\mathfrak{f} ; \varepsilon = \mathfrak{p}$.

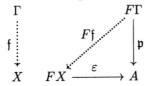

EXAMPLES 7·1·8:

(a) Coclosures, universal quantifiers ($\forall$), necessity ($\Box$) and other right adjoint monotone functions, analogously to Examples 7.1.2.

(b) Limits, $X = \lim A_i$, the duals of Examples 7.1.3. The co-unit ε is the family of projections (π_i).

(c) $((-) \wedge \phi) \dashv (\phi \Rightarrow (=))$ in a Heyting semilattice (Proposition 3.6.14).

(d) Let S be a cartesian closed category and $F = (Y \times (-)) : S \to S$. Then $\varepsilon : FX \to A$ is universal iff $X = A^Y$ and ε is evaluation.

(e) For the forgetful functor $F : Set \to Rel$, $X = \mathcal{P}(A)$ and $\varepsilon(a) = \{a\}$.

(f) For $F : Set \to Pfn$, $X = A_\perp$ and $\varepsilon(a) = \text{lift}\, a$ (Definition 3.3.7).

(g) The co-reflection of Gp into Mon gives the group of units.

(h) Set, again with the discrete structure, is co-reflective in Bin, Sp, $Graph$, $Preord$, Gpd and Cat. Co-universality of $\varepsilon : FX \to A$ says that X is A with the **indiscriminate** structure (also called indiscrete or chaotic), which cannot distinguish between points. So $x \mapsto y$, $x \leqslant y$, $X(x, y) = \{\star\}$, etc. for all x, y. Classically, the indiscriminate topology only has $\varnothing$ and X as open sets.

(i) For any set ("alphabet") G, the set Γ of G-**streams** carries a map read : $\Gamma \to \Gamma \times G$, so it is a coalgebra for the functor $(-) \times G$ (Lemma 6.1.8). Let F be the underlying set functor. Then $FX \to 1$ is co-universal from F to 1 (*i.e.* X is the terminal coalgebra) iff $X = G^{\mathbb{N}} \cong G^{\mathbb{N}} \times G$, this structure being induced by $\mathbb{N} + 1 \cong \mathbb{N}$.

There are also examples of symmetrically adjoint contravariant functors, generalising Galois connections (Proposition 3.8.14). Recall in particular that negation, $(-) \Rightarrow \perp$, is symmetrically self-adjoint (Exercise 3.50).

EXAMPLES 7·1·9:

(a) The Lineland Army is a monoid equipped with a symmetric self-adjunction (Example 1.2.7 and Exercise 7.8).

(b) Let Σ be any object of a cartesian closed category S, such as 2 in Set. Then $X \mapsto \Sigma^X$, as a functor $S \to S^{op}$, is symmetrically adjoint to itself on the right.

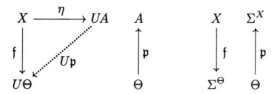

(c) Let Σ be a field, say $\mathbb{R}$. For a vector space V, the **dual space** V^* consists of the linear maps $V \to \Sigma$. Then $(-)^* : S \to S^{op}$ defines a self-adjoint functor on the category of vector spaces, whose unit was the original natural transformation (Example 4.8.8(b)).

EXAMPLES 7·1·10:

(a) The inclusion $\mathcal{BA} \subset \mathcal{HSL}$ of Boolean algebras in the category of Heyting *semi*lattices has both a reflection and a co-reflection, and these functors are the same. Exercise 7.12 shows that this situation always arises from a natural idempotent, in this case $\neg\neg$.

(b) The inclusion $\mathcal{Gp} \subset \mathcal{Mon}$ is also both reflective and co-reflective, but now the two adjoints are different, as they are for the inclusion of any complete sublattice (Remark 3.8.9).

Classifying categories. We have constructed the preorder or category of contexts and substitutions $\mathsf{Cn}_{\mathcal{L}}^{\square}$ for various fragments $\square$ of logic.

REMARK 7·1·11: There is an analogy between $\mathbb{Z}[\underline{x}]$ and $\mathsf{Cn}_{\mathcal{L}}^{\square}$:

(a) Extensionally, each of them is the system of well formed formulae built up from some indeterminates ($\underline{x}$ or $\mathcal{L}$) using certain operations, namely addition and multiplication, or the logical connectives of the fragment $\square$. (Recall also that the hom-set $\mathsf{Cn}_{\mathcal{L}}^{\times}([\underline{x} : X], Y)$ is the free algebra on generators $\underline{x}$, *modulo* the laws.)

(b) Any sequence $\underline{a} \in R$ of elements of another ring, or any model $\mathcal{M}$ of $\mathcal{L}$ in a semantic world $\mathcal{E}$, may be substituted for the indeterminates, and the syntactic formulae "evaluated" using structural recursion.

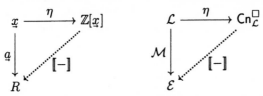

(c) Intensionally, this is a universal property, because $\mathbb{Z}[\underline{x}]$ and $\mathsf{Cn}_{\mathcal{L}}^{\square}$ have the same structure as R or $\mathcal{E}$ and $[\![-]\!]$ is the unique homomorphism of this structure taking the generic object to the concrete one.

EXAMPLES 7·1·12: This analogy is precise for propositional logic.

(a) Let $\square$ be unary propositional logic. Then $\mathcal{L} \to \mathsf{Cn}_{\mathcal{L}}$ is inclusion of (a set with) a binary relation (unary closure condition, Section 3.8) in its reflexive–transitive closure. This is the universal map from $\mathcal{L}$ to the forgetful functor $U : \mathcal{Preord} \to \mathcal{Bin}$; in fact $\mathcal{Preord}$ is reflective in $\mathcal{Bin}$ (an object of which is a set with a binary endorelation).

(b) Let $\square$ be Horn logic (propositional algebraic); then U is the forgetful functor from $\mathcal{SLat}$. For propositional geometric and intuitionistic logic, the category $\mathcal{A}$ is $\mathcal{Frm}$ or $\mathcal{HSL}$ (Section 3.9).

(c) Let $\square$ be classical propositional logic, so $\mathcal{A} = \mathcal{BA}$ (Boolean algebras), and let X be a finite set. Then $\mathsf{Cn}_X^{\text{boole}} = 2^{2^X}$ is the set of disjunctive or conjunctive normal forms (Remark 1.8.4) in the set X of atomic propositions, together with their negations.

(d) $U : \mathcal{CSLat} \to \mathcal{Pos}$. Then $A = \mathsf{sh}(X) = [X^{\text{op}} \to 2]$, the lattice of lower sets, ordered by inclusion (Definition 3.1.7), with $\eta(x) = X \downarrow x$ and $p(I) = \bigvee_\Theta \{f(x) \mid x \in I\}$ by Proposition 3.2.7(b).

It also works at the level of types for unary algebraic theories.

(e) The classifying monoid for a free single-sorted unary theory gives the universal map to $U \equiv \mathsf{List} : \mathcal{Mon} \to \mathcal{Set}$. In Section 2.7 we wrote $[\![\ell]\!] = \mathsf{fold}(e, m, f, \ell)$.

(f) A many-sorted free unary theory is described by an oriented graph; the classifying category is composed of paths ($U : \mathcal{Cat} \to \mathcal{Graph}$).

(g) Finally, elementary sketches present equational many-sorted unary theories, and the classifying category is free on the sketch.

REMARK 7·1·13: The *intuition* that $\mathsf{Cn}_{\mathcal{L}}^{\square}$ has a polynomial structure is a valuable one, but there are three technical problems.

(a) For an "adjunction" $\mathsf{Cn}_{(-)}^{\square} \dashv \mathsf{L}^{\square}$ we need a *category* of languages, but there seems to be no convincing notion of map $\mathcal{L}_1 \to \mathcal{L}_2$, even for elementary sketches. In Sections I §10 and II §14 of [LS86], Joachim Lambek and Philip Scott define language morphisms to match their semantic interpretation, $\mathsf{Cn}_{\mathcal{L}_1}^{\square} \to \mathsf{Cn}_{\mathcal{L}_2}^{\square}$. However, this is not consistent with our use of $\mathcal{Bin}$ and $\mathcal{Graph}$ above, and so loses the *dynamic* aspect of the language which was the theme of Sections 3.7–3.9 and also of Chapter VI. See also Remark 7.5.2.

(b) The semantic structure corresponding to the type theory (products, exponentials, *etc.*) is itself characterised categorically by universal properties. In defining the classifying category, should we require $[\![-]\!]$ to preserve them on the nose (and to be unique up to equality), as we expect from syntax, or only up to isomorphism, as is the case for semantics? This question itself is the subject of Section 7.6. If we take the semantic option, then the universal property of the classifying category is more complicated than Definition 7.1.1: the interpretation functor $[\![-]\!]$ is only unique up to unique isomorphism — if it is defined at all, as some Choice is to be made.

(c) The axiom-scheme of replacement is needed to construct $[\![-]\!]$.

The second question also arises when we define limits and colimits of categories; we shall return to this point at the end of the next section.

REMARK 7·1·14: Historically, these intuitions emerged from algebraic geometry in the form of a classifying *topos*. The fragment of logic which is classified by toposes (known as **geometric logic**) includes products, equalisers and arbitrary pullback-stable colimits, the coproducts being disjoint and the quotients effective (Chapter V). In particular it allows existential quantification and infinitary disjunction.

By analogy with polynomials, the classifying topos for, say, groups was written $\mathcal{S}et[\mathsf{G}]$. Given a group G in another topos $\mathcal{E}$, $[\![-]\!] : \mathcal{S}et[\mathsf{G}] \to \mathcal{E}$ evaluates type-expressions by substituting the particular group G for the generic G. Being a homomorphism of the categorical structure, $[\![-]\!]$ preserves finite limits and has a right adjoint, written $\mathfrak{p}^* \equiv [\![-]\!] \dashv \mathfrak{p}_*$. Such an adjoint pair is called a **geometric morphism**, and is the analogue for toposes of the inverse and direct image operations on open sets that arise from a continuous function $\mathfrak{p}$ between spaces. Then groups in $\mathcal{E}$ correspond to continuous functions $\mathfrak{p} : \mathcal{E} \to \mathcal{S}et[\mathsf{G}]$; in particular the "points" of the topos $\mathcal{S}et[\mathsf{G}]$ are ordinary groups, since the topos $\mathcal{S}et$ denotes a singleton space. Hence $\mathcal{S}et[\mathsf{G}]$ is thought of as the "space of groups", not to be confused with the category $\mathcal{G}p$ — homomorphisms are in fact the "specialisation order" (Example 3.1.2(i)) on this space.

This analogy between model theory and topology explains why both subjects depend so heavily on Choice: it is necessary to *find* points, models or prime ideals with certain properties. However, propositional geometric logic is only special in so far as its model theory has this familiar points-and-open-sets form, whereas the categorical model theory of full first order logic has not yet been fully worked out. Classifying gadgets may be constructed in a uniform way for any fragment of logic, although we have considered simpler ones; besides, we have been interested in toposes as models of higher order logic, the relevant homomorphisms being logical functors, *i.e.* those that preserve $\mathfrak{2}$.

7.2 ADJUNCTIONS

Now we shall transform universal maps from their logically quantified statement into a purely algebraic form. This involves making a Choice (Exercise 3.26) of universal and co-universal maps, which provide the components of two *natural transformations*. Although they are unique up to unique isomorphism, these play a crucial role, and are just as much part of the definition as are the functors U and F. In particular we shall show that the laws which they obey express the β- and η-rules in type theory (Sections 2.3 and 2.7), and (as in Definition 4.7.2(c)) the naturality condition handles substitution or continuation.

In this section we drop the Convention 4.1.2 for brackets (which arose from Currying, Convention 2.3.2): FUA means $F(UA)$, etc.

DEFINITION 7·2·1:   An **adjunction** $F \dashv U$ consists of

(a) the **right** or **upper adjoint** functor $U : \mathcal{A} \to \mathcal{S}$,

(b) the **left** or **lower adjoint** functor $F : \mathcal{S} \to \mathcal{A}$,

(c) the **unit** natural transformation, $\eta : \mathrm{id}_{\mathcal{S}} \to U \cdot F$, sometimes called the **front adjunction**, and

(d) the **co-unit** natural transformation, $\varepsilon : F \cdot U \to \mathrm{id}_{\mathcal{A}}$, also called the **back adjunction**,

satisfying the following **triangular laws**:

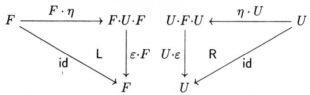

cf. Lemma 3.6.2 for posets. The letters L and R will be used to identify the triangles in the same way as N and Z for naturality in Definition 4.8.1.

THEOREM 7·2·2:   The following are equivalent:

(a) $(F, U, \eta, \varepsilon)$ form an adjunction.

(b) A natural isomorphism $\lambda : \mathcal{A}(F(-), (=)) \cong \mathcal{S}((-), U(=))$, called the **adjoint transposition** (*cf.* Definition 3.6.1)

$$\frac{FX \xrightarrow{\ \mathsf{p}\ } \Theta \ \text{in } \mathcal{A}}{X \xrightarrow{\ \mathsf{f}\ } U\Theta \ \text{in } \mathcal{S}} \quad F \dashv U$$

is given between functors $\mathcal{S}^{\mathrm{op}} \times \mathcal{A} \rightrightarrows \mathit{Set}$. (It is actually enough for λ to be natural in one argument for each fixed object as the other.)

(c) There is an assignment $\eta_X : X \to UA_X$ of universal maps from each object $X \in \mathrm{ob}\,\mathcal{S}$ to the functor U.

(d) There is an assignment $\varepsilon_A : FX_A \to A$ of co-universal maps from the functor F to each $A \in \mathrm{ob}\,\mathcal{A}$.

The components are derived from one another by

$$
\begin{aligned}
FX &= A_X & UA &= X_A \\
F\mathsf{u} &= \lambda^{-1}(\mathsf{u}\,;\eta Y) & U\mathsf{z} &= \lambda(\varepsilon A\,;F\mathsf{z}) \\
\eta X &= \lambda(\mathrm{id}_{FX}) & \varepsilon A &= \lambda^{-1}(\mathrm{id}_{UA}) \\
\lambda(\mathsf{p}) &= \eta X\,;U\mathsf{p} & \lambda^{-1}(\mathsf{f}) &= F\mathsf{f}\,;\varepsilon A
\end{aligned}
$$

We shall break with our usual custom by deferring the proof to page 380.

REMARK 7·2·3: Naturality of λ means that the following square in *Set* commutes, which, for $\mathfrak{p} : FX \to B$, says that $\lambda(Fu\,;\mathfrak{p}\,;\mathfrak{z}) = u\,;\lambda(\mathfrak{p})\,;U\mathfrak{z}$.

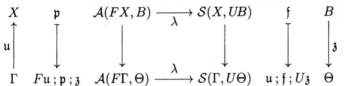

Hence if the square on the left below commutes in $\mathcal{A}$ then so does that on the right in $\mathcal{S}$ (and conversely since λ is bijective).

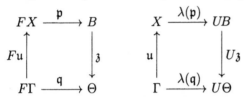

Put $\mathfrak{z} = \mathsf{id}$ and compare these diagrams with Definition 4.7.2(c).

Applications. Recall from Sections 2.3 and 2.7 that the introduction, elimination, equality, β- and η-rules for the product, sum, function-type and List constructions are summed up by adjoint correspondences.

REMARK 7·2·4: Consider the common situation in which a universal property is used to define a new construction in terms of an old functor. Indeed the definition of universal maps, unlike that of adjunctions, was phrased in just this way.

(a) The unit η provides the operations for the introduction rules, if these are direct (ν_0, ν_1, 0, +, empty list and append), and the co-unit ε gives the direct operations for elimination (π_0, π_1, ev).

(b) The triangle law on the old side is the β-rule.

(c) The triangle law on the new side is extensionality or the η-rule. (It is a pity that the letter η has established meanings for two different parts of the anatomy of an adjunction.)

(d) The adjoint transposition λ is the meta-operation or indirect rule (λ-abstraction, pairing, case analysis and recursion).

(e) The naturality condition on the new side of the adjunction defines the effect of the new construction on morphisms.

(f) Product and exponential are right adjoints, and naturality on the old side (the left) states the substitution-invariance of $\langle -, - \rangle$ and λ.

(g) For sum and List, which are left adjoints, old (right) naturality gives the continuation rules; in these cases substitution-invariance must be expressed by an extra condition.

EXAMPLE 7·2·5: For products, $F : \mathcal{S} \to \mathcal{S} \times \mathcal{S}$ is the diagonal functor and $U = \times$. Then λ is $\langle -, - \rangle$, the co-unit ε is the family of projections (π_i) and the unit η is the diagonal map $\Delta : X \to X \times X$. The product functor is defined on maps,

$$\frac{\Gamma' \xrightarrow{u} \Gamma \xrightarrow{a} X \xrightarrow{\jmath 0} X' \qquad \Gamma' \xrightarrow{u} \Gamma \xrightarrow{b} Y \xrightarrow{\jmath 1} Y'}{\Gamma' \xrightarrow{u} \Gamma \xrightarrow{\langle a, b \rangle} X \times Y \xrightarrow{\jmath 0 \times \jmath 1} X' \times Y'}$$

by naturality on the right (the new side). The triangle laws are

$$\Delta \, ; \pi_i = \mathrm{id} \quad \text{and} \quad \Delta \, ; (\pi_0 \times \pi_1) = \mathrm{id}$$

or, type-theoretically, $\pi_0\langle x, x \rangle = x$, $\pi_1\langle x, x \rangle = x$ and $\langle \pi_0(z), \pi_1(z) \rangle = z$. We must use naturality of ε to put the β-rules in the form $\pi_0\langle x, y \rangle = x$.

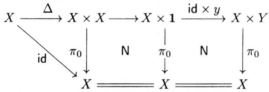

Notice that $y : \mathbf{1} \to Y$ is a *global* element: if it is only partially defined then commutativity of the square on the right becomes an inequality, so strict naturality fails in p-categories (Exercise 5.15).

EXAMPLE 7·2·6: For colimits, λ^{-1} is case analysis $[\ , \ ]$, the unit η is the family of inclusions (ν_i) and the co-unit ε is the codiagonal.

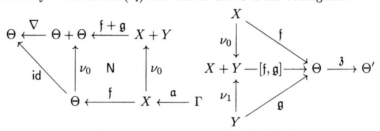

The naturality condition on the left (the new side) defines $+$ on maps. On the right (old) it states invariance under *continuation* $\jmath$ (Remark 2.3.13). This says that the $(+\mathcal{E})$-box is open-ended, *cf.* Remark 1.6.5.

EXAMPLE 7·2·7: For function-types, $F = (-) \times Y$ and $U = (=)^Y$. The adjoint transposition (λ) is λ-abstraction and the co-unit ε is evaluation. Naturality on the left (the old side) states substitution-invariance of λ (Definition 4.7.2(c)), and on the right (new) is postcomposition. The naturality of ε is $(-)^X$ on morphisms; and that of $\eta : x \mapsto \lambda y. \langle x, y \rangle$ is substitution under λ.

The left triangle law is the essence of the β-rule, but in the special case $p = \langle x, y \rangle$, $(\lambda y'. \langle x, y' \rangle)y = \langle x, y \rangle$. To see it at work we must pre- and postcompose with the argument and the function, and use naturality, *cf.* Remark 2.3.9.

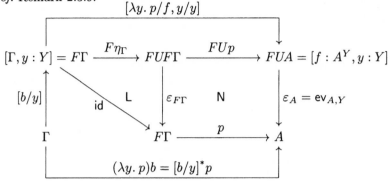

The right triangle law is the η-rule in a more direct way.

EXAMPLE 7·2·8: List : $\mathcal{S}et \to \mathcal{M}on$ is left adjoint to the forgetful functor. The unit is the singleton list, and the co-unit $\mathsf{List}(A) \to A$ multiplies out a list of elements of the monoid A. The adjoint transposition λ^{-1} is (roughly) what we called fold in Proposition 2.7.5 and $[\![-]\!]$ in Section 4.6 and elsewhere. Naturality on the left (the new side) for $u : X \to Y$ is the action $\mathsf{map}\, u$ of the functor (Example 2.7.6(d)). On the right it is the action of a monoid homomorphism $\mathfrak{z}$ (Remark 2.7.12).

Example 7.4.6 shows how $\varepsilon_A : \mathsf{List}(A) \to A$ captures the structure of the algebra A.

REMARK 7·2·9: This account is connected with the *binary* rules for List (append and fold) in Definition 2.7.4ff, whereas recursion in Section 6.1 was based on the unary form (cons and listrec). The unary and binary theories employ universal properties in different ways, which are related to the different roles of the alphabet X.

(a) The adjunction $\mathcal{M}on \rightleftarrows \mathcal{S}et$ describes *all* lists whatever, and we make use of the *free* monoid on any set of generators.

(b) Definition 6.1.1 built X into the functor T, so we only use the *initial* T-algebra for recursion.

Proof of the equivalence. It is surprising how little structure need exist in advance: we need naturality of λ on *one* side, for which we must know what the morphisms are. The rest of the structure is derivable: the naturality and triangle laws are the essential part.

PROOF OF THEOREM 7·2·2: We must show that the eight expressions are functorial, natural, mutually inverse or universal as appropriate.

[a⇒b] Bijectivity of λ uses naturality and the two triangular laws.

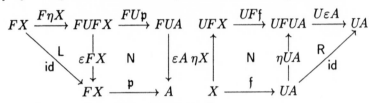

Naturality uses naturality (what else?).

$$X \xrightarrow{\eta X} UFX \xrightarrow{U\mathfrak{p}} UA \qquad FX \xrightarrow{F\mathfrak{f}} FUA \xrightarrow{\varepsilon A} A$$

$$u \uparrow \quad Z \quad UFu \uparrow \qquad \downarrow U\mathfrak{z} \quad Fu \uparrow \qquad FU\mathfrak{z} \downarrow \quad Z \quad \downarrow \mathfrak{z}$$

$$Y \xrightarrow{\eta Y} UFY \xrightarrow{U\mathfrak{q}} UB \qquad FY \xrightarrow{F\mathfrak{f}'} FUB \xrightarrow{\varepsilon B} B$$

[b⇒c] Putting λ for $B = FX$, $\mathfrak{p} = \mathrm{id}_{FX}$, $u = \mathrm{id}_X$ in Remark 7.2.3 gives

$$X \xrightarrow{\lambda(\mathrm{id}_{FX}) = \eta X} UFX$$
$$\Big\| \qquad\qquad\qquad \Big\downarrow U\mathfrak{z}$$
$$X \xrightarrow{\lambda(\mathfrak{z})} UB'$$

so $\mathfrak{z}$ mediates for $\lambda(\mathfrak{z})$ and is unique because λ is bijective (*cf.* the proof of the Yoneda Lemma, Theorem 4.8.12).

[c⇒a] First we must define F functorially on maps and show that η is natural. $Fu = \lambda^{-1}(u\,;\eta Y) : FX \to FY$ is the unique map making the square commute, but this square states naturality of η:

$$UFX \xrightarrow{UFu} UFY \xrightarrow{UF\mathfrak{v}} UFZ$$
$$\eta X \uparrow \quad N \quad \eta Y \uparrow \quad N \quad \uparrow \eta Z$$
$$X \xrightarrow{u} Y \xrightarrow{\mathfrak{v}} Z$$

For $u = \mathrm{id}$ only $Fu = \mathrm{id}$ will do, and similarly $F(u\,;\mathfrak{v}) = Fu\,;F\mathfrak{v}$ both fill in for the composite along the top (*cf.* Proposition 4.5.13).

Putting $\varepsilon A = \lambda^{-1}(\mathrm{id}_{UA})$ immediately satisfies law R, but we have to show that it satisfies L and is natural.

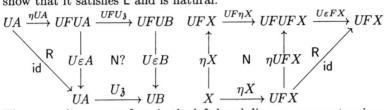

The naturality square for ε in the left-hand diagram commutes since both routes serve for $\lambda^{-1}(U_3)$ using universality of ηUA. The right hand diagram (the L law) commutes by naturality of η with respect to ηX and by the definition of εFX; the top composite $(F\eta X; \varepsilon FX)$ and id both serve for $\lambda^{-1}(\eta X)$, so L holds.

The proofs of [b⇒d] and [d⇒a] are dual, *i.e.* they are obtained from [b⇒c⇒a] by interchanging the parameters and reversing the arrows. □

Reflections and representables. The equivalence amongst these four presentations explains how various other terminologies arise. Recall from Theorem 4.8.12(b) that the **Yoneda embedding** identifies $X \in \mathrm{ob}\,\mathcal{S}$ with $\mathcal{H}_X = \mathcal{S}(-, X)$ in $\mathit{Set}^{\mathcal{S}^{\mathrm{op}}}$, cf. $\mathcal{S} \downarrow x$ in Definition 3.1.7.

COROLLARY 7·2·10:

(a) $F : \mathcal{S} \to \mathcal{A}$ has a right adjoint iff for each $A \in \mathrm{ob}\,\mathcal{A}$, the presheaf (contravariant functor, Example 4.4.2(g)) $\mathcal{A}(F(-), A) : \mathcal{S}^{\mathrm{op}} \to \mathit{Set}$ is *representable*, by $X = UA$ (Definition 4.8.13). Cf. the representable lower set $\{X \mid F(X) \leqslant A\}$ in the proof of Theorem 3.6.9.

(b) $U : \mathcal{A} \to \mathcal{S}$ is the inclusion of a reflective subcategory iff it is part of some adjunction whose *co-unit* is invertible (being mono suffices), cf. closure operations (Section 3.7).

(c) U is an equivalence functor iff it is part of a strong equivalence.

PROOF: The first part is immediate and the other two are similar to Lemma 7.1.5. Recall the characterisations of

▸ reflectivity, by ηUA being invertible for each $A \in \mathrm{ob}\,\mathcal{A}$;

▸ weak equivalence, by ηX being invertible for each $X \in \mathrm{ob}\,\mathcal{S}$;

▸ strong equivalence, by ηX and εA both being invertible, for each $X \in \mathrm{ob}\,\mathcal{S}$ and $A \in \mathrm{ob}\,\mathcal{A}$ respectively (Definition 4.8.9(b)).

By the R law ηUA is invertible iff $U\varepsilon A$ is; indeed the latter being mono suffices. This holds if ε is mono because U, being a right adjoint, preserves limits (Theorem 7.3.5) and hence monos (Proposition 5.2.2(d)). Conversely if U is the inclusion of a reflective subcategory then it is full and faithful, so ε is invertible. □

The natural bijection is a simple and symmetrical way of presenting an adjunction and remembering (in a subject with plenty of traps for the dyslexic) which is left and which is right. Although an "intuitively" natural construction *usually* turns out to be natural in the formal sense, Theorem 7.6.9 shows that naturality may be the point at issue. We saw in Remark 7.2.3 how checking it can be built into the idiom.

The formulation using representables shows how to generalise the notion of universal property, for example to that of $Cn_{\mathcal{L}}^{\square}$.

REMARK 7·2·11:   Let $\mathfrak{A}$ and $\mathfrak{C}$ be 2-categories.   The *isomorphism* $\mathfrak{A}(F(-), (=)) \cong \mathfrak{C}(-, U(=))$ may be replaced by either a strong or a weak *equivalence* of hom-categories (Definition 4.8.9). The various forms are illustrated by $\mathfrak{Cat}$: this has limits, exponentials and coproducts in the "isomorphism" sense (*cf.* Exercise 4.49), but other colimits can only be defined by an adjoint correspondence which is an equivalence functor. These different situations are distinguished by referring to *2-limits*, but *bi-colimits*. Definition 7.3.8 also gives a *new* kind of 2-limit. An even more general situation is where $\mathfrak{A}$ and $\mathfrak{C}$ are *enriched*, *i.e.* $\mathfrak{A}(A, B)$ and $\mathfrak{C}(X, Y)$ belong to some other category $\mathcal{V}$ such as $\mathcal{Ab}$ [Kel82].

Adjunctions are more common than the newcomer to category theory might expect, and individual cases are correspondingly less remarkable. Not unusually they are produced like a rabbit out of a hat, following an argument in which the more canny reader has watched the conjurer stuffing them in, as free algebras. The point is that, if the categories ($\mathcal{S}$ and $\mathcal{A}$) and the functors (F and U) are artificial ($\mathcal{A} = \mathcal{S} \times \mathcal{S}$ for the product, to give a mild example), then the significance of the adjunction $F \dashv U$ is not particularly cogent. The universal property in the style of Definition 7.1.1 expresses the important facts more directly. It also avoids the question of Choice, which is, frankly, a red herring: this is a side-effect of imposing one-dimensional algebraic notation on a logical situation which is naturally a little subtler.

7.3   GENERAL LIMITS AND COLIMITS

Limits and colimits are perhaps the most important cases of universal properties, so we devote the next three sections to them and to how they relate to adjunctions in general. Even if you learn no other theorem in category theory, the most important result is that left adjoints preserve colimits and right adjoints limits: the next section consists entirely of applications of this fact. For the converse, which (unlike the version for posets) needs an extra hypothesis, we make use of a simple example of a 2-limit, but this will repay ample dividends in Section 7.7.

The original case for which a definition like the one which follows was needed was that of a chain or ω-sequence, where $\mathrm{ob}\,\mathcal{I} = \mathbb{N}$ and there is an arrow $(n + 1) \to n$ in the limit case (*projective* or *inverse limits*) and $n \to (n + 1)$ in the colimit case (*inductive* or *direct limits*). When it was realised that products, pullbacks, equalisers and their duals fit the same pattern, it became customary to define a diagram-shape as a category. More recently, since the identities and composition play no role in the definition of (co)cones and (co)limits — and often get in the way[1] — authors' habits have reverted to regarding them as graphs. Each convention has its uses, and the notion of elementary sketch covers both.

DEFINITION 7·3·1: A **diagram-shape** is an elementary sketch by yet another name (Definition 4.2.5ff), usually with only a *set* of nodes. (See Definition 3.2.9 for the poset analogues.)

(a) A **diagram** (of shape $\mathcal{I}$) in a category $\mathcal{C}$ is an interpretation of the sketch, *i.e.* an assignment of an object $X_i \in \mathrm{ob}\,\mathcal{C}$ to each node $i \in \mathcal{I}$ and a morphism $X_{\mathfrak{u}} : X_i \to X_j$ to each arrow $\mathfrak{u} : i \to j$ in such a way that the given polygons (laws) commute. So if $\mathcal{I}$ is given as a category, a diagram is a functor $X_{(-)} : \mathcal{I} \to \mathcal{C}$.

(b) A **cone** with **vertex** Γ over this diagram assigns a map $\mathfrak{a}_i : \Gamma \to X_i$ to each node $i \in \mathcal{I}$ such that for each ("base") arrow $\mathfrak{u} : i \to j$ in $\mathcal{I}$, the ("cross-section") triangle $\mathfrak{a}_i \,;\, X_{\mathfrak{u}} = \mathfrak{a}_j$ commutes.

(c) A **limit** is a universal cone $(L, (\pi_i)_{i\in\mathcal{I}})$, so that for any other cone $(\Gamma, (\mathfrak{a}_i)_{i\in\mathcal{I}})$ there is a unique mediating map $\mathfrak{h} : \Gamma \to L$ with $\mathfrak{h}; \pi_i = \mathfrak{a}_i$ for each $i \in \mathcal{I}$.

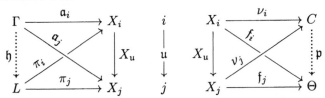

(d) A **cocone** with **covertex** Θ *under* the diagram is an assignment of a morphism $\mathfrak{f}_i : X_i \to \Theta$ to each node $i \in \mathcal{I}$ such that for each arrow

---

[1] For example, in domain theory there is a result known as the limit–colimit coincidence. This is valid for general filtered diagrams, but most accounts only state and use it for chains; as it was discovered during the "category" phase, these treatments are cluttered with notation like ϕ_{ij} and ψ_{ji}. Functors were introduced to get rid of such subscripts, *cf.* the comments after Lemma 6.3.11. My name *bilimit* has been adopted in domain theory for the coincident limit and colimit, although this conflicts with the bi-colimit terminology mentioned in Remark 7.2.11.

$\mathfrak{u} : i \to j$ in $\mathcal{I}$, the law $X_{\mathfrak{u}}$; $\mathfrak{f}_j = \mathfrak{f}_i$ holds.

(e) A **colimit** is a universal cocone $(C, (\nu_i)_{i \in \mathcal{I}})$, so that for any cocone $(\Theta, (\mathfrak{f}_i)_{i \in \mathcal{I}})$ there is a unique mediating map $\mathfrak{p} : C \to \Theta$ with $\nu_i ; \mathfrak{p} = \mathfrak{f}_i$ for each $i \in \mathcal{I}$.

(f) A category which has all set-indexed limits or colimits is called **complete** or **cocomplete** respectively, *cf.* Definition 3.6.12 for lattices.

(g) See Definition 4.5.10 regarding preservation and creation of limits and colimits. Some authors say that a functor is (*co*)*continuous* when it preserves (co)limits, but we reserve this word for *Scott*-continuity, *i.e.* preservation of filtered colimits.

As for all universal properties, limits and colimits, where they exist, are unique up to unique isomorphism. They are written

$$\lim_{i \in \mathcal{I}} X_i \text{ or } \underrightarrow{\lim}_{\mathcal{I}} X_i \qquad \text{and} \qquad \operatorname*{colim}_{i \in \mathcal{I}} X_i \text{ or } \underleftarrow{\lim}_{\mathcal{I}} X_i$$

respectively. The projections π_i and inclusions ν_i are an essential part of the definition.

EXAMPLES 7·3·2:

(a) If $\mathcal{C}$ is a preorder then a cone is a lower bound and a cocone is an upper bound; the arrows of $\mathcal{I}$ are redundant. Limits and colimits are meets and joins respectively (Definition 3.2.4).

(b) For $\mathcal{I} = \emptyset$, a (co)cone is simply an object, the (co)vertex; then a limit is a terminal object and a colimit is an initial object.

(c) If $\mathcal{I}$ is a discrete graph (with no arrows), or a discrete category (with only identity maps), then a diagram is a family of objects, a limit is its product and a colimit its coproduct.

(d) For the graph on the left below, a limit is an equaliser and a colimit is a coequaliser.

(e) For the graph on the right, a limit is a pullback, but the colimit is simply the value of the diagram at the corner. For the opposite of this diagram-shape, a colimit is a pushout.

(f) The equaliser of a parallel pair consisting of an endomap $\mathfrak{i} : X \to X$ and the identity is the set of fixed points. If $\mathfrak{i}$ is idempotent then the coequaliser is the image, and these two objects are isomorphic; the mono and epi split the idempotent $\mathfrak{i}$ (Definition 1.3.12).

(g) Section 6.4 used similar diagrams (without the requirement that $\mathfrak{i}$ be idempotent) to study **while** loops.

(h) If $\mathcal{I}$ has a terminal object then we call it a **wide pullback**; such diagrams arose in Exercise 3.34 and Lemma 5.7.8. The term is also applied to a diagram of such a shape or its limit.

(i) If the graph $\mathcal{I}$ is connected (in an unoriented sense, *cf.* Lemma 1.2.4) then we similarly refer to **connected limits**. The epithet refers to the *shape* of the defining diagram, and does not mean that the object L is connected, for example as a topological space. The empty diagram is *not* connected; wide pullbacks are, as are equalisers, which are not wide pullbacks. For simple connectedness, see Exercise 8.13.

(j) A category in which every finite diagram has a cocone (which need not be colimiting) is called **filtered**; this generalises directedness for posets (Definition 3.4.1). The forgetful functor $U : \mathcal{M}od(\mathcal{L}) \to \mathcal{S}et$ for a finitary theory $\mathcal{L}$ creates filtered colimits (*i.e.* of diagrams whose shape is filtered), *cf.* Theorem 3.9.4 and Definition 6.6.14.

(k) (Freyd) Assuming excluded middle (so $2 = \mathbf{2}$), if a small category has arbitrary limits then it is a preorder. Let $\mathfrak{a}, \mathfrak{b} : \Gamma \rightrightarrows X$ be distinct and $I = \mathsf{mor}\mathcal{C}$ be the set of all morphisms; then, by the definition of I-fold product, $\mathbf{2}^I \subset \mathcal{C}(\Gamma, X)^I \cong \mathcal{C}(\Gamma, X^I) \subset I$, contradicting Cantor's Theorem (Proposition 2.8.8). This result remains true in any sheaf topos over the classical category of sets.

(l) Martin Hyland, Edmund Robinson and Giuseppe Rosolini [HRR90] showed that the effective topos has a reflective, and so complete, subcategory (whose objects are known as *modest sets*) which is *weakly* equivalent to the internal ("small") category of PERs (Exercise 5.10). Because of the problem of Choosing amongst isomorphic objects, the latter does not have limits of *parametric* diagrams.

LEMMA 7·3·3: Let $X_{(-)} : \mathcal{I} \to \mathcal{C}$ be a diagram in a category with equalisers. If $\mathcal{C}$ has products indexed by $\mathsf{ob}\,\mathcal{I}$ and $\mathsf{mor}\mathcal{I}$ then $\lim_i X_i$ exists. Moreover if $F : \mathcal{C} \to \mathcal{D}$ preserves these equalisers and products then it preserves the limit.

PROOF: The limit of $X_{(-)} : \mathcal{I} \to \mathcal{C}$ is given by the equaliser

$$
(s_i)_{i \in \mathsf{ob}\,\mathcal{I}} \longmapsto \left(X_u(s_{\mathsf{src}\,u}) \right)_{i \xrightarrow{u} j}
$$

$$
\underset{\mathcal{I}}{\mathsf{Lim}}\, X_i \lhook\joinrel\longrightarrow \prod_{i \in \mathsf{ob}\,\mathcal{I}} X_i \rightrightarrows \prod_{u \in \mathsf{mor}\mathcal{I}} X_{\mathsf{tgt}\,u}
$$

$$
(s_i)_{i \in \mathsf{ob}\,\mathcal{I}} \longmapsto \left(s_{\mathsf{tgt}\,u} \right)_{i \xrightarrow{u} j}
$$

EXAMPLES 7·3·4:

(a) The lemma was found by abstracting the construction of general limits in *Set*, namely the subset consisting of those $(\mathsf{ob}\,\mathcal{I})$-tuples (s_i)

which satisfy $X_u(s(i)) = s(j)$ for each $u : i \to j$ in $\mathcal{I}$.

(b) Equalisers, binary products and a terminal object suffice to construct all finite limits (see also Example 5.1.3(c)).

(c) Similarly, the *co*limit C of any diagram $X_{(-)} : \mathcal{I} \to \mathcal{S}et$ is the *co*equaliser of a pair of functions *into* the *co*product C_0. By Lemma 5.6.11, this coequaliser is the quotient $C = C_0/\sim$ by an equivalence relation, namely that generated by the relation $\langle i, x \rangle \sim \langle j, y \rangle$ if there is an arrow $u : i \to j$ in $\mathcal{I}$ such that $y = X_u(x)$.

Although it is useful for showing that limits and colimits exist in certain semantic categories, this *decomposition* into products and equalisers is misleading for type theory, as we shall see in Section 8.3. (See [ML71, p. 109] for a more detailed proof of the lemma.)

Adjoints preserve (co)limits. The most important result.

THEOREM 7·3·5: Let $F \dashv U$. Then F preserves any colimits that exist and U any limits, *cf.* Proposition 3.6.8.

PROOF: Let $X_{(-)} : \mathcal{I} \to \mathcal{S}$ be any diagram and let $\nu_i : X \to C$ be a colimiting cocone under it in $\mathcal{S}$. Then

(a) By functoriality, $F\nu_i : FX \to FC$ is also a cocone under $FX_{(-)}$.

(b) Let $\mathfrak{p}_i : FX_i \to \Theta$ be a cocone under the diagram $FX_{(-)} : \mathcal{I} \to \mathcal{A}$. By the adjoint correspondence there are maps $\mathfrak{f}_i : X_i \to U\Theta$, which, by naturality with respect to X_u, form a cocone under the diagram $X_{(-)} : \mathcal{I} \to \mathcal{S}$.

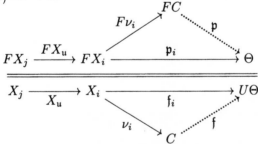

Hence there is a unique mediator $\mathfrak{f} : C \to U\Theta$ in $\mathcal{S}$.

(c) Taking this back along the adjoint correspondence, there is a map $\mathfrak{p} : FC \to \Theta$, which is a mediator for the cocone $\mathfrak{p}_i$ by naturality. It is unique because the argument is reversible.

The result for limits is the same with the arrows reversed.   □

REMARK 7·3·6: Colimits are preserved by left adjoints because they are themselves left adjoints, and a diagram of left adjoints commutes iff the

right adjoints do, *cf.* Remark 3.6.11. However, to state this precisely we
have to formulate the analogue of Lemma 3.6.4ff (see Exercise 7.27ff).

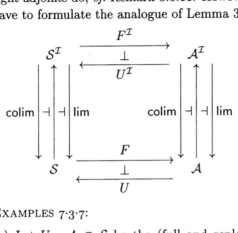

EXAMPLES 7·3·7:

(a) Let $U : \mathcal{A} \subset \mathcal{S}$ be the (full and replete) inclusion of a reflective
    subcategory, with $F \dashv U$, and let $A_{(-)} : \mathcal{I} \to \mathcal{A}$ be a diagram. Then

$$F \operatorname{colim}^{\mathcal{S}} UA_I = \operatorname{colim}^{\mathcal{A}} FUA_I \cong \operatorname{colim}^{\mathcal{A}} A_I$$

by Corollary 7.2.10(b), *cf.* Proposition 3.7.3 for posets. As an easy
special case of Proposition 7.5.6, the inclusion U not only preserves
but *creates* limits (Definition 4.5.10(c)).

(b) In a cartesian closed category, where $X \times (-) \dashv (=)^X$,
    - ► $X \times (-)$ preserves colimits;
    - ► in particular, products distribute over sums (Definition 5.5.1);
    - ► $(-)^X$ preserves limits;
    - ► the contravariant functor $\Sigma^{(-)}$ sends colimits to limits, because
      it is symmetrically self-adjoint on the right (Example 7.1.9(b));
      in particular $\Sigma^{X+Y} \cong \Sigma^X \times \Sigma^Y$;

    cf. Proposition 3.6.14 for Heyting lattices.

(c) Since the forgetful functors $\mathcal{S}et \to \mathcal{P}fn$ and $\mathcal{S}et \to \mathcal{R}el$ have right
    adjoints (namely the lifting and covariant powerset functors), the
    colimits with respect to total functions work for partial functions
    and for relations too.

(d) The forgetful functor $U : \mathcal{M}od(\mathcal{P}) \to \mathcal{S}et$ from the category of alge-
    bras for the covariant powerset functor creates (small) limits, but
    has no left adjoint, since the initial $\mathcal{P}$-algebra would be $\mathcal{P}(A) \cong A$,
    contradicting Cantor's theorem (Propositions 2.8.8 and 6.1.4(b)).

Because of Example 7.3.2(k) we cannot reasonably ask for categories to
have and functors to preserve class-indexed (co)limits. Theorem 7.3.12
shows what alternative condition suffices to obtain the right adjoint for
a colimit-preserving functor. The forgetful functor from the category of

complete Boolean algebras to *Set* is a more popular counterexample, but it takes rather more work to show that $F\mathbb{N}$ doesn't exist [Joh82, p. 57].

The next section gives some further applications of Theorem 7.3.5.

Comma categories. The next construction is a new kind of limit which arises in 2-categories, just as equalisers appeared when we moved from posets to categories. Section 7.7 makes a powerful application of this simple idea to the equivalence of syntax and semantics.

DEFINITION 7·3·8: Let $\mathcal{C} \xrightarrow{F} \mathcal{S} \xleftarrow{U} \mathcal{A}$ be functors. Then the *comma category* $F \downarrow U$ has

(a) *objects* (X, A, α) where $X \in \mathrm{ob}\,\mathcal{C}$, $A \in \mathrm{ob}\,\mathcal{A}$ and $\alpha : FX \to UA$ in $\mathcal{S}$ (we commonly abbreviate this to the map $\alpha : F(X) \to U(A)$ alone, bracketing the objects for emphasis if necessary; such an object will be called *tight* if the map α is an isomorphism),

(b) *morphisms* the pairs $\phi : X \to Y$, $\mathrm{u} : A \to B$ making the square on the left commute:

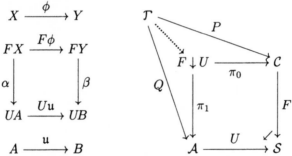

The square of functors on the right doesn't commute; the little arrow is a natural transformation $\alpha : F \cdot \pi_0 \to U \cdot \pi_1$ that has a categorical universal property:

PROPOSITION 7·3·9: Let $\varGamma$ be another category with functors $P : \varGamma \to \mathcal{C}$ and $Q : \varGamma \to \mathcal{A}$ and a natural transformation $\phi : F \cdot P \to U \cdot Q$. Then there is a functor $\langle P, Q, \phi \rangle : \varGamma \to F \downarrow U$, unique up to equality, such that $P = \pi_0 \cdot \langle P, Q, \phi \rangle$, $Q = \pi_1 \cdot \langle P, Q, \phi \rangle$ and $\phi = \alpha \cdot \langle P, Q, \phi \rangle$.

In particular, if the square from $\varGamma$ commutes up to isomorphism then the mediator factors through the *pseudo-pullback*, *i.e.* the full subcategory of $F \downarrow U$ consisting of the tight objects.   □

EXAMPLES 7·3·10: The notation can be specialised in several ways.

(a) If $\mathcal{A} = \{\star\}$, $U\star = A$, $\mathcal{S} = \mathcal{C}$, $F = \mathrm{id}_{\mathcal{C}}$, we have the *slice* $\mathcal{C} \downarrow A$ (5.1.8).

(b) Dually, for $\mathcal{C} = \{\star\}$, $F\star = X$, $\mathcal{S} = \mathcal{A}$, $U = \mathrm{id}_{\mathcal{A}}$, the *coslice* $X \downarrow \mathcal{A}$.

(c) If $\mathcal{C}$, $\mathcal{S}$ and $\mathcal{A}$ are discrete categories then $F \downarrow U$ is the pullback in *Set* of their sets of objects (Example 5.1.4(b)).

(d) If they are preorders then $F \downarrow U = \{\langle x, a \rangle \mid x \in \mathcal{C}, a \in \mathcal{A}, Fx \leqslant Ua\}$ (see footnote 1 on page 129).

(e) An initial object of $X \downarrow U$ (where $\mathcal{C} = \{\star\}$ and $F(\star) = X$) is a universal map from the object X to the functor U (Definition 7.1.1).

(f) The functor U is called **final**[2] if for every $X \in \mathrm{ob}\,\mathcal{S}$, $X \downarrow U$ is a connected category. See Exercise 7.19 for filtered final functors.

(g) A terminal object of $F \downarrow A$ (where $\mathcal{A} = \{\star\}$ and $U(\star) = A$) is a co-universal map from the functor F to the object A (Lemma 4.5.16 and Definition 7.1.7ff);

(h) (Lawvere) $F \dashv U$ iff there is an isomorphism of categories

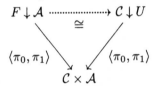

such that the triangle of functors commutes.

(i) For $\mathcal{S} = \mathcal{C}$, $F = \mathrm{id}_{\mathcal{C}}$ and $U : \mathcal{A} \to \mathcal{S}$ any functor, $\mathcal{S} \downarrow U$ is called the **gluing construction** (Section 7.7).

Equivalent colimits. We now prove Proposition 3.2.10 for categories. Let $X_{(-)} : \mathcal{I} \to \mathcal{C}$ be a diagram. Then any functor $U : \mathcal{J} \to \mathcal{I}$ between diagram-shapes induces a "restriction" of cocones $\mathfrak{g}_I : X_I \to \Theta$ under the diagram of shape $\mathcal{I}$ to $\mathfrak{f}_J = \mathfrak{g}_{UJ}$ under the composite $\mathcal{J} \to \mathcal{I} \to \mathcal{C}$.

PROPOSITION 7·3·11: If U is final then this is a bijection: every cocone $\mathfrak{f}_J : X_{UJ} \to \Theta$ for $\mathcal{J}$ extends uniquely to a cocone $\mathfrak{g}_I$ for $\mathcal{I}$. So whenever either of them exists, $\mathrm{colim}_{I \in \mathcal{I}}\, X_I \cong \mathrm{colim}_{J \in \mathcal{J}}\, X_{UJ}$.

PROOF: Let $\mathfrak{f}_J : X_{UJ} \to \Theta$ be a cocone in $\mathcal{C}$ under $\mathcal{J}$. By finality, for $I \in \mathrm{ob}\,\mathcal{I}$ the comma category $I \downarrow U$ is connected. Let $\mathfrak{a} : I \to UJ$ be

---

[2]The prefix "co-" in the original word cofinal (Definition 7.3.10(f)) carried the usual Latin–English meaning of "together", rather than the meaning of dualisation inherited from (co)homology (and maybe trigonometry before that). Although final functors are the analogue, not the dual, of cofinal monotone functions, the prefix was dropped in [ML71, p. 213] as it was considered inappropriate. I feel that it was unnecessary to introduce this confusion, as Proposition 7.3.11 associates them with *colimits* (but *cf.* Exercise 3.32). Even so, the definitions are not the same: any surjective function between discrete posets is cofinal and they give rise to the same joins, but to different coproducts. This difference is attributable to the hidden existential quantifier mentioned in footnote 1 on page 129.

any object of it and put $\mathfrak{g}_I = X_\mathfrak{a} \, ; \mathfrak{f}_J$, as we must for this to be a cocone under $\mathcal{I}$ with respect to $\mathfrak{a}$, since $\mathfrak{f}_J = \mathfrak{g}_{UJ}$.

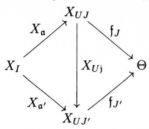

Any other object $\mathfrak{a}' : I \to UJ'$ of $I \downarrow U$ is linked to this one by a zig-zag, of which (for induction on its length) we need only consider one step $\mathfrak{j} : J \to J'$. Since this is a morphism of $I \downarrow U$ and $\mathfrak{f}_{(-)}$ was a cocone, the triangles commute, so $\mathfrak{g}_I = X_{\mathfrak{a}'} \, ; \mathfrak{f}_{J'}$.

We may then take $\mathfrak{g}_I$ as the value at I of a cocone over $\mathcal{I}$, because for $\mathfrak{i} : I' \to I$, $I \downarrow U \hookrightarrow I' \downarrow U$ as a subcategory, so $\mathfrak{g}_{I'} = X_\mathfrak{i} \, ; \mathfrak{g}_I$. $\square$

The general adjoint functor theorem. We would like to show that if $\mathcal{S}$ has and $F : \mathcal{A} \to \mathcal{S}$ preserves colimits then F has a right adjoint.

By Example 7.3.10(g), it suffices to show that, for each $A \in \mathrm{ob}\,\mathcal{A}$, the category $F \downarrow A$ has a terminal object. Since the terminal object of any category $\mathcal{X}$ is the colimit of the diagram $\mathrm{id}_\mathcal{X} : \mathcal{I} \to \mathcal{X}$ with $\mathcal{I} \equiv \mathcal{X}$,

$$UA = \mathrm{colim}\{X \mid (FX \overset{\mathfrak{f}}{\to} A) \in F \downarrow A\},$$

as in the proof of Theorem 3.6.9. (This is an example of a **left Kan extension**, the categorical analogue $\langle F\rangle\mathrm{id}(A)$ of the modal operator $\Diamond$ in Section 3.8.)

Unfortunately, $F \downarrow A$ is a large category, so by Example 7.3.2(k) we cannot ask for its colimit directly. There may, however, be a smaller diagram which has the same colimit, in the sense of Proposition 7.3.11.

THEOREM 7·3·12: (Peter Freyd, 1963) Let $\mathcal{S}$ be a category which has and $F : \mathcal{S} \to \mathcal{A}$ a functor which preserves all small colimits. Let $A \in \mathrm{ob}\,\mathcal{A}$ and suppose that the **solution-set condition** holds:

there are a small category $\mathcal{I}_A$ and a final functor $\mathcal{I}_A \to F \downarrow A$.

Then there is a co-universal map $\varepsilon_A : FX \to A$ from the functor F to the object A. Making an assignment of these, there is a right adjoint functor $F \dashv U$. $\square$

If we are already in possession of the co-universal map, the singleton subdiagram consisting of this alone suffices, so the solution-set condition is trivially necessary.

Preservation of limits is normally excellent heuristic evidence justifying the *search* for an adjunction: it is well worth making a habit of checking whether functors preserve products, coproducts and a few other cases. Frequently a few minutes' work will identify the essential features of any adjunction which holds, and the quickest way of proving adjointness is the universal property. We have also seen in the preceding sections that the effect of the two functors on morphisms, the unit, η, co-unit, ε, transposition, λ, *and their naturality* all shed light on the phenomena under study. So, although this is technically "redundant" information, they should always be investigated and described.

The general adjoint functor theorem has been cited [ML88] as the first *theorem* of category theory itself, but the solution-set condition seriously limits its value. Which is more useful, an *ad hoc* construction verifying this condition, or a presentation of the adjoint, which is an invariant? Clearly the latter, because subsequent researchers will want to know as much as possible about what it does, and maybe study it in its own right. Describing the adjoint is elementary (in the technical sense) and usually simpler than testing preservation of all (co)limits *properly*.

Although the theory of general limit diagrams is important to bringing together products, equalisers, pullbacks and inverse chains, it begs a number of questions, in particular *where such* (infinite) *diagrams come from*. We shall discuss this and the existence of *arbitrary* limits and colimits in *Set* in the final section of the book.

7.4 FINDING LIMITS AND FREE ALGEBRAS

Simply as applications of Theorem 7.3.5, we shall show how to find limits and colimits in topology and algebra, and conversely how coequalisers may be used to find initial algebras. As in Chapter V, this approach quickly identifies many of the basic features of an area of mathematics.

Limits and colimits in topology and order theory. We *do not* ask that the spaces be T_0, the preorders antisymmetric or the groupoids skeletal, as this destroys the adjunction (γ) — a quotient such as that in Proposition 3.1.10 must be applied to the colimits.

THEOREM 7.4.1: These categories have limits and colimits.

(a) Any limit of preorders is the limit of the underlying sets, equipped with some order; similarly the vertex classes of a limit of relations, graphs, categories or groupoids (from β).

(b) Moreover the hom-sets are also calculated as limits, and the order relation as a conjunction (from δ).

(c) Any limit of spaces is the limit of the underlying sets, equipped with some topology (from β).

(d) The limit of a diagram of discrete relations, *etc.* is discrete (from α). This is not so for spaces, as it is not possible to define the connected components of an arbitrary topological space.

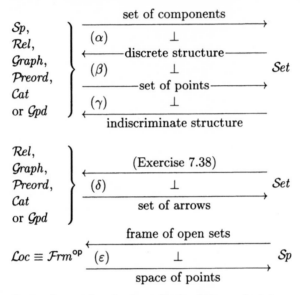

(e) A colimit of preorders is the colimit of the underlying sets, equipped with some *preorder*; also the vertex classes of graphs, *etc.* (from γ).

(f) A colimit of spaces is the colimit of the underlying sets, equipped with some topology or *preorder* (from γ).

(g) $\Omega : \mathcal{S}p \to \mathcal{F}rm^{\mathrm{op}}$ and pts $: \mathcal{F}rm \to \mathcal{S}p^{\mathrm{op}}$ (Exercise 4.14) are symmetrically adjoint on the right, so they send colimits to limits. The frame of open sets of a colimit is the limit of the frames of open sets of the spaces; in particular the Aleksandrov topology (whose open sets are the upper sets) on a colimit of preorders and the Scott topology (Proposition 3.4.9) on dcpos are found as limits (from ε).

(h) A colimit of discrete spaces is discrete (from β).

(i) The set of connected components of a colimit of graphs *etc.* is the colimit of those of the graphs in the diagram (from α).

(j) The space of points of a limit of locales is the limit of the spaces of points of the locales themselves (from ε).   □

REMARK 7·4·2: It remains to find the topology on a limit.

(a) As we know the underlying set of a limit of spaces, identifying the topology is reduced to a lattice-theoretic problem: it is the coarsest that makes the projections continuous.

(b) Similarly, a colimit carries the finest topology making the inclusions continuous. (In fact these properties may also be expressed in pure category theory, by observing that the points functor $Sp \to Set$ is a bifibration, Definition 9.2.6(e).)

(c) Limits of locales (colimits of frames) are not in general the same as the corresponding limits of spaces; they must be found using generators and relations, as below. Nuclei (Example 3.9.10(a)) give the most efficient technique; see [Joh82] for more detail. □

REMARK 7·4·3: Van Kampen's Theorem 5.4.8 may also be formulated as the preservation of colimits by a left adjoint. The right adjoint takes a groupoid or category to its **nerve**, which is a simplicial complex whose 0- and 1-cells are the objects and morphisms, and whose higher n-cells are composable sequences of length n (so Theorem 4.2.12 described the two-dimensional skeleton).

Generators and relations. We can exploit the relationship between left adjoints and colimits to construct free algebras using coequalisers and *vice versa*.

THEOREM 7·4·4: Every finitary equational algebraic theory $\mathcal{L}$ has a free algebra FX/R on any set X of generators for any set R of relations.

PROOF: We can only really sketch this complicated construction.

(a) Forget the sorts and laws of $\mathcal{L}$ and the relations R;

(b) add the generators X as nullary operation-symbols to the theory;

(c) Proposition 6.1.11 and Example 6.2.7 give equationally free algebras;

(d) the minimal subalgebra of an equationally free algebra, consisting of raw terms, is well founded and parsable; recall that it may also be constructed either as the colimit $\bigcup T^n(\varnothing)$, or as the union of all extensional well founded T-coalgebras (Section 6.3);

(e) it satisfies the recursion scheme by Theorem 6.3.13;

(f) the sorts are restored by restricting to the well formed formulae (Proposition 6.2.6); let $F_0 X$ be the free algebra for the theory $\mathcal{L}_0$ (known as the *signature*) obtained by forgetting the laws of $\mathcal{L}$;

(g) the congruence K is generated from the laws and relations R;

(h) the required free algebra FX/R is the quotient $Q = F_0 X/K$ by this equivalence relation.

Let us spell out the last step; notice that it treats "general" laws and "particular" relations in the same way, so without loss of generality R includes the laws of $\mathcal{L}$. Each member of R relates two raw terms, so we have a parallel pair of functions $R \rightrightarrows F_0X$. Let K be the congruence generated from R in step (g), and Q its quotient in $\mathcal{S}et^\Sigma$.

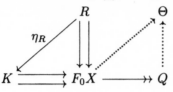

For Theorem 5.6.9, the theory must be finitary in order to define the operations on the quotient, making it a coequaliser in $\mathcal{M}od(\mathcal{L})$. Any algebra Θ for the equational theory $\mathcal{L}$ is *a fortiori* a model of the free theory $\mathcal{L}_0$, and so it has a unique homomorphism $F_0X \to \Theta$. That it satisfies the laws is exactly to say that the composites with $R \rightrightarrows F_0X$ are equal. The composite homomorphisms with $K \rightrightarrows F_0X$ are also equal because K is the congruence-closure of R. Hence there is a mediating function $Q \to \Theta$, and this is a homomorphism.   □

We shall discuss the way in which K is generated shortly.

EXAMPLE 7·4·5:   Recall from Example 4.6.3(i) that a category with a given set O of objects is an algebra for a theory with O^2 sorts, $O + O^3$ operation-symbols and $O^2 + O^2 + O^4$ laws. Theorem 6.2.8(a) constructed the free such algebra on an oriented graph (with the same set O of nodes). By Theorem 7.4.4 we now have the free category on any elementary sketch (Theorem 4.2.12).   □

EXAMPLE 7·4·6:   Every algebra presents itself by its *multiplication table*. It is generated by its own elements $a \in A$, subject to the relations that each depth 1 expression $r(\underline{a}) \in TA$ (Definition 6.1.1) is identified with its value $r_A(\underline{a}) \equiv \mathrm{ev}_A(r(\underline{a})) \in A$ (*cf.* the ambiguity in the usage of the word *law* mentioned in Remarks 1.2.2 and 6.6.4).

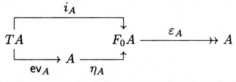

This is the **canonical language** for the algebra A. It has the same operation-symbols as the single-sorted theory for which A is an algebra, plus a constant for each element of A, and a law for each $r_A(\underline{a})$. We shall generalise the canonical language to larger fragments of logic in Section 7.6.   □

Daniel Kan's original example of an adjoint derived the tensor product $M \otimes (-)$ *à la* Remark 7.2.4 from the internal hom $(=) \otimes M$. As such it is constructed from generators and relations.

EXAMPLE 7·4·7: The tensor product of two modules M and N for a commutative ring R is the free R-module generated by symbols $\ulcorner(m,n)\urcorner$ (for $m \in M$, $n \in N$) subject to relations such as $r\ulcorner(m,n)\urcorner = \ulcorner(rm,n)\urcorner$ and $\ulcorner(m,n_1+n_2)\urcorner = \ulcorner(m,n_1)\urcorner + \ulcorner(m,n_2)\urcorner$ (*cf.* Example 3.9.10(e)). □

Computing colimits. Conversely, we may exploit the self-presentation of algebras for finitary algebraic theories. If we had a convincing notion of morphism of languages (*cf.* Remark 7.1.13(a)), these results would simply be applications of the fact that $Cn^{\times}_{(-)}$, being a left adjoint, preserves colimits. Indeed the first is plausibly the coproduct of theories.

LEMMA 7·4·8: Let A and B be generated by X and Y respectively, subject to the relations R and S.

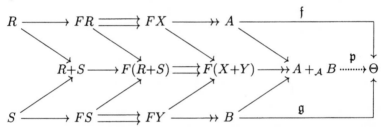

Then the algebra-coproduct $A +_{\mathcal{A}} B$ is generated by the disjoint union $X +_S Y$ modulo $R +_S S$. □

LEMMA 7·4·9: With similar notation, the algebra-coequaliser of $B \rightrightarrows A$ is generated by X subject to relations $R + P$.

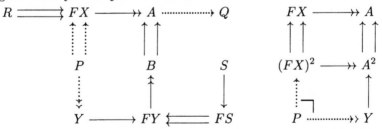

PROOF: The relations S for B are irrelevant: Q is also the coequaliser of $FY \rightrightarrows A$, where FY is the algebra *freely* generated by the given generators of B. Then Y may be viewed as an additional system of relations, but these are on A, not FX as required, so let $P \to (FX)^2$ be the pullback of $Y \to A^2$. (See Exercise 5.53 for why this is not just the pullback of $Y \rightrightarrows A$.) □

Some of these manipulations of colimits have been implemented in ML by David Rydeheard [RB88].

Treating relations as another theory. These constructions are not very tractable: they mix the fundamentally incompatible techniques of structural recursion and quotients by equivalence relations. We cannot expect any better (such as from the General Adjoint Functor Theorem 7.3.12), because **while** programs can be interpreted with coequalisers. From the point of view of machine representation, we have to accept generators and relations as a legitimate way of naming objects. We must also resort to these methods to find coproducts of groups, and so the fundamental groupoids of spaces obtained by surgery *à la* van Kampen (Theorem 5.4.8).

Nevertheless some unification is possible, by treating both the creation of raw terms and their identifications under the laws in the same way, as generative processes, but at different levels. (We shall have to do this for generalised algebraic theories in Remark 8.4.2.) We should take the principle of interchangeability more seriously, adding a categorical dimension *below* that which we're used to considering in mathematics: individuals can only be represented by tokens, with interchange arrows between them, *cf.* the comments in Definition 1.2.12 and Example 2.4.8.

REMARK 7·4·10: The congruence K may be generated in various ways:

(a) It is the closure of (the image of) $R \subset A^2 = (F_0 X)^2$ under

$$\{(a_1, b_2), (a_2, b_2), \ldots, (a_k, b_k)\} \triangleright (r_A(\underline{a}), r_A(\underline{b})) \qquad (r \in \Omega)$$

$$\otimes \triangleright (a, a) \qquad \{(a, b)\} \triangleright (b, a) \qquad \{(a, b), (b, c)\} \triangleright (a, c).$$

It is not possible to separate this into two processes of generating an equivalence relation and generating an $\mathcal{L}$-subalgebra.

(b) It is the free algebra on R for a more complicated theory than $\mathcal{L}$, namely with *operation-symbols* for each of the closure conditions above, including reflexivity, symmetry and transitivity.

(c) This theory may be chosen to have all of the laws induced by being a congruence on this particular algebra A, or may instead have no laws at all (*cf.* Lemma 7.4.9), making it a *free* theory. Another option is the theory of groupoids in $\mathcal{M}od(\mathcal{L}_0)$.   □

Definition 8.1.11 and Example 9.2.4(h) suggest a framework in which such a theory can be formulated, using dependent types and indexed categories. The next section uses self-presentation to give a concise functorial description of algebraic theories. Section 7.6 defines the self-presentation or *canonical language* of any semantic category using the type theory corresponding to the structure it has.

7.5 MONADS

Monads are the view of adjunctions which we get by looking at them from one end, generalising closure operations from Section 3.7. They describe (single-sorted) finitary equational theories, but also characterise many apparently unalgebraic categories as categories of infinitary algebras.

Let $F \dashv U$ be an adjunction and put $M = U \cdot F$. Then the natural transformation $\mu = \varepsilon \cdot F$ (the **multiplication**) makes the following diagrams commute, by the triangle laws (Definition 7.2.1) and naturality of ε.

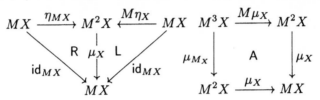

DEFINITION 7.5.1: A triple consisting of a functor $M : \mathcal{S} \to \mathcal{S}$ and natural transformations $\eta : \mathrm{id}_{\mathcal{S}} \to M$ and $\mu : M^2 \to M$ that obey these laws is called a **monad**.

REMARK 7.5.2: Algebras for functors have arisen in informatics, and algebras for monads in categorical algebra, so the uncritical reader of abstract accounts of these separate topics is in danger of making inappropriate value-judgements. Functors and monads can both be used to code the *same* (free) theory, but in different ways:[3]

(a) The functor T is the analogue of Lemma 3.7.10, which coded a system of closure *conditions*. It is *dynamic*: we can *watch* the genesis of the free algebra from its well founded coalgebras (Section 6.3), and there is an associated notion of recursion.

(b) The monad (M, η, μ) generalises the closure *operation* and is *static*: the construction of the free algebra is already finished.

In particular, MX is much bigger than TX, and may be a proper class if, for example, $T = \mathcal{P}$.

We shall only consider single-sorted theories (as before, for many-sorted theories we must work in Set^{Σ}), and start with a free algebraic theory in the sense of Section 6.1, presented as a functor T. Let $F_0 X$ be the free T-algebra on X, with inclusion $\eta_X : X \to M_0 X$, where $M_0 = U \cdot F_0$. Then (M_0, η, μ) is a monad, for a certain natural transformation μ.

---

[3]Note that [BW85, Section 9.4] uses T for the monad and R for our T.

For example, let $T = \{1\} + (-)^2$ and $X = \{a, b\}$. Then $M_0 X$ contains

$$1 \quad a \quad b \quad ab \quad ba \quad a1 \quad 1a \quad 11 \quad (ab)a \quad a(ba) \quad \ldots,$$

in which we use (round) parentheses in the usual way to disambiguate non-associative operations. Then $M_0^2 X$ consists of terms in generators from $M_0 X$, which we may think of as [square] *bracketed* terms:

$$1 \quad [1] \quad [a] \quad [ab] \quad [ba] \quad [b][a] \quad [a]1 \quad [a1] \quad [ab]a \quad [a]([b][ab]).$$

Then $\mu_X \equiv \varepsilon_{F_0 X}$ removes the square brackets to give plain terms.

A monad is needed to code laws. For a T-algebra $\mathrm{ev}_A : TA \to A$, the map $\varepsilon_A : M_0 A \to A$ constructed in Example 7.4.6 evaluates terms of arbitrary depth, whereas ev_A only handles those of depth 1. The more comprehensive structure is needed to test whether A satisfies the laws of an equational theory of which T is the signature, as these laws may be between terms of any depth. But even for a free theory, ε_A must satisfy certain conditions if it is given by some ev_A.

The Kleisli and Eilenberg–Moore categories. All monads arise from adjunctions, in fact in two different ways, both found in 1965.

PROPOSITION 7·5·3: Let (M, η, μ) be a monad on a category $\mathcal{S}$.

(a) The **Kleisli category** $\mathsf{Kl}(M, \eta, \mu)$ has the same objects as $\mathcal{S}$, but its maps $X \to Y$ are the $\mathcal{S}$-maps $X \to MY$; the Kleisli identity on X is η_X and the composite of $\mathfrak{f} : X \to MY$ with $\mathfrak{g} : Y \to MZ$ is $\mathfrak{f}; M\mathfrak{g}; \mu_Z$ (*cf.* Exercise 3.38 for preorders). There is an adjunction,

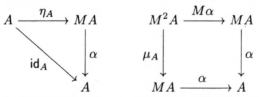

with co-unit $\varepsilon_X = \mathrm{id}_{MX}$, and this induces the given monad.

(b) An **algebra** is an $\mathcal{S}$-map $\alpha : MA \to A$ such that

$$
\begin{array}{ccc}
A \xrightarrow{\;\eta_A\;} MA & \qquad M^2 A \xrightarrow{\;M\alpha\;} MA \\
\end{array}
$$

cf. fixed points for closure operations (Section 3.7). Although this is a more complicated notion of algebra than that in Section 6.1, the definition of homomorphism $\mathfrak{f} : (A, \alpha) \to (B, \beta)$ is the same: an $\mathcal{S}$-map $\mathfrak{f} : A \to B$ such that $\alpha ; \mathfrak{f} = M\mathfrak{f} ; \beta$. They constitute the

Eilenberg–Moore category, $\mathcal{M}od(M,\eta,\mu)$. The forgetful functor U_{EM} has a left adjoint, $F_{EM}: X \mapsto (MX, \mu_X)$, and $\varepsilon_{(A,\alpha)} = \alpha$ is the co-unit. These also induce the given monad.

(c) Let $U: \mathcal{A} \to \mathcal{S}$ with $F \dashv U$ be any adjunction (with co-unit ε and transposition λ) giving rise to this monad. Then there are unique functors making the triangles commute:[4]

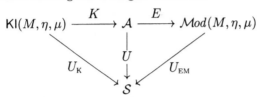

K takes the Kleisli morphism $\mathfrak{g}: X \to MY$ to $\lambda^{-1}(\mathfrak{g}): FX \to FY$, and, for $A \in \text{ob}\,\mathcal{A}$, $E(A)$ is the algebra $\varepsilon_A: UFUA \to UA$. $\qquad\square$

DEFINITION 7·5·4: An adjunction $F \dashv U$ (or just the functor U) for which the functor E is a weak equivalence is said to be ***monadic***.

EXAMPLES 7·5·5:

(a) $\mathcal{V}sp$ is the Eilenberg–Moore category for the monad on $\mathcal{S}et$ induced by the adjunction in Example 7.1.4(b). The Kleisli category consists of those vector spaces that have bases (which is all of them, given the axiom of choice).

(b) $\mathcal{R}el \hookrightarrow \mathcal{CSL}at$ are (equivalent to) the Kleisli and Eilenberg–Moore categories for the covariant powerset monad $(\mathcal{P}, \{-\}, \bigcup)$ on $\mathcal{S}et$.

(c) The following Kleisli, co-Kleisli and Eilenberg–Moore adjunctions arise from lifting in domain theory (Remark 3.4.5, Definition 3.3.7 and Example 3.9.8(c)), where $(-)_\perp$-homomorphisms are continuous functions that also preserve $\perp$.

$$\mathcal{P}fn \xrightarrow{(-)_\perp} (-)_\perp\text{-algebras} \underset{K^{op}\top EM}{\overset{(-)_\perp}{\rightleftarrows}} \mathcal{IPO}$$

$$\mathcal{S}et \underset{\text{discrete}}{\overset{\top}{\rightleftarrows}} \mathcal{D}cpo$$

Classically, every $(-)_\perp$-algebra (or ipo) A is $X_\perp$, for $X = A \smallsetminus \{\perp\}$, so the middle adjunction above is also Kleisli.

(d) Any single-sorted finitary algebraic theory $\mathcal{L}$ gives rise to a monad on $\mathcal{S}et$, for which the Eilenberg–Moore category is $\mathcal{M}od(\mathcal{L})$. The

---

[4]K is called L and E is called K in [ML71, Chapter VI].

Kleisli category consists of the *free* algebras on any set of generators. Restricting to finite such sets, we obtain $(\mathrm{Cn}_{\mathcal{L}}^{\times})^{\mathrm{op}}$, *cf.* Corollary 3.9.5.

(e) Exercises 4.27 and 7.41 show how some cartesian closed categories of domains arise as co-Kleisli categories of monads on symmetric monoidal closed categories.

Beck's theorem. The importance of monadic adjunctions lies in the fact that universal constructions with algebras may be computed for the carriers, so long as the two structures commute. Jon Beck's theorem was presented at a conference in 1966 but he never wrote it up for publication.

PROPOSITION 7·5·6: The forgetful functor $\mathcal{M}od(M, \eta, \mu) \to \mathcal{S}et$ *creates all small limits* (Definition 4.5.10), *and whatever colimits M preserves.*

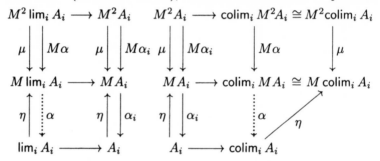

PROOF: The structure map for the (co)limit algebra is the mediator α to $\lim_i A_i$ or from $\mathrm{colim}_i M A_i$ shown dotted. Similarly the Eilenberg–Moore equations $\eta\,;\alpha = \mathrm{id}$ and $M\alpha\,;\alpha = \mu\,;\alpha$ hold because both sides are mediators to $\lim_i A_i$ or from $\mathrm{colim}_i M A_i$ or $\mathrm{colim}_i M^2 A_i$.   □

This property is characteristic, as we can see from a special case:

REMARK 7·5·7: (Bob Paré) Consider *contractible* coequalisers (Exercise 5.2). Such coequalisers exist in any category $\mathcal{S}$ where idempotents split, and they are preserved by *all functors* out of $\mathcal{S}$, in particular by M.

An algebra for the monad (by definition) makes the two squares below commute and the rows identities (recall that Z indicates a naturality square). It is a contractible coequaliser.

$$MA \xhookrightarrow{\eta_{MA}} M^2 A \xrightarrow{\mu_A} A$$

The contraction η_{MA} is not a homomorphism: the coequaliser diagram for the algebras only *becomes* contractible when we apply the forgetful functor U. Such a parallel pair of homomorphisms $\mathfrak{p}, \mathfrak{r} : C \rightrightarrows B$ for which there is some $\mathcal{S}$-map $\mathfrak{n} : B \to C$ with $\mathfrak{n}\,;\mathfrak{p} = \mathrm{id}_B$ and $\mathfrak{p}\,;\mathfrak{n}\,;\mathfrak{r} = \mathfrak{r}\,;\mathfrak{n}\,;\mathfrak{r}$ is called a *U-**contractible coequaliser***.

Notice in particular that the structure map of any algebra $\alpha : FA \to A$ is the coequaliser of $F\alpha$ and μ_A not only in $\mathcal{S}$ but also *in the category of algebras and homomorphisms*, so it is a self-presentation.

EXAMPLE 7·5·8: For the discrete $\dashv$ points adjunction (Theorem 7.4.1), the maps $F\alpha$ and μ_A are both the identity on the underlying set of A equipped with the *discrete* topology, so the original topology on the space A is not recovered as a coequaliser.

THEOREM 7·5·9: (Jon Beck) Let $F \dashv U$ between categories in which idempotents split (Exercise 4.16). Then the following are equivalent:

(a) the adjunction is monadic,

(b) U creates whatever colimits M preserves: for any diagram $\mathcal{I} \to \mathcal{A}$, if M preserves the colimit of $\mathcal{I} \to \mathcal{A} \to \mathcal{C}$, then U creates the colimit of $\mathcal{I} \to \mathcal{A}$;

(c) U creates U-contractible coequalisers.

The comparison functor is full and faithful (and in particular reflects invertibility) iff every ε_A is a self-presentation.

PROOF: To show that E is full and faithful, let $\mathfrak{g} : UA \to UB$ in $\mathcal{S}$. There is a unique $\mathfrak{f}$ with $\mathfrak{g} = U\mathfrak{f}$ iff *the top row is a coequaliser* in $\mathcal{A}$:

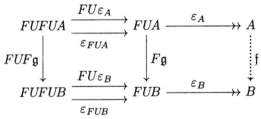

For essential surjectivity, let $\alpha : UFX \to X$ be an algebra. Then $F\alpha$ and ε_{FX} form a U-contractible pair in $\mathcal{A}$, whose coequaliser A (created by U) gives rise to the algebra $E(A) \cong (X, \alpha)$. $\qquad\square$

EXAMPLES 7·5·10:

(a) Reflective subcategories are always monadic, with invertible μ: we call the monad ***idempotent***. (By Corollary 7.2.10(b), ε is invertible and the forgetful functor is full and faithful, so the contraction in $\mathcal{S}$ is already a morphism of the category $\mathcal{A}$).

(b) $U : \mathcal{M}od(T) \to \mathcal{S}et$ creates whatever colimits T preserves, by the same argument as for Proposition 7.5.6, so it is monadic. Hence the

functor and monad have the same algebras, and we also justified this
name in terms of multiplication tables in Definition 6.1.1.

(c) When does $\alpha : M_0 A \to A$ satisfy the laws of an equational theory $\mathcal{L}$?
As in Theorem 7.4.4, they can be stated as the equality of composites
$R \rightrightarrows M_0 A \to A$. However, we already know the coequaliser of this
pair: it is the free $\mathcal{L}$-algebra FA on A. Hence $\alpha : MA \to A$ is an
M-algebra iff $M_0 A \twoheadrightarrow MA \to A$ is an $\mathcal{L}_0$-algebra satisfying the
laws.

(d) (Bob Paré) $Set^{\mathrm{op}} \simeq \mathcal{M}od(M, \eta, \mu)$, where this monad arises from the
symmetric self-adjunction of Example 7.1.9(b) with $\Sigma = 2$, *i.e.* the
contravariant powerset. Monadicity in this case is a consequence
of the Beck-Chevalley condition for the quantifiers (Remarks 9.3.7
and 9.4.3). From the definition of an elementary topos as having
finite *limits* and powersets, it follows immediately that it has finite
colimits, though the resulting constructions are more complicated
than those in Section 2.1. Indeed toposes have products and sums
of the same shapes, Proposition 9.6.13 and [BW85, Section 5.1].

Applications. We came to monads from finitary algebraic theories.
The types analogue of Theorem 3.9.4 and Exercise 3.37 is

PROPOSITION 7·5·11:   Every *finitary* monad on *Set*, *i.e.* for which M
preserves filtered colimits, is isomorphic to that given by the free algebras
for some single-sorted finitary algebraic theory $\mathcal{L}$, and then $\mathcal{M}od(\mathcal{L})$ is
externally locally finitely presentable (Definition 6.6.14(c)).

PROOF: We shall describe the Lawvere theory (Exercise 4.29). The set
of k-ary operation-symbols is $\mathsf{Cn}^{\times}_{\mathcal{L}}(\mathbf{k}, \mathbf{1}) = M\mathbf{k}$. The composition

$$(M\mathbf{n})^{\mathbf{k}} \times (M\mathbf{k}) = \mathsf{Cn}^{\times}_{\mathcal{L}}(\mathbf{n}, \mathbf{k}) \times \mathsf{Cn}^{\times}_{\mathcal{L}}(\mathbf{k}, \mathbf{1}) \to \mathsf{Cn}^{\times}_{\mathcal{L}}(\mathbf{n}, \mathbf{1}) = M\mathbf{n},$$

which determines how to apply a k-ary operation-symbol to arguments,
is the effect of the adjoint transposition $f \mapsto p$:

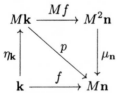

The multiplication μ thereby captures the laws of $\mathcal{L}$ (*cf.* saturation for
closure conditions). For a finite set $\mathbf{n}$, the value at $\mathbf{n}$ of the monad derived
from the theory $\mathcal{L}$ is the carrier of the free algebra on n generators, which
is $\mathsf{Cn}^{\times}_{\mathcal{L}}(\mathbf{n}, \mathbf{1}) = M\mathbf{n}$ as required, and similarly for functions between
finite sets. Since every set is a filtered colimit of finite(ly presented) sets
(Exercise 7.21) and M preserves filtered colimits, M is determined up

to unique isomorphism by its values $M\mathbf{n}$, so it agrees with the monad arising from $\mathcal{L}$. Again since M preserves filtered colimits, U creates them; F preserves finite presentability, and $\mathcal{M}od(\mathcal{L})$ is LFP. $\square$

As we have repeatedly pointed out, infinitary algebraic theories with arbitrary laws present problems when we reject the Axiom of Choice. Fred Linton developed monads as a useful alternative, showing that the symbolic and diagrammatic notions are equivalent in the presence of Choice [Lin69]. Monads can give an unexpected algebraic perspective on topology: for example the category of compact Hausdorff spaces is monadic over *Set*, the left adjoint being the space of ultrafilters. See [Man76] for this and a monadic treatment of algebra.

REMARK 7·5·12: The infinitary operations of most interest are meets, joins, limits and colimits. Proposition 3.2.7(b) and Theorem 3.9.7 showed in particular how to add joins to a poset, retaining certain specified joins. If a poset A is able to carry an algebra for a join-adding monad, then this structure α is unique, and $\alpha \dashv \eta_A$; such monads may be recognised by the fact that $\mu_X \dashv \eta_{MX}$. They were investigated by Anders Kock [Koc95] and V. Zöbelein. For meet- or limit-adding monads, the adjunctions are reversed. Alan Day found such a monad over *Sp* (or *Dcpo*) whose algebras are continuous lattices and whose homomorphisms preserve $\bigwedge$ and $\bigvee$ [GHK+80]. This was generalised (to smaller classes of meets) in [Tay90] and [Sch93].

REMARK 7·5·13: The Transfinite Recursion Theorem 6.7.4 relates the algebras for the covariant powerset *monad* (complete join-semilattices) equipped with an endofunction, to the well founded *coalgebras* for the *functor* alone, which we discussed in Example 6.3.3 and Remark 6.7.14. In fact the coalgebras for any functor which happens to be part of a monad carry partial "successor" and "union" operations (Exercise 7.45) [CP92, JM95, Tay96b].

REMARK 7·5·14: Eugenio Moggi has argued that certain monads should be regarded as notions of computation [Mog91]. To each type X we associate the type MX of *computations* (whose ultimate results would be) of type X. For example, with the lift monad (Definition 3.3.7), morphisms $\Gamma \to X_\perp$ are *partial* (possibly non-terminating) programs. Additional structure, known as a *strength*, is needed for this, as for parametric recursion in Remark 6.1.6 and Exercise 6.23. Moggi gave a symbolic form (the "**let**" calculus) for this piece of category theory.

REMARK 7·5·15: Jon Beck himself studied monads to unify homological algebra [BB69]. Applying the functor M repeatedly to an object X, the natural transformations $M^{n-i-1}\mu_{M^iX}$ provide the boundary operations

(S-morphisms) of a *simplicial complex* (Exercise 4.15). On the other hand the "ordinals" $M^n\mathbb{Q}$ just mentioned provide an abstract system of *simplices* (point, interval, triangle, tetrahedron, ...), so we may use maps $M^n\mathbb{Q} \to X$ to investigate the homotopy of any object X.

Echoing what we said about the utility of adjoints at the end of Section 7.3, whenever you have a construction which yields an object of the same category as its data, it is well worth looking for natural transformations making it a monad. The opportunities for mathematical investigation from such a simple thing are quite striking: the algebras for the monad often turn out to form important categories in their own right, and the iterates of its functor provide detailed information about the abstract topology and recursion theory of the category.

7.6   FROM SEMANTICS TO SYNTAX

When we introduced categories in Chapter IV we saw how they capture both semantic things such as topological spaces, and also the syntactic category of contexts and substitutions. The latter has chosen products, function-types *etc.* simply because these have symbolic names, with rules which say just this. In order to give specific values to the interpretation functor $[\![-]\!] : \mathsf{Cn}_{\mathcal{L}}^{\square} \to S$, we therefore have to *choose* products and function-spaces in any category S in which we want to define semantics. On the other hand, mathematical intuition is that there is no God-given choice of the product of two topological spaces: these objects may always be interchanged with isomorphic copies.

Consequently there are strong and conflicting opinions about whether, when we say that a category *has* products *etc.*, we mean them to be chosen or merely to exist. The second point of view is important even in formal reasoning, as a systematic way of discarding type-theoretic detail, *cf.* Example 2.4.8. This conflict, to which we have referred in Sections 1.2, 4.5 and 7.1, is the price we have to pay for the versatility of category theory, but now we intend to *resolve* it, showing that it amounts to the difference between strong and weak equivalences.

Logicians, when they have considered semantics at all, have traditionally called it *complete* if there are enough models in a fixed universe of sets to distinguish the syntax (Remark 1.6.13). That is, if $\mathcal{L} \vDash \phi$, *i.e.* some property ϕ holds for all models of a theory $\mathcal{L}$, then it is provable ($\mathcal{L} \vdash \phi$). We have already achieved this goal (for the fragments of logic we have discussed) by opening out the universe to encompass models built from the syntax itself. Now we want to go further, and treat syntax and semantics as equals — literally — requiring completeness of the syntax for the semantics as well as *vice versa*.

We aim to construct a language $\mathcal{L}$ from any suitable *given* category $\mathcal{C}$, such that $\mathcal{C} \simeq \mathsf{Cn}_{\mathcal{L}}^{\square}$. Theorem 4.2.12 did this for the unary case, where the structure $\square$ just consisted of identities and compositions: the canonical elementary language $\mathcal{L} = \mathsf{L}(\mathcal{C})$ has a type $\ulcorner X \urcorner$ for each object of $\mathcal{C}$ and an operation-symbol $\ulcorner \alpha \urcorner$ for each $\mathcal{C}$-morphism.[5] $\mathcal{L}$ also names as laws *all* of the equations which hold in $\mathcal{C}$, so $\ulcorner - \urcorner$ preserves the structure $\square$, *i.e.* it is a functor, and $\mathsf{Cn}_{\mathcal{L}} \cong \mathcal{C}$ by Theorem 6.2.8(c).

Recall that we also called $\mathcal{L}$ an elementary *sketch*. Although we shall describe the language for products *etc.* symbolically, many of the ideas of this section come from sketch theory (and others from sheaves).

Encoding operations. We shall consider those connectives $\square$ of type theory which are characterised by universal properties: products, sums, exponentials, quantifiers, $\mathsf{List}\,(-)$ and $\mathcal{P}(-)$, but not tensor products.

REMARK 7·6·1: Suppose that the objects X and Y have a product P and a sum S in a category $\mathcal{C}$. This means that there are $\mathcal{C}$-maps

$$X \xleftarrow{\;\;\pi_0\;\;} P \xrightarrow{\;\;\pi_1\;\;} Y \qquad X \xrightarrow{\;\;\nu_0\;\;} S \xleftarrow{\;\;\nu_1\;\;} Y$$

satisfying certain universal properties, but we shall ignore the latter for the time being. Adding product and sum types to the canonical elementary language, together with the pairing and case analysis meta-operations, in the syntactic category $\mathsf{Cn}^{\times +}$ there are morphisms

$$\ulcorner P \urcorner \xrightarrow{\;\langle \ulcorner \pi_0 \urcorner, \ulcorner \pi_1 \urcorner \rangle\;} \ulcorner X \urcorner \times \ulcorner Y \urcorner \qquad \ulcorner S \urcorner \xleftarrow{\;[\ulcorner \nu_0 \urcorner, \ulcorner \nu_1 \urcorner]\;} \ulcorner X \urcorner + \ulcorner Y \urcorner .$$

Although they are interpreted as identities in the *intended* semantics, these maps are not invertible in the syntax (or in arbitrary models of it) as it stands. So if we want P to *name* the product we must add

$$x : \ulcorner X \urcorner, y : \ulcorner Y \urcorner \vdash \mathsf{pair}_{X,Y,P}(x,y) : \ulcorner P \urcorner$$

as a new symbol (***encoding operation***) of the language, with β-rules

$$\ulcorner \pi_0 \urcorner(\mathsf{pair}(x,y)) \rightsquigarrow x \qquad \ulcorner \pi_1 \urcorner(\mathsf{pair}(x,y)) \rightsquigarrow y$$

and an η-rule $\mathsf{pair}(\ulcorner \pi_0 \urcorner(z), \ulcorner \pi_1 \urcorner(z)) = z$, which force pair to be the inverse of $\langle \ulcorner \pi_0 \urcorner, \ulcorner \pi_1 \urcorner \rangle$. The definable operation for the sum goes in the opposite direction, so the encoding operation is

$$z : \ulcorner S \urcorner \vdash \mathsf{split}(z) : \ulcorner X \urcorner + \ulcorner Y \urcorner .$$

---

[5] We temporarily use Greek letters for $\mathcal{C}$-maps to distinguish them from those of $\mathsf{Cn}_{\mathcal{L}}^{\square}$, which are written with $\mathfrak{German}$ letters; the term calculus, such as e, is in *italics*.

This difference is attributable to the fact that they are given by right and left adjoints, respectively. In the generic case we shall write

$$e : \Box_{\mathcal{L}} \ulcorner \underline{X} \urcorner \xrightarrow{\;\cong\;} \ulcorner \Box_{\mathcal{C}} \underline{X} \urcorner$$

for the encoding operation, in the "product" direction.

REMARK 7·6·2: P need not be the *chosen* product. Let $\alpha : Q \cong P$ be a semantic isomorphism, so we have another product diagram in $\mathcal{C}$ (*cf.* Theorem 4.5.6), and the composite $\ulcorner \alpha \urcorner \circ \mathsf{pair}_{X,Y,P}$ obeys the laws for $\mathsf{pair}_{X,Y,Q}$. Since inverses are unique, if this operation had also been named it would be provably equal to the derived form. For a semantically given category, therefore, it is harmless to name *all* of the diagrams (or as many as we please) which satisfy the universal property.

On the other hand, some diagrams may be products "accidentally", not because we intended them to be. If we omit the corresponding pair encoding, we get a *new* product (that is, the syntactic product will not be isomorphic to the unwanted semantic one). Including its encoding operation in the language is how we give our approval to a particular semantic product, and this will ensure that the functor $\ulcorner - \urcorner$ preserves it.

EXAMPLE 7·6·3: The theory of monoids was given in Example 4.6.3(f) by a ***finite product sketch***. A traditional syntactic account would say that multiplication has *two* (separate) arguments, but in this section all operations are treated as *unary*. We introduce a *new* object $M^{(2)}$ to be the source, and have to give additional information to force $M^{(2)}$ to be $M \times M$. This is what the upper parts of the diagrams in Example 4.6.3(f) say. An ***interpretation*** of this finite product sketch in any category $\mathcal{S}$ with finite products is an interpretation of the underlying elementary sketch for which $M \leftarrow M^{(2)} \rightarrow M$ becomes a product cone in $\mathcal{S}$. We have just described the syntax which is needed to do the same thing.

EXAMPLE 7·6·4: The canonical elementary language of a poset $(\Sigma, \leqslant)$ lists the elements of Σ and all instances of the order relation $\leqslant$ between them; a general language $\mathcal{L}$ of this kind is simply a set Σ with any binary relation $<$, and $\mathsf{Cn}_{\mathcal{L}}$ is the reflexive–transitive closure (Section 3.8).

Analogously to $\mathsf{Cn}_{\mathcal{L}}^{\times}$ we have the classifying semilattice $\mathsf{Cn}_{\mathcal{L}}^{\wedge}$ of a Horn theory $(\Sigma, \triangleright)$ constructed in Theorem 3.9.1. By Proposition 3.9.3, any meet-semilattice arises in this way, where, if we want to specify that $p = x \wedge y$, we must add this fact to the language $\mathcal{L}$, by writing

$$\{x, y\} \triangleright p \qquad \text{or} \qquad \mathsf{pair} : \ulcorner x \urcorner \wedge \ulcorner y \urcorner \leqslant \ulcorner p \urcorner$$

in the notation of the two sections. Otherwise we get Example 3.2.8.

Canonical language. The tradition of sketches is a minimalist one: to include *only* enough objects to determine the theory. We also want to show that categories which have *all* finite products, specified or not, arise as classifying categories. To deal with each of these applications, we give three versions of the definition.

The equational laws ought be treated in the same way as the additional structure such as products and function-types. However, it is technically simpler to assume in part (a) below that the unary part of the structure, which includes all of the equational laws amongst expressions, has been dealt with already (we did this in Theorem 3.9.7 for the lattice case too). The underlying elementary sketch is therefore a category.

To this we add encoding operations such as pair, split and (for function-types) abs, together with the β- and η-rules that force them to be inverse to the corresponding definable operation in Remark 7.6.1.

DEFINITION 7·6·5: A language $\mathcal{L}$ for a small category $\mathcal{C}$ is

(a) *a subcanonical* language for $\mathcal{C}$ if it consists of (the whole of) the elementary language $\mathsf{L}(\mathcal{C})$, together with some encoding operations for constructions which obey the appropriate universal properties in $\mathcal{C}$ (there need only be a few of these, or maybe none at all, but it is important that we only nominate products *etc.* for the language if they really were products in the semantics, *cf.* Definition 3.9.6(b));

(b) *the weakly canonical $\square$-language* $\mathsf{L}^{\square}(\mathcal{C})$ if it has *at least* one encoding operation for each tuple of sub-types;

(c) *a strongly canonical $\square$-language* for $\mathcal{C}$ if it has *exactly* one object and one encoding operation for each tuple of sub-types. That is, it makes an *assignment* of $\square$-structure, which it registers by including these particular encoding operations in the language. (The word "canonical" is carrying two senses here: the plain English one that it is making a *choice*, and the one from sheaf theory, Definition 3.9.6, that it accounts exactly for *all* of the semantic structure.)

In the last case the interpretation $[\![-]\!]$ of $\mathcal{L}$ in $\mathcal{S}$ is defined, by Remarks 4.6.5, 4.7.4 and 5.5.2. Besides the operations and meta-operations of $\square$, it must define the elementary language (*i.e.* the objects, maps and composites) of $\mathcal{C}$, and the encoding operations. These must be

$$[\![x : \ulcorner X \urcorner]\!] = X \qquad [\![\ulcorner \alpha \urcorner(x)/y]\!] = \alpha \qquad [\![e]\!] = \mathsf{id},$$

where e is the encoding operation for the *chosen* product, *etc.*

The quoting operation $\ulcorner - \urcorner$ preserves just that structure of $\mathcal{C}$ which $\mathcal{L}$ names. In particular it is a functor because the identities and composites $\ulcorner \alpha \urcorner ; \ulcorner \beta \urcorner = \ulcorner \alpha ; \beta \urcorner$ are named as laws of $\mathcal{L}$.

REMARK 7·6·6: Let $X \leftarrow P \rightarrow Y$ be any cone in $\mathcal{C}$.

(a) The functor $\ulcorner - \urcorner$ sends it to a product cone in $\mathsf{Cn}_{\mathcal{L}}^{\square}$ iff there is some definable term **pair** in the syntax such that pair : $\ulcorner X \urcorner \times \ulcorner Y \urcorner \cong \ulcorner P \urcorner$. Either **pair** is an encoding operation obeying the β- and η-rules above, or there is some other definable term which does this job.

(b) In this case $\ulcorner - \urcorner$ is full and faithful for morphisms $\Gamma \rightarrow P$ iff P was a product in the semantics, since otherwise $\ulcorner P \urcorner$ has mediators in $\mathsf{Cn}_{\mathcal{L}}^{\square}$ which either did not exist or were not unique in $\mathcal{C}$ (*cf.* subcanonical coverages, Definition 3.9.6).

PROPOSITION 7·6·7:   The role of categorical equivalence.

(a) Let $(F, U, \eta, \varepsilon) : \mathcal{C} \simeq \mathcal{D}$ be a strong equivalence of categories, where $\mathcal{D}$ has *assigned* categorical structure of some kind, such as products. Then the equivalence transfers it to $\mathcal{C}$.

(b) Suppose that products *etc. exist* as a *property* in $\mathcal{D}$ and that there is an equivalence functor $F : \mathcal{C} \rightarrow \mathcal{D}$. Then $\mathcal{C}$ also has products.

PROOF: Let $P = FX \times^{\mathcal{D}} FY$.

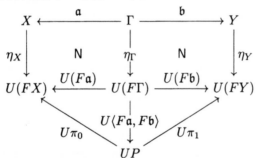

However, the unit η is also needed — and it must be natural — to show that $UP = X \times^{\mathcal{C}} Y$. $\square$

DEFINITION 7·6·8:   The structure $\square$ is said to be ***conservative*** if, for any category $\mathcal{C}$ and subcanonical $\square$-language $\mathcal{L}$, the functor

$$\ulcorner - \urcorner : \mathcal{C} \rightarrow \mathsf{Cn}_{\mathcal{L}}^{\square}$$

is full and faithful. That is, every map $\mathfrak{a} : \Gamma \equiv [x : \ulcorner X \urcorner] \rightarrow [y : \ulcorner Y \urcorner]$ is of the form $[\ulcorner \alpha \urcorner(x)/y]$ for a unique $\alpha : X \rightarrow Y$ in $\mathcal{C}$. If the interpretation is defined then $\alpha = [\![\mathfrak{a}]\!]$, so the issue is uniqueness, *i.e.* to show that $\mathfrak{a} = \ulcorner [\![\mathfrak{a}]\!] \urcorner$; in particular there is nothing more to do in the poset case.

Conservativity is a theorem that we must prove for each fragment $\square$ of structure, along with giving its interpretation. Traditionally, this term means the *relative* property of the *extension* of a theory by some new connective, for example adding function-types to algebra. According to our definition, $\square$ is conservative if the extension to the *full* $\square$-structure is conservative relative to *every* intermediate structure $\mathcal{L}$.

The equivalence.

THEOREM 7·6·9: Let $\square$ be a conservative structure, $\mathcal{C}$ a small category which has this structure and $\mathcal{L}$ a subcanonical $\square$-language for $\mathcal{C}$.

(a) Suppose $\mathcal{C}$ has and $\mathcal{L}$ names a *choice* of all $\square$-structure; that is, it is a strongly canonical $\square$-language (Definition 7.6.5(c)). Then the corresponding $\square$-type theory has an interpretation $[\![-]\!]$ in $\mathcal{C}$, which is a $\square$-preserving functor, and

$$\text{the syntax of the canonical language } \mathsf{Cn}_{\mathcal{L}}^{\square} \; \underset{\text{quoting } \ulcorner - \urcorner}{\overset{\text{interpretation } [\![-]\!]}{\underset{\simeq}{\rightleftarrows}}} \; \mathcal{C} \text{ semantics}$$

is a *strong equivalence* of categories (Definition 4.8.9(b)).

(b) Suppose $\mathcal{C}$ merely has (all) $\square$-structure as a *property*, and $\mathcal{L}$ is the weakly canonical $\square$-language (Definition 7.6.5(b)). (So for example if $\square = \times$ it must include at least one pair operation for each pair of objects.) Then $\ulcorner - \urcorner$ is an *equivalence functor* (Definition 4.8.9(c)) but $[\![-]\!]$ need not be defined.

(c) For a subcanonical language, $\ulcorner - \urcorner$ is full and faithful.

PROOF: [a] The functors $\ulcorner - \urcorner$ and $[\![-]\!]$ are given: we must find natural isomorphisms $\mathfrak{e}_\Gamma : \Gamma \cong \ulcorner [\![\Gamma]\!] \urcorner$ and $\varepsilon_X : [\![x : \ulcorner X \urcorner]\!] \cong X$ such that

$$\mathfrak{e}_{\ulcorner X \urcorner} \, ; \ulcorner \varepsilon_X \urcorner = \mathsf{id}_{\ulcorner X \urcorner} \qquad [\![\mathfrak{e}_\Gamma]\!] \, ; \varepsilon_{[\![\Gamma]\!]} = \mathsf{id}_{[\![\Gamma]\!]}$$

(normally we would write η for $\mathfrak{e}$ but we use the 𝔊erman letter to make it clear that these are morphisms of the syntactic category). The base case of the structural recursion defining the interpretation $[\![-]\!]$ is that $[\![x : \ulcorner X \urcorner]\!] = X$ and $[\![\ulcorner \alpha \urcorner(x)/y]\!] = \alpha$, so $\varepsilon_X = \mathsf{id}_X$ and ε is natural. The steps of that recursion made $[\![-]\!]$ a structure-preserving functor.

The first law now reduces to $\mathfrak{e}_{\ulcorner X \urcorner} = \mathsf{id}_{\ulcorner X \urcorner}$, which is the beginning of the recursive definition of $\mathfrak{e}_\Gamma$. A typical step in this definition is

$$
\begin{array}{ccc}
[U \to V] & \overset{\mathfrak{e}_{[U \to V]}}{\dashrightarrow} & \ulcorner [U \to V] \urcorner \\[4pt]
{\scriptstyle [U \to \mathfrak{e}_V]} \big\downarrow & & \big\uparrow {\scriptstyle \mathsf{abs}_{\ulcorner U \urcorner, \ulcorner V \urcorner}} \\[4pt]
[U \to \ulcorner [\![V]\!] \urcorner] & \underset{[\mathfrak{e}_U^{-1} \to \ulcorner [\![V]\!] \urcorner]}{\longrightarrow} & [\ulcorner [\![U]\!] \urcorner \to \ulcorner [\![V]\!] \urcorner]
\end{array}
$$

and so involves structural operations like $\ulcorner \mathsf{ev} \urcorner$ as well as the encoding operations $e = \mathsf{abs}$ of the type sub-expressions. By induction $\mathfrak{e}_\Gamma$ is invertible and $[\![\varepsilon_\Gamma]\!] = \mathrm{id}$, since this was given to be the case for the connectives of the canonical language, so the second law, $[\![\varepsilon_\Gamma]\!] \, ; \varepsilon_{[\![\Gamma]\!]} = \mathrm{id}_{[\![\Gamma]\!]}$, holds.

Naturality of $\varepsilon_{(-)}$ is not trivial: it is equivalent to conservativity.

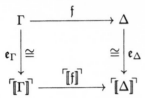

There is a unique morphism $\alpha : [\![\Gamma]\!] \to [\![\Delta]\!]$ of $\mathcal{C}$ with $\ulcorner \alpha \urcorner = \varepsilon_\Gamma^{-1} \, ; f \, ; \varepsilon_\Delta$. Then $\alpha = [\![\ulcorner \alpha \urcorner]\!] = \mathrm{id} \, ; [\![f]\!] \, ; \mathrm{id}$ since $[\![\varepsilon]\!] = \mathrm{id}$, so the square commutes.

The weaker result [b] follows by application of a meta-theorem analogous to Remark 3.3.2. A *particular* type may be given a (non-canonical) interpretation by *choosing* structure for each of finitely many instances of the connectives. Hence for each $\Gamma \in \mathrm{ob}\,\mathsf{Cn}_{\mathcal{L}}^{\square}$ there is some $X \in \mathrm{ob}\,\mathcal{C}$ such that $\Gamma \cong [x : \ulcorner X \urcorner]$ in $\mathsf{Cn}_{\mathcal{L}}^{\square}$, *i.e.* $\ulcorner - \urcorner : \mathcal{C} \to \mathsf{Cn}_{\mathcal{L}}^{\square}$ is essentially surjective. By conservativity it is full and faithful, so an equivalence functor. $\square$

REMARK 7·6·10:   Going from a category $\mathcal{C}$ with, for example, products (not necessarily specified) to its canonical language $\mathsf{L}^\times(\mathcal{C})$ and then back to the category of contexts and substitutions $\mathsf{Cn}_{\mathsf{L}^\times(\mathcal{C})}^\times$, we obtain a category with *specified* products. This is in a sense a cheat, because it has *far more* objects (the contexts) than the original category: the products are constructed not explicitly in the original category but *formally* as lists. Compare this with the meets in Proposition 3.9.3.

> We do not attempt to make global choices
> of products in semantic categories.

The introduction of variables (the number of which we have refused to specify) also proliferates the isomorphic copies of the objects. The old and new products are isomorphic *via* the pair operation, whilst no new maps are added between existing objects, or old ones identified, so the categories are equivalent.

REMARK 7·6·11:   There is a certain principle of interchangeability for *objects* (and their universal constructions such as products and function-types) in semantic categories. Syntactic categories do not obey it, but are more convenient for symbolic reasoning. By making use of a principle of interchangeability for *categories* we may transfer the advantages of the syntactic calculus to the semantic situation, *cf.* using symmetry to simplify problems such as Remark 1.6.9(c), so long as we only consider those internal properties of the categories which are invariant under weak equivalence.

Conservativity by normalisation. We still have to prove that the familiar connectives are conservative. Let us consider very briefly how this can be done syntactically. We have to show that any morphism $\mathfrak{a} : [x : \ulcorner X \urcorner] \to [y : \ulcorner Y \urcorner]$ in $\mathsf{Cn}_{\mathcal{L}}^{\square}$ is a substitution $[\ulcorner \alpha \urcorner(x)/y]$ for some unique $\alpha : X \to Y$ in $\mathcal{C}$.

REMARK 7·6·12: Without knowing anything about $\square$, the normal form theorem of *the direct declarative language* (Theorem 4.3.9) says that

$$\mathfrak{a} = [a/y],$$

where a is a (unique) term. By structural recursion, this may be

(a) $a = x$, in which case $\alpha = \mathrm{id}_X$,

(b) $a = \ulcorner \gamma \urcorner(b)$ with $b = \ulcorner \beta \urcorner(x)$ by the induction hypothesis, so $\alpha = \beta\,;\gamma$,

(c) the result of a meta-operation of $\square$ such as $\langle\,,\,\rangle$, λ or $[\,,\,]$,

(d) or an encoding operation (pair, abs or split) applied to a sub-term.

Cases (c) and (d) have to be handled for each connective $\square$ in turn. However, conservativity only asks us to consider terms of ground type, whereas $\langle\,,\,\rangle$, λ and split have results of product, exponential or sum type. So we only have to deal with pair, abs and case analysis.

THEOREM 7·6·13: Algebra is conservative over the unary language, *cf.* Exercise 4.25, and so every category with finite products is the classifying category for some algebraic theory.

PROOF: Suppose that the outermost operation-symbol is

$$\mathsf{pair}_{U,V,P} : \ulcorner U \urcorner \times \ulcorner V \urcorner \cong \ulcorner P \urcorner.$$

Since it is binary it must have two sub-terms, which are of (base) types $\ulcorner U \urcorner$ and $\ulcorner V \urcorner$. By the induction hypothesis, they are $\ulcorner \beta_i \urcorner(x)$. Then

$$\mathsf{pair}(\ulcorner \beta_0 \urcorner(x), \ulcorner \beta_1 \urcorner(x)) = \mathsf{pair}(\ulcorner \pi_0 \urcorner(\ulcorner \alpha \urcorner(x)), \ulcorner \pi_1 \urcorner(\ulcorner \alpha \urcorner(x))) = \ulcorner \alpha \urcorner(x),$$

where $\alpha = \langle \beta_0, \beta_1 \rangle : X \to U \times V = P$ in $\mathcal{C}$. $\square$

COROLLARY 7·6·14: In algebra, every term

$$\Gamma \equiv [x_1 : \ulcorner X_1 \urcorner, x_2 : \ulcorner X_2 \urcorner, \ldots, x_n : \ulcorner X_n \urcorner] \vdash a : \ulcorner Y \urcorner$$

equals $\ulcorner \alpha \urcorner(\mathsf{tuple}(x_1, \ldots, x_n))$ for some $\alpha : X_1 \times \cdots \times X_n \to Y$ in $\mathcal{C}$. $\square$

We shall ignore tuple from now on, just as we took the identities and composition for granted in Definition 7.6.5.

THEOREM 7·6·15: The λ-calculus is conservative over algebra, so every cartesian closed category is the classifying category of some λ-theory.

PROOF: Let $\mathsf{abs} : \ulcorner U \urcorner \to \ulcorner V \urcorner \to \ulcorner F \urcorner$. We show that any term

$$\Gamma \equiv [x_1 : \ulcorner X_1 \urcorner, x_2 : \ulcorner X_2 \urcorner, \ldots, x_n : \ulcorner X_n \urcorner] \vdash a = \mathsf{abs}(f) : \ulcorner F \urcorner$$

is provably equal to $\ulcorner \alpha \urcorner(x_1, \ldots, x_n)$. Now the arity of abs does not itself force $f = \lambda z.\, p$, but this follows from the normalisation theorem *of the λ-calculus* (Fact 2.3.3 and Exercise 2.23). The assumption that the terms are normal is to be made at the beginning of the argument.

Applying the corollary to p, for some $\beta : X_1 \times \cdots \times X_n \times U \to V$,

$$x_1 : \ulcorner X_1 \urcorner, \ldots, x_n : \ulcorner X_n \urcorner, z : \ulcorner U \urcorner \vdash p = \ulcorner \beta \urcorner(\underline{x}, z) : \ulcorner V \urcorner.$$

Then (in the context $x_1 : \ulcorner X_1 \urcorner, \ldots, x_n : \ulcorner X_n \urcorner$)

$$
\begin{aligned}
a &= \mathsf{abs}\left(\lambda z.\, \ulcorner \beta \urcorner(\underline{x}, z)\right) \\
&= \mathsf{abs}\left(\lambda z.\, \ulcorner \mathsf{ev} \urcorner(\ulcorner \alpha \urcorner(\underline{x}), z)\right) = \ulcorner \alpha \urcorner(\underline{x}),
\end{aligned}
$$

where $\alpha = \tilde{\beta}$. This is justified by the β- and η-rules for abs,

$$\ulcorner \mathsf{ev} \urcorner(\mathsf{abs}\, f, z) \rightsquigarrow fz \qquad \mathsf{abs}(\lambda z.\, \ulcorner \mathsf{ev} \urcorner(p, z)) \rightsquigarrow p$$

for $f : \ulcorner U \to V \urcorner$, $p : \ulcorner U \urcorner \to \ulcorner V \urcorner$ and $z : \ulcorner U \urcorner$. □

In the case of the sum we must simplify $[f, g](c)$, where c is a closed term. Now we want c to be provably equal to either $\nu_0(a)$ or $\nu_1(b)$, so that this expression is just fa or gb. This is the **disjunction property**.

For algebra there is probably no alternative to the syntactic method, as in any case we have to prove the correctness of the canonical language, which is necessarily rather complicated. For generalised algebraic theories, this is the subject of Section 8.4. Once we have the language in categorical form, category theory provides a powerful method to prove conservativity.

7.7   GLUING AND COMPLETENESS

To complete the equivalence between syntax and semantics, it remains to prove confluence, strong normalisation for the λ-calculus and the disjunction property for intuitionistic logic. The conceptual content of these results, when proved syntactically, is drowned in a swamp of symbolic detail which cannot be transferred to new situations. The remarkable construction which we use illustrates how much can be discovered simply by playing with adjoints and pullbacks.

The origin of the name *gluing* is that this is how to recover a topological space from an open set and its complementary closed set (Exercise 3.71). The construction for Grothendieck toposes was first set out in [AGV64, Exposé IV, §9.5]. Considered as inverse images, functors between toposes with the rich properties of $(\pi_0, \pi_1) : S{\downarrow}U \to S \times \mathcal{A}$ or $\pi_1 : S{\downarrow}U \to \mathcal{A}$ are called **surjections** and **open inclusions** respectively (geometrically, $S \times \mathcal{A}$ is the disjoint union of S and $\mathcal{A}$.)

The gluing construction. Recall from Example 7.3.10(i) that, for any functor $U : \mathcal{A} \to S$, the **gluing construction** is the category $S{\downarrow}U$ whose objects consist of $I \in \mathrm{ob}\,S$, $\Gamma \in \mathrm{ob}\,\mathcal{A}$ and $\mathfrak{f} : I \to U\Gamma$ in S, and whose morphisms are illustrated by the diagram below. We shall say that $(I, \Gamma, \mathfrak{f})$ is **tight** if $\mathfrak{f}$ is an isomorphism.

$$
\begin{array}{cccccc}
I & I & \xrightarrow{\;\mathfrak{f}\;} & U\Gamma & \Gamma \\
{\scriptstyle \xi}\big\downarrow & {\scriptstyle \xi}\big\downarrow & & {\scriptstyle U\mathfrak{u}}\big\downarrow & \big\downarrow{\scriptstyle \mathfrak{u}} \\
J & J & \xrightarrow{\;\mathfrak{g}\;} & U\Delta & \Delta
\end{array}
$$

Since it preserves everything in sight, π_1 is also called a **logical functor**.

PROPOSITION 7·7·1: Let $U : \mathcal{A} \to S$ be any functor. (We emphasise the case where it preserves finite products and maybe pullbacks, and S is a topos; the dotted lines signify even stronger assumptions which we do not wish to make. See Exercise 3.70 for posets.) Then

$$
\begin{array}{ccccc}
S{\downarrow}U & (0 \to U\Gamma) & (I \xrightarrow{\mathfrak{f}} U\Gamma) & (U\Gamma \xrightarrow{\mathrm{id}} U\Gamma) & (I \xrightarrow{\mathfrak{f}} U\Gamma) \\[2pt]
& E\big\uparrow \quad \dashv & \pi_1 \updownarrow \quad \dashv & A\big\uparrow \quad \dashv & \vdots \\[2pt]
\mathcal{A} & \Gamma & \Gamma & \Gamma & \Gamma \times_{RU\Gamma} RI
\end{array}
$$

(a) $\pi_1 : S {\downarrow} U \to \mathcal{A}$ is an op-fibration, and also a fibration if S has pullbacks (Definition 9.2.6);

(b) if S has an initial object 0_S then π_1 also has a full and faithful left adjoint, E, so π_1 creates limits and E preserves colimits;

(c) E identifies $\mathcal{A}$ with the full (co-reflective) subcategory of $S {\downarrow} U$ in which $I = 0_S$;

(d) if 0_S is strict (Definition 5.5.1) and $\mathcal{A}$ has a terminal object $1_\mathcal{A}$ then this subcategory of $S {\downarrow} U$ is the *slice* by $(0_S, 1_\mathcal{A}, 0_S \to U1_\mathcal{A})$, so E is a discrete fibration and creates non-empty limits;

(e) in this case, moreover,

$$
E(\Gamma \times \pi_1(J \xrightarrow{\mathfrak{f}} U\Delta)) = (0 \to U(\Gamma \times \Delta)) = E(\Gamma) \times (J \xrightarrow{\mathfrak{f}} U\Delta)
$$

holds and the co-unit $E \cdot \pi_1 \to \mathrm{id}$ is a cartesian transformation (Remark 6.3.4, *cf.* the Frobenius law, Lemma 1.6.3, Corollary 9.3.9);

(f) π_1 has a right adjoint, A, so π_1 preserves colimits and A limits;

(g) A identifies $\mathcal{A}$ with the full subcategory of $\mathcal{S} \downarrow U$ consisting of tight objects, which is therefore reflective; it is also an exponential ideal (Exercise 7.11) if, as we shall show, $\mathcal{S} \downarrow U$ is cartesian closed;

(h) $U = \pi_0 \cdot A$ and $\pi_1 \cdot A = \mathrm{id}$;

(i) A preserves whatever colimits U does (but one of our main goals is to show that this happens); if U has a right adjoint $R : \mathcal{S} \to \mathcal{A}$ and $\mathcal{A}$ has pullbacks then A also has a right adjoint;

$$\mathcal{S} \downarrow U \quad (I \xrightarrow{\eta_I} UFI) \quad (I \xrightarrow{f} U\Gamma) \quad (I \to U\mathbf{1})$$

with the diagram showing $\mathcal{S}$, I, I, I below, and adjunctions $\dashv$ with π_0 and T.

(j) π_0 is a fibration;

(k) π_0 preserves whatever limits U does; indeed if $F \dashv U$ with unit η then π_0 has a left adjoint, identifying $\mathcal{S}$ with the full subcategory of $\mathcal{S} \downarrow U \simeq F \downarrow \mathcal{A}$ which consists of universal maps (Definition 7.1.1);

(l) if $\mathcal{A}$ and $\mathcal{S}$ have and U preserves $\mathbf{1}$ then π_0 has a right adjoint T, so π_0 preserves colimits and T limits;

(m) T identifies $\mathcal{S}$ with the full subcategory of $\mathcal{S} \downarrow U$ in which $\Gamma = 1_{\mathcal{A}}$;

$$\mathcal{S} \times \mathcal{A} \quad (I, \Gamma) \quad (I, \Gamma) \quad (I, \Gamma)$$

with the diagram showing adjunctions via (π_0, π_1) and V, and below:

$$\mathcal{S} \downarrow U \quad (I \to U\Delta) \quad (I \xrightarrow{f} U\Gamma) \quad (J \xrightarrow{\pi_1} U\Gamma)$$
$$\Delta = (FI + \Gamma) \qquad J = (I \times U\Gamma)$$

(n) (π_0, π_1) creates colimits, and has a right adjoint V if $\mathcal{S}$ has binary products;

(o) (Gavin Wraith) $\mathcal{S} \downarrow U$ is the category of coalgebras for the comonad (*cf.* Definition 7.5.4) on $\mathcal{S} \times \mathcal{A}$ induced by $(\pi_0, \pi_1) \dashv V$;

(p) (π_0, π_1) creates any limits which U preserves, and has a left adjoint if U does and $\mathcal{A}$ has binary coproducts. □

COROLLARY 7·7·2: Assuming only that U preserves finite limits, if $\mathcal{A}$ and $\mathcal{S}$ have the following structure, so does $\mathcal{S} \downarrow U$, and π_1 preserves it:

(a) finite limits;

(b) stable disjoint sums (Section 5.5);

(c) regularity (Section 5.8);

(d) effective regularity (Barr-exactness);

(e) the structure of a prelogos;

(f) that of a (countably) complete prelogos;

(g) being a pretopos;

(h) $\mathbb{N}$ (Example 6.4.13) and $\mathsf{List}(X)$ (Exercise 6.30).

If U preserves this structure then so does π_0. □

Implication and the function-type are considered in Exercise 3.72 and Proposition 7.7.12; Exercises 7.50–7.51 deal with factorisation systems (and so the existential quantifier, Section 9.3) and 2 (higher order logic).

Conservativity. Let $\mathcal{C}$ be a category and $\mathcal{L}$ a subcanonical □-language for it (Definition 7.6.5(a)), *i.e.* $\mathcal{L}$ names *all* of the objects and maps of $\mathcal{C}$, stating all of the laws which hold between them, possibly with encoding operations for *some* □-structure. Recall from Definition 7.6.8 that □ is called **conservative** if the functor $\ulcorner - \urcorner : \mathcal{C} \to \mathsf{Cn}_{\mathcal{L}}^{\square}$ is full and faithful.

NOTATION 7·7·3:

(a) Define a functor $U : \mathsf{Cn}_{\mathcal{L}}^{\square} \to \mathcal{S} \equiv \mathcal{S}et^{\mathcal{C}^{op}}$ by

$$\Gamma \mapsto \mathsf{Cn}_{\mathcal{L}}^{\square}(\ulcorner - \urcorner, \Gamma),$$

which preserves any limits that $\mathsf{Cn}_{\mathcal{L}}^{\square}$ has;

(b) and another functor $Q : \mathcal{C} \to \mathcal{S} \downarrow U$ by

$$X \mapsto (\mathcal{H}_X, \ulcorner X \urcorner, \mathsf{q}_X),$$

where $\mathsf{q}_X : \mathcal{H}_X \to F \ulcorner X \urcorner$ is a morphism of $\mathcal{S} \equiv \mathcal{S}et^{\mathcal{C}^{op}}$, *i.e.* a natural transformation between presheaves. It has components

$$\mathsf{q}_{X,Z} : \mathcal{H}_X(Z) \equiv \mathcal{C}(Z, X) \to U \ulcorner X \urcorner(Z) \equiv \mathsf{Cn}_{\mathcal{L}}^{\square}(\ulcorner Z \urcorner, \ulcorner X \urcorner)$$

given by quoting $\ulcorner - \urcorner$ of $\mathcal{C}$-maps $Z \to X$ (which are the operation-symbols of $\mathcal{L}$). In Proposition 7.3.9, Q is the mediator to the comma category from the lax square consisting of the Yoneda embedding $\mathcal{H}_{(-)} : \mathcal{C} \hookrightarrow \mathcal{S}et^{\mathcal{C}^{op}}$, quoting $\ulcorner - \urcorner : \mathcal{C} \to \mathsf{Cn}_{\mathcal{L}}^{\square}$ and $\mathsf{q} : \mathcal{H}_{(-)} \to U \ulcorner - \urcorner$.

EXAMPLE 7·7·4: Let $\mathcal{C} = \{1\}$ and suppose that $\mathcal{L}$ says that $\ulcorner 1 \urcorner$ is indeed the terminal object (so there is a nullary encoding operation $\vdash \star : \ulcorner 1 \urcorner$ with one η-rule $x : \ulcorner 1 \urcorner \vdash x = \star$). Then

$$U \equiv \mathsf{Cn}_{\mathcal{L}}^{\square}(1, -) : \mathsf{Cn}_{\mathcal{L}}^{\square} \to \mathcal{S} \equiv \mathcal{S}et$$

gives the set of global elements of a context, *i.e.* its closed terms, or proofs under no hypotheses (Remark 4.5.3). In sheaf theory this functor is traditionally called Γ, which of course conflicts with the notation of this book. When $\mathcal{A}$ is a topos (instead of $\mathsf{Cn}_{\mathcal{L}}^{\square}$), $\widehat{\mathcal{A}} \equiv \mathcal{S}et \downarrow U$ is called the **Freyd cover** or **scone** (Sierpiński cone) of $\mathcal{A}$, the (closed) vertex being $\mathcal{S}et$ *quâ* the one-point topos (*cf.* lifting a domain, Definition 3.3.7 and Exercise 3.71). Also, $Q(1) = (\mathsf{id} : 1 \to U1)$. □

LEMMA 7·7·5: Q is full and faithful.

PROOF: An $(\mathcal{S} \downarrow U)$-map $QX \to QY$ is a pair $(\xi, \mathfrak{a})$ making the square below commute, but by the Yoneda Lemma (Theorem 4.8.12(a)) any $\mathcal{S}$-map (natural transformation) $\xi : \mathcal{C}(-, X) \to \mathcal{C}(-, Y)$ is of the form $\mathrm{post}(\alpha)$ for some unique (semantic) map $\alpha = \xi_X(\mathrm{id}_X) : X \to Y$ in $\mathcal{C}$.

$$
\begin{array}{ccccc}
X & \mathcal{C}(-, X) \xrightarrow{\ulcorner = \urcorner_{Z,X}} \mathrm{Cn}_{\mathcal{L}}^{\Box}(\ulcorner - \urcorner, \ulcorner Y \urcorner) & \ulcorner X \urcorner \\[4pt]
\alpha \downarrow \quad & \xi = \mathrm{post}(\alpha) \downarrow \qquad \mathrm{post}(\mathfrak{a}) \downarrow & \downarrow \mathfrak{a} \\[4pt]
Y & \mathcal{C}(-, Y) \xrightarrow{\ulcorner = \urcorner_{Z,Y}} \mathrm{Cn}_{\mathcal{L}}^{\Box}(\ulcorner - \urcorner, \ulcorner Y \urcorner) & \ulcorner Y \urcorner
\end{array}
$$

The other map, $\mathfrak{a}$, belongs to $\mathrm{Cn}_{\mathcal{L}}^{\Box}$, *i.e.* it is syntactic, a substitution. We must show that $\mathfrak{a} = \ulcorner \alpha \urcorner$ without assuming the theorem we're trying to prove, that $\ulcorner - \urcorner$ is full and faithful. In fact this follows from the fact that the square commutes at $\mathrm{id}_X \in \mathcal{C}(X, X)$. Hence $(\xi, \mathfrak{a}) = Q\alpha$.   □

THEOREM 7·7·6:   Suppose that

(a) $\mathcal{S} \downarrow U$ has and π_1 preserves $\Box$-structure,

(b) $\mathcal{S}$ satisfies the axiom-scheme of replacement, and

(c) Q is a model of $\mathcal{L}$, *i.e.* Q preserves any $\Box$-structure which $\mathcal{L}$ specifies.

Then $\ulcorner - \urcorner$ is full and faithful, *i.e.* $\Box$ is conservative.

In practice, π_1 preserves $\Box$-structure *on the nose*, and $[\![-]\!]$ is defined by structural recursion so that $[\![\Gamma]\!] = ([\![\Gamma]\!]_0, \Gamma, q_\Gamma)$, where $[\![-]\!]_0 = \pi_0 [\![-]\!]$.

The functor $[\![-]\!]$ reflects the *existence* of isomorphisms: if $[\![\Gamma]\!] \cong [\![\Delta]\!]$ *somehow* then $\Gamma \cong \Delta$. (This is the categorical analogue of an injective function between posets.) For higher order logic, $[\![-]\!]$ need not be full.

PROOF: Since $\mathrm{Cn}_{\mathcal{L}}^{\Box}$ is the classifying category, and using the axiom-scheme of replacement to justify the recursion, the model Q extends to a $\Box$-preserving functor $[\![-]\!] : \mathrm{Cn}_{\mathcal{L}}^{\Box} \to \mathcal{S} \downarrow U$ making the upper triangle commute, uniquely up to unique isomorphism:

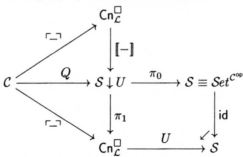

The lower $\mathsf{Cn}_{\mathcal{L}}^{\square}$ is also a model, for which both id and $\pi_1[\![-]\!]$ serve as the mediator from the classifying category (since $[\![-]\!]$ and π_1 preserve $\square$-structure), so they are isomorphic, $\pi_1[\![\Gamma]\!] \cong \Gamma$. Hence $[\![-]\!]$ is faithful, whilst Q is full and faithful, so $\ulcorner-\urcorner$ is also full and faithful. $\qquad\square$

EXAMPLES 7·7·7: The first clause of the theorem is applicable to the structures $\square$ listed in Corollary 7.7.2 (and more). The difficulty lies in condition 7.7.6(c), *i.e.* what structure is named by the language $\mathcal{L}$.

(a) $\mathcal{L}$ may be just the canonical *elementary* language of $\mathcal{C}$, with no extra structure.

(b) $\mathcal{L}$ may also include some tuple maps, or, more generally, encoding operations for some finite limits (Definition 7.6.5). Then $\mathcal{C}$, $\mathsf{Cn}_{\mathcal{L}}^{\square}$, $\mathcal{S}$ and $\mathcal{S}{\downarrow}U$ have these limits and $\mathcal{H}_{(-)}$, $\ulcorner-\urcorner$, U and Q preserve them.

(c) $\mathcal{L}$ also includes some stable colimits, encoded by a Grothendieck topology $\mathcal{J}$. The category $\mathcal{S} = \mathcal{S}et^{\mathcal{C}^{op}}$ of presheaves must be replaced by the category $\mathsf{Sh}(\mathcal{C}, \mathcal{J})$ of sheaves, which freely adjoins colimits, but keeping those in $\mathcal{J}$, *cf.* Theorem 3.9.7 for posets.

(d) Corollary 7.7.13 shows that Q and A also preserve exponentials.

The construction relies on the fact that $\mathcal{S}$, which is a topos, has all of the extra structure $\square$ (plus the axiom-scheme of replacement) and the Yoneda embedding is full and faithful and preserves it. However, $\mathcal{S}$ does not *freely* adjoin this structure (except in the case of arbitrary stable colimits), and the question is whether the embedding into the free category $\mathsf{Cn}_{\mathcal{L}}^{\square}$ is full and faithful.

REMARK 7·7·8: In the case $\mathcal{C} = \{1\}$, an object $I \xrightarrow{f} U\Gamma \equiv \mathsf{Cn}_{\mathcal{L}}^{\square}(1, \Gamma)$ of $\mathcal{S}{\downarrow}U$ is a family of closed terms or proofs of Γ, indexed by I. More generally, it is a cocone $f_i : \ulcorner X_i \urcorner \to \Gamma$ of such proofs under a certain diagram $X_{(-)} : \mathcal{I} \to \mathcal{C}$ of base types or hypotheses. This diagram is the discrete fibration corresponding to the sheaf $I : \mathcal{C}^{op} \to \mathcal{S}et$ by Proposition 9.2.7.

Notation 7.7.3(b) provided a *specific* sheaf of closed terms q_X of each base type $X \in \mathrm{ob}\,\mathcal{C}$, so we shall call it the **realisation** of X. Theorem 7.7.6 showed that this is an isomorphism (the realisation is *tight*) for base types, and extended the notation to general contexts. We already know that the full subcategory of tight objects in $\mathcal{S}{\downarrow}U$ is closed under definable limits, so the same is true of the class of tightly realised contexts. We shall see that this extends to colimits and exponentials, so if $\square$ consists only of this (first order) structure then A is already the interpretation $[\![-]\!]$, and this is full as well as faithful.

For higher order logic the realisation is no longer tight. Andre Scedrov and Philip Scott [SS82] trace the method back in the symbolic tradition to Stephen Kleene's realisability methods (1962), and link it to the categorical construction. Peter Freyd found this after hearing the presentation of Scott's work with Joachim Lambek [LS80] at a conference, and not at first believing the theorem below which their results implied.

Existence and disjunction properties. Recall that $[-]_0 \equiv \pi_0 \cdot [-]$.

LEMMA 7·7·9: The realisation $q : [-]_0 \to U$ is naturally split epi.

PROOF: First observe that, by the Yoneda lemma (Theorem 4.8.12(c)),

$$\mathcal{S}([\ulcorner-\urcorner]_0, [\Gamma]_0) = \mathcal{S}(\mathcal{H}_{(-)}, [\Gamma]_0) \cong [\Gamma]_0.$$

The section $\mathsf{m} : \mathcal{U} \to [-]_0$ (called u in [AHS95]) is the action of $[-]_0$,

$$\mathsf{Cn}_{\mathcal{L}}^{\square}(\ulcorner-\urcorner, \Gamma) \xrightarrow{\;([-]_0)^{\ulcorner-\urcorner,\Gamma}\;} \mathcal{S}([\ulcorner-\urcorner]_0, [\Gamma]_0) \cong [\Gamma]_0$$

naturally in ($-$ and) Γ. We have to show that $q_\Gamma(\mathsf{m}_\Gamma(\mathsf{u})) = \mathsf{u}$ for each $\mathsf{u} \in U\Gamma \equiv \mathsf{Cn}_{\mathcal{L}}^{\square}(\ulcorner-\urcorner, \Gamma)$, along the bottom row of the next diagram. Evaluating this equation at each $X \in \mathrm{ob}\,\mathcal{C}$, we use naturality with respect to $\mathsf{u}_X : \ulcorner X \urcorner \to \Gamma$.

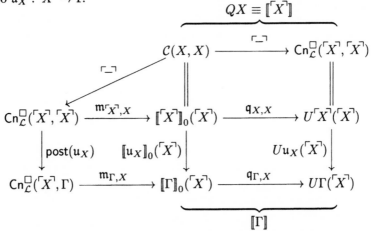

Then the diagram in Set commutes and the required law follows from its effect on $\mathrm{id}_X \in \mathcal{C}(X, X)$.   □

THEOREM 7·7·10: (Freyd) $U : \mathsf{Cn}_{\mathcal{L}}^{\square} \to \mathcal{S}$ preserves colimits named in $\square$, but not necessarily those named in $\mathcal{L}$. In particular, the global sections functor (Example 7.7.4) preserves

(a) the initial object, so there is no closed term $\vdash \mathbf{0}$ in $\mathsf{Cn}^{\square}$;

(b) coproducts, so there are just two closed terms $\vdash \mathbf{2}$;

(c) regular epis, so $\mathbf{1}$ is projective (Remark 5.8.4(e));

(d) coproducts and coequalisers (Example 6.4.13), so the closed terms $\vdash \mathbb{N}$ are the numerals.   □

PROOF: We have just shown that U is a retract of $[\![-]\!]_0$, which preserves whatever colimits are in $\square$, and hence so does U (Exercise 7.13). $\square$

COROLLARY 7·7·11: In terms of proof theory, the fragment $\square$

(a) is consistent;

(b) has the disjunction property: if $\vdash \phi \vee \psi$ then either $\vdash \phi$ or $\vdash \psi$;

(c) has the existence property: if $\vdash \exists x.\ \phi[x]$ then $\vdash \phi[a]$ for some a;

(d) has standard arithmetic. $\square$

$\mathcal{S}$ can prove consistency of $\square$ because it has been strengthened with the axiom-scheme of replacement, to which we return in Section 9.6.

Exponentials. Recall that $[I \to J](X) = \mathcal{S}(\mathcal{H}_X \times I, J)$ by Exercise 4.41, whilst $U\Gamma(X) = \mathsf{Cn}_{\mathcal{L}}^{\square}(\ulcorner X \urcorner, \Gamma)$ by Notation 7.7.3(a).

PROPOSITION 7·7·12: Suppose $\mathcal{A}$ and $\mathcal{S}$ are cartesian closed, $\mathcal{S}$ has pullbacks and U preserves finite products. Then $\mathcal{S} \downarrow U$ is cartesian closed and π_1 and A (but not π_0 or U) preserve exponentials. (See Exercise 3.72 for the version for Heyting semilattices.)

PROOF: Given $(I \xrightarrow{\mathfrak{f}} U\Gamma)$ and $(J \xrightarrow{\mathfrak{g}} U\Delta)$ in $\mathcal{S} \downarrow U$, we form an internal version of the hom-set $(\mathcal{S} \downarrow U)(\mathfrak{f}, \mathfrak{g})$, namely the pullback

$$
\begin{array}{ccc}
H & \xrightarrow{\hspace{5cm}} & [I \to J] \\
\Big\downarrow{\scriptstyle \mathfrak{h}} & {\scriptstyle \lrcorner} & \Big\downarrow{\scriptstyle [I \to \mathfrak{g}]} \\
U[\Gamma \to \Delta] \longrightarrow [U\Gamma \to U\Delta] & \xrightarrow{[\mathfrak{f} \to U\Delta]} & [I \to U\Delta]
\end{array}
$$

in $\mathcal{S}$, where the lower left map is the exponential transpose of

$$
U\Gamma \times U[\Gamma \to \Delta] \cong U\big(\Gamma \times [\Gamma \to \Delta]\big) \xrightarrow{U\,\mathsf{ev}} U\Delta.
$$

Then $(H \xrightarrow{\mathfrak{h}} U[\Gamma \to \Delta])$ is the required exponential, with λ-abstraction

$$
\begin{array}{ccc}
C \times I \dashrightarrow J & \qquad & C \dashrightarrow H \xrightarrow{\hspace{2cm}} [I \to J] \\
{\scriptstyle c \times \mathfrak{f}}\Big\downarrow \quad \Big\downarrow{\scriptstyle \mathfrak{g}} \quad \mathfrak{g} \xmapsto{\ \lambda\ } c & \quad {\scriptstyle c}\Big\downarrow \quad {\scriptstyle \mathfrak{h}}\Big\downarrow \quad {\scriptstyle \lrcorner} \quad \Big\downarrow{\scriptstyle [I \to \mathfrak{g}]} \\
U(\Xi \times \Gamma) \dashrightarrow U\Delta & \quad U\Xi \dashrightarrow U[\Gamma \to \Delta] \longrightarrow [I \to U\Delta]
\end{array}
$$

Notice that if $\mathfrak{g}$ is an isomorphism, then so is $\mathfrak{h}$. $\square$

COROLLARY 7·7·13: Q preserves any exponentials that are named in $\mathcal{L}$. If Δ has tight realisation then so does $[\Gamma \to \Delta]$ for any Γ whatever.

PROOF: In computing $Q(X \to Y)$, the edges of the pullback square above are all invertible, the vertices being isomorphic to $\mathcal{H}_{X \to Y}$. $\square$

THEOREM 7·7·14: (Yves Lafont, [Laf87, Annexe C]) The λ-calculus is a conservative extension of algebra. ☐

This is as much as is needed for the equivalence between semantics and syntax in Theorem 7.6.9. We haven't proved the normalisation theorem as such, but by a variation on this technique every term is provably equal to a normal form [MS93, AHS95, ˜CDS98]. It seems likely that a purely categorical proof will be found for strong normalisation itself, handling reduction paths in the fashion of Exercise 4.34. We return to consistency and the axiom of replacement in Section 9.6.

EXERCISES VII

1. Find the co-units and adjoint transformations of the examples in Section 7.1.

2. Define the Cauchy and Dedekind completions as functors on suitable categories and prove their universal properties.

3. Show that the functor which assigns the set of components to a graph (Example 7.1.6(c)) preserves finite products but not equalisers or pullbacks. Explain why the axiom of choice is needed to extend this to infinite products.

4. Let $T : Set \to Set$ be a functor coding an infinitary free theory as in Chapter VI. What is the categorical structure $\square$-Cat which corresponds to this fragment of logic ($\square$)? [Hint: restrict the arity to κ and consider κ-ary products.] Describe the classifying category $Cn_T^\square$. Find a category $\mathcal{C}$ of languages (whose objects are functors such as T) so that this classifying category is the universal map from the object T to the forgetful functor $U : \square$-$Cat \to \mathcal{C}$.

5. Explain how naturality of λ and ε defines postcomposition and the effect of the functor $(-)^X$ on maps.

6. Show that any adjunction between groups must be a strong equivalence, the two functors being isomorphisms whose composition is a conjugacy (inner automorphism) (*cf.* Exercise 4.36).

7. Let $U = \mathcal{S}(1, -) : \mathcal{S} \to Set$ be the global sections functor of a topos $\mathcal{S} = Set^{\mathcal{I}^{op}}$. Show that U calculates limits of presheaves considered as diagrams in Set, and that $K \dashv U$, where $KX(I) = X$ for $X \in ob\, Set$ and $I \in ob\, \mathcal{I}$. The functor K itself has a left adjoint: what is it?

8. Let $\mathcal{S}$ be the monoid of parades in the Lineland Army (Example 1.2.7). Show that the operation $\mathcal{S}^{op} \hookrightarrow \mathcal{S}$ which reverses a parade and promotes everyone by one grade is a monoid homomorphism which is

symmetrically adjoint to itself on the right. Show that this is the free monoid-with-a-monad.

9. Given "$F \dashv U$" with $\varepsilon : F \cdot U \cong \mathrm{id}_D$ naturally, and a family of isomorphisms $\eta_X : X \cong U(FX)$, show that U is full and faithful iff η is natural (*cf.* the proof of Theorem 7.6.9).

10. Let $\mathcal{A}$ be a cartesian closed category and $F : \mathcal{A} \to \mathcal{S}$ a functor which preserves binary products and has a right adjoint U. Show that $[X \to_\mathcal{S} UA] \cong U[FX \to_\mathcal{A} A]$ for $X \in \mathrm{ob}\,\mathcal{S}$ and $A \in \mathrm{ob}\,\mathcal{A}$ (*cf.* Exercise 3.69).

11. Deduce that any reflective subcategory $\mathcal{A} \subset \mathcal{S}$ of a cartesian closed category is an **exponential ideal** ($[X \to_\mathcal{S} UA] \in \mathrm{ob}\,\mathcal{A}$ and is the exponential there) iff the reflection preserves products. In this case $\mathcal{A}$ is also cartesian closed.

12. Let $U : \mathcal{A} \subset \mathcal{S}$ be a full replete subcategory. Show that there exists a functor $F : \mathcal{S} \to \mathcal{A}$ such that both $F \dashv U$ and $U \dashv F$ iff each object $X \in \mathrm{ob}\,\mathcal{S}$ carries a natural split idempotent $\alpha_X : X \to X$ (Definition 1.3.12) such that $X \in \mathrm{ob}\,\mathcal{A} \Leftrightarrow \alpha_X = \mathrm{id}_X$.

13. Let $F : \mathcal{C} \to \mathcal{D}$ be a (co)limit-preserving functor and $\alpha : F \to F$ a natural idempotent. Suppose that each $\alpha_X : FX \to FX$ has a splitting GX in $\mathcal{D}$. Show that $G : \mathcal{C} \to \mathcal{D}$ is also a (co)limit-preserving functor.

14. Treating diagrams as functors, explain how (co)cones are natural transformations.

15. Construct finite wide pullbacks from binary pullbacks.

16. Construct (finite) connected limits from equalisers and (finite) wide pullbacks.

17. Show that $\mathrm{colim}_\mathcal{I}\,\mathrm{colim}_\mathcal{J}\,X_{(I,J)} \cong \mathrm{colim}_\mathcal{J}\,\mathrm{colim}_\mathcal{I}\,X_{(I,J)}$. (For joins, see Lemma 3.5.4.)

18. Suppose that $\mathcal{S}$ has (co)limits of shape $\mathcal{I}$, and let $\mathcal{C}$ be any (small) category. Show that $[\mathcal{C} \to \mathcal{S}] \equiv \mathcal{S}^\mathcal{C}$ also has (co)limits of shape $\mathcal{I}$, and that they are constructed pointwise, *cf.* Lemma 3.5.7.

19. Let $\mathcal{I}$ be a filtered category and $U : \mathcal{I} \to \mathcal{J}$ a final functor. Show that $\mathcal{J}$ is also filtered. [Hint: use filteredness to simplify the definition of finality first.] Show that if U is also full and faithful then filteredness of $\mathcal{J}$ implies that of $\mathcal{I}$ (*cf.* Exercise 3.15). In this case show that, to test finality, it suffices that the object class be "cofinal" (Proposition 3.2.10): $\forall X : \mathrm{ob}\,\mathcal{S}.\ \exists A : \mathrm{ob}\,\mathcal{A}.\ \exists f : X \to UA$.

20. Prove the converse of Proposition 7.3.11. [Hint: consider the Yoneda embedding $X_{(-)} : \mathcal{I} \to \mathcal{C} = \mathcal{S}et^{\mathcal{I}^{op}}$ and let $\Theta = \mathcal{H}_I$.]

21. Show that any set X is finitely presentable iff $(-)^X$ preserves filtered colimits, and that every set is a filtered colimit of finitely presented sets, *cf.* Proposition 6.6.13ff. [Hint: consider $\mathcal{I} = \mathcal{S}_{\text{fp}} \downarrow X$.]

22. Collect and compare properties (such as Definition 6.6.14) of the form "$(-)^X$ preserves colimits of shape $\mathcal{I}$" for various classes of diagrams $\mathcal{I}$.

23. We say that X is **finitely related** (*cf. stable* in [Joh77, p. 233]) if, in any pullback of the form

with A and B finitely generated, K is also finitely generated. In other words, for any two lists of elements, there is (in the internal sense) some list of coincidences between them. Show that $X \in \text{ob}\,Set$ is finitely related iff it has decidable equality.

24. Formulate the definitions of finitely presentable, generable and related for objects of $\mathcal{Mod}(\mathcal{L})$, where $\mathcal{L}$ is a finitary algebraic theory. Show that X is finitely generated iff $\mathcal{C}(X, -)$ preserves directed unions, finitely related iff this functor preserves filtered colimits of surjections, and finitely presented iff all filtered colimits are preserved. Hence show that it is finitely presented iff it is finitely generated and finitely related.

25. Let $\mathcal{L}$ be a *disjunctive* theory such as trichotomous orders, coherence spaces or fields (Section 5.5). Show that the forgetful functor $\mathcal{Mod}(\mathcal{L}) \to Set^\Sigma$ creates connected limits. For the theory of fields having roots of specified polynomials, show that the forgetful functor creates wide pullbacks but not equalisers.

26. Let $F : \mathcal{C} \to \mathcal{A}$, $U : \mathcal{A} \to \mathcal{C}$, $B : \mathcal{A} \to \mathcal{K}$ and $Y : \mathcal{C} \to \mathcal{K}$ be functors. Show that $F \dashv U$ if there is a natural bijection

$$\frac{B \longrightarrow Y \cdot F}{B \cdot U \longrightarrow Y}$$

for all B and Y, *i.e.* in the "opposite" sense to Theorem 7.2.2.

27. Given adjunctions

$$\mathcal{A} \underset{F_1}{\overset{U_1}{\underset{\top}{\rightleftarrows}}} \mathcal{I} \underset{F_2}{\overset{U_2}{\underset{\top}{\rightleftarrows}}} \mathcal{S}$$

show that $U_2 \cdot U_1 \dashv F_1 \cdot F_2$ with unit $\eta_2 ; (U_2 \cdot \eta_1 \cdot F_2)$, co-unit $(F_1 \cdot \varepsilon_2 \cdot U_1) ; \varepsilon_1$ and adjoint transposition $\lambda_1 \circ \lambda_2$. Show also that this notion of composition is associative (up to equality) and has a unit.

28. Let $(F_1, U_1, \eta_1, \varepsilon_1)$ and $(F_2, U_2, \eta_2, \varepsilon_2)$ be adjunctions with the same source and target categories. Find a natural bijection between natural transformations $v : U_1 \to U_2$ and $\phi : F_2 \to F_1$. Show that it respects identities and (vertical) composition of v and ϕ, and is preserved by composition on either side with another adjunction, as in Exercise 7.27.

29. Deduce that if a diagram of left adjoints commutes up to isomorphism then so does the corresponding diagram of right adjoints.

30. Explain how the triangle laws (Definition 7.2.1) give a meaning to the notion of adjunction within *any* 2-category $\mathfrak{C}$ (Definition 4.8.15). Show how to define the 2-category $\mathfrak{C}^{\dashv}$ which has

(a) the same objects (0-cells) as $\mathfrak{C}$;

(b) as 1-cells, adjunctions, composition being given by Exercise 7.27;

(c) as 2-cells, natural transformations as in Exercise 7.28.

31. Show that the forgetful 2-functor $\mathfrak{C}^{\dashv} \to \mathfrak{C}$ which extracts the left (or right) part of the adjunction and natural transformations is full and faithful at the 2-level. Deduce that the 2-category of left adjoints is equivalent to that of right adjoints, in the weakest sense of Definition 4.8.9(d). Explain why they are not directly related.

32. For any 2-category $\mathfrak{C}$, $(\mathfrak{C}^{\dashv})^{\dashv}$ has the same objects (0-cells), but the 1-cells $\mathcal{E} \to \mathcal{F}$ consist of four 1-cells $A : \mathcal{E} \to \mathcal{F}$, $B_1, B_2 : \mathcal{F} \rightrightarrows \mathcal{E}$ and $C : \mathcal{E} \to \mathcal{F}$ and eight 2-cells of $\mathfrak{C}$ with $A \dashv B_1$ and $B_2 \dashv C$. By looking at various adjoint transpositions, show that $B_1 \cong B_2$ canonically and hence that $(\mathfrak{C}^{\dashv})^{\dashv}$ is strongly 2-equivalent to the 2-category whose 1-cells are adjoint triples in $\mathfrak{C}$. Describe the 2-cells. By applying the result for $\mathfrak{C}$ to $\mathfrak{C}^{\dashv}$, show that $((\mathfrak{C}^{\dashv})^{\dashv})^{\dashv}$ is strongly 2-equivalent to the 2-category of adjoint sequences of length 4 and so on. (Arbitrarily long chains of adjunctions exist by Exercise 3.61 and its categorical analogue.)

33. What does the solution-set condition in the General Adjoint Functor Theorem 7.3.12 mean in the case of Proposition 6.1.11, and how do the conjunctive interpretation and equationally free algebras show that it is satisfied?

34. Formulate a solution-set condition for a prefactorisation system $(\mathcal{E}, \mathcal{M})$ in which $\mathcal{M}$-maps need not be mono, and use it to factorise maps (*cf.* Proposition 5.7.11). [Hint: consider the category whose objects are factorisations of the given map into an $\mathcal{E}$ and an arbitrary map.]

35. Let $F : \mathcal{C}^{\mathrm{op}} \times \mathcal{C} \to \mathcal{C}$ be a mixed variance functor and $\Gamma \in \mathrm{ob}\,\mathcal{C}$. A *wedge from* Γ *to* F is a dinatural transformation (Exercises 4.44ff), and the *end* $\int^X F(X, X)$ is the final wedge, just as a cone is a natural transformation from an object to a diagram and its limit is the final

cone. Show that $A = \int^X \left[[A \to X] \to X \right]$ for any object A of a cartesian closed category. Deduce the other parts of Remark 2.8.11.

36. Let $U : \mathcal{A} \to \mathcal{C}$ be a functor. A $\mathcal{C}$-map $e : \Gamma \to UA$ is called a **candidate** if, in any commutative square of the form on the left,

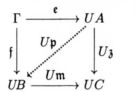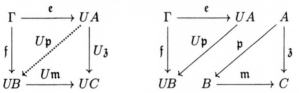

there is a *unique* $\mathfrak{p} : A \to B$ such that *both* triangles commute (without U, i.e. $\mathfrak{p};\mathfrak{m} = \mathfrak{z}$ and not just $U\mathfrak{p};U\mathfrak{m} = U\mathfrak{z}$). Suppose that *every* map $\Gamma \to TX$ in $\mathcal{C}$ factors uniquely as $e ; U\mathfrak{m}$, with e a candidate. Show that U preserves wide pullbacks. (*Cf.* factorisation systems, Definition 5.7.1.)

37. Show that any functor $T : \mathcal{Set} \to \mathcal{Set}$ of the form $\coprod_{n \in \mathbb{N}} c_n \times X^n$ satisfies the conditions of the previous exercise, where the candidates with $\Gamma = \mathbf{1}$ and $A = \mathbf{n}$ correspond bijectively to elements of c_n. André Joyal has called such a functor **analytic** [Joy87, Tay89].

38. Describe the left adjoint of $\mathrm{mor} : \mathcal{Cat} \to \mathcal{Set}$. [Hint: *not* List.]

39. Show that $\mathcal{IPO} \subset \mathcal{Dcpo}$ is the full subcategory of objects which are able to support an algebra structure for the lift monad, and that this is what the co-Kleisli category on the category of algebras is always like.

40. Show that an adjunction $F \dashv U$ is of Kleisli type (Proposition 7.5.3(a)) iff F is essentially surjective and every ε_A is a self-presentation (*cf.* Theorem 7.5.9).

41. (Robert Seely) Let $\mathcal{A}$ be a symmetric monoidal closed category (so it has a tensor product $\otimes$ and a mixed-variance functor $\multimap$ satisfying $(-)\otimes X \dashv X \multimap (=)$) which also has finite products and is equipped with a comonad $!$ such that $!(A \times B) \cong !A \otimes !B$ and $!\mathbf{1} = I$ (the tensor unit). Show that the co-Kleisli category is cartesian closed, with $[A \to B] = !A \multimap B$. (Exercises 4.27ff give a concrete example.)

42. A function $f : A \multimap B$ between L-domains is called *linear* if it preserves locally least upper bounds *i.e.* if a is a locally least upper bound of a_1 and a_2 in A then so is $f(a)$ of $f(a_1)$ and $f(a_2)$ in B. In particular f is Scott-continuous, but this is weaker than the condition needed for a right adjoint unless A is a complete semilattice (*cf.* Exercise 3.33ff). Let $A \multimap B$ be the dcpo of linear functions, with the pointwise order, which agrees with Exercise 4.27 for complete semilattices. There we found $(-) \otimes A \dashv A \multimap (=)$. Use the fact that $A = \bigcup_a A \downarrow a$ for any L-domain A (Exercise 3.34), to deduce the existence of $A \otimes B$ for L-domains from the

interaction of colimits and left adjoints. Describe $A \otimes B$ for boundedly complete domains (Exercise 3.21).

43. Show that M_0 in Remark 7.5.2 is the *free* monad on the functor T, formulating a suitable 2-category in which this is so.

44. By analogy with Exercise 3.36 and Proposition 7.5.11, explain how *any* (infinitary) monad can be seen as an infinitary single-sorted algebraic theory with a proper class of operation-symbols and laws.

45. Let (M, η, μ) be a monad on $\mathcal{S}$ and suppose that ev : $X \to MX$ is a final *coalgebra* for the *functor* M. Show that for each $\Gamma \in \mathrm{ob}\,\mathcal{S}$ there is a unique map $f : \Gamma \to X$ such that $f \,;\, \eta_X = f \,;\, \mathrm{ev}$. [Hint: η_Γ.]

46. Formulate the results of Section 7.6 as a reflection of the 2-category of categories with □-structure *as a property* and functors preserving it (and natural transformations) into the 2-subcategory where this structure is canonical and preserved on the nose.

47. Prove Proposition 7.7.1.

48. Investigate the effect of a natural transformation $\phi : U \to U'$ on $\mathcal{S} \downarrow U$. Consider in particular the case where ϕ is *cartesian*, *i.e.* its naturality squares are pullbacks [Tay88].

49. Describe the effect of the functor Q in Notation 7.7.3(b) on morphisms.

50. Let $U : \mathcal{A} \to \mathcal{S}$ be a functor between categories, each equipped with a factorisation system, such that U takes "monos" of one kind to "monos" of the other. (For example *all* maps in $\mathcal{S}$ could be called "monos".) Define a factorisation system on $\mathcal{S} \downarrow U$ which is preserved by π_0 and π_1. Show that if the given factorisation systems are stable, then so is the resulting one, so long as U preserves pullbacks.

51. Let $U : \mathcal{A} \to \mathcal{S}$ be a functor between toposes that preserves finite limits. In particular it preserves monos, so there is a semilattice homomorphism $p : U\Omega_A \to \Omega_S$.

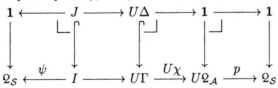

Show that $(\Omega_S \downarrow p, \Omega_A, \pi_1)$ is the subobject classifier of $\mathcal{S} \downarrow U$.

Algebra with Dependent Types

M ATHEMATICAL REASONING IN THE LARGE (as we have been *doing* it) involves the introduction and manipulation of symbols for individuals and structures which successively depend on one another. For example we may introduce a category C, some objects X and Y of C, a morphism f of the hom-set *from X to Y*, and (if C is a concrete category) elements of some pullback whose construction *involves* f. Similar idioms of presentation may be found in geometry, algebra, analysis and so on. Simple type theory (reasoning as we have *studied* it) does not allow for this dependency: we must consider more complicated calculi.

We studied *propositions* dependent on terms in Section 1.4, where they were called predicates (but type-theorists prefer the word proposition). The main story in a proof in the predicate calculus is told by the logical formulae which are asserted at each step. The proof boxes which we used in Sections 1.4–1.7 were designed to reflect this: variables were consigned to the margin, and we hardly mentioned types as they and any function-symbols which we used were understood to come from a simply typed algebraic theory (maybe even a free one).

A dependent-type proof is much busier, since we also have to show that the types and terms we use are well formed. This goes on in the same current of reasoning, attention passing amongst these different aspects. Nikolas de Bruijn devised ways of expressing successive dependencies in AUTOMATH, with various conventions for saying that the context of the previous line was to be repeated, augmented or diminished, or that the current line was asserted in a global or some other context. As before, here we shall use the box or sequent style according to the emphasis we wish to place on the changes of context.

For some authors, the phrase "dependent type theory" means the study of the universal quantifier $\forall$ or dependent product Π, which generalises the function-type $\rightarrow$. That is the subject of the final chapter: this one sets up the correspondence between type theory and category theory *at the algebraic level*. The symbolic rules and universal properties of the quantifiers are then directly related. If you wish to go straight to

Chapter IX, you should just observe the way in which contexts are used as vertices of commutative diagrams, and that this notation is sometimes abbreviated to a juxtaposition of letters denoting contexts or types.

Examples of *types* dependent on terms that we have already met include the hom-set $C[x, y]$ for two objects of any category, and the arity ar$[r]$ of an operation-symbol in any infinitary free theory. In each of these cases, an alternative presentation mor$C \longrightarrow (\text{ob}\,C)^2$ or $\kappa \longrightarrow \Omega$ is useful for some purposes (Remark 4.1.10 and Definition 6.1.1). The targets of these maps are the types of the independent variables, and the source is the disjoint union $\coprod_x Y[x]$, *displaying* the type $Y[a]$ over its index $a \in X$. Diagrams (expressing limits or colimits) in categories may also be seen as dependent types, but the additional arrow information makes the situation more complicated: it is handled by the *Grothendieck construction*, Proposition 9.2.7.

The notion of generalised algebraic theory, *i.e.* with dependent types, is powerful enough to serve as a general meta-language. Not only is the theory of categories an example of it, but so is the theory of generalised algebraic theories itself, as are even stronger notions such as cartesian closed categories and toposes. (*Cf.* that a few basic styles of symbolic reasoning such as the ruled lines and sequent notation for deductions suffice to set out the rules of complex calculi.)

Although any practising mathematician can formulate any *particular* dependent-type argument quite fluently, a very complicated recursive construction is needed to describe the *generality* — which is what this chapter is about. After describing the calculus, we shall construct the classifying category using the techniques developed in Sections 4.2–4.6.

Recall that for simply typed algebraic theories this was a category with products; by Section 7.6 every such category arises in this way. The analogue for dependent types is a category with a class of *display* maps, such as $\kappa \longrightarrow \Omega$. Syntactically, displays drop one typed variable $y : Y[x]$ from a context $[\Gamma, y : Y]$, whilst substitution of a term a for the variable x gives rise to a pullback square whose vertical maps are displays and which has the substitution morphism $[a/x]$ along the bottom. The class of displays must therefore be closed under pullback against any map.

According to set theory, the passage from a dependent type $Y[x]$ to its display involves the *axiom of replacement*. Semantically, the general notation $Y[x]$ is meaningless in itself: we understand it only *via* the interpretation of this syntax in terms of displays as given in this chapter, so the correspondence is the *definition* of $Y[x]$, rather than a theorem or axiom. However, the particular case $Y[n] = T^n(U)$ of a (possibly transfinitely) iterated functor does depend for its existence on a substantive axiom, as does the interpretation functor $[\![-]\!]$ (Section 9.6).

Earlier investigations of this subject made the perhaps more obvious generalisation from products to *pullbacks* (Remark 5.2.9). This is the extreme case of our formulation, in which *arbitrary* functions between sets play the role of displays. It has the effect of building *equality types* into the syntax, but this is not necessarily appropriate when the objects under study are computational (Proposition 8.3.4ff).

Our approach is based on a philosophical and idiomatic point to which John Cartmell [Car86] first drew attention. He called the relationship between $[\Gamma, x : X]$ and Γ *analytic* in the sense of Immanuel Kant. For example, to speak of a "sentence" one must *presuppose* some particular language, since this is implicit in the nature of a sentence. By contrast, general functional relationships like "the official language of a country" are *synthetic*. The point of view we took in Remark 4.1.10 regarding the morphisms of a category was similarly based on the idea that when f and g compose it is not merely because $\mathsf{tgt}\, f = \mathsf{src}\, g$ *accidentally*.

Generalised algebraic theories can be interpreted in other categories besides sets. On the semantic side, the investigation of what morphisms may arise as displays can involve non-trivial issues in topology, order theory and other disciplines about which we shall just give a few hints in Section 8.3. Similarly, Chapter IX does some calculations about quantifiers and even the type of types.

This subject is very much in its infancy: it is notationally complicated, most algebraists concern themselves with simply typed theories such as groups, rings, modules, *etc.*, and logicians are more interested in the quantifiers. The account in this chapter is perhaps overly influenced by type theory (though the practitioners of that subject would, on the contrary, find it inadequate), as it allows for a high degree of interdependence amongst types. This is needed for comprehension (*cf.* Exercises 2.17 and 8.3ff) and for the equivalence with display categories.

There are, however, numerous examples from algebra whose dependency is "stratified", with *simply* typed theories at each level (as in Section 4.6). We saw such a stratification in Sections 5.6 and 7.4 for the *congruences* for simply typed theories. Another is the theory of categories. This means that we must expect all of the difficulties associated with the category of categories (two-dimensional limits, pseudo-colimits, failure of regular epis to compose) to beset generalised algebraic theories too.

If the intended laws can be separated from the operation-symbols by stratification, then it is no longer a severe restriction to concentrate on *free* theories (with no laws within levels). Then there would a stratified theory of structural recursion, unification and resolution. All of these issues demand further research, but are beyond the scope of this book.

The aims of the present account are to identify how an object-theory contributes types, operation-symbols and laws to the language, and how the arguments are supplied by (the category composed of) substitutions. Section 8.2 describes this category, and Remark 8.4.1 the object-theory. Section 8.1 leads *informally* towards these from the vernacular.

Instead of defining the objects and maps of the *category* of contexts and substitutions *directly*, we do this by means of an elementary sketch as in Section 4.3. I began work on this chapter from the *exercise* of showing that the context-morphisms defined recursively in [Pit95] (essentially Remark 8.2.12) do indeed satisfy the axioms of a category. This proved to be unreasonably laborious, being a highly convoluted version of Proposition 2.7.5, and led me to the approach *via* sketches. This profoundly influenced the rewriting of Chapters IV and I.

It is well known that syntactic substitution is characterised by pullback, but Section 8.2 is probably the first explicit proof of this in a category fashioned directly from language. From this careful analysis we may determine what coherence conditions pullbacks along composites must satisfy, namely *none* so long as on-the-nose equalities are never asserted between type-expressions involving different outermost type-symbols.

In Section 8.4 we repeat the discussion in Section 7.6 about whether the notion of a class of displays in a category is a structure or a property: again this dispute is *resolved* by means of the distinction between strong and weak equivalences of categories.

8.1   THE LANGUAGE

In the mathematical vernacular (Section 1.6), *let* and *suppose* introduce hypothetical things and properties. Formally, they expand the context in the sequent style, or open nested boxes in the box presentation. Things introduced by "let" are denoted by typed variables. Hypothetical properties may of course depend on entities which have already been introduced in the argument; now the types of the variables may also be dependent on what has gone before. This means that, as a precondition of opening the box or expanding the context, these types must themselves be shown to be well formed.

DEFINITION 8·1·1:   The steps in an argument are called **judgements**. Amongst the direct algebraic steps (those within a fixed context Γ), we distinguish the following forms:

(a) **Term formation** ($\Gamma \vdash a : X$): the term a is well formed and of type X.

(b) **Truth** ($\Gamma \vdash \phi$ holds): that ϕ is true in the context, *i.e.* that there is a well formed proof p of (type) ϕ, where p is implicit in the history of the deduction. This is a special case of the previous form, where we omit proof(-term)s because no distinction is made between them.

(c) **Term equality** ($\Gamma \vdash a = b : X$): that a and b are equal *quâ* terms of type X. What notion of "equality" we mean needs some discussion.

(d) **Proof equality.**

(e) **Type formation** ($\Gamma \vdash X$ type or $\Gamma \vdash \phi$ prop): that X is a well formed type (or ϕ a proposition).

(f) **Type equality** ($\Gamma \vdash X = Y$ or $\Gamma \vdash \phi = \psi$): that these propositions or types are intensionally equal.

For the postulates at the beginning of an argument we need

(g) **Context formation**:

$$\frac{\Gamma \vdash X \text{ type} \qquad x \notin \Gamma}{[\Gamma, x : X] \text{ ctxt}} \qquad \frac{\Gamma \vdash \phi \text{ prop} \qquad x \notin \Gamma}{[\Gamma, x : \phi] \text{ ctxt}}$$

Having shown that a type or proposition is well formed in a certain context, we may expand the context by a new variable. The rule above belongs to the sequent style; in the box idiom we open a nested box which begins with a term- or proof-formation judgement saying that the new variable x is a well formed term of type X or ϕ.

We discuss context formation in the next section, where it will become clear that we must also say when two contexts are equal, using equality of their types. The provision of arguments of types and operation-symbols also needs formation and equality rules for substitutions.

The idioms which *close* boxes or make the contexts smaller arise from the quantifiers (Remark 9.1.6 and the whole of the final chapter). For the time being, therefore, boxes never get closed. Despite the boxes and sequents, this chapter generalises, not the predicate calculus as we saw it in Sections 1.4–1.5, but just resolution of Horn clauses involving *atomic* predicates, as in PROLOG (Remark 1.7.2ff).

At the algebraic level the only distinction between the behaviour of types and propositions is that elements (terms) of the same set (type) are distinguished, but proofs are anonymous. By giving names to the Horn clauses, proof-terms may be assigned to propositions (Remark 6.2.10); the main point of Section 2.4 was to do the same for implicational logic, using λ-terms. Alternatively, we may simply assert that *any* two proofs of the same proposition in the same context are equal.

The classification of judgements also applies to the rules which justify them. These are the subject of the rest of this section.

Terms. In this chapter, a term is an algebraic expression: just as in Definition 1.1.3, it is either a variable x belonging to the context, or an operation-symbol r applied to (zero or more) sub-terms.

REMARK 8·1·2: Let r be an operation-symbol, from the object-theory. It is applied to arguments,

$$\frac{a_1 : X_1 \quad a_2 : X_2[a_1] \quad a_3 : X_3[a_1, a_2] \quad \cdots \quad a_k : X_k[a_1, \ldots, a_{k-1}]}{r(a_1, \ldots, a_k) : Y[a_1, a_2, \ldots, a_k]}$$

where now the *types* of the second and subsequent sub-expressions may depend on the preceding *terms*, and the type of the result may depend on all of them. If there is no such dependency, as in Section 4.6, we can allow the sub-expressions to have been born simultaneously — at any rate there is no restriction on their order of formation. If X_2 really does depend on a_1 then we must have had a fragment of proof like

　₁ X_1 type

　₂ $a_1 : X_1$

　₃ $X_2[a_1]$ type

　₄ $a_2 : X_2[a_1]$

and so on (with intermediate steps). The operation-symbol r might for example be composition, the types being the set of objects and hom-sets of a category. For propositions r names a Horn clause:

$$\alpha_1, \alpha_2, \ldots, \alpha_k \vdash \phi.$$

The *arity* of an operation-symbol r is given by listing the types on the left and right of the turnstile. Now, in order to express the dependency of the later types, the earlier ones must be accompanied by variables. In the dependent-type situation it is convenient to regard the operation-symbol not as the letter r alone, but as applied to a list of variables:

$$x_1 : X_1, x_2 : X_2[x_1], \ldots, x_k : X_k[x_1, \ldots, x_{k-1}] \vdash r(\underline{x}) : Y[\underline{x}].$$

The informal notation $r(\underline{a})$ and $X[\underline{a}]$ quickly becomes inadequate, so we shall develop a formalism in which the arguments are delivered to r by *substitution* of $\underline{a}$ for $\underline{x}$, writing

$$[\underline{a}/\underline{x}]^* r \quad \text{for} \quad r(\underline{a}).$$

This notation allows substitution into expressions ($[\underline{a}/\underline{x}]^* t$) as well as operation-symbols, repeated substitutions and weakening ($\hat{x}^* t$).

A lot of work remains to be done in the next section to define general substitutions u and their action u^*t on terms — the calculus is highly recursive, and we have to break into the circle *somewhere* — but, in anticipation of this, here are the first of the formal rules.

DEFINITION 8·1·3: The **term-formation rules** are as follows:

(a) **resolution** (*cf.* Remark 1.7.6) of an operation-symbol r,

$$\frac{u : \Gamma \to \Delta \qquad \Delta \vdash r : Y}{\Gamma \vdash u^*r : u^*Y}$$

by substitution $u = [q/x]$ of arguments q, and

(b) variables considered as terms,

$$\frac{[\Gamma, x : X, \Psi] \ \mathsf{ctxt}}{\Gamma, x : X, \Psi \vdash x : X}$$

of which the **identity axiom** $(x : X \vdash x : X)$ is a special case.

Both of these rules incorporate **weakening**, *cf.* Definition 1.4.8 and Remark 2.3.8. We study the structural rules in the next section.

Equality of terms. The notion of equality needed in the foundations of dependent type theory is that from algebra: congruence. The judgement $\Gamma \vdash a = b$ means that, for any valid judgement $\Gamma, \Psi \vdash \mathcal{J}$ in which a is a sub-expression, replacing it with b gives another valid judgement.

DEFINITION 8·1·4: The **intensional term-equality rules** are:

(a) the laws of the object-theory, as in Definition 4.6.1(d);

(b) the rules for an equivalence relation (Definition 1.2.3),

$$\frac{a : X}{a = a : X} \qquad \frac{a = b : X}{b = a : X} \qquad \frac{a = b : X \qquad b = c : X}{a = c : X}$$

(c) and pre- and postsubstitution, including in particular congruence for each operation-symbol r,

$$\frac{a_1 = b_1 : X_1 \quad a_2 = b_2 : X_2[a_1] \quad \cdots \quad a_k = b_k : X_k[a_1, \ldots, a_{k-1}]}{r(a_1, \ldots, a_k) = r(b_1, \ldots, b_k) : Y[a_1, \ldots, a_k].}$$

The formal rule, *cf.* Definition 8.1.3(a), is

$$\frac{u = v : \Gamma \rightrightarrows \Delta \qquad \Delta \vdash a = b : X}{\Gamma \vdash u^*a = v^*b : X}$$

The anonymity of proofs is expressed by an indiscriminate equality rule:

$$\frac{\Gamma \vdash \phi \ \mathsf{prop} \qquad \Gamma \vdash a : \phi \qquad \Gamma \vdash b : \phi}{\Gamma \vdash a = b : \phi} \ \text{anon}$$

REMARK 8·1·5: Intensional equality $\Gamma \vdash a = b : X$ remains at the level of judgements: it does not (within the basic calculus) provide us with a term of some propositional type eq[a, b]. We discuss extensional equality briefly in Section 8.3. If we want to do something *conditionally* on the equality of a and b, it is extensional equality that we need.

Dependent types. The dependency of types on terms (in predicates, hom-sets, arities, *etc.*) is expressed in a similar way to the application of operation-symbols to arguments, but there are no type variables.

DEFINITION 8·1·6: The ***direct type formation rule*** is

$$\frac{a_1 : X_1 \qquad a_2 : X_1[a_1] \qquad \cdots \qquad a_k : X_k[a_1, \ldots, a_{k-1}]}{Y[a_1, \ldots, a_k] \text{ type}} Y\mathcal{F}$$

where Y is a dependent type-symbol in the object-theory. As we did for the operation-symbols (Definition 8.1.3(a)), it is convenient to regard the primitive form of the dependent type as having variables for its arguments: $\underline{x} : \underline{X} \vdash Y[\underline{x}]$. The formal type-formation rule is then

$$\frac{u : \Gamma \to \Delta \qquad \Delta \vdash Y \text{ type}}{\Gamma \vdash u^*Y \text{ type}}$$

Equality of types. The instantiation of a dependent type $Y[\underline{x}]$ at equal terms $\underline{a} = \underline{b}$ gives rise to equal types. We have not addressed this phenomenon before — indeed we have gone to some lengths to exclude it — but how can we deny that Factors[9×4] = Factors[6×6]? There must at least be a canonical isomorphism between the two, but if we chose to make this explicit we would be obliged to introduce formation and equality rules for it, which would have to obey further coherence rules with respect to other substitutions. Predicates at equal subjects also give rise to equal types; again there is a canonical way of translating a proof of one into a proof of the other, which must also obey coherence.

Rather than enter this labyrinth, we accept that types can be *intensionally* equal, *i.e.* if they have a common history of formation. Whereas set theory allows independently given types to be tested for equality or inequality, we do not. Section 9.2, however, does look at some of the categorical consequences of replacing equality by isomorphism.

DEFINITION 8·1·7: The ***type equality rules*** are

(a) reflexivity, symmetry, transitivity,

$$\frac{X \text{ type}}{X = X} \qquad \frac{X = Y}{Y = X} \qquad \frac{X = Y \qquad Y = Z}{X = Z}$$

(b) and congruence,

$$\frac{a_1 = b_1 : X_1 \qquad a_2 = b_2 : X_2[a_1] \qquad \cdots \qquad a_k = b_k : X_k[a_1, \ldots, a_{k-1}]}{Y[a_1, \ldots, a_k] = Y[b_1, \ldots, b_k]}$$

for which the formal rule is

$$\frac{\mathfrak{u} = \mathfrak{v} : \Gamma \rightrightarrows \Delta \qquad \Delta \vdash Y \text{ type}}{\Gamma \vdash \mathfrak{u}^* Y = \mathfrak{v}^* Y}$$

(c) But the most important consequence of type equality is that terms may acquire new types:

$$\frac{a : X \qquad X = Y}{a : Y} \text{ coerce} \qquad \frac{a = b : X \qquad X = Y}{a = b : Y}$$

Notice that the congruence rules in Definition 8.1.4(c) *don't even make sense* without this ability to transfer types of sub-terms.

For the sake of giving a little more thought to the all-important coercion rule, we pause to consider its one-way version.

REMARK 8·1·8:   **Subtyping** generalises equality between types to a *non-symmetric* relation(-judgement) $\Gamma \vdash U \subset X$ satisfying

$$\frac{U = X}{U \subset X} \qquad \frac{U \subset V \qquad V \subset W}{U \subset W} \qquad \frac{U \subset X \qquad V \subset Y}{(U \times V) \subset (X \times Y)}$$

$$\frac{X \supset U \qquad V \subset Y}{(X \to V) \subset (U \to Y)} \qquad \frac{t : U \qquad U \subset X}{t : X} \qquad \frac{u = v : U \qquad U \subset X}{u = v : X}$$

in which $\to$ is contravariant in its first argument. Comprehension gives the most familiar example (Remark 2.2.4). In principle terms may be equal with respect to one type but not another, for example because the second consists of "tokens" for the first, this having finer equality rules (*cf.* PERs, Exercise 5.10).

Subtyping also arises in **object-oriented programming languages**, in which complex types are developed from simpler ones by the addition of constructors and properties. As a mathematical analogy to this style, we may define a field as an Abelian group, with multiplication as an extra operation and the field axioms as extra conditions.

Terms of the narrower type **inherit** the covariant properties (positive, in the sense of Remark 1.5.9) associated with the wider one to which we **coerce** them; negative properties pass the other way. The coercion functions, like forgetful functors, need not be injective, but if they are suppressed from the notation then there must be at most one between any two types. (This has non-trivial consequences for function-types.)

REMARK 8·1·9:   Some authors allow axioms stating equality of types, so that Heyting semilattices can be treated in the same framework. We forbid them, because interchangeability of objects should be expressed as isomorphisms, *i.e.* by means of two operation-symbols and two laws between their composites (*cf.* Section 7.6). See Exercise 4.7 regarding the quotient of a category by a system of canonical isomorphisms. As we commented before Proposition 3.2.11, the antisymmetry law for posets is a side-effect of the imposition of algebraic notation ($\wedge$, $\vee$, $\Rightarrow$), and is not an intrinsic feature of logic.

The restriction drastically simplifies the issue of type equality, as type-expressions can be equal *only* as a result of making equal substitutions into the same "outermost" type-symbol. This is essential to the validity of the structural recursion used in the interpretation (Section 8.4).

Definitional equality need not be excluded: it is of course very useful to define $\mathbb{R} = \{(L, U) : \mathcal{P}(\mathbb{Q}) \mid \ldots\}$ as in Remark 2.1.1. This is harmless to the interpretation of dependent type theory, as we may simply replace the left hand side of the definition by the right. Inter-provability as a notion of equality of propositions will be discussed in Remark 9.5.6.

The object-language. The pure theory cannot prove the existence of anything apart from the empty context [ ], so an object-theory is needed. As in simply typed algebra, it has types, operation-symbols and laws.

DEFINITION 8·1·10: A *generalised algebraic theory* $\mathcal{L}$ is given by

(a) type-symbols $\Delta \vdash X$ type and proposition-symbols $\Delta \vdash \phi$ prop, each defined in a context Δ;

(b) operation-symbols, $\Delta \vdash r : X$, which are typed and in context;

(c) laws between terms $\Delta \vdash a = b : X$ of the same type in context.

In order to give meaning to the "contexts" which occur in these data, we have to generate a small part of the language — as we did to state laws for an algebraic theory in Section 4.6. However, the ubiquitous dependencies must not be allowed to become circular, so we require that

the types and operation-symbols which occur in Δ
must have been declared in advance.

Abstractly, the presentation defines a *relation* between each new symbol being defined and the type-symbols and operation-symbols that are used in the formation of its defining context Δ (there are operation-symbols in the terms that are the arguments of the type-expressions). This relation must be *well founded*.

DEFINITION 8·1·11: A *stratified algebraic theory* is one which obeys a stronger well-foundedness condition, that all of the operation-symbols $r : X$ and laws $a = b : X$ must be declared before variables $x : X$ may be used as arguments of further type-symbols $Y[x]$. We shall not impose this condition in this chapter, as it is violated by the canonical language in Definition 8.4.6. See Exercises 8.1–8.5 and Examples 9.2.4.

EXAMPLE 8·1·12: The theory of categories, with $c(f,g)$ for $f \, ; g$.

$$\vdash \quad O \text{ type}$$
$$x,y:O \quad \vdash \quad H[x,y] \text{ type}$$
$$x,y:O, f:H[x,y] \quad \vdash \quad \text{src}(f) = x:O$$
$$x,y:O, f:H[x,y] \quad \vdash \quad \text{tgt}(f) = y:O$$
$$x:O \quad \vdash \quad \text{id}(x):H[x,x]$$
$$x,y,z:O, f:H[x,y], g:H[y,z]$$
$$\vdash \quad c(f,g):H[x,z]$$
$$x,y:O, f:H[x,y] \quad \vdash \quad c(\text{id}(x),f) = f:H[x,y]$$
$$x,y:O, f:H[x,y] \quad \vdash \quad c(f,\text{id}(y)) = f:H[x,y]$$
$$w,x,y,z:O, f:H[w,x], g:H[x,y], h:H[y,z]$$
$$\vdash \quad c(c(f,g),h) = c(f,c(g,h)):H[w,z]$$

In Remark 4.1.10 we said that the maps of a category $\mathcal{C}$ with $O = \text{ob}\,\mathcal{C}$ could be collected together and presented by $\langle \text{src}, \text{tgt}\rangle : (\text{mor}\,\mathcal{C}) \longrightarrow O \times O$ instead of using the dependent type $H[x,y]$. In fact $\langle \text{src}, \text{tgt}\rangle$ will serve as the display map $\widehat{f}$ that corresponds to this dependent type, which we introduce syntactically and semantically in the next two sections.

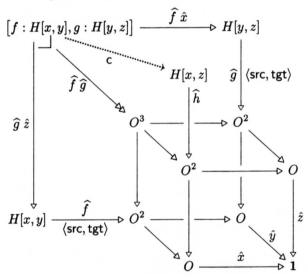

Before composition can be defined, its support — the set of composable pairs (f,g) with $\text{tgt}(f) = \text{src}(g)$ — must be constructed. There is a *natural* syntactic description of this set, namely the context of the rule which introduces c. It is also the pullback of tgt and src, as marked in the diagram above, *cf.* transitivity of a kernel pair (Proposition 5.6.4).

This chapter shows how to translate between the symbolic and diagrammatic idioms. The next section constructs the classifying category for $\mathcal{L}$, *i.e.* the category in which this diagram is drawn, and shows that squares like the one marked really are pullbacks. In fact pullback performs all of the substitutions u^* in Definitions 8.1.3(a), 8.1.4(c), 8.1.6 and elsewhere. In Section 8.4 we shall give the interpretation of the language, for which we shall need a sharper form of the induction which generates it.

8.2   THE CATEGORY OF CONTEXTS

We have presented the formation and equality rules for terms and types using generalised substitutions $u : \Gamma \to \Delta$. These form a category, which is generated by the structural rules (weakening and cut) subject to the laws for substitution. This presentation is an elementary sketch: it is the culmination of the approach to semantics of formal languages which we began in Section 4.2. As we have some heavy symbol-pushing to do, the diagrammatically minded reader may find it helpful to read the next section first, as it presents the corresponding semantic arguments.

If you *prefer* symbols, remember that commutative diagrams simply show the types of the symbols. Giving a list of successively equal terms does not show the type information (which, after all, is what *dependent type* theory is all about), and makes it difficult to follow the reduction rules. In view of the fact that types contain terms as sub-expressions, the practice of annotating terms with their types is inadequate:

EXAMPLE 8·2·1: $f : H[\mathsf{src}(f), \mathsf{tgt}(f)]$.

After defining equality of contexts we discuss the structural rules. These are *derived* in the sense that *if the premise is deducible using the rules of the previous section then so is the conclusion*. It is left to the reader to demonstrate this. The calculus has a "term model" in which

$$\mathsf{Cn}(\Gamma) = \{(\Psi, \mathcal{J}) \mid \Gamma, \Psi \vdash \mathcal{J}\}$$

represents[1] the context Γ, where we shall use $\mathcal{J}$ for the right hand side of an arbitrary judgement. The structural rules translate elements of one such set to another: they define an *action* on $\mathsf{Cn}(-)$ which is essentially the Cayley–Yoneda action introduced in Section 4.2. This satisfies the laws of the Extended Substitution Lemma (Proposition 1.1.12).

We write $x \in \Gamma$ to mean that $x : X$ is one of the variables which are listed in this context. Note that for a *context* Γ, if $x \in \mathsf{FV}(Y)$ for the type

---

[1]Beware that this is not a *model* of the theory, because it does not preserve the type constructors such as function-space, or even product, *cf.* term "models" of the λ-calculus.

(expression) Y of one of the variables $y \in \Gamma$ then $x \in \Gamma$ itself. However, for a context *extension* $[\Gamma, \Psi]$ we write $x \in \mathsf{FV}(\Psi)$ if $x \in \mathsf{FV}(Y)$ for some type in Ψ, where x may belong to Γ rather than Ψ (*cf.* Exercise 1.12).

Objects. As before, these are contexts, but type dependency means that the class of well formed contexts must be defined recursively. Indeed the terms, types and contexts must be generated in a single simultaneous process (as remarked in Section 6.2). The starting point is the *empty context* [ ], which will be interpreted as the terminal object **1**, but we shall find that a general context Γ is always present in the background.

DEFINITION 8·2·2: The *hypothesis* or *context formation* rule is

$$\frac{\Gamma \vdash X \text{ type} \qquad x \notin \Gamma}{\Gamma, x : X \text{ ctxt}} \text{ hyp}$$

$$\begin{array}{|ccc|} \hline & X \text{ type} & \\ \hline x : X & & \text{hyp} \\ \vdots & & \\ \hline \end{array}$$

where (following Convention 1.1.8) the variable name x does not already occur in Γ. It is what we use at the beginning of a function definition $(\to\mathcal{I})$, universal proof $(\forall\mathcal{I})$ or, without a variable, implication $(\Rightarrow\mathcal{I})$.

The objects and maps of the category will be contexts and multiple substitutions. Morphisms may only compose if the target of one matches the source of the other, *i.e.* they are *equal objects* of the category. This means we have to give rules defining equality of contexts which allow substitution of a term of one type for a variable of an equal type. These simply extend the type equality rules (Definition 8.1.7).

DEFINITION 8·2·3: The *context equality* rule is

$$\frac{\Gamma = \Delta \qquad \Gamma \vdash X = Y}{[\Gamma, x : X] = [\Delta, x : Y]}$$

With the axiom for the empty context ($[\,] = [\,]$) we have all the rules for the additional forms of judgement for formation and equality of contexts, because reflexivity, symmetry and transitivity are derivable from those of type equality, and hence equality of substitution (into a type-symbol). Briefly, contexts are well formed and equal iff the corresponding types are, in the preceding context, and the variable names agree.

This satisfies $\Gamma = \Delta \Rightarrow \mathsf{Cn}(\Gamma) = \mathsf{Cn}(\Delta)$, because the rule

$$\frac{\Gamma \vdash X = Y \qquad \Gamma, x : X, \Psi \vdash \mathcal{J}}{\Gamma, x : Y, \Psi \vdash \mathcal{J}} \text{ coerce}$$

is derivable, by inserting coercion rules (Definition 8.1.7(c)).

REMARK 8·2·4: Our convention is to retain the variable names — in the analogy of Remark 4.3.14, we use coloured wires where, *e.g.*, [Pit95] numbers the pins. In accounts of the latter kind, $[\Gamma, x : X]$ and $[\Gamma, y : X]$ are equal contexts. Nevertheless for us the rule

$$\frac{\Gamma, x : X, \Psi \vdash \mathcal{J}}{\Gamma, y : X, [y/x]^* \hat{y}^* \Psi \vdash [y/x]^* \hat{y}^* \mathcal{J}}$$

is derivable: Remark 8.2.8 gives a canonical isomorphism between any two contexts which differ only in these names (open α-equivalence).

Coloured wires, unlike numbered pins, may be permuted, but now our freedom to do this is restricted by the type dependency. Of course, it is *necessary* that the two types be well formed in the same context,

$$\frac{\Gamma \vdash X \text{ type} \qquad \Gamma \vdash Y \text{ type} \qquad \Gamma, x : X, y : Y, \Psi \vdash \mathcal{J}}{\Gamma, y : Y, x : X, \Psi \vdash \mathcal{J}} \text{ exchange}$$

but this is also sufficient.

It would be nice to treat permutation as equality. A context would be no longer a list but a *partially* ordered set, subject to the order relation implicit in Definition 1.5.4, namely that the variables occurring freely in any type must be mentioned before it in the listing of the context.

Unfortunately the explicit listing is required for the semantics, since, even in a category with specified products (or pullbacks), the choice usually cannot be made commutatively or associatively (Remark 8.3.1). Contexts which differ only by permutation are nevertheless canonically isomorphic, by a substitution which is the identity on each variable. They also share the same clone, so to make the isomorphism explicit the order of the variables must be specified.

Display maps. As in Definition 4.3.11(b), the arrow class is generated by display maps and single substitutions. The former arise from

DEFINITION 8·2·5: The **weakening rule** is

$$\frac{\Gamma \vdash X \text{ type} \qquad x \notin \Gamma \cup \Psi \qquad \Gamma, \Psi \vdash \mathcal{J}}{\Gamma, x : X, \Psi \vdash \mathcal{J}} \text{ weaken}$$

i.e. the addition of an unused variable at *any* valid position in a context, but by the permutations above it may be taken to be the last. This rule was given in Definitions 1.4.8 and Remark 2.3.8 for propositions and for types, respectively.

Weakening, like coercion, is a derived rule, since in all of the rules of the previous section variables may be added to the context. Indeed only Definition 8.1.3(b) mentioned the context at all, but we added passive Γ and Ψ to it to incorporate weakening.

The action on clones is the inclusion $\hat{x}^* : \mathsf{Cn}(\Gamma) \hookrightarrow \mathsf{Cn}([\Gamma, x : X])$ since

$$\{(\Psi, \mathcal{J}) \mid \Gamma, \Psi \vdash \mathcal{J}\} \subset \{(\Psi, \mathcal{J}) \mid \Gamma, x : X, \Psi \vdash \mathcal{J}\},$$

and we abstract this as (the contravariant action of) a ***display map***:

$$\frac{\Gamma \vdash X \text{ type} \qquad x \notin \Gamma \cup \Psi \qquad [\Gamma, \Psi] \text{ ctxt}}{[\Gamma, x : X, \hat{x}^*\Psi] \xrightarrow{\quad \hat{x}_\Psi \quad} [\Gamma, \Psi]}$$

The hat was introduced in Notation 1.1.11 and used in Sections 1.5, 2.3 and 4.3. Display maps are marked with open triangle arrowheads. The composite of *zero or more* displays will be indicated by a double triangle arrowhead ($\longrightarrow\!\!\!\triangleright$) and $\widehat{\Psi}$. The subscript on $\hat{x}_\Psi$ refers to the extension Ψ of the context Γ in which $\hat{x}$ is defined.

Cuts. The other class of generating maps consists of the substitutions of one term for a variable of the same type in the same context. As the types may now have free variables, Definition 1.1.10 must be extended to deal with substitution of terms into types and contexts. This is routine, but once again we see the need for type-equality rules.

DEFINITION 8·2·6: The ***cut rule*** is
$$\frac{\Gamma \vdash X \text{ type} \qquad \Gamma \vdash a : X \qquad \Gamma, x : X, \Psi \vdash \mathcal{J}}{\Gamma, [a/x]^*\Psi \vdash [a/x]^*\mathcal{J}} \text{ cut}$$

Like coercion and weakening, it is a derived rule, because all of the rules are invariant under substitution.

Cut acts on terms, types, judgements and derivations by substitution,

$$(\Psi, \mathcal{J}) \mapsto ([a/x]^*\Psi, [a/x]^*\mathcal{J}),$$

i.e. $[a/x]^* : \mathsf{Cn}([\Gamma, x : X]) \to \mathsf{Cn}(\Gamma)$ by $\mathcal{J}(x) \mapsto \mathcal{J}(a)$.

The associated generating map is of the form

$$\frac{\Gamma \vdash a : X \qquad [\Gamma, x : X, \Psi] \text{ ctxt}}{[\Gamma, [a/x]^*\Psi] \xrightarrow{\quad [a/x]_\Psi \quad} [\Gamma, x : X, \Psi]}$$

which Definition 8.3.2 characterises semantically.

The context Γ of global parameters is needed in the cut rule in order to compose cuts. Let $\Gamma \vdash X$ type and $\Gamma, x : X \vdash Y$ type; then

$$\frac{\Gamma \vdash b : [a/x]^*Y \qquad \dfrac{\Gamma \vdash a : X \qquad \Gamma, x : X, y : Y, \Psi \vdash \mathcal{J}}{\Gamma, y : [a/x]^*Y, [a/x]^*\Psi \vdash [a/x]^*\mathcal{J}}}{\Gamma, [b/y]^*[a/x]^*\Psi \vdash [b/y]^*[a/x]^*\mathcal{J}}$$

where, by the first cut, $[a/x]^*Y$ is a well formed type in the context Γ.

Notice that the double substitution is by $[b/y]$; $[a/x]$ in that order.

Laws. In the Extended Substitution Lemma (Proposition 1.1.12) we must now say whether the variables x and y may or may not occur free in the type-expressions, as well as in the terms. In the following, $y : Y$ is later in the context than $x : X$, and Y may in general depend on x.

REMARK 8·2·7: The variable y never occurs free, and $x \notin \mathsf{FV}(a, X)$.

$$\hat{x} \, ; \hat{y} \quad\quad\quad \rightsquigarrow \quad \hat{y} \, ; \hat{x} \quad\quad\quad\quad\quad x \notin \mathsf{FV}(Y) \quad (\mathsf{P})$$

$$[a/x] \, ; \hat{x} \quad\quad \rightsquigarrow \quad \mathsf{id} \quad\quad\quad\quad\quad\quad\quad\quad (\mathsf{T})$$

$$[a/x] \, ; \hat{y} \quad\quad \rightsquigarrow \quad \hat{y} \, ; [a/x] \quad\quad\quad\quad\quad\quad (\hat{\mathsf{S}})$$

$$\hat{x} \, ; [b/y] \quad\quad \rightsquigarrow \quad [b/y] \, ; \hat{x} \quad\quad\quad x \notin \mathsf{FV}(Y, b) \quad (\check{\mathsf{S}})$$

$$[a/x] \, ; [b/y] \quad\quad \rightsquigarrow \quad \left[[a/x]^* b/y\right] \, ; [a/x] \quad\quad\quad\quad (\mathsf{W})$$

$$[x/y] \, ; \hat{x} \, ; [y/x] \, ; \hat{y} \quad = \quad \mathsf{id} \quad\quad\quad\quad X \equiv Y \quad (\mathsf{R})$$

The type Y may depend on x in $(\hat{\mathsf{S}})$ but not in $(\check{\mathsf{S}})$. We shall see in Lemma 8.2.10 that these laws play different roles, with opposite senses as reduction rules. In fact $(\check{\mathsf{S}})$ may be derived (as an unoriented law) from $(\hat{\mathsf{S}})$ and (R).

Of course the object-language also contributes laws, but we shall regard them as part of the term calculus (so a denotes an equivalence class *modulo* such laws) rather than of the sketch for the category composed of substitutions.

Now that the types may be dependent it is essential to include them in the statement of the laws. We must in particular ensure that the terms and the intermediate types are well formed. As usual, commutative diagrams provide the clearest mode of expression; the little arrows in the corners of the squares indicate the sense of the reduction rules.

The general forms of the laws, which we shall mark with a star, allow for an extension Ψ to the context which may depend on y. For example the general form of (T^*) is that the composite

$$[\Gamma, \Psi] = \left[\Gamma, [b/y]^* \hat{y}^* \Psi\right] \xrightarrow{\;[b/y]\;} \left[\Gamma, y : Y, \hat{y}^* \Psi\right] \xrightarrow{\;\hat{y}\;} [\Gamma, \Psi]$$

is the identity, but to show that this is well formed we need the version without Ψ first.

Similarly we have to know that $\hat{y}^* \hat{x}^* \Psi = \hat{x}^* \hat{y}^* \Psi$ for

$$
\begin{array}{ccc}
[\Gamma, x : X, \hat{x}^*\Phi, y : \hat{x}^* Y, \hat{x}^* \hat{y}^* \Psi] & \xrightarrow{\;\;\hat{x}\;\;} & [\Gamma, \Phi, y : Y, \hat{y}^* \Psi] \\
\Big\downarrow{\hat{y}} & \mathsf{P}^* & \Big\downarrow{\hat{y}} \\
[\Gamma, x : X, \hat{x}^*\Phi, \hat{x}^*\Psi] & \xrightarrow{\;\;\hat{x}\;\;} & [\Gamma, \Phi, \Psi]
\end{array}
$$

The following diagrams show how the (S*) and (W*) laws including the context extension Ψ reduce to the simpler forms, where PP and $\widehat{SS}$ stand for as many squares as there are variables in Ψ.

$$[\Gamma, [a/x]^*\Phi, y : [a/x]^*Y, \hat{y}^*[a/x]^*\Psi] \xrightarrow{\;[a/x]^*\hat{y}^*\Psi\;\;[a/x]_{y\Psi}\;} [\Gamma, x : X, \Phi, y : Y, \hat{y}^*\Psi]$$

$$\widehat{\Psi} \qquad \widehat{SS} \qquad \widehat{\Psi}$$

$$[\Gamma, [a/x]^*\Phi, y : [a/x]^*Y] \xrightarrow{\;[a/x]_y\;} [\Gamma, x : X, \Phi, y : Y]$$

$$\hat{y} \quad PP \quad \hat{y} \qquad \hat{S} \qquad \hat{y} \quad PP \quad \hat{y} \quad (\hat{S}^*)$$

$$[\Gamma, [a/x]^*\Phi] \xrightarrow{\;[a/x]\;} [\Gamma, x : X, \Phi]$$

$$\widehat{\Psi} \qquad \widehat{SS} \qquad \widehat{\Psi}$$

$$[\Gamma, [a/x]^*\Phi, [a/x]^*\Psi] \xrightarrow{\;[a/x]_\Psi\;} [\Gamma, x : X, \Phi, \Psi]$$

$$[\Gamma, x : X, \hat{x}^*\Phi, \hat{x}^*[b/y]^*\Psi] \xrightarrow{\;[b/y]^*\hat{x}^*\Psi\;\;\hat{x}^*[b/y]_\Psi\;} [\Gamma, x : X, \hat{x}^*\Phi, y : \hat{x}^*Y, \hat{x}^*\Psi]$$

$$\widehat{\Psi} \qquad \widehat{SS} \qquad \widehat{\Psi}$$

$$[\Gamma, x : X, \hat{x}^*\Phi] \xrightarrow{\;\hat{x}^*[b/y]\;} [\Gamma, x : X, \hat{x}^*\Phi, y : \hat{x}^*Y]$$

$$\hat{x} \quad PP \quad \hat{x} \qquad \check{S} \qquad \hat{x} \quad PP \quad \hat{x} \quad (\check{S}^*)$$

$$[\Gamma, \Phi] \xrightarrow{\;[b/y]\;} [\Gamma, \Phi, y : Y]$$

$$\widehat{\Psi} \qquad \widehat{SS} \qquad \widehat{\Psi}$$

$$[\Gamma, \Phi, [b/y]^*\Psi] \xrightarrow{\;[b/y]_\Psi\;} [\Gamma, \Phi, y : Y, \Psi]$$

The (W*) law is the Substitution Lemma itself:

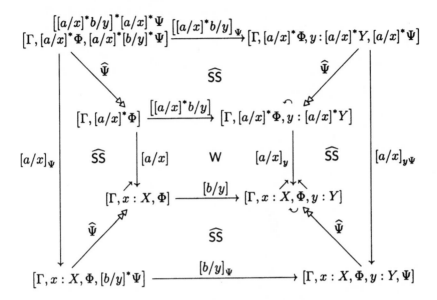

REMARK 8·2·8: The final law justifies **renaming** and **permutation** (**exchange**) of variables. It uses **contraction**, *i.e.* the special case of cut where the term is a new variable of the same type:

$$\frac{\Gamma, x : X, \Phi, y : X, \Psi \vdash \mathcal{J}}{\Gamma, x : X, \Phi, [x/y]^*\Psi \vdash [x/y]^*\mathcal{J}} \qquad \frac{\Gamma, x : X, \Phi, y : X, \Psi \vdash \mathcal{J}}{\Gamma, \Phi, y : X, [y/x]^*\Psi \vdash [y/x]^*\mathcal{J}}$$

Of course $[x/y]$ is a section of the display $\hat{y}$, but its composite with $\hat{x}$ performs renaming, so the parallel pairs make $[\Gamma, x : X] \cong [\Gamma, y : X]$:

$$[\Gamma, x : X] \underset{\hat{y}}{\overset{[x/y]}{\rightleftarrows}} [\Gamma, x : X, y : X] \underset{[y/x]}{\overset{\hat{x}}{\rightleftarrows}} [\Gamma, y : X] \qquad \text{(R)}$$

Again this simpler law, applied to Ψ and with $x \notin \mathrm{FV}(\Phi)$, is needed to show that the general case is well formed.

$$\begin{array}{ccc} [\Gamma, x : X, \Phi, \Psi] & \xrightarrow{\ [x/y]\ } & [\Gamma, x : X, \Phi, y : X, \hat{x}^*[y/x]^*\hat{y}^*\Psi] \\ \hat{y}\uparrow & \text{R}^* & \downarrow\hat{x} \\ [\Gamma, x : X, \Phi, y : X, \hat{y}^*\Psi] & \xleftarrow{\ [y/x]\ } & [\Gamma, \Phi, y : X, [y/x]^*\hat{y}^*\Psi] \end{array}$$

Hence the renaming and exchange rules (Remark 8.2.4) are derivable, since we have shown how to turn x into y and move it past Φ (which may be empty). $\qquad \square$

Normal forms for morphisms. We shall now prove Theorem 4.3.9 and Corollary 4.3.13 for dependent types: each map in the category can be written uniquely as a multiple substitution in a certain normal form. As usual the most natural way of expressing this involves a ground context Γ. Essentially, this corresponds to the slice category $\mathsf{Cn}_{\mathcal{L}}^{\times}\!\downarrow\Gamma$ (Definition 5.1.8), but the appropriate notion is actually a full subcategory, *cf.* $\mathsf{Sub}_{\mathcal{C}}(X) \subset \mathcal{C}\!\downarrow\! X$ (Remark 5.2.5), which is the slice relative to monos.

DEFINITION 8·2·9: The ***relative slice*** $\mathsf{Cn}_{\mathcal{L}}^{\times}\!\downarrow\Gamma$ is the category whose

(a) *objects* are contexts $[\Gamma, \Phi]$ extending Γ, so that the structure map $\widehat{\Phi} : [\Gamma, \Phi] \longrightarrow\!\!\!\!\!\twoheadrightarrow \Gamma$ is always a composite of displays, and

(b) *morphisms* $[\Gamma, \Phi] \to [\Gamma, \Delta]$ are generated by displays and cuts which leave the context Γ untouched, subject to the laws above.

By Exercise 8.9, such maps are exactly the commutative triangles

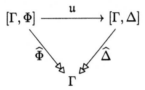

LEMMA 8·2·10: Any map $u : [\Gamma, \Phi] \to [\Gamma, \Delta]$ in $\mathsf{Cn}_{\mathcal{L}}^{\times}\!\downarrow\Gamma$ may be expressed uniquely (up to the choice of new names Δ') in the form

$$\Phi \xrightarrow{\;v\;} [\Delta', \Phi] \xrightarrow{\;\widehat{\widehat{\Phi}}\;}\!\!\!\!\!\twoheadrightarrow \Delta' \xrightarrow[\cong]{\;e\;} \Delta$$

where
$$
\begin{aligned}
\Delta &= [y_1 : Y_1, \cdots, y_m : Y_m] \\
\Delta' &= [y_1' : Y_1, \cdots, y_m' : Y_m] \\
v &= [b_m/y_m'] ; \cdots ; [b_2/y_2'] ; [b_1/y_1'] \\
b_i &= u^*y_i : u^*Y_i
\end{aligned}
$$

and $e : [\Gamma, \Delta'] \cong [\Gamma, \Delta]$ renames $y_1', \ldots, y_m'$ as $y_1, \ldots, y_m$.

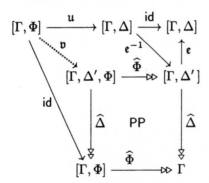

Indeed $\mathfrak{v}$ is the unique map such that $\mathfrak{v} ; \widehat{\Phi} ; \mathfrak{e} = \mathfrak{u}$ and $\mathfrak{v} ; \widehat{\Delta} = \mathrm{id}$. If Φ and Δ have no variables in common, the renaming is not needed.

PROOF: We reduce a composite of displays and cuts to normal form by induction on its length. For the base case, the renaming law (R) expresses id_Δ as a normal form with

$$\Phi = \Delta, \qquad \mathfrak{v} = [y_m/y'_m] ; \cdots ; [y_1/y'_1],$$

where we use the P, $\widehat{S}$ and W laws to commute distinct variables.

For a general map $\mathfrak{u} = \mathfrak{u} ; \mathrm{id}_\Delta$, the first part $\mathfrak{v}$ will remain a composite of m substitutions $[b_i/y'_i]$ and the last part will always be $\mathfrak{e}$. However, the terms b_i and the list $\widehat{\Phi}$ will vary as we append generating maps to the *front*; at first, $b_i = y_i$ and $\widehat{\Phi} = \widehat{\Delta}$. We have to consider the cases of prepending displays $\hat{x}$ and single substitutions $[a/x]$ to a normal form. (To save space we abbreviate the notation, writing Ψ' for $[a/x]^*\Psi$.)

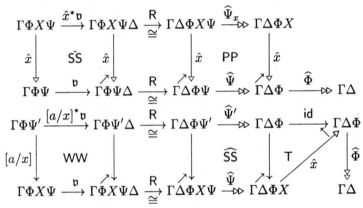

The $\check{S}$ and W squares do the real work, replacing b_i by $\hat{x}^*b_i$ or $[a/x]^*b_i$; the rest of the diagram maintains the list of displays. Notice that $\widehat{S}$ and $\check{S}$ have opposite senses as reduction rules, and that only unstarred reductions are needed, by use of the mediating (R)-permutations.

Now suppose that $\mathfrak{v}'$ satisfies the condition, *i.e.* $\mathfrak{v}' ; \widehat{\Delta} = \mathrm{id}_{[\Gamma,\Phi]}$ and $\mathfrak{v}' ; \widehat{\Delta} ; \mathfrak{e} = \mathfrak{u}$. Applying the construction to $\mathfrak{v}'$ with $\Gamma' = [\Gamma, \Phi]$, $\Phi' = [\ ]$ and Δ we obtain a sequence of single substitutions and *no displays*:

$$\mathfrak{v}' = [b'_m/y'_m] ; \cdots ; [b'_2/y'_2] ; [b'_1/y'_1].$$

But $\Gamma, \Phi \vdash b_i = \mathfrak{u}^*y_i = b'_i : \mathfrak{u}^*Y_i$, so $\underline{b}$ and $\mathfrak{v}$ are uniquely determined.

Finally, if Φ and Δ have no variables in common then $\widehat{\Phi}$ commutes with $\mathfrak{i}$. Then $\mathfrak{v} ; \mathfrak{i}$ is a sequence of assignments to the original $\underline{y}$. $\square$

Notice that the previous result is a special case of the statement that the P square is a pullback (Lemma 8.2.15(a) below). Cuts and displays *almost* form a factorisation system, but the diagonal fill-in is not unique (Exercise 8.8). In future we shall assume that the variables are disjoint, disregarding the renaming part. As in Section 4.3, we write $[\underline{b}/\underline{y}]_\Gamma$ for the normal form. Using this result we can recover the types and terms from the category; this will be our handle on semantic categories, and so the means to define the interpretation and canonical language in Section 8.4.

COROLLARY 8·2·11:   Analogously to Corollary 4.3.13,

(a) types X in context Γ correspond to displays $\hat{x} : [\Gamma, x : X, \Psi] \relbar\joinrel\twoheadrightarrow \Gamma$ *modulo* renaming of variables;

(b) terms $\Gamma \vdash a : X$ correspond to sections of the displays: if $\mathfrak{a} \,;\hat{x} = \mathrm{id}_\Gamma$, where $\mathfrak{a} : \Gamma \to [\Gamma, x : X]$, then $\mathfrak{a} \equiv [a/x]$ with $\Gamma \vdash a \equiv \mathfrak{a}^* x : X$;

(c) in particular, variables *quâ* terms correspond to contractions (Remark 8.2.8) $[x/x'] : [\Gamma, x : X, \Psi] \to [\Gamma, x : X, \Psi, x' : X]$, which are diagonals for the pullback in Lemma 8.2.15(a) below;

(d) the action on clones, indeed on variables, is faithful.   □

REMARK 8·2·12:   The category may be defined directly using normal forms [Pit95]. This is technically more difficult because *recursive* methods do not work easily for *associative* operations (*cf.* append, Definition 2.7.4); in particular a *double induction* is needed to show that the identity id_Γ is well formed. As we noted for the rational numbers in Example 1.2.1, the construction must perform the normalisation process explicitly, rather than leaving it to work for itself at the end.

There are two new forms of judgement, whose rules we have derived:

(a) **Substitution formation** $(\Gamma \vdash \mathfrak{u} : \Phi \to \Delta)$: that the morphism $\mathfrak{u} : [\Gamma, \Phi] \to [\Gamma, \Delta]$ in $\mathsf{Cn}_{\mathcal{L}}^\times \!\downarrow\! \Gamma$ is well formed.

$$\frac{\Gamma \vdash \mathfrak{u} : \Phi \to \Delta \quad \Gamma, \Delta \vdash X \text{ type} \quad \Gamma, \Phi \vdash a : \mathfrak{u}^* X \quad x \notin \Gamma \cup \Delta}{\Gamma \vdash [\mathfrak{u}, a/x] : \Phi \to [\Delta, x : X]}$$

The order of the terms in this notation for morphisms is rather misleading since $[\mathfrak{u}, a/x] = [a/x] \,; \mathfrak{u}_x$, where $\mathfrak{u}_x$ is the same sequence of substitutions as $\mathfrak{u}$, but in the context of an additional variable.

(b) **Substitution equality** $(\Gamma \vdash \mathfrak{u} = \mathfrak{v} : \Phi \to \Delta)$: that the morphisms $\mathfrak{u}, \mathfrak{v} : [\Gamma, \Phi] \rightrightarrows [\Gamma, \Delta]$ are equal (in $\mathsf{Cn}_{\mathcal{L}}^\times \!\downarrow\! \Gamma$).

$$\frac{\Gamma \vdash \mathfrak{u} = \mathfrak{v} : \Phi \rightrightarrows \Delta \quad \Gamma, \Phi \vdash a = b : \mathfrak{u}^* X}{\Gamma \vdash [\mathfrak{u}, a/x] = [\mathfrak{v}, b/x] : \Phi \rightrightarrows [\Delta, x : X]}$$

There are also axioms for the empty substitution $[\ ]_\Gamma \equiv \widehat{\Phi} : [\Gamma, \Phi] \to \Gamma$, which is the terminal projection in the slice $\mathsf{Cn}_{\mathcal{L}}^\times \!\downarrow\! \Gamma$.

It doesn't matter whether we say that two maps are equal globally or in the context Γ, *i.e.* the forgetful functor $\widehat{\Gamma}_! : \mathsf{Cn}^\times_{\mathcal{L}} \!\downarrow\! \Gamma \to \mathsf{Cn}^\times_{\mathcal{L}}$ is faithful and reflects invertibility (it is not in general full).

DEFINITION 8·2·13: The **generalised cut rule**, which includes the weakening rule as well as multiple cuts (Definitions 8.2.5 and 8.2.6) is

$$\frac{\mathsf{u} : \Gamma \to \Delta \qquad \Delta, \Psi \vdash \mathcal{J}}{\Gamma, \mathsf{u}^*\Psi \vdash \mathsf{u}^*\mathcal{J}}$$

The bijection between morphisms $[\mathsf{u}, a/x] : \Gamma \to [\Delta, x : X]$ and terms $\Gamma \vdash a : \mathsf{u}^*X$ is a recurrent idiom, because it is the syntactic form of the generalised elements (arbitrary morphisms into $[\Delta, x : X]$) which we introduced categorically in Remark 4.5.3. It is a special case of the fact that this (PŠ) square is a pullback, and also says how to append a display *after* a normal form.

LEMMA 8·2·14: Let $\mathsf{u} : \Gamma \to [\Delta, y : Y]$, such that $(\mathsf{u}; \hat{y}) = (\mathsf{v}; \widehat{\Gamma}) : \Gamma \to \Delta$ in normal form, where $\mathsf{v} : \Gamma \to [\Gamma, \Delta]$ with $\mathsf{v} \,; \widehat{\Delta} = \mathrm{id}_\Gamma$. Then the normal form of u is $[b/y] \,; \mathsf{v}_y \,; \widehat{\Gamma}$, where $\Gamma \vdash b : \mathsf{v}^*Y$.

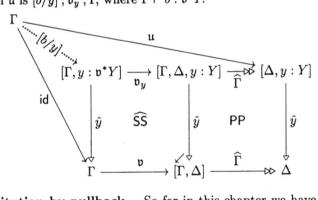

Substitution by pullback. So far in this chapter we have used the upper star notation *only* in the sense of substitution. Now we shall justify the other use (Section 5.1), that u^* is given by pullback along u. Notice that the commutative squares have already been provided by the syntax, and our job is to show that they have the *universal property*: there is no assumption about forming or choosing pullbacks categorically.

LEMMA 8·2·15: The P, $\widehat{\mathsf{S}}$, Š and W squares are pullbacks.

PROOF: We aim to use Lemma 5.1.2 to compose and cancel pullback rectangles as much as possible to cut down the symbolic manipulation. The other tool is the *uniqueness* of normal forms. The choice of maps in a test quadrilateral is not a free one: we work out what their normal forms must be to commute, and then read off the mediator.

(a) Comparing P with another commutative square from Γ to Δ, let $u\,;\widehat{\Gamma}$ be the normal form of the common composite.

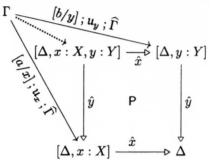

By Lemma 8.2.14, the normal forms of the maps to $[\Delta, x : X]$ and $[\Delta, y : Y]$ are as marked, with $\Gamma \vdash a : u^*X, b : u^*Y$. Then by the same argument the pullback mediator must be $[a/x]\,;[b/y]\,;u_{xy}\,;\widehat{\Gamma}$.

(b) ($\check{\mathsf{S}}$) follows from (P) by Lemma 5.1.2, as the rows are identities.

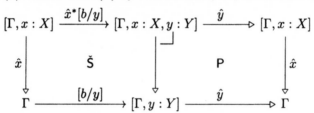

(c) We consider the form of the $\widehat{\mathsf{SS}}$ square in which *all* of the additional variables Φ are deleted, and Ψ is empty, so the letterings $\Gamma X \Phi Y \Psi$ and $\Gamma X \Phi \Psi$ in the diagram on page 442 become $\Delta X \Phi$ and ΔX here.

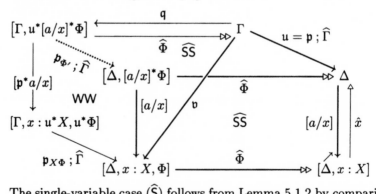

The single-variable case ($\widehat{\mathsf{S}}$) follows from Lemma 5.1.2 by comparing Φ with $[\Phi, y : Y]$. We also abbreviate $\Phi' = [a/x]^*\Phi$. Consider a test quadrilateral from Γ, so $u = \mathfrak{p}\,;\widehat{\Gamma} = \mathfrak{v}\,;\widehat{\Phi}\,;\hat{x}$. (This and the claimed pullback are drawn above in bold.) Using a and $\mathfrak{p}$, the WW and $\widehat{\mathsf{SS}}$ parallelograms shown commute.

Now let the normal form of $\mathfrak{v}$ be $\mathfrak{q};[a'/x];\mathfrak{p}'_{X\Phi};\widehat{\Gamma}$. For commutativity $\Gamma \to \Delta X$, we must have $\mathfrak{p}' = \mathfrak{p}$ and $a' = \mathfrak{p}^*a$ by uniqueness of normal forms. Hence the pullback mediator is $\mathfrak{q};\mathfrak{p}_{\Phi'};\widehat{\Gamma}$, and it is unique by considering the normal form of any other candidate.

(d) W (but not W^*) may be shown to be a pullback by Lemma 5.1.2, since the composites along the top and bottom are identities:

$$
\begin{array}{ccccc}
[\Gamma,[a/x]^*\Phi] & \xrightarrow{\;[[a/x]^*b/y]\;} & [\Gamma,[a/x]^*\Phi,y:[a/x]^*Y] & \xrightarrow{\;\hat{y}\;} & [\Gamma,[a/x]^*\Phi] \\
{\scriptstyle [a/x]}\downarrow & \mathsf{W} & {\scriptstyle [a/x]}\downarrow & \widehat{\mathsf{S}} & \downarrow{\scriptstyle [a/x]} \\
[\Gamma,x:X,\Phi] & \xrightarrow{\;[b/y]\;} & [\Gamma,x:X,\Phi,y:Y] & \xrightarrow{\;\hat{y}\;} & [\Gamma,x:X,\Phi]
\end{array}
$$

Finally, by their decompositions displayed on pages 442–3, the general (P^*), $(\check{\mathsf{S}}^*)$, $(\check{\mathsf{S}}^*)$ and (W^*) squares are also pullbacks, by Lemma 5.1.2. $\square$

THEOREM 8·2·16: Let $\Delta \vdash a : X$ and $\mathsf{u} : \Gamma \to \Delta$ be any map. Then the following squares (are well formed, commute and) are pullbacks in the category of contexts and substitutions.

$$
\begin{array}{ccc}
[\Gamma,x:\mathsf{u}^*X,\mathsf{u}^*\Psi] \xrightarrow{\;\mathsf{u}_{x\Psi}\;} [\Delta,x:X,\Psi] & \quad & [\Gamma,[a/x]^*\mathsf{u}^*\Psi] \xrightarrow{\;\mathsf{u}_\Psi\;} [\Delta,[a/x]^*\Psi] \\
{\scriptstyle \hat{x}}\downarrow \quad \widehat{\mathsf{SP}} \quad \downarrow{\scriptstyle \hat{x}} & & {\scriptstyle [\mathsf{u}^*a/x]}\downarrow \quad \mathsf{W}\check{\mathsf{S}} \quad \downarrow{\scriptstyle [a/x]} \\
[\Gamma,\Psi] \xrightarrow{\;\mathsf{u}_\Psi\;} [\Delta,\Psi] & & [\Gamma,x:\mathsf{u}^*X,\mathsf{u}^*\Psi] \xrightarrow{\;\mathsf{u}_{x\Psi}\;} [\Delta,x:X,\Psi]
\end{array}
$$

PROOF: Writing u as a composite of displays and single substitutions, these squares similarly decompose into (P), (S) and (W) squares, which we have shown to be pullbacks. $\square$

In the syntactic setting the pullback functor is defined, and preserves all of the syntactic structure, up to equality. Moreover pullback along composable maps in turn (*cf.* Lemma 5.1.2) is equal to pullback along their composite, by Proposition 1.1.12.

In Section 8.4 we shall show that this simple categorical property — that the class of displays is closed under pullback — characterises the verbosity of the syntactic presentation of generalised algebraic theory.

8.3   DISPLAY CATEGORIES AND EQUALITY TYPES

This section investigates the semantic structure needed to interpret the syntax of the previous two sections. We shall see that there are many naturally occurring examples of it in general topology and elsewhere, so one of the aims in setting out these ideas is to encourage topos theorists and others to adopt the notation of dependent type theory.

The structure requires less than that the semantic category have all finite limits, and we show that this shortfall corresponds exactly to the addition of equality types to the language. In preparation for the interpretation of and equivalence with syntax in the next section, we show how to present any *semantic* category equipped with a class of display maps as a sketch analogous to that used to construct the category of contexts and substitutions in Section 8.2.

Display maps. In Section 4.6 the interpretation of a (simply typed) algebraic theory $\mathcal{L}$ in a category $\mathcal{C}$ was a *product*-preserving functor $\mathsf{Cn}_\mathcal{L}^\times \to \mathcal{C}$. In the dependent type case the functor preserves (certain) *pullbacks*, but there are a lot more of these, so how can the extension to type dependency be conservative?

REMARK 8·3·1:   The P pullback is simply the binary product $X \times Y$ in the slice $\mathsf{Cn}_\mathcal{L}^\times \downarrow \Gamma$, as in simple type theory. This is commutative and associative, but only up to isomorphism.

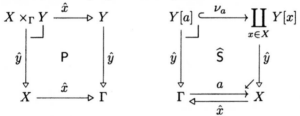

The second square was not mentioned before, as it was automatically a pullback by Lemma 5.1.2. This is because when $x \notin \mathsf{FV}(Y)$, as also for (Š), the X-indexed union becomes $X \times Y$ and the map marked ν_a is a section of $\hat{x} = \pi_1 : X \times Y \longrightarrow Y$ (*cf.* Lemma 8.2.15(b)).

In the dependent case, ν_a is no longer a (generalised) *element* but the ath inclusion into a sum indexed by X. Being a pullback expresses stability of this sum, as in Section 5.5, and ν_a is a regular mono, but not necessarily split. We also see the semantic reason why the components of the normal form of a map to $[\Gamma, x : X, y : Y]$ come in the order $[b/y];[a/x]$: the element $b \in Y[a]$ must be selected before it is included in the sum.

Putting these together, to abstract Theorem 8.2.16 we need the pullback of any display map $\hat{y}$ against arbitrary maps in the category.

DEFINITION 8·3·2:   Let $\mathcal{D} \subset \mathsf{mor}\mathcal{C}$ be a class of maps of any category, which we write $\longrightarrow\!\!\triangleright$.

Then we call $\mathcal{D}$

(a) a **display structure**, if for each $\eth : X \longrightarrow \Delta$ in $\mathcal{D}$ and $u : \Gamma \to \Delta$ in $\mathcal{C}$ there is a *given*[2] pullback square, in which $u^*\eth \in \mathcal{D}$;

(b) a **class of displays** if every such square *exists*, with $u^*\eth \in \mathcal{D}$, and the class $\mathcal{D}$ is closed under composition with isomorphisms.

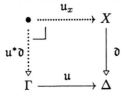

An **interpretation** of $(\mathcal{C}, \mathcal{D})$ considered as a generalised algebraic theory is a functor from $\mathcal{C}$ which preserves displays and these pullbacks.

We shall say that a $\mathcal{C}$-map is a **cut** if, like ν_a above, it can be expressed as a pullback of a section of a display along a sequence of displays.

The class $\mathcal{D}$ need not be a subcategory, *i.e.* include all identities and be closed under composition. In fact these properties of $\mathcal{D}$ say that there are *base types* isomorphic to singletons and dependent sums (Remark 9.3.1). But the semantic classes of displays frequently are subcategories, and in practice it is convenient and harmless to make this assumption.

Even when $\mathcal{D}$ is a subcategory, it is "closed under pullbacks" only in the sense that if the right hand side of a pullback square in $\mathcal{C}$ lies in $\mathcal{D}$ then so does the left. The category $\mathcal{D}$ need not *have* pullbacks: we have not required pullback *mediators* to lie within it (Exercise 8.13). In particular it need not be the $\mathcal{M}$-class for a factorisation system. Indeed if the cancellation property $(u \,;\, \eth)$, $\eth \in \mathcal{D} \Rightarrow u \in \mathcal{D}$ (*cf.* Lemma 5.7.6(e)) were required, this would defeat the point of Definition 8.3.8 below.

Next we consider the extreme cases between which display structures interpolate.

---

[2]In the syntax the substitution action is *functorial*, as John Cartmell required in his definition [Car86], *i.e.* $\mathrm{id}^*\hat{x} = \hat{x}$ and $(\mathfrak{v} \,;\, u)^*\hat{x} = \mathfrak{v}^*(u^*\hat{x})$ up to *equality*. The left-associated product also has this property because it only depends on X, not on u. However, we shall find that the interpretation makes no use of this, because the pullbacks which are used are always "anchored" at the interpretation of the type-*symbol*. This is because we forbade laws between types as axioms in the theory (Remark 8.1.9).

The map u is determined, in the syntax, by a list of terms, where the terms are *equivalence classes* of raw terms *modulo* laws. The assignment of $u^*\eth$ to u must depend only on the morphism, *i.e.* on the equivalence classes, not on the choice of raw terms to represent it.

Indeed we only need pullbacks against the *interpretations* of syntactic morphisms, but since our aim is a completeness theorem, *i.e.* to give a syntactic name to every semantic map, we do not exploit this weakening of the definition.

Products and equality types.

EXAMPLE 8·3·3:   Let C be a category with specified terminal object
1 and specified binary products. Then the class D of maps that are
specified left projections, $\pi_0 : \Delta \times X \longrightarrow \Delta$, forms a display structure.
The resulting interpretation of the context $[x_1 : X_1, \ldots, x_n : X_n]$ is the
left-associated product $((\cdots(((1 \times X_1) \times X_2) \times X_3) \times \cdots) \times X_n)$ which
was defined in Remark 4.5.15.

PROOF: Given any map $u : \Gamma \to \Delta$, let the specified product projection
$\pi_0 : \Gamma \times X \longrightarrow \Gamma$ serve as the choice of pullback. The "dependent" types
in this example are in fact constant.   □

PROPOSITION 8·3·4:   Let C be a category with a class of displays. If all
product projections and pullback diagonals are displays, then *every* map
is (isomorphic to) a composite of displays, and C has all finite limits.

PROOF: Consider the graph of $u : X \to Y$ (Remark 5.1.7):

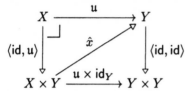

Then the pullback of u against any $v : Z \to Y$ must exist.   □

REMARK 8·3·5:   The diagonal is the ***equality type***,

$$\hat{=} : \{\langle y_0, y_1 \rangle \mid y_0 = y_1\} \longrightarrow [y_0, y_1 : Y],$$

so Proposition 8.3.4 says that the classifying category of a generalised
algebraic theory has all finite limits (and all of its maps are isomorphic
to displays) iff the theory has all equality types.

The presence or absence of equality types influences the quantifiers. In
the syntax, the quantifiers are always associated with a (bound) *variable*,
and we shall find in Sections 9.3 and 9.4 that they are the two adjoints
to the *weakening* functor for that variable. Bill Lawvere, however, who
first had this insight, described them as adjoint to pullback v^* against
arbitrary maps v [Law69], and emphasised this by discussing the diagram
above [Law70, p. 8]. Quantification along a general function f in *Set*
gives the ***guarded quantifiers*** (Remarks 1.5.2 and 3.8.13(b)),

$$\begin{aligned}
(\exists_f \phi)[y] &\equiv \exists x. (y = f(x)) \wedge \phi[x] \\
(\forall_f \phi)[y] &\equiv \forall x. (y = f(x)) \Rightarrow \phi[x],
\end{aligned}$$

in which the effect of the graph $\{\langle x, y \rangle \mid y = f(x)\}$ is clearly visible.

Consider in particular quantification along the diagonal map $\hat{=}_X$:

$$(\exists_{\hat{=}} \top)[x_0, x_1] \Leftrightarrow (x_0 = x_1) \qquad (\forall_{\hat{=}} \bot)[x_0, x_1] \Leftrightarrow (x_0 \neq x_1).$$

Perhaps the equality type itself may be acceptable in a computational setting, but the inequality type $(x_0 \neq x_1)$ begins to raise doubts. To be able to distinguish *positively* between x_0 and x_1 suggests a "Hausdorff" condition on the type, *i.e.* that there is some computation $f : X \to \mathbf{2}$ that terminates for these arguments (but not necessarily everywhere) with $f(x_0) = 0$ and $f(x_1) = 1$. In fact $\forall_{\hat{=}}$ exists for $X \in \operatorname{ob} Sp$ (classically) iff it is locally Hausdorff (Example 9.4.11(e)). On the other hand, $\exists_{\hat{=}}$ exists iff X is a discrete space (Remark 9.3.13).

Display maps in topology and elsewhere. Since Definition 8.3.2 is a unary closure condition (Section 3.8) there are lots of examples.

EXAMPLES 8·3·6: The following are semantic classes of displays:

(a) Product projections (*i.e.* all legs of spans which have the universal property of a product) in a category with finite products.

(b) *All* maps in a lex category.

(c) All instances of the order relation in a meet-semilattice.

(d) Monos in *Set*. If we think of subobjects as subsets with *canonical* inclusions then these provide a corresponding display *structure*. The pullback in this case is called inverse image.

(e) A map $\partial : X \to \Delta$ in any category is called **carrable** if the pullback $u^*\partial$ along every map $u : \Gamma \to \Delta$ exists.

(f) The carrable maps in $\mathcal{IPO}$ (domains *with bottom*) are exactly the projections (Example 3.6.13(b)).

(g) Inclusions of normal subgroups form a class of displays in $\mathcal{Gp}$ which is not closed under composition.

(h) Replete functors (Definition 4.4.8(d)) form a class of displays in the 1-category of categories and functors.

(i) Subspaces, local homeomorphisms and open surjections (but not general continuous surjections) of topological spaces. Example 9.3.10 shows how to use open maps to interpret the existential quantifier. These and other classes may also be defined for toposes.

(j) We may form the closure under pullbacks of any class whatever of carrable maps, to give a class of displays.

(k) More usefully, given a class of "grue" maps (for example closed maps between topological spaces), we say that a morphism is **stably grue** if all pullbacks of it exist (*i.e.* it is carrable) and are grue. (In the modal logic of Section 3.8, stably means $\Box$, and part (j) said $\Diamond$.)

Many of the examples of classes of displays of toposes may be described as an internal set, locale or other structure in the target topos.

EXAMPLES 8·3·7:

(a) D. Lazard (*c.* 1950) defined a *sheaf* as a local homeomorphism $\eth : X \longrightarrow \Delta$ between two topological spaces. The fibres $X[u] \equiv \eth^{-1}(u)$ for $u \in \Delta$ are discrete and may be regarded as the values of a "variable" *set* as u ranges over the space Δ. The relative slice $Sp \downarrow \Delta \simeq Sh(\Delta)$ is a topos, so it obeys the same (intuitionistic) logic as *Set*. The (global) sections $\Delta \to X$ of $\eth$ are the global elements of the set X, but these are inadequate to characterise it: we need to consider arbitrary maps (generalised elements) $\Gamma \to X$ (Definition 8.2.13).

(b) The category Loc of locales has all finite limits, so that *arbitrary* continuous functions $\eth : X \to \Delta$ may be used as displays. They correspond to internal locales in the topos $Sh(\Delta)$.

Intermediate pullback-stable classes of continuous functions or geometric morphisms correspond to notions of space over Δ lying between the discrete and the general cases.

(c) Algebraic lattices (Sections 3.4 and 3.9) provide the simplest notion of "domain". Recall that they are of the form $Idl(\mathcal{C}^{op})$, where $\mathcal{C}$ is a meet-semilattice, and the Scott topology is the frame of monotone functions $2^{\mathcal{C}^{op}}$. Martin Hyland and Andrew Pitts showed how to make these domains "variable" [HP89], by allowing $\mathcal{C}$ to be an internal semilattice in $Sh(\Delta)$. The topology of the domain is then $A^{\mathcal{C}^{op}}$, the topology of Δ being A; the topos display is $\eth : Sh(\Delta)^{\mathcal{C}^{op}} \longrightarrow Sh(\Delta)$, where $\eth^*$ takes $F \in ob\, Sh(\Delta)$ to $\lambda X:ob\,\mathcal{C}.\ F$.

(d) They also generalised this idea from propositions to types, replacing semilattices by lex categories, *i.e.* the classifying categories of finitary essentially algebraic theories internal to $Sh(\Delta)$. A similar construction could be based on our generalised algebraic theories (classified by display categories) instead.

In the next section we show that for each class of display maps there is a certain generalised algebraic theory. The terms of the corresponding type in context are exactly the points of the variable space, and the type theory allows us to reason about it as if it were a set. Unlike classical logic, no assumption is built in that structures are determined by their points: they may have *none* globally. The "points" provided by type theory are terms or generalised elements (Remark 4.5.3). In this way dependent type theory is applicable in general topology, topos theory and domain theory to justify more "synthetic" styles of argument.

Relative slices. From the semantics we shall now move gradually back towards syntax, starting with the analogue of Definition 8.2.9.

DEFINITION 8·3·8: For $\Gamma \in \mathrm{ob}\,\mathcal{C}$, the **relative slice category** $\mathcal{C}\!\downarrow\!\Gamma$ has

(a) as *objects* Δ the composable sequences of $\mathcal{D}$-maps

$$\Gamma \longleftarrow X_1 \longleftarrow X_2 \longleftarrow \cdots \longleftarrow X_n$$

(b) and as *morphisms* the commuting triangles in $\mathcal{C}$ of the form

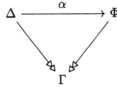

where $\alpha : \Delta \to \Phi$ is any $\mathcal{C}$-map.

So the forgetful functor $\mathrm{src} : \mathcal{C}\!\downarrow\!\Gamma \to \mathcal{C}$ is faithful and reflects invertibility. In particular, if $\mathcal{C}$ has a terminal object, $\mathcal{C}\!\downarrow\!1 \to \mathcal{C}$ is full and faithful. Don't confuse $\mathcal{C}\!\downarrow\!1$ with $\mathcal{C}\!\downarrow\!1$, which is trivially isomorphic to $\mathcal{C}$.

REMARK 8·3·9: Every context may be reduced to [ ] by successively omitting variables, so every object of $\mathsf{Cn}^\times_{\mathcal{L}}$ has a canonical sequence of displays down to the terminal object. Since it preserves displays, the interpretation only makes use of a particular semantic object if it too has such a sequence. Cartmell [Car86] focused on the sequences of displays, defining a **contextual category** to have a tree-structure on its object-class (and functorial assignment of pullbacks). However, there may be isomorphic objects with entirely different paths in the tree structure. Our relative slice category $\mathcal{C}\!\downarrow\!1$ has this tree structure, but $\mathcal{C}$ need not.

If every terminal projection $X \to 1$ in $\mathcal{C}$ is a display, then $\mathcal{C}\!\downarrow\!1$ is strongly equivalent to $\mathcal{C}$. More generally, we say that $(\mathcal{C}, \mathcal{D})$ is **rooted** if every $X \to 1$ is a composite of isomorphisms and displays ([HP89] calls this the **display condition**). Then $\mathcal{C}\!\downarrow\!1 \simeq \mathcal{C}$, either strongly or weakly according as the composite is given or just exists.

No further hypothesis is needed concerning the existence of products in $\mathcal{C}\!\downarrow\!\Gamma$, because they are given by pullbacks of displays over Γ and so are guaranteed by the display axioms.

Using the contextual structure of the semantic category $\mathcal{C}\!\downarrow\!1$, its maps can be expressed in "normal form" and this category is presented by a sketch in the same way as that used to define the syntactic category $\mathsf{Cn}^\times_{\mathcal{L}}$. The proof is much easier than for the corresponding results of the previous section, because this time we know in advance that P and Š are pullbacks, whereas before we had to work up to this from the special case in Lemma 8.2.10.

LEMMA 8·3·10:   Every morphism $\beta : \Delta \to \Phi$ of $\mathcal{C} \!\downarrow\! \Gamma$ may be expressed as a composite of a sequence of cuts (in the sense of Definition 8.3.2) of the same length as Φ, followed by displays corresponding to Δ. This is unique up to unique isomorphism of the intermediate objects.

PROOF: The pullback on the left gives $\beta = \gamma \,;\, \widehat{\Delta}$ with $\gamma \,;\, \widehat{\Phi} = \mathrm{id}_\Delta$:

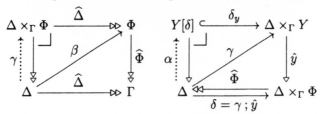

We decompose γ into a sequence of cuts by induction on (the length of) Φ, the base case [ ] being trivial. The second diagram is the induction step, adding one display $Y \longrightarrow \Phi$, where $\delta = \gamma \,;\, \hat{y}$ is already in normal form. The extra cut α is found as shown, and $\gamma = \alpha \,;\, \delta_y$. Finally, as in Lemma 8.2.10, the number of displays involved is fixed by the source and target of the given map, and the intermediate objects are determined (as pullbacks) up to isomorphism.   □

PROPOSITION 8·3·11: $\mathcal{C} \!\downarrow\! \Gamma$ is given by a sketch with laws analogous to those called (P), (T), $(\widehat{\mathsf{S}})$, $(\check{\mathsf{S}})$ and (W) in Remark 8.2.7.

PROOF: These laws are needed to take a composite which is the wrong way round and rearrange it into normal form.   □

We have presented the syntactic and semantic categories by sketches of the same form. Now we shall turn this into a categorical equivalence.

8.4   INTERPRETATION

Now we shall interpret the language in a category with displays, and show that any such category arises up to equivalence as $\mathrm{Cn}_{\mathcal{L}}^{\times}$ for some generalised algebraic theory $\mathcal{L}$, as we did in Sections 4.6 and 7.6.

Derivation histories in normal form. Before giving the interpretation itself we must clarify the well founded structure over which it is defined, because Example 8.2.1 shows that this is delicate. We must take (equivalence classes of) histories, rather than the strings of operation-symbols and variables, as the terms *etc.* in $\mathrm{Cn}_{\mathcal{L}}^{\times}$.

REMARK 8·4·1:   The rules of Section 8.1 may be reorganised according to the use they make of the object-language (Definition 8.1.10).   Let

$u : \Gamma \to \Delta$ be any morphism (substitution), $[\Gamma, \Psi]$ an extended context and x a variable which is not in Γ, Δ or Ψ.

(a) Each type-symbol $(\Delta \vdash X \text{ type}) \in \mathcal{L}$ provides the following features:
- ▸ the type-expression $\Gamma \vdash u^* X$, whose arguments are given by the components of u,
- ▸ the extended context $[\Gamma, x : u^* X, \Psi]$,
- ▸ weakening $\hat{x}_\Psi : [\Gamma, x : u^* X, \Psi] \dashrightarrow [\Gamma, \Psi]$,
- ▸ contraction $[x/x']_\Psi : [\Gamma, x : u^* X, \Psi] \to [\Gamma, x : u^* X, x' : u^* X, \Psi]$,
- ▸ and the variable $\Gamma, x : u^* X, \Psi \vdash x : u^* X$ *quâ* term.

In the presentation of the syntax these were listed under different headings (8.1.6, 8.2.2, 8.2.5, 8.2.8 and 8.1.3(b) respectively), but they are obtainable from one another *immediately*, *i.e.* by the use of a single rule of derivation. Types are equal (8.1.7) iff they result from equal substitutions into the same type-symbol; similarly with context equality and coercion (8.2.3).

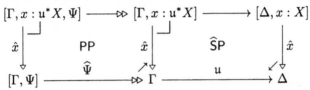

These are all aspects of the same thing, for which it is convenient to take the display $\hat{x} : [\Delta, x : X] \dashrightarrow \Delta$ as the primitive form. The other, substituted, forms are obtained from the display by pullback.

(b) The type of an operation-symbol may in general be a substitution instance $\mathfrak{p}^* X$ of a type-symbol, given by the previous part. Each operation-symbol $(\Delta \vdash r : \mathfrak{p}^* X) \in \mathcal{L}$ provides
- ▸ the term $\Gamma, \Psi \vdash u^* r : (u \,;\, \mathfrak{p})^* X$ and
- ▸ the cut $[u^* r / x]_\Psi : [\Gamma, [u^* r / x]^* \Psi] \to [\Gamma, x : (u \,;\, \mathfrak{p})^* X, \Psi]$

(Definitions 8.1.3(a) and 8.2.6). We shall use the section $[r/x]$ of $\mathfrak{p}^* \hat{x}$ as the primitive form of the operation-symbol.

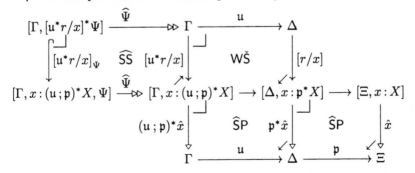

(c) Each law $(\Delta \vdash \mathfrak{f}^*r = \mathfrak{g}^*s : \mathfrak{h}^*X) \in \mathcal{L}$, where $\mathfrak{h} = \mathfrak{f} ; \mathfrak{p} = \mathfrak{g} ; \mathfrak{q}$, relates
two terms of the same type (Definition 8.1.4(a));

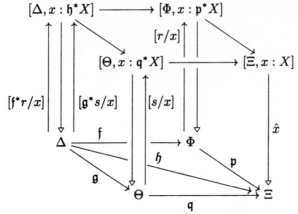

it extends to $\Gamma, \Psi \vdash (\mathfrak{u} ; \mathfrak{f})^*r = (\mathfrak{u} ; \mathfrak{g})^*s : (\mathfrak{u} ; \mathfrak{h})^*X$ (congruence,
8.1.4(c)), but there is insufficient space to show this.

(d) The (reflexive, symmetric and) transitive laws of term equality are
also needed (8.1.4(b)), as are the substitution laws (8.1.4(c) and
8.2.7). All of them feature in these diagrams, apart from (R), which
disappears along with the variables in the interpretation anyway.

(e) Type and context equality and coercion are derivable.

REMARK 8·4·2: The typical formation step is therefore simply
$$\frac{\Gamma \text{ ctxt} \quad [\Gamma, \Psi] \text{ ctxt} \quad \Delta \text{ ctxt} \quad \mathfrak{u} : \Gamma \to \Delta \quad (\Delta \vdash \mathcal{J}) \in \mathcal{L}}{\Gamma, \Psi \vdash \mathfrak{u}^* \mathcal{J}}$$
where $\mathcal{J}$ is a type-symbol, operation-symbol or law, *cf.* the generalised
cut rule (Definition 8.2.13). (If we laid out the derivation as a tree in the
way suggested by this rule there would be a great deal of redundancy;
the box style is more natural.)

To show that a context $\Gamma \equiv [\underline{x} : \underline{X}]$ is well formed we must prove
$$x_1 : X_1, \ldots, x_{n-1} : X_{n-1} \vdash X_n \text{ type} \qquad (n \leqslant \mathsf{len}(\Gamma)).$$
Similarly, for a map $\mathfrak{u} \equiv [\underline{a}/\underline{y}]$ in the form of Remark 8.2.12(a) we need

$$\frac{\begin{array}{c} \Gamma \xrightarrow{\mathfrak{v}} \Delta' \\ \mathfrak{f} \downarrow \qquad \downarrow \mathfrak{g} \\ \Phi \xrightarrow{\mathfrak{p}} \Xi \qquad \Xi \vdash Y \text{ type} \qquad \Phi \vdash r : \mathfrak{p}^*Y \end{array}}{\Gamma \vdash a_n : \mathfrak{v}^*Y_n}$$

where $\mathfrak{v} \equiv [a_1/y_1, \ldots, a_{n-1}/y_{n-1}] : \Gamma \to \Delta' \equiv [y_1 : Y_1, \ldots, y_{n-1} : Y_{n-1}]$,
$a_n \equiv \mathfrak{f}^*r$ and $Y_n \equiv \mathfrak{g}^*Y$. We have to prove commutativity of the square,

i.e. that the maps are composable and their composites are equal, in order to justify the typing. This uses the laws (which must be part of the derivation tree) and the normal form theorem (Lemma 8.2.10): the derivations are of *raw* terms, but the justification of such formations may rely on certain laws, the two sides of which themselves each require derivation sub-trees.

Comparing this with the recursive paradigm (Definition 2.5.1 and Section 6.2), the sub-expressions to be considered in practice are the types of the context Γ and the terms used in the substitution $u : \Gamma \to \Delta$. The latter are the arguments of the type- and operation-symbols as presented informally in Section 8.1. The well-foundedness condition mentioned in Definition 8.1.10 is needed for the existence of derivations.

THEOREM 8·4·3: This is the normal form for derivations.

PROOF: We have shown throughout Remark 8.4.1 how the rules (for the generalised algebraic fragment) set out informally in Section 8.1 fit into the normal form. Weakening and cut (Definitions 8.2.5ff) commute with the last rule in the normal form, increasing the height of the derivation history by at most that of the substituted term, so histories are strongly normalising.[3] □

The induction in this and later results is over the derivation tree, not just on the number of variables in Γ, since the target Δ of u may be a longer context than Γ. This allows for the possibility that nested operation-symbols may have greater arity than the outermost one. The base case is the empty context [ ].

In Section 7.4 we found the free model for equational algebraic theories with many but simple types as the quotient of the associated absolutely free model by the congruence generated by the laws. Because of the effect the laws have on types, this is not possible in the dependent case. We have to generate the well formed instances of equality as syntactic entities in themselves, along with the types, terms, contexts and substitutions, allowing coercions arising from such equalities. At the end these will form a congruence, and the terms may finally be treated as equivalence classes. The derivation histories form a recursive cover (Definition 6.2.2), but there is no canonical choice of history for any given term or type.

---

[3]I shall leave the matter of confluence to some reader with a better stomach for such manipulation. I have to confess that I am not particularly confident of this result, but I feel it would be better to study stratified algebraic theories (Definition 8.1.11) than to pursue this complication for its own sake. Then we should ask what role meta-operations like comprehension play, and *exactly how* they break the stronger well-foundedness condition.

Interpretation.   To present the data for a model of a simply typed algebraic theory in Definition 4.6.2, we needed only each object A_X itself which was to be the denotation of the sort X, together with some products of these objects to serve as the sources of the denotations of the operation-symbols. The dependent type situation is much more complicated. Now, for the sort $\Delta \vdash X$, the object A_X is the source of some semantic display map whose target is the interpretation $[\![\Delta]\!]$. We must begin the *proof* of the following theorem, *i.e.* the construction of $[\![-]\!] : \mathrm{Cn}_{\mathcal{L}}^{\times} \to \mathcal{S}$ by structural recursion, before completing its *statement*.

THEOREM 8·4·4:   Let $\mathcal{L}$ be a generalised algebraic theory and let $\mathcal{C}$ be a category with a display structure $\mathcal{D}$. Then the interpretations of $\mathcal{L}$ correspond bijectively to functors $\mathrm{Cn}_{\mathcal{L}}^{\times} \to \mathcal{C} \!\downarrow\! 1 \to \mathcal{C}$ which preserve displays and the P and S pullbacks. The choice of pullbacks of displays defines the interpretation up to equality. Homomorphisms correspond to natural transformations as in Example 4.8.2(e).

PROOF: Using structural recursion over the derivations in Remark 8.4.2 the interpretation given in Remark 4.6.5 extends to dependent types:

(a) The base case is the empty context $[\,]$, which is interpreted as the terminal object 1 of $\mathcal{C}$.

(b) The (displays corresponding to) type-symbols $\Delta \vdash X$ type are given.

(c) Substitution along $u : \Gamma \to \Delta$ and $\widehat{\Psi} : [\Gamma, \Psi] \dashrightarrow \Gamma$ uses the chosen P and S pullbacks of display maps against arbitrary maps in $\mathcal{C}$; cut and weakening are sound.

(d) The (sections of displays corresponding to) operations are given.

(e) The laws of $\mathcal{L}$ are given. Any provable equality has a derivation tree, by induction over which the terms have equal interpretations.

(f) The laws T, P, S, W and R (the Extended Substitution Lemma) hold by Proposition 8.3.11. $\square$

COROLLARY 8·4·5:   To interpret a *particular* context, type, term or morphism only *finitely many* choices of pullback are used. Hence such fragmentary interpretations exist even when no global choice of pullbacks is provided, and are unique up to unique isomorphism (*cf.* the proof of Theorem 7.6.9(b)). Also, when the interpretation of a type in a context has been chosen, the meaning of its terms is fixed up to equality. $\square$

Canonical language.   Conversely, we shall show that every category with a rooted class of displays is equivalent to the category of contexts and substitutions for some generalised algebraic theory. This may be read off from the sketch in Proposition 8.3.11, but we need encoding operations (Section 7.6) to say that the P and S squares are pullbacks.

As in Sections 4.2, 7.6 and 8.3, we shall temporarily use Greek letters for maps of C, the semantic category. German letters still denote syntactic morphisms. We shall not make a notational distinction between the objects of $C \downarrow 1$ and the type-symbols and contexts which they name; type *expressions* are flagged by the presence of a substitution action.

DEFINITION 8·4·6: Let C be a small category with a class of displays $\mathcal{D}$. The **canonical language** $\mathsf{L}(C, \mathcal{D})$ is the following theory:

(a) Essentially, each display map $X \longrightarrow \Delta$ gives a type-symbol which we shall call just X (though this is an abuse of notation).
More precisely, each sequence Φ of displays, *i.e.* each object of $C \downarrow 1$,

$$1 \longleftarrow X_1 \longleftarrow X_2 \longleftarrow \cdots \longleftarrow X_n \longleftarrow X_{n+1},$$

names a **context-symbol**

$$\ulcorner \Phi \urcorner \equiv [x_1 : X_1, x_2 : X_2, \ldots, x_n : X_n, x_{n+1} : X_{n+1}],$$

in which all of the types are type-*symbols* with arguments exactly the preceding variables. The variable names are chosen arbitrarily.
Remark 8.4.1(a) put no restriction on the defining context Δ of a type-symbol: its types could be any substitution instances. Here we only introduce type-symbols in contexts which themselves consist of type-*symbols* alone. Since no substitution operations are used, no pullbacks are needed to construct the interpretation of these contexts and type-symbols in C; in fact we recover $[\![\ulcorner \Phi \urcorner]\!] = \Phi$.

There are two classes of operation-symbols.

(b) Each section of a display names an operation-symbol:

$$\Delta \overset{\alpha}{\underset{\longleftarrow}{\longrightarrow}} X \qquad \Delta \vdash \ulcorner \alpha \urcorner : X$$

with $[\![\ulcorner \alpha \urcorner / x]\!] = \alpha$. Again this is a simplification of Remark 8.4.1(b), *viz.* that the type of an operation-symbol is a type-symbol ($\mathfrak{p} = \mathsf{id}$), and they are defined in a context $\Delta \equiv \Xi$ that itself consists only of type-symbols.

(c) Let $\square$ be any semantic pullback square (on the left), in which $\mathfrak{v}$ is a *syntactic* morphism; so there are type-symbols Y and Z, defined in contexts Δ and Φ respectively that consist of type-symbols alone.

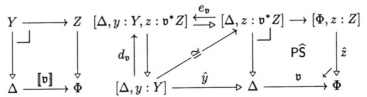

In each such situation we introduce encoding operations $d_\mathfrak{v}$ and $e_\mathfrak{v}$

$$\Delta, y : Y \vdash d_\mathfrak{v}(y) : \hat{y}^*\mathfrak{v}^*Z \qquad \Delta, z : \mathfrak{v}^*Z \vdash e_\mathfrak{v}(z) : \hat{z}^*Y,$$

noting that neither of them is a section of a type-symbol. It will be convenient to name them according to the substitution $\mathfrak{v}$ involved, ignoring the objects Y and Z.

In particular, the generating cases (Remark 8.3.1) are

(P) $\Delta = [\Phi, x : X] \xrightarrow{\mathfrak{v}=\hat{x}} \Phi$ with $Y \underset{e}{\overset{d}{\underset{\cong}{\rightleftarrows}}} Z \times_\Phi X$ over Δ, so, in

Remark 7.6.12, d corresponds to $\langle \ulcorner\pi_0\urcorner, \ulcorner\pi_1\urcorner\rangle$ and e to pair, and

(S) $\Delta \xrightarrow{\mathfrak{v}=[a/x]} \Phi = [\Delta, x : X]$ with $Y \underset{e}{\overset{d}{\underset{\cong}{\rightleftarrows}}} Z[a] \xhookrightarrow{\nu_a} \sum_{x \in X} Z[x]$.

Recall the distinctions made in Definition 7.6.5, between the *weak* canonical language of (in this case) a *semantic* class of displays and a *strong* language which picks out the display *structure*. (The *choice* of pullback $Y = [\mathfrak{v}]^*Z$ is registered by including *its* encoding operations $d_\mathfrak{v}$ and $e_\mathfrak{v}$ in the language.) We require the latter to define the interpretation $[\![-]\!]$, and then $[\![d_\mathfrak{v}]\!] = [\![e_\mathfrak{v}]\!] = \mathrm{id}$. For the former, $d_\mathfrak{v}$ and $e_\mathfrak{v}$ are defined for *every* pullback square $\square$. These operations obey the weaker well-foundedness condition in Definition 8.1.10, but they violate the stronger one for a stratified theory (Definition 8.1.11).

There are four classes of laws.

(d) For each pullback-diagonal α arising from a display $\hat{x}$,

$$\Delta \xleftarrow{\hat{x}} X \underset{\hat{y}}{\overset{\alpha}{\rightleftarrows}} X \times_\Delta X,$$

we have $\Delta, x : X \vdash \ulcorner\alpha\urcorner = x : X$, that is, $[\ulcorner\alpha\urcorner(x)/y] = [x/y]$.

(e) The operations $d_\mathfrak{v}$ and $e_\mathfrak{v}$ are mutually inverse,

$$\Delta, y : Y \vdash e_\mathfrak{v}(d_\mathfrak{v}(y)) = y \qquad \Delta, z : \mathfrak{v}^*Z \vdash d_\mathfrak{v}(e_\mathfrak{v}(z)) = z,$$

(f) and coherent with respect to identities and composites of pullbacks,

$$\Gamma, x : X \qquad \vdash \quad d_{u;\mathfrak{v}} = [d_\mathfrak{v}/y]^* u_y^* d_u : \hat{x}^*(u ; \mathfrak{v})^*Z$$
$$\Gamma, z : (u ; \mathfrak{v})^*Z \quad \vdash \quad e_{u;\mathfrak{v}} = [u_z^* e_\mathfrak{v}/y]^* \hat{z}^* e_u : \hat{z}^* X,$$

or, regarding these operations as (iso)morphisms in relative slices,

$$d_u ; u^* d_\mathfrak{v} = d_{u;\mathfrak{v}} \qquad u^* e_\mathfrak{v} ; e_u = e_{u;\mathfrak{v}}$$

(For identity squares, *i.e.* $X = Y$ and $u = \mathrm{id}$, $e_{\mathrm{id}} = d_{\mathrm{id}} = \mathrm{id}_X$. We could just include $d_\mathfrak{v}$ and $e_\mathfrak{v}$ in the language when $\mathfrak{v}$ is a display or cut, with coherences with respect to the Extended Substitution Lemma.)

(g) Terms (substituted operation-symbols f^*r) must be related to the operation-symbols with the same denotation. For each semantic square $\square$ as above, where $\alpha\,;\hat{x} = \mathrm{id}_\Gamma$ and $\beta\,;\hat{y} = \mathrm{id}_\Delta$, the diagram on the right must commute in the syntax if that on the left does so semantically:

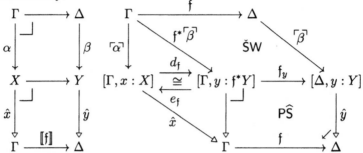

Again, in the cases $\square = \mathsf{P}$ and S,

(P, $\hat{x}$) $\ulcorner\alpha\urcorner = \hat{x}^{*}\ulcorner\beta\urcorner$ is $\ulcorner\beta\urcorner$ plus an unused argument x, and

(S, $[a/x]$) $\ulcorner\alpha\urcorner = [a/x]^{*}\ulcorner\beta\urcorner$ is a special case of $\ulcorner\beta\urcorner$,
              with its x-argument fixed at a.

REMARK 8·4·7: The canonical language of $\mathsf{Cn}_{\mathcal{L}}^{\times}$ is the *clone*: it has a new (type- or operation-) symbol for each equivalence class of expressions in $\mathcal{L}$; *cf.* the comments following Theorems 4.6.7 and 7.6.9.

Completeness. The next two results state conservativity of dependent type theory, the first in a syntactic idiom, and the second in the way in which we defined it in Definition 7.6.8.

LEMMA 8·4·8: Every term $\Gamma \vdash t : u^{*}Y$ defined in a context consisting only of type-symbols is of the form $t = \ulcorner\alpha\urcorner/x]^{*}d_u$, where $\alpha = [\![e_u(t)]\!]$.

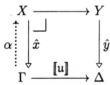

In particular, every term $\Gamma \vdash a : X$ whose type is and context consists of type-symbols is provably equal to a unique operation-symbol $\ulcorner\alpha\urcorner$ naming a section $\alpha = [\![a]\!]$ of the corresponding display (Definition 8.4.6(b)).

(This does not make the $e_{\mathfrak{v}}$ encoding operations redundant, because the contexts in which they were defined include substituted types.)

PROOF: Put $\Gamma \vdash a = [t/y]^{*}e_u : X$, or $e_u(t)$ in the informal notation, so $d_u(a) \rightsquigarrow t$ by Definition 8.4.6(e). Note that t is a sub-term of a and not *vice versa*. Since the type of a is the type-symbol X, the interpretation

$\alpha = [\![a]\!]$ is uniquely determined as a section of the display $X \longrightarrow \Gamma$ (Corollary 8.4.5). We shall show that $a = \ulcorner[\![a]\!]\urcorner$ is provable, by structural induction on the history of the derivation of t (Remark 8.4.2).

The idea is to compare the type-expression $\mathfrak{u}^*Y$ with the type of the outermost operation-symbol of t, so $\mathfrak{u} = \mathfrak{f}\,;\mathfrak{p}$ in Remark 8.4.1(b).

(a) If $t = y$ then $y : \mathfrak{u}^*Y$ must belong to the context Γ, which consists only of type-symbols, so $\mathfrak{u} = \mathrm{id}$, and α is a pullback-diagonal. Then $t = a = y = \ulcorner\alpha\urcorner$ by Definition 8.4.6(d).

(b) If $t = \mathfrak{f}^*\ulcorner\beta\urcorner$ then $\mathfrak{f} = \mathfrak{u}$ since the type of $\ulcorner\beta\urcorner$ is a type-symbol ($\mathfrak{p} = \mathrm{id}$ in Definition 8.4.6(b)). Then $t = d_\mathfrak{u}(\ulcorner\alpha\urcorner)$ by Definition 8.4.6(g).

The outermost operation-symbol of t may instead be one of the encoding operations $d_\mathfrak{v}$ or $e_\mathfrak{v}$ of Definition 8.4.6(c). We identify a component c of the arguments substituted into $d_\mathfrak{v}$ or $e_\mathfrak{v}$ by $\mathfrak{f}$, so c is a sub-term of t and the induction hypothesis applies. The reason why we need to consider $\mathfrak{u}$ in the general result is that the type of c need not be a type-symbol.

(c) If $t = \mathfrak{f}^*e_\mathfrak{v}$ then $\mathfrak{p} = \hat{z}$ in Remark 8.4.1(b), and $\mathfrak{u} = \mathfrak{f}\,;\hat{z}$.

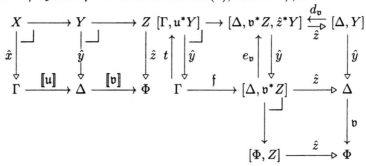

So $\mathfrak{f} = [c/z]\,;\mathfrak{u}_z$ for some unique $\Gamma \vdash c : (\mathfrak{u}\,;\mathfrak{v})^*Z$, by Lemma 8.2.14. Put $\Gamma \vdash b \equiv [c/z]^*e_{\mathfrak{u};\mathfrak{v}} : X$, so $\Gamma \vdash [b/x]^*d_{\mathfrak{u};\mathfrak{v}} = c : (\mathfrak{u}\,;\mathfrak{v})^*Z$ as for a and t. Then

$$\begin{aligned}
a &= [t/y]^*e_\mathfrak{u} = [\mathfrak{f}^*e_\mathfrak{v}/y]^*e_\mathfrak{u} = [[c/z]^*\mathfrak{u}_z^*e_\mathfrak{v}/y]^*[c/z]^*\hat{z}^*e_\mathfrak{u} \\
&= [c/z]^*[\mathfrak{u}_z^*e_\mathfrak{v}/y]^*\hat{z}^*e_\mathfrak{u} = [c/z]^*e_{\mathfrak{u};\mathfrak{v}} = b
\end{aligned}$$

by the Substitution Lemma and coherence (Definition 8.4.6(f)). But the induction hypothesis applies to b, so $a = b = \ulcorner[\![b]\!]\urcorner = \ulcorner[\![a]\!]\urcorner$.

(d) The remaining case, $t = \mathfrak{f}^*d_\mathfrak{v}$, is similar, with $\mathfrak{p} = \hat{z}\,;\mathfrak{v}$ and $\mathfrak{u} = \mathfrak{f}\,;\hat{z}\,;\mathfrak{v}$, but beware that y and z are interchanged in the notation. $\qquad\square$

COROLLARY 8·4·9:   Dependent type theory is conservative, *i.e.* the quoting functor $\ulcorner-\urcorner : \mathcal{C}{\downarrow}1 \to \mathrm{Cn}_\mathcal{L}^\times$ is full and faithful, indeed on slices.

PROOF: Quoting ($\ulcorner-\urcorner$) has so far only been defined for displays and cuts (Definition 8.4.6), so we use Proposition 8.3.11 to extend it to a functor.

The laws $(\mathsf{T},\mathsf{P},\mathsf{S},\mathsf{W})$ of the sketch hold because they do syntactically. Comparing the syntactic and semantic normal forms (Lemmas 8.2.10 and 8.3.10), we only have to establish a bijection between terms and semantic cuts, indeed sections of displays, as Lemma 8.4.8 does. $\square$

Finally we demonstrate the equivalence between syntax and semantics.

THEOREM 8·4·10: Let $\mathcal{C}$ be a small category with a class $\mathcal{D}$ of displays.

(a) $\mathsf{Cn}^{\times}_{\mathsf{L}(\mathcal{C},\mathcal{D})}$ is weakly equivalent to $\mathcal{C}\!\downarrow\!\mathbf{1}$ and to a full subcategory of $\mathcal{C}$.

(b) If $(\mathcal{C},\mathcal{D})$ is rooted then $\mathcal{C}\!\downarrow\!\mathbf{1}\simeq\mathcal{C}$, strongly if every terminal projection is itself a display or has a canonical decomposition into displays.

$$\mathsf{Cn}^{\times}_{\mathsf{L}(\mathcal{C},\mathcal{D})} \xleftarrow[{[-]}]{\ulcorner-\urcorner\;\simeq\;} \mathcal{C}\!\downarrow\!\mathbf{1} \xrightarrow{\;\mathsf{src}\;} \mathcal{C}$$

(c) If pullbacks are canonical then the first equivalence is strong.

PROOF: We already know that $\ulcorner-\urcorner$ and src are full and faithful, whilst rootedness by definition makes src an equivalence functor.

In order to show that $[-]$ and $\ulcorner-\urcorner$ form a strong equivalence (on slices) with $e_{\ulcorner\Phi\urcorner}=\mathsf{id}$ we must construct $e_{\Gamma}:\Gamma\cong\ulcorner[\Gamma]\urcorner$; the interpretation gives $\varepsilon:[\ulcorner-\urcorner]=\mathsf{id}$. (As in Section 7.6 we use $\mathfrak{e}$ instead of η for this syntactic morphism.) By the simplifications in the construction of the canonical language, induction on the length of Γ suffices, starting at $\mathfrak{e}_{[]}:[]=\ulcorner\mathbf{1}\urcorner$.

We must extend $\mathfrak{e}$ from Γ to $[\Gamma,y:u^{*}Y]$. The arity Δ of the type-symbol Y is a context-symbol, *i.e.* $\Delta=\ulcorner[\Delta]\urcorner$ (Definition 8.4.6(a)). With

$$\Gamma'=\ulcorner[\Gamma]\urcorner \qquad \mathfrak{v}=\mathfrak{e}_{\Gamma}^{-1}\,;\mathsf{u} \qquad \Gamma'\vdash X=\ulcorner[\mathfrak{v}^{*}Y]\urcorner$$

the diagram shows the passage from syntax (the back of the cube) to semantics (the square on the right) and to syntax again (front):

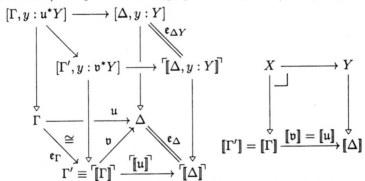

The bottom of the cube commutes (and, in general, the isomorphism $\mathfrak{e}$ is natural) because $\ulcorner-\urcorner$ is full and faithful, just as in Theorem 7.6.9.

The semantic square is a canonical pullback, and defines $e_{\Gamma Y}$ as

$$[\Gamma, y : \mathfrak{u}^*Y] \xrightarrow[\cong]{\hat{y}^*e_\Gamma} [\Gamma', y : \mathfrak{v}^*Y] \underset{d_\mathfrak{u}}{\overset{e_\mathfrak{u}}{\underset{\cong}{\rightleftarrows}}} [\Gamma', x : X] \equiv \ulcorner[\Gamma, y : \mathfrak{u}^*Y]\urcorner.$$

Since only finitely many pullbacks are used to interpret any particular Γ, $\ulcorner\!\urcorner$ is an equivalence functor even if pullbacks are not specified.   □

Type-theoretic notation in categories.

REMARK 8·4·11:   The equivalence between type theory and categories is very important. It may be exploited in two ways:

(a) Since $\mathsf{Cn}^\times_{\mathcal{L}}$ is a category, we may investigate universal properties in it and relate them to the syntactic names of its objects and morphisms. This is the subject of the final chapter.

(b) Conversely, for any semantic category $\mathcal{C}$ with a pullback-stable class $\mathcal{D}$ of maps, there is a generalised algebraic theory $\mathcal{L}$ with $\mathcal{C} \simeq \mathsf{Cn}^\times_{\mathcal{L}}$ and the same class of displays. The objects and maps of $\mathsf{Cn}^\times_{\mathcal{L}}$ have syntactic names and its pullbacks are canonical, but $\mathcal{C}$ enjoys any property which we can prove with the aid of the canonical structure as long as it is invariant under weak equivalence.

So we may use *whichever notation best fits the circumstances.* Category theory teaches the importance of maps and universal properties. Type theory provides their names and the idiom for using them. It is foolish to see this as an excuse to avoid either diagrams or formal languages.

The full context notation and the following abbreviation of it are closer to ordinary mathematical practice than the categorical notation that has been used for type theory and fibrations.

NOTATION 8·4·12:   The symbol $\hat{x}$ is an *unanalysed* name (*cf.* single keys for é and ü on French and German keyboards) to give to morphisms in a class $\mathcal{D}$ of displays in an arbitrary category.

$$\begin{array}{ccc} \Delta X & \xrightarrow{\;u_x\;} & \Gamma X \\ {\scriptstyle \hat{x}_\Delta}\Big\downarrow & \lrcorner & \Big\downarrow{\scriptstyle \hat{x}_\Gamma} \\ \Delta & \xrightarrow[\;u\;]{} & \Gamma \end{array}$$

Moreover if Γ is the target of such a map then we shall call its source ΓX, again unanalysed. The pullbacks of $\hat{x}$ will also be called $\hat{x}$, though we sometimes distinguish them by using the target as a subscript (unfortunately, this conflicts with Definition 8.2.5). Similarly the pullback of an arbitrary map (substitution) u against a display $\hat{x}$ will be sometimes named by placing x, or more generally the passive part of the context, as a subscript. The association amongst $\hat{x}$, u_x and ΓX is, in the semantic setting, purely a typographical one.

Having dealt with the direct (algebraic) rules of type theory, we shall now turn to the indirect ones: the quantifiers. We shall see that they are the two adjoints to substitution along display maps. When a map is a display, its name $\hat{x}$ according to the Notation incorporates that of a variable x, so the adjoints may be written

$$\exists x \dashv \hat{x}^* \dashv \forall x.$$

In this way the quantifier symbols (Σ, Π, $\exists$, $\forall$ and λ) may be introduced into categorical notation. We can only use this for displays — because non-displays do not admit quantification.

Exercises VIII

1. Show that, when src and tgt are omitted, the theory of categories in Example 8.1.12 is stratified (Definition 8.1.11). Extend this theory to categories with products, coproducts and exponentials by transcribing the universal properties.

2. Describe the stratified algebraic theory of 2-categories.

3. Why, with only O and H as types, is the theory of categories with pullbacks *not* a generalised algebraic theory as defined in Section 8.1? Describe such a theory using an additional type of commutative squares or triangles. This is not a stratified theory — do you think that any theory of pullbacks could be?

4. Formulate the notions of plain, distributive, positive, division and power allegory in [FS90] as stratified algebraic theories. Why is the theory of tabular allegories *not* stratified?

5. Formulate the generalised algebraic theory of generalised algebraic theories (*cf.* Remark 8.4.1). Extend this to the theory of a theory together with a model.

6. Show that coercion (mentioned in Definition 8.2.3), weakening (8.2.5) and cut (8.2.6) are derived rules of the calculus of Section 8.1, in the external sense that if there is a valid deduction of the premise then there is also one of the conclusion. Explain what one must check in order to show that extensions of the calculus retain these properties, and verify them for the sum and product rules.

7. Deduce the starred laws in Remark 8.2.7 from their unstarred forms using Lemma 8.2.15. [Hint: writing $\mathfrak{h} : \Xi \to \Delta$ for the unstarred

commutative square, and $\mathfrak{f}, \mathfrak{g} : [\Xi, \mathfrak{h}^*\Psi] \rightrightarrows [\Delta, \Psi]$ for the two sides of the starred one, show that $\mathfrak{f} = \mathrm{id}_{\mathfrak{h}^*\Psi}$; $\mathfrak{h}_\Psi = \mathfrak{g}$.]

8. Let $\mathfrak{e} = [a/x]$ and $\mathfrak{m} = \hat{y}$ where $\Gamma \vdash a : X$ and $\Delta \vdash y : Y$. Show that the diagonal fill-in for the orthogonality property (Definition 5.7.1) exists. [Hint: use Lemmas 5.7.10 and 8.2.10.]

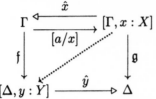

Show that if $\mathfrak{e} \perp \hat{y}$, *i.e.* the fill-in is unique, for all displays $\hat{y}$ then $\mathfrak{e}$ is an isomorphism, *i.e.* $X \cong \mathbf{1}_\Gamma$. [Hint: let $\hat{y} : [\Gamma, x : X, y : X] \longrightarrow [\Gamma, x : X]$, $\mathfrak{f} = [a/x, a/y]$ and $\mathfrak{g} = \mathrm{id}$ and show that both $[x/y]$ and $[a/y]$ are fill-ins.]

9. Show that every map $\mathfrak{f} : [\Gamma ; \Phi] \to [\Gamma, \Delta]$ such that $\mathfrak{f} ; \widehat{\Delta} = \widehat{\Phi}$ (*i.e.* the triangle in Definition 8.2.9 commutes *in the whole category*) can be expressed as a composite of displays and cuts which each leave Γ untouched.

10. Show that $(\Delta, \mathcal{J}) \mapsto ([x : X, \Delta], \mathcal{J})$ defines a functor $\hat{x}_! : \mathsf{Cn}^\times_{\mathcal{L}} \!\downarrow [\Gamma, x : X] \to \mathsf{Cn}^\times_{\mathcal{L}} \!\downarrow \Gamma$ with $\hat{x}_! \dashv \hat{x}^*$.

11. Show that pullback along any map $u : \Gamma \to \Delta$ extends to a functor $u^* : \mathcal{C} \!\downarrow \Delta \to \mathcal{C} \!\downarrow \Gamma$.

12. Prove the normal form theorem for (semantic) relative slice categories (Proposition 8.3.11).

13. Let $\mathcal{D} \subset \mathcal{C}$ be a class of display maps. Let $X_{(-)} : \mathcal{I} \to \mathcal{D}$ be a diagram where $\mathcal{I}$ is a finite oriented graph which, *quâ* unoriented graph, is simply connected. Show that this diagram has a limit in $\mathcal{C}$ and that the limiting cone consists of $\mathcal{D}$-maps. Assuming that $\mathcal{D}$ is a subcategory closed under cofiltered limits, extend the result to arbitrary simply connected diagrams.

14. Draw the diagram which introduces the variable *quâ* term of an arbitrary instance of a type.

15. Explain how the definition of a generalised algebraic theory reduces to Definition 4.6.2ff in simple type theory. Similarly compare the notions of interpretation and homomorphism, reconciling the presentation of laws as commutative polygons and as pullbacks. How does the canonical language reduce to Section 7.6 in simple type theory?

16. Show that, once the interpretation of the display map $\hat{x}$ is fixed, changing those of the auxiliary objects by isomorphisms does not affect the interpretation of a.

The Quantifiers

G ENTZEN LIBERATED the quantifiers from their old metaphysical interpretation as infinitary truth-functions and expressed them as elementary sequent rules. In Chapter I we showed that his *natural* deduction really does agree with the mathematical vernacular. Dag Prawitz, Nikolas de Bruijn and Per Martin-Löf developed calculi which their followers have used to formalise many mathematical arguments, and even for programming. These help us to understand very clearly the extent to which equality, the quantifiers, powersets and the other components of logic are actually employed.

We have claimed that universal properties in category theory deserve to be called "foundational" like the logical connectives. In a tradition which began with the axiomatisation of Abelian categories, Chapter V showed that sum types and relational algebra also have an intimate connection with the idioms of mathematics. Employing a technology from algebraic geometry, Bill Lawvere saw that the rules for the quantifiers say that they too are characterised by universal properties. By similar methods, Jean Bénabou reduced the infinitary limits and colimits in category theory to elementary form. Robert Seely concluded this phase of development by extending the logic of Remark 5.2.8 to locally cartesian closed categories.

The introduction by Jean-Yves Girard and Thierry Coquand of calculi that have quantification over all types and even universal types forced a rethink of Lawvere's ideas. These calculi do not have direct set-theoretic interpretations, so many sorts of *domain*, sometimes themselves categories, were investigated in the search for their semantics. As a result, we now have very efficient ways of calculating quantifiers in semantic categories, which we shall illustrate for Sp (such a quantifier was found in 1965). Together with the previous one, this chapter makes the link between categories and syntax precise, to the point that we cease to distinguish between them notationally.

The study of these powerful calculi has given us ways to tackle other "big" questions in logic. In category theory, what does it mean to treat categories such as Set and functors between them as single entities?

Are the axiom of replacement and large cardinals in set theory of relevance to mathematical constructions? How can we understand Gödel's incompleteness theorem on the one hand, and the proofs of consistency afforded by the gluing construction on the other?

Although this book has kept its promise to remain within or equivalent to Zermelo-Fraenkel set theory, our study of the roles of propositions and types in the quantifiers enables us to *change* the rules of engagement between them. Semantically, we may look for pullback-stable classes of maps in topology and elsewhere which obey the rules for the quantifiers, and then start to use the notation of logic to describe geometric and other constructions. Syntactically, the blanket assumption of such a clumsy framework as ZF for mathematical reasoning can be fine-tuned to capture exactly the arguments which we want to express, and thereby apply them in novel situations. It seems to me that the quantifiers and equalities on which Cantor's diagonalisation argument and the Burali-Forti paradox rely are stronger than are justified by logical intuition. We now have both the syntax and the semantics to weaken them.

Such a revolution in mathematical presentation may be another century away: excluded middle and Choice were already on the agenda in 1908, but the consensus of that debate has yet to swing around to the position on which such developments depend. On the other hand, the tide of technology will drive mathematicians into publishing their arguments in computer-encoded form. Theorems which provide nirvana will lose out to those that are programs and do calculations for their users on demand. A new philosophical and semantic basis is needed to save mathematics from being reduced yet again to programming.

9.1   THE PREDICATE CONVENTION

This section, which is an extended introduction to the chapter, sets up certain conventions for the way in which we shall use the familiar notation of the predicate calculus to explain a much more general and abstract theory. This overview is closer than is the rest of the chapter to Bill Lawvere's treatment of the semantics and Per Martin-Löf's of the syntax of type theory. We shall also show how the quantifiers fit into the general recursive scheme for constructing and interpreting type theory.

Throughout the book we have stressed the formal analogy between types and propositions. The only distinction between them in Chapter VIII, which set up the algebraic formalism for dependent types and its relationship with category theory, was the rather superficial one that proofs of propositions are not distinguished, so their displays are mono.

REMARK 9·1·1: The theme of this chapter is not the *extensional* difference between all maps and monos, but the separate *roles* which sets and propositions play in the quantifiers, comprehension and powerset.

(a) Elements (terms) of the same set (type) are distinguished, but proofs are *anonymous*. (In Section 2.4 we mentioned the possibility that, by adding proof-terms, one might extract programs from proofs.)

(b) Predicates $\phi[x]$ depend on set-variables but not *vice versa*, so we may *rearrange any context* to put the set-variables first and the propositions dependent on them afterwards. This is an important technical simplification, which we shall exploit in Section 9.2.

(c) Some forms of *quantification*, $\Sigma x{:}X.\ Y[x]$, $\exists x{:}X.\ \phi[x]$, $\Pi x{:}X.\ Y[x]$ and $\forall x{:}X.\ \phi[x]$, are allowed (Sections 9.3 and 9.4), but not others such as $\exists y{:}\phi.\ X[y]$, because of (b),

(d) and we may extract a *witness* $x : X$ from a term of type $\Sigma x{:}X.\ Y[x]$ but, by (a), not from the provability of $\exists x{:}X.\ \phi[x]$.

(e) We may form the *comprehension* $\{x : X \mid \phi[x]\}$ of a predicate on a set, giving a (sub)set, and

(f) there is a set 2 whose elements name propositions (Section 9.5).

(g) There may even be a *type of all types* — if for "type" we allow a domain with fixed points instead of a set with equality (Section 9.6) — but there is no proposition of propositions or proposition of sets.

We say that sets and propositions are types of two **kinds** (some authors say *sort* or *order*). These differences and the generality of which they are examples are sometimes presented in an extremely abstract way. Henk Barendregt uses $*$, $\square$ and like symbols for kinds in describing his *pure type systems* [Bar92]. We end up proposing the use of such a formalism for *semantic* reasons, but we will keep the familiar terminology of the predicate calculus for the sake of motivation.

CONVENTION 9·1·2: However, we employ the words "proposition" and "type" as *variables*. In each section, these kinds are only required to obey the conditions of that particular section. They may in fact stand for the same kind — ϕ may be a type like X — but we use this **predicate convention** to show where they are *potentially* different. What we say in propositional notation applies *mutatis mutandis* to types — in fact simply by *transliteration* of X for ϕ etc.

Like characters in a Greek tragedy, prop, type, $*$ and $\square$ act out dramas about the essential interactions of things, in which their own identity is not relevant. In particular, the principle (a) of proof-anonymity is only used at one point in this chapter, namely in the discussion of the η-rule for the powerset (Remark 9.5.6). Assuming proof-anonymity would in fact make the second half of Section 9.4 pointless.

Classes of display maps. Chapter VIII set up the interpretation of types in context $(\Gamma \vdash X \text{ type})$ as pullback-stable "display maps" $X \longrightarrow \Gamma$. The partition of the class of all types into various kinds is handled by equipping the category with two or more classes of displays.

NOTATION 9·1·3: Following the predicate convention, we shall write

$$\left([\Gamma, x : X] \overset{\hat{x}}{\longrightarrow} \Gamma\right) \in \mathcal{D} \quad \text{and} \quad \left([\Gamma, \phi] \hookrightarrow \Gamma\right) \in \mathcal{M}$$

for sets and propositions respectively. But *this is only a convention*:

$\hookrightarrow$ *does not necessarily signify an injective function,*

because this would mean that proofs were anonymous, which, as we have said, we shall usually not assume. There are, for example, realisability models of the predicate calculus in which proofs are distinguished and displays for propositions are not monos.

Note that nothing that we did in Chapter VIII mixes up the two classes, since there we considered displays one at a time. This unary closure condition gave rise to the □-modality "stably" in Example 8.3.6(k). It is the *union* of the kinds to which Theorem 8.4.10 refers.

The same semantic display map may belong to several classes, and be used as the interpretation of types of different kinds. In particular, we shall set out the theory of the existential quantifier and the type of propositions in detail, and deduce that of the dependent sum and type of types from it by substituting $\mathcal{M} = \mathcal{D}$ and prop = type. Proof-anonymity is itself the parameter which distinguishes $\exists$ from Σ as we understand them semantically, so of course we are careful not to *assume* it.

Fibrations. We shall continue to work in the category of contexts and substitutions developed in the previous chapter, but now these contexts consist of types *and* propositions.

REMARK 9·1·4: Do not confuse $[x : X, y : \phi[x]]$ with $\{x : X \mid \phi[x]\}$, which is the subset formed by comprehension. Compare the "virtual objects" consisting of program-variables and midconditions introduced in Remark 5.3.3: semantically, a context is not just a set but a subset of a *particular* "ambient" set. The semantics of the category of contexts is not *Set* but the comma category $\mathcal{M} \downarrow Set$ (Definition 7.3.8).

Until Section 9.5, where we consider comprehension as a type-forming operation, this means that the essential logical information is in a sense duplicated in the category. Exercise 9.9 explains how the results in Section 9.3 about factorisation systems in $\mathcal{M} \downarrow \mathcal{S}$ relate to $\mathcal{S}$ itself.

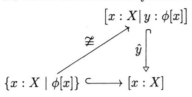

REMARK 9·1·5: $\mathsf{Cn}^{\times}_{\mathsf{type}}$, which consists of set-only contexts, is called the *base category*. For each set (or each context Γ consisting only of sets), there is a subcategory of this big category that consists of the predicates $\phi[\underline{x}]$ over Γ; it is called the *fibre* $\mathcal{P}(\Gamma)$. For each function $\mathsf{u} : \Gamma \to \Delta$, there is a substitution or inverse image functor $\mathsf{u}^{*} : \mathcal{P}(\Delta) \to \mathcal{P}(\Gamma)$; the assignment $\mathsf{u} \mapsto \mathsf{u}^{*}$ provides a functor $\mathcal{P}(-) : \left(\mathsf{Cn}^{\times}_{\mathsf{type}}\right)^{\mathsf{op}} \to \mathcal{Cat}$, which is known as an *indexed category*. The big category $\mathsf{Cn}^{\times}_{\mathsf{type|prop}}$ whose contexts consist of both types and propositions collects all of these fibres together. There's a proposition-erasing functor $P : \mathsf{Cn}^{\times}_{\mathsf{type|prop}} \longrightarrow \mathsf{Cn}^{\times}_{\mathsf{type}}$, called a *fibration*. Beware that P and $\mathcal{P}$ are in different typefaces!

Bill Lawvere and Jean Bénabou used fibrations to study the quantifiers and infinitary limits and colimits. We show in Section 9.2 that fibrations capture the independence of types from propositions (Remark 9.1.1(b)), but this is not relevant to the quantifiers (it *is* to comprehension).

The quantifier formation rules. As for the predicate and λ-calculi (Sections 1.5 and 2.3), we present the syntactic rules for the quantifiers in both the box and sequent styles.

REMARK 9·1·6: The **quantifier formation rule** binds a variable, so the box begins with a context-formation (Definition 8.2.2).

$$\frac{\Gamma \vdash X \text{ type} \quad \Gamma, x : X \vdash \phi \text{ prop}}{\Gamma \vdash Qx{:}X.\ \phi \text{ prop}} \, Q\mathcal{F}$$

X type

$x : X$	hyp
$\vdots$	
$\phi[x]$ prop	

$Qx{:}X.\ \phi[x]$ prop $Q\mathcal{F}$

Beware that this only says, for example, that $\exists x.\ \phi[x]$ is *well formed*, not that it is *true*. The truth of $\exists x{:}X.\ \phi$ is shown by a *direct introduction rule*, from a witness $a : X$ and its evidence $b : \phi[a]$, which we put together with an operation-symbol ve. It is the *elimination* rule which is direct for universal quantification and comprehension, the operation-symbol being called ev. As these are operation-symbols, the corresponding term equality rules are just instances of congruence.

The *indirect* $(\forall\mathcal{I})$- and $(\exists\mathcal{E})$-rules define terms

$$\lambda x.\, p \qquad \text{and} \qquad \text{let } (x, y) \text{ be } c \text{ in } f$$

which bind the variables x and y, so they are subject to α-equivalence, as are the type formation rules. Corresponding to the formation rules, there are also type and term *equality* rules, such as

$$\frac{\Gamma \vdash X = Y \qquad \Gamma, x : X \vdash \phi = \psi \qquad y \notin \Gamma}{\Gamma \vdash (Qx{:}X.\ \phi) = (Qy{:}Y.\ [y/x]^*\psi)}$$

There is no necessary connection between the use of the name x in the quantifier $Qx.\ \phi$ and in its terms: they may be renamed separately. That we don't is one way in which we break Convention 1.1.8 about reusing the name x. It is also broken by the way in which we express the rules for the quantifiers, especially in the adjoint form; this is inevitable, as they are all about passing from a world with x to one without, and back.

As in Section 1.6, types and terms may be imported into boxes from above, and exported if wrapped in the quantifier or indirect operation.

Adjointness in foundations. These are symbolic trivia: what we really want to know is how the indirect, β- and η-rules correspond to the universal properties of ve and ev.

REMARK 9·1·7: Bill Lawvere finally brought symbolic logic into the heart of mathematics by recognising the bijective correspondences

$$\frac{\Gamma, x : X, \qquad \phi[x] \vdash \hat{x}^*\vartheta}{\Gamma, \qquad \exists x.\, \phi[x] \vdash \quad \vartheta} \qquad\qquad \frac{\Gamma, x : X, \hat{x}^*\psi \vdash \qquad \phi[x]}{\Gamma, \qquad\qquad \psi \vdash \forall x.\, \phi[x]}$$

He said that the quantifiers are adjoint to *substitution*. This functor in the middle is traditionally invisible, but, unifying category theory with type theory in our notation, we now see that it is not substitution but *weakening*:

$$\exists x \dashv \hat{x}^* \dashv \forall x,$$

where $\hat{x} : [\Gamma, x : X] \longrightarrow \Gamma$ (Remark 1.5.5).

REMARK 9·1·8: Since universal properties describe objects in category theory only up to isomorphism, we shall understand notation such as $\exists x.\ \phi$ in the same way: it means *any* object ψ which satisfies the introduction, elimination, β- and η-rules for an existential quantifier for the predicate $\phi[x]$ over the type X. This need not be the string "$\exists x.\ \phi$", just as $X \to Y$ in Section 4.7 did not have to be literally an arrow. In fact the symbolic rules also characterise ψ *only* up to isomorphism.

Following the preference stated in Chapter VII, we shall discuss the universal properties of the quantifiers diagrammatically, instead of using Lawvere's adjunctions.

Substitution and the Beck–Chevalley condition. We are used to substituting directly under the quantifiers: Definition 1.1.10(d) said that $u^*\exists x{:}X.\ \phi$ is *equal* to $\exists x{:}u^*X.\ u^*\phi$. We *don't* state such a rule here because to assert equalities between types may conflict with their possibly different histories of formation.

REMARK 9·1·9: Instead of asserting a substitution rule directly for the *types*, we give stronger indirect rules for the *terms*. Then, as for all universal properties, we prove that $u^*Qx{:}X.\ \phi \cong Qx{:}u^*X.\ u^*\phi$, in a unique way which commutes with the structure.

Categorically, the contexts involved form a pullback as shown on the left,

$$
\begin{array}{ccc}
[\Gamma, x:u^*X] & \xrightarrow{\ u_x\ } & [\Delta, x:X] \\
\hat{x}_\Gamma \Big\uparrow \quad\ \ \lrcorner & \widehat{\mathsf{PS}} & \Big\downarrow \hat{x}_\Delta \\
\Gamma & \xrightarrow[\ u\ ]{} & \Delta
\end{array}
\qquad
\begin{array}{ccc}
\mathcal{P}(\Gamma, x:u^*X) & \xleftarrow{\ u_x^*\ } & \mathcal{P}(\Delta, x:X) \\
Qx_\Gamma \Big\downarrow & \cong & \Big\downarrow Qx_\Delta \\
\mathcal{P}(\Gamma) & \xleftarrow[\ u^*\ ]{} & \mathcal{P}(\Delta)
\end{array}
$$

and we want the diagram on the right to commute, in the sense that a certain natural transformation (provided by the universal properties of Qx_Γ and Qx_Δ) is invertible. This equation between types is known as the **Beck–Chevalley condition**, although it was Lawvere and Bénabou who identified it in categorical logic, respectively attributing it to Jon Beck and Claude Chevalley because of analogous properties in their work in the descent theory of algebraic geometry.

This property is often presented as an additional burden: something *extra* to be checked after the already heavy labour of the construction of the concrete objects to be used in the interpretation. Of course, if you choose to define the quantifiers categorically as the adjoints to substitution, then this condition does need verification. But we give another characterisation of $\forall$ with the condition built in, so *it can do the work for us*: we choose the values of quantifiers of the form $\forall x{:}X.\ \phi$ where X and ϕ are type- and proposition-*symbols*, and then the substituted forms $\forall x{:}u^*X.\ \mathsf{p}^*\phi$ are *derived* from the Beck–Chevalley condition.

The recursive definition of interpretations. Theorem 8.4.4 showed how to use the history of formation of types and terms to interpret generalised algebraic theories in categories with display maps.

REMARK 9·1·10: The quantifiers and their terms contribute new cases to this structural recursion, just as the simply typed λ-calculus in Remark 4.7.4 extended the interpretation of algebra in categories with products. By the recursion hypothesis, we already have

$$[\phi] \lhook\joinrel\longrightarrow [X] \longrightarrow [\Gamma]$$

and have to find $[\![Qx\!:\!X.\ \phi]\!]$ and $[\![\mathsf{ve}]\!]$ or $[\![\mathsf{ev}]\!]$. We shall express the syntactic rules diagrammatically as universal properties, so it remains to mimic these in the semantics. This fixes the interpretation uniquely, at least up to isomorphism, by saying that $[\![-]\!]$ preserves these properties. They must be stable in the semantics as well as the syntax, since $[\![-]\!]$ also preserves pullbacks.

REMARK 9·1·11:   For the sake of making $[\![-]\!]$ into a *bona fide* functor, let us consider briefly how to choose a *particular* object from the isomorphism class which the universal property provides. In Zermelo type theory (Remark 2.2.4) this may be done by *fiat* for $\exists$, as comprehension names *canonical* subsets for the image factorisation.

The formal rules as we give them say that the Beck–Chevalley condition only holds up to unique isomorphism, even in the syntax. Without some trick, $\exists x\!:\!X.\ \phi$ must be regarded as a new proposition-*symbol*, even when X and ϕ are type-*expressions*: $\mathfrak{f}^*\exists x\!:\!X.\ \phi$ is a substitution-instance of it, but $\exists x\!:\!\mathfrak{f}^*X.\ \mathfrak{f}^*\phi$ is not.

We would like to find some way of prescribing interpretations to make the Beck–Chevalley condition hold up to equality. Although we have introduced it semantically, this is really a symbolic problem: does cut or substitution commute with the type formation rules? If it does then once again we need only choose the results of quantification at type-*symbols*. Recall that Theorem 8.4.4 did this for unquantified type-expressions: given a choice of semantic displays at type-symbols, it used pullbacks *anchored* there to interpret type-expressions.

Since $\mathfrak{f}^*C(\mathfrak{p}^*\phi) = (\mathfrak{f};\mathfrak{p})^*C\phi$, there is no difficulty with comprehension (or the powerset). In the cases of $\exists$ and $\forall$, substitutions may be embedded in *two* places: for the range and the body. Corollary 9.4.15 shows that

$$Qx\!:\!\mathsf{u}^*X.\ \mathfrak{p}^*\phi \cong [\mathsf{u}, \lambda x.\ \mathfrak{p}]^*Qx\!:\!X.\ \mathsf{ev}^*\phi,$$

where λ and ev belong to a certain dependent *product*, so u and $\mathfrak{p}$ can be brought outside together and combined with the extra substitution $\mathfrak{f}$. Thus the problem can be solved for $\forall$, and for $\exists$ if $\forall$ is also present.

9.2   INDEXED AND FIBRED CATEGORIES

In the familiar predicate calculus, propositions may depend on elements of sets, but sets do not depend on the actual proofs of propositions. The technology used to take advantage of this separation principle pre-dates the formal study of syntax: it was introduced for algebraic geometry in 1960 by Alexander Grothendieck [Gro64, Exposé VI], to handle the way in which an arbitrary mathematical structure (rather than simply a predicate) varies over an indexing space. John Gray developed the category theory [Gra66]. In 1970, Bill Lawvere applied it to what we would now call type theory, and Jean Bénabou used it to account for the infinitary aspects of category theory.

Separating propositions from types. Instead of beginning, as is usual, with the formal definition of fibrations, let us first consider the feature of the predicate calculus which lies behind it (Remark 9.1.1(b)); Bart Jacobs called this a ***propositional situation*** [Jac90].

DEFINITION 9·2·1: Let $\Sigma = \Sigma_{\text{type}} + \Sigma_{\text{prop}}$ be a partition of the class of types of a generalised algebraic theory $\mathcal{L}$ (Definition 8.1.10), the members of the second class being styled "propositions". Then $\mathcal{L}$ is said to admit a ***division of contexts*** if

(a) the *types* do not depend on propositional variables,

$$\frac{\Gamma \vdash \phi \text{ prop} \qquad \Gamma, y : \phi \vdash X \text{ type}}{\Gamma, y : \phi \vdash X = \hat{y}^* X' \qquad \text{for some } \Gamma \vdash X' \text{ type}}$$

(b) the type *terms* do not depend on propositional variables,

$$\frac{\Gamma \vdash X \text{ type} \qquad \Gamma, y : \phi \vdash a : \hat{y}^* X}{\Gamma, y : \phi \vdash a = \hat{y}^* a' : \hat{y}^* X \qquad \text{for some } \Gamma \vdash a' : X}$$

(c) and likewise the *laws* for type terms,

$$\frac{\Gamma \vdash X \text{ type} \qquad \Gamma, y : \phi \vdash \hat{y}^* a = \hat{y}^* b : \hat{y}^* X}{\Gamma \vdash a = b : X}$$

So certain *converses of weakening* hold. Using the *exchange rule*,

$$\frac{\Gamma, y : \phi, x : X, \Psi \vdash \mathcal{J}}{\Gamma, x : X, y : \hat{x}^* \phi, \Psi \vdash \mathcal{J}}$$

the type variables in a context may be listed first, with the propositions following. This is often indicated by a vertical bar:

$$x : X, y : Y, z : Z, \ldots \mid p : \phi, q : \psi, r : \vartheta, \ldots \vdash \cdots$$

or, briefly, $\Gamma \mid \Phi \vdash \mathcal{J}$ and $[\Gamma \mid \Phi]$.

The point is that every jumbled context is isomorphic to a divided one.

We shall study the full subcategory $\mathsf{Cn}^\times_{\text{type}|\text{prop}} \subset \mathsf{Cn}^\times_{\mathcal{L}}$ of divided contexts. The conditions say that this inclusion is an equivalence, so our results about divided contexts actually apply to the whole category.

We shall write $\longrightarrow$ for the display maps corresponding to the types and $\hookrightarrow$ for the propositions (Notation 9.1.3). According to the *predicate convention*, the latter are not necessarily monos.

LEMMA 9·2·2: Every pair of displays $\Gamma\phi X \xrightarrow{\hat{x}} \Gamma\phi \overset{\hat{y}}{\hookrightarrow} \Gamma$ is part of a (P-) pullback,

$$
\begin{array}{ccc}
[\Gamma, y : \phi, x : X] & \xrightarrow{\hat{x}} & [\Gamma, y : \phi] \\
\cong [\Gamma, x : X', y : \hat{x}^*\phi] & & \\
\downarrow \quad \text{P} & & \downarrow \hat{y} \\
[\Gamma, x : X'] & \dashrightarrow[\hat{x}] & \Gamma
\end{array}
\qquad
\begin{array}{ccc}
[\Gamma, x : X, \hat{x}^*\phi] & \xleftarrow{[a/x]} & [\Gamma, \phi] \\
& \xrightarrow{\hat{x}_\phi} & \\
\hat{y} \downarrow & \text{Š} & \downarrow \hat{y} \\
& [a'/x] & \\
[\Gamma, x : X] & \xleftarrow{} & \Gamma \\
& \xrightarrow{\hat{x}} &
\end{array}
$$

and every section of $\hat{x}_\phi$ is the (Š) pullback of a unique section of $\hat{x}$.

PROOF: $\Gamma\phi X \cong \Gamma X'\phi$ by (a) and the R equation (Remark 8.2.8). The second diagram follows from (b) and uniqueness from (c).   □

Parts (b) and (c) of Definition 9.2.1 say that, given a *morphism* of divided contexts, we may *erase all propositional information*: this is well defined as a functor, the **fibration**. So the tree structure in John Cartmell's contextual categories (Remark 8.3.9) may be pruned to the types. We write $\longrightarrow\!\!\!\!\!\cdot$ for the fibration, as it "displays" propositions over types.

PROPOSITION 9·2·3:   Let $\mathrm{Cn}^\times_\mathsf{type}$ be the full subcategory consisting of those contexts which only involve types. Then there are adjoint functors,

$$
\begin{array}{ccc}
\mathrm{Cn}^\times_\mathsf{type|prop} & [\Gamma \mid u^*\Phi] \xrightarrow{u_\Phi} [\Delta \mid \Phi] & [\Gamma \mid u^*\Phi] \xleftarrow{u^*} [\Delta \mid \Phi] \quad [\Gamma \mid] \\
P \downarrow \dashv \uparrow T & \hat{\Phi} \downarrow\!\downarrow \quad\quad \downarrow\!\downarrow \hat{\Phi} & \uparrow T \\
\mathrm{Cn}^\times_\mathsf{type} & \Gamma \xrightarrow{u} \Delta & \Gamma \xrightarrow{u} \Delta \quad \Gamma
\end{array}
$$

where P forgets the propositional part of a divided context and T gives an empty propositional part to a type-only context. $\mathrm{Cn}^\times_\mathsf{type|prop}$ is called the **total category** and P the **fibration**.

For each object Γ of $\mathrm{Cn}^\times_\mathsf{type}$ (known as the **base category**), the **fibre** is the relative slice $\mathrm{Cn}^\times \!\!\downarrow\!\! \Gamma$, whose objects are contexts of the form $[\Gamma \mid \Phi]$ and whose morphism act as the identity on Γ. $T\Gamma \equiv [\Gamma \mid]$ is the terminal object of the fibre over Γ.

(a) Syntactic substitution defines a functor $u^* : \mathrm{Cn}^\times \!\!\downarrow\!\! \Gamma \to \mathrm{Cn}^\times \!\!\downarrow\!\! \Delta$ for $u : \Delta \to \Gamma$; moreover $(u ; v)^*\phi = u^*(v^*\phi)$ and $\mathrm{id}^*\phi = \phi$.

(b) Semantically, the analogous operation is pullback (inverse image), but this is only defined up to unique isomorphism.

PROOF: Notice that P cannot be defined for types which depend on the omitted propositional variables. The other two parts of Definition 9.2.1 are needed to define P on morphisms involving type operation-symbols, such that the laws are respected. The substitution functors were defined in Section 8.2 and TT is terminal in the relative slice by Lemma 8.2.10. $\square$

EXAMPLES 9·2·4:

(a) *Sets and predicates.* The fibre over Γ is the Lindenbaum algebra of predicates with free variables in Γ with respect to the provability order; TT is the constantly true predicate. Semantically, the fibre is the powerset $\mathcal{P}(\Gamma)$, consisting of subsets of $[\![\Gamma]\!]$. The substitution functor u^* is the inverse image u^{-1}; notice that it preserves $\rightarrow$, $\bigvee$ and $\bigwedge$, and has adjoints on both sides (Remark 3.8.13(b)).

(b) In declarative programs, the types are those of the *run-time program-variables* but the propositions or *midconditions* only appear in the analysis, which may perhaps be carried out by a proof-checking compiler (Remark 5.3.3).[1] The fibration P erases the midconditions (on objects), and the correctness proofs from programs (morphisms). The substitution functors give the weakest precondition interpretation of Remark 4.3.5ff.

(c) If the types of *both* kinds are independent of each other's terms then $Cn^\times_{type|prop} = Cn^\times_{type} \times Cn^\times_{prop}$, the two projections being fibrations. The normal form theorem (Lemma 8.2.10) expresses each morphism of the product as a commuting pair, as in Proposition 4.8.3.

(d) Three or more kinds obeying the analogous exchange rules give rise to three-part contexts $[\Gamma \,|\, \Phi \,|\, \Lambda]$, and to a composition of fibrations.

(e) Let $\mathcal{L} = (\Sigma, \triangleright)$ be a propositional Horn theory (Section 3.7). Suppose that $\Sigma = \Sigma_{type} + \Sigma_{prop}$ such that Definition 9.2.1(b) holds, *i.e.* if $K \triangleright t \in \Sigma_{type}$ then already $K_{type} \equiv K \cap \Sigma_{type} \triangleright t$. Let $\mathcal{L}_{type}$ be Σ_{type} with the restriction of $\triangleright$. Then there is a division of the contexts in the classifying semilattice of $\mathcal{L}$ (Theorem 3.9.1).

(f) The classifying category $Cn^\times_{type}$ for the theory of rings has lists of polynomials as morphisms. Contexts for the theory $\mathcal{L}$ of rings-with-modules divide into two parts, of which the first refers purely to rings (Exercise 4.23). The dependency of modules on rings is not at the type level, but is due to the term (action) $R \times M \rightarrow M$.

(g) The theory of categories (Example 8.1.12) admits a division into objects (O) and morphisms $(H[x,y])$, *if the operation-symbols* src *and* tgt *are omitted from the theory* (*cf.* Example 8.2.1 and Exercise 8.1). Similarly the theory of 2-categories is fibred over the theory of categories, dividing natural transformations from functors.

---

[1]For imperative programs, the maps of the base category (programs with particular source and target types) are partial functions. I cannot see how to make this situation into a fibration, but *cf.* Exercise 9.46.

(h) Let $\mathcal{L}_{\mathsf{type}}$ consist of the sorts and operation-symbols of an algebraic
    theory. Then the laws of the given theory, together with reflexivity,
    transitivity, symmetry and congruence with respect to each of the
    operation-symbols, may be formulated as $\mathcal{L}_{\mathsf{prop}}$ (Remark 7.4.10).

(i) More generally, let $\mathcal{L}$ be a ***conservative extension*** of a gener-
    alised algebraic theory $\mathcal{L}_{\mathsf{type}}$: the type-symbols of $\mathcal{L}_{\mathsf{type}}$ are defined
    in type-only contexts, as are all operation-symbols and laws of $\mathcal{L}$
    whose types are in $\mathcal{L}_{\mathsf{type}}$. The levels of a stratified algebraic theory
    (Definition 8.1.11) are successive conservative extensions.

(j) Let $\mathcal{C}$ be a category with two classes $\mathcal{D}$ and $\mathcal{M}$ of displays (Def-
    inition 8.3.2) satisfying Lemma 9.2.2. Then there is a fibration
    $\mathcal{C}\!\downarrow_{\forall\mathcal{D}|\mathcal{M}} 1 \longrightarrow \mathcal{C}\!\downarrow_{\forall\mathcal{D}} 1$ of relative slices (Definition 8.3.8).

We take one example separately from the rest, since for Jean Bénabou
it was the paradigm.

EXAMPLE 9·2·5:   An *I-indexed family of objects* $(X_i)_{i\in I}$ is a context

$$\big[i : I | x : X[i]\big].$$

The division is not intrinsic but imposed by us *arbitrarily*, between the
indices and what they index. The fibration forgets everything but the
indices, and the substitution functors perform "re-indexing".

Jean Bénabou explained how the dependent sum $\Sigma i{:}I.\ X[i]$ and product
$\Pi i{:}I.\ X[i]$, which we study in the next two sections, give an *elementary*
axiomatisation of coproducts and products of infinitary indexed families
of sets or other mathematical objects. We regard $X[i]$ and $\Sigma i.\ X[i]$ as
idiomatic notation for the associated display map. The previous chapter
developed this notation and its formal interpretation.

In practice, more than one such suffix $(i : I,\ j : J,\ ...)$ is often needed,
so it is better to write $[\Gamma \mid x : X]$ than to collect I, J, ... as a single type
$I \times J$ or $\Sigma i{:}I.\ J[i]$ (category theory exists to eliminate suffixes from
mathematics). There may also be several I-indexed families (X_i), (Y_i),
..., from which we may for example need to select tuples, so these too
we keep together in the many-type divided context $[\Gamma \mid \Phi]$.

I find this example very confusing as the main use used to demonstrate
the theory of fibrations, for one thing because the category of sets is too
special to illustrate many of the difficulties, such as the Beck–Chevalley
condition and quantification over equality (Remark 8.3.5). But, in terms
of our definition, lemma and proposition above, how can the principle
of independence possibly apply, when the types on both sides of the

division are of the same kind (sets)? Making such a division cannot be expressing any semantic fact about the theory of sets: it is a *convention*:

we don't use elements of *indexed* sets as *indices*.

In the process of inventing a formal language with divided contexts that uses displays of sets on both sides, we "taint" the displays *by their use in the two roles*: we take the same class twice, or one class and a subclass $\mathcal{M}$ of it, and mark the copies as "types" and "propositions". Then the formal language (*i.e.* the arguments we permit ourselves to write in it), is artificially restricted according to the rules in Definition 9.2.1. The axiom of comprehension forgets this restriction (Section 9.5).

Fibrations. The "substitution" functors have a universal property, which (like adjunctions) we may capture without making choices of them.

DEFINITION 9·2·6: Let $P : \mathcal{C} \longrightarrow S$ be any functor.

(a) For any object $\Gamma \in \mathrm{ob}\,S$, a map $\mathfrak{g} : \Phi \to \Psi$ with $P\Phi = P\Psi = \Gamma$ and $P\mathfrak{g} = \mathrm{id}_\Gamma$ in $\mathcal{C}$ is called **vertical**. The subcategory $P^{-1}(\Gamma)$ is known as the **fibre over** Γ; its objects are those $\Phi \in \mathrm{ob}\,\mathcal{C}$ with $P\Phi = \Gamma$ and its maps are the vertical ones. A morphism that is merely made invertible by P (not necessarily an identity) is called **pseudo-vertical**.

(b) A morphism $\mathfrak{p} : \Psi \to \Theta$ in $\mathcal{C}$ is said to be **horizontal, prone** or **cartesian** if it has the universal property illustrated in the *right* hand diagram below, in which $\mathfrak{v} = P\mathfrak{p} : \Delta = P\Psi \to \Xi = P\Theta$: given any map $\Phi \to \Theta$ whose image factors into $\mathfrak{v}$ in the sense of forming a commutative triangle in S as illustrated, there is a unique fill-in $\Phi \to \Psi$ such that the upper triangle commutes *and* the image is the given map $\Gamma \to \Delta$.

(c) The functor P is a **fibration** if for every object $\Theta \in \mathrm{ob}\,\mathcal{C}$ and map $\mathfrak{v} : \Delta \to \Xi = P\Theta$ in S there is a prone **lifting** $\mathfrak{p} : \Psi \to \Theta$ with $P\Psi = \Delta$ and $P\mathfrak{p} = \mathfrak{v}$. In this case the source, $\mathcal{C}$, and target, S, of P are called the **total category** and **base category** respectively.

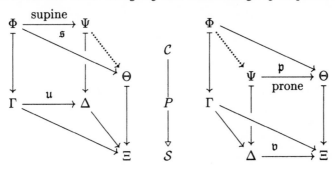

(d) Dually, $\mathfrak{s} : \Phi \to \Psi$ is called **op-horizontal, supine**[2] or **cocartesian** if it has the property shown on the left; P is an **op-fibration** if each base map $\mathfrak{u} : \Gamma \to \Delta$ has a supine lifting at each object Φ over Γ. So $P : \mathcal{C} \longrightarrow \mathcal{S}$ is an op-fibration iff $P : \mathcal{C}^{\mathrm{op}} \longrightarrow \mathcal{S}^{\mathrm{op}}$ is a fibration.

(e) A functor which is both a fibration and an op-fibration is known as a **bifibration**. This is the case iff the substitution functors are adjunctible, the unit and co-unit being the comparisons between prone and supine maps whose targets and sources agree, respectively.

(f) A **hyperdoctrine** is a fibration $P : \mathcal{C} \longrightarrow \mathcal{S}$ where $\mathcal{S}$ and the fibres $\mathcal{P}(\Gamma)$ for $\Gamma \in \mathrm{ob}\,\mathcal{S}$ are locally cartesian closed, and the substitution functors $\mathfrak{u}^* : \mathcal{P}(\Delta) \to \mathcal{P}(\Gamma)$ for $\mathfrak{u} : \Gamma \to \Delta$ in $\mathcal{S}$ have adjoints on both sides obeying the Beck–Chevalley conditions. (As we do not follow Lawvere's approach, we don't use the word hyperdoctrine.)

By the same argument as for Theorem 4.5.6, prone liftings are unique up to unique isomorphism; indeed pseudo-vertical and prone maps form a cartesian factorisation system (Definition 5.7.2 and Exercise 9.5). The fibration is recovered from the fibres and substitution functors by the following **Grothendieck construction**.

PROPOSITION 9·2·7: Let $\mathcal{P}(-) : \mathcal{S}^{\mathrm{op}} \to \mathcal{C}at$ be any functor, where we write $\mathfrak{u}^* : \mathcal{P}(\Delta) \to \mathcal{P}(\Gamma)$ instead of $\mathcal{P}(\mathfrak{u})$ for the action of maps in $\mathcal{S}$. Then the following define a fibration $P : \mathcal{C} \longrightarrow \mathcal{S}$:

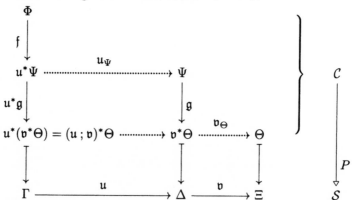

(a) the *objects* of $\mathcal{C}$ are pairs, which we write as $[\Gamma \mid \Phi]$, consisting of $\Gamma \in \mathrm{ob}\,\mathcal{S}$ and $\Phi \in \mathrm{ob}\,\mathcal{P}(\Gamma)$; and then $P[\Gamma \mid \Phi] = \Gamma$;

---

[2]Prone means horizontal and face down, supine means horizontal but face up; they are new words in this topic, chosen to reflect the different ways in which these maps are orthogonal to the vertical maps (Exercise 9.5). Prone has another meaning of being likely to do something, and supine also means being dis-inclined to action. The mnemonic is that s̲u̲pine maps are related to s̲u̲ms, although prone maps are not in fact directly concerned with products.

(b) the *morphisms* $[\Gamma \mid \Phi] \rightarrow [\Delta \mid \Psi]$ are also pairs $[u \mid f]$, where $u : \Gamma \rightarrow \Delta$ and $f : \Phi \rightarrow u^* \Psi$ in $\mathcal{P}(\Gamma)$, and then $P[u \mid f] = u$;

(c) *vertical morphisms* are those of the form $[\mathrm{id} \mid f]$, and the fibre over Γ is isomorphic to the category $\mathcal{P}(\Gamma)$;

(d) the *prone lifting* of $\mathfrak{v} : \Delta \rightarrow \Xi$ at $[\Xi \mid \Theta]$ is

$$\mathfrak{v}_\Theta \equiv [\mathfrak{v} \mid \mathrm{id}_{\mathfrak{v}^* \cdot \Theta}] : [\Delta \mid \mathfrak{v}^* \Theta] \rightarrow [\Xi \mid \Theta]$$

(e) the composite $[u \mid f] ; [\mathfrak{v} \mid g]$ is $[u ; \mathfrak{v} \mid f ; u^* g]$ (*cf.* Definition 8.4.6(f)).

PROOF: Associativity of composition in $\mathcal{C}$ depends on functoriality of $\mathcal{P}(-)$. The mediator for $[u ; \mathfrak{v} \mid f]$ in the universal property of the prone map $[\mathfrak{v} \mid \mathrm{id}]$ is $[\mathfrak{v} \mid g]$. Conversely if $[\mathfrak{v} \mid g]$ is prone then g is invertible, since $[u \mid \mathrm{id}]$ is also prone. Everything else in the definition of fibration is straightforward.   □

EXAMPLES 9·2·8: As the data for Proposition 9.2.7 consist merely of an assignment of a category to each object of the base category (in a functorial way), examples of fibrations are not difficult to find.

(a) For sets and predicates, the total category is the comma category $\mathcal{M} \downarrow \mathcal{S}et$ (Remark 9.1.4), of which an object is a set with a subset.

(b) For indexed families (Example 9.2.5), tgt : $\mathcal{S}et^{\rightarrow} \dashrightarrow \mathcal{S}et$ is the fibration. (Section 9.5 takes up these two examples.)

(c) Any fibration is a replete functor (Definition 4.4.8(d)), so its strict pullback $U^* P$ against any functor $U : \mathcal{A} \rightarrow \mathcal{S}$ is equivalent to the pseudo-pullback (Definition 7.3.9). This is also a fibration, whose fibres over $A \in \mathrm{ob}\,\mathcal{A}$ are isomorphic to those over $UA \in \mathrm{ob}\,\mathcal{S}$. For example, $\pi_1 = U^*$ tgt in the gluing construction (Proposition 7.7.1).

(d) A presheaf $\mathcal{S}^{\mathrm{op}} \rightarrow \mathcal{S}et$ is an indexed *discrete* category. It gives rise to a **discrete fibration**, characterised by the fact that all vertical maps are identities, *e.g.* Remark 7.7.8. *Any* indexed category $\mathcal{S}^{\mathrm{op}} \rightarrow \mathcal{C}at$ is a (pre)sheaf of categories, or an internal category in $\mathsf{Sh}(\mathcal{S})$.

(e) In an indexed *groupoid*, all maps are prone, and all vertical maps are isomorphisms. The restriction of the fibration to each slice (*sic*, not fibre) $\mathcal{C} \downarrow \Theta \rightarrow \mathcal{S} \downarrow P\Theta$ is an equivalence of categories; any functor with the latter property is called an **isotomy** and is weakly equivalent to a fibration with groupoid fibres.

(f) A groupoid *homomorphism* is a fibration iff it is an op-fibration iff it is replete. The substitution functors are equivalences.

(g) Any continuous function $f : X \rightarrow Y$ between spaces induces a homomorphism of their fundamental groupoids $\pi_1(f) : \pi_1(X) \rightarrow \pi_1(Y)$. This is a fibration iff the *weak path lifting property* holds: for every point $x \in X$ and path $q : I \equiv [0,1] \rightarrow Y$ with $f(x) = q(0)$, there is some path $p : I \to X$ with $p(0) = x$ whose image $f \circ p$ is homotopic (relative to the endpoints) to q.

Models. In line with the theme of the book, we have so far considered fibrations of *classifying* categories $\mathsf{Cn}_{\mathcal{L}}^{\times}$, but these underlie much more familiar fibrations of categories of *models*, for which there is a richer structure. The typical example is the theory of rings-with-modules, in which the fibre over each ring is its category of modules.

PROPOSITION 9·2·9:   Let $\mathcal{L}_{\mathsf{type}} \subset \mathcal{L}$ be generalised algebraic theories, one a conservative extension of the other. Then there is a bifibration of their categories of models, which has adjoints on both sides.

$$
\begin{array}{ccc}
\mathcal{M}od(\mathcal{L}) & M \in \mathcal{M}od(R) \xrightarrow[\text{restriction}]{\overset{\text{induction}}{\underset{\perp}{\longrightarrow}}} \mathcal{M}od(S) \ni N \\
\text{initial} \Big\uparrow \dashv \dashv \Big\uparrow \text{final} & \\
\mathcal{M}od(\mathcal{L}_{\mathsf{type}}) & R \xrightarrow{\quad u \quad} S
\end{array}
$$

Let $u : R \to S$ be a homomorphism in the sub-theory $(\mathcal{L}_{\mathsf{type}})$.

(a) In the *final* algebra for the larger theory $(\mathcal{L})$ in the fibre over R, the interpretation of each "proposition" is a singleton.

(b) For an algebra N over S, the *restriction* along $R \to S$ has the same interpretation of "types" as R, but the "propositions" and actions are obtained from those of S by substitution.

(c) The *initial* algebra over R is that generated by R together with the "propositional" operation-symbols, *modulo* "propositional" laws,

(d) and *induced* algebras are computed in a similar way. $\square$

EXAMPLES 9·2·10:

(a) Let $u : R \to S$ be a homomorphism of commutative rings and M, N be modules for R and S respectively. Then

$$R \times N \xrightarrow{(u, \mathsf{id})} S \times N \longrightarrow N$$

is the action of the **restriction** of N to R, and the left adjoint $M \mapsto S \otimes_R M$ is known as **induction** (not in the logical sense).

(b) Let $\Sigma = \Sigma_{\mathsf{type}} + \Sigma_{\mathsf{prop}}$ be a divided Horn theory, $M, N \subset \Sigma$ two closed subsets and $u : R = M \cap \Sigma_{\mathsf{type}} \subset S = N \cap \Sigma_{\mathsf{type}}$. Then the initial algebra in the fibre over R is the closure of R with respect to the larger theory; the final one is $R + \Sigma_{\mathsf{prop}}$. The restriction of N is $R + N \cap \Sigma_{\mathsf{prop}}$, and the induction of M is the closure of $M \cup S$.

For finitary theories, the restriction functors preserve filtered colimits. Restriction and induction maps between algebraic lattices are called **projections** and **embeddings** respectively (Exercises 9.10ff). □

Coherence issues. For any ("semantically given") fibration, a *choice* of prone liftings $\mathfrak{p} : \Psi \to \Theta$ for each $\mathfrak{v}$ and Θ is called a **cleavage**, and extends to "substitution" functors $\mathfrak{v}^*$ between fibres. A **split fibration** is one for which these act functorially, as they do for syntactic categories. In general, the natural isomorphisms between $(\mathfrak{u} \,;\, \mathfrak{v})^*$ and $\mathfrak{u}^* \cdot \mathfrak{v}^*$ in a cleavage must be specified, and obey a system of coherence laws.

These isomorphisms are defined by universal properties, so are uniquely determined by certain equations. However, as Jean Bénabou rather forcefully pointed out [Bén85], there is a casual tendency just to call them "canonical" — as if this were an intrinsic property like continuity. The problem is that if isomorphisms are not kept under strict control, they conspire to form non-trivial *groups*.

PROPOSITION 9·2·11: A *group* homomorphism $P : \mathcal{C} \dashrightarrow S$ is a fibration iff it is surjective, and the fibre over the unique object of S is the kernel,

$$ K \longhookrightarrow \mathcal{C} \xrightarrow{\;P\;} S. $$

The fibration P is split iff this is a **split extension** in the sense of group theory, *i.e.* S may be embedded as a subgroup of $\mathcal{C}$ such that every element of $\mathcal{C}$ is uniquely $\mathfrak{u} \,;\, \mathfrak{f}$ with $\mathfrak{u} \in S$ and $\mathfrak{f} \in K$. Then each $\mathfrak{u} \in S$ acts on K by conjugation $\mathfrak{u}^* : \mathfrak{f} \mapsto \mathfrak{f}^{\mathfrak{u}} \equiv \mathfrak{u} \,;\, \mathfrak{f} \,;\, \mathfrak{u}^{-1}$ within $\mathcal{C}$. □

EXAMPLES 9·2·12:

(a) Every product projection in $\mathcal{G}p$ is a split extension, for example the Klein 4-group $(\mathbb{Z}/(2))^2 \twoheadrightarrow \mathbb{Z}/(2)$, where $\mathbb{Z}/(2)$ is cyclic of order 2.

(b) $\mathsf{S}_3 \twoheadrightarrow \mathbb{Z}/(2)$ is a split extension with kernel $\mathbb{Z}/(3)$, but not a product, where S_3 is the non-Abelian group of order 6.

(c) $\mathbb{Z}/(4) \twoheadrightarrow \mathbb{Z}/(2)$ is not a split extension, since it cannot be described by conjugation, all three groups being Abelian.

(d) The squaring map $S^1 \twoheadrightarrow S^1$ on the unit circle $S^1 \subset \mathbb{C}$ (Examples 2.4.8 and 6.6.7) is a non-split extension of topological groups.

(e) There is a double cover $\mathsf{SU}_2 \twoheadrightarrow \mathsf{SO}_3$, where SO_3 is the group of rotations of the sphere $S^2 = \{(\mathrm{i}x, y + \mathrm{i}z) \mid x^2 + y^2 + z^2 = 1\} \subset \mathbb{C}^2$ and SU_2 consists of the complex matrices $\left(\begin{smallmatrix} a & b \\ c & d \end{smallmatrix}\right)$ with $ac^* + bd^* = 0$ and $|a|^2 + |b|^2 = |c|^2 + |d|^2 = 1$, which acts by conjugation. This has a manifestation in quantum mechanics called the Pauli exclusion principle: electrons have a non-geometrical property called "spin" which changes sign if the particle is rotated through 360°, whereas it stays the same for photons.

REMARK 9·2·13: Let $\bar{u} \in C$ be a cleavage, *i.e.* a choice of pre-images (prone liftings) for each $u \in S$. This is also known as a **transversal** for the normal subgroup $K \lhd C$. Then

$$\bar{u} \, ; \bar{v} = \delta(u, v) \, ; \overline{u \, ; v} \qquad \text{for some unique } \delta(u, v) \in K,$$

where $\delta(u, \text{id}) = \text{id} = \delta(\text{id}, u)$. By associativity in C,

$$\delta(u, v) \, ; \delta((u \, ; v), w) = u^* \delta(v, w) \, ; \delta(u, (v \, ; w)),$$

which is called the **cocycle condition**. The set $H^1(S, K)$ of functions $\delta : S^2 \to K$ satisfying these three laws is called the **first cohomology**; it is a group by pointwise multiplication in K. Then

$$[u \, | \, f] \, ; [v \, | \, g] = [u \, ; v \, | \, f \, ; u^* g \, ; \delta(u, v)]$$

defines an extension of K by S, using a version of the Grothendieck construction with cleavages.   □

Example 9.2.12(d) is a *topological* group. In this case, the base S^1 can be *covered* by open subsets (in the sense of Section 3.9) for each of which the two square roots may be distinguished continuously. The indexation is now over the frame $\Omega(S^1)$ of open subsets of the circle, but it is also necessary to say how these (and things defined over them) are pasted together with respect to the coverage. A fibration which respects this pasting is called a **stack**, or **champs** in French. (The French for a field of numbers is *corps*, but beware that Bénabou has used the word *corpus* in yet another sense!) For an account of the Grothendieck school's approach to fibred categories, see [Gir71], in which Jean Giraud studies non-Abelian cohomology of topological groups up to dimension 3.

Stacks arise in type theory too, for example to say that $S^2 \simeq S \downarrow 2$, respecting the coproduct $2 = 1 + 1$, *cf.* Exercise 9.20. They were also used by Hyland, Robinson and Rosolini [HRR90] to state precisely the completeness of the category of modest sets (Example 7.3.2(l)).

In type theory, Bénabou's criticism has been (mis-understood and) a source of confusion — for *syntactically defined* indexed structures, the substitutions really are functorial *on the nose*, and the group-theoretic issues do not arise. The more difficult technology of fibrations has been employed where the simpler indexed one would have been enough. Even for semantics, we saw in Sections 7.6 and 8.4 that there is an *equivalent* syntactic category: Exercise 9.17 provides a similar construction for fibred *versus* indexed categories. One should not try to *choose* product, substitution functors, *etc.* but work instead in the syntactically constructed category, taking the results back along the weak equivalence.

9.3 SUMS AND EXISTENTIAL QUANTIFICATION

Remark 5.8.5 used relational algebra to illustrate the link between the existential quantifier (with proof-anonymity) and stable factorisation (into regular epis and monos). In this section we shall demonstrate this directly from syntax. Recall that we used the notation $\hookrightarrow$ and $\twoheadrightarrow$ in Section 5.7 for factorisation systems, without assuming that these maps really were mono and epi; we are using the same Convention 9.1.2 in this chapter, calling the body and result of the quantifier "propositions", without assuming proof-anonymity.

In the Lawvere presentation, the existential quantifier is the left adjoint of the substitution functor. So $P : \mathsf{Cn}^{\times}_{\mathsf{type}|\mathsf{prop}} \longrightarrow \mathsf{Cn}^{\times}_{\mathsf{type}}$ is a *bi*fibration (Definition 9.2.6(e)), but the Beck–Chevalley condition, for invariance under substitution, must be stated separately (Exercise 9.19). The dependent sum arises in the same way from arbitrarily divided contexts (Example 9.2.5), the Beck–Chevalley condition now being automatic.

We shall use display maps instead of the fibred technology of the previous section, and derive diagrammatic universal properties directly from the syntactic introduction and elimination rules. The quantifiers are only naturally defined when the "substitution" is weakening, so $\exists x \dashv \hat{x}^{*}$.

Dependent sums and composition. We begin with the case where the range, body and result types are of the same kind. This quantifier, the dependent sum Σ, exists when the (single) class of display maps is closed under composition (*cf.* Definition 8.3.2).

REMARK 9·3·1: We defined a display map to be one which arises in the category of contexts from forgetting a *single* variable, so if the class is to be closed under composition, every context $[\Gamma, x : X, y : Y]$ with two (extra) variables must be the same as some $[\Gamma, z : Z]$ with just one.

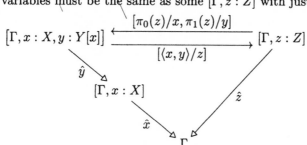

In this case we have an isomorphism over Γ making Z the dependent sum $\Sigma x{:}X.\, Y[x]$. As the notation suggests, this situation reduces to a binary product in the case where $Y[x]$ does not in fact depend on $x : X$ (indeed, ΓXY is then the (P-)pullback of ΓX and ΓY, Remark 8.3.1).

But it is asymmetrical in X and Y: like **if ϕ and ψ then** in imperative programming languages, the x-component is "read" first.

The projection operations have arity

$$\Gamma, z : Z \vdash \pi_0(z) : X \qquad \Gamma, z : Z \vdash \pi_1(z) : Y[\pi_0(z)].$$

Notice that $\pi_0(z)$ is used twice; in this sense the $(\exists \mathcal{E})$-rules which we shall discuss below are computationally simpler.

Similarly, if an identity (or more generally an isomorphism) is a display then it must be possible to omit the corresponding type, and conversely to introduce its value in a unique way. This type is therefore a singleton.

Hence dependent sums are freely added to a type theory by closing its class of displays under composition.

Sum-introduction. We cannot recover $a : X$ and $b : \phi[a]$ from a proof of $\exists x.\ \phi[x]$, so the projections π_i are lost when we quantify one kind over another. But we still have the pairing map. This operation has not previously been given a name in type theory: we do so to stress that it need not be an isomorphism, and that it is to $\exists$ as evaluation is to $\forall$.

DEFINITION 9·3·2:   The term-formation rule introducing sum types is

$$\Gamma, x : X, y : \phi \vdash \text{ve} : \exists x{:}X.\ \phi. \tag{$\exists \mathcal{I}$}$$

This operation takes a witness $a : X$ and its evidence $b : \phi[a]$, and yields a proof of $\exists x.\ \phi[x]$, so we shall call it the **verdict**, $\text{ve}(a, b)$. Spelling this out, the right rule in the sequent calculus is

$$\frac{\Gamma \vdash X \text{ type} \quad \Gamma \vdash a : X \quad \Gamma, x : X \vdash \phi \text{ prop} \quad \Gamma \vdash b : [a/x]^* \phi}{\Gamma \vdash \text{ve}(a, b) : \exists x{:}X.\ \phi}$$

The verdict operation is a map $[\Gamma, x : X, y : \phi] \twoheadrightarrow [\Gamma, z : \exists x.\ \phi]$. We use the double arrowhead because, if the propositional displays $\hookrightarrow$ are monos, then the verdict maps are stable regular epis, as we shall see.

For the binary sum (Remark 2.3.10) we wrote $\nu_x(y)$ for $\text{ve}(x, y)$. Note that $x, y \notin \text{FV}(\exists x.\ \phi)$; we ought perhaps to write $\hat{y}^* \hat{x}^* \exists x'{:}X.\ [x'/x]^* \phi$, but we shan't. Also, we never meet the quantifier $\exists y{:}\phi.\ X$; this is because its verdict operation $\text{ve}(y, x)$ would be a *type*-term depending on a *proof* $y : \phi$, violating Remark 9.1.1(b) and Definition 9.2.1(b).

Unsubstituted weak sums and the adjunction.

REMARK 9·3·3:   The simplest **weak sum elimination rule** is

$$\frac{\Gamma \vdash \vartheta \text{ prop} \qquad \Gamma, x : X, y : \phi \vdash f : \vartheta}{\Gamma, z : \exists x{:}X.\ \phi \vdash \text{let } (x, y) \text{ be } z \text{ in } f : \vartheta} \exists^-_{\text{id}} \mathcal{E}$$

which reduces to $(\exists \mathcal{E})$ as in Definition 1.5.1 if we erase the proofs (terms of propositional type). "$\Gamma \vdash \vartheta$ prop" is the more precise type-theoretic way of saying that ϑ is a well formed proposition with $x \notin \mathsf{FV}(\vartheta)$. With the *variable* z, this is actually the "left rule" in the sequent calculus, *cf.* Remark 1.4.9: the elimination rule in natural deduction gives a value c to z. This term might involve parameters from another context, but we shall ignore these at first.

The **let** syntax, though ugly, is needed to match the argument z or c against the *pattern* $\mathsf{ve}(x, y)$, binding the variables x and y (*cf.* ν_i in $(+\mathcal{E})$, Remark 2.3.10), whereas $(\lambda x.\, p)a$ assigns its argument to just one variable. The β-rule for sums (cut-elimination) executes the match:

$$\cfrac{\cfrac{\Gamma \vdash a : X \quad \Gamma \vdash b : [a/x]^* \phi}{\Gamma \vdash \mathsf{ve}(a, b) : \exists x{:}X.\, \phi}\; \exists\mathcal{I} \qquad \cfrac{\Gamma, x : X, y : \phi \vdash f : \vartheta}{\Gamma, z : \exists x{:}X.\, \phi \vdash \mathbf{let} : \vartheta}\; \exists^-_{\mathsf{id}}\mathcal{E}}{\Gamma \vdash \big(\mathbf{let}\ (x, y)\ \mathbf{be}\ \mathsf{ve}(a, b)\ \mathbf{in}\ f\big) \leadsto [a/x, b/y]^* f : \vartheta}\; \exists^-\beta$$

On the face of it, we associate the value a to the variable x and b to y within the proof f. But the verdict need not be the pairing function, and many different pairs $\langle a, b\rangle$ may be associated with the same proof $c = \mathsf{ve}(a, b)$ of $\exists x.\, \phi[x]$. So there is an equivalence relation amongst such pairs, and the **let** expression corresponds to the familiar idiom of "choosing" a member of the equivalence class, *cf.* Remark 1.6.7 — *i.e.* to using an $(\exists\mathcal{E})$-rule. This indeterminacy is why we say **let** instead of **put**.

The box idiom is

The η-rule is

$$\cfrac{\Gamma, z : \exists x.\, \phi \vdash p : \vartheta}{\Gamma, z' : \exists x.\, \phi \vdash \big(\mathbf{let}\ (x, y)\ \mathbf{be}\ z'\ \mathbf{in}\ [\mathsf{ve}(x, y)/z]^* p\big) = [z'/z]^* p : \vartheta}\; \exists^-\eta$$

Like λ-abstraction, **let** respects equality:

$$\cfrac{\Gamma \vdash \vartheta\ \mathrm{prop} \quad \Gamma, x : X, y{:}\phi \vdash f = g : \vartheta}{\Gamma, z : \exists x{:}X.\, \phi \vdash \big(\mathbf{let}\ (x, y)\ \mathbf{be}\ z\ \mathbf{in}\ f\big) = \big(\mathbf{let}\ (x, y)\ \mathbf{be}\ z\ \mathbf{in}\ g\big) : \vartheta}\; \mathbf{let} =$$

PROPOSITION 9·3·4: ∃ and ve satisfy the unsubstituted weak sum rules iff in the square (commutative since the rules happen in the context Γ),

$$[\Gamma, x : X, y : \phi] \xrightarrow{\ \epsilon = [\mathsf{ve}(x,y)/z]\ } [\Gamma, z : \exists x.\, \phi]$$

with maps f, p, $\hat{z}$, $[c/z]$, $\widehat{m}$, to $[\Gamma, m : \vartheta]$ and Γ

there is a unique fill-in which makes the triangles commute.

PROOF: The sum elimination rule $(\exists^{-}_{\mathsf{id}}\mathcal{E})$ gives such a map; it is a term of type ϑ, so the lower triangle commutes. The upper one is the $(\exists^{-}\beta)$-rule in the unsubstituted form (*cf.* Example 7.2.7)

$$
\begin{aligned}
[x'/x, y'/y]^{*} f &= [\mathsf{ve}(x',y')/z]^{*}(\mathbf{let}\ (x,y)\ \text{be}\ z\ \text{in}\ f)\\
&= \mathbf{let}\ (x,y)\ \text{be}\ \mathsf{ve}(x',y')\ \text{in}\ f.
\end{aligned}
$$

Finally, the $(\mathbf{let} =, \exists^{-}\eta)$-rules say that p is unique. □

REMARK 9·3·5: ***Existential quantification is the left adjoint of weakening.*** Indeed $[\mathsf{ve}(x,y)/z]$ is a universal map (Definition 7.1.1) in the fibre category $\mathcal{P}\big([\Gamma, x : X]\big)$ from the object $[\Gamma, x : X, y : \phi]$, *i.e.* the proposition ϕ, to the functor $\hat{x}^{*} : \mathcal{P}(\Gamma) \to \mathcal{P}\big([\Gamma, x : X]\big)$. Then

$$[\Gamma, x : X, y : \phi] \xrightarrow{\ \eta = [\mathsf{ve}(x,y)/z]\ } [\Gamma, x : X, z : \hat{x}^{*}\exists x.\, \phi]$$

and

$$[\Gamma, z : \exists x.\, \hat{x}^{*}\vartheta] \xrightarrow{\ \varepsilon = [\mathbf{let}\ (x,y)\ \text{be}\ z\ \text{in}\ y/m]\ } [\Gamma, m : \vartheta]$$

are respectively the unit and co-unit of the adjunction $\exists x \dashv \hat{x}^{*}$. □

Substitution and the Beck–Chevalley condition. The adjunction is very neat, but it is not the whole story, because we have overlooked substitution for the parameters which might occur in the proof c of $\exists x.\, \phi$.

These parameters are provided by the substitution $\mathsf{u} : \Gamma \to \Delta$. We make the convention that *the types X, $\phi[x]$ and $\exists x.\, \phi$ are defined in Δ* (*cf.* type-*symbols* being defined over Δ in the previous chapter), although for reasons of space we often omit these from the rules. Then

$$\Gamma \vdash c : \mathsf{u}^{*}\exists x.\, \phi[x],$$

corresponds by Lemma 8.2.14 to a morphism

$$[\mathsf{u}, c/z] : \Gamma \to [\Delta, z : \exists x.\, \phi].$$

As Γ, u and c are arbitrary, this is a generalised element of $[\Delta, z : \exists x.\, \phi]$.

DEFINITION 9·3·6: The (substituted) **weak sum elimination rule** is

$$\frac{\Delta \vdash X \text{ type} \quad \Delta, x : X \vdash \phi \text{ prop} \quad u : \Gamma \to \Delta}{\Gamma \vdash c : u^* \exists x.\, \phi \quad \Gamma \vdash \vartheta \text{ prop} \quad \Gamma, x : u^* X, y : u^* \phi \vdash f : \vartheta}{\Gamma \vdash \text{let } (x, y) \text{ be } c \text{ in } f : \vartheta} \quad \exists^- \mathcal{E}$$

which turns the diagram in Proposition 9.3.4 into

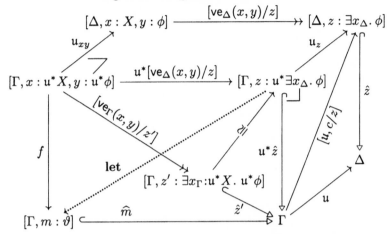

REMARK 9·3·7: The comparison (marked $\cong$ in the diagram)

$$\Gamma, z' : \exists x{:}u^* X.\, u^* \phi \vdash \text{let}_\Gamma(x, y) \text{ be } z' \text{ in } u^*\, \mathsf{ve}_\Delta(x, y) : u^* \exists x.\, \phi$$

arises from the *unsubstituted* rules — the introduction rule (unit, ve) of $\exists x_\Delta \dashv \hat{x}_\Delta$ and the elimination rule (co-unit, let) of $\exists x_\Gamma \dashv \hat{x}_\Gamma$:

$$\exists x_\Gamma \cdot u_x^* \xrightarrow{\eta_\Delta} \exists x_\Gamma \cdot u_x^* \cdot \hat{x}_\Delta^* \cdot \exists x_\Delta = \exists x_\Gamma \cdot \hat{x}_\Gamma^* \cdot u^* \cdot \exists x_\Delta \xrightarrow{\varepsilon_\Gamma} u^* \cdot \exists x_\Delta.$$

In the case $u = \text{id}$, this is the identity, by the triangular laws given in Definition 7.2.1. The fully substituted form of $(\exists^- \mathcal{E})$ provides the inverse of the comparison for any substitution u, as may be seen by putting $\exists x{:}u^* X.\, u^* \phi$ for ϑ in the diagram. This is the **Beck–Chevalley condition for sums**, *cf.* the context diagrams in Remark 9.1.9.

THEOREM 9·3·8: The verdict obeys the weak sum rules iff $\exists x \dashv \hat{x}^*$ and the Beck–Chevalley condition holds.   □

COROLLARY 9·3·9: With $u : [\Delta, \psi] \hookrightarrow \Delta$, we have the Frobenius law

$$\exists x.\, (\phi[x] \wedge \psi) \cong (\exists x.\, \phi[x]) \wedge \psi.$$

because the two factorisations of $[\Delta, X, \phi \wedge \psi] \hookrightarrow [\Delta, X] \longrightarrow \Delta$ in the diagram overleaf must be isomorphic.   □

Our Frobenius Law follows a *corollary* of Beck–Chevalley because, unlike Lawvere and Bénabou, we include substitution for propositional as well as type variables in the change of base u.

Putting $\phi = \top$, the co-unit of $\exists x \dashv \hat{x}^*$ is a cartesian transformation (*cf.* Remark 6.3.4 and Proposition 7.7.1(e)).

$$[\Delta, \psi, \exists x_{\Delta\psi}.\,\phi] \xrightarrow[\cong]{\text{Beck–Chevalley}} [\Delta, \psi, \exists x_\Delta.\,\phi] \cong [\Delta, \psi \wedge \exists x_\Delta.\,\phi]$$

Frobenius $\cong$

$$[\Delta, x : X, \phi, \psi] \cong [\Delta, x : X, \phi \wedge \psi] \longrightarrow [\Delta, \exists x_\Delta.\,\phi \wedge \psi] \hookrightarrow \Delta$$

The box proof (*cf.* Lemma 1.6.3) is

$\quad$ 1 $d : \psi$

$\quad$ 2 $c : \exists x.\,\phi$

$\exists\mathcal{E}$	3 $x : X$	witn
	4 $y : \phi[x]$	witn
	5 $d : \psi$	(1)
	6 $\langle y, d \rangle : \phi \wedge \psi$	$\wedge\mathcal{I}(4,5)$
	7 $\mathsf{ve}(x, \langle y, u \rangle) : \exists x.\,\phi \wedge \psi$	$\exists\mathcal{I}(3,6)$

$\quad$ 8 **let** (x, y) **be** c **in** $\mathsf{ve}(x, \langle y, u \rangle) : \exists x.\,\phi \wedge \psi \qquad \exists\mathcal{E}$

An idiomatic form of this proof would use "**let** (x, y) **be** c **in**" to open the box, after which we may say $c = \mathsf{ve}(x, y)$. However, when the box is ultimately closed (at the end of the proof or before closing a surrounding box), the term has to be exported in its **let** form.

Semantics and open maps. The existential quantifier adds a case to the structural recursion in Remark 9.1.10. Given displays $[\![\hat{y}]\!]$ and $[\![\hat{x}]\!]$,

$$[\![\Delta, x : X, y : \phi]\!] \dashrightarrow^{[\![\mathsf{ve}]\!]} [\![\Delta, z : \exists x.\,\phi]\!]$$

$[\![\hat{y}]\!]$ $\qquad\qquad\qquad$ $[\![\hat{z}]\!]$

$$[\![\Delta, x : X]\!] \xrightarrow{\;\;[\![\hat{x}]\!]\;\;} [\![\Delta]\!]$$

we have to find denotations for ve and $\hat{z}$ making the square commute. For soundness of the **let** syntax, these must have *the same* universal property in the semantics as that described in Proposition 9.3.4 for the syntax. So the functor $[\![-]\!]$ preserves the factorisation system.

EXAMPLE 9·3·10: Let $\mathcal{S}$ be $\mathcal{S}p$ or $\mathcal{L}oc$ and $\mathcal{M}$ the class of open inclusions; by the definition of continuity, this is closed under pullback.

For any $\Gamma \in \mathrm{ob}\,\mathcal{S}$, the relative slice $\mathcal{P}(\Gamma) \equiv \mathcal{S} \!\downarrow\! \Gamma$ is the frame of open subsets of the space Γ, and for any continuous map $\mathsf{u} : \Gamma \to \Delta$, the substitution functor $\mathsf{u}^* : \mathcal{P}(\Delta) \to \mathcal{P}(\Gamma)$ is the inverse image (frame homomorphism) of the same name. By Remark 9.3.5, if u (deserves to be called a display $\hat{x}$ and) admits an existential quantifier $\exists x$, then this must be the left adjoint $u_! \dashv u^*$ with the *Frobenius law*

$$u_!(U \cap u^*V) = u_!(U) \cap V \qquad \text{for all } U \in \mathcal{P}(\Gamma) \text{ and } V \in \mathcal{P}(\Delta).$$

Such a u is called an **open map**. This usage is consistent with the term open inclusion, because any mono which satisfies the Frobenius law is the inclusion of an open subset. Then the class $\mathcal{D}$ of open maps

(a) contains all isomorphisms,

(b) is closed under composition,

(c) is stable under pullback against arbitrary continuous maps,

(d) satisfies the Beck–Chevalley condition for such pullbacks, and

(e) obeys the cancellation law that, if e is a surjection (*i.e.* e^* is full) and e ; u is open, then u is also open.

Since the class $\mathcal{E}$ of open surjections also has these properties, and in particular the factorisation of *open* maps into open surjections and open inclusions is stable under pullback, we have a sound interpretation of $\exists$ (**geometric logic**), where $\mathcal{D}$ provides display maps for the types. $\qquad \square$

Strong sums. So far we have only defined the adjunction $\exists x \dashv \hat{x}^*$ between the categories whose objects are a *single* type ϕ over $[\Gamma, x : X]$ and ϑ over Γ. If the kind of propositions admits dependent "sums" over itself, *i.e.* the class $\mathcal{M}$ of displays is closed under composition, then this is enough to deal with contexts (*lists* of propositions) Φ and Θ (*cf.* Exercises 5.47 and 9.27).

Otherwise we must use a more complicated version of $(\exists \mathcal{E})$ for the context $\Theta = [\vartheta_1, \vartheta_2]$, and in particular for the case where $\vartheta_1 = \exists x.\ \phi$:

$$\frac{[\Gamma, x : X, \phi] \xrightarrow{\ [\mathsf{ve}, f]\ } [z : \exists x.\ \phi, m : \vartheta]}{[\Gamma, z : \exists x.\ \phi] \xrightarrow{\ [\mathsf{id}, p]\ } [z : \exists x.\ \phi, m : \vartheta]} \exists \mathcal{E}$$

The orthogonality condition for a factorisation system (Definition 5.7.1) provides the stronger rule, in which ϑ may depend on $z : \exists x.\ \phi$.

Theorem 9·3·11: (Martin Hyland, Andrew Pitts [HP89]) A verdict

$$\Gamma, x : X, y : \phi \vdash \mathsf{ve}(x, y) : \exists x{:}X.\ \phi$$

is orthogonal to all propositional displays $\widehat{m} : [\Theta, m : \vartheta] \hookrightarrow \Theta$ iff it satisfies the **strong sum elimination rule**,

$$\frac{\Gamma, z : \exists x{:}X.\ \phi \vdash \vartheta \text{ prop} \quad \Gamma, x : X, y : \phi \vdash f : [\mathsf{ve}(x, y)/z]^*\vartheta}{\Gamma, z : \exists x{:}X.\ \phi \vdash \big(\text{\bf let } (x, y) \text{ \bf be } z \text{ \bf in } f\big) : \vartheta} \exists \mathcal{E}$$

where ϑ also becomes $[\mathsf{ve}(x, y)/z]^*\vartheta$ in the conclusion of $(\exists \beta)$ and $[z'/z]^*\vartheta$ in $(\exists \eta)$. Hence strong sums with range $\mathcal{D}$ and body $\mathcal{M}$ exist iff there is another class $\mathcal{E}$ (consisting of all maps which are isomorphic to verdicts) such that $\mathcal{E} \perp \mathcal{M}$ and $\mathcal{M}\,;\mathcal{D} \subset \mathcal{E}\,;\mathcal{M}$, this factorisation ($\hat{y}\,;\hat{x} = \mathsf{ve}\,;\hat{z}$) being stable under pullback.

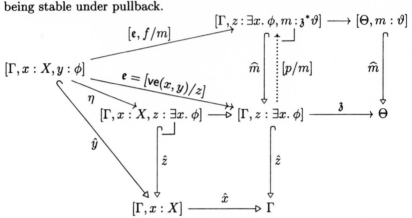

PROOF: According to the original definition, we should show that ve is orthogonal to $[\Theta, \vartheta] \hookrightarrow \Theta$ with respect to any $\mathfrak{z} : [\Gamma, \exists x.\,\phi] \longrightarrow \Theta$. However, Lemma 5.7.10 lets us consider "epi" and "mono" maps with the same target $[\Gamma, z : \exists x.\,\phi]$, ignoring $\mathfrak{z}$. Recall that the proof of this lemma used pullback along $\mathfrak{z}$ (*cf.* Exercise 9.30). As for the weak sum in Proposition 9.3.4, the fill-in p is given by the elimination rule, the β-rule says that it makes the triangle commute, and it is unique by the η- and equality rules. Remark 9.3.7 has already shown that the pullback along u of a verdict map must be another such. $\qquad\square$

REMARK 9·3·12: As a special case, a single kind admits strong sums over itself ($\mathcal{M} = \mathcal{D}$) iff it is closed under composition (Exercise 9.26). In this case the verdict maps are isomorphisms, so the Beck–Chevalley condition and stability under pullback are automatic. The formulation with "product" projections (Remark 9.3.1) relies on the kinds being the same, otherwise we would have witnesses for existential quantification.

REMARK 9·3·13: The strong sum generalises the *test* proposition ϑ to a context extension, but there seems to be no type-theoretic notation for $\exists x.\,\Phi$, where Φ is a list of propositions involving x. This is because

$$\exists x.\,\phi[x] \wedge \psi[x] \vdash \big(\exists y.\,\phi[y]\big) \wedge \big(\exists z.\,\psi[z]\big)$$

is irreversible without "co-operation" between the witnesses of ϕ and ψ. By Lemma 5.7.6(e), the comparison map

$$[\Gamma, \exists x.\,(\phi, \psi)] \dashrightarrow [\Gamma, \exists x.\,\phi] \hookrightarrow \Gamma$$

is in $\mathcal{E}^{\perp}$, so maybe it should be regarded as a proposition (in $\mathcal{M}$), but this is not clear. If we do adopt this view, then the factorisation problem reduces to two cases, of which the first is called the **support** of X:

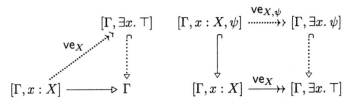

Even then, type theory does not require *all* maps to be factorisable. In particular the pullback diagonal (contraction, variable *quâ* term)

$$[\Gamma, x : X] \xrightarrow{\ [x/y]\ } [\Gamma, x : X, y : X],$$

which is of course a split mono, does not have to be (factorisable into a verdict and) a propositional display. For example, this map is open (Example 9.3.10) with $\Gamma = 1$ iff the space X is discrete, *cf.* Remark 8.3.5.

Finally, notice that it is the *factorisation* $\mathsf{e}\,;\mathsf{m}$ rather than the *class* $\mathcal{E}$ which is required to be stable under pullback, since the target of the substitution $\mathsf{u} : \Gamma \to \Delta$ is that of m, not e. (Before Definition 9.3.6 we did mention a morphism to $[\Delta, \exists x.\ \phi]$, but its extra component c did not contribute to the pullback.)

9.4 DEPENDENT PRODUCTS

The rules for the dependent product are those for the quantifier $\forall$ in Section 1.5, with the addition of terms which come from the λ-calculus. The dependent product Πx and universal quantifier $\forall x$ are *right* adjoint to weakening, with a Beck–Chevalley condition, but this is where the similarity with Σx and $\exists x$ ends. As usual, we relate the syntactic rules to various diagrammatic forms.

Whereas the previous section was very much about the *quantifier* (albeit directly applicable to categories in which the "propositional" displays need not be mono), this one quickly leaves the predicate calculus behind, and has much more of the flavour of *function-types*. However, following Convention 9.1.2, we retain the predicate notation, although its *only* purpose is to distinguish the positive role of the body $\phi[x]$ from the negative range $x : X$ of the quantifier (Remark 1.5.9). Far from assuming the display of ϕ to be mono, in Lemma 9.4.10ff we consider the special case in which ϕ is a fixed object and the *type* display is mono. You may prefer to start with Definition 9.4.8, skipping the type theory, as there are applications in geometric topology, as well as to free algebraic theories.

Application and abstraction.

DEFINITION 9·4·1:   The **application** map ($\forall$-elimination) is

$$\Delta, x : X, f : \forall x{:}X.\ \phi \vdash \mathsf{ev}(f, x) : \phi, \qquad\qquad (\forall\mathcal{E})$$

the corresponding left rule in the sequent calculus being

$$\frac{\Delta \vdash a : X \qquad \Delta, p : [a/x]^*\phi \vdash \jmath : \Theta}{\Delta, f : \forall x.\ \phi \vdash [\mathsf{ev}(f,a)/p]^*\jmath : \Theta}$$

The term-formation rule introducing the $\forall$-type is λ-**abstraction**:

$$\frac{\Delta \vdash X\ \mathsf{type} \quad \Delta, x : X \vdash \phi\ \mathsf{prop} \quad u : \Gamma \to \Delta \quad \Gamma, x : u^*X \vdash p : u^*\phi}{\Gamma \vdash \lambda x.\ p : u^*\forall x{:}X.\ \phi}\ \forall\mathcal{I}$$

combining the $(\forall\mathcal{I})$- and $(\to\mathcal{I})$-rules of Sections 1.5 and 2.3, together with the substitution u which we have discussed in detail for $\exists$. There is also an equality rule $(\lambda{=})$ saying that if $p = q$ then $\lambda x.\ p = \lambda x.\ q$. The β-rule incorporates the substitution u into Definition 2.3.7,

$$\frac{\Delta \vdash X\ \mathsf{type} \qquad \Delta, x : X \vdash \phi\ \mathsf{prop}}{\Gamma \vdash \mathsf{ev}(\lambda x.\ p, a) \rightsquigarrow [a/x]^*p : [a/x]^*u^*\phi}\ \forall\beta$$

with the premises $u : \Gamma \to \Delta \qquad \Gamma \vdash a : u^*X \qquad \Gamma, x : u^*X \vdash p : u^*\phi$

and finally the η-rule is

$$\frac{\Delta \vdash f : \forall x{:}X.\ \phi}{\Delta \vdash (\lambda x{:}X.\ fx) = f : \forall x{:}X.\ \phi}\ \forall\eta$$

We have already devoted Sections 2.3 and 4.7 to simple type theory, where the dependent product reduces to the function-type $X \to Y$. The diagram in Definition 4.7.9 is obtained from that opposite by replacing $\Gamma\Psi$, ϕ and $\forall x.\ \phi$ by Γ, Y and F respectively, and otherwise deleting Γ, Δ and u. In particular, the ground context Γ was previously the terminal object and was not drawn. See also Example 7.2.7.

THEOREM 9·4·2:   The octagon interprets the rules for the dependent product iff for every dotted map as the lower left oblique edge,

$$\Gamma, x : u^*X, \Psi \vdash p : u^*\phi$$

there is a unique map for each of the upper diagonals

$$\Gamma, \Psi \vdash f : u^*\forall x{:}X.\ \phi$$

making the diagram commute. We write $f = \lambda x.\ p$ and $F = \forall x.\ \phi$.

PROOF: As before, $(\forall\mathcal{I})$ provides the map, $(\forall\beta)$ makes the diagram opposite commute and $(\lambda{=}, \forall\eta)$ say that the fill-in is unique. The obtuse dotted triangles define bijections $\tilde{\mathsf{p}} \leftrightarrow [\mathsf{u}, f]$ and $\mathsf{p} \leftrightarrow [\mathsf{u}, p]$ by Lemma 8.2.14, and the parallelograms (but not the kites) are pullbacks — there is no room for the right-angle symbol any more! $\square$

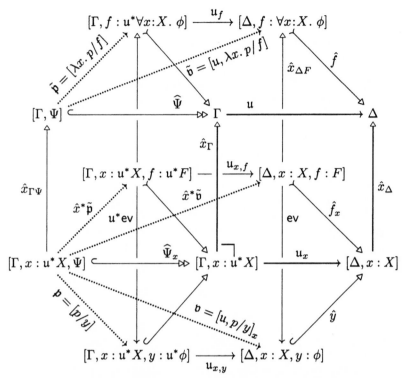

The composite $\widehat{\Psi}\,;\mathsf{u}:[\Gamma,\Psi]\longrightarrow\Delta$ above in fact corresponds to the u in the syntactic rules; Ψ has only been included to state the next result.

The adjunction, Beck–Chevalley condition and other forms.

REMARK 9·4·3: ***Universal quantification is the right adjoint of weakening.*** With $\mathsf{u}=\mathrm{id}$, consider Ψ not as a context extension but as an object of the relative slice $\mathcal{S}\!\downarrow\!\Gamma$. Then there is a natural bijection

$$\frac{\Gamma,x:X\vdash\hat{x}^*\Psi\xrightarrow{\ p\ }\phi}{\Gamma\qquad\vdash\quad\Psi\xrightarrow{\ f\ }\forall x.\,\phi}$$

for which the unit and co-unit of $\hat{x}^*\dashv\forall x$ are

$$[\Gamma,z:\psi]\xrightarrow{\ \eta=[\lambda x.\,z/f]\ }[\Gamma,f:\forall x.\,\hat{x}^*\psi]$$

$$[\Gamma,x:X,f:\hat{x}^*\forall x.\,\phi]\xrightarrow{\ \varepsilon=[\mathsf{ev}(f,x)/y]\ }[\Gamma,x:X,y:\phi]$$

cf. Theorem 9.3.8 for the weak sum. Currying (Example 4.7.3(c)) and Lemma 9.4.13 essentially show how to do this for a *list* Φ instead of a single proposition ϕ.

Now consider the general substitution $u : \Gamma \to \Delta$, which *imports proofs into the box* in the language of Lemma 1.6.3ff. In that section we only saw the consequences of importation for $\exists$, not for $\forall$, because the range of quantification (X) was a simple type.

The separate adjunctions $\hat{x}_\Gamma^* \dashv \forall x_\Gamma$ and $\hat{x}_\Delta^* \dashv \forall x_\Delta$ give a comparison

$$\Gamma, f : u^*\forall x{:}X.\ \phi \vdash \lambda x_\Gamma.\ \mathsf{ev}_\Delta(f, x) : \forall x{:}u^*X.\ u^*\phi.$$

In categorical notation this is

$$\forall x_\Gamma \cdot u_x^* \xleftarrow{\ \varepsilon_\Delta\ } \forall x_\Gamma \cdot u_x^* \cdot \hat{x}_\Delta^* \cdot \forall x_\Delta = \forall x_\Gamma \cdot \hat{x}_\Gamma^* \cdot u^* \cdot \forall x_\Delta \xleftarrow{\ \eta_\Gamma\ } u^* \cdot \forall x_\Delta,$$

but from the fully substituted $(\forall\mathcal{E})$-rule we obtain

$$\frac{\Gamma, f : \forall x{:}u^*X.\ u^*\phi, x : u^*X \vdash \mathsf{ev}_\Gamma(f, x) : u^*\phi}{\Gamma, f : \forall x{:}u^*X.\ u^*\phi \vdash \lambda x_\Delta.\ \mathsf{ev}_\Gamma(f, x) : u^*\forall x{:}X.\ \phi}$$

so that $u^*\forall x_\Delta.\ \phi \cong \forall x_\Gamma.\ u_x^*\phi$. This is the **Beck–Chevalley condition** for products, using the context diagrams in Remark 9.1.9. $\qquad\square$

REMARK 9·4·4: There are two complications in this formulation:

(a) the Beck–Chevalley condition has to be stated separately, and

(b) the bijection $p \leftrightarrow f$ is quantified over a *class of displays* such as $\widehat{\Psi}$.

In Section 9.3 the analogous object to Ψ, testing the adjunction, was called ϑ. There the choice between types and propositions *for ϑ* (rather than for the other participating objects X, $\phi[x]$ and $\exists x.\ \phi$) was the crucial difference between the dependent sum and existential quantifier.

The next result, due to Thomas Streicher [Str91], shows that, for the product, these two complications actually cancel each other out. Hence dependent products are *absolute*: they are defined independently of the choice of the classes $\mathcal{M}$ and $\mathcal{D}$ of displays.

THEOREM 9·4·5: Let $\Delta X \xrightarrow{\ \hat{x}\ } \Delta$, $\Delta X\phi \xrightarrow{\ \hat{y}\ } \Delta X$ and $\Delta F \rightarrowtail^{\hat{f}} \Delta$ be displays, and $\mathsf{ev} : \Delta XF \to \Delta X\phi$. Then F and ev satisfy the rules for the dependent product $\forall x{:}X.\ \phi$ over Δ iff there is a natural bijection

$$\frac{[\Gamma, x : u^*X] \xrightarrow{\ [u, p/y]_x\ } [\Delta, x : X, y : \phi] \text{ in } \mathcal{C}\!\downarrow\![\Delta, x : X]}{\Gamma \xrightarrow{\ [u, f]\ } [\Delta, f : F] \text{ in } \mathcal{C}\!\downarrow\!\Delta}$$

for *all* $u : \Gamma \to \Delta$ *whatever*. Notice too that we have reverted to the ordinary slice $\mathcal{C}\!\downarrow\!\Delta$, since Γ is not a context extension of Δ. There is no further Beck–Chevalley condition.

PROOF: Again, this follows from the octagonal diagram. $\qquad\square$

Local cartesian closure. Dependent products may also be described by restricting to slices or fibres, as the octagonal diagram resides there.

Together with Chapter V, this is the usual categorical treatment of full first order type theory as it was done in the 1970s.

PROPOSITION 9·4·6: The following are equivalent for any category C which has a terminal object.[3]

(a) C has all dependent products;

(b) every (ordinary) slice $C \downarrow \Gamma$ is cartesian closed;

(c) C has pullbacks and every pullback functor u^* has a right adjoint.

In particular the Beck–Chevalley condition is automatic and all colimits are pullback-stable. Then C is said to be *locally cartesian closed*. $\square$

In a locally cartesian closed category all maps are treated as displays, so it has equality types and dependent sums (Remarks 8.3.5 and 9.3.1). Taking the class of monos for the propositional displays, Exercise 9.34 shows that this is closed under dependent products along all maps, so we may interpret universal quantification. Existential quantification may be interpreted as in Section 5.8 on the additional assumption that equivalence relations have quotients (which are automatically stable).

However, the syntax may lack equality types, for the reasons discussed in Section 8.3. In semantic categories, such as those consisting of domains or (locally compact) topological spaces, not all finite limits need exist. Hence we need to consider products whose ranges belong to a restricted class, which we have called type displays ($\longrightarrow\!\triangleright$).

DEFINITION 9·4·7: A *relatively cartesian closed category* C has a single rooted class $\mathcal{D}$ of displays, for which all dependent products exist (with the range, body and result in $\mathcal{D}$). Equivalently, every relative slice $C \downarrow \Gamma$ is cartesian closed and pullback preserves exponentials, *i.e.* $u^* ev_\Delta$ has the universal property of ev_Γ. Such categories were introduced by me, together with an essentially syntactic example based on retracts of a model of the untyped λ-calculus [Tay86a]. Dependent products of domains were constructed in my thesis [Tay86b], and this notion was studied further in [HP89].

---

[3]The terminal object is irrelevant here. It stands for the empty context, where there are no hypotheses, but (philosophically) we are never in such a situation: as Johann Lambert remarked, "no two concepts are so completely dissimilar that they do not have a common part". The definition should simply be that all slices are cartesian closed. The same point can be made for lex, extensive and regular categories, pretoposes and toposes. This suggestion has been made by other authors, but with the *proviso* that binary products be nevertheless required, to interpret concatenations of contexts. I reject this too, on the grounds that we never in fact concatenate two contexts *wholesale*, but either transfer single hypotheses and types from the second to the first, or merge the two (*i.e.* form their pullback) relative to an *a priori* context.

Partial products. Examples of a particular case of dependent product were identified in geometric topology by Boris Pasynkov [Pas65], before Lawvere had studied the quantifiers categorically, and long before the modern type-theoretic account of them had been formulated and applied in informatics. Partial products are more complicated than ordinary function-spaces, but not much, so it is quite feasible to investigate them in semantic categories such as in topology.

Susan Niefield [Nie82] characterised those continuous functions which admit partial products for the Sierpiński space $\Phi = S$, and hence for all $\Phi \in \text{ob}\,Sp$. She also studied the categories of uniform spaces and affine varieties. Building on this, the relationship between partial products and a notion of exponentiability was formulated by Roy Dyckhoff and Walter Tholen [DT87], although none of these authors discussed the Beck–Chevalley condition. In fact we shall see that *all* dependent products (*with* this condition) can be derived from partial products.

DEFINITION 9·4·8:   Let $\hat{x} : \Delta X \longrightarrow \Delta$ a carrable map in a category $\mathcal{C}$, *i.e.* one for which all pullbacks exist (Example 8.3.6(e)), so it is legitimate to call it a display. Also, let Φ be any object of $\mathcal{C}$. Then the sub-diagram in bold below (where the parallelogram is a pullback) is called a *partial product* if it is the universal such figure, *i.e.* given $u : \Gamma \to \Delta$ and $\mathfrak{p} : [\Gamma, u^* X] \to \Phi$ there is a unique map $[u, \mathfrak{f}] : \Gamma \to \Delta\Phi^X$ making the diagram commute.

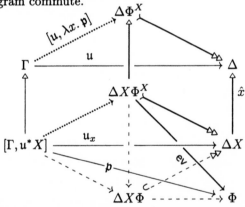

Treating Φ as a constant type (defined in the empty context), and using the binary product $(\Delta X) \times \Phi$, shown dashed, this is another special case of the octagonal diagram in Theorem 9.4.2. Hence the partial product $\Delta\Phi^X$ is the dependent product $\forall x.\, \hat{x}^*\Phi$ in the context Δ.

When $\Delta = \mathbf{1}$, this is the ordinary function-type Φ^X (Definition 4.7.9).

The map $\Delta X \longrightarrow \Delta$ is *exponentiable* if partial products $\Delta\Phi^X$ exist for *all* $\Phi \in \text{ob}\,\mathcal{C}$.

EXAMPLES 9·4·9:

(a) In *Set*, elements of $\Delta\Phi^X$, *i.e.* maps $\Gamma = \{\star\} \to \Delta\Phi^X$, are pairs (u, p), where $u \in [\![\Delta]\!]$ and $p : X[u] \to \Phi$, so

$$\Delta\Phi^X = \sum_{u\in[\![\Delta]\!]} \Phi^{X[u]} \xrightarrow{\hat{p}} \Delta.$$

(b) The functor T that codes a free algebraic theory (Definition 6.1.1) is of the form $\Delta(-)^X$, where $X \longrightarrow \Delta$ is $\kappa \longrightarrow \Omega$ and $X[r] = \mathsf{ar}[r]$.

(c) In the general dependent product in *Set* for displays,

$$\coprod_{u\in[\![\Delta]\!]} \coprod_{x\in X[u]} \phi[u, x] \xrightarrow{\hat{y}} \coprod_{u\in[\![\Delta]\!]} X[u] \xrightarrow{\hat{x}} \Delta,$$

the elements of $F[u]$ are the sections of $\coprod_x \phi[u, x] \longrightarrow X[u]$.

(d) In *Pos*, the elements of $\Delta\Phi^X$ are again pairs (u, p), where now p is monotone. As in Theorem 4.7.13, to determine the order on the function-space we consider $f : \Gamma = \{0 < 1\} \to \Delta\Phi^X$; then

$$(u_0, p_0) \leqslant (u_1, p_1) \iff u_0 \leqslant_\Delta u_1 \land$$
$$\forall x \in X[u_0].\, \forall y \in X[u_1].\, x \leqslant_X y \Rightarrow p_0(x) \leqslant_\Phi p_1(y).$$

For this relation to be transitive on $\Delta\Phi^X$, it would suffice that

$$\forall u_0 \leqslant u_1 \leqslant u_2.\, \forall x \in X[u_0].\, \forall z \in X[u_2].\, \exists y \in X[u_1].\, x \leqslant y \leqslant z,$$

but this condition is also necessary for $\Phi = \{\bot < \top\}$, $p_0 : x \mapsto \top$, $p_2 : z \mapsto \bot$. General dependent products in *Pos* are sets of sections as for *Set*, equipped with this modified pointwise order.

(e) F. Conduché found the analogous characterisation of exponentiable functors in *Cat*, which is discussed in [Joh77, p. 57]. In particular, any fibration *or* op-fibration is exponentiable, but if $\Delta X \longrightarrow \Delta$ belongs to one class then $\Delta\Phi^X \rightarrowtail \Delta$ is in the other. Also, all replete functors between groupoids are exponentiable.

We have stressed that $\hookrightarrow$ for the body of the quantifier is merely a notational convention. For many of the interesting examples of partial products, it is actually the *type* display $\Delta X \longrightarrow \Delta$ that is mono.

LEMMA 9·4·10: If $\Delta X \longrightarrow \Delta$ is mono and the partial product $\Delta\Phi^X$ exists, then $\Delta X \Phi^X \cong (\Delta X) \times \Phi$, *i.e.* the dashed vertical morphism in the diagram opposite is invertible.

PROOF: Given $\Phi \xleftarrow{p'} \Gamma' \xrightarrow{u'} \Delta X$, put $u : \Gamma = \Gamma' \longrightarrow \Delta X \longrightarrow \Delta$, so $\Gamma X = \Gamma'$ is the inverse image, and let $p = p'$. Then $[u, f] : \Gamma \to \Delta\Phi^X$ factors through ΔX by construction, so the pullback mediator is the required product mediator $\Gamma \to \Delta X \Phi^X$. $\square$

EXAMPLES 9·4·11: In the following examples, S denotes the Sierpiński space whilst $S^n \subset \mathbb{R}^{n+1}$ is a circle or sphere.

(a) Lifting is a partial product, since the test is a partial map $\Gamma \rightharpoonup \Phi$. $\Phi_\perp$ is the final solution to forming a pullback square containing $\Phi \longrightarrow 1 \rightharpoonup 2$; it is called a *pullback complement* in [DT87].

(b) See Exercise 3.71 for the same construction in $\mathcal{L}oc$; Example 7.7.4, the Freyd cover, gives it for toposes and geometric morphisms.

(c) Since every open inclusion $U \hookrightarrow \Gamma$ is the pullback of the open point of the Sierpiński space along the classifying map $\Gamma \to S$ of U, and the universal property of the (topological) lift $\Phi_\perp$ is stable by the Beck–Chevalley condition, $U \hookrightarrow \Gamma$ is exponentiable in Sp.

(d) Similarly[4] for a *closed* subset $A \subset \Gamma$. By composition, any *locally closed inclusion* $A \cap U \subset \Gamma$ is exponentiable in Sp. Niefield showed that this characterises exponentiability for subspace inclusions.

(e) The diagonal $\hat{=}\; : X \hookrightarrow X \times X$ is (locally) closed iff X is (locally) Hausdorff, *cf.* Remark 8.3.5. If Y is locally compact and X locally Hausdorff, then any map $Y \to X$ is exponentiable.

(f) Pasynkov's original example [Pas65, p. 181] applied (c) to the interior of the ball $B^n \subset \mathbb{R}^n$. The sphere S^2 can be seen as the partial product of either the interval B^1 by the circle S^1, or the disc B^2 by two points S^0. Similarly for higher dimensions:

$$
\begin{array}{ccc}
S^{n+m} = \{(\underline{x}, \underline{y}) \mid \sum_1^n x_i^2 + \sum_0^m y_j^2 = 1\} & \longrightarrow & \{\underline{x} \mid \sum_1^n x_i^2 \leqslant 1\} = \overline{B^n} \\
\Big\uparrow & & \Big\uparrow \\
B^n \times S^m = \{(\underline{x}, \underline{y}) \mid \sum_1^n x_i^2 < 1, \sum_0^m y_j^2 = 1\} & \longrightarrow & \{\underline{x} \mid \sum_1^n x_i^2 < 1\} = B^n
\end{array}
$$

(g) Using the universal property, a path $[\mathsf{u}, \mathsf{f}] : \Gamma = [0,1] \to \Delta\Phi^X$ is determined by the path u in Δ, together with open segments p in Φ defined on the inverse image of the interior $\Delta X \subset \Delta$. This suggests a way of computing the fundamental groupoid $\pi_1(\Delta\Phi^X)$.

---

[4]This relies, of course, on assuming that the closed point of the Sierpiński space classifies closed subsets. Whether this is so intuitionistically depends on how we formulate Definition 3.4.10. It seems to me that S *should* be defined to have this property, with an underlying set that is strictly between **2** and **2** [Tay98].

EXAMPLES 9·4·12: In practice, the result of a quantifier need not lie in the same class of maps as its (range or) body.

(a) The function-space $\mathbb{N}^{\mathbb{N}}$ exists in *Sp* (it is called **Baire space**), but it is not locally compact, so $S^{\mathbb{N}^{\mathbb{N}}}$ does not exist.

(b) Equality and the order relation on $\{\perp < \top\}$ consist respectively of two and three points of the four-point Boolean algebra. The order can also be thought of as implication, and is definable from lifting. However, by Example 9.4.9(d), a full subposet $\Delta X \longrightarrow \Delta$ is exponentiable iff it is convex, which these examples are not.

(c) In the analogous topological situation for the Sierpiński space S, these subspaces are not locally closed (Example 9.4.11(e)).

Facts such as these about a category could be presented by identifying certain *triples* of classes of display maps ($\longrightarrow, \hookrightarrow, \rightarrowtail$) such that we may form quantifiers whose range, body and result belong to the respective classes. In other words, we have a type theory with many kinds $(\mathcal{K}_1, \mathcal{K}_2, \mathcal{K}_3)$, and the usual introduction, elimination, β- and η-rules, but a restriction on the *formation*-rule for the dependent product:

$$\frac{\Gamma \vdash X : \mathcal{K}_1 \qquad \Gamma, x : X \vdash \phi : \mathcal{K}_2}{\Gamma \vdash (\forall x{:}X.\ \phi) : \mathcal{K}_3} \ (\mathcal{K}_1, \mathcal{K}_2, \mathcal{K}_3)\text{-}\forall\mathcal{F}$$

Henk Barendregt [Bar92, §5.4] has developed just such a formalism, but with the quite different motivation of unifying various syntactic calculi, including Girard's System F (Definition 2.8.10) and Thierry Coquand's Calculus of Constructions [CH88].

Partial products suffice. Partial products are easier to calculate semantically than general dependent products, but we shall now show that, together with pullbacks, they are enough. The trick, categorically, is to consider naturality with respect to Φ.

The reason for writing capital Φ and $\hookrightarrow\!\!\!\!\to$ with a double head above is that we shall later use Φ as the defining *context* of the proposition-symbol ϕ, a substitution-instance of which occurs as the body of some dependent product to be calculated. We shall now write

$$[\Phi, y : \phi] \stackrel{\hat{y}}{\hookrightarrow} \Phi \stackrel{\hat{\Phi}}{\hookrightarrow\!\!\!\to} [\,],$$

assuming that these belong to a rooted class of displays. Comparing their universal properties, there is a unique mediator

$$\Delta(\Phi\phi)^X \rightarrowtail\!\!\!\dashrightarrow \Delta\Phi^X \equiv [\Delta, \mathfrak{f} : \Phi^X] \rightarrowtail\!\!\!\to \Delta$$

in the *third* class. We now show that this corresponds to the proposition $\forall x.\ \phi[\mathfrak{f}x]$. This method may be adapted to finding dependent products of contexts $\forall x.\ \Phi$, *cf.* Remark 9.3.13 for $\exists x.\ \Phi$.

LEMMA 9·4·13: Suppose that $\Delta\Phi^X$ (at the top of the diagram below) is the partial product of Φ along $\hat{x}$ with evaluation map ev_Φ, as in the parallelogram on the right, and let $\hat{y} : [\Phi, y : \phi] \hookrightarrow \Phi$ be a propositional display. Then the following are equivalent:

(p) $\Delta(\Phi\phi)^X$ is the partial product of $\Phi\phi$ along $\hat{x}$ with evaluation $ev_{\Phi\phi}$, shown as the big bold rectangle,

(d) $\Delta(\Phi\phi)^X \rightarrowtail \Delta\Phi^X$ is the dependent product $\Delta, f : \Phi^X \vdash \forall x.\ \phi[fx]$, whose evaluation map ev_ϕ is shown dashed.

The subscripts Φ and $\Phi\phi$ on ev and λ refer to the *partial* products $\Delta\Phi^X$ and $\Delta(\Phi\phi)^X$, whilst ϕ indicates the *dependent* product $\forall x.\ \phi$.

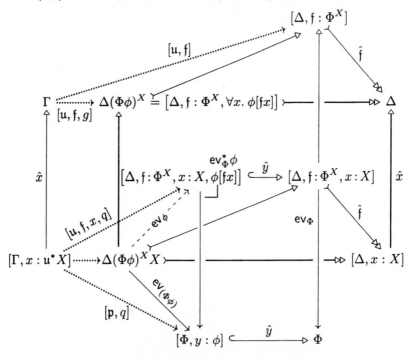

PROOF: For both universal properties, consider a test object Γ.

[p⇒d] We are given $[u, f] : \Gamma \to \Delta\Phi^X$ and $[u, f, x, q] : \Gamma X \to \Delta\Phi^X X\phi$, so by composition of the equilateral triangle put $p = ev_\Phi(f, x)$ to get $[p, q] : \Gamma X \to \Phi\phi$, and $[u, f, g] = \lambda_{\Phi\phi}x.\ [p, q] : \Gamma \to \Delta(\Phi\phi)^X$ using the partial product. This is the dependent product by Theorem 9.4.5, whose u, Δ and ϕ correspond to $[u, f]$, $\Delta\Phi^X$ and $ev_\Phi^*\phi \equiv \phi[fx]$ here. The slender triangle at the top left commutes by universality of $\Delta\Phi^X$, and the triangle involving ev_ϕ commutes using the pullback $ev_\Phi^*\phi$.

[d⇒p] Given $u : \Gamma \to \Delta$ and $[\mathfrak{p}, q] : \Gamma X \to \Phi\phi$, use the dependent product to put $[u, \mathfrak{f}] = \lambda_\Phi x.\ \mathfrak{p} : \Gamma \to \Delta\Phi^X$, define $[u, \mathfrak{f}, x, q]$ using the pullback $ev^*_\Phi\phi$ and let $[u, \mathfrak{f}, \lambda_\phi x.\ q] : \Gamma \to \Delta\Phi^X \forall x.\ \phi[\mathfrak{f}x]$. □

THEOREM 9·4·14: Let $\mathcal{C}$ be a category with a carrable map $\hat{x}$, a rooted ("propositional") display structure, and (choices of) partial products of all objects along $\hat{x}$. Then all dependent products along $\hat{x}$ exist.

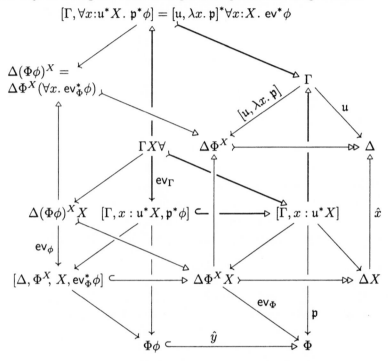

$$[\Gamma, \forall x{:}u^*X.\ \mathfrak{p}^*\phi] = [u, \lambda x.\ \mathfrak{p}]^* \forall x{:}X.\ ev^*\phi$$

PROOF: We want the dependent product $\forall x{:}u^*X.\ \mathfrak{p}^*\phi$ in Γ (shown in bold), where $u : \Gamma \to \Delta$ and $\mathfrak{p} : \Gamma X \to \Phi$. The Beck–Chevalley condition says that the universal property of $\forall x.\ ev^*_\Phi\phi$ is stable, so the required product is given by pullback along $[u, \lambda x.\ \mathfrak{p}] : \Gamma \to \Delta\Phi^X$ as shown, since $[u, \lambda x.\ \mathfrak{p}]_x$; $ev_\Phi = \mathfrak{p}$ using the partial product $\Delta\Phi^X$. □

COROLLARY 9·4·15: The interpretation of dependent products can be defined to make the Beck–Chevalley condition hold up to equality.

PROOF: The binary dependency of $\forall x{:}u^*X.\ \mathfrak{p}^*\phi$ on u and $\mathfrak{p}$ is replaced by a unary one on $[u, \lambda x.\ \mathfrak{p}]$ and the pullback is anchored at $\Delta\Phi^X$, where X and ϕ are type-symbols (*cf.* Remark 9.1.11). □

9.5 COMPREHENSION AND POWERSET

To complete the category-theoretic interpretation of Zermelo type theory (Section 2.2), we need to discuss how to form a subtype from a predicate, and also to investigate the type 2 which names all of the propositions.

Comprehension. The formation and elimination rules of $\{x : X \mid \phi[x]\}$ resemble those of an existential quantifier, yielding a set instead of a proposition, and it behaves like Σx on its range. But the introduction rule for comprehension is different from that for sums. Comprehension turns a proposition into a type, so the effect is *to move it across the division of the context* in Section 9.2. It is like the single-kind Example 9.2.5, where the division was imposed arbitrarily.

Although $\{x : X \mid \phi[x]\}$ is usually called a sub*set*, it is a common idiom to put two or more variables on the left of the divider, or none at all (Remark 2.2.7). In fact the formal rules also suggest that we should view comprehension as an operation on *contexts*. Then $\{\Delta \mid \phi\}$ is the context Δ extended by a new type $C\phi$, which is what the proposition ϕ becomes after its move across the division.

There is a very simple account of comprehension in terms of fibrations, which was, of course, found by Bill Lawvere [Law70]. It was rediscovered by Thomas Ehrhard [Ehr89] and considered further by Bart Jacobs [Jac90] and Duško Pavlović [Pav90].

DEFINITION 9·5·1:   In a generalised algebraic theory with divided contexts, the **extent** of a proposition ϕ is the type $C\phi$ with type-formation and *indirect* term-formation (type-introduction) rules

$$\frac{\Delta \mid \vdash \phi \;\mathsf{prop}}{\Delta \mid \vdash C\phi \;\mathsf{type}}\; C\mathcal{F} \qquad \frac{\Delta \mid \vdash p : \phi}{\Delta \mid \vdash \kappa.p : C\phi}\; C\mathcal{I}$$

(compare $C\phi$ with $\{\star \mid \phi\}$ in Remark 2.2.7.) There are also type- and term-equality rules as you would expect. The elimination rule (that "an element provides its own evidence") is

$$\Delta, y : C\phi \mid \; \vdash \mathsf{ev}(y) : \phi \quad (C\mathcal{E})$$

Note that κ *is not a function-symbol*, so we cannot deduce $\Delta \vdash \phi \cong C\phi$. As we shall see, it is an adjoint transposition like λ-abstraction, which is why we put a dot after it, but it does not bind any variables, so no α-equivalence need be stated. The β- and η-rules for comprehension are

$$\frac{\Delta \mid \vdash p : \phi}{\Delta \mid \vdash \mathsf{ev}(\kappa.p) = p : \phi}\; C\beta \qquad \Delta, y : C\phi \mid \; \vdash y = \kappa.\mathsf{ev}(y) : C\phi \quad (C\eta).$$

The context Δ in these rules has no propositional part; this ensures that the condition for the propositional situation (Definition 9.2.1) is preserved. In fact Δ *may* be extended (weakened) by a propositional part Ψ, but ϕ and p must not depend on it.

REMARK 9·5·2: The **comprehension** $\{\Delta \mid \phi\}$ is the context $[\Delta, C\phi\mid]$. The introduction rule (combined with cut) is then

$$\frac{\mathfrak{a} : \Gamma \to \Delta \qquad \Gamma \mid \,\vdash p : \mathfrak{a}^*\phi}{[\mathfrak{a}, \kappa.p] : \Gamma \to \{\Delta \mid \phi\} \equiv [\Delta, C\phi]} \; C\mathcal{I}$$

and the β- and η-rules are summed up by the diagram

In particular, for $\Delta = [x : X]$ and $\Gamma = [\,]$,

$$\frac{a : X \qquad p : \phi[a]}{[a, \kappa.p] : \{x : X \mid \phi[x]\}} \quad \text{or just} \quad \frac{a : X \qquad \phi[a]}{a : \{x : X \mid \phi[x]\}}$$

with proof-anonymity, *cf.* Definition 2.2.3.

THEOREM 9·5·3: Let $P : \mathcal{C} \longrightarrow \mathcal{S}$ be a fibration and $P \dashv T$ with co-unit $P \cdot T = \mathrm{id}_{\mathcal{S}}$ (Proposition 9.2.3). Then P interprets comprehension iff there is an adjoint $T \dashv C$, with co-unit ev and transposition κ.

In other words, the way in which comprehension turns propositions into types is by a *co-reflection* (*cf.* the support of a type, which is its *reflection* into propositions, Remark 9.3.13). Substitution-invariance is automatic.

PROOF: From the above diagram, there is an adjoint correspondence:

$$\frac{T\Gamma \equiv [\Gamma\mid] \xrightarrow{\;[\mathfrak{a},p]\;} [\Delta\mid\phi]}{\Gamma \xrightarrow{\;[\mathfrak{a},\kappa.p]\;} \{\Delta \mid \phi\}}$$

By Corollary 7.2.10(b), T is full and faithful and the second unit is also a natural isomorphism $\mathrm{id}_{\mathcal{S}} \cong C \cdot T$. $\qquad\square$

EXAMPLES 9·5·4:

(a) In the predicate calculus, where the class $\mathcal{M}$ of propositional displays consists of monos in *Set*, the total category $\mathrm{Cn}^{\times}_{\mathsf{type|prop}}$ is $\mathcal{M} \downarrow \mathit{Set}$. The maps of this category are commutative squares whose verticals are mono. Such squares are prone iff they are pullbacks (hence the alternative name *cartesian*). The fibres are $\mathcal{P}(\Gamma) = 2^\Gamma$.

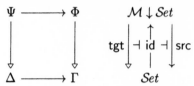

(b) In Example 9.2.5 the division of the context is an arbitrary one between an indexing set I and the sets $X[i]$ which it indexes. This can be described in the same way with $\mathcal{M} = Set$, so *any* four maps may form the commutative square. The comma category $Set \downarrow Set$ is the same as the functor category $\mathcal{S}^{\rightarrow}$, the fibration functor P being **tgt**, so this is known as the **codomain** (target) **fibration**. The fibre over Γ is $Set \downarrow \Gamma$ and the adjoints in this case (and the previous one) are given by $X \mapsto \mathrm{id}_X$ and **src**. The slice $Set \downarrow \Gamma$ is equivalent to the functor category Set^Γ, *cf.* extensive categories if $\Gamma = \mathbf{2}$ (Section 5.5).

(c) Let $\mathcal{S} = Cat$ and $\mathcal{P}(\Gamma) = Set^\Gamma$; then the comprehension functor gives discrete op-fibrations [Law70, SW73].

(d) Not all comprehension fibrations are like this. The fibration P of the category of rings-with-modules over rings (Example 9.2.10(a)) has comprehension, but it is trivial: $C = P$ (since T gives the zero module, we have $T \dashv P$ as well as $P \dashv T$).   □

Not all subobjects in the base category need arise by comprehension. Those that do form a class of support maps (Definition 5.2.10); indeed the notion of a class of supports corresponds to that of a fragment of logic in the propositional kind.

This account of comprehension, pretty though it is, does not explain its role in Zermelo type theory as a way of *creating* sets beside those definable with $\mathbf{1}$, $\times$, $\mathbb{N}$ and $\mathcal{P}$. Exercises 9.45ff describe three approaches.

The type of propositions. In higher order logic, proposition- and type-expressions are handled as if they were terms. In a purely syntactic investigation such as [Bar92], the colon notation $a : U$ between terms and types can be extended to types and kinds U : type or ϕ : prop, and even used to say things like prop : type or type : type.

With categorical interpretations in mind, we prefer to keep the term:type relationship a two-level one, and distinguish the *substance* of a type (to which its terms belong) from its *name* considered as a term belonging to the type of types; the new types Prop and Type classify the kinds or classes of display maps prop and type respectively. (Per Martin-Löf attributes these two approaches to higher order logic to Bertrand Russell and Alfred Tarski respectively.)

DEFINITION 9·5·5: A *generic proposition* $\omega[z]$ involves translation of

(a) names into propositions, using a special dependent proposition

$$z : \mathsf{Prop} \vdash \omega[z] \text{ prop,} \qquad\qquad (\mathsf{Prop}\ \mathcal{E})$$

dependent on a new *type*

$$\vdash \mathsf{Prop} \text{ type}$$

of names of propositions;

(b) and propositions into names:

for each well formed $\Gamma \vdash \phi$ prop,

there is some $\Gamma \vdash a : \mathsf{Prop}$ with $\Gamma \vdash \phi \cong \omega[a]$, $\qquad (\mathsf{Prop}\ \mathcal{I})$

where the isomorphism is as usual expressed by operation-symbols

$$\Gamma, x : \phi \vdash i(x) : \omega[a] \qquad \Gamma, y : \omega[a] \vdash j(y) : \phi$$

and laws (Prop β)

$$\Gamma, x : \phi \vdash j(i(x)) = x : \phi \qquad \Gamma, y : \omega[a] \vdash i(j(y)) = y : \omega[a].$$

See Exercise 9.51 for another version of this rule.

Recall from Theorem 8.2.16 that the substituted type $\omega[a]$ is given by a canonical pullback, as in the diagram on the left:

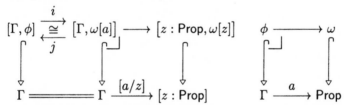

More concisely, for any propositional display $\phi \hookrightarrow \Gamma$ there is some *characteristic map* $a : \Gamma \to \mathsf{Prop}$ which makes the square on the right a (not necessarily canonical) pullback.

How does a depend on ϕ? Since ϕ is a proposition (or type) and not a term, there can be no operation-symbol $\phi \mapsto a$, so the rule (b) must be a *scheme*: the existence of a is asserted individually for each proposition ϕ.

REMARK 9·5·6: Is the characteristic map $a : \Gamma \to \mathsf{Prop}$ at least uniquely determined by the display $\phi \hookrightarrow \Gamma$? If it is then any *isomorphic* display $\psi \hookrightarrow \Gamma$ must correspond to *the same* map a (*up to equality*), since we have discarded the isomorphism (i, j) in the translation. This is an *extensionality* rule (*cf.* Definition 2.2.5 and Remark 2.8.4),

$$\frac{\Gamma \vdash \phi \cong \psi}{\Gamma \vdash \chi.\phi = \chi.\psi : \mathsf{Prop}} \text{ ext}$$

where we write $\chi.\phi$ for a (*cf.* $\kappa.p$ in Definition 9.5.1), so

$$z : \mathsf{Prop} \vdash z = \chi.\omega[z] \qquad\qquad\qquad\qquad (\mathsf{Prop}\ \eta).$$

The type Prop is then a skeletalisation of the relative slice category $\mathsf{Cn}\!\!\int\Gamma$, *cf.* Exercise 4.37 and Remark 5.2.5. This only makes sense for propositions, with anonymous proofs, *i.e.* where this slice is a preorder.

THEOREM 9·5·7: Prop is a support classifier (Definition 5.2.10) iff it satisfies comprehension, proof-anonymity and extensionality.

PROOF: We must replace the proposition ϕ by its extent, the type $C\phi$ (Definition 9.5.1), as it remains to show that $\{z : \mathsf{Prop} \mid \omega[z]\} \cong \mathbf{1}$. Consider any map $\Gamma \to \{z : \mathsf{Prop} \mid \omega[z]\}$, which corresponds to

$$\Gamma \Vdash a : \mathsf{Prop} \qquad \Gamma \Vdash b : C\omega[a]$$

by Lemma 8.2.10. But $b = \kappa.p$ by Definition 9.5.1 ($C\eta$), for some unique $\Gamma \Vdash p : \omega[a]$. By proof-anonymity there is only one such p. Then

$$\frac{\Gamma \Vdash \omega[a] \cong \top}{\Gamma \Vdash a = \chi.\omega[a] = \chi.\top}\ \mathsf{ext}$$

so a is also unique ($\Gamma \Vdash \omega$ and $\Gamma \Vdash \omega \cong \top$ are interchangeable judgements without extensionality, since $X \cong \mathbf{1}^X \times X^\mathbf{1}$, Exercise 2.21). $\qquad\square$

Observe that *any* single display map generates a class of displays which satisfies Definition 9.5.5(a), by Example 8.3.6(j). Conversely, if a kind has a generic display $\omega \hookrightarrow \mathsf{Prop}$ then we may state the type-theoretic properties of the kind in terms of this instead of the whole *class*.

REMARK 9·5·8:   Using the uniquely determined characteristic map, the connectives may be transferred from the propositions to their names, *i.e.* they define algebraic operations on Prop as in Remark 2.8.5. We write 2 for Prop, as in an elementary topos, when it classifies *all* monos (Definition 5.2.6); this happens exactly when the internal connectives and quantifiers for all types X exist ([Tay98], Exercise 9.52).

$$
\begin{array}{ccc}
\omega[x] \wedge \omega[y] \longrightarrow \omega & \omega[x] \vee \omega[y] \longrightarrow \omega & \omega[x] \Rightarrow \omega[y] \longrightarrow \omega \\
\Big\downarrow \qquad\qquad \Big\downarrow & \Big\downarrow \qquad\qquad \Big\downarrow & \Big\downarrow \qquad\qquad \Big\downarrow \\
[x,y:2] \xrightarrow{\ \mathsf{and}\ } 2 & [x,y:2] \xrightarrow{\ \mathsf{or}\ } 2 & [x,y:2] \xrightarrow{\ \mathsf{implies}\ } 2
\end{array}
$$

The displays on the left of each square are derived from the universal properties or proof rules of the type-theoretic connectives $\wedge$, $\vee$ and $\Rightarrow$: they are the product, coproduct and exponential in the context $[x, y : 2]$. Then and, or and $\mathsf{implies}$ are *defined* by these diagrams, *i.e.*

$$\mathsf{and}(x,y) \overset{\mathrm{def}}{=\!=} \chi.\big(\omega[x] \wedge \omega[y]\big).$$

There is a similar correspondence between predicates $\Gamma, x : X \vdash \phi$ prop and terms of type $X \to$ Prop, so this type is the **powerset** $\mathcal{P}(X)$ or 2^X. The internalised quantifiers (Remark 2.8.5) some, all $: 2^X \to 2$ are defined by $\mathsf{some}_X(f) = \chi.\big(\exists x.\, \omega[fx]\big)$ and $\mathsf{all}_X(f) = \chi.\big(\forall x.\, \omega[fx]\big)$. $\qquad\square$

REMARK 9·5·9: We may do the same for other support classifiers such as the Sierpiński space (Definition 3.4.10, Example 5.2.11(d)). In this case the kind of propositions consists of open inclusions in $\mathcal{S}p$, for which we discussed existential quantification along open maps in Example 9.3.10. Although $\big(\omega[x] \Rightarrow \omega[y]\big) \subset S \times S$ exists as a subspace, it is not open, or even locally closed (Example 9.4.12(c)). However, Barendregt's calculus allows us to juggle the kinds so that when $\forall$ is applied to a geometric proposition it is called something else (not G_δ, but something similar) and so is no longer a candidate for classification by S.

REMARK 9·5·10: There is a certain awkwardness in the type-theoretic rules for the powerset. In symbolic logic (where proofs and formulae alike are expressions) it is natural to repeat the term:type relationship as type:kind. This is also the practice in that tradition of category theory which is founded in algebraic topology (though not the one from logic), and it is unnatural to force isomorphisms into equalities. These considerations do not arise in first order logic, as equality is a matter between two terms *relative to some type*. But the extensionality law brings us back to the questions of equality *versus* interchangeability of mathematical objects in Section 1.2, so let's review its uses.

(a) Suppose we have a type of proof-anonymous propositions that does not necessarily satisfy (Prop η). Then there is an extensional such type 2 iff every equivalence relation on any type has an effective quotient (*cf.* Example 2.1.5 and Proposition 5.6.8).

(b) Proposition 3.1.10 used equivalence relations to reduce any preorder such as (Prop, $\vdash$) to a poset. We went on from there to discuss the Yoneda embedding, but from Chapter IV onwards we used this quite successfully for *non-skeletal* categories.

(c) The extensional lattice of ideals of a ring (Example 3.2.5(f)) must often be replaced by its non-skeletal category of modules, and for algebraic rather than logical reasons. For example, the ring may have an automorphism such as $x \mapsto x^p$ that does not act faithfully on the lattice.

(d) Let $x = (L, U)$ and $y = (M, V)$ be Dedekind cuts (Example 2.1.1) bearing the same relation to all rationals, *i.e.* $\forall q{:}\mathbb{Q}.\ q \in L \Leftrightarrow q \in M$ and similarly for U and V. If $f : \mathbb{R}_D \to \Theta$ is continuous then its values on $\mathbb{Q}$ force $f(x) = f(y)$, *without* extensionality $(x = y)$.

Extensionality was needed in older mathematical constructions because Zermelo type theory has very few connectives. Nowadays function- and list-types are used in functional programming to do many jobs previously done by the powerset (Exercise 9.55). However, Exercises 2.54 and 9.57 do depend on extensionality.

Set theory treats higher order logic metaphysically, model theory pretends that it doesn't exist, type theory bureaucratises it and category theory has no view which is identifiably its own. I feel that it does play an important part in mathematics, at least in the weak case of the Sierpiński space, and that this needs to be explained [Tay98].

9.6   UNIVERSES

The real test of foundations is of course how they support the edifice above. We have already demonstrated this throughout the book, but without going beyond its proper scope, we may ask what our logic has to say about its own construction: meta-mathematics as an example of mathematics. This does put the theory in jeopardy as the main question is consistency.

The description of the object language is in two parts: in these last two chapters we have said *what it is to be* a structure for a certain fragment of logic, and Chapters VI and VII constructed the *free* such structure.

REMARK 9·6·1:   Example 8.1.12 gave the generalised algebraic theory of categories, which may be summed up semantically by the display

$$\mathsf{mor} \equiv \big[x, y : O, f : H[x, y]\big] \xrightarrow{\langle \mathsf{src}, \mathsf{tgt} \rangle} O^2$$

together with operation-symbols id and compose satisfying the axioms for a category. Then we may add constants unit, empty $\in O$ and operations

$$\mathsf{product, \ coproduct, \ exponential} : O \times O \longrightarrow O$$

to this language, together with those for the corresponding direct and indirect operations, β- and η-rules, *cf.* Exercise 8.1 and Remark 9.5.8. With a little more work the class of monos can also be encoded, and hence the subobject classifier Ω and powersets (Exercise 9.58). In this way we can speak of an ***internal topos*** or other logical structure.

Gödel's incompleteness theorem. Consider first the *free* structure. Although Cantor's theorem is valid inside, from the outside the objects (contexts) and maps (substitutions of terms) are defined syntactically. So, by the techniques of Section 6.2, there is a recursive cover of each hom-set, in particular $\mathbb{N} \twoheadrightarrow H[\mathsf{unit}, \mathsf{P}]$, where P is the internal version of

$\mathcal{P}(\mathbb{N})$ (as long as the sorts, operation-symbols and axioms are recursively enumerable). This is the Skolem paradox, page 79.

In 1931 Kurt Gödel used powers of primes to describe the enumeration, but recent authors seem to forget that any modern technology that they may employ to write books about his argument works with Gödel numbers as a matter of course. (These spectacularly infeasible calculations do, however, illustrate the need for exponentials, in both the logical and arithmetical senses.) Instead of numbers, it is more natural to use *texts*, *i.e.* terms of type $\mathsf{List}(A)$, where the alphabet A contains everything used in the syntax of $\square$, including variables and the meta-notation for substitution and proofs. It might be the set of distinct symbols used in this book, including my TEX macros for proof trees and boxes, which specify the two-dimensional arrangement of formulae using a linear stream of tokens. We also need a quoting function $\ulcorner - \urcorner : \mathsf{List}(A) \to \mathsf{List}(A)$ such as

$$\ulcorner [3,0,5,7] \urcorner = [0,4,1,6,8,0], \text{ relative to some } A \cong \mathbb{N}.$$

THEOREM 9·6·2: Let $\square$ be a consistent fragment of logic that is recursively axiomatisable and adequate for arithmetic. Then $\square$ cannot prove its own consistency.

PROOF: Using primitive recursion, it is a decidable property of a triplet (p,Γ,ϕ) of texts whether p is a well formed proof in $\square$ whose last line is $\Gamma \vdash \phi$ (*cf.* the proof of Proposition 6.2.6). This property can itself be expressed as a text $ok \in \mathsf{List}(A)$ containing (symbols in A for) variables x, y and z. Using the informal notation $ok[\ulcorner p \urcorner, \ulcorner \Gamma \urcorner, \ulcorner \phi \urcorner]$ to indicate substitution for x, y and z, it satisfies

(a) an *introduction rule* for each sequent rule r of $\square$,

$$\frac{\Gamma_1 \vdash \phi_1 \quad \Gamma_2 \vdash \phi_2 \quad \cdots \quad \Gamma_k \vdash \phi_k}{\Delta \vdash \vartheta} \, r$$

that $ok[\ulcorner p_1 \urcorner, \ulcorner \Gamma_1 \urcorner, \ulcorner \phi_1 \urcorner], \ldots, ok[\ulcorner p_k \urcorner, \ulcorner \Gamma_k \urcorner, \ulcorner \phi_k \urcorner] \vdash ok[\ulcorner r(p) \urcorner, \ulcorner \Delta \urcorner, \ulcorner \vartheta \urcorner]$,

(b) and an *elimination rule* that

$$ok[\ulcorner p \urcorner, \ulcorner \Gamma \urcorner, \ulcorner \phi \urcorner] \vdash ok \left[\ulcorner \bar{p} \urcorner, \ulcorner [\,] \urcorner, \ulcorner \exists q. \, ok[\ulcorner q \urcorner, \Gamma, \ulcorner \phi \urcorner] \urcorner \right],$$

where $\bar{p}$ and q are obtained from p by structural recursion.

In this notation, inconsistency of $\square$ is the statement

$$\psi \equiv \exists p. \, ok[\ulcorner p \urcorner, \ulcorner [\,] \urcorner, \ulcorner \bot \urcorner].$$

We shall show that this is equivalent to the Gödel sentence

$$\vartheta \equiv \exists p. \, ok\left[\ulcorner p \urcorner, \ulcorner [y,z] \urcorner, \ulcorner \neg \exists q. \, ok[\ulcorner q \urcorner, y, z] \urcorner \right].$$

Note that y and z are not free variables of ϑ — they are not variables at all, being quoted. Using the introduction rule (a) for $r = (\perp\mathcal{E})$, and for weakening by y and z, we have $\psi \vdash \vartheta$. Conversely, from ϑ we deduce

$$\exists q.\ ok\left[\ulcorner q\urcorner,\ \ulcorner[\ ]\urcorner,\ \ulcorner\vartheta\urcorner\right]$$

by the elimination rule (b), and also

$$\exists p.\ ok\left[\ulcorner r(p)\urcorner,\ \ulcorner[\ ]\urcorner,\ \ulcorner\neg\vartheta\urcorner\right]$$

by (a), the introduction rule, where r is the cut that substitutes

$$\ulcorner[y, z]\urcorner \text{ for } y \quad \text{and} \quad \ulcorner\neg\exists p.\ ok\left[\ulcorner p\urcorner, y, z\right]\urcorner \text{ for } z.$$

But then, using the introduction rule for $r = (\neg\mathcal{E})$, we have $\vartheta \vdash \psi$. Hence if $\square$ is consistent, $\neg\psi$ and $\neg\vartheta$ are true, but $\neg\vartheta$ actually says that $\neg\vartheta$ is unprovable. Notice that each $\phi[x] \equiv \neg ok\left[x, \ulcorner[\ ]\urcorner, \ulcorner\perp\urcorner\right]$ is provable in $\square$, but $\forall x.\ \phi[x]$ is not, as this is $\neg\psi$. $\square$

One might suppose that Gödel's theorem says that the concept of truth is metaphysical, and needs simply to be replaced by provability. On the contrary, André Joyal (1973) considered the free model $\mathsf{Cn}_{\mathcal{L}}^{\square}$ instead of *Set* for the *outer* world, and then the *internal* free model within that. As the objects, maps and equality in $\mathsf{Cn}_{\mathcal{L}}^{\square}$ are given by the syntax of $\square$, *their truth is our provability*, and the internal notion is different again.

In particular, the property ψ of inconsistency says that there is a map unit $\to$ empty, *i.e.* that *the global sections functor* $U \equiv H[\mathsf{unit}, -]$ *does not preserve the initial object* (*cf.* Theorem 7.7.10(a)). The subobject classified by the Gödel sentence ϑ is then the non-empty equaliser

$$\varnothing \underset{\neq}{\subsetneqq} \{\star \mid \vartheta\} \underset{\neq}{\subsetneqq} 1 \xrightarrow[\text{no}]{\text{yes}} U[\mathsf{coproduct}\,(\mathsf{unit}, \mathsf{unit})].$$

Full internal subcategories. This book was originally supposed to provide the basis for categorical domain theory, in which I wanted to speak of *Set* as if it were an ipo. Functor categories like [*Set* $\to$ *Set*] were to be made legitimate using a Grothendieck universe (*cf.* Remarks 4.1.8 and 4.8.11). In modern parlance, we want the "small sets" to form an *internal* topos $\mathcal{S}$, *i.e.* for which the "large sets" $O = \mathrm{ob}\,\mathcal{S}$ and $\mathrm{mor}\,\mathcal{S}$ are types to which the ordinary constructors of logic are applicable.

A Grothendieck universe is more than an internal topos, as we want each small set $x \in O$ to serve as a large set too. As in the previous section, we call this $U[x]$ to retain the distinction between the name x and substance $U[x]$. Since we intend this correspondence to make the structures of the two categories agree as far as possible, $U[-]$ is in particular to be a full and faithful functor that preserves the terminal object (unit $\in O$), so

$$U[x] \cong Set\,(\mathbf{1}, U[x]) \xleftarrow[\cong]{U} \mathcal{S}(\mathsf{unit}, x) \equiv H[\mathsf{unit}, x],$$

i.e. $U : \mathcal{S} \to Set$ is once again the global sections functor.

LEMMA 9·6·3: If *Set* is relatively cartesian closed, the **full internal subcategory** S is determined up to isomorphism by the display of $U[x]$.

PROOF: $H[x,y] \equiv S(x,y) \xrightarrow[\cong]{U} Set(U[x], U[y]) = U[y]^{U[x]}$. □

PROPOSITION 9·6·4: Type $\equiv O$ and $U[x]$ satisfy Definition 9.5.5 *modulo* Convention 9.1.2: prop becomes the kind of indexed families of S-types.

$$
\begin{array}{ccc}
X \cong U[a] & \cdots\cdots\rightarrow & V \equiv [x : O, U[x]] \\
\downarrow \, \llcorner & & \downarrow \\
\Gamma & \xrightarrow[\quad a \quad]{} & \mathsf{Type} \equiv [x : O]
\end{array}
$$

Proof-anonymity and extensionality no longer hold, so the characteristic map a is not unique. Although we have previously abused the name of the type X as the source of the display $[\Gamma, x : X] \equiv \sum_{u \in [\Gamma]} X[u] \relbar\joinrel\rightarrow \Gamma$, we will later need to distinguish carefully between $U[-]$ and V. □

REMARK 9·6·5: Now we can compare the rest of the structure in S and *Set*. Limits are always preserved by a global sections functor, so, as in Remark 9.5.8, consider exponential : Type $\times$ Type $\to$ Type such that

$$
U[\text{exponential}(x,y)] \; \underset{\mathsf{abs}}{\overset{\widetilde{\mathsf{ev}}}{\underset{\longleftarrow}{\cong \; \longrightarrow}}} \; (U[x] \to U[y]),
$$

where $\widetilde{\mathsf{ev}}$ is the application of closed function-terms to closed arguments in S. Its inverse abs is essentially an encoding operation like those in Section 7.6; it gives names in the internal model to *all* functions between small sets in the external world, so there is no longer a recursive cover $\mathbb{N} \twoheadrightarrow U[\mathsf{P}] \cong \mathcal{P}(\mathbb{N})$ of the internal powerset P.

Analogous structure accompanies the binary operation sum and the constants empty and N. If Type is also closed under comprehension and powerset (or just Ω), it is a **full internal topos**, or, as the set theorists call it, a **limit power cardinal**. In this case, every subclass of a small set is small (*cf.* finiteness, Examples 6.6.3). But there is also a constraint on the logic of the *large* sets, that its class Ω of truth values is small — of course this goes unnoticed in classical logic, where **2** is always small.

Various closure conditions like these on a full internal *category*, rather than properties of a set, would seem to be the way to understand (large) cardinals in set theory. Although they didn't use the word "cardinal", André Joyal and Ieke Moerdijk took an approach like this in [JM95], and Ross Street has also advocated the use of full internal subcategories as a notion of size [Str80].

EXAMPLES 9·6·6:

(a) The display $\mathsf{tgt} : (<) \longrightarrow \mathbb{N}$ classifies finite sets, with $U[n] = \mathbf{n}$.
(b) If (O, ϵ) is an internal model of some set theory then $\mathsf{tgt} : (\epsilon) \longrightarrow O$ is a full internal topos, where $U[x] = \{y \mid y \, \epsilon \, x\}$.   □

Notice that these conditions force the global sections functor U to be the same as the standard interpretation $[\![-]\!]$.

REMARK 9·6·7: Suppose that all types are given syntactically, as Per Martin-Löf has described. Then the (external) history of formation may be transcribed internally as a term of some type Type equipped with operation-symbols for each of the constructors of the given type theory. There may be base types, so long as these are named by constants in Type. So far, this is essentially the *free* internal model, *cf.* Theorem 9.6.2.

What makes Type into a *universe* is a new type-symbol $U[-]$, along with encoding operations and **reflection principles** for each type connective that re-enact the history of the types from the transcript. (Martin-Löf writes V for Type and $T[-]$ for our $U[-]$, whereas we shall need to use T for a functor.)

Russell's paradox. Why can't *all* types be classified by $V \longrightarrow$ Type? Jean-Yves Girard showed that Martin-Löf's *original* type theory, which required this, was inconsistent by proving the Burali-Forti paradox in it. It is rather easier to encode Russell's $\{x \mid x \notin x\}$.

REMARK 9·6·8: Recall that Cantor's Theorem 2.8.8 showed that there is no injection $m : 2^V \hookrightarrow V$, where 2 classifies propositions. But Type classifies *types* like Type^V as $U[w]$ for some $w : \mathsf{Type}$, so $m : \mathsf{Type}^V \hookrightarrow V$ by the wth inclusion into $\Sigma u : \mathsf{Type}.\, U[u]$. (The propositional analogue of V in Theorem 9.5.7 was $\mathbf{1}$.)

Is m split? As in the proof of Cantor's theorem we might define p by

$$U[p(x)(y)] \cong \Pi f : \mathsf{Type}^V.\, I_V[x, m(f)] \to U[f(y)],$$

where I_V is the equality type for V (Remark 8.3.5), or, more crudely,

$$p(x) = \begin{cases} x & \text{if } x = \langle w, f \rangle \text{ for some } f \in \mathsf{Type}^V \cong U[w] \\ \text{unit} & \text{otherwise} \end{cases}$$

since Type is essentially syntax and it is not unreasonable to compare *names* for decidable equality. As we are about to see, neither of these is sound, but when handling a paradoxical argument it is better to leave a gap for some future constructive interpretation than to be stubbornly destructive about a particular formal system.

Recall from Exercise 2.28 that a **reflexive** type X is one for which there are some type R and a retraction $X^R \lhd R$, and that in this case every endofunction of X has a fixed point.

PROPOSITION 9·6·9: Suppose the full internal category S is relatively cartesian closed. If Type is reflexive then so is every type it classifies.

PROOF: Let Θ be any type. We shall treat it like falsity, *cf.* the joker $\bot$ (Remark 1.4.4), so $R \to \Theta$ is the "negation" of R. This is internalised as $n :$ Type $\to$ Type with $U[n(u)] \cong (U[u] \to \Theta)$.

There are two conceptual differences between our formalism and the one that Russell demolished. Type is the universe of *types*, to which (the names of) all types *belong*, whereas V is a universe of *elements*, and all types are *subtypes* of V. We also replace the *fact* that $x \in X$ in Cantor's theorem by its *type* $U[fx]$ of proofs, where $f : V \to$ Type codes X. Then

$$\left(U[pxx] \to \Theta\right) \cong U[n(pxx)] = U[r(x)],$$

where $r = \lambda x.\, n(pxx)$ codes the Russell set. Now consider R, the type of proofs that the element $x = m(r)$ belongs to this set.

$$
\begin{aligned}
R \to \Theta &= U[r(mr)] \to \Theta && \text{definition of } R \\
&= U[p(mr)(mr)] \to \Theta && p \circ m = \text{id at } r \in \text{Type}^V \\
&\cong U[n(p(mr)(mr))] && \text{definition of } n \\
&= U[r(mr)] = R && \text{definition of } r \text{ and } R
\end{aligned}
$$

so $\Theta^R \cong R$. $\qquad\qquad\qquad\square$

EXAMPLES 9·6·10: Such a structure cannot therefore be a *set* theory, but there are categories of *domains* that do have a type of types.

(a) Dana Scott studied the category of countably based algebraic lattices and Scott-continuous functions [Sco76]. Every such object is the image of a closure operation on $\mathcal{P}(\mathbb{N})$ (Theorem 3.9.4); the closure operations themselves also form an algebraic lattice (Exercise 3.44). In the notion of type-dependency that this classifies, if $x \leqslant y$ then $U[y]$ is the image of a closure operation on $U[x]$.

(b) For a more flexible notion of type-dependency, we let Type be the *category* of embedding–projection pairs, or, more generally, adjoint pairs of Scott-continuous functions ($S^{\dashv}$). Although the category-domain Type $\equiv S^{\dashv}$ has as objects the types in the internal category S that it classifies, the morphisms are quite different. The filtered colimits in $S^{\dashv}$ as an "ipo" are both limits of the right and colimits of the left adjoints in S. It is possible to "cover" this category with an ipo, but it is more natural to extend S, the notion of domain, to include categories. Hyland and Pitts gave such a model [HP89], with an interpretation of Coquand's calculus of constructions.

(c) Thierry Coquand and others had previously found a model of the second order λ-calculus (Remark 2.8.10), in which Type is the *range* of a quantifier but, being a category-domain, is not itself one of the types, which are posets.

The results have not been particularly satisfactory. For example, the model in [GLT89, Appendix A], gives $\Pi\vartheta.\vartheta \to \vartheta \to \vartheta$ an extra element called "intersection". Other types and other models are much worse. See [Tay89] for a model that makes non-trivial use of permutation groups.

The axiom-scheme of replacement. Zermelo–*Fraenkel* set theory has an axiom we haven't considered. It was formulated independently by Abraham Fraenkel, Nels Lennes and Thoralf Skolem around 1922.[5]

REMARK 9·6·11: In the formulation of set theory as an untyped first order theory with one binary relation $\in$, **replacement** is a *scheme* of axioms, one for each functional relation R, that

$$\forall x.\, \exists y.\, \forall j. \quad j \in y \iff \exists i.\, i \in x \wedge iRj.$$

This seems to assert the existence of the image y of the set x under R. However, Notation 1.3.4 insisted that the source and target be given along with any relation or function, and the image is a subset of the target. Even without this, Lemma 5.7.3ff found the image as the quotient of the kernel of the function: $y \cong x/{\sim}$, where $i_1 \sim i_2$ if $\exists j.\, i_1 R j \wedge i_2 R j$. Accounts of set theory often draw other trivial or redundant corollaries of Replacement, without giving any indication of its real power.

This confusingly named postulate can easily slip through our fingers into tautology (Exercise 9.61). Indeed Fraenkel, at first, made this mistake by asking for R to be definable. It is a little clearer to say that

if x and y_i are sets for each $i \in x$ then $\bigcup_{i \in x} y_i$ is also a set,

which is equivalent in Zermelo set theory *via* the union axiom (Definition 2.2.9(e)). Beware that these are *a priori* different notions of "union" as the axiom is again vacuous if there is already a set z with $\forall i.\, i \in x \Rightarrow y_i \subset z$.

EXAMPLE 9·6·12: The ordinals 0, 1, ..., ω, $\omega + 1$, ..., $\omega + n$, ... are definable in Zermelo set theory. However, $\omega 2$ does not exist in the model in Exercise 2.20 *if you insist that its well founded relation be* $\in$ (this classical ordinal is the order-type of the even natural numbers followed by the odd ones). All definable sets have rank $< \omega 2$ (Definition 6.7.5).

---

[5]Cantor wrote in 1899 that "two equivalent [=bijective] multiplicities are either both 'sets' or both inconsistent" [vH67, pp. 113ff]. This statement, *verbatim*, is equivalent to Replacement *in the 1920s conception of set theory*, but it seems to me that, considering his frame of mind, Cantor could not have appreciated that it made the difference between two formal theories, or that it was such a strong axiom. To a categorist, Cantor's words carry no force at all.

Infinite limits and colimits. Although the set-theoretic union above is meaningless under our principle of interchangeability (Section 1.2), it is but a small step from here to infinitary coproducts. As is common in 1960s category theory, Section 7.3 took these for granted.

PROPOSITION 9·6·13: Let $\mathcal{S}$ be a full internal topos and $X_{(-)} : I \to \mathcal{S}$ be an indexed family (discrete diagram), where $I = U[s]$ for some $s \in O$. Then (a)$\Leftrightarrow$(b)$\Rightarrow$(c):

(a) there are an object U in $\mathcal{S}$ and monos $m_i : X_i \hookrightarrow U$;

(b) the coproduct $\coprod_i X_i$ exists in $\mathcal{S}$;

(c) the product $\prod_i X_i$ exists in $\mathcal{S}$.

Conversely, the coproduct may be obtained from the product $\prod_i \mathcal{P}(X_i)$. By Lemma 7.3.3, $\mathcal{S}$ then has limits and colimits of all small diagrams, *i.e.* whose vertices and arrows form small sets.

PROOF: The first part of the proof is very often given without comment as the construction of the coproduct or disjoint union of sets.

[a$\Rightarrow$b] The coproduct is $C = \{\langle i, x \rangle \mid x \in X_i\}$, or, more precisely,

$$C = \{\langle i, u \rangle \mid i \in I, u \in U, \exists! x{:}X_i.\ u = m_i(x)\}.$$

[b$\Rightarrow$a] As $\mathcal{S}$, being a topos, is locally cartesian closed, all colimits are stable under pullback. The inclusion ν_i is the pullback of the mono $i : 1 \to I$ against the display $\Sigma i.\ X_i \longrightarrow I$ (which is the coproduct mediator for the cocone $X_i \to \{i\} \hookrightarrow I$), so ν_i is also mono.

[b$\Rightarrow$c] The product is the set of sections of the display map $\pi_0 : C \to I$, *cf.* Example 9.4.9(a).

[c$\Rightarrow$a] Exercise 2.14. □

To ensure that $P \neq \varnothing$ when $X_i \neq \varnothing$, we need exactly the Axiom of Choice (Definition 1.8.8) as Russell and Whitehead stated it.

So $\mathcal{S}$ is a **self-complete** full internal category in the sense that it has limits and colimits *indexed by the externalisations of its objects*. In set theory this is called a **regular cardinal**, and a self-complete full internal topos is a **strongly inaccessible cardinal**.

These operations on $\mathcal{S}$ are "infinitary" in the same sense as in Section 6.1 (and Definition 3.6.12). This is meaningful *only* for an *internal* category: it relies on the enclosing universe of classes or large sets. What *is* an indexed family of sets? What is it to be a cocone which tests the infinitary coproduct? We have only succeeded in defining these notions in terms of each other.

Avoiding replacement. Chapter VIII was devoted to indexed families.

REMARK 9·6·14: We encoded $X_{(-)}$ or $X[-]$ as the display

$$\coprod_{i \in I} X_i \xrightarrow{\ i\ } I$$

of its coproduct over the diagram shape. The subscripted or dependent notation, which is simply *undefined* in category theory (without an enclosing universe), was re-introduced *by convention* on page 427 to mean this display.

Jean Bénabou characterised infinitary limits and colimits essentially as in Sections 9.3–9.4, using fibrations. The Grothendieck construction (Proposition 9.2.7) does for diagrams (category-indexed families of categories) what we have just said for sets. According to Bénabou's notion of limit, *every* topos is complete. Nevertheless, there are intuitive limiting constructions that cannot be performed in it.

Transfinite iteration and fixed points of functors. As Replacement is clothed in so much obfuscation, it is tempting to ask whether it is needed at all. If it is, we would like to put it in an *elementary* form, like the subobject classifier in an elementary topos, without universes. Ideally we would also like to apply it directly to other situations (Abelian categories, domain theory, *etc.*), rather than *via* some set theory.

EXAMPLES 9·6·15: How are infinite things defined without prior reference to infinity? The first infinite thing, $\mathbb{N}$, embodies *iteration*.

(a) How can we make the large/small distinction *relative*, so that any set X can be considered small ($X \cong U[x]$) and thereby benefit from an enclosing universe? Forming it either as a coproduct or colimit,

$$\mathcal{P}^\omega(X) \stackrel{\text{def}}{=\!=} \bigcup_{n \in \mathbb{N}} \mathcal{P}^n(X)$$

carries the structure of a full internal topos (Exercise 2.20). With $X = \varnothing$, it is the smallest such, so (by Cantor's Theorem 2.8.8) we have a topos with no full internal subtopos. Hence their existence is not provable in Zermelo type theory.

(b) Taking unions at limit ordinals in the same way,

$$\aleph_{\text{succ}\,\alpha} = \omega_{\text{succ}\,\alpha} = \mathfrak{H}(\aleph_\alpha) \quad \text{and} \quad \beth_{\text{succ}\,\alpha} = \mathcal{P}(\beth_\alpha)$$

are defined by transfinite iteration of the powerset and Hartogs' construction (Lemma 6.7.11, which is functorial on monos), where $\omega_0 = \omega$ begins a sequence of ordinals and $\aleph_0 = \beth_0 = \mathbb{N}$ begin two sequences of (underlying) sets. $\aleph$ (aleph) and $\beth$ (beth) are the first two letters of the Hebrew alphabet. $\beth_1$ (also called $\mathfrak{c}$) is quite familiar, since classically $\beth_1 \cong \mathbb{R}$. The **continuum hypothesis** says that $\aleph_1 \cong \beth_1$, and the **generalised continuum hypothesis** that the $\aleph$ and $\beth$ sequences coincide forever. Here $\cong$ means bijection, not isomorphism: *it destroys the* (different) *structure* on $\aleph_\alpha$ and $\beth_\alpha$.

(c) In domain theory, any ipo X is a retract of some model Ω of the untyped λ-calculus: $X \lhd \Omega \cong \Omega^\Omega$, where Ω is the bilimit

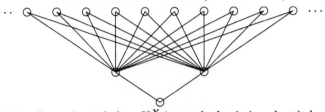

Now let X be the infinite L-domain (Exercise 3.34)

Achim Jung showed that X^X is an algebraic ipo, but it has a pair of compact elements (Definition 3.4.11) with $2^{\mathbb{N}}$ minimal upper bounds. Similarly the nth term (classically) has cardinality at least $\beth_n$, so Ω, which is at least $\beth_\omega$, is not definable in Zermelo type theory. $\square$

The theme of these examples is that Replacement provides a full internal model of some fragment of logic within a world where a stronger logic is valid. Whereas Proposition 9.6.13 stated Replacement for the *inner* model by measuring its completeness from the outside, we are now demonstrating the power of the *outer* world by its ability to probe the logic of an internal structure.

REMARK 9·6·16: Recall from Remark 6.7.14 that an **ordinal** is a well founded extensional coalgebra $\mathsf{parse} : \alpha \to \mathsf{sh}(\alpha)$ for the functor

$$\mathsf{sh} : \mathcal{P}os \to \mathcal{P}os \text{ by } \alpha \mapsto \{ U \subset \alpha \mid \forall \beta, \gamma \in \alpha.\ \gamma \leqslant \beta \in U \Rightarrow \gamma \in U \}.$$

In order to formulate Replacement in an elementary way *à la* Bénabou, we must consider not the *individual* iterates $X[\beta] = T^\beta(X_0)$ but the α-indexed family of them. Together with the maps $X[\gamma] \to X[\beta]$ for $\gamma \leqslant \beta$, this family forms a *discrete op-fibration* in $\mathcal{P}os$. As usual in the Bénabou approach, the functor $T : \mathcal{S}et \to \mathcal{S}et$ must be extended to act on families (discrete op-fibrations), not just on individual sets.

$$
\begin{array}{ccc}
[\beta : \alpha, X[\beta]] & \cdots\cdots\rightarrow & \left[U : \mathsf{sh}(\alpha),\ \underset{\gamma \in U}{\mathrm{colim}}\, T X[\gamma] \right] \\
\Big\downarrow & & \Big\downarrow \\
\alpha & \xrightarrow{\ \mathsf{parse} : \beta \mapsto \{ \gamma \mid \gamma \prec \beta \}\ } & \mathsf{sh}(\alpha)
\end{array}
$$

Each $U \in \mathsf{sh}(\alpha)$ is a diagram-shape, with arrows for each $\delta \leqslant \gamma$, and we form the colimit of this diagram as shown (this is a parametric colimit, defined like the parametric dependent sums in Section 9.3). By its usual substitution interpretation, this square is a pullback iff

$$X[\beta] \cong \operatorname*{colim}_{\gamma \prec \beta} TX[\gamma],$$

i.e. we have found the transfinite iterates as required (*cf.* Lemma 6.7.2ff).

Our *formulation* of this property is elementary, and no longer depends on an enclosing universe. The *existence* of such a family $X[\beta]$ for each ordinal α and indexed functor T is the **categorical axiom of iterative replacement**. It can be stated for any $\mathcal{P}os$-indexed category, not just $\mathcal{S}et$, and for coalgebras for other functors besides sh.

REMARK 9·6·17: Recall from Lemma 3.7.10 and Definition 6.1.1 the way in which T arises: not as a *single* construction but as an amalgam of *all* of the operations under discussion. Unscrambling this and the associated theory of ordinals (with a *single* successor), we see that Replacement is just what we need to justify the recursion-scheme over types,

$$\frac{1 : \mathcal{S} \quad \times : \mathcal{S} \times \mathcal{S} \to \mathcal{S} \quad + : \mathcal{S} \times \mathcal{S} \to \mathcal{S} \quad \mathcal{P} : \mathcal{S} \to \mathcal{S} \quad \cdots}{[\![-]\!] : \mathsf{Cn}^\square \to \mathcal{S}}$$

i.e. Martin-Löf's reflection principles or universe-elimination. He proved strong normalisation for his type theory with a sequence of universes by syntactic methods (1973). For us, this recursion scheme lies at the heart of categorical type theory, and in particular the application of the gluing construction to such questions (Section 7.7).

There is a naïve proof of consistency, which says that the connectives have a standard interpretation, whilst the rules preserve truth, so falsity never enters the system. The role of Replacement or universe-elimination is to *form* that standard interpretation (as a whole) and do *recursion* over it. Unfortunately, *Replacement is not conservative*: we cannot observe consistency without disturbing the system under study. Although the logic $\square$ from which the internal model $\mathcal{S} = \mathsf{Cn}^\square$ is built stays the same, Gödel's Theorem says that the properties of the model within a world that has the *same* logic are different from those in a world that also satisfies Replacement. In particular, the global sections functor $U : \mathcal{S} \to \mathcal{S}et$ preserves $\aleph$ in the latter case but not the former.

For the same reason Grothendieck universes are also not conservative.

By harnessing type theory to category theory, we now have the tools to formalise a logical foundation for mathematics of whatever strength may in future be deemed appropriate. Here and in Section 7.7 we have seen how powerful this method is: we have a *uniform* proof of consistency of the chosen fragment by adjoining some part of the axiom-scheme of replacement. Neither category theory nor the Bourbaki programme has grasped the significance of Replacement, but then that is because set theory and proof theory muddled it up with Cantor's megalomania.

Although a new categorical analysis of high-powered logic would be enlightening, for the foundations of both mainstream mathematics and programming something considerably simpler than Zermelo–Fraenkel set theory is needed.

EXERCISES IX

1. Investigate the naturality of $\exists x \dashv \hat{x}^* \dashv \forall x$ as in Section 7.2.

2. Definition 9.2.1 associates a divided context to any jumbled one. Construct the isomorphism between them.

3. In the first part of Lemma 9.2.2, show that this is the *initial* solution to forming a pullback square with the given top and right sides.

4. Show that any fibration $P : C \longrightarrow S$ preserves pullbacks. Give a semilattice example in which it does not preserve $\top$, but show that if there is a functor T with $P \cdot T = \mathrm{id}_S$ then P does preserve all finite limits. Conversely, show that if C and S have and $P : C \to S$ preserves pullbacks, and $P \dashv T$ with $P \cdot T = \mathrm{id}_S$, then P is a fibration.

5. Show that the pseudo-vertical and prone maps for any fibration $P : C \longrightarrow S$ form a factorisation system (Definition 5.7.2) in C. [Hint: the prone part of a C-morphism $\mathfrak{f}$ is the lifting of $P\mathfrak{f}$; use the universal property again to show that the fill-in property holds.] Also show that every square consisting of parallel vertical and prone maps is a pullback. We call this situation a **cartesian factorisation system**.

6. Show that every fibration in which the base category and fibres have all finite limits arises uniquely from a conservative extension of Horn (or essentially algebraic) theories: the two classes of displays consist of the prone and vertical maps respectively.

7. Show that $\pi_0 : S \downarrow U \to S$ in Proposition 7.7.1 is a fibration, but is not comprehensive or an op-fibration. What is the corresponding conservative extension of theories?

8. In any fibration $P : C \longrightarrow S$, show how to find limits in C, given limits in S and the fibres. Show that $\mathsf{pts} : Sp \to Set$ and $\mathsf{ob} : Cat \to Set$ are bifibrations (Definition 9.2.6(e)), and use this to describe limits and colimits in Sp and Cat, cf. Section 7.4.

9. Let $\mathcal{E} \perp \mathcal{M}$ be a factorisation system in a category C. Verify that the supine, vertical and prone morphisms for the codomain fibration

tgt $= \pi_1 : \mathcal{M} \downarrow \mathcal{C} \longrightarrow \mathcal{C}$ are as shown below, and that if $\mathcal{E} \perp \mathcal{M}$ is stable then so is the supine–vertical factorisation.

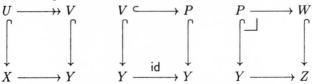

Conversely, suppose that $\mathcal{M} \subset \mathcal{C}$ is such that the class $\mathcal{M}'$ of squares like those in the middle, but with isomorphisms at the bottom, is part of a (stable) factorisation system; show that $\mathcal{M}$ is too.

10. Prove Proposition 9.2.9 in detail for Horn theories. Let $S = \bigcup R_i$ in the base category (algebraic lattice) $\mathcal{M}od(\mathcal{L}_{\text{type}})$. Prove the limit–colimit coincidence, that the projections $\mathcal{M}od(S) \to \mathcal{M}od(R_i)$ form a limiting cone, and the embeddings a colimiting cocone in the category of algebraic lattices and Scott-continuous functions. Show that every fibration of algebraic lattices with these properties arises from a conservative extension of Horn theories.

11. Formulate the divided theory whose type part has one sort and whose propositional part is the theory of groups. Use this and Proposition 9.2.9 to compute limits and colimits in $\mathcal{G}p$. More generally, describe the theory which divides the laws of an algebraic theory from its sorts and operation-symbols (*cf.* Remark 7.4.10 and Example 9.2.4(h)), and use it to compute free algebras.

12. Formulate the sense in which the total category $\mathcal{C}$ of a fibration $P : \mathcal{C} \to \mathcal{S}$ is the **lax colimit** of the corresponding indexed category, considered as a diagram of shape $\mathcal{S}$ in the 2-category $\mathfrak{C}$at.

13. Use the Grothendieck construction for a diagram $\mathcal{I} \to \mathcal{S}et$ to compute its (strict) colimit. [Hint: consider the connected components of the total category.]

14. (For group theorists.) Use the fact that fibrations of groups are surjective homomorphisms and that this class is closed under pullbacks to prove the Jordan–Hölder theorem.

15. Repeat Exercise 4.41, that $\mathcal{S}et^{\mathcal{C}^{op}}$ is cartesian closed, using the Grothendieck construction to replace presheaves by discrete fibrations.

16. Let $P : \mathcal{C} \longrightarrow \mathcal{S}$ be a fibration. Define the morphisms of the category $\mathcal{D}$, whose objects are triples $(\Gamma, \Phi, u : \Gamma \to P\Phi)$, such that $\pi_0 : \mathcal{D} \longrightarrow \mathcal{S}$ is a split fibration and $\Phi \mapsto (P\Phi, \Phi, \text{id}_\Phi)$ is a weak equivalence. [Hint: this is *not* the comma category $\mathcal{C} \downarrow P$.] Describe the groupoid that arises in this way from $\mathbb{Z}/(4) \twoheadrightarrow \mathbb{Z}/(2)$.

17. Let $P : \mathcal{C} \longrightarrow \mathcal{S}$ be any fibration. Suppose that there is a choice of re-indexing functors $u^* : \mathcal{P}(\Delta) \to \mathcal{P}(\Gamma)$ for each $u : \Gamma \to \Delta$ in $\mathcal{S}$.

Formulate the coherence conditions which relate id^* to the identity re-indexing functor and $(\mathsf{u};\mathsf{v})^*$ to $\mathsf{u}^* \cdot \mathsf{v}^*$. Conversely, given assignments $\mathcal{P}(\Gamma)$ and u^* satisfying these conditions, adapt the Grothendieck construction (Proposition 9.2.7) to recover the fibration.

18. Investigate the formation and coherence rules for explicit isomorphisms instead of equalities of types in Definition 8.1.7.

19. In terms of bifibrations, show that the Beck–Chevalley condition for a particular pullback in the base category says that the prone and supine liftings going two ways around the square "join up".

20. Let $\mathcal{C}$ be a category with products and $\mathbf{2} = \mathbf{1} + \mathbf{1}$ such that the inclusions $\mathbf{1} \to \mathbf{2}$ are carrable. Show that $\mathcal{C}$ is extensive (Section 5.5) iff there is a class $\mathcal{D}$ of displays with the following type-theoretic property: for any two types Y, N there is a dependent type $i : \mathbf{2} \vdash X[i]$ type such that $Y \cong X[1]$, $N \cong X[0]$, and $\Sigma i{:}\mathbf{2}.\, X[i]$ is their coproduct. In this case show also that $\Pi i{:}\mathbf{2}.\, X[i]$ is their product. (*Cf.* Exercise 5.35.)

21. Show that $(\exists x{:}X.\ \bot) \cong \bot$,

$$\exists x{:}\mathbf{1}.\ \phi \cong \phi[\star] \qquad \exists x{:}\emptyset.\ \phi \cong \bot \qquad \Sigma x{:}X.\ \mathbf{1} \cong X$$
$$\forall x{:}\mathbf{1}.\ \phi \cong \phi[\star] \qquad \forall x{:}\emptyset.\ \phi \cong \mathbf{1} \qquad \Pi x{:}X.\ \mathbf{1} \cong \mathbf{1}$$

and $\exists x{:}X.\ \exists y{:}Y[x].\ \phi[x,y] \cong \exists z{:}(\Sigma x{:}X.\ Y[x]).\ \phi[\pi_0(z), \pi_1(z)]$.

22. Consider a generalised cut of a substitution $\mathsf{u} : \Gamma \to \Delta$ with the $(\exists^- \mathcal{E})$-rule. Show that the substituted form of the rule as we gave it in Definition 9.3.6 is what is needed to commute these sequent rules.

23. What is the type of $\lambda z.\ (\mathbf{let}\ (x,y)\ \mathbf{be}\ z\ \mathbf{in}\ d)$? Considering this type as a proposition, what is its relationship to that of $\lambda x, y.\ d$? [Hint: Exercise 1.22.] How is this relationship expressed by the ve operation?

24. Formulate the canonical $\exists$-language (Section 7.6) for a stable factorisation system, and use gluing (Section 7.7) to prove conservativity and equivalence.

25. Show that the Beck–Chevalley comparison maps as given in Definition 9.3.6ff for the existential quantifier are mutually inverse.

26. Show that the Beck–Chevalley condition is automatic for dependent sums, *i.e.* composition of displays. Using Exercise 7.29, deduce the same condition for locally cartesian closed categories (Proposition 9.4.6).

27. (Jacobs, Moggi and Streicher) Show that if $\mathcal{M}$ is closed under composition, *i.e.* it admits strong sums over itself, then weak sums over $\mathcal{D}$ are sufficient to derive the strong version (*cf.* Exercise 5.47).

28. Show how to define

$$\frac{[\Gamma, x : X, \phi] \xrightarrow{\ [f_1, f_2, f_3]\ } [\Gamma, \vartheta_1, \vartheta_2, \vartheta_3]}{[\Gamma, \exists x{:}X.\ \phi] \xrightarrow{\ [p_1, p_2, p_3]\ } [\Gamma, \vartheta_1, \vartheta_2, \vartheta_3]} \exists \mathcal{E}$$

[Hint: define $p_1 : \exists x{:}X.\ \phi \to \vartheta_1$ with the weak sum, then use the strong sum for $p_2 : \exists x{:}X.\ \phi \to p_1^*\vartheta_2$ and again for p_3.]

29. Adapt Section 9.3 to sums $\Sigma x{:}X.\ \phi$ in which the result is of a different kind ($\mathcal{M}$) from the body ($\mathcal{P}$). [Hint: Theorem 9.3.11 becomes $\mathcal{E} \perp \mathcal{M}$ and $\mathcal{P}\,;\mathcal{D} \subset \mathcal{E}\,;\mathcal{M}$ stably.]

30. For the continuation rule for the strong sum type, the top right square in Theorem 9.3.11 need not be a pullback. What is the effect of this in terms of the universal property of a factorisation system, and why is this automatic? Formulate the corresponding type-theoretic rule for the strong and weak sums and also discuss it in the idioms of Section 7.2.

31. Let $\mathcal{E}$ and $\mathcal{M}$ be two classes of maps satisfying the conditions of Theorem 9.3.11. Suppose that either (a) all maps (in particular verdicts) factorise, or (b) all $\mathcal{M}$-maps are mono. Show that $\mathcal{E}$ is pullback-stable.

32. Formulate the cut-elimination step which expresses $\Pi\beta$ in the sequent calculus. (The right rule is essentially the same as $(\forall\mathcal{I})$, and the left rule was given in Definition 9.4.1.)

33. Show how to extend the universal quantifier to $\forall x.\ \Phi$, where Φ is a list of propositions dependent on $x : X$. [Hint: adapt Lemma 9.4.13.]

34. Deduce from Exercise 7.36 that $\Delta(-)^X$ preserves monos, so the class of monos is closed under universal quantification in the sense needed for the remarks following Proposition 9.4.6.

35. Show that the free algebra FG on a set G for a finitary free theory is the partial product $\Pi x.G$ of G along the map $X \longrightarrow F\mathbf{1}$ that is itself defined as the pullback of $\mathrm{tgt} : (<) \longrightarrow \mathbb{N}$ along the interpretation $[\![-]\!] : F\mathbf{1} \to \mathbb{N}$ defined in Exercise 6.3.

36. Formulate the canonical language for partial products and prove conservativity for dependent products as in Sections 7.6 and 7.7.

37. Prove Theorem 9.4.14, that dependent products can be reduced to partial products, type-theoretically.

38. Prove Lemma 9.4.10, that $\Delta X\Phi^X \cong (\Delta X) \times \Phi$ if $\Delta X \longrightarrow \Delta$ is mono, using type theory.

39. Let Φ be a *propositional* context and suppose that the partial products of Φ along the displays $\Delta X\psi \hookrightarrow \Delta X \longrightarrow \Delta$ exist. Prove, both type-theoretically and by diagram-chasing, that the partial product along $\Delta(\exists x.\ \psi) \hookrightarrow \Delta$ also exists. [Hint: $\forall z{:}(\exists x.\ \psi).\ \Phi \cong \forall x.\ \forall y{:}\psi.\ \Phi$.]

40. Deduce that any local homeomorphism is exponentiable in $\mathcal{S}p$.

41. Find a Scott-continuous function between dcpos which satisfies the conditions of Example 9.4.9(d) for being exponentiable in $\mathcal{P}os$, but

which is not exponentiable in $\mathcal{D}cpo$. [Hint: the bilimit of the fibres over some ascending sequence fails to be the fibre over its directed join.]

42. How can Pasynkov's expression of spheres as partial products be used to calculate $\pi_1(S^1)$, the fundamental group of the circle?

43. Assuming proof-anonymity, explain in terms of Theorem 9.5.3 why the results of comprehension are monos.

44. Show how to calculate the comprehension $\{\Gamma \mid \Phi\}$ of a list of propositions.

45. Give the equational formulation in a 2-category or bicategory (*cf.* Exercise 7.30) of the notion of a fibration with lex fibres, and of the same with comprehension. In this sense show that tgt : $C^\vee \longrightarrow C$ freely adds comprehension to $P : C \longrightarrow S$, where the objects of $C^\vee$ are the vertical morphisms of C; type-theoretically, this inserts an arbitrary division in the propositional part of contexts. [Hint: for uniqueness of the meditator between the total categories, you need to know that

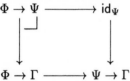

is a pullback in $C^\vee$, where $\Gamma = TP\Phi = TP\Psi$.] Although having comprehension is a *property* of the fibration, *viz.* that an adjoint exists, this construction is not idempotent; explain why this is. In Section 5.3 we wanted to use fibrations and comprehension to add partial maps; does the construction achieve this goal?

46. What does it mean for a fibration $P : C \longrightarrow S$ to have equality *predicates* corresponding to the *types* given by product diagonals in S (Section 8.3)? Assume that the fibres are all preorders. Construct the category of partial maps (*cf.* Propositions 5.3.5) and a fibration with comprehension whose objects are the virtual objects of the allegory (*cf.* footnote 1 on page 479).

47. Assuming that $P : C \longrightarrow S$ admits existential quantification, construct the allegory of relations (Proposition 5.8.7), and recover the comprehensive fibration using Freyd's tabular allegories [FS90, §2.166].

48. Let S be the category of types in Zermelo type theory generated by 1, $\mathbb{N}$, $\times$ and $\mathcal{P}$ (*cf.* Exercise 2.17), and all functions between them, so S is a full subcategory of Set which is closed under these operations but not under pullback. Carry out and contrast the constructions of the preceding three exercises for the predicate fibration $P : C \longrightarrow S$.

49. Discuss the commutation of comprehension and powerset with substitution (*cf.* Remark 9.1.11) and with the connectives and quantifiers

(*cf.* Exercise 2.17). Show that all proposition-symbols except ω can be eliminated, and that if types do not depend on proofs (Section 9.2) then the fibres $\mathcal{P}(\Gamma)$ are simply typed.

50. Describe the types $\omega[x] \times \omega[y]$ *etc.* more explicitly and show that $\wedge$, $\vee$ and $\Rightarrow$ make 2 into a Heyting lattice.

51. Reformulate the type-theoretic rules in Definition 9.5.5 so that, categorically, propositions over Γ are classified by spans $\Gamma \leftarrow\!\!\!\leftarrow \Delta \to \mathsf{Prop}$ (with two pullbacks) instead of just maps $\Gamma \to \mathsf{Prop}$. Relate this to Remark 6.6.5 and Example 9.3.10(e). Rework Sections 9.5 and 9.6 with this definition, which is the one used in [JM95].

52. Let $\top : 1 \to \Sigma$ in $\mathcal{S}$ be a support classifier (Definition 5.2.10). Suppose that Σ admits $\Rightarrow$, Σ^X exists for each $X \in \mathrm{ob}\,\mathcal{E}$, the Leibniz principle holds (Proposition 2.8.7) and the quantifiers $\exists : \Sigma^X \to \Sigma$ and $\forall : \Sigma^{\Sigma^X} \to \Sigma$ exist. Show that $\mathcal{S}$ is a topos and $\Sigma = 2$ [Tay98].

53. Show that the inverse image map $!^* : S \to S^X$, where $! : X \to 1$ in Sp, has a *continuous* right adjoint (all_X) iff X is compact.

54. What are the assumptions behind Example 9.5.10(a)?

55. Implement Proposition 6.1.11 in a functional programming language with list- and function-types, showing that any free theory has an equationally free model without assuming propositional extensionality (Remark 9.5.10). Assume that Prop is a semilattice with structure $\mathsf{and} : \mathsf{Prop} \to \mathsf{Prop} \to \mathsf{Prop}$ and $\mathsf{true} : \mathsf{Prop}$, and the type Ω of operation-symbols has an equality function $\mathsf{eq} : \Omega \to \Omega \to \mathsf{Prop}$.

56. Prove the second order representations of the propositional connectives and Leibniz' principle (Proposition 2.8.6ff), making use of proof-anonymity but not extensionality.

57. For any support classifier Σ (satisfying extensionality), show that $a \wedge f(a) = a \wedge f(\top)$ for $\Gamma \vdash a : \Sigma$ and $\Gamma, x : \Sigma \vdash f : \Sigma$.

[Hint: consider the subobject classified by $[a] \hookrightarrow \Gamma \underset{\langle \mathsf{id}, \top \rangle}{\overset{\langle \mathsf{id}, a \rangle}{\rightrightarrows}} \Gamma \times \Sigma \xrightarrow{f} \Sigma$.]

58. Show how to encode the class of monos of an internal category and hence its subobject classifier 2 (Remark 9.6.1). If $U \subset |X| = \mathcal{S}(1, X)$ then $U = |V|$ for some V, *i.e.* a subclass of a small set is a small set.

59. Characterise precisely the displays $X \longrightarrow\!\!\!\!\!\bullet\; \Gamma$ of algebraic lattices that are classified by Scott's type of types in Example 9.6.10(a).

60. Make Proposition 9.6.13 meaningful.

61. Let $Y[x]$ be a type-expression Zermelo type theory (Section 2.2), Show that there are a type Z in which x is not free, and injective functions $\nu_x : Y[x] \subset Z$.

62. Consider the category whose objects are discrete fibrations in
$\mathcal{P}os$ and whose morphisms are monotone functions. Adapt Section 6.3
and Remark 6.7.14 to coalgebras in this category for the "indexed" functor described in Remark 9.6.16. Show that such coalgebras are well
founded iff the underlying $\alpha \to \mathsf{sh}(\alpha)$ is well founded in $\mathcal{P}os$. Taking
as the notion of "mono" the prone maps in Exercise 9.9, show that
the solutions of the iteration equation (*i.e.* pullbacks) are "extensional".
Deduce that, in this generalisation, Example 7.1.6(g) *does* depend on
Replacement.

63. For each of the fragments of logic which we have considered
in this book (unary, algebraic, λ-calculus, *etc.*), formulate the corresponding replacement-scheme or reflection principle in Remark 9.6.17,
for example that for algebra is a certain partial product.

64. Investigate the analogue for these replacement-schemes of Corollary 7.7.2 in the gluing construction.

65. Identify the fallacy in this argument. Let $\Box_0$ be Zermelo type
theory. For each n, let $\Box_{n+1}$ be $\Box_n$ plus as much of the axiom-scheme
of replacement as is needed to justify the gluing construction that shows
that $\Box_{n+1} \vdash$ "$\Box_n$ is consistent". Now let $\Box_\omega = \bigcup_n \Box_n$. If $\Box_\omega \vdash \bot$ then
$\Box_n \vdash \bot$ for some n, but $\Box_\omega \vdash$ "$\Box_n$ is consistent", so $\Box_\omega$ proves its own
consistency, contradicting Gödel's theorem. But $\Box_\omega$ has a standard non-
trivial interpretation in Zermelo–Fraenkel set theory, which is therefore
inconsistent.

Bibliography

Further bibliographical information and access to those papers that are available on-line may be found *via* the World Wide Web at http://www.dcs.qmw.ac.uk/~pt/Practical_Foundations/

[AJ94] Samson Abramsky and Achim Jung. Domain theory. In Samson Abramsky et al., editors, *Handbook of Logic in Computer Science*, volume 3, pages 1–168. Oxford University Press, 1994.

[Acz88] Peter Aczel. *Non-well-founded Sets*. Number 14 in Lecture Notes. Center for the Study of Language and Information, Stanford University, 1988.

[AR94] Jiři Adámek and Jiři Rosický. *Locally Presentable and Accessible Categories*. Number 189 in London Mathematical Society Lecture Notes. Cambridge University Press, 1994.

[Age92] Pierre Ageron. The logic of structures. *Journal of Pure and Applied Algebra*, 79:15–34, 1992.

[AHS95] Thorsten Altenkirch, Martin Hofmann, and Thomas Streicher. Categorical reconstruction of a reduction-free normalisation proof. In Peter Johnstone, David Pitt, and David Rydeheard, editors, *Category Theory and Computer Science VI*, number 953 in Lecture Notes in Computer Science, pages 182–199. Springer-Verlag, 1995.

[AC98] Roberto Amadio and Pierre-Louis Curien. *Domains and Lambda-Calculi*. Number 46 in Cambridge Tracts in Theoretical Computer Science, Cambridge University Press, 1998.

[App92] Andrew Appel. *Compiling with Continuations*. Cambridge University Press, 1992.

[AGV64] Michael Artin, Alexander Grothendieck, and Jean-Louis Verdier, editors. *Séminaire de Géometrie Algébrique, IV: Théorie des Topos*, numbers 269–270 in Lecture Notes in Mathematics. Springer-Verlag, 1964. Second edition, 1972.

[Bae97] John Baez. An introduction to *n*-categories. In Eugenio Moggi and Giuseppe Rosolini, editors, *Category Theory and Computer Science VII*, number 1290 in Lecture Notes in Computer Science, pages 1–33. Springer-Verlag, 1997.

[Bar81] Henk Barendregt. *The Lambda Calculus: its Syntax and Semantics*. Number 103 in Studies in Logic and the Foundations of Mathematics. North-Holland, 1981. Second edition, 1984.

[Bar92] Henk Barendregt. Lambda calculi with types. In Samson Abramsky et al., editors, *Handbook of Logic in Computer Science*, volume 2, pages 117–309. Oxford University Press, 1992.

[BHPRR66] Yehoshua Bar-Hillel, E. I. J. Poznanski, M. O. Rabin, and Abraham Robinson, editors. *Essays of the Foundations of Mathematics*. Magnes Press, Hebrew University, 1966. Distributed by Oxford University Press.

[BB69] Michael Barr and Jon Beck. Homology and standard constructions. In Eckmann [Eck69], pages 245–335.

[BGvO71] Michael Barr, Pierre Grillet, and Donovan van Osdol, editors. *Exact Categories and Categories of Sheaves*. Number 236 in Lecture Notes in Mathematics. Springer-Verlag, 1971.

[Bar79] Michael Barr. **-Autonomous Categories*. Number 752 in Lecture Notes in Mathematics. Springer-Verlag, 1979.

[BW85] Michael Barr and Charles Wells. *Toposes, Triples, and Theories*. Number 278 in Grundlehren der mathematischen Wissenschaften. Springer-Verlag, 1985.

[BW90] Michael Barr and Charles Wells. *Category Theory for Computing Science*. International Series in Computer Science. Prentice-Hall, 1990. Second edition, 1995.

[Bar91] Michael Barr. *-Autonomous categories and linear logic. *Mathematical Structures in Computer Science*, 1:159–178, 1991.

[Bar77] Jon Barwise, editor. *Handbook of Mathematical Logic*. Number 90 in Studies in Logic and the Foundations of Mathematics. North-Holland, 1977.

[Bee80] Michael Beeson. *Foundations of Constructive Mathematics: Metamathematical Studies*. Number 6 in Ergebnisse der Mathematik und ihrer Grenzgebiete. Springer-Verlag, 1980. Second edition, 1985.

[Bel88] John Lane Bell. *Toposes and Local Set Theories: an Introduction*. Number 14 in Logic Guides. Oxford University Press, 1988.

[Bén85] Jean Bénabou. Fibred categories and the foundations of naïve category theory. *Journal of Symbolic Logic*, 50:10–37, 1985.

[BP64] Paul Benacerraf and Hilary Putnam, editors. *Philosophy of Mathematics: Selected Readings*. Prentice-Hall, 1964. Second edition, Cambridge University Press, 1983.

[Ben64] Paul Benacerraf. What numbers could not be. In Benacerraf and Putnam [BP64], pages 272–294. Second edition, Cambridge University Press, 1983.

[Ber35] Paul Bernays. Sur le platonisme dans les mathématiques. *Enseignement Mathématique*, 34:52–69, 1935. English translation, "Platonism in Mathematics" in [BP64], pages 258–271.

[Bib82] Wolfgang Bibel. *Automated Theorem Proving*. Friedrich Vieweg & Sohn, Braunschweig, 1982. Second edition, 1987.

[BML41] Garrett Birkhoff and Saunders Mac Lane. *A Survey of Modern Algebra*. MacMillan, New York, 1941. Fourth edition, 1977.

[Bir76] Garrett Birkhoff. The rise of modern algebra. In Jan Dalton Tarwater, John White, and John Miller, editors, *Men and*

Institutions in American Mathematics, pages 41–86. Texas Technical University, 1976.

[BB85] Errett Bishop and Douglas Bridges. *Constructive Analysis*. Number 279 in Grundlehren der mathematischen Wissenschaften. Springer-Verlag, 1985.

[Bla33] Max Black. *The Nature of Mathematics, a Critical Survey*. International Library of Psychology. Kegan Paul, 1933.

[Boc51] Josef Bochenski. *Ancient Formal Logic*. Studies in Logic and the Foundations of Mathematics. North-Holland, 1951.

[Boe52] Philotheus Boehner. *Medieval Logic: an Outline of its Development from 1250 to c.1400*. Manchester University Press, 1952.

[Boe58] Philotheus Boehner. *Collected articles on Ockham*. Franciscan Institute, 1958. Edited by Elio Marie Buytaert.

[Bol51] Bernard Bolzano. *Paradoxien des Undlichen*. 1851. English translation, "Paradoxes of the Infinite" by Fr. Prihonsky, published by Routledge, 1950.

[BJ74] George Boolos and Richard Jeffrey. *Computability and Logic*. Cambridge University Press, 1974. Third edition, 1989.

[Boo93] George Boolos. *The Logic of Provability*. Cambridge University Press, 1993.

[Boo98] George Boolos. *Logic, Logic and Logic*. Harvard University Press, 1998. Edited by Richard Jeffrey.

[Bor94] Francis Borceux. *Handbook of Categorical Algebra*. Number 50 in Encyclopedia of Mathematics and its Applications. Cambridge University Press, 1994. Three volumes.

[Bou57] Nicolas Bourbaki. *Eléments de Mathématique XXII: Théories des Ensembles, Livre I, Structures*. Number 1258 in Actualités scientifiques et industrielles. Hermann, 1957. English translation, "Theory of Sets", 1968.

[Boy68] Carl Boyer. *A History of Mathematics*. Wiley, 1968. Revised edition by Uta Merzbach, Wiley, 1989.

[BM88] Robert Boyer and J. Strother Moore. *A Computational Logic Handbook*. Number 23 in Perspectives in Computing. Academic Press, 1988.

[Bro75] Jan Brouwer. *Collected Works: Philosophy and Foundations of Mathematics*, volume 1. North-Holland, 1975. Edited by Arend Heyting.

[Bro81] Jan Brouwer. *Brouwer's Cambridge Lectures on Intuitionism*. Cambridge University Press, 1981. Edited by Dirk van Dalen.

[Bro76] Felix Browder, editor. *Mathematical Developments Arising from Hilbert Problems*, number 28 in Proceedings of Symposia in Pure Mathematics. American Mathematical Society, 1976.

[Bro87] Ronald Brown. From groups to groupoids: a brief survey. *Bulletin of the London Mathematical Society*, 19:113–134, 1987.

[Bro88] Ronald Brown. *Topology: a Geometric Account of General Topology, Homotopy Types and the Fundamental Groupoid*.

Mathematics and its Applications. Ellis Horwood, 1988. First
edition "Elements of modern topology", 1968.

[BS81] Stanley Burris and H. P. Sankappanavar. *A Course in Universal Algebra*. Number 78 in Graduate Texts in Mathematics. Springer-Verlag, 1981.

[Bur81] Albert Burroni. Algèbres graphiques. *Cahiers de Topologie et Géométrie Différentielle*, XXII, 1981.

[Caj93] Florian Cajori. *A History of Mathematics*. MacMillan, 1893. Fifth edition, Chelsea, N.Y., 1991.

[Caj28] Florian Cajori. *A History of Mathematical Notations*. Open Court, 1928. Reprinted by Dover, 1993.

[Cam98] Peter Cameron. *Introduction to Algebra*. Oxford University Press, 1998.

[Can15] Georg Cantor. *Contributions to the Founding of the Theory of Transfinite Numbers*. Open Court, 1915. Translated and edited by Philip Jourdain; reprinted by Dover, 1955.

[Can32] Georg Cantor. *Gesammelte Abhandlungen mathematischen und philosophischen Inhalts*. Springer-Verlag, 1932. Edited by Ernst Zermelo; reprinted by Olms, Hildesheim, 1962.

[CPR91] Aurelio Carboni, Maria-Cristina Peddicchio, and Giuseppe Rosolini, editors. *Proceedings of the 1990 Como Category Theory Conference*, number 1488 in Lecture Notes in Mathematics. Springer-Verlag, 1991.

[CLW93] Aurelio Carboni, Steve Lack, and Robert Walters. Introduction to extensive and distributive categories. *Journal of Pure and Applied Algebra*, 84:145–158, 1993.

[Car34] Rudolf Carnap. *Logische Syntax der Sprache*. Vienna, 1934. English translation by Amethe Smeaton, "The Logical Syntax of Language", Kegan Paul, 1937.

[CE56] Henri Cartan and Sammy Eilenberg. *Homological Algebra*. Princeton University Press, 1956.

[Car86] John Cartmell. Generalised algebraic theories and contextual categories. *Annals of Pure and Applied Logic*, 32:209–243, 1986.

[CK73] Chen Chung Chang and Jerome Keisler. *Model Theory*. Number 73 in Studies in Logic and the Foundations of Mathematics. North-Holland, 1973. Third edition, 1990.

[CR92] Jon Chapman and Frederick Rowbottom. *Relative Category Theory and Geometric Morphisms: a Logical Approach*. Number 16 in Logic Guides. Oxford University Press, 1992.

[Chu56] Alonso Church. *Introduction to Mathematical Logic*. Princeton University Press, 1956.

[Coc93] Robin Cockett. Introduction to distributive categories. *Mathematical Structures in Computer Science*, 3:277–307, 1993.

[CCS98] A. M. Cohen, H. Cuypers, and H. Sterk, editors. *Some Tapas of Computer Algebra*, number 4 in Algorithms and Computation in Mathematics. Springer-Verlag, 1998.

[Coh66] Paul Cohen. *Set Theory and the Continuum Hypothesis.* W.A. Benjamin, 1966.

[Coh77] Paul Cohn. *Algebra*, volume 2. Wiley, 1977.

[Coh81] Paul Cohn. *Universal Algebra.* Number 6 in Mathematics and its Applications. Reidel, 1981. Originally published by Harper and Row, 1965.

[Con71] John Horton Conway. *Regular Algebra and Finite Machines.* Chapman and Hall, 1971.

[Con76] John Horton Conway. *On Numbers and Games.* Number 6 in London Mathematical Society Monographs. Academic Press, 1976.

[CH88] Thierry Coquand and Gérard Huet. The calculus of constructions. *Information and Computation*, 76:95–120, 1988.

[Coq90a] Thierry Coquand. Metamathematical investigations of a calculus of constructions. In Odifreddi [Odi90], pages 91–122.

[Coq90b] Thierry Coquand. On the analogy between propositions and types. In Gérard Huet, editor, *Logical Foundations of Functional Programming*, pages 399–418. Addison-Wesley, 1990.

[Coq97] Thierry Coquand. Computational content of classical logic. In Pitts and Dybjer [PD97], pages 33–78.

[Cos79] Michel Coste. Localisation, spectra and sheaf representation. In Fourman et al. [FMS79], pages 212–238.

[Cou05] Louis Couturat. *Les Principes des Mathématiques, avec un Appendice sur le Philosophie de Kant.* 1905.

[CP92] Roy Crole and Andrew Pitts. New foundations for fixpoint computations: FIX-hyperdoctrines and the FIX-logic. *Information and Computation*, 98:171–210, 1992.

[Cro93] Roy Crole. *Categories for Types.* Cambridge Mathematical Textbooks. Cambridge University Press, 1993.

[CDS98] Djordje Čubrić, Peter Dybjer, and Philip Scott. Normalisation and the Yoneda embedding. *Mathematical Structures in Computer Science*, 8:153–192, 1998.

[Cur86] Pierre-Louis Curien. *Categorical Combinators, Sequential Algorithms, and Functional Programming.* Pitman, 1986. Second edition, Birkhäuser, Progress in Theoretical Computer Science, 1993.

[CF58] Haskell Curry and Robert Feys. *Combinatory Logic I.* Studies in Logic and the Foundations of Mathematics. North-Holland, 1958. Volume II, with Jonathan Seldin, 1972.

[Cur63] Haskell Curry. *Foundations of Mathematical Logic.* McGraw–Hill, 1963. Republished by Dover, 1977.

[CSH80] Haskell Curry, Jonathan Seldin, and Roger Hindley, editors. *To H.B. Curry: Essays on Combinatory Logic, Lambda Calculus and Formalism.* Academic Press, 1980.

[Dau79] Joseph Warren Dauben. *Georg Cantor: his Mathematics and Philosophy of the Infinite.* Harvard University Press, 1979.

[DST88] James Davenport, Y. Siret, and E. Tournier. *Computer Algebra: Systems and Algorithms for Algebraic Computation*. Academic Press, 1988. Translated from French; third edition 1993.

[DP90] B. A. Davey and Hilary Priestley. *Introduction to Lattices and Order*. Cambridge University Press, 1990.

[Dav65] Martin Davis. *The Undecidable. Basic Papers on Undecidable, Unsolvable Problems and Computable Functions*. Raven Press, Hewlett, N.Y., 1965.

[dB80] Nikolas de Bruijn. A survey of the project Automath. In Curry et al. [CSH80], pages 579–606.

[Ded72] J. W. Richard Dedekind. *Stetigkeit und irrationale Zahlen*. Braunschweig, 1872. Reprinted in [Ded32], pages 315–334; English translation, "Continuity and Irrational Numbers" in [Ded01].

[Ded88] J. W. Richard Dedekind. *Was sind und was sollen die Zahlen?* Braunschweig, 1888. Reprinted in [Ded32], pages 335–391; English translation, "The Nature and Meaning of Numbers" in [Ded01].

[Ded01] J. W. Richard Dedekind. *Essays on the theory of numbers*. Open Court, 1901. English translations by Wooster Woodruff Beman; republished by Dover, 1963.

[Ded32] J. W. Richard Dedekind. *Gesammelte mathematische Werke*, volume 3. Vieweg, Braunschweig, 1932. Edited by Robert Fricke, Emmy Noether and Øystein Ore; republished by Chelsea, New York, 1969.

[Det86] Michael Detlefsen. *Hilbert's Program: an Essay on Mathematical Instrumentalism*. Number 182 in Synthese Library. Reidel, 1986.

[Die77] Jean Alexandre Dieudonné. *Panorama des Mathématiques Pures: la Choix Bourbachique*. Gauthier-Villars, 1977. English translation, "A panorama of pure mathematics, as seen by N. Bourbaki" by I. G. Macdonald, Academic Press, Pure and applied mathematics, 97, 1982.

[Die88] Jean Alexandre Dieudonné. *A History of Algebraic and Differential Topology 1900–1960*. Birkhäuser, 1988.

[Dol72] Albrecht Dold. *Lectures on Algebraic Topology*. Number 200 in Grundlehren der mathematischen Wissenschaften. Springer-Verlag, 1972.

[Dum77] Michael Dummett. *Elements of Intuitionism*. Logic Guides. Oxford University Press, 1977.

[Dum78] Michael Dummett. *Truth and Other Enigmas*. Duckworth, London, 1978.

[DT87] Roy Dyckhoff and Walter Tholen. Exponentiable maps, partial products and pullback complements. *Journal of Pure and Applied Algebra*, 49:103–116, 1987.

[Eck69] Beno Eckmann, editor. *Seminar on Triples and Categorical Homology Theory*, number 80 in Lecture Notes in Mathematics. Springer-Verlag, 1969.

[Ehr84] Charles Ehresman. *Œuvres complètes et commentées*. Amiens, 1980–84. Edited by Andrée Charles Ehresmann; published as supplements to volumes 21–24 of *Cahiers de topologie et géométrie différentielle*.

[Ehr88] Thomas Ehrhardt. Categorical semantics of constructions. In Yuri Gurevich, editor, *Logic in Computer Science III*, pages 264–273. IEEE Computer Society Press, 1988.

[Ehr89] Thomas Ehrhardt. Dictoses. In Pitt et al. [PRD⁺89], pages 213–223.

[ES52] Sammy Eilenberg and Norman Steenrod. *Foundations of Algebraic Topology*. Princeton University Press, 1952.

[EHMLR66] Sammy Eilenberg, D. K. Harrison, Saunders Mac Lane, and Helmut Röhrl, editors. *Categorical Algebra (La Jolla, 1965)*. Springer-Verlag, 1966.

[EK66] Sammy Eilenberg and Max Kelly. Closed categories. In Eilenberg et al. [EHMLR66].

[EE70] Sammy Eilenberg and Calvin Elgot. *Recursiveness*. Academic Press, 1970.

[EML86] Sammy Eilenberg and Saunders Mac Lane. *Eilenberg–Mac Lane, Collected Works*. Academic Press, 1986.

[FG87] John Fauvel and Jeremy Gray. *The History of Mathematics, a Reader*. Macmillan and the Open University, 1987.

[Fen71] Jens Erik Fenstad, editor. *Second Scandinavian Logic Symposium*, number 63 in Studies in Logic and the Foundations of Mathematics. North-Holland, 1971.

[FF69] Richard Feys and Frederic Fitch. *Dictionary of Symbols of Mathematical Logic*. Studies in Logic and the Foundations of Mathematics. North-Holland, 1969.

[FJM⁺96] Marcelo Fiore, Achim Jung, Eugenio Moggi, Peter O'Hearn, Jon Riecke, Giuseppe Rosolini, and Ian Stark. Domains and denotational semantics: History, accomplishments and open problems. *Bulletin of the EATCS*, 59:227–256, 1996.

[Fit52] Frederic Benton Fitch. *Symbolic Logic: an Introduction*. Ronald Press, New York, 1952.

[Fit69] Melvin Fitting. *Intuitionistic Logic, Model Theory and Forcing*. Studies in Logic and the Foundations of Mathematics. North-Holland, 1969.

[Flo67] Robert Floyd. Assigning meaning to programs. In J. T. Schwartz, editor, *Mathematical Aspects of Computer Science*, number 19 in Proceedings of Symposia in Applied Mathematics, pages 19–32. American Mathematical Society, 1967.

[FMS79] Michael Fourman, Chris Mulvey, and Dana Scott, editors. *Applications of Sheaves*, number 753 in Lecture Notes in Mathematics. Springer-Verlag, 1979.

[FJP92] Michael Fourman, Peter Johnstone, and Andrew Pitts, editors. *Applications of categories in computer science*, number 177 in

London Mathematical Society Lecture Notes. Cambridge University Press, 1992.

[Fow87] David Fowler. *The Mathematics of Plato's Academy: a New Reconstruction.* Oxford University Press, 1987.

[FBH58] Abraham Fraenkel and Yehoshua Bar-Hillel. *Foundations of Set Theory.* Studies in Logic and the Foundations of Mathematics. North-Holland, 1958.

[Fre60] Gottlob Frege. *Translations from the Philosophical Writings of Gottlob Frege.* Blackwell, 1960. Edited by Peter Geach and Max Black; third edition, 1980.

[Fre84] Gottlob Frege. *Collected Papers on Mathematics, Logic and Philosophy.* Blackwell, 1984. Edited by Brian McGinness.

[Fre64] Peter Freyd. *Abelian Categories: an Introduction to the Theory of Functors.* Harper and Row, 1964.

[Fre66] Peter Freyd. The theory of functors and models. In John Addison, Leon Henkin, and Alfred Tarski, editors, *Theory of Models*, Studies in Logic and the Foundations of Mathematics, pages 107–120. North-Holland, 1966.

[Fre72] Peter Freyd. Aspects of topoi. *Bulletin of the Australian Mathematical Society*, 7:1–76 and 467–480, 1972.

[FK72] Peter Freyd and Max Kelly. Categories of continuous functors, I. *Journal of Pure and Applied Algebra*, 2:169–191, 1972.

[FS90] Peter Freyd and Andre Scedrov. *Categories, Allegories.* Number 39 in Mathematical Library. North-Holland, 1990.

[Fre91] Peter Freyd. Algebraically complete categories. In Carboni et al. [CPR91], pages 95–104.

[GU71] Peter Gabriel and Fritz Ulmer. *Lokal präsentierbare Kategorien.* Number 221 in Lecture Notes in Mathematics. Springer-Verlag, 1971.

[Gal38] Galileo Galilei. *Two New Sciences.* 1638. Translated by Stillman Drake, University of Wisconsin Press, 1974; Re-published by Wall & Thompson, 1989.

[Gal86] Jean Gallier. *Logic for Computer Science: Foundations of Automated Theorem Proving.* Computer Science and Technology Series. Harper and Row, 1986. Republished by Wiley, 1987.

[Gan56] Robin Gandy. On the axiom of extensionality. *Journal of Symbolic Logic*, 21:36–48 and 24:287–300, 1956.

[Gen35] Gerhard Gentzen. Untersuchungen über das Logische Schliessen. *Mathematische Zeitschrift*, 39:176–210 and 405–431, 1935. English translation in [Gen69], pages 68–131.

[Gen69] Gerhard Gentzen. *The Collected Papers of Gerhard Gentzen.* Studies in Logic and the Foundations of Mathematics. North-Holland, 1969. Edited by M. E. Szabo.

[GHK+80] Gerhard Gierz, Karl Heinrich Hoffmann, Klaus Keimel, Jimmie Lawson, Michael Mislove, and Dana Scott. *A Compendium of Continuous Lattices.* Springer-Verlag, 1980.

[Gil82] Donald Gillies. *Frege, Dedekind and Peano on the Foundations of Arithmetic*. Number 2 in Methodology and Science Foundation. Van Gorcum, 1982.

[Gir71] Jean-Yves Girard. Une extension de l'interpretation de Gödel à l'analyse, et son application à l'élimination des coupures dans l'analyse et la théorie des types. In Fenstad [Fen71], pages 63–92.

[Gir87a] Jean-Yves Girard. Linear logic. *Theoretical Computer Science*, 50:1–102, 1987.

[Gir87b] Jean-Yves Girard. *Proof Theory and Logical Complexity*, volume 1. Bibliopolis, 1987.

[GLT89] Jean-Yves Girard, Yves Lafont, and Paul Taylor. *Proofs and Types*. Number 7 in Cambridge Tracts in Theoretical Computer Science. Cambridge University Press, 1989.

[Gir71] Jean Giraud. *Cohomologie non-abélienne*. Number 179 in Grundlehren der mathematischen Wissenschaften. Springer-Verlag, 1971.

[Gir72] Jean Giraud. Classifying topos. In Lawvere [Law72], pages 43–56.

[Göd31] Kurt Gödel. Über formal unentscheidbare Sätze der Principia Mathematica und verwandter Systeme I. *Monatshefte für Mathematik und Physik*, 38:173–198, 1931. English translations, "On Formally Undecidable Propositions of 'Principia Mathematica' and Related Systems" published by Oliver and Boyd, 1962 and Dover, 1992; also in [vH67], pages 596–616 and [Dav65], pages 5–38.

[Göd80] Kurt Gödel. *Kurt Gödel: Collected Works*. Oxford University Press, 1980. Edited by Solomon Feferman and others.

[God58] Roger Godement. *Topologie Algebrique et Theorie des Faisceaux*. Hermann, 1958.

[Gol79] Robert Goldblatt. *Topoi: The Categorial Analysis of Logic*. Number 98 in Studies in Logic and the Foundations of Mathematics. North-Holland, 1979. Third edition, 1983.

[GG80] Ivor Grattan-Guinness, editor. *From the Calculus to Set Theory, 1630–1910: an Introductory History*. Duckworth, London, 1980.

[GG97] Ivor Grattan-Guinness. *The Fontana History of the Mathematical Sciences: the Rainbow of Mathematics*. Fontana, 1997.

[Grä68] George Grätzer. *Universal Algebra*. Van Nostrand, 1968.

[Gra66] John Gray. Fibred and cofibred categories. In Eilenberg et al. [EHMLR66], pages 21–83.

[Gra79] John Gray. Fragments of the history of sheaf theory. In Fourman et al. [FMS79], pages 1–79.

[GS89] John Gray and Andre Scedrov, editors. *Categories in Computer Science and Logic*, number 92 in Contemporary Mathematics. American Mathematical Society, 1989.

[Gro64] Alexander Grothendieck, editor. *Séminaire de Géometrie Algébrique, I (1960/1)*, number 224 in Lecture Notes in Mathematics. Springer-Verlag, 1964.

[Gun92] Carl Gunter. *Semantics of Programming Languages: Structures and Techniques.* Foundations of Computing. MIT Press, 1992.

[Hal60] Pál Halmos. *Naïve Set Theory.* Van Nostrand, 1960. Reprinted by Springer-Verlag, Undergraduate Texts in Mathematics, 1974.

[Har40] G. H. Hardy. *A Mathematician's Apology.* Cambridge University Press, 1940. Reprinted 1992.

[Hat82] William Hatcher. *The Logical Foundations of Mathematics.* Foundations and Philosophy of Science and Technology. Pergamon Press, 1982.

[Hau14] Felix Hausdorff. *Mengenlehre.* 1914. English translation, "Set Theory", published by Chelsea, 1962.

[Hec93] André Heck. *Introduction to Maple.* Springer-Verlag, 1993. Second edition, 1996.

[Hei86] Gerhard Heinzmann, editor. *Poincaré, Russell, Zermelo et Peano: Textes de la Discussion (1906–1912) sur les Fondements des Mathématiques: des Antinomies à la Prédicativité.* Albert Blanchard, Paris, 1986.

[HM75] Matthew Hennessey and Robin Milner. Algebraic laws for non-determinism and concurrency. *Journal of the ACM*, 32:137–161, 1975.

[Hen88] Matthew Hennessy. *Algebraic Theory of Processes.* Foundations of Computing. MIT Press, 1988.

[Hen90] Matthew Hennessy. *The Semantics of Programming Languages: an Elementary Introduction using Structural Operational Semantics.* Wiley, 1990.

[Hey56] Arend Heyting. *Intuitionism, an Introduction.* Studies in Logic and the Foundations of Mathematics. North-Holland, 1956. Revised edition, 1966.

[HA28] David Hilbert and Wilhelm Ackermann. *Grundzüge der theoretischen Logik.* Springer-Verlag, 1928. Republished 1972; English translation by Lewis Hammond et al., "Principles of Mathematical Logic", Chelsea, New York, 1950.

[HB34] David Hilbert and Paul Bernays. *Grundlagen der Mathematik.* Number 40 in Grunlagen der Mathematischen Wissenschaften. Springer-Verlag, 1934.

[Hil35] David Hilbert. *Gesammelte Abhandlungen*, volume 3. Springer-Verlag, 1935. Reprinted, 1970.

[HS71] Peter Hilton and Urs Stammbach. *A Course in Homological Algebra.* Number 4 in Graduate Texts in Mathematics. Springer-Verlag, 1971. Second edition, 1997.

[HS86] Roger Hindley and Jonathan Seldin. *Introduction or Combinators and Lambda Calculus.* Number 1 in London Mathematical Society Student Texts. Cambridge University Press, 1986.

[Hoa69] Tony Hoare. An axiomatic basis for computer programming. *Communications of the ACM*, 12:576–580 and 583, 1969.

[Hod93] Wilfrid Hodges. *Model Theory*. Number 42 in Encyclopedia of Mathematics and its Applications. Cambridge University Press, 1993.

[Hof95] Martin Hofmann. On the interpretation of type theory in locally cartesian closed categories. In Leszek Pacholski and Jerzy Tiuryn, editors, *Computer Science Logic VIII*, number 933 in Lecture Notes in Computer Science, pages 427–441. Springer-Verlag, 1995.

[Hof79] Dougals Hofstadter. *Gödel, Escher, Bach, and Eternal Golden Braid*. Harvester, 1979. Reprinted by Penguin, 1980.

[How80] William Howard. The formulae-as-types notion of construction. In Curry et al. [CSH80], pages 479–490.

[Hue73] Gérard Huet. The undecidability of unification in third order logic. *Information and Control*, 22(3):257–267, April 1973.

[Hue75] Gérard Huet. A unification algorithm for typed lambda calculus. *Theoretical Computer Science*, 1:27–57, 1975.

[HJP80] Martin Hyland, Peter Johnstone, and Andrew Pitts. Tripos theory. *Mathematical Proceedings of the Cambridge Philosophical Society*, 88:205–232, 1980.

[Hyl81] Martin Hyland. Function spaces in the category of locales. In Bernhard Banachewski and Rudolf-Eberhard Hoffman, editors, *Continuous Lattices*, number 871 in Lecture Notes in Mathematics, pages 264–281. Springer-Verlag, 1981.

[Hyl82] Martin Hyland. The effective topos. In Troelstra and van Dalen [TvD82], pages 165–216.

[Hyl88] Martin Hyland. A small complete category. *Annals of Pure and Applied Logic*, 40:135–165, 1988.

[HP89] Martin Hyland and Andrew Pitts. The theory of constructions: Categorical semantics and topos-theoretic models. In Gray and Scedrov [GS89], pages 137–199.

[HRR90] Martin Hyland, Edmund Robinson, and Giuseppe Rosolini. The discrete objects in the effective topos. *Proceedings of the London Mathematical Society*, 60:1–36, 1990.

[Jac90] Bart Jacobs. *Categorical Type Theory*. PhD thesis, Universiteit Nijmegen, 1990.

[JMS91] Bart Jacobs, Eugenio Moggi, and Thomas Streicher. Relating models of impredicative type theories. In Pitt et al. [PCA⁺91], pages 197–218.

[Jac93] Bart Jacobs. Comprehension categories and the semantics of type dependency. *Theoretical Computer Science*, 107(2):169–207, 1993.

[Jás34] Stanisław Jáskowski. On the rules of suppositions in formal logic. *Studia Logica*, 1, 1934. Reprinted in [McC67], pages 232–258.

[Jec78] Thomas Jech. *Set Theory*. Number 79 in Pure and Applied Mathematics. Academic Press, 1978. Second edition, 1997.

[Joh77] Peter Johnstone. *Topos Theory*. Number 10 in London Mathematical Society Monographs. Academic Press, 1977.

[JPR⁺78] Peter Johnstone, Robert Paré, Robert Roseburgh, Steve Schumacher, Richard Wood, and Gavin Wraith. *Indexed Categories and their Applications.* Number 661 in Lecture Notes in Mathematics. Springer-Verlag, 1978.

[Joh82] Peter Johnstone. *Stone Spaces.* Number 3 in Cambridge Studies in Advanced Mathematics. Cambridge University Press, 1982.

[Joh85] Peter Johnstone. When is a variety a topos? *Algebra Universalis,* 21:198–212, 1985.

[Joh90] Peter Johnstone. Collapsed toposes and cartesian closed varieties. *Journal of Algebra,* 129:446–480, 1990.

[JT84] André Joyal and Myles Tierney. An extension of the Galois theory of Grothendieck. *Memoirs of the American Mathematical Society,* 51(309), 1984.

[Joy87] André Joyal. Foncteurs analytiques et espèces de structures. In Gilbert Labelle and Pierre Leroux, editors, *Combinatoire énumerative,* number 1234 in Lecture Notes in Mathematics, pages 126–159. Springer-Verlag, 1987.

[JM95] André Joyal and Ieke Moerdijk. *Algebraic Set Theory.* Number 220 in London Mathematical Society Lecture Notes. Cambridge University Press, 1995.

[Jun90] Achim Jung. Cartesian closed categories of algebraic CPO's. *Theoretical Computer Science,* 70:233–250, 1990.

[KR91] Hans Kamp and Uwe Reyle. *From Discourse to Logic: Introduction to Model-theoretic Semantics of Natural Language, Formal Logic and Discourse Representation Theory.* Reidel, 1991. Re-published by Kluwer, Studies in Linguistics and Philosophy, 42, 1993.

[Kan58] Daniel Kan. Adjoint functors. *Transactions of the American Mathematical Society,* 87:294–329, 1958.

[Kel55] John Kelley. *General Topology.* Van Nostrand, 1955. Reprinted by Springer-Verlag, Graduate Texts in Mathematics, 27, 1975.

[Kel69] Max Kelly. Monomorphisms, epimorphisms and pull-backs. *Journal of the Australian Mathematical Society,* 9:124–142, 1969.

[Kel74] Max Kelly, editor. *Proceedings of the Sydney Category Theory Seminar 1972-3.* Number 420 in Lecture Notes in Mathematics. Springer-Verlag, 1974.

[Kel82] Max Kelly. *Basic Concepts of Enriched Category Theory.* Number 64 in London Mathematical Society Lecture Notes. Cambridge University Press, 1982.

[Kle52] Stephen Kleene. *Introduction to Metamathematics.* Number 1 in Bibliotheca mathematica. North-Holland, 1952. Revised edition, Wolters-Noordhoff, 1971.

[KV65] Stephen Kleene and Richard Vesley. *The Foundations of Intuitionistic Mathematics, Especially in relation to Recursive Functions.* North-Holland, 1965.

[Kle67] Stephen Kleene. *Mathematical Logic.* John Wiley and Sons, 1967.

[Knu68] Donald Knuth. *The Art of Computer Programming.* Addison-Wesley, 1968. Three volumes published out of seven planned; second edition, 1973.

[KB70] Donald Knuth and Peter Bendix. Simple word problems in universal algebra. In John Leech, editor, *Computational Problems in Abstract Algebra*, pages 263–297. Pergamon Press, 1970.

[Knu74] Donald Knuth. *Surreal Numbers*. Addison-Wesley, 1974.

[Koc81] Anders Kock. *Synthetic Differential Geometry*. Number 51 in London Mathematical Society Lecture Notes. Cambridge University Press, 1981.

[Koc95] Anders Kock. Monads for which structures are adjoint to units. *Journal of Pure and Applied Algebra*, 104:41–59, 1995.

[Kol25] Andrei Kolmogorov. On the principle of excluded middle. *Matematičeskii Sbornik*, 32:646–667, 1925. In Russian; English translation in [vH67], pages 414–437.

[Koy82] C. P. J. Koymans. Models of the lambda calculus. *Information and Control*, 52:206–332, 1982.

[Kre58] Georg Kreisel. Mathematical significance of consistency proofs. *Journal of Symbolic Logic*, 23:155–182, 1958.

[Kre67] Georg Kreisel. Informal rigour and completeness proofs. In Imre Lakatos, editor, *Problems in the Philosophy of Mathematics*. North-Holland, 1967.

[Kre68] Georg Kreisel. A survey of proof theory. *Journal of Symbolic Logic*, 33:321–388, 1968.

[Kre71] Georg Kreisel. A survey of proof theory II. In Fenstad [Fen71], pages 109–170.

[KKP82] Norman Kretzmann, Anthony Kenny, and Jan Pinborg, editors. *The Cambridge history of later medieval philosophy: from the rediscovery of Aristotle to the disintegration of scholasticism, 1100-1600*. Cambridge University Press, 1982.

[KM66] Kazimierz Kuratowski and Andrzej Mostowski. *Teoria mnogosci*. Polish Scientific Publishers, 1966. English translation, "Set Theory" by M. Maczynski, North-Holland, Studies in Logic and the Foundations of Mathematics, number 86, 1968; second edition, 1976.

[Laf87] Yves Lafont. *Logiques, Catégories et Machines*. PhD thesis, Université de Paris 7, 1987.

[LS91] Yves Lafont and Thomas Streicher. Game semantics for linear logic. In *Logic in Computer Science VI*, pages 43–50. IEEE Computer Society Press, 1991.

[Lai83] Christian Lair. Diagrammes localement libres, extensions de corps et théorie de Galois. *Diagrammes*, 10, 1983.

[Lak63] Imre Lakatos. Proofs and refutations: the logic of mathematical discovery. *British Journal for the Philosophy of Science*, 14:1–25, 1963. Edited by John Worrall and Elie Zahar, Cambridge University Press, 1976.

[Lak86] George Lakoff. *Women, Fire, and Dangerous Things: What Categories Reveal about the Mind*. University of Chicago Press, 1986.

[Lam58] Joachim Lambek. The mathematics of sentence structure. *American Mathematical Monthly*, 65:154–170, 1958.

[Lam68] Joachim Lambek. Deductive systems and categories I: Syntactic calculus and residuated categories. *Mathematical Systems Theory*, 2:287–318, 1968.

[Lam69] Joachim Lambek. Deductive systems and categories II: Standard constructions and closed categories. In Peter Hilton, editor, *Category Theory, Homology Theory and their Applications*, number 86 in Lecture Notes in Mathematics, pages 76–122. Springer-Verlag, 1969.

[Lam72] Joachim Lambek. Deductive systems and categories III: Cartesian closed categories, intuitionist propositional calculus, and combinatory logic. In Lawvere [Law72], pages 57–82.

[LS80] Joachim Lambek and Philip J. Scott. Intuitionist type theory and the free topos. *Journal of Pure and Applied Algebra*, 19:215–257, 1980.

[LS86] Joachim Lambek and Philip Scott. *Introduction to Higher Order Categorical Logic*. Number 7 in Cambridge Studies in Advanced Mathematics. Cambridge University Press, 1986.

[Lam89] Joachim Lambek. Multicategories revisited. In Gray and Scedrov [GS89], pages 217–240.

[Lan70] Serge Lang. *Introduction to linear algebra*. Addison-Wesley, 1970. Third edition, Springer-Verlag, Undergraduate Texts in Mathematics, 1987.

[Law63] Bill Lawvere. Functorial semantics of algebraic theories. *Proceedings of the National Academy of Sciences of the United States of America*, 50(1):869–872, 1963.

[Law64] Bill Lawvere. An elementary theory of the category of sets. *Proceedings of the National Academy of Sciences of the United States of America*, 52:1506–1511, 1964.

[Law66] William Lawvere. The category of categories as a foundation for mathematics. In Eilenberg et al. [EHMLR66], pages 1–20.

[Law68a] Bill Lawvere. Diagonal arguments and cartesian closed categories. In Peter Hilton, editor, *Category Theory, Homology Theory and their Applications II*, number 92 in Lecture Notes in Mathematics, pages 134–145. Springer-Verlag, 1968.

[Law68b] Bill Lawvere. Some algebraic problems in the context of functorial semantics of algebraic structures. In Saunders Mac Lane, editor, *Reports of the Midwest Category Seminar II*, number 61 in Lecture Notes in Mathematics, pages 41–61. Springer-Verlag, 1968.

[Law69] Bill Lawvere. Adjointness in foundations. *Dialectica*, 23:281–296, 1969.

[Law70] Bill Lawvere. Equality in hyperdoctrines and the comprehension schema as an adjoint functor. In Alex Heller, editor, *Applications of*

Categorical Algebra, number 17 in Proceedings of Symposia in Pure Mathematics, pages 1–14. American Mathematical Society, 1970.

[Law71] Bill Lawvere. Quantifiers and sheaves. In *Actes du Congrès International des Mathématiciens*, volume 1, pages 329–334. Gauthier-Villars, 1971.

[Law72] Bill Lawvere, editor. *Toposes, Algebraic Geometry, and Logic*, number 274 in Lecture Notes in Mathematics. Springer-Verlag, 1972.

[Law73] Bill Lawvere. Metric spaces, generalised logic, and closed categories. In *Rendiconti del Seminario Matematico e Fisico di Milano*, volume 43. Tipografia Fusi, Pavia, 1973.

[LMW75] Bill Lawvere, Christian Maurer, and Gavin Wraith, editors. *Model Theory and Topoi*, number 445 in Lecture Notes in Mathematics. Springer-Verlag, 1975.

[LS97] Bill Lawvere and Stephen Schanuel. *Conceptual Mathematics: a First Introduction to Categories*. Cambridge University Press, 1997.

[LS81] Daniel Lehmann and Michael Smyth. Algebraic specifications of data types: a synthetic approach. *Mathematical Systems Theory*, 14:97–139, 1981.

[Lei90] Daniel Leivant. Contracting proofs to programs. In Odifreddi [Odi90], pages 279–327.

[Lew18] Clarence Lewis. *A Survey of Symbolic Logic*. University of California Press, 1918. Republished by Dover, 1960.

[Lin71] Carl Linderholm. *Mathematics made Difficult*. Wolfe, London, 1971.

[Lin69] Fred Linton. An outline of functorial semantics. In Eckmann [Eck69], pages 7–52.

[Luk51] Jan Łukasiewicz. *Aristotle's Syllogistic from the Standpoint of Modern Formal Logic*. Oxford University Press, 1951. Second edition, 1963.

[Luk63] Jan Łukasiewicz. *Elements of Mathematical Logic*. Pergamon Press, 1963. Translated from Polish by Olgierd Wojtasiewicz.

[Luk70] Jan Łukasiewicz. *Selected Works*. Studies in Logic and the Foundations of Mathematics. North-Holland, 1970. Edited by Ludwig Berkowski.

[Luo92] Zhaohui Luo. A unifying theory of dependent types: the schematic approach. In Anil Nerode and Mikhail Taitslin, editors, *Logical Foundations of Computer Science (Logic at Tver '92)*, number 620 in Lecture Notes in Computer Science, pages 293–304. Springer-Verlag, 1992.

[MW91] Malcolm MacCallum and Francis Wright. *Algebraic Computing with REDUCE: lecture notes from the First Brazilian School on Computer Algebra*. Oxford University Press, 1991.

[MLB67] Saunders Mac Lane and Garrett Birkhoff. *Algebra*. MacMillan, New York, 1967. Second edition, 1979.

[ML71] Saunders Mac Lane. *Categories for the Working Mathematician*. Number 5 in Graduate Texts in Mathematics. Springer-Verlag, 1971.

[ML81] Saunders Mac Lane. Mathematical models: a sketch for the philosophy of mathematics. *American Mathematical Monthly*, 88:462–472, 1981.

[ML79] Saunders Mac Lane. *Selected Papers*. Springer-Verlag, 1979. Edited by Irving Kaplansky.

[ML86] Saunders Mac Lane. *Mathematics, Form and Function*. Springer-Verlag, 1986.

[ML88] Saunders Mac Lane. Categories and concepts in perspective. In Peter Duren, Richard Askey, and Uta Merzbach, editors, *A Century of Mathematics in America*, volume 1, pages 323–365. American Mathematical Society, 1988. Addendum in volume 3, pages 439–441.

[MLM92] Saunders Mac Lane and Ieke Moerdijk. *Sheaves in Geometry and Logic: a First Introduction to Topos Theory*. Universitext. Springer-Verlag, 1992.

[MR77] Michael Makkai and Gonzalo Reyes. *First Order Categorical Logic: Model-Theoretical Methods in the Theory of Topoi and Related Categories*. Number 611 in Lecture Notes in Mathematics. Springer-Verlag, 1977.

[Mak87] Michael Makkai. Stone duality for first order logic. *Advances in Mathematics*, 65:97–170, 1987.

[MP87] Michael Makkai and Andrew Pitts. Some results on locally finitely presentable categories. *Transactions of the American Mathematical Society*, 299:473–496, 1987.

[MP90] Michael Makkai and Robert Paré. *Accessible Categories: the Foundations of Categorical Model Theory*. Number 104 in Contemporary Mathematics. American Mathematical Society, 1990.

[Mak93] Michael Makkai. The fibrational formulation of intuitionistic predicate logic. *Notre Dame Journal of Formal Logic*, 34:334–7 and 471–498, 1993.

[Mak96] Michael Makkai. Avoiding the axiom of choice in category theory. *Journal of Pure and Applied Algebra*, 108:109–173, 1996.

[Mak97] Michael Makkai. First order logic with dependent sorts. ftp.math.mcgill.ca, 1997.

[Mal71] Anatolii Mal'cev. *The Metamathematics of Algebraic Systems. Collected Papers 1936–67*. Number 66 in Studies in Logic and the Foundations of Mathematics. North-Holland, 1971. Edited by Benjamin Wells.

[Man98] Paolo Mancosu. *From Brouwer to Hilbert: the Debate on the Foundations of Mathematics in the 1920s*. Oxford University Press, 1998.

[Man76] Ernest Manes. *Algebraic Theories*. Number 26 in Graduate Texts in Mathematics. Springer-Verlag, 1976.

[Mar98] Francisco Marmolejo. Continuous families of coalgebras. *Journal of Pure and Applied Algebra*, 130:197–215, 1998,

[ML75] Per Martin-Löf. An intuitionistic theory of types: Predicative part. In Harvey Rose and John Sheperdson, editors, *Logic Colloquium*

'73, number 80 in Studies in Logic and the Foundations of Mathematics, pages 73–118. North-Holland, 1975.

[ML84] Per Martin-Löf. *Intuitionistic Type Theory*. Bibliopolis, Naples, 1984.

[MR73] Adrian Mathias and Hartley Rogers, editors. *Cambridge Summer School in Mathematical Logic*. Number 337 in Lecture Notes in Mathematics. Springer-Verlag, 1973.

[McC67] Storrs McCall. *Polish Logic, 1920-1939*. Oxford University Press, 1967.

[MF87] Ralph McKenzie and Ralph Freese. *Commutator Theory for Congruence Modular Varieties*. Number 125 in London Mathematical Society Lecture Notes. Cambridge University Press, 1987.

[MMT87] Ralph McKenzie, George McNulty, and Walter Taylor. *Algebras, Lattices, Varieties*. Wadsworth and Brooks, 1987.

[McL92] Colin McLarty. *Elementary Categories, Elementary Toposes*. Number 21 in Logic Guides. Oxford University Press, 1992.

[MNPS91] Dale Miller, Gopalan Nadathur, Frank Pfenning, and Andre Scedrov. Uniform proofs as a foundation for logic programming. *Annals of Pure and Applied Logic*, 51:125–137, 1991.

[Mit65] Barry Mitchell. *Theory of Categories*. Number 17 in Pure and applied mathematics. Academic Press, 1965.

[MS93] John Mitchell and Andre Scedrov. Notes on sconing and relators. In Egon Börger, Gerhard Jäger, Hans Büning, and Michael Richter, editors, *Computer Science Logic '92*, number 702 in Lecture Notes in Computer Science, pages 352–378. Springer-Verlag, 1993.

[Mit96] John Mitchell. *Foundations for Programming Languages*. MIT Press, 1996.

[Mog91] Eugenio Moggi. Notions of computation and monads. *Information and Computation*, 93:55–92, 1991.

[Mon66] Richard Montague. Fraenkel's addition to the axioms of Zermelo. In Bar-Hillel et al. [BHPRR66], pages 91–114.

[Moo82] Gregory Moore. *Zermelo's Axiom of Choice: its Origins, Development, and Influence*. Number 8 in Studies in the History of Mathematics and Physical Science. Springer-Verlag, 1982.

[MS55] John Myhill and John Shepherson. Effective operations on partial recursive functions. *Zeitschrift für Mathematische Logik und Gründlagen der Mathematik*, pages 310–317, 1955.

[NST93] Peter Neumann, Gabrielle Stoy, and Edward Thompson. *Groups and Geometry*. Oxford University Press, 1993.

[Nie82] Susan Niefield. Cartesianness: Topological spaces, uniform spaces and affine varieties. *Journal of Pure and Applied Algebra*, 23:147–167, 1982.

[Noe83] Emmy Noether. *Gesammelte Abhandlungen*. Springer-Verlag, 1983. Edited by Nathan Jabobson.

[NPS90] Bengt Nordström, Kent Petersson, and Jan Smith. *Programming in Martin-Löf's Type Theory: an Introduction*. Number 7 in

International Series of Monographs on Computer Science. Oxford University Press, 1990.

[Obt89] Adam Obtułowicz. Categorical and algebraic aspects of Martin-Löf type theory. *Studia Logica*, 48:299–318, 1989.

[Odi89] Piergiorgio Odifreddi. *Classical Recursion Theory: the Theory of Functions and Sets of Natural Numbers*. Number 125 in Studies in Logic and the Foundations of Mathematics. North-Holland, 1989.

[Odi90] Piergiorgio Odifreddi, editor. *Logic and Computer Science*. Number 31 in APIC Studies in Data Processing. Academic Press, 1990.

[Osi74] Gerhard Osius. Categorical set theory: a characterisation of the category of sets. *Journal of Pure and Applied Algebra*, 4:79–119, 1974.

[Par76] David Park. The Y-combinator in Scott's lambda-calculus models. Research Report CS-RR-013, Department of Computer Science, University of Warwick, June 1976. Revised, 1978.

[Pas65] Boris Pasynkov. Partial topological products. *Transactions of the Moscow Mathematical Society*, 13:153–271, 1965.

[Pau87] Lawrence Paulson. *Logic and Computation: Interactive proof with Cambridge LCF*. Number 2 in Cambridge Tracts in Theoretical Computer Science. Cambridge University Press, 1987.

[Pau91] Lawrence Paulson. *ML for the Working Programmer*. Cambridge University Press, 1991. Second edition, 1996.

[Pau92] Lawrence Paulson. Designing a theorem prover. In Samson Abramsky et al., editors, *Handbook of Logic in Computer Science*, pages 415–475. Oxford University Press, 1992.

[Pau94] Lawrence Paulson. *Isabelle: a Generic Theorem Prover*. Number 828 in Lecture Notes in Computer Science. Springer-Verlag, 1994.

[Pav90] Duško Pavlović. *Predicates and Fibrations*. PhD thesis, Rijksuniversiteit Utrecht, 1990.

[Pav91] Duško Pavlović. Constructions and predicates. In Pitt et al. [PCA+91], pages 173–197.

[Pea89] Giuseppe Peano. *Arithmetices Principia, Nova Methodo Exposita*. Fratres Bocca, Turin, 1889. English translation, "The Principles of Arithmetic, presented by a new method", in [vH67], pages 20–55.

[Pea73] Giuseppe Peano. *Selected Works of Giuseppe Peano*. Toronto University Press, 1973. Translated and edited by Hubert Kennedy.

[Pei33] Charles Sanders Peirce. *Collected Papers*. Harvard University Press, 1933. Edited by Charles Hartshorne and Paul Weiss.

[Pie91] Benjamin Pierce. *Basic Category Theory for Computer Scientists*. Foundations of Computing. MIT Press, 1991.

[PRD+89] David Pitt, David Rydeheard, Peter Dybjer, Andrew Pitts, and Axel Poigné, editors. *Category Theory in Computer Science III*, number 389 in Lecture Notes in Computer Science. Springer-Verlag, 1989.

[PCA+91] David Pitt, Pierre-Louis Curien, Samson Abramsky, Andrew Pitts, Axel Poigné, and David Rydeheard, editors. *Category Theory in*

Computer Science IV, number 530 in Lecture Notes in Computer Science. Springer-Verlag, 1991.

[Pit89] Andrew Pitts. Non-trivial power types can't be subtypes of polymorphic types. In *Logic in Computer Science IV*, pages 6–13. IEEE Computer Society Press, 1989.

[PT89] Andrew Pitts and Paul Taylor. A note on Russell's Paradox in locally cartesian closed categories. *Studia Logica*, 48:377–387, 1989.

[Pit95] Andrew Pitts. Categorical logic. Technical Report 367, University of Cambridge Computer Laboratory, May 1995.

[PD97] Andrew Pitts and Peter Dybjer, editors. *Semantics and Logics of Computation*, Publications of the Newton Institute. Cambridge University Press, 1997.

[Plo77] Gordon Plotkin. LCF considered as a programming language. *Theoretical Computer Science*, 5:223–255, 1977.

[Plo81] Gordon Plotkin. Domain theory. Post-graduate lecture notes, known as the Pisa Notes; ftp.lfcs.ed.ac.uk, 1981.

[Poh89] Wolfram Pohlers. *Proof Theory: an Introduction*. Number 1407 in Lecture Notes in Mathematics. Springer-Verlag, 1989.

[Pol45] George Polya. *How to Solve It: a New Aspect of Mathematical Method*. Princeton University Press, 1945. Re-published by Penguin, 1990.

[Pra65] Dag Prawitz. *Natural Deduction: a Proof-Theoretical Study*. Number 3 in Stockholm Studies in Philosophy. Almquist and Wiskell, 1965.

[Pra77] Dag Prawitz. Meanings and proofs: on the conflict between classical and intuitionistic logic. *Theoria*, 43:2–40, 1977.

[Qui60] Willard van Orman Quine. *Word and Object*. Studies in Communication. MIT Press, 1960.

[Ram31] Frank Ramsey. *Foundations of Mathematics*. Kegan Paul, 1931.

[RS63] Helena Rasiowa and Roman Sikorski. *The Mathematics of Metamathematics*. Number 41 in Monogrfie Matematyczne. Polish Scientific Publishers, 1963.

[Ras74] Helena Rasiowa. *An Algebraic Approach to Non-classical Logics*. Number 78 in Studies in Logic and the Foundations of Mathematics. North-Holland, 1974.

[Rey83] John Reynolds. Types, abstraction and parametric polymorphism. In Richard Mason, editor, *Information Processing*, pages 513–524. North-Holland, 1983.

[Rey84] John Reynolds. Polymorhism is not set-theoretic. In Gilles Kahn, David MacQueen, and Gordon Plotkin, editors, *Semantics of Data Types*, number 173 in Lecture Notes in Computer Science, pages 145–156. Springer-Verlag, 1984.

[RP93] John Reynolds and Gordon Plotkin. On functors expressible in the polymorphic lambda calculus. *Information and Computation*, 105:1–29, 1993.

[Rey98] John Reynolds. *Theories of Programming Languages*. Cambridge University Press, 1998.

[Ric56] Gordon Rice. Recursive and recursively enumerable orders. *Transactions of the American Mathematical Society*, 83:277, 1956.

[Rob66] Abraham Robinson. *Non-standard Analysis*. Studies in Logic and the Foundations of Mathematics. North-Holland, Amsterdam, 1966.

[Rob82] Derek Robinson. *A Course in the Theory of Groups*. Number 80 in Graduate Texts in Mathematics. Springer-Verlag, 1982. Second edition, 1996.

[RR88] Edmund Robinson and Giuseppe Rosolini. Categories of partial maps. *Information and Computation*, 79:95–130, 1988.

[Rus03] Bertrand Russell. *The Principles of Mathematics*. Cambridge University Press, 1903.

[Rus08] Bertrand Russell. Mathematical logic based on the theory of types. *American Journal of Mathematics*, 30:222–262, 1908. Reprinted in [vH67], pages 150–182.

[RW13] Bertrand Russell and Alfred North Whitehead. *Principia Mathematica*. Cambridge University Press, 1910–13.

[RB88] David Rydeheard and Rod Burstall. *Computational Category Theory*. Prentice-Hall, 1988.

[Sam48] Pierre Samuel. On universal mappings and free topological groups. *Bulletin of the American Mathematical Society*, 54:591–598, 1948.

[SS82] Andre Scedrov and Philip Scott. A note on the Friedman slash and Freyd covers. In Troelstra and van Dalen [TvD82], pages 443–452.

[Sch93] Andrea Schalk. Domains arising as algebras for power space constructions. *Journal of Pure and Applied Algebra*, 1993.

[Sch67] Joseph Schoenfield. *Mathematical Logic*. Addison-Wesley, 1967.

[Sco66] Dana Scott. More on the axiom of extensionality. In Bar-Hillel et al. [BHPRR66], pages 115–139.

[Sco70a] Dana Scott. Constructive validity. In Michel Laudet, D. Lacombe, L. Nolin, and M. Schützenberger, editors, *Automatic Demonstration*, number 125 in Lecture Notes in Mathematics, pages 237–275. Springer-Verlag, 1970.

[Sco70b] Dana Scott. Outline of a mathematical theory of computation. In *Information Sciences and Systems*, pages 169–176. Princeton University Press, 1970.

[Sco76] Dana Scott. Data types as lattices. *SIAM Journal on Computing*, 5:522–587, 1976.

[Sco79] Dana Scott. Identity and existence in intuitionistic logic. In Fourman et al. [FMS79], pages 660–696.

[SS70] J. Arthur Seebach and Lynn Arthur Steen. *Counterexamples in Topology*. Holt, Rinehart and Winston, 1970. Republished by Springer-Verlag, 1978 and by Dover, 1995.

[See84] Robert Seely. Locally cartesian closed categories and type theory. *Mathematical Proceedings of the Cambridge Philosophical Society*, 95:33–48, 1984.

[See87] Robert Seely. Categorical semantics for higher order polymorphic lambda calclus. *Journal of Symbolic Logic*, 52:969–989, 1987.

[See89] Robert Seely. Linear logic, ∗-autonomous categories and cofree algebras. In Gray and Scedrov [GS89], pages 371–382.

[Ser71] Jean-Pierre Serre. *Repésentations Linéaires des Groupes Finis*. Hermann, 1971. English translation, "Linear representations of finite groups" by Leonard Scott, Springer-Verlag, Graduate Texts in Mathematics 42, 1976, reprinted 1986.

[Sha91] Stewart Shapiro. *Foundations without Foundationalism: a Case for Second-order Logic*. Number 17 in Logic Guides. Oxford University Press, 1991.

[Sko22] Thoralf Skolem. Einige Bemerkungen zur axiomatischen Begründung der Mengenlehre. In *Skandinaviska matematikenkongressen*, pages 217–232. Akademiska Bokhandeln, 1922. English translation, "Some Remarks on Axiomatized Set Theory" in [vH67], pages 290–301.

[Sko70] Thoralf Skolem. *Selected Works in Logic*. Universitetsforlaget, Oslo, 1970. Edited by Jens Erik Fenstad.

[Ste67] Norman Steenrod. A convenient category of topological spaces. *Michigan Mathematics Journal*, 14:133–152, 1967.

[Sto37] Marshall Stone. Applications of the theory of Boolean rings to general topology. *Transactions of the American Mathematical Society*, 41:375–481, 1937.

[SW73] Ross Street and Robert Walters. The comprehensive factorization of a functor. *Bulletin of the American Mathematical Society*, 79:936–941, 1973.

[Str80] Ross Street. Cosmoi of internal categories. *Transactions of the American Mathematical Society*, 258:271–318, 1980.

[Str91] Thomas Streicher. *Semantics of Type Theory: Correctness and Completeness of a Categorical Semantics of the Calculus of Constructions*. Progress in Theoretical Computer Science. Birkhäuser, 1991. His 1988 Passau Ph.D. thesis.

[Str69] Dirk Struik. *A Source Book in Mathematics, 1200–1800*. Harvard University Press, 1969.

[Sty69] N. I. Styazhkin. *History of Mathematical Logic from Leibniz to Peano*. MIT Press, 1969. Translated from Russian, originally published by Nauka, Moscow, 1964.

[Tai75] William Tait. A realizability interpretation of the theory of species. In R. Parik, editor, *Logic Colloquium*, pages 240–251. Springer-Verlag, 1975.

[Tak75] Gaisi Takeuti. *Proof Theory*. Number 81 in Studies in Logic and the Foundations of Mathematics. North-Holland, 1975. Second edition, 1987.

[Tar56] Alfred Tarski. *Logic, Semantics, Metamathematics*. Oxford University Press, 1956. Edited by J. H. Woodger.

[Tay86a] Paul Taylor. Internal completeness of categories of domains. In David Pitt, editor, *Category Theory and Computer Programming*, number 240 in Lecture Notes in Computer Science, pages 449–465. Springer-Verlag, 1986.

[Tay86b] Paul Taylor. *Recursive Domains, Indexed Category Theory and Polymorphism.* PhD thesis, Cambridge University, 1986.

[Tay87] Paul Taylor. Homomorphisms, bilimits and saturated domains — some very basic domain theory. `ftp.dcs.qmw.ac.uk`, 1987.

[Tay88] Paul Taylor. The trace factorisation of stable functors. 1988.

[Tay89] Paul Taylor. Quantitative domains, groupoids and linear logic. In Pitt et al. [PRD+89], pages 155–181.

[Tay90] Paul Taylor. An algebraic approach to stable domains. *Journal of Pure and Applied Algebra*, 64:171–203, 1990.

[Tay91] Paul Taylor. The fixed point property in synthetic domain theory. In Gilles Kahn, editor, *Logic in Computer Science 6*, pages 152–160. IEEE Computer Society Press, 1991.

[Tay96a] Paul Taylor. Intuitionistic sets and ordinals. *Journal of Symbolic Logic*, 61:705–744, 1996.

[Tay96b] Paul Taylor. On the general recursion theorem. 1996.

[Tay98] Paul Taylor. An abstract stone duality, I: Geometric and higher order logic. 1998. In preparation.

[Ten81] Robert Tennent. *Principles of Programming Languages.* Prentice-Hall, 1981.

[Ten91] Robert Tennent. *Semantics of Programming Languages.* International Series in Computer Science. Prentice-Hall, 1991.

[Tie72] Miles Tierney. Sheaf theory and the continuum hypothesis. In Lawvere [Law72], pages 13–42.

[Tro69] Anne Sjerp Troelstra. *Principles of Intuitionism.* Number 95 in Lecture Notes in Mathematics. Springer-Verlag, 1969.

[Tro77] Anne Sjerp Troelstra. *Choice Sequences: a Chapter of Inttuitionistic Mathematics.* Logic Guides. Oxford University Press, 1977.

[TvD82] Anne Sjerp Troelstra and Dirk van Dalen, editors. *L. E. J. Brouwer Centenary Symposium,* number 110 in Studies in Logic and the Foundations of Mathematics. North-Holland, 1982.

[TvD88] Anne Sjerp Troelstra and Dirk van Dalen. *Constructivism in Mathematics, an Introduction.* Number 121 and 123 in Studies in Logic and the Foundations of Mathematics. North-Holland, 1988.

[TS96] Anne Sjerp Troelstra and Helmut Schwichtenberg. *Basic Proof Theory.* Number 43 in Cambridge Tracts in Theoretical Computer Science. Cambridge University Press, 1996.

[Tur35] Alan Turing. On computable numbers with an application to the Entscheidungsproblem. *Proceedings of the London Mathematical Society (2),* 42:230–265, 1935.

[vD80] Dirk van Dalen. *Logic and Structure.* Universitext. Springer-Verlag, 1980. Second edition, 1983.

[vdW31] Bartel van der Waerden. *Moderne Algebra.* Ungar, 1931. Fifth edition, Springer-Verlag, 1960; English translation by Fred Blum and John Schulenberger, "Algebra", Springer-Verlag, 1971.

[vH67] Jan van Heijenoort, editor. *From Frege to Gödel: a Source Book in Mathematical Logic, 1879–1931.* Harvard University Press, 1967. Reprinted 1971, 1976.

[Vic88] Steven Vickers. *Topology via Logic.* Number 5 in Cambridge Tracts in Theoretical Computer Science. Cambridge University Press, 1988.

[vN61] John von Neumann. *Collected Works.* Pergamon Press, 1961. Edited by A. H. Taub.

[Web80] Judson Chambers Webb. *Mechanism, Mentalism and Metamathematics: an Essay on Finitism.* Number 137 in Synthese Library. Reidel, 1980.

[Wel96] Charles Wells. *The Handbook of Mathematical Discourse,* http://www-math.cwru.edu/~cfw2/abouthbk.htm

[Wey19] Hermann Weyl. Der Circulus vitiosus in der heutigen Begründung der Analysis. *Jahrbericht der deutschen Matematiker-Vereinigung,* 28:85–92, 1919. English translation, "The Continuum: a Critical Examination of the Foundations of Analysis" by Stephen Pollard and Thomas Bole, published by Thomas Jefferson University Press, 1987, reprinted by Dover, 1993.

[Wey68] Hermann Weyl. *Gesammelte Abhandlungen.* Springer-Verlag, 1968. Edited by K. Chandrasekharan.

[ZS58] Oscar Zariski and Pierre Samuel. *Commutative Algebra.* Van Nostrand, 1958. Reprinted by Springer-Verlag, Graduate Texts in Mathematics, numbers 28–9, 1975.

[Zer08a] Ernst Zermelo. Neuer Beweis für die Möglichkeit einer Wohlordnung. *Mathematische Annalen,* 65:107–128, 1908. English translation, "New proof that every set can be well ordered" in [vH67], pages 183–198.

[Zer08b] Ernst Zermelo. Untersuchungen über die Grundlagen der Mengenlehre I. *Mathematische Annalen,* 65:261–281, 1908. English translation, "Investigations in the foundations of set theory" in [vH67], pages 199–215.

Index

Printed in the United Kingdom
by Lightning Source UK Ltd.
135925UK00001B/114/A

UNIVERSITIES AI MEDWAY LIBRARY